Fundamentals and Applications in Aerosol Spectroscopy

Fundamentals and Applications in Aerosol Spectroscopy

Edited by

Ruth Signorell ▪ Jonathan P. Reid

CRC Press
Taylor & Francis Group
Boca Raton   London   New York

CRC Press is an imprint of the
Taylor & Francis Group, an **informa** business

MATLAB® is a trademark of The MathWorks, Inc. and is used with permission. The MathWorks does not warrant the accuracy of the text or exercises in this book. This book's use or discussion of MATLAB® software or related products does not constitute endorsement or sponsorship by The MathWorks of a particular pedagogical approach or particular use of the MATLAB® software.

CRC Press
Taylor & Francis Group
6000 Broken Sound Parkway NW, Suite 300
Boca Raton, FL 33487-2742

First issued in paperback 2017

© 2011 by Taylor & Francis Group, LLC
CRC Press is an imprint of Taylor & Francis Group, an Informa business

No claim to original U.S. Government works

ISBN-13: 978-1-4200-8561-7 (hbk)
ISBN-13: 978-1-138-11394-7 (pbk)

Visit the Taylor & Francis Web site at
http://www.taylorandfrancis.com

and the CRC Press Web site at
http://www.crcpress.com

Contents

Preface ix
Editors xiii
Contributors xv

SECTION I   Infrared Spectroscopy

Chapter 1   Infrared Spectroscopy of Aerosol Particles 3

Thomas Leisner and Robert Wagner

Chapter 2   Vibrational Excitons: A Molecular Model to Analyze Infrared
Spectra of Aerosols 25

George Firanescu, Thomas C. Preston, Chia C. Wang, and Ruth Signorell

Chapter 3   Aerosol Nanocrystals of Water Ice: Structure, Proton Activity, Adsorbate
Effects, and H-Bond Chemistry 49

J. Paul Devlin

Chapter 4   Infrared Extinction and Size Distribution Measurements
of Mineral Dust Aerosol 79

Paula K. Hudson, Mark A. Young, Paul D. Kleiber, and Vicki H. Grassian

Chapter 5   Infrared Spectroscopy of Dust Particles in Aerosols for Astronomical
Application 101

Akemi Tamanai and Harald Mutschke

SECTION II   Raman Spectroscopy

Chapter 6   Linear and Nonlinear Raman Spectroscopy of Single Aerosol Particles 127

N.-O. A. Kwamena and Jonathan P. Reid

Chapter 7   Raman Spectroscopy of Single Particles Levitated by an Electrodynamic
Balance for Atmospheric Studies 155

Alex K. Y. Lee and Chak K. Chan

Chapter 8   Micro-Raman Spectroscopy for the Analysis of Environmental Particles 193

Sanja Potgieter-Vermaak, Anna Worobiec, Larysa Darchuk,
and Rene Van Grieken

Chapter 9   Raman Lidar for the Characterization of Atmospheric Particulate Pollution 209

Detlef Müller

SECTION III   VIS/UV Spectroscopy, Fluorescence, and Scattering

Chapter 10 UV and Visible Light Scattering and Absorption Measurements on
Aerosols in the Laboratory 243

Zbigniew Ulanowski and Martin Schnaiter

Chapter 11 Progress in the Investigation of Aerosols' Optical Properties Using
Cavity Ring-Down Spectroscopy: Theory and Methodology 269

Ali Abo Riziq and Yinon Rudich

Chapter 12 Laser-Induced Fluorescence Spectra and Angular Elastic Scattering Patterns
of Single Atmospheric Aerosol Particles 297

R. G. Pinnick, Y. L. Pan, S. C. Hill, K. B. Aptowicz, and R. K. Chang

Chapter 13 Femtosecond Spectroscopy and Detection of Bioaerosols 321

Luigi Bonacina and Jean-Pierre Wolf

Chapter 14 Light Scattering by Fractal Aggregates 341

C. M. Sorensen

SECTION IV   UV, X-ray, and Electron Beam Studies

Chapter 15 Aerosol Photoemission 367

Kevin R. Wilson, Hendrik Bluhm, and Musahid Ahmed

Chapter 16 Elastic Scattering of Soft X-rays from Free Size-Selected Nanoparticles 401

Harald Bresch, Bernhard Wassermann, Burkhard Langer, Christina Graf, and
Eckart Rühl

Chapter 17 Scanning Transmission X-ray Microscopy: Applications in
Atmospheric Aerosol Research 419

Ryan C. Moffet, Alexei V. Tivanski, and Mary K. Gilles

Chapter 18 Electron Beam Analysis and Microscopy of Individual Particles 463

Alexander Laskin

Index 493

Preface

This book is intended to provide an introduction to aerosol spectroscopy and an overview of the state-of-the-art of this rapidly developing field. It includes fundamental aspects of aerosol spectroscopy as well as applications to atmospherically and astronomically relevant problems. Basic knowledge is the prerequisite for any application. However, in aerosol spectroscopy, as in many other fields, there remain crucial gaps in our understanding of the fundamental processes. Filling this gap can only be a first step, with the challenge then remaining to develop instruments and methods based on those fundamental insights, instruments that can easily be used to study aerosols in planetary atmospheres as well as in space. With this in mind, this book also touches upon some of the aspects that need further research and development. As a guideline, the chapters in this book are arranged in the order of decreasing wavelength of light/electrons, starting with infrared spectroscopy and concluding with x-ray and electron beam studies.

Infrared spectroscopy is one of the most important aerosol characterization methods in laboratory studies, for field measurements, for remote sensing, and in space missions. It provides a wealth of information about aerosol particles ranging from properties such as particle size and shape to information on their composition and chemical reactivity. The analysis of spectral information, however, is still a challenge. In Chapter 1, Leisner and Wagner provide a detailed description of the most widely used method to analyze infrared extinction spectra, namely classical scattering theory in combination with continuum models of the optical properties of aerosol particles. The authors explain how information such as number concentration, size distribution, chemical composition, and shape can be retrieved from infrared spectra, and outline where pitfalls could occur. Theoretical considerations are illustrated with experiments performed in the large cloud chamber, aerosol interaction and dynamics in the atmosphere (AIDA).

Classical scattering theory and continuum models for optical properties are not always suitable for a detailed analysis of particle properties. Available optical data are often not accurate enough, and for small particles, where the molecular structure becomes important, these methods fail altogether. In Chapter 2, Firanescu, Preston, Wang, and Signorell discuss a molecular model that allows a detailed analysis of particle properties on the basis of the band shapes observed in infrared extinction spectra. In particular, this approach explains why and when infrared spectra of molecular aerosols are determined by particle properties such as shape, size, or architecture. After a description of the approach, the authors illustrate its application by means of a variety of examples.

Water and ice are the most important components of aerosols in our Earth's atmosphere. They play a crucial role in many atmospheric processes. Water ice is also ubiquitous beyond our planet and solar system. In Chapter 3, Devlin uses infrared spectroscopy to characterize this important type of particle and shows how the structural properties of pure and mixed ice nanocrystals can be unraveled by this technique. Special consideration is given to the nature of the surface of these particles, the role it plays, and how it is influenced by adsorbates. The formation and transformation of numerous naturally occurring hydrates are discussed. These studies reveal the exceptional properties of water ice surfaces.

Chapters 4 and 5 are devoted to the infrared spectroscopy of dust particles. The infrared radiative effects of mineral dust aerosols in the Earth's atmosphere are investigated by Hudson, Young, Kleiber, and Grassian in Chapter 4. Remote sensing studies using infrared data from satellites provide the source of information to determine the radiative effects of these particles. Such data are commonly analyzed using Mie theory, which treats all particles as spheres. The authors discuss the errors associated with this assumption and demonstrate that the proper treatment of particle shape is crucial in retrieving reliable information about the radiative effect of mineral dust particles from remote sensing. The properties of dust grains occurring in astrophysical environments are the subject of Chapter 5 by Tamanai and Mutschke. Dust grains of different composition with sizes in the micrometer range are widely distributed throughout space. Ground-based as well as satellite-based telescopes are used for infrared studies of these dust particles. Tamanai and Mutschke discuss infrared laboratory studies of astrophysically relevant dust grains and their application to the interpretation of astronomical spectra. While the wide variety of dust properties makes spectral analysis a difficult task, the authors demonstrate that important information can be obtained from such measurements about the conditions under which dust grains exist and evolve in astronomical environments.

Raman spectroscopy has proved to be a versatile tool for examining aerosol particles in controlled laboratory measurements, allowing the unambiguous identification of chemical species, the determination of particle composition, and even the determination of particle size and temperature. Although Raman scattering is inherently a weak process, measurements have been routinely performed on droplet trains using pulsed laser and continuous-wave laser techniques, on aerosol particles isolated in optical or electrodynamic traps, and on particles deposited on substrates. Section II begins with a general introduction to the fundamentals of both linear and nonlinear Raman scattering from aerosol particles. In particular, Kwamena and Reid highlight the considerable accuracy (<1 nm) that can be achieved in the determination of droplet size from the unique fingerprint of enhanced Raman scattering that occurs at discrete wavelengths commensurate with whispering gallery modes, also referred to as morphology-dependent resonances. Before reviewing some recent applications of Raman spectroscopy for characterizing aerosol, they introduce some of the key experimental considerations that must be remembered when designing a Raman instrument for aerosol studies. Lee and Chan describe the coupling of Raman spectroscopy with an electrodynamic balance in Chapter 7, outlining how information gained from Raman measurements can complement that from other methods, including light scattering for probing particle size and morphology, or tracking evolving particle mass. In particular, they review recent studies of hygroscopicity and heterogeneous chemistry. They demonstrate that resolving Raman line shapes can provide important insights into intermolecular interactions between solvent and solute molecules within the condensed aerosol phase, particularly important for understanding the properties of metastable supersaturated states accessed at high solute concentrations.

Raman analysis can provide an important tool for characterizing particulate matter of atmospheric origin as well as for probing particles in controlled laboratory measurements. Potgieter-Vermaak, Worobiec, Darchuk, and Van Grieken review the application of micro-Raman spectroscopy for the analysis of environmental particles in Chapter 8. They begin by reviewing the methods available for ambient sampling and the importance of choosing suitable substrates, before discussing the advantages and challenges of utilizing the technique on a stand-alone basis. The practicalities of coupling micro-Raman measurements with other techniques, such as scanning electron microscopy coupled with energy-dispersive x-ray spectrometric detection, are also described and assessed.

Key uncertainties remain in the direct and indirect impact of aerosols on climate, and coordinated monitoring of the temporal variability of global aerosol distribution is a basic requirement of climate research. In Chapter 9, Müller describes the application of Raman LIDAR (light detection and ranging) in the characterization of atmospheric pollution. After a description of the basic principles of Raman LIDAR, methods for deriving the optical and microphysical properties of particulate pollution are introduced. This is followed by an illustration of the potential of modern Raman LIDARs, particularly when measurements are made with a network of systems on a continental scale.

Elastic light scattering by particles in the visible and UV parts of the electromagnetic spectrum provides the basis for many conventional and routine techniques for determining particle size and concentration. More recently, it has been shown that resolving the light scattering from single particles may lead to the development of new instruments for assessing particle size and shape.

In addition, fluorescence spectroscopy is becoming an increasingly applied technique for identifying particle composition. Ulanowski and Schnaiter begin Section III with a discussion of light scattering and absorption measurements on aerosols in the laboratory. Following an introduction to key parameters that must be typically measured, they review some of the common methods for performing extinction spectroscopy, using an optical extinction cell, and absorption spectroscopy, specifically photoacoustic spectroscopy, and applications of these instruments in laboratory and chamber measurements. Resolving the angular dependence of light scattering has a long history in the field of particle analysis, and recent developments have concentrated on the measurement and analysis of complex morphologies recorded at the single-particle level, allowing the categorization of sampled particles into distinct classes.

In Chapter 11, Riziq and Rudich describe the information that can be gained by measuring light extinction from ensembles of accumulation mode aerosol particles using cavity ring-down spectroscopy (CRD-S). CRD-S is widely used for performing highly sensitive measurements of gas-phase composition and is now becoming more extensively used in both field and laboratory-based aerosol measurements. The authors introduce the underlying principles of CRD-S, before describing pulsed and continuous-wave implementations of the technique, and the sensitivity that can be achieved. The chapter concludes with a review of recent applications, particularly focusing on the retrieval of aerosol optical properties.

The application of laser-induced fluorescence (LIF) spectroscopy for identifying and classifying biological aerosol particles is described by Pinnick, Pan, Hill, Aptowicz, and Chang in Chapter 12. Although many compounds have similar fluorescence spectra with relatively broad and indistinguishable features, unlike those that occur in Raman or IR spectra, single-particle LIF measurements can provide clear and distinguishable signatures for different classes of biological and anthropogenic aerosol. Further classification of particle type/morphology can be achieved by two-dimensional angular optical scattering (TAOS), complementing and expanding on the discussion of this technique provided by Ulanowski and Schnaiter in Chapter 10. Bonacina and Wolf describe the improved specificity of bioaerosol detection that can be achieved using ultrafast laser techniques, including time-resolved pump–probe fluorescence spectroscopy, femtosecond laser-induced break down spectroscopy, and coherent optimal control in Chapter 13. In particular, they show that the application of an ultrafast double-pulse excitation scheme can induce strong fluorescence depletion from biological samples such as bacteria-containing droplets, allowing discrimination from possible interferents, such as polycyclic aromatic compounds, which otherwise have similar spectroscopic properties.

In many optical studies of aerosols, particles can be assumed to be spherical in shape, allowing the application of Mie scattering theory. In many cases, this only provides an approximate picture and the application of more rigorous treatments that describe the nonspherical morphology of a particle must be considered. Sorensen explores the complexity apparent in scattering measurements from fractal aggregates in Chapter 14, concentrating on diffusion-limited cluster aggregates. The theoretical treatment of such particles is based on the Rayleigh–Debye–Gans (RDG) approximation, which assumes that the monomeric units forming the aggregate scatter light independently. Once the fundamental concepts describing scattering in such complex systems have been introduced, the absolute scattering and differential cross-sections are defined, and the methods used in the analysis of data recorded from polydisperse systems are described.

Section IV deals with VUV, x-ray, and electron beam studies of aerosols. All these techniques constitute fairly new ways of characterizing aerosols, many aspects of which have been developed in recent years by the authors of these chapters. This book contains a unique overview of the different aspects and prospects of these methods. Photoelectron spectroscopy as applied to aerosol science is the subject of Chapter 15 by Wilson, Bluhm, and Ahmed, who provide a comprehensive overview of the techniques, the history, and the literature in the field. The use of photoelectric charging to probe surface composition and chemical as well as physical properties of aerosols is demonstrated by various examples in the second part of their chapter. The third part demonstrates, with many examples, how synchrotron-based aerosol photoemission can be used to unravel chemical information on the interfaces and properties of biological nanoparticles. Bresch, Wassermann, Langer, Graf, and Rühl demonstrate in Chapter 16 how x-ray light scattering allows them to obtain information on aerosol properties such as surface properties or size. The use of tunable x-rays for the aerosol scattering experiment is an exciting new approach. The authors present novel experimental results and developments for the proper analysis of the observed scattering patterns.

New approaches to characterize aerosols by scanning x-ray transmission microscopy and electron microscopy are presented in Chapter 17 by Moffet, Tivanski, and Gilles and in Chapter 18 by Laskin. Chapter 17 provides a unique introduction to scanning transmission x-ray microscopy and the latest developments in this field. This is the first and so far only comprehensive overview of this promising technique to become available in the literature. The power of this technique for the characterization of atmospherically relevant aerosols is illustrated by applying the method to aerosol samples collected from various sources in different field campaigns. The authors outline how information on aerosol morphology, surface coating, mixing state, and atmospheric processing can be extracted from such measurements. Following this overview of scanning x-ray transmission microscopy, Laskin gives a similarly unique review of electron beam microscopy studies of aerosols and complementary microspectroscopic methods in Chapter 18. Besides many other particle properties, the microanalysis of aerosol particles allows one to retrieve information on the lateral distribution of chemical species within individual particles. In one of his examples, the author shows how chemical information is extracted from studies of field-collected particles. In another, he reports on the use of electron microscopy to study the hygroscopic properties and ice nucleation of individual particles.

Our special thanks go to all authors who have contributed their time and expertise to this overview of the spectroscopy of aerosols. We hope that the result is as enjoyable as it is informative, not only for aerosol scientists but also for students and other readers interested in the field.

MATLAB® is a registered trademark of The MathWorks, Inc. For product information, please contact:

The MathWorks, Inc.
3 Apple Hill Drive
Natick, MA 01760-2098 USA
Tel: 508 647 7000
Fax: 508-647-7001
E-mail: info@mathworks.com
Web: www.mathworks.com

Ruth Signorell
Jonathan P. Reid

Editors

Ruth Signorell received undergraduate and postgraduate degrees from ETH Zürich in Switzerland before moving to a postdoctoral fellowship at the University of Göttingen in Germany where she became assistant professor in 2002. Since 2005, she has been professor in physical and analytical chemistry at the University of British Columbia in Canada. She has been awarded the ETH Medal in 1999 for her PhD thesis, the 2005 Werner Award of the Swiss Chemical Society, an A. P. Sloan Fellowship from the United States in 2007, the 2009 Thermo Fisher Scientific Spectroscopy Award from the Canadian Society for Analytical Sciences and Spectroscopy, and the 2010 Keith Laidler Award from the Canadian Society for Chemistry. Her research interests focus on infrared and extreme ultraviolet studies of aerosols.

Jonathan P. Reid received undergraduate and postgraduate degrees from the University of Oxford (MA, DPhil) before moving to a postdoctoral fellowship at JILA, University of Colorado. In 2000, he took up a lectureship at the University of Birmingham, United Kingdom, before moving to the University of Bristol, United Kingdom, in 2004. He is currently professor in physical chemistry and a Leadership Fellow of the Engineering and Physical Sciences Research Council. He was awarded the 2001 Harrison Memorial Prize and the 2004 Marlow Medal by the Royal Society of Chemistry. His research interests focus on developing new techniques to characterize and manipulate aerosol particles using light.

Contributors

Musahid Ahmed
Chemical Sciences Division
Lawrence Berkeley National Laboratory
Berkeley, California

K. B. Aptowicz
Department of Physics
West Chester University
West Chester, Pennsylvania

Hendrik Bluhm
Chemical Sciences Division
Lawrence Berkeley National Laboratory
Berkeley, California

Luigi Bonacina
University of Geneva—GAP-Biophotonics
Rue de l'Ecole de Medecine
Geneva, Switzerland

Harald Bresch
Physikalische Chemie
Freie Universität Berlin
Berlin, Germany

Chak K. Chan
Division of Environment
Hong Kong University of Science and
    Technology
Kowloon, Hong Kong, China

R. K. Chang
Department of Applied Physics
Yale University
New Haven, Connecticut

Larysa Darchuk
Department of Chemistry
University of Antwerp (Campus Drie Eiken)
Universiteitsplein, Wilrijk-Antwerpen, Belgium

J. Paul Devlin
Department of Chemistry
Oklahoma State University
Stillwater, Oklahoma

George Firanescu
Department of Chemistry
University of British Columbia
Vancouver, British Columbia, Canada

Mary K. Gilles
Chemical Sciences Division
Lawrence Berkeley National Laboratory
Berkeley, California

Christina Graf
Physikalische Chemie
Freie Universität Berlin
Berlin, Germany

Vicki H. Grassian
Department of Physics and Astronomy
University of Iowa
Iowa City, Iowa

S. C. Hill
U.S. Army Research Laboratory
Adelphi, Maryland

Paula K. Hudson
Center for Global and Regional Environmental
    Research
University of Iowa
Iowa City, Iowa

Paul D. Kleiber
Department of Physics and Astronomy
University of Iowa
Iowa City, Iowa

Nana Kwamena
School of Chemistry
University of Bristol
Bristol, United Kingdom

Burkhard Langer
Physikalische Chemie
Freie Universität Berlin
Berlin, Germany

Alexander Laskin
W. R. Wiley Environmental Molecular Science
    Laboratory
Pacific Northwest National Laboratory
Richland, Washington

Alex K. Y. Lee
Department of Chemical and Biomolecular
    Engineering
Hong Kong University of Science and
    Technology
Kowloon, Hong Kong, China

Thomas Leisner
Karlsruhe Institute of Technology
Institute for Meteorology and Climate
    Research
Hermann-von-Helmholtz-Platz
Eggenstein-Leopoldshafen, Germany

Ryan C. Moffet
Chemical Sciences Division
Lawrence Berkeley National Laboratory
Berkeley, California

Detlef Müller
Atmospheric Remote Sensing Laboratory
Gwangju Institute of Science and
    Technology
Gwangju, Republic of Korea

and

Department of Physics
Leibniz Institute for Tropospheric
    Research
Leipzig, Germany

Harald Mutschke
Astrophysical Institute and University
    Observatory
Friedrich-Schiller-University Jena
Schillergäßchen, Jena, Germany

Y. L. Pan
U.S. Army Research Laboratory
Adelphi, Maryland

R. G. Pinnick
U.S. Army Research Laboratory
Adelphi, Maryland

Sanja Potgieter-Vermaak
Department of Chemistry
University of Antwerp
    (Campus Drie Eiken),
Universiteitsplein, Wilrijk-Antwerpen,
    Belgium

Thomas C. Preston
Department of Chemistry
University of British Columbia
Vancouver, British Columbia, Canada

Jonathan P. Reid
School of Chemistry
University of Bristol
Bristol, United Kingdom

Ali Abo Riziq
Department of Environmental Sciences
Weizmann Institute
Rehovot, Israel

Yinon Rudich
Department of Environmental
    Sciences
Weizmann Institute
Rehovot, Israel

Eckart Rühl
Physikalische Chemie
Freie Universität Berlin
Berlin, Germany

Martin Schnaiter
Karlsruhe Institute of Technology
Institute for Meteorology and Climate
    Research
Hermann-von-Helmholtz-Platz
Eggenstein-Leopoldshafen,
    Germany

C. M. Sorensen
Department of Physics
Kansas State University
Manhattan, Kansas

Akemi Tamanai
Astrophysical Institute and University
    Observatory
Friedrich-Schiller-University Jena
Schillergäßchen, Jena, Germany

Alexei V. Tivanski
Chemical Sciences Division
Lawrence Berkeley National Laboratory
Berkeley, California

Zbigniew Ulanowski
Centre for Atmospheric and Instrumentation
    Research
University of Hertfordshire, Hatfield
Herts, United Kingdom

Rene Van Grieken
Department of Chemistry
University of Antwerp (Campus Drie Eiken)
Universiteitsplein, Wilrijk-Antwerpen,
    Belgium

Robert Wagner
Karlsruhe Institute of Technology
Institute for Meteorology and Climate Research
Hermann-von-Helmholtz-Platz
Eggenstein-Leopoldshafen, Germany

Chia C. Wang
Department of Chemistry
University of British Columbia
Vancouver, British Columbia, Canada

Bernhard Wassermann
Physikalische Chemie
Freie Universität Berlin
Berlin, Germany

Kevin R. Wilson
Chemical Sciences Division
Lawrence Berkeley National Laboratory
Berkeley, California

Jean-Pierre Wolf
University of Geneva—GAP-
    Biophotonics
Rue de l'Ecole de Medecine
Geneva, Switzerland

Anna Worobiec
Department of Chemistry
University of Antwerp
    (Campus Drie Eiken)
Universiteitsplein, Wilrijk-Antwerpen,
    Belgium

Mark A. Young
Department of Chemistry
University of Iowa
Iowa City, Iowa

Section I

Infrared Spectroscopy

1 Infrared Spectroscopy of Aerosol Particles

Thomas Leisner and Robert Wagner

CONTENTS

1.1 Introduction 3
1.2 Theory..5
    1.2.1 Particle Sizes Small Compared to the Wavelength (Rayleigh Regime) 5
        1.2.1.1 General Equations and Comparison with Bulk Absorption Measurements 5
        1.2.1.2 Influence of Particle Shape 7
        1.2.1.3 Derivation of Optical Constants from the Absorption Spectra of Small Particles 8
    1.2.2 Infrared Extinction Spectra of Wavelength-Sized Particles (Mie Regime) 10
        1.2.2.1 Dependence of the Spectral Habitus on the Particle Size..........................10
        1.2.2.2 Influence of Particle Shape 13
        1.2.2.3 Size Distribution Retrieval..16
1.3 Examples..17
    1.3.1 Typical Infrared Spectral Habitus of Large Cloud Particles 17
    1.3.2 Solution Ambiguity of the Size Distribution Retrieval for Aspherical Ice Particles 19
1.4 Concluding Remarks 20
References..22

1.1 INTRODUCTION

Mid-infrared extinction spectroscopy has been established as an important tool to derive microphysical properties such as size, shape, and phase of aerosols and individual aerosol particles and to monitor multiphase processes, both in laboratory measurements as well as in remote sensing applications.[1,2] The extinction of an incident infrared beam is the sum of light absorption in the particles and light scattering by the particles. Absorption is the dominant contribution for particle sizes small compared to the wavelength of the incident light. At mid-infrared wavelengths, this holds for particle diameters below approximately 200 nm. In its absorption contribution, the infrared spectrum is susceptible to the distinctive bands of organic and inorganic functional groups inherent in molecularly structured aerosol particles and can thus be a powerful tool for chemical characterization. Recent examples include the analysis of the chemical evolution of secondary organic aerosol in a smog chamber and the unique discrimination between different types of polar stratospheric cloud particles in satellite infrared measurements.[3,4] Moreover, infrared spectroscopy is ideally suited to investigate the deliquescent and efflorescent behavior of aerosol particles, identifying the phase transition by the appearance and disappearance of the broad liquid water absorption band at around 3300 cm^{-1}.[5–9] Exploiting the different spectral habitus of the absorption bands of liquid water droplets and ice crystals, infrared measurements are also a common experimental tool in studies on the ice-freezing behavior of supercooled aqueous solution droplets.[10–14]

In the limit of small particles whose interaction with light can be described by Rayleigh theory,[15] the absorption spectrum only depends on the volume of the particles but not on the details of the aerosol size distribution. Only for particle diameters below approximately 20 nm, pronounced size-dependent phenomena might appear in the absorption spectra, in particular for particles composed of equivalent molecules and vibrational bands with a strong molecular transition dipole.[16] The size-dependent spectral habitus of such transitions can be modeled quite accurately with the quantum–mechanical vibrational exciton model by taking into account the resonant transition dipole coupling between the molecules in the aerosol particle. As shown, for example, with the asymmetric stretching mode of CO_2 (see Figure 3 in Sigurbjörnsson et al.[16]), the modulation of these intermolecular interactions by the particle boundaries becomes important for particle diameters below 20 nm, provoking that each particle size below this threshold exhibits a unique fine structure in the absorption spectrum.

For particle diameters larger than 200 nm, the scattering contribution to the infrared extinction spectra begins to manifest itself in slanted baselines in nonabsorbing spectral regimes and in dispersion features superimposing and distorting the absorption bands.[17] In contrast to the Rayleigh limit, scattering is sensitive to the particle size and in principle allows a retrieval of the aerosol size distribution. This involves a least-squares minimization procedure between a measured and a calculated infrared spectrum, using the size-distribution vector of the aerosol sample as the optimization parameter. Most frequently, the classical scattering theory is used to compute the extinction spectrum of aerosol particles, including Mie theory for spheres[15] and the T-matrix approach[18] or the discrete dipole approximation (DDA)[19] for nonspherical particles. The quantitative applicability of this approach relies on accurate frequency-dependent optical constants, that is, the real (n) and imaginary (k) parts of the complex refractive index N ($N = n + ik$) that are used as input values in the calculations. Over the past decade, a significant portion of the publications on aerosol infrared spectroscopy has been devoted to improving the database of optical constants of atmospherically relevant aerosol particles, see also the discussion on the indices of refraction tabulated in the recent HITRAN 2008 database of spectroscopic parameters.[20] In particular, the pronounced temperature dependence of the infrared refractive indices, which is apparent for many substances, has been systematically investigated for the first time, including, for example, the $H_2SO_4/H_2O/HNO_3$ system,[21–24] supercooled water,[25,26] and ice.[25]

For spherical particles, it was shown that even bi- or multimodal aerosol size distributions can be retrieved with good accuracy from measured infrared extinction spectra.[1] Important aerosol constituents such as solid sodium chloride and ammonium sulfate crystals, mineral dust particles, and ice crystals, however, partly reveal highly irregular morphologies. In such cases, the spectral analysis might be affected by severe size/shape ambiguities: different sets of shape–size distributions might satisfy the same optical data, thereby impeding the unique retrieval of both the size and the shape of the aerosol particles.[27] In such cases, an *a priori* information or an independent reference measurement of either the size distribution or the particle morphology is indispensable. Note that particle shape does not only influence the magnitude of the scattering contribution for larger aerosol particles, but might also strongly affect the spectral habitus of the absorption bands in the Rayleigh limit. For certain values of the optical constants n and k (e.g., $n \approx 0$ and $k \approx \sqrt{2}$ for a sphere), shape–dependent resonances (surface modes) might provoke that a small-particle spectrum strongly deviates from the corresponding bulk absorption spectrum and shows a high sensitivity to the particle shape.[15]

In the present contribution, we want to give a concise survey of size and shape effects on the infrared extinction spectra of aerosol particles within the framework of classical continuum models (Section 1.2). We will briefly address the strategies for the derivation of optical constants from the extinction spectra of airborne particles and point to the uncertainties associated with the size distribution retrieval for nonspherical particles. Selected applications of aerosol infrared spectroscopy to retrieve particle properties and to analyze aerosol multiphase processes are shown in Section 1.3 for measurements on particle ensembles in the large aerosol and cloud chamber AIDA of the Karlsruhe Institute of Technology.

1.2 THEORY

1.2.1 PARTICLE SIZES SMALL COMPARED TO THE WAVELENGTH (RAYLEIGH REGIME)

1.2.1.1 General Equations and Comparison with Bulk Absorption Measurements

For infrared optical depth measurements on airborne particles, the Lambert–Beer equation can be written under the single scattering criterion in discrete form as

$$\tau(\tilde{v}_j) = -\log \frac{I(\tilde{v}_j)}{I_0(\tilde{v}_j)} = \frac{l}{\ln 10} \sum_{i=1}^{N} n(D_i) C_{\text{ext}}(D_i, \tilde{v}_j) \quad j = 1\dots M. \tag{1.1}$$

It relates the measured optical depth $\tau(\tilde{v}_j)$ at a specific wave number $\tilde{v}_j$ to the optical path length l, the number concentration $n(D_i)$ of particles in a particular size bin D_i of width ΔD, and the size-bin averaged extinction cross-section $C_{\text{ext}}(D_i, \tilde{v}_j)$, given by

$$C_{\text{ext}}(D_i, \tilde{v}_j) = \frac{1}{\Delta D} \int_{D_i - \frac{\Delta D}{2}}^{D_i + \frac{\Delta D}{2}} C_{\text{ext}}(D, \tilde{v}_j) \, dD. \tag{1.2}$$

The extinction cross-section C_{ext} is the sum of the absorption cross-section C_{abs} and the scattering cross-section C_{sca}:

$$C_{\text{ext}}(D, \tilde{v}_j) = C_{\text{abs}}(D, \tilde{v}_j) + C_{\text{sca}}(D, \tilde{v}_j). \tag{1.3}$$

In the Rayleigh approximation, the absorption cross-section of a small sphere for transmission measurements in air (refractive index of the medium ≈ 1) is written as

$$C_{\text{abs}}(D, \tilde{v}_j) = 6\pi \tilde{v}_j V(D) \text{Im}\left(\frac{N^2(\tilde{v}_j) - 1}{N^2(\tilde{v}_j) + 2} \right), \tag{1.4}$$

with $V(D)$ denoting the volume of the sphere of diameter D and $N(\tilde{v}_j)$ symbolizing the complex refractive index of the particle with $N(\tilde{v}_j) = n(\tilde{v}_j) + ik(\tilde{v}_j)$. Note that Equation 1.4 is derived under the assumption that $x = \pi D \tilde{v}_j \ll 1$ and $x|N(\tilde{v}_j)| \ll 1$.[15] Further assuming that the scattering contribution to extinction can be neglected, Equation 1.1 reduces to

$$\tau(\tilde{v}_j) = \frac{6\pi \tilde{v}_j l V_{\text{tot}}}{\ln 10} \text{Im}\left(\frac{N^2(\tilde{v}_j) - 1}{N^2(\tilde{v}_j) + 2} \right)$$

$$= \frac{6\pi \tilde{v}_j l V_{\text{tot}}}{\ln 10} \left(\frac{6n(\tilde{v}_j) k(\tilde{v}_j)}{\left(n(\tilde{v}_j)^2 - k(\tilde{v}_j)^2 + 2 \right)^2 + \left(2n(\tilde{v}_j) k(\tilde{v}_j) \right)^2} \right). \tag{1.5}$$

In this expression, the recorded optical depth only depends on the total particle-volume concentration V_{tot} and not on the details of the aerosol number size distribution $n(D_i)$, that is, different size distributions with the same overall particle-volume concentration give rise to identical infrared absorption spectra. It is interesting to compare Equation 1.5 with the absorption spectrum of the same substance in the bulk phase. For transmission measurements of a small film of thickness d, the optical depth is directly proportional to the imaginary part of the complex refractive index:

$$\tau(\tilde{v}_j) = \frac{4\pi\tilde{v}_j d}{\ln 10} k(\tilde{v}_j). \qquad (1.6)$$

The ratio of the optical depths measured for a small-particle and a thin-film spectrum thereby becomes proportional to $n(\tilde{v}_j)/((n(\tilde{v}_j)^2 - k(\tilde{v}_j)^2 + 2)^2 + (2n(\tilde{v}_j)k(\tilde{v}_j))^2)$. For spectral regimes with less intense absorption bands ($k < 0.3$), which only provoke a small amplitude of the anomalous dispersion feature in the corresponding n spectrum, the proportionality factor reduces to $n(\tilde{v}_j)/(n(\tilde{v}_j)^2 + 2)^2$ when assuming $n(\tilde{v}_j) \gg k(\tilde{v}_j)$ over the considered wave number region. With $n(\tilde{v}_j)$ only revealing small-amplitude dispersion features, the small-particle absorption spectrum will not be considerably different from that of the bulk phase. This is demonstrated in panels a and b of Figure 1.1 with aqueous sulfuric acid (25 wt% $H_2SO_4^{28}$) as an example. On the other hand,

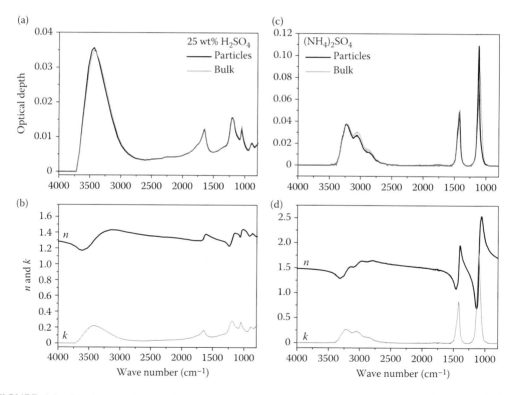

FIGURE 1.1 Panel a: Small-particle absorption spectrum of aqueous sulfuric acid with 25 wt% H_2SO_4 (black line), as computed from Equation 1.5 with $V_{tot} = 1000$ $\mu m^3/cm^3$ and $l = 100$ m based on the optical constants from Palmer and Williams[28] (shown in panel b). The same refractive index data set was used to compute the corresponding thin-film absorption spectrum (gray line) from Equation 1.6 with $d = 0.085$ μm. Panel c: Small-particle absorption spectrum of crystalline ammonium sulfate spheres (black line), as computed from Equation 1.5 with $V_{tot} = 1000$ $\mu m^3/cm^3$ and $l = 100$ m based on the optical constants from Earle et al.[29] (shown in panel d, data set for $T = 298$ K). Comparison with a calculated thin-film absorption spectrum (gray line) for $d = 0.080$ μm (Equation 1.6).

strong absorption bands with $k > 1$ and concomitantly high-amplitude anomalous dispersions in the n spectrum might provoke that certain spectral regimes approximately fulfill the resonance condition that is inherent in Equation 1.5. For wave numbers with $n \approx 0$ and $k \approx \sqrt{2}$, there will be an enhanced cross-section in the absorption spectra of small spheres, provoking that the spectral habitus of an absorption band (including band intensity and peak position) might strongly differ from the corresponding bulk absorption feature. As an example, panel c of Figure 1.1 compares the small-particle and bulk absorption spectrum of crystalline ammonium sulfate spheres.[29] Just in the regime of the intense $\nu_3(SO_4^{2-})$ absorption band at 1100 cm^{-1}, the small-particle absorption band is shifted to higher wave numbers and shows an increased intensity. Thus, the absorption maximum is shifted from $\tilde{\nu}(k_{max})$, that is, the bulk peak wave number, to a position where the corresponding optical constants n and k (Figure 1.1d) better fulfill the resonance condition of Equation 1.5.

1.2.1.2 Influence of Particle Shape

As already indicated in the introduction, the spectral habitus of intense small-particle absorption bands might also strongly depend on the particle shape. As a simple case study, we want to summarize the results for needle- and disk-like spheroids, representing two subgroups of a general ellipsoidal particle. For an exhaustive discussion of the shape effects, the reader is referred to the textbook of Bohren and Huffman.[15] For ellipsoidal particles, the geometrical factor L has to be introduced in the expression for the absorption cross-section:

$$C_{abs}(D_V, \tilde{\nu}_j, L) = 6\pi \tilde{\nu}_j V(D_V) \text{Im}\left(\frac{N^2(\tilde{\nu}_j) - 1}{3L\left[N^2(\tilde{\nu}_j) - 1\right] + 3} \right). \tag{1.7}$$

The particle diameter D_V may now be interpreted as the diameter of the sphere with the same volume as the nonspherical particle. For each of the three principal axes of an ellipsoid, there is a distinct value for the geometrical factor L (L_1, L_2, and L_3) and the average absorption cross-section for randomly oriented ellipsoids can be obtained from the arithmetic mean of the three principal cross-sections. For a sphere with $L_1 = L_2 = L_3 = \frac{1}{3}$, Equation 1.7 just reduces to Equation 1.5. For needle- and disk-like spheroids, there are two distinct geometrical factors with $L_1 = 0$, $L_2 = L_3 = 0.5$ (needle) and $L_1 = L_2 = 0$, $L_3 = 1$ (disk). Therefore, two distinct resonances might be observed for these particle shapes instead of the single absorption band for a sphere. The intensity of these bands and their spectral splitting, however, depends on whether the actual spectral variation of the n and k values over the considered wave number range is sufficient to cover both resonance conditions.[30] As an example, Figure 1.2a compares the small-particle absorption spectra of randomly oriented crystalline ammonium sulfate needles and disks with the corresponding sphere computation from Figure 1.1c in the regime of the $\nu_3(SO_4^{2-})$ vibration. And indeed, both spheroidal shapes reveal a band splitting, featuring one common mode at 1090 cm^{-1} from the principal component with $L = 0$. This band gains a higher intensity for the disk-like shape due to its duplicate contribution to the averaged cross-section, see Figure 1.2b, c. On the other hand, the 1120 cm^{-1} needle absorption band and the shoulder at 1140 cm^{-1} for ammonium sulfate disks are due to the resonances for $L = 0.5$ and $L = 1$, respectively. For these two bands, the needle-like shape gives rise to a higher intensity, both due to the doubled weight of the $L = 0.5$ principal cross-section and a better match of the resonance condition compared to the geometrical factor $L = 1$.

In a set of recent publications (see Sigurbjörnsson et al.[16] and references therein), also the quantum–mechanical vibrational exciton model was successfully applied to reproduce the shape effects in the infrared absorption spectra of Rayleigh-sized particles. In these analyses, the calculations were explicitly done for particle radii from 10 to 100 nm, that is, size and shape effects that occur in nanosized particles, as addressed in the introduction, were excluded. From the observation that strong shape effects are only evident for intense vibrational bands with a high molecular transition

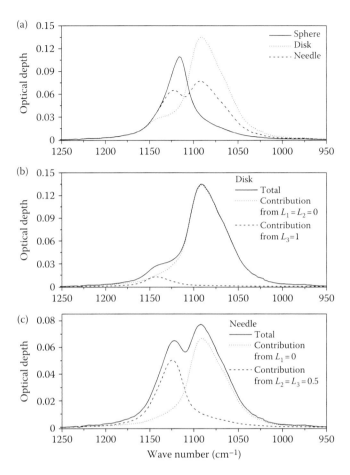

FIGURE 1.2   Panel a: Small-particle absorption spectra of crystalline ammonium sulfate spheres, needles, and disks, as computed from Equation 1.7 with $V_{tot} = 1000$ μm^3/cm^3 and $l = 100$ m based on the optical constants from Earle et al.[29] Panels b and c elucidate the contributions from the two distinct geometrical factors for the disk- and needle-like particles. See text for details.

dipole, it was concluded that the strong resonant intermolecular transition dipole coupling provides the microscopic explanation of the shape effects in small-particle absorption spectra. The exciton coupling leads to a delocalization of the excitation energy over the whole particle, which in turn gives rise to the shape sensitivity of the absorption bands.

Apart from crystalline ammonium sulfate, other important aerosol constituents, which feature pronounced shape-dependent infrared absorption bands in certain spectral regimes include mineral dust (which, if containing silicates, exhibits a prominent Si–O stretch resonance at around 1050 cm^{-1}),[31] nitric acid dihydrate (in the nitrate absorption regime between 1500 and 1000 cm^{-1}),[32] and ammonia aerosols (v_2 N–H bending mode at 1060 cm^{-1}).[33]

1.2.1.3   Derivation of Optical Constants from the Absorption Spectra of Small Particles

Two different approaches are usually applied to determine the frequency-dependent complex refractive indices from infrared optical measurements. On the one hand, the spectra of the optical constants n and k can be approximated by a set of Lorentz damped harmonic oscillators, with each oscillator characterized by its peak wave number, band width (damping constant), and intensity.[34] Starting from an *a priori* guess for the band parameters, their values can be optimized in an inversion scheme by minimizing the summed-squared residuals between measured and calculated infrared spectra. The other approach exploits the Kramers–Kronig relation between the real and imaginary parts of the complex refractive index,[15]

$$n(\tilde{v}_0) - 1 = \frac{2}{\pi} P \int_0^\infty \frac{k(\tilde{v})\tilde{v}}{\tilde{v}^2 - \tilde{v}_0^2}\, d\tilde{v}, \tag{1.8}$$

from which the value for n at a specific wave number $\tilde{v}_0$ can be computed from the spectrum of k over the entire frequency range. Equation 1.8 can be directly applied to the analysis of thin-film absorption spectra,[23] given that $k(\tilde{v})$ is directly obtained from the transmission measurements, see Equation 1.6. The experimental data, however, are often limited to mid-infrared wavelengths and thus do not cover the complete wave-number range to evaluate the integral in Equation 1.8. Therefore, suitable extensions of the $k(\tilde{v})$ spectrum beyond the experimentally accessible wave-number range (e.g., at UV–VIS and far-IR wavelengths) have to be introduced to avoid potentially severe truncation errors in the Kramers–Kronig transformation.[35] It is sometimes proposed to employ the so-called subtractive Kramers–Kronig integration to minimize the effect of truncation errors.[36,37] In this approach, the real part of the refractive index has to be known at some specific wave-number position $\tilde{v}_x$, preferentially located within the measured frequency range. Using $n(\tilde{v}_x)$ as a so-called anchor point in the evaluation of the Kramers–Kronig integral, that is,

$$n(\tilde{v}_0) = n(\tilde{v}_x) + \frac{2(\tilde{v}_0^2 - \tilde{v}_x^2)}{\pi} P \int_0^\infty \frac{k(\tilde{v})\tilde{v}}{(\tilde{v}^2 - \tilde{v}_0^2)(\tilde{v}^2 - \tilde{v}_x^2)}\, d\tilde{v}, \tag{1.9}$$

may reduce the weight of the unknown frequency behavior of $k(\tilde{v})$ for wave numbers far above or below the anchor point by introducing the additional factor $(\tilde{v}^2 - \tilde{v}_x^2)$ in the denominator of the Kramers–Kronig integral. A Kramers–Kronig relation also exists between the reflectivity and phase shift for reflection and can be used to obtain complex refractive indices from infrared reflection spectra of bulk materials.[38,39]

Concerning infrared transmission measurements of airborne particles, Rouleau and Martin[40] have emphasized that the Kramers–Kronig relation not only holds for $N(\tilde{v}_j)$, but also for the composite function $f = (N^2(\tilde{v}_j) - 1)/(N^2(\tilde{v}_j) + 2)$:

$$\mathrm{Re}\{f\}(\tilde{v}_0) = \frac{2}{\pi} P \int_0^\infty \frac{\mathrm{Im}\{f\}(\tilde{v})\tilde{v}}{\tilde{v}^2 - \tilde{v}_0^2}\, d\tilde{v}. \tag{1.10}$$

Equation 1.10 thereby offers the most direct approach to deduce the optical constants from transmission spectroscopy of particles, provided that, (1), the particle sizes are small enough to fulfill the requirements for Equation 1.5, (2), the particles are of (near) spherical shape, and (3), the overall particle volume concentration can be measured with high accuracy by supplementary methods (e.g., analyses of filter samples or size distribution measurements). Then, the imaginary part of f can be directly obtained from the measured optical depth, and, together with its real part, computed from the Kramers–Kronig integral (Equation 1.10) with proper extension for the unmeasured spectral range, allows the calculation of n and k (see Segal–Rosenheimer et al.[41] for a recent example). This procedure may also be applied with sufficient accuracy to the infrared spectra of particles with a small scattering contribution, manifesting itself in a slightly slanted baseline at nonabsorbing wave numbers. Then, the scattering part can be subtracted from the extinction spectrum by assuming a Rayleigh-like $C_{sca}(\tilde{v}) \propto \tilde{v}^4$ behavior (see, e.g., Figure 4 in Norman et al.[22]) and the residuum absorption contribution can be treated with Equation 1.5. If it is not possible to experimentally prepare particle sizes which fall into the regime of Equation 1.5, the retrieval of the optical constants becomes much more laborious. The inversion schemes are then based on Mie theory and usually involve an iterative adjustment of n and k together with the parameters of the underlying particle size distribution.[42,43] Small differences between the data sets of optical constants obtained from different studies (see, e.g., a comparison between two recently derived n and k data sets for supercooled water[25,26]) reflect the less stringent and approximate nature of these iterative inversion strategies.

A significant part of the atmospheric aerosol is composed of inhomogeneous particles, ranging from comparatively simple core–shell structures (e.g., a particle containing a solid nucleus like soot and a liquid organic or inorganic coating layer) to complex aggregates such as mineral dust, featuring a mixture of various minerals whose infrared refractive indices might strongly vary from mineral to mineral.[44] In the case of dust samples, only the so-called effective or average optical constants can be deduced from (infrared) optical measurements when performing the spectra analysis as if the particles were homogeneous.[15] Clearly, such data sets have a limited range of applicability, given that each individual sample features a diverse mineralogical composition and aggregate structure. On the modeling part, different mixing rules are proposed to calculate the refractive indices of inhomogeneous, multicomponent particles such as mineral dust aggregates from the data sets of the individual components. In the Maxwell–Garnet approximation, the composite particle is treated as a homogeneous matrix with embedded inclusions, implying that a clear distinction between the inclusions and the host matrix can be made. On the contrary, the Bruggeman theory applies to a random inhomogeneous medium where the distinction between inclusion and host becomes unnecessary and both components can be treated symmetrically.[15,44]

1.2.2 Infrared Extinction Spectra of Wavelength-Sized Particles (Mie Regime)

1.2.2.1 Dependence of the Spectral Habitus on the Particle Size

In the Rayleigh limit, the cross-sections for extinction and scattering are obtained by imposing that the particles at each instant are exposed to an electromagnetic field that is uniform over their dimension. The scattered field is then described by the electric dipole radiation of an oscillating dipole. For particle sizes comparable to the wavelength of the infrared light, that is, in the framework of Mie theory for spheres, the scattered electromagnetic field is written as an infinite sum of normal modes of the spherical particles weighted by the scattering coefficients $a_n(\tilde{v}, D, N)$ and $b_n(\tilde{v}, D, N)$, yielding the following expression for the extinction cross-section:

$$C_{\text{ext}}(\tilde{v}, D, N) = \frac{1}{2\pi\tilde{v}^2} \sum_{n=1}^{\infty} (2n+1)\text{Re}[a_n(\tilde{v}, D, N) + b_n(\tilde{v}, D, N)]. \tag{1.11}$$

Guidelines for the computation of the scattering coefficients a_n and b_n, including the number of terms that are required to obtain convergence in the series of Equation 1.11, are given for example, in the textbook by Bohren and Huffman.[15]

In the following, we give a brief overview about extinction features in the framework of Mie theory, taking ice as an example. Based on these results for spherical ice particles, Section 1.2.2.2 addresses the influence of particle asphericity on the absorption and scattering cross-sections. Panel a of Figure 1.3 illustrates the evolution of the spectral habitus of the infrared extinction spectrum of ice spheres when going from submicron to wavelength-sized particles. The Mie calculations were done for a log-normal number size distribution with a common mode width of $\sigma_g = 1.5$ and count median diameters (*CMD*) ranging from 0.1 to 13 µm; the employed optical constants are shown in panel b. The lowermost extinction spectrum for the 0.1 µm-sized ice spheres is solely governed by the absorption contribution, with the most prominent absorption bands located at around 3250 cm^{-1} (O−H molecular stretching mode) and 800 cm^{-1} (intermolecular vibration). The increasing scattering contribution for larger particle sizes first manifests itself in slightly slanted baselines at non-absorbing wave numbers greater than 3600 cm^{-1}, without provoking in the first part a significant distortion of the spectral habitus of the extinction bands at 3250 and 800 cm^{-1} (spectra for *CMD* of

0.3 and 0.5 μm). For particle sizes between 1 and 4 μm, the prominent extinction bands gradually adopt the spectral shape of the anomalous dispersion feature that is inherent in the spectrum of the real part of the complex refractive index. Toward even larger sphere diameters, the infrared spectra are characterized by a quite constant optical depth over the entire wave number range, except for two pronounced extinction minima at around 3500 and 950 cm^{-1}. These minima are caused by the Christiansen effect and reflect the reduced scattering cross-sections of the ice spheres in these wave number regimes, because the value for the real part of the refractive index approaches unity, that is, corresponds to the value of the surrounding medium.[45–47] A detailed view of the Christiansen band at 950 cm^{-1} is shown in panel c of Figure 1.3. It becomes obvious that the wavelength position of minimal optical depth does not exactly correspond to the minimum of the n spectrum (panel d), because the minimized scattering contribution is counterbalanced by a large absorption contribution due to the high value for the imaginary index k. Instead, the extinction minimum is shifted to larger wave numbers where absorption by the ice spheres is reduced. For smaller particle diameters

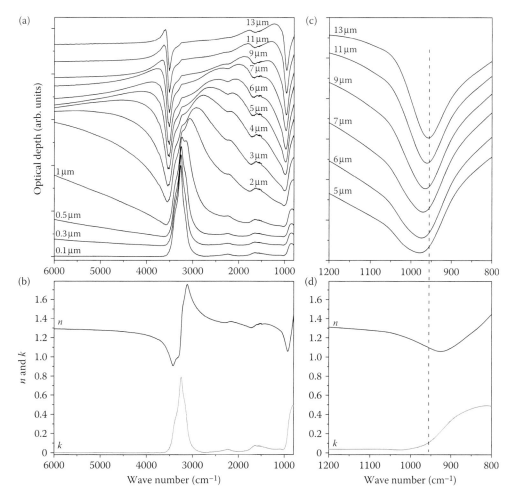

FIGURE 1.3   Panel a: Infrared extinction spectra of ice spheres computed with Mie theory for a log-normal number distribution of particle sizes with a mode width of $\sigma_g = 1.5$ and varying values for the count medium diameter, as indicated in the figure panel. All spectra are normalized to unity and are offset for clarity. The employed complex refractive indices for ice were taken from Zasetsky et al.[25] (data set for 210 K) and are shown in panel b. Panel c: Subset of the computed extinction spectra for the larger particle diameters in the wave number regime of the Christiansen minimum at 950 cm^{-1}. The corresponding part of the spectrum of the optical constants n and k is shown in panel d. See text for details.

with a reduced scattering contribution to overall extinction, the need for minimizing absorption gradually becomes more important, thereby explaining the continuous shift of the Christiansen minimum to higher wave numbers.[47]

For the interpretation of the results from Mie computations, it is convenient to plot the size dependency of the extinction, scattering, and absorption efficiencies that are obtained by dividing the corresponding cross-sections through the projected area of the sphere. In panel a of Figure 1.4, the extinction efficiencies Q_{ext} of ice spheres are shown for two selected wave numbers, (1), 4457 cm^{-1} (nonabsorbing wave number with $n = 1.26$ and $k < 10^{-3}$), and (2), 3251 cm^{-1} (center of the intense O—H stretching mode with $n = 1.22$ and $k = 0.78$). For the nonabsorbing wave number, the trace of the extinction efficiency bears two fine structures, referred to as interference and ripple structure.[15] The interference structure manifests itself in a series of broad and regularly spaced extinction

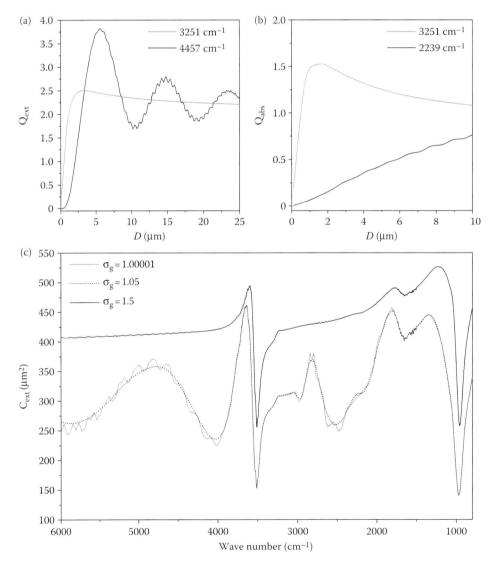

FIGURE 1.4 Mid-infrared extinction (panel a) and absorption (panel b) efficiencies for ice spheres as a function of the sphere diameter at different wave numbers. (c) Size-averaged extinction cross-sections for log-normally distributed ice spheres with a count medium diameter of 13 μm and different values for the mode width σ_g.

minima and maxima, which arise from the interference of incident and forward-scattered light. The oscillations gradually converge to a value of two, that is, the limiting value for Q_{ext} for particles much larger than the wavelength. According to the descriptive analysis by Bohren and Huffman,[15] the maxima and minima of the extinction cross-sections are determined by the term $\sin[x(N-1)]$, which appears in the numerators of approximated expressions for the scattering coefficients a_n and b_n. The first maximum of this function is reached for a particle diameter of

$$D = \frac{1}{2\tilde{v}(N-1)}. \tag{1.12}$$

The first interference peak is therefore shifted to smaller diameters as the wave number and the refractive index increase.

Superimposed on the broad-scale interference structure, the ripple structure is characterized by sharp and irregular peaks that are due to resonances in the Mie scattering functions a_n and b_n (Equation 1.11).[48] The interference and ripple structures are mapped onto the infrared extinction spectrum of an ensemble of large ice spheres at nonabsorbing wave numbers, provided that the mode width of the size distribution is extremely narrow. The gray trace in panel c of Figure 1.4 shows the size-averaged extinction cross-sections of almost mono-sized ice spheres with a count median diameter of 13 μm and a mode width of 1.00001. The oscillations due to the ripple and interference structure can be clearly seen in the spectral regimes characterized by low values for k, in particular between 6000 and 4000 cm^{-1} as well as 2800 and 2400 cm^{-1}. With only a slight increase in the mode width ($\sigma_g = 1.05$), the ripple structure completely disappears, leaving behind only the broad-scale extinction oscillations due to the interference structure. The latter is also flattened out when further increasing the mode width to $\sigma_g = 1.5$. As shown by the gray trace in Figure 1.4a, the ripple and interference structure also disappear when the value for the imaginary part of the complex refractive index, that is, particle absorption, is sufficiently large.

The size dependency of the absorption efficiency Q_{abs} is plotted in panel b of Figure 1.4 for a weak absorption band at 2239 cm^{-1} and for the intense O−H stretching mode at 3251 cm^{-1}. At 2239 cm^{-1}, Q_{abs} almost linearly increases with increasing sphere diameter over the considered particle size range, implying, in other words, that the absorption cross-section C_{abs} is proportional to the particle volume. For the strong absorption band at 3251 cm^{-1}, this linear relationship between Q_{abs} and D only holds for diameters below 1 μm. For larger sizes, Q_{abs} starts to level off, indicating that C_{abs} becomes proportional to the surface area of the particle instead of the particle volume: the photons are predominantly absorbed in the surface layer of the particle so that the interior becomes unimportant in the absorption process. The trace for the absorption efficiency also features a pronounced resonance region where Q_{abs} exceeds 1.0, provoked by the above-edge contribution of photons beyond the physical cross-section of the particle.[49]

1.2.2.2 Influence of Particle Shape

The morphologies of ice crystals growing in the atmosphere can vary from pristine hexagonal columns and plates to complex bullet rosettes and aggregates. Therefore, ice particles are a suitable candidate to analyze the influence of particle asphericity on the infrared absorption and scattering cross-sections.[50] In contrast to Section 1.2.1.2, we will now focus on the shape effect on the extinction spectra of micron-sized particles. Note that shape-induced spectral distortions can also be expected for the absorption cross-sections of Rayleigh-sized ice crystals in the regime of the intense O−H stretching mode at 3250 cm^{-1}.

The shape dependency of the extinction, scattering, and absorption cross-sections of micron-sized ice crystals at mid-infrared wavelengths is shown in Figure 1.5 for three selected equal-volume sphere diameters D_V of 3.1, 5.05, and 6.7 μm. The cross-sections were calculated with a combination of the T-matrix and the DDA approach, see Wagner et al.[51] for the computational details. The ice crystals were modeled as circular cylinders in random orientation with aspect ratios ϕ (ratio of the diameter

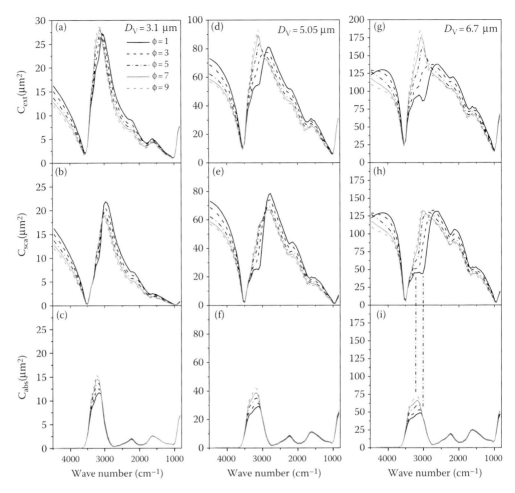

FIGURE 1.5 T-matrix and DDA computations of the mid-infrared extinction, scattering, and absorption cross-sections of cylindrical ice crystals in random orientation as a function of the aspect ratio φ for equal-volume sphere diameters D_V of 3.1 μm (panels a–c), 5.05 μm (panels d–f), and 6.7 μm (panels g–i). Note that the cross-section—axes in the three subpanels for a given particle size are identically scaled to underline the relative weight of the absorption and scattering contributions to overall extinction.

to the length of the cylinder) ranging from φ = 1 (compact cylinder) to φ = 9 (plate-like cylinder). Circular cylinders have proven to be an appropriate surrogate for pristine hexagonal ice crystals in extinction calculations at infrared wavelengths.[52] The cross-sections computed from Mie theory only slightly differ from those for φ = 1 and are therefore not shown. Figure 1.5 underlines that a high degree of particle asphericity (φ = 9) leads to a significant distortion of the spectral habitus of the extinction spectrum compared to compact ice crystals of φ = 1. The shape-induced spectral differences in extinction can be traced back to changes in both the scattering and absorption cross-sections, with the scattering contribution, however, as the dominant part responsible for the general habitus of the extinction spectrum. For wave-number regimes where light absorption is low (2800–1000 cm⁻¹) or even completely negligible (>3600 cm⁻¹), there is a continuous decrease in the extinction and scattering cross-section with increasing particle asphericity. Following the analysis from Zakharova and Mishchenko,[53,54] the scattering cross-sections of particles with extreme aspect ratios are more sensitive to the size parameter $x = \pi D \tilde{v}_j$ along the smallest particle dimension. The thickness of ice disks with, for example, D_V = 5.05 μm and φ = 9 only amounts to about 1 μm, thereby explaining their reduced scattering cross-sections compared to those for spheres and compact ice cylinders.

In contrast to the purely scattering regimes of the extinction spectrum, the wavelength range between 3600 and 2800 cm^{-1}, containing the intense O−H stretching absorption mode, displays the opposite shape-related trend in the extinction cross-sections, that is, increasing cross-sections with increasing degree of particle asphericity. Partly, this effect can be explained by the higher absorption cross-sections of the plate-like ice cylinders around 3250 cm^{-1}. For this intense absorption band, as already indicated in the discussion of Figure 1.4b, C_{abs} becomes proportional to the surface area of the particle. As the average cross-sectional area of ice cylinders with a given D_V strongly increases with increasing aspect ratio ϕ, the absorption spectra shown in panels c, f, and i reveal an increasing intensity around 3250 cm^{-1} with increasing particle asphericity. For the less intense absorption bands, the absorption cross-sections remain proportional to the particle volume over the considered size range and therefore, for a given D_V, no shape effect can be made out in the other regimes of the absorption spectrum. At a closer look, the absorption cross-sections around 3250 cm^{-1} are not exclusively a function of the particles' average cross-sectional area but also depend on their shape. Panel a of Figure 1.6 underlines that the absorption resonance with $Q_{abs} > 1$ is reduced for strongly aspherical particles and even completely disappears for $\phi = 9$. Note that this effect of

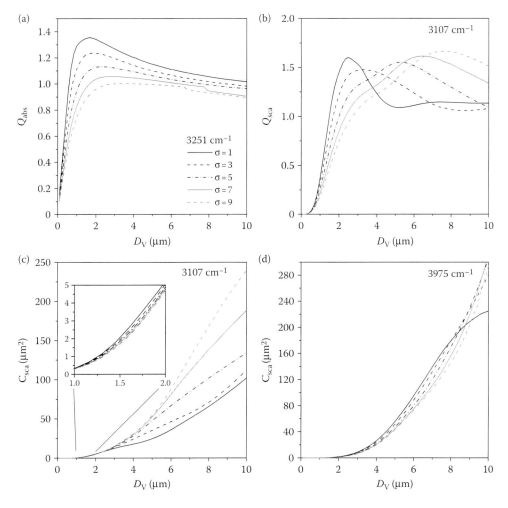

FIGURE 1.6 Shape-dependent trends in the absorption (panel a) and scattering (panel b) efficiencies as well as scattering cross-sections (panel c and d) for randomly oriented ice cylinders at selected wave numbers. Note that Q_{abs} and Q_{sca} were obtained by dividing the respective cross-sections through the average cross-sectional area of the particles, projected on a plane perpendicular to the incident beam.

reduced absorption efficiencies in the resonance region for, e.g., ice cylinders with $D_V = 3.1$ and $\phi = 9$ compared with $\phi = 1$ cannot compensate for the higher cross-sectional area of the more aspherical particles, meaning that the actual absorption cross-sections show an increase with increasing degree of particle asphericity (Figure 1.5c).

In addition to the increase in the absorption cross-sections between 3600 and 2800 cm^{-1}, also the scattering cross-sections are increasing for more aspherical particles in this particular wave-number regime, thereby inverting the trend from the nonabsorbing regions of the spectrum. As underlined by the two vertical lines between panels h and i of Figure 1.5, the maximum deviation between C_{sca} for $\phi = 1$ and $\phi = 9$ does not coincide with the center of the O—H stretching absorption mode. Instead, it is slightly shifted to lower wave numbers and thereby falls into the regime with maximum values (up to about 1.75) for the real part n of the complex refractive index for ice (see Figure 1.3b). Following from Equation 1.12, this high value for n shifts the position of the first interference maximum to smaller particle sizes compared to, for example, the wave number regime between 6000 and 4000 cm^{-1} with $n < 1.3$. As shown in panel b of Figure 1.6, the first interference maximum of compactly shaped ice cylinders of $\phi = 1$ at 3107 cm^{-1} ($n = 1.75$) is located at about $D_V = 2.5$ μm. Note that the interference structure is slightly damped, because the imaginary part of the complex refractive index is not equal to zero at this wave number. The first interference peak, however, is still clearly conserved. Up to this interference maximum, the scattering cross-sections reveal the "usual" trend, that is, decreasing values for C_{sca} with increasing particle asphericity, as shown by the insert in Figure 1.6c. For larger particles, the magnitude of the scattering cross-sections is influenced by the habitus of the oscillations due to the interference structure. As shown in Figure 1.6b, the regular series of scattering maxima and minima, which is typical for compact particle shapes is severely distorted for more aspherical ice cylinders. For $\phi = 1$, particle sizes from $D_V = 4$ to 7 μm fall into the regime of reduced scattering efficiencies after the first interference maximum has been surpassed. On the contrary, for $\phi = 7$ and $\phi = 9$, the scattering efficiencies have their maxima in just this size range, thereby explaining the opposite trend in the shape dependency of the scattering cross-sections for large ice crystals in the spectral regime characterized by high values for n. As a comparison, at a wave number of 3975 cm^{-1} with $n = 1.22$, the first interference peak is shifted to much larger particle sizes due to the lower value for n (Figure 1.6d). The size range from $D_V = 4–7$ μm is thereby unaffected by the shape-dependent distortions in the interference structure and features the common trend of diminishing scattering cross-sections with increasing ϕ.

The preceding discussion has revealed that the magnitude and the sign (increase or diminution) of the shape-dependent variations in the infrared extinction spectra of micron-sized particles crucially depend on the particle size and values for the optical constants. It is therefore indispensable to carry out a detailed shape analysis for each chemical species on its own. Shape-induced spectral distortions also clearly impede the accurate retrieval of the particle size distribution from infrared extinction measurements, as discussed in the following section.

1.2.2.3 Size Distribution Retrieval

As already indicated in Section 1.1, infrared extinction spectra of wavelength-sized particles can be inverted to provide the unknown particle size distribution in situations where direct particle sampling is not amenable (e.g., in remote-sensing applications). In matrix notation, Equation 1.1 can be written as

$$\boldsymbol{\tau} = \mathbf{An}, \tag{1.13}$$

where $\boldsymbol{\tau}$ is an $N \times 1$ vector whose component τ_j is the measured optical depth at the wave number $\tilde{v}_j$, $\mathbf{n}$ is an $M \times 1$ vector whose component n_i is the number concentration of particles in the size bin D_i, and $\mathbf{A}$ is an $N \times M$ matrix of the size-bin averaged extinction cross-sections at each wave-number position (scaled by $l/\ln 10$). The inversion problem thereby seems to be straightforward, because apparently only the matrix Equation 1.13 would have to be solved. As a major problem, however, the matrix **A** is often ill-conditioned, that is, some of the M equations in Equation 1.1 can be approximated as a linear combination of the other equations.[55] This in turn leads to a solution ambiguity in the size distribution retrieval, meaning that many different size distributions may satisfy the same optical data. To reduce the solution ambiguity, *a priori* constraints are imposed on the solution, for example, non-negativity and smoothness constraints.[27] The performance of a specific inversion procedure is then ideally tested in laboratory experiments where the particle size distribution can be independently measured, as for example, recently done by Zasetsky et al.[1] The authors showed that for spherical or near-spherical particles even complicated multimodal size distributions could be accurately deduced. This, however, makes high demands on the accuracy of the employed optical constants. As convincingly demonstrated by Liu et al.[27] with liquid water as an example, even slight perturbations of the complex refractive indices can provoke severe retrieval errors, particularly toward the small particle end of the size distribution.

The same problem is encountered in the size distribution retrieval for aspherical particles. When computing the extinction cross-sections inherent in the base matrix **A**, it is almost inevitable to employ an idealized, only approximate shape representation for the broad diversity of particle habits that can be observed for certain aerosol species such as ice crystals or mineral dust grains. The limited quality of the adopted shape representation directly provokes errors in the retrieved size distribution, yielding, for example, a multimodal size distribution with spurious modes of smaller and larger particles even if the true size distribution is unimodal.[27] A descriptive example of the serious solution ambiguity and the size distribution retrieval for ice crystals is discussed in Section 1.3.2. Particularly for aspherical particles, it might therefore be necessary to reinforce the *a priori* constraints on the desired size distribution, for example, by imposing a parametric form for the unknown size distribution (e.g., a uni- or bimodal log-normal size distribution).[56–58] The retrieval then reduces to find a unique solution for the limited parameter set of the distribution function by minimizing the summed-squared residuals between measured and calculated infrared extinction spectra.

1.3 EXAMPLES

In this section, we want to illustrate some of the theoretical considerations from Section 1.2 by suited examples of infrared measurements of cloud particles inside the aerosol and cloud chamber AIDA of the Karlsruhe Institute of Technology. The AIDA aerosol vessel (temperature range 333–183 K, pressure range 1100–0.01 hPa) can be used to perform controlled expansion cooling experiments which mimic the expansion cooling of rising air parcels in the atmosphere.[59] The formation of supercooled liquid water and/or ice crystal clouds is triggered when the relative humidity inside the chamber has exceeded a threshold value whose magnitude depends on the type of the preadded seed aerosol particles and the temperature. The generated cloud particles can be detected with a variety of techniques, including *in situ* infrared extinction measurements with an internal White-type multiple reflection cell, see Figure 1.7.[60,61]

1.3.1 TYPICAL INFRARED SPECTRAL HABITUS OF LARGE CLOUD PARTICLES

Underlining the explanations from Section 1.2.2.1, Figure 1.8 compares selected infrared extinction spectra of large cloud particles, recorded during three different expansion–cooling experiments. In experiment A ($T = 270$ K, seed aerosol: mineral dust), a cloud of supercooled liquid water droplets has formed. All nuclei have grown to droplets of roughly the same diameter, that is, the AIDA chamber has just operated as a giant condensation nuclei counter in this experiment. The narrow droplet size distribution leads to a pronounced Mie interference structure. As the oscillations in the interference structure are very sensitive to the droplet diameter, the size distribution of the water cloud can be retrieved with high accuracy. The fit parameters for the calculated spectrum are $CMD = 24.7$ μm and $\sigma_g = 1.05$. Trace B of Figure 1.8 represents a rare occasion where a distinct interference structure can also be made out in the infrared extinction spectrum of an ice cloud.

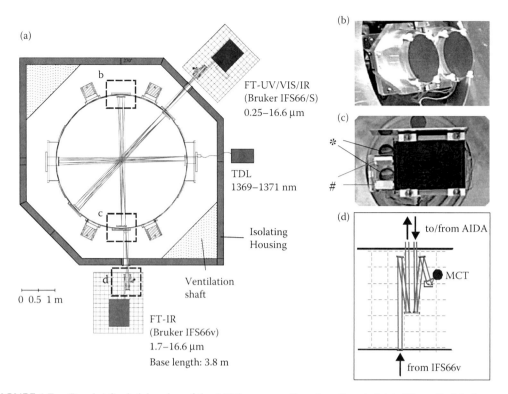

FIGURE 1.7 (Panel a) Scaled drawing of the AIDA cross-section at medium height of the cylindrical aerosol chamber.[60] The FTIR spectra are recorded *in situ* using an internal White-type multiple reflection cell with a base length of 3.8 m. The field mirror (panel c) and its opposite double mirrors (panel b) are mounted on the inside of two flanges which are enclosed in the chamber walls. All mirrors are gold-coated and underlain with heating foils to prevent them from icing. The infrared beam exciting the FTIR spectrometer (type IFS66v, Bruker) is directed into the White cell via two focusing mirrors (panel d), passing the chamber walls through a flange equipped with heated BaF$_2$ cell windows (marked by * in panel c). In addition to the basic White-type arrangement, a pair of diagonal mirrors (marked by # in panel c) can be used to double the number of cell transversals by displacing the beam vertically and redirecting it to the double mirrors for another set of traversals. The beam exciting the White cell is focused onto a liquid-nitrogen cooled MCT detector (panel d). As shown in panel a, two additional multiple reflection cells are used for *in situ* water vapor measurements by tunable diode laser (TDL) absorption spectroscopy and for extinction spectroscopy at visible wavelengths.

In this particular experiment, the homogeneous freezing of supercooled sulfuric acid solution droplets was studied at $T = 211$ K. The measured spectrum can only be accurately fitted by a narrow size distribution of compactly shaped particles (ice cylinders with $\phi = 1$), yielding $CMD_V = 12.7$ μm and $\sigma_g = 1.12$ (see Figure 12 in Wagner et al.[60] for details). The most commonly observed spectral habitus for clouds of large aspherical ice crystals (trace C, $T = 244$ K, seed aerosol: mineral dust), however, reveals a flattened extinction profile, only interrupted by the two pronounced Christiansen minima. Here, a set of different size–shape distributions of ice crystals can reproduce the spectral signature, leading to ambiguous retrieval results. For compact particle shapes, the interference structure can only be leveled out by imposing a broad size distribution. The gray trace for experiment C shows the best-fitted spectrum for $\phi = 1$, yielding $CMD_V = 7.6$ μm and $\sigma_g = 2.0$. Equally good-fit results, however, could be obtained for various other scenarios that also would provoke a flattening of the interference structure, including, (1), an ice cloud composed of a broad variety of differently shaped crystals or increasingly aspherical particles, even when prescribing a narrow size distribution, and (2), an appropriately chosen bi- or multimodal size distribution. In such a case, the broad diversity of potential solutions to the retrieval problem necessitates independent reference

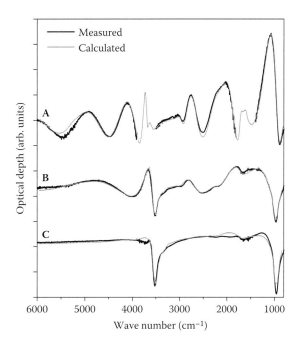

FIGURE 1.8 Comparison between measured and best-fitted infrared extinction spectra of cloud particles generated by three different expansion cooling experiments in the AIDA chamber. Trace A: Cloud of super-cooled water droplets at $T = 270$ K. Traces B and C: Ice crystal clouds at $T = 211$ and 244 K, respectively. For the calculated spectra, a log-normal number distribution of particle sizes was prescribed. The low-temperature refractive index data sets for supercooled water (trace A) and ice (traces B and C) from Zasetsky et al.[25] were used as input in the computations. See text for details.

measurements of either the number size distribution or the particle shape to corroborate the infrared retrieval results. A quantitative example of the solution ambiguity is given in Section 1.3.2.

1.3.2 SOLUTION AMBIGUITY OF THE SIZE DISTRIBUTION RETRIEVAL FOR ASPHERICAL ICE PARTICLES

Panels a and b of Figure 1.9 show two sets of infrared extinction spectra of ice crystal clouds that were monitored at time steps of 20 s during two different expansion cooling experiments in AIDA, see Wagner et al.[51] for a detailed discussion. In experiment A ($T = 210$ K, seed aerosol: secondary organic aerosol particles), the evolution of the spectral habitus of the growing ice crystals, featuring a continuous shift of the extinction maximum in the O–H stretching regime toward lower wave numbers, can be accurately mimicked by compact particle shapes ($\phi = 1$, panel c). By contrast, the spectra series from experiment B ($T = 210$ K, seed aerosol: illite clay mineral with organic coating) bears a distorted spectral habitus, which can be approximated by the T-matrix computations for plate-like ice cylinders with $\phi = 8$ (panel d). With an appropriate distortion of the true number size distribution, however, the infrared spectra from series B can also be accurately reproduced by compactly shaped ice crystals. This is demonstrated in Figure 1.10a, showing a selected measured spectrum from experiment B together with the best-fitted spectrum for ice cylinders of $\phi = 1$ when prescribing a log-normal size distribution. The fit yields a broad size distribution with a median equal-volume sphere diameter of about 1 μm and an overall ice particle number concentration of about 200 cm^{-3}. The latter is one order of magnitude larger than the actual number concentration that was independently measured with an optical particle counter. When forcing the ice particle number concentration to be close to the measured value (20 cm^{-3}) and only allowing the mode width and median diameter of the size distribution to be optimized in the retrieval (panel b of Figure 1.10),

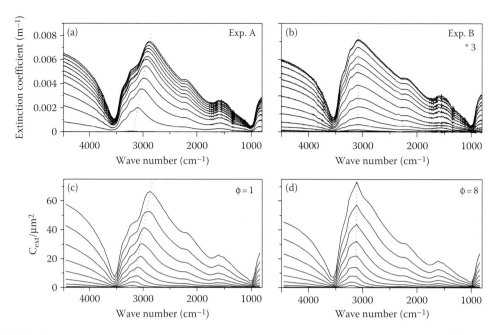

FIGURE 1.9   (Panels a and b) Series of infrared extinction spectra (extinction coefficient $= (\tau(\tilde{\nu}) \cdot \ln 10)/l)$, recorded at time steps of 20 s during two different AIDA expansion cooling experiments at $T = 210$ K with secondary organic aerosol particles (experiment A) as well as coated illite clay mineral particles (experiment B) as seed aerosol. With respect to their temporal order, the spectra are to be read from bottom to top. (Panels c and d) Computed extinction cross-sections of circular ice cylinders in random orientation of aspect ratios $\phi = 1$ and 8 for equal-volume sphere diameters ranging from 1 μm (bottom spectra) to 4.6 μm (top spectra) in 0.45 μm size steps.

the fit for $\phi = 1$, now yielding a narrow distribution of larger ice crystals, clearly fails to reproduce the measured spectral habitus. Only for the fit with $\phi = 8$ (panel c), both the computed spectral habitus and the deduced number concentration of ice crystals are simultaneously in satisfactory agreement with the measurements. This analysis underlines that the shape-induced distortions in the infrared extinction cross-sections between compact and plate-like ice crystals (see also Figure 1.5) can be efficiently balanced by deforming the true size distribution, meaning that for example, a narrow size distribution of about 5–7 μm sized plate-like ice crystals and a broad size distribution of compactly shaped ice crystals with a much smaller median diameter yield very similar infrared extinction spectra. As shown in Figure 16 of Wagner et al.[51] the infrared extinction cross-sections of a unimodal distribution of ice crystal plates can also be approximated very well by a bimodal size distribution of compactly shaped particles, with one mode of larger and one mode of smaller ice crystals compared to the median size of the unimodal distribution. Therefore, independent measurements are needed for a correct interpretation of the infrared retrieval results. Concerning experiment B from Figure 1.9 (panels b and d), the assumption that strongly aspherical ice crystals have formed is also backed up by supplementary *in situ* laser light scattering measurements in AIDA. For this particular ice cloud, a very low backscattering linear depolarization ratio was detected which is typical for needle- or plate-like particles.[53,54]

1.4   CONCLUDING REMARKS

Infrared spectroscopy can be a powerful tool to analyze not only the composition and size distribution of aerosol particles, but also to gain shape information of micron-sized particles. In order to tap the full potential of aerosol particle analysis by infrared extinction spectroscopy, however, a careful

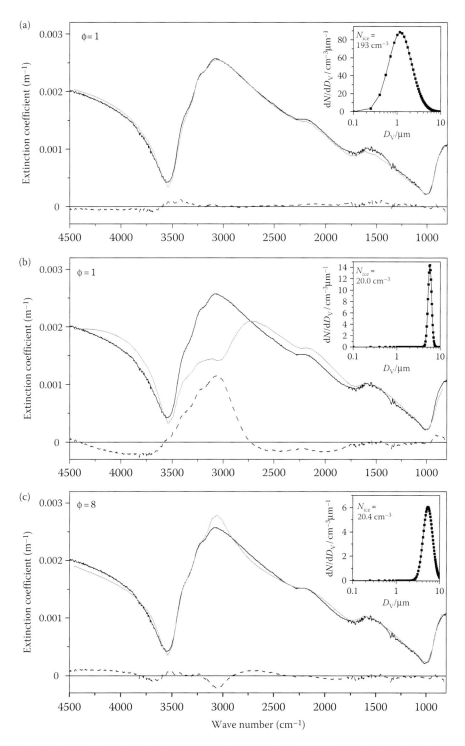

FIGURE 1.10 Selected infrared extinction spectrum from experiment B in Figure 1.9 (black line) in comparison with the best-fit results for cylindrical ice crystals of various shapes (gray lines). The spectral residuals between measurement and computation are shown as dashed black lines. The retrieved ice crystal number size distribution (forced to be of log-normal type) as well as the overall ice particle number concentration (N_{ice}) are displayed in the top right corner of panels a, b, and c. For the fits shown in panels a and c, N_{ice}, σ_g, and CMD_V were independently optimized; fit example b, N_{ice} was prescribed to 20 cm^{-3} and only σ_g and CMD_V were allowed to vary.

examination of the size- and shape-related changes on the spectral habitus is indispensable. This has to be carried out specifically for each substance, as the size- and shape-dependency of any infrared spectrum crucially depends on the magnitude of the real and imaginary parts of the refractive index. Therefore, it is not advisable to transfer the exemplary computational results for Rayleigh-sized ammonium sulfate particles (Section 1.2.1.2) and micron-sized ice crystals (Section 1.2.2.2) to different chemical species. When aspherical particle habits are involved, only favorable conditions (see, e.g., Figure 1.8, trace B) allow a unique determination of the particle number concentration, size, and shape from infrared extinction spectra without the need of confining one or more retrieval parameters by auxiliary measurements. To illustrate this, an example of a severe size–shape ambiguity in the spectral analysis was presented (Figure 1.10) where an independent reference measurement of the aerosol number concentration was required to constrain the infrared retrieval results.

REFERENCES

1. Zasetsky, A. Y., Earle, M. E., Cosic, B. et al. 2007. Retrieval of aerosol physical and chemical properties from mid-infrared extinction spectra. *J. Quant. Spectrosc. Radiat. Transfer* 107:294–305.
2. Zasetsky, A. Y., Khalizov, A. F., and Sloan, J. J. 2004. Characterization of atmospheric aerosols from infrared measurements: Simulations, testing, and applications. *Appl. Opt.* 43:5503–11.
3. Sax, M., Zenobi, R., Baltensperger, U., and Kalberer, M. 2005. Time resolved infrared spectroscopic analysis of aerosol formed by photo-oxidation of 1,3,5-trimethylbenzene and alpha-pinene. *Aerosol. Sci. Tech.* 39:822–30.
4. Höpfner, M., Luo, B. P., Massoli, P. et al. 2006. Spectroscopic evidence for NAT, STS, and ice in MIPAS infrared limb emission measurements of polar stratospheric clouds. *Atmos. Chem. Phys.* 6:1201–19.
5. Cziczo, D. J. and Abbatt, J. P. D. 1999. Deliquescence, efflorescence, and supercooling of ammonium sulfate aerosols at low temperature: Implications for cirrus cloud formation and aerosol phase in the atmosphere. *J. Geophys. Res. (Atmos.)* 104:13781–90.
6. Cziczo, D. J., Nowak, J. B., Hu, J. H., and Abbatt, J. P. D. 1997. Infrared spectroscopy of model tropospheric aerosols as a function of relative humidity: Observation of deliquescence and crystallization. *J. Geophys. Res. (Atmos.)* 102:18843–50.
7. Onasch, T. B., Siefert, R. L., Brooks, S. D. et al. 1999. Infrared spectroscopic study of the deliquescence and efflorescence of ammonium sulfate aerosol as a function of temperature. *J. Geophys. Res. (Atmos.)* 104:21317–26.
8. Schlenker, J. C. and Martin, S. T. 2005. Crystallization pathways of sulfate–nitrate–ammonium aerosol particles. *J. Phys. Chem. A* 109:9980–5.
9. Najera, J. J. and Horn, A. B. 2009. Infrared spectroscopic study of the effect of oleic acid on the deliquescence behaviour of ammonium sulfate aerosol particles. *Phys. Chem. Chem. Phys.* 11:483–94.
10. Clapp, M. L., Niedziela, R. F., Richwine, L. J., Dransfield, T., Miller, R. E., and Worsnop, D. R. 1997. Infrared spectroscopy of sulfuric acid water aerosols: Freezing characteristics. *J. Geophys. Res. (Atmos.)* 102:8899–907.
11. Prenni, A. J., Wise, M. E., Brooks, S. D., and Tolbert, M. A. 2001. Ice nucleation in sulfuric acid and ammonium sulfate particles. *J. Geophys. Res. (Atmos.)* 106:3037–44.
12. Bertram, A. K., Patterson, D. D., and Sloan, J. J. 1996. Mechanisms and temperatures for the freezing of sulfuric acid aerosols measured by FTIR extinction spectroscopy. *J. Phys. Chem.* 100:2376–83.
13. Chelf, J. H. and Martin, S. T. 2001. Homogeneous ice nucleation in aqueous ammonium sulfate aerosol particles. *J. Geophys. Res. (Atmos.)* 106:1215–26.
14. Hung, H. M. and Martin, S. T. 2002. Infrared spectroscopic evidence for the ice formation mechanisms active in aerosol flow tubes. *Appl. Spectrosc.* 56:1067–81.
15. Bohren, C. F. and Huffman, D. R. 1983. *Absorption and Scattering of Light by Small Particles*. New York: John Wiley & Sons, Inc.
16. Sigurbjörnsson, O. F., Firanescu, G., and Signorell, R. 2009. Intrinsic particle properties from vibrational spectra of aerosols. *Annu. Rev. Phys. Chem.* 60:127–46.
17. Weritz, F., Simon, A., and Leisner, T. 2002. Infrared microspectroscopy on single levitated droplets. *Environ. Sci. Pollut. Res.* 92–9.
18. Mishchenko, M. I. and Travis, L. D. 1998. Capabilities and limitations of a current FORTRAN implementation of the T-matrix method for randomly oriented, rotationally symmetric scatterers. *J. Quant. Spectrosc. Radiat. Transfer* 60:309–24.

19. Draine, B. T. and Flatau, P. J. 1994. Discrete-dipole approximation for scattering calculations. *J. Opt. Soc. Am. A* 11:1491–9.

20. Rothman, L. S., Gordon, I. E., Barbe, A. et al. 2009. The HITRAN 2008 molecular spectroscopic database. *J. Quant. Spectrosc. Radiat. Transfer* 110:533–72.

21. Niedziela, R. F., Norman, M. L., DeForest, C. L., Miller, R. E., and Worsnop, D. R. 1999. A temperature- and composition-dependent study of H_2SO_4 aerosol optical constants using Fourier transform and tunable diode laser infrared spectroscopy. *J. Phys. Chem. A* 103:8030–40.

22. Norman, M. L., Qian, J., Miller, R. E., and Worsnop, D. R. 1999. Infrared complex refractive indices of supercooled liquid HNO_3/H_2O aerosols. *J. Geophys. Res. (Atmos.)* 104:30571–84.

23. Biermann, U. M., Luo, B. P., and Peter, T. 2000. Absorption spectra and optical constants of binary and ternary solutions of H_2SO_4, HNO_3, and H_2O in the mid infrared at atmospheric temperatures. *J. Phys. Chem. A* 104:783–93.

24. Myhre, C. E. L., Grothe, H., Gola, A. A., and Nielsen, C. J. 2005. Optical constants of HNO_3/H_2O and $H_2SO_4/HNO_3/H_2O$ at low temperatures in the infrared region. *J. Phys. Chem. A* 109:7166–71.

25. Zasetsky, A. Y., Khalizov, A. F., Earle, M. E., and Sloan, J. J. 2005. Frequency dependent complex refractive indices of supercooled liquid water and ice determined from aerosol extinction spectra. *J. Phys. Chem. A* 109:2760–4.

26. Wagner, R., Benz, S., Möhler, O., Saathoff, H., Schnaiter, M., and Schurath, U. 2005. Mid-infrared extinction spectra and optical constants of supercooled water droplets. *J. Phys. Chem. A* 109:7099–112.

27. Liu, Y. G., Arnott, W. P., and Hallett, J. 1999. Particle size distribution retrieval from multispectral optical depth: Influences of particle nonsphericity and refractive index. *J. Geophys. Res. (Atmos.)* 104:31753–62.

28. Palmer, K. F. and Williams, D. 1975. Optical constants of sulfuric acid: Application to the clouds of Venus? *Appl. Opt.* 14:208–19.

29. Earle, M. E., Pancescu, R. G., Cosic, B., Zasetsky, A. Y., and Sloan, J. J. 2006. Temperature-dependent complex indices of refraction for crystalline $(NH_4)_2SO_4$. *J. Phys. Chem. A* 110:13022–8.

30. Bonnamy, A., Jetzki, M., and Signorell, R. 2003. Optical properties of molecular ice particles from a microscopic model. *Chem. Phys. Lett.* 382:547–52.

31. Hudson, P. K., Gibson, E. R., Young, M. A., Kleiber, P. D., and Grassian, V. H. 2008. Coupled infrared extinction and size distribution measurements for several clay components of mineral dust aerosol. *J. Geophys. Res. (Atmos.)* 113:doi:10.1029/2007JD008791.

32. Wagner, R., Möhler, O., Saathoff, H., Stetzer, O., and Schurath, U. 2005. Infrared spectrum of nitric acid dihydrate—influence of particle shape. *J. Phys. Chem. A* 109:2572–81.

33. Clapp, M. L. and Miller, R. E. 1993. Shape effects in the infrared-spectrum of ammonia aerosols. *Icarus* 105:529–36.

34. Thomas, G. E., Bass, S. F., Grainger, R. G., and Lambert, A. 2005. Retrieval of aerosol refractive index from extinction spectra with a damped harmonic-oscillator band model. *Appl. Opt.* 44:1332–41.

35. Segal-Rosenheimer, M. and Linker, R. 2009. Impact of the non-measured infrared spectral range of the imaginary refractive index on the derivation of the real refractive index using the Kramers–Kronig transform. *J. Quant. Spectrosc. Radiat. Transfer* 110:1147–61.

36. Ahrenkiel, R. K. 1971. Modified Kramers–Kronig analysis of optical spectra. *J. Opt. Soc. Am.* 61:1651–5.

37. Milham, M. E., Frickel, R. H., Embury, J. F., and Anderson, D. H. 1981. Determination of optical constants from extinction measurements. *J. Opt. Soc. Am.* 71:1099–106.

38. Yamamoto, K. and Masui, A. 1995. Complex refractive-index determination of bulk materials from infrared reflection spectra. *Appl. Spectrosc.* 49:639–44.

39. Boer, G. J., Sokolik, I. N., and Martin, S. T. 2007. Infrared optical constants of aqueous sulfate–nitrate–ammonium multi-component tropospheric aerosols from attenuated total reflectance measurements—Part I: Results and analysis of spectral absorbing features. *J. Quant. Spectrosc. Radiat. Transfer* 108:17–38.

40. Rouleau, F. and Martin, P. G. 1991. Shape and clustering effects on the optical-properties of amorphous-carbon. *Astrophys. J.* 377:526–40.

41. Segal-Rosenheimer, M., Dubowski, Y., and Linker, R. 2009. Extraction of optical constants from mid-IR spectra of small aerosol particles. *J. Quant. Spectrosc. Radiat. Transfer* 110:415–26.

42. Clapp, M. L., Miller, R. E., and Worsnop, D. R. 1995. Frequency-dependent optical constants of water ice obtained directly from aerosol extinction spectra. *J. Phys. Chem.* 99:6317–26.

43. Dartois, E. and Bauerecker, S. 2008. Infrared analysis of CO ice particles in the aerosol phase. *J. Chem. Phys.* 128:154715.

44. Sokolik, I. N. and Toon, O. B. 1999. Incorporation of mineralogical composition into models of the radiative properties of mineral aerosol from UV to IR wavelengths. *J. Geophys. Res. (Atmos.)* 104:9423–44.

45. Carlon, H. R. 1980. Aerosol spectrometry in the infrared. *Appl. Opt.* 19:2210–8.

46. Carlon, H. R. 1979. Christiansen effect in IR-spectra of soil-derived atmospheric dusts. *Appl. Opt.* 18:3610–4.

47. Arnott, W. P., Dong, Y. Y., and Hallett, J. 1995. Extinction efficiency in the infrared (2–18 μm) of laboratory ice clouds–observations of scattering minima in the Christiansen bands of ice. *Appl. Opt.* 34:541–51.

48. Chýlek, P. 1976. Partial-wave resonances and ripple structure in mie normalized extinction cross-section. *J. Opt. Soc. Am.* 66:285–7.

49. Baran, A. J., Francis, P. N., Havemann, S., and Yang, P. 2001. A study of the absorption and extinction properties of hexagonal ice columns and plates in random and preferred orientation, using exact T-matrix theory and aircraft observations of cirrus. *J. Quant. Spectrosc. Radiat. Transfer* 70:505–18.

50. Baran, A. J. 2004. On the scattering and absorption properties of cirrus cloud. *J. Quant. Spectrosc. Radiat. Transfer* 89:17–36.

51. Wagner, R., Benz, S., Möhler, O., Saathoff, H., Schnaiter, M., and Leisner, T. 2007. Influence of particle aspect ratio on the midinfrared extinction spectra of wavelength-sized ice crystals. *J. Phys. Chem. A* 111:13003–22.

52. Lee, Y. K., Yang, P., Mishchenko, M. I. et al. 2003. Use of circular cylinders as surrogates for hexagonal pristine ice crystals in scattering calculations at infrared wavelengths. *Appl. Opt.* 42:2653–64.

53. Zakharova, N. T. and Mishchenko, M. I. 2000. Scattering properties of needlelike and platelike ice spheroids with moderate size parameters. *Appl. Opt.* 39:5052–7.

54. Zakharova, N. T. and Mishchenko, M. I. 2001. Scattering by randomly oriented thin ice disks with moderate equivalent-sphere size parameters. *J. Quant. Spectrosc. Radiat. Transfer* 70:465–71.

55. Kandlikar, M. and Ramachandran, G. 1999. Inverse methods for analysing aerosol spectrometer measurements: A critical review. *J. Aerosol Sci.* 30:413–37.

56. Gonda, I. 1984. On inversion of aerosol size distribution data. *Aerosol. Sci. Tech.* 3:345–6.

57. Dellago, C. and Horvath, H. 1993. On the accuracy of the size distribution information obtained from light extinction and scattering measurements. 1. Basic considerations and models. *J. Aerosol Sci.* 24:129–41.

58. Walters, P. T. 1980. Pratical applications of inverting spectral turbidity data to provide aerosol size distributions. *Appl. Opt.* 19:2353–65.

59. Möhler, O., Büttner, S., Linke, C. et al. 2005. Effect of sulphuric acid coating on heterogeneous ice nucleation by soot aerosol particles. *J. Geophys. Res. (Atmos.)* 110:doi:10.1029/2004JD005169.

60. Wagner, R., Benz, S., Möhler, O., Saathoff, H., and Schurath, U. 2006. Probing ice clouds by broadband mid-infrared extinction spectroscopy: Case studies from ice nucleation experiments in the AIDA aerosol and cloud chamber. *Atmos. Chem. Phys.* 6:4775–800.

61. Wagner, R., Linke, C., Naumann, K. H. et al. 2009. A review of optical measurements at the aerosol and cloud chamber AIDA. *J. Quant. Spectrosc. Radiat. Transfer* 110:930–49.

2 Vibrational Excitons
A Molecular Model to Analyze Infrared Spectra of Aerosols

George Firanescu, Thomas C. Preston, Chia C. Wang, and Ruth Signorell

CONTENTS

2.1 Introduction 25
2.2 Theory... 27
    2.2.1 The Vibrational Exciton Model 27
    2.2.2 Local Environment Effects on Transitions Dipoles and Transitions Frequencies...... 31
    2.2.3 Numerical Implementation 32
    2.2.4 Correlating Spectral and Structural Properties—Excitation Density 34
2.3 Description of the Experimental Setup 36
2.4 Some Examples.. 37
    2.4.1 The Influence of Particle Size and Internal Structure on IR Spectra......... 37
    2.4.2 The Influence of the Particle Shape on IR Spectra.................................... 40
    2.4.3 Spectral Features of Multicomponent Particles.. 40
2.5 Conclusion 46
Acknowledgments... 46
References.. 46

2.1 INTRODUCTION

The interest in molecular aerosols—large molecular aggregates with sizes ranging from subnanometers to many microns—has increased significantly over the past few years because of their critical role in atmospheric and industrial processes, in air pollution where they pose health risks, and in astrophysical and astrochemical processes. To improve our understanding of the impact of aerosols in these various fields, it is crucial to study their physical (size, shape, architecture, phase behavior), and chemical (composition, reactivity) properties, and their formation processes (aggregation, chemical reaction, agglomeration, and coagulation).

Infrared (IR) spectroscopy is one of the most important methods for the characterization of molecularly structured matter—including molecular aerosols—in laboratory studies, in field measurements, for remote sensing, and as part of space missions. Since the intramolecular and the intermolecular vibrational dynamics are governed by the particles' properties, IR spectra contain a wealth of information about these very properties. Figure 2.1 illustrates this for various carbon dioxide ice aerosol particles in the region of the antisymmetric stretching vibration of carbon dioxide around 2350 cm^{-1}. It is striking how diverse the appearance of a single vibrational band can be for particles of varying size (traces a and b), shape (traces c and d), composition (trace e), and architecture (trace f).

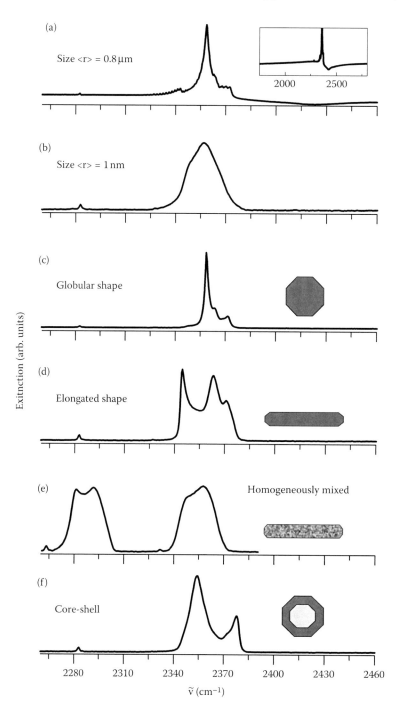

FIGURE 2.1 Experimental IR extinction spectra of $^{12}CO_2$ aerosol particles in the region of the antisymmetric stretching vibration. The spectra are determined by properties. Size: (a) Spectra of large particles (mean radius 0.8 μm) are characterized by elastic scattering features (slanted baseline and dispersion shape of absorption bands as highlighted in the inset). (b) Spectra of small particles are dominated by direct and indirect size effects (see Section 2.4.1). Shape: (c) Globular and (d) elongated crystalline particles have different spectra because of the different shapes. Composition: (e) Homogenously mixed $^{12}CO_2$–$^{13}CO_2$ particles are characterized by broad unstructured IR bands. Architecture: (f) A $^{12}CO_2$ shell grown on an SF_6 core causes a characteristic double band.

Spectral differences could hardly be more obvious than those shown in Figure 2.1, but it remains to be explained how information on particle properties is gained from the experimentally observed band structures. This step turns out to be the major challenge and is often the limiting factor in using IR spectra for particle characterization. On a molecular level aerosol particles are many-dimensional highly complex systems that are extremely challenging to model. Comparatively simple continuum approaches in combination with classical scattering theory are thus commonly used in a first attempt [1–3] with Mie theory as probably the most popular variant [1,2,4]. All these classical approaches solve Maxwell's equations for the particles, assuming their optical data which are required as input, are known beforehand. Such calculations may be able to provide limited information about the particle size and sometimes also about the particle shape. Any more sophisticated analysis of spectra such as those shown in Figure 2.1 is beyond the scope of classical models (see [5]) and requires modeling on a molecular level instead.

As an example of such a microscopic approach, this chapter summarizes the theory behind the so-called vibrational exciton model [6–13]. The dominating interaction of this molecular model is dipole–dipole coupling between all molecules in an aerosol particle. As further outlined in the subsequent section, it is mainly this dipole interaction which determines the band structure of IR bands for systems with strong molecular transition dipoles, such as the antisymmetric stretching band of carbon dioxide shown in Figure 2.1. Section 2.4 illustrates through various examples how this molecular model can be used to uncover information on particle properties hidden in IR spectra. The examples we have chosen focus on weakly bound molecular aerosols with relevance to planetary atmospheres.

2.2 THEORY

2.2.1 THE VIBRATIONAL EXCITON MODEL

The vibrational exciton model is a quantum mechanical model which allows the prediction of IR spectra of large molecular aggregates. The model was initially used in the pioneering work of Fox and Hexter [6] to describe the vibrational dynamics of molecular bulk crystals. A comprehensive discussion on this subject can be found in *Molecular Vibrations in Crystals* by Decius and Hexter [14]. The first attempt to apply the vibrational exciton model to finite systems was made by Cardini et al. [7], who tried to reproduce the experimental IR spectra of CO_2 clusters measured by Barnes and Gough [15]. Without definite information on the experimental conditions to allow a thorough comparison—particle size and temperature were only estimated in Ref. [15]—the suitability of the model in describing molecular aggregates remained uncertain. Later attempts to use the exciton model made by Ewing and coworkers [8–10] were also only partially successful owing to computational limitations regarding particle size. The best agreement with experimental results was obtained for thin molecular slabs, where the effectively reduced dimensionality of the problem leads to a faster convergence of the exciton spectra as a function of system size, in this case for 450–1800 molecules [10]. The importance of exciton coupling for IR spectra of large molecular aggregates, however, remained unclear.

In recent years, our research group has been able to demonstrate unambiguously that resonant transition dipole coupling (exciton coupling) is indeed the dominant interaction that controls whether or not IR bands are determined by intrinsic particle properties, such as shape, size, or architecture [5,11–13,16–22]. Furthermore, our combination of experiment with modeling also allows us to apply the exciton model to the treatment of large aerosol particles. As mentioned, the exciton model describes the vibrational bands of aerosols in terms of resonant transition dipole coupling. The sketch in Figure 2.2 emphasizes the differences between systems without dipole coupling (left) and systems where this interaction is dominant (right). On the left-hand side, no coupling occurs between the vibrating molecules (symbolized by the short strong springs) in an aerosol particle. As a consequence, their vibrational eigenstates remain degenerate and the vibrational band of the particle remains structureless. By contrast, strong dipole coupling (right; symbolized by the long soft springs)

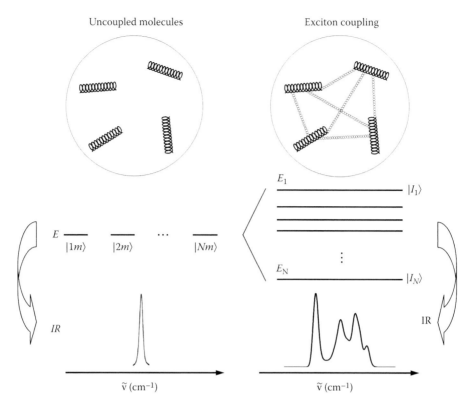

FIGURE 2.2   Schematic representation of the exciton coupling. Left: The hypothetical case without coupling between the different molecules (shown for four molecules represented by the four springs) in aerosol particles. The energies (E) of the vibrationally excited states of the four molecules are degenerate. Accordingly, they all show the same IR transition and thus a single band in the IR spectrum. Right: If a vibrational mode is strong (transition dipole >0.1 D) dipole–dipole coupling (represented by the small springs between the molecules) lifts the degeneracy of the molecular vibrational states and leads to concerted vibrations that extend over the entire particle acting as a sensor for its properties. There are now many vibrational transitions with different energies, which leads to the widths and band structure observed in the IR spectrum.

lifts the degeneracy of the molecular vibrational eigenstates producing vibrational modes that are delocalized over the entire particle, which are called vibrational excitons. Their delocalized nature makes exciton coupling an excellent probe for intrinsic particle properties. As a consequence of the coupling, the IR bands become structured. The aerosol particles in Figure 2.2 are oversimplified for clarity, composed of only four molecules, each with a single vibrational mode represented by short strong springs and the couplings represented by long soft springs (right). Resonant transition dipole coupling, however, extends not only to neighboring, but to all molecules in an aerosol particle. Restricting dipole coupling to nearest neighbors produces completely misleading results. This is a crucial point that is often misunderstood. It obviously makes the theoretical treatment challenging in particular for large aerosol particles. Research on various model systems in our group [5,11–13,16–22] further revealed that for vibrations with molecular transition dipoles larger than about 0.1 D, resonant transition dipole coupling dominates the appearance of the corresponding IR bands. These vibrations are therefore ideally suited to study intrinsic particle properties. Although dipole–dipole interactions are clearly dominant for such vibrational transitions, induced dipole effects can be significant for molecules with high polarizability, as indicated by results obtained for the SF_6 dimer [23,24] where induced dipole contributions account for approximately 10% of the frequency splitting in the IR spectrum of the ν_3 vibrational mode. Therefore, we have also included dipole-induced dipole coupling in the exciton model. Restricted to dipole–dipole coupling ($\hat{H}_{DD}$) and dipole-induced dipole interactions ($\hat{H}_{DID}$) (the 1st and 2nd term, respectively, in A_{ij} of Equation 2.1), the vibrational Hamiltonian is greatly simplified, so that the calculation of IR spectra for particles consisting of tens of thousands of molecules become tractable.

$$\hat{H} = \hat{H}_0 + \hat{H}_{DD} + \hat{H}_{DID}$$

$$= \hat{H}_0 - \frac{1}{2}\sum_{i,j}\mu_i^+\lambda_{ij}\mu_j - \frac{1}{2}\sum_k\left(\alpha_k\sum_i\lambda_{ki}\mu_i\right)^+\left(\sum_j\lambda_{kj}\mu_j\right)$$

$$= \hat{H}_0 + \sum_{i,j}\mu_i^+ A_{ij}\mu_j \tag{2.1}$$

with

$$A_{ij} = -\frac{1}{2}\left(\lambda_{ij} + \sum_k\lambda_{ik}\alpha_k\lambda_{jk}\right).$$

$\hat{H}_0$ is the vibrational Hamiltonian of the uncoupled molecules. The sum extends over pairs of molecules, μ_i is the dipole moment vector and α_i the molecular polarizability tensor of molecule i ($^+$denotes the transpose) and λ_{ij} is a scaled projection matrix:

$$\lambda_{ij} = \frac{(1-\delta_{ij})}{4\pi\varepsilon_0 R_{ij}^3}\left(3e_{ij}e_{ij}^+ - 1\right), \tag{2.2}$$

where δ_{ij} is Kronecker's delta, R_{ij} is the distance between molecules i and j, and e_{ij} is the unit vector pointing from the center of mass of molecule i to that of molecule j.

$\hat{H}$ is expressed in a direct product basis of single molecular oscillator eigenfunctions $|n_{im}\rangle$ with n quanta in mode m of molecule i. For the purposes of this work, the basis is restricted to a near-resonant single oscillator, single quantum excitations $|i_m\rangle$ represented by the product function with vibrational mode m excited on molecule i and all other oscillators in the ground state. $|0\rangle$ denotes the overall ground state. We further approximate matrix elements by assuming harmonic oscillators and linear dipole functions. Within this double-harmonic approximation, the Hamiltonian is completely defined by the transition energy in wave numbers $\tilde{\nu}_{im}$, the molecular transition dipole moments $\mu_{im} = \langle 0|\mu_i|1_{im}\rangle$, and α_i, the molecular polarizability tensor. These input parameters are derived independently, for example, from independent experimental data or from explicit potential and dipole functions as explained in Section 2.2.2.

To determine the Hamilton matrix elements, we insert the series expansion of the molecular dipole moment operators μ_j up to first order in normal coordinates q_{jm}:

$$\mu_j = \mu_j^0 + \sum_m \mu_j^m q_{jm}, \quad \text{with } \mu_j^m = \left(\frac{\partial\mu_j}{\partial q_{jm}}\right)_{q_{jm}=0} \tag{2.3}$$

into the Hamiltonian (Equation 2.1) and rearrange the terms by normal modes q_{jm}:

$$\hat{H} = \hat{H}_0 + \sum_{i,j}\left[(\mu_i^0)^+ A_{ij}\mu_j^0 + \sum_m q_{im}(\mu_i^m)^+ A_{ij}\mu_j^0 + \sum_n (\mu_i^0)^+ A_{ij}\mu_j^n q_{jn}\right.$$

$$\left. + \sum_{m,n} q_{im}(\mu_i^m)^+ A_{ij}\mu_j^n q_{jn}\right] \tag{2.4}$$

The zeroth-order terms in q contribute the same constant value to the diagonal terms $\langle j_m | \hat{H} | j_m \rangle$ and $\langle 0 | \hat{H} | 0 \rangle$ and can therefore be disregarded. Furthermore, the first-order terms do not couple (near) resonant levels, to which our exciton model is restricted. Neglecting zeroth- and first-order terms in the dipole interaction yields

$$\hat{H} = \hat{H}_0 + \sum_{i,m,j,n} q_{im} (\mu_i^m)^+ A_{ij} \mu_j^n q_{jn}. \tag{2.5}$$

$(\mu_i^m)^+ A_{ij} \mu_j^n$ is a scalar, and the matrix elements can be written as

$$\langle r_s | \hat{H} | t_u \rangle = \delta_{rt} \delta_{su} hc\tilde{v}_{rs} + \sum_{i,m,j,n} (\mu_i^m)^+ A_{ij} \mu_j^n \langle r_s | q_{im} q_{jn} | t_u \rangle \tag{2.6}$$

with Kronecker's symbol $\delta_{rt} = 1$ if $r = t$ and $\delta_{rt} = 0$ otherwise, Planck's constant h, speed of light c, and molecular transition wave numbers $\tilde{v}_{rs}$. Within the basis of single-harmonic excitations only the following terms are nonzero:

$$\langle 0_{im} | q_{im} | 1_{im} \rangle = \frac{1}{\sqrt{2}}, \quad \langle 0_{im} | q_{im}^2 | 0_{im} \rangle = \frac{1}{2}, \quad \langle 1_{im} | q_{im}^2 | 1_{im} \rangle = \frac{3}{2} \tag{2.7}$$

Applying Equations 2.7 to Equation 2.6 leaves only the following terms:

$$\langle r_s | \hat{H} | t_u \rangle = \delta_{rt} \delta_{su} hc\tilde{v}_{rs} + \Delta \sum_{i,m,j,n} (\mu_i^m)^+ A_{ij} \mu_j^n \langle r_s | q_{im} q_{jn} | t_u \rangle \tag{2.8}$$

with $\Delta = \delta_{ij} \delta_{mn} (1 + 2\delta_{ir} \delta_{ms}) + \delta_{ir} \delta_{ms} \delta_{jt} \delta_{nu} + \delta_{it} \delta_{mu} \delta_{jr} \delta_{ns}$.
This gives for diagonal elements

$$\langle r_s | \hat{H} | r_s \rangle = hc\tilde{v}_{rs} + (\mu_r^s)^+ A_{rr} \mu_r^s + \sum_{i,m} \frac{1}{2} (\mu_i^m)^+ A_{ii} \mu_i^m \tag{2.9}$$

and for nondiagonal elements

$$\langle r_s | \hat{H} | t_u \rangle = \frac{1}{2} (\mu_r^s)^+ A_{rt} \mu_t^u + \frac{1}{2} (\mu_t^u)^+ A_{tr} \mu_r^s = (\mu_r^s)^+ A_{rt} \mu_t^u. \tag{2.10}$$

The sum in Equation 2.9 is the same for all basis functions, in particular for the ground state $\langle 0 | \hat{H} | 0 \rangle$ energy. Thus, shifting the zero energy arbitrarily but without any loss of generality to the uncoupled ground state energy and using $\mu_{im} = \langle 0_{im} | \mu_i | 1_{im} \rangle = (1/\sqrt{2}) \mu_i^m$, the matrix elements of the Hamiltonian are cast into the following simple form:

$$\langle r_s | \hat{H} | t_u \rangle = \delta_{rt} \delta_{su} hc\tilde{v}_{rs} + 2\mu_{rs}^+ A_{rt} \mu_{tu}. \tag{2.11}$$

The overall computational effort in setting up this matrix scales to the third power with the number of oscillators in the system. The treatment of induced dipole effects is the most expensive part of the calculations. Considering resonant dipole coupling alone, the calculations scale only quadratically with the system size.

2.2.2 LOCAL ENVIRONMENT EFFECTS ON TRANSITION DIPOLES AND TRANSITION FREQUENCIES

In its simplest form, the vibrational exciton model uses as input the same transition wave number $\tilde{v}_{im}$ and transition dipole moment $\mu_{im} = \langle 0|\mu_i|_{im}\rangle$ for all molecules. This assumption, however, is only correct in highly symmetric environments, such as some crystalline phases, where all molecules can be considered equivalent. Variations in the structure of a particle, such as the disorder in the amorphous phase or the different environment experienced by molecules in the particle surface, lead to different local potentials for individual molecules with corresponding changes of transition dipoles and frequencies. These local variations are accounted for by deriving individual transition frequencies and transition dipoles for each molecule within the field of all others using explicit potential and dipole functions. The electrostatic interactions in the intermolecular potential function are closely related to the molecular transition dipoles. Therefore the dipole function should ideally be consistent with the potential instead of being defined independently [12]. Under this condition, the calculated intensities provide an independent check on the quality of the intermolecular potential, because physically the strength of interaction and the absorption strengths are correlated within the exciton approach. The transition frequencies and transition dipoles are obtained in the double-harmonic approximation by performing a local normal mode analysis, that is, for each molecule the corresponding block F_{ii} in the overall Hessian matrix F is calculated and diagonalized separately to provide the local transition frequency.

$$F = \begin{pmatrix} F_{11} & \cdots & F_{1N} \\ \vdots & \ddots & \vdots \\ F_{N1} & \cdots & F_{NN} \end{pmatrix}, \quad \text{with } F_{ii} = \begin{pmatrix} \dfrac{\partial^2 V}{m_1 \partial x_{1i}^2} & \cdots & \dfrac{\partial^2 V}{\sqrt{m_1 m_n}\, \partial x_{1i} \partial x_{ni}} \\ \vdots & \ddots & \vdots \\ \dfrac{\partial^2 V}{\sqrt{m_n m_1}\, \partial x_{ni} \partial x_{1i}} & \cdots & \dfrac{\partial^2 V}{m_n \partial x_{ni}^2} \end{pmatrix}, \tag{2.12}$$

where V denotes the potential, $(x_{1i} \ldots x_{ni})$ the Cartesian atomic coordinates, and m_i the atomic masses of molecule i. This approach considers, within the harmonic approximation, all interactions between each molecule and the rest of the particle, save for the coupling between the normal modes of different molecules, represented by the off-diagonal F_{ij} blocks. Then the derivatives of the overall dipole moment are calculated with respect to all atomic Cartesian coordinates and converted to the local normal mode representation, which yields the local transition dipoles.

The Hamiltonian defined so far still needs the molecular positions and orientations to be specified. For crystalline particles, experimental bulk crystal structures can be used although deviations near the surface may become important especially for very small particles. The choice is less clear-cut for noncrystalline phases, that is, partially disordered or even completely amorphous particles. Model amorphous configurations in this work are generated starting from a fully crystalline structure cut from the crystalline bulk. In a first step, all molecules within the particle region designated as amorphous are randomly shifted and rotated [12]. The resulting structure contains physically implausible, that is to say very high-energy configurations, such as two atoms being very close, since the randomization process does not include any energetic or structural bias. These unfavorable configurations are removed by simulated annealing using classical molecular dynamics (MD) [25–27]. As energy is repeatedly transferred into the system and then gradually removed, the artificial heating and cooling of the particle allows the relaxation of strained configurations. The simulated annealing process should not be mistaken for an attempt to describe the physical annealing processes undergone by experimentally generated particles. In the context of the present work, MD simply serves as an effective means to sample the physically relevant configuration space, rather than to describe the detailed temporal evolution. Its application to a large variety of complex systems is well established [25–27], which makes classical MD the method of choice to treat the large molecular aggregates we are interested in.

2.2.3 Numerical Implementation

The particles of interest in this work cover a size range of approximately 1–100 nm, that is, from several hundred to billions of molecules. The computational problem is simplified by the fact that the absorption spectra of particles converge as a function of size at diameters around 10 nm, that is, the band positions and band structures do not change if the size of the particle is increased further. It is thus sufficient to exceed the size convergence threshold in the simulations. However, this still leaves tens of thousands of degrees of freedom to be dealt with.

For such systems, the direct diagonalization of the Hamiltonian (Equations 2.1 and 2.11) becomes increasingly impractical even when simulating a single nondegenerate vibrational mode. It should be noted that the full set of eigenvalues and eigenvectors of the exciton Hamiltonian is required. When the complexity of the system increases, an efficient numerical implementation of the exciton model becomes essential to calculate particle spectra in a reasonable time frame. A few examples where direct diagonalization is no longer a viable method are when larger systems are required to achieve spectral convergence or when dealing with molecules with several degenerate or near-resonant vibrational modes.

The time-dependent numerical approach used throughout this work calculates the absorption spectra directly from the electric dipole autocorrelation function. If $\{E_I\}$ is the set of eigenvalues of $\hat{H}$ with corresponding eigenvectors $|I\rangle$ and transition moments $M_I = \langle 0|\mu|I\rangle$ (overall dipole function $\mu = \sum_i \mu_i$), then the absorption spectrum is proportional to

$$\sigma(E) = \sum_I \left| M_I \right|^2 * f(E - E_I),$$

(2.13)

where * denotes a convolution and $f(E)$ is an appropriate line shape, for example, a Gaussian or a Lorentzian function. Calculating the Fourier transform of $\sigma(E)$ gives

$$\tilde{\sigma}(t) = g(t) \sum_I \left| M_I \right|^2 \int e^{-iEt/\hbar} \delta(E - E_I)\, \mathrm{d}E/\hbar$$

(2.14)

$$= g(t) \sum_I \langle 0|\mu|I\rangle e^{-iE_I t/\hbar} \langle I|\mu|0\rangle = g(t)\langle 0|\mu e^{-i\hat{H}t/\hbar}\mu|0\rangle$$

(2.15)

$$= g(t) \sum_{i,m,j,n} \langle 0|\mu|i_m\rangle \langle i_m|e^{-i\hat{H}t/\hbar}|j_n\rangle \langle j_n|\mu|0\rangle$$

(2.16)

$$= g(t) \sum_{i,m,j,n} \langle i_m|\mu_{im}e^{-i\hat{H}t/\hbar}\mu_{jn}|j_n\rangle = g(t)C(t),$$

(2.17)

where $g(t)$ is the Fourier transform of $f(E)$ and $C(t)$ is the dipole autocorrelation function.

The time propagation of the dipole-weighted wave function in Equation 2.17 is performed using a second-order scheme, noting that $C(-t) = C^*(t)$. Inserting the Taylor expansion of the time evolution operator

$$e^{-i\hat{H}\Delta t/\hbar} = 1 - i\hat{H}\Delta t/\hbar + \cdots$$

(2.18)

where Δt is the propagation time step, into

$$\left|\psi_{t+\Delta t}\right\rangle - \left|\psi_{t-\Delta t}\right\rangle = (e^{-i\hat{H}\Delta t/\hbar} - e^{i\hat{H}\Delta t/\hbar})\left|\psi_t\right\rangle \tag{2.19}$$

and rearranging Equation 2.19 gives

$$\left|\psi_{t+\Delta t}\right\rangle \approx -i\frac{2\Delta t}{\hbar}\hat{H}\left|\psi_t\right\rangle + \left|\psi_{t-\Delta t}\right\rangle. \tag{2.20}$$

Within this numerical implementation of the exciton model, based on Equations 2.11, 2.17, and 2.20, the calculation of particle spectra at a resolution of ~1 cm⁻¹ requires approximately 3 h of computer time on a 3 GHz Pentium 4 processor for the largest systems (30,000 degrees of freedom), once the Hamilton matrix is set up. This is about a factor of 100 faster than full diagonalization. The time propagation depends quadratically on the system size and is scalable as a function of two parameters: the propagation step Δt and the total number of time steps, N_t. Together they define the resolution of the calculated spectra: Δt is related to the Nyquist frequency and thus defines the frequency interval, while N_t defines the number of grid points within the interval. The Nyquist frequency

$$\omega_{\text{Nyquist}} = \frac{1}{2\Delta t} \tag{2.21}$$

is the largest frequency value that can still be represented correctly by a Fourier transform—higher frequencies are convoluted back into the $[0, \omega_{\text{Nyquist}}]$ interval leading to artifacts. Ideally, Δt should be chosen such that $h \cdot \omega_{\text{Nyquist}}$ equals or just exceeds the spectral range of the Hamiltonian, with the minimum number of propagation steps N_t required to obtain the desired spectral resolution. Vibrational transitions, however, lie between several hundreds and a few thousands of wave numbers in frequency, imposing a rather strict limitation on Δt. Since, in our case, the ground state is uncoupled from excited states, it is possible to circumvent this problem by subtracting the average value $\bar{E}$ of the diagonal elements of the Hamiltonian from its diagonal. This roughly corresponds to the contribution of the uncoupled molecular vibrations ($\hat{H}_0$). This constant shift is added back to the spectrum after time evolution and Fourier transformation. Thus, the Nyquist frequency for a time step of 0.002 ps is 8333 cm⁻¹ only needs to be comparable to the width of the vibrational band, in most cases less than 100 cm⁻¹, reducing the number of propagation steps at equal resolution by up to two orders of magnitude. The time step is also limited by the numerical accuracy of the propagation scheme. The first sign of numerical error is an increasing broadening of the spectrum, nearly negligible at $\bar{E}$ and more pronounced further away. To illustrate this effect, we chose a small particle composed of a CHF₃ spherical core ($r = 1$ nm) surrounded by a SF₆ cubic shell ($5 \times 5 \times 5$ unit cells) and simulate the ν_3 band of SF₆ and the ν_5 and ν_2 bands of CHF₃. Thus, the vibrational eigenvalues span ~300 cm⁻¹ making the effect more pronounced than for a single band. We calculate the particle spectrum with a time step of 0.002 ps ($\omega_{Ny} \cong 833$ cm⁻¹) over $N_t = 41{,}840$ steps (full line) and a second time with a time step of 0.02 ps ($\omega_{Ny} \cong 833$ cm⁻¹) over $N_t = 4184$ steps (dashed line) to introduce a small numerical error as shown in Figure 2.3. In both cases ω_{Nyquist} exceeds the width of the band in question. The number of time steps was changed simultaneously to keep the simulation time constant, that is, maintain the same frequency grid. Comparing the slightly erroneous calculation ($\Delta t = 0.02$ ps) against the accurate result ($\Delta t = 0.002$ ps), shows that the further the spectral features are away from $\bar{E} = 1050.4$ cm⁻¹ (for definition see text after Equation 2.21) the stronger their broadening, clearly visible for both bands. The error is due solely to the propagation, since a particle spectrum calculated from every tenth value of the autocorrelation function

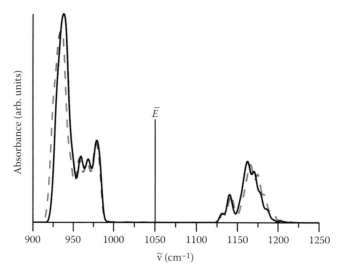

FIGURE 2.3 Calculated absorption spectrum at constant resolution, using different time steps, illustrating the onset of numerical instability on the example of a small particle composed of a CHF_3 spherical core ($r = 1$ nm) surrounded by an SF_6 cubic shell ($5 \times 5 \times 5$ unit cells). Solid line: Accurate spectrum calculated with $\Delta t = 0.002$ ps, $N_t = 41{,}840$. Dashed line: Spectrum calculated for $\Delta t = 0.02$ ps, $N_t = 4184$ to demonstrate the effect of a slight numerical error. The vertical line at 1050.4 cm^{-1} indicates the mean energy $\bar{E}$.

calculated with the 0.002 ps time step is identical to the reference spectrum. Consequently, the time step should always be chosen as large as possible, while still ensuring numerical stability and keeping calculated frequencies below the Nyquist frequency. The number of simulation steps can then be adjusted to obtain the desired spectral resolution.

2.2.4 Correlating Spectral and Structural Properties—Excitation Density

Locating where spectroscopic features originate in a particle is an important step toward a better understanding of intrinsic particle properties. Such a correlation would allow one to determine whether a contribution observed in a vibrational absorption spectrum can be attributed to a particle's surface or core, to amorphous or crystalline regions, or whether features characteristic of a given shape are localized in key regions. The excitation density is introduced as a tool to analyze the correlation between the spectroscopic and structural properties of the particles.

The excitation density can be regarded as an intensity-weighted state density [12,13]. Dividing a particle into volume elements ΔV_k, it represents the contributions of these individual volume elements to the total spectrum. With the overall dipole moment $\mu = \Sigma_i \mu_i$ and eigenvectors $\left| I \right\rangle = \Sigma_{i,m} c_{im,I} \left| i_m \right\rangle$ the exciton transition moments become

$$M_I = \left\langle 0 \middle| \mu \middle| I \right\rangle = \sum_{i,m} c_{im,I} \left\langle 0 \middle| \mu_i \middle| i_m \right\rangle. \tag{2.22}$$

A local transition moment is then defined at coordinates (R, Ω) (radius and solid angle, respectively)

$$m_I(R, \Omega) = \sum_{i,m} c_{im,I} \left\langle 0 \middle| \mu_i \delta(R - R_i)\delta(\Omega - \Omega_i) \middle| i_m \right\rangle. \tag{2.23}$$

Inserting $m_I(R, \Omega)$ in Equation 2.13 defines the excitation density as

$$\sigma(E, R, \Omega) = \sum_I (m_I(R, \Omega) M_I^*) * f(E - E_I), \tag{2.24}$$

which can be regarded as the contribution to the spectrum arising from the infinitesimal volume element at (R, Ω) interfering with the entire particle. As mentioned above, the excitation densities in the discussion section refer to integrals over finite volumes ΔV_k. For example, for the most commonly used division into spherical shells, $\sigma(E, R, \Omega)$ is integrated over the angular component Ω. As the size of different ΔV_k may vary, the excitation densities $\sigma(E, \Delta V_k)$ are always normalized by dividing by the size of the corresponding volume elements.

The above definition of the excitation density requires the eigenvectors of the Hamiltonian. When the IR spectrum of the particle is calculated from the dipole correlation function, the excitation density must be derived in a different but mathematically equivalent way. To this end, we write the time evolution operator matrix representation in the molecular oscillator product basis $\{|i_m\rangle\}$ (all molecular oscillators are in the ground state except for mode m excited on molecule i):

$$U_{im,jn} = \langle i_m | e^{-i\hat{H}t/\hbar} | j_n \rangle = \sum_I c_{im,I}^* \exp\left(-\frac{iE_I t}{\hbar}\right) c_{jn,I}. \tag{2.25}$$

Inserting $U_{im,jn}$ into the correlation function (2.16) gives

$$\tilde{\sigma}(t) = g(t) \sum_{i,m,j,n} \langle 0 | \mu | i_m \rangle U_{im,jn} \langle j_n | \mu | 0 \rangle, \tag{2.26}$$

where $g(t)$ is once again the Fourier transform of the line shape $f(E)$, μ is the overall dipole moment operator, and $|0\rangle$ is the ground state. The contribution of the vibrational mode m of molecule i interfering with all other modes can then be written as

$$\tilde{\sigma}_{im}(t) = g(t) \langle 0 | \mu | i_m \rangle \sum_{j,n} U_{im,jn}(t) \langle j_n | \mu | 0 \rangle \tag{2.27}$$

Applying the inverse Fourier transform one obtains

$$\begin{aligned}
\sigma_{im}(E) &= \int e^{iEt/\hbar} \tilde{\sigma}_{im}(t) \, dt \\
&= \langle 0 | \mu | i_m \rangle \sum_I c_{im,I} \int g(t) e^{-i(E-E_I)t/\hbar} \, dt \sum_{j,n} c_{jn,I} \langle j_n | \mu | 0 \rangle \\
&= \sum_I c_{im,I} \langle 0 | \mu | i_m \rangle \langle I | \mu | 0 \rangle \delta(E - E_I) * f(E - E_I)
\end{aligned} \tag{2.28}$$

The summation over all molecules i and all modes m yields the spectrum as defined in Equation 2.13. This demonstrates that the total correlation function $\tilde{\sigma}(t)$ can be decomposed into partial correlation functions $\tilde{\sigma}_{im}(t)$ whose Fourier transforms give the local contributions of the molecules to the overall spectrum. Thus for any partitioning of a particle into volume elements ΔV_k, the excitation density can be obtained by Fourier transform of the local partial correlation functions $\tilde{\sigma}_k(t)$, with

$$\tilde{\sigma}_k(t) = \sum_{i \in \Delta V_k} \tilde{\sigma}_{im}(t). \tag{2.29}$$

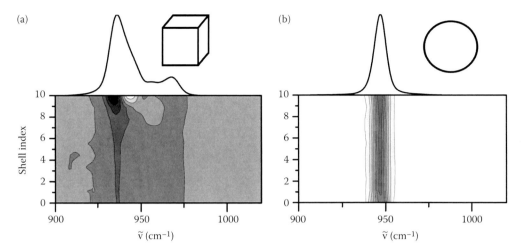

FIGURE 2.4   Calculated IR spectra (upper traces) and corresponding excitation densities (lower panels) for a crystalline SF_6 (a) cube and (b) sphere in the region of v_3 SF stretching mode. Both particles are partitioned into 10 spherical shells which are labeled with a shell index ranging from 1 to 10. A shell index of 1 corresponds to the center region of the particle and a shell index of 10 corresponds to the surface region of the particle. The excitation density (Equation 2.24), increasing from light to dark, was integrated over each shell and normed by the number of molecules in the shell.

This does not involve any additional numerical work. Strictly speaking, the excitation density is not a true density, as the interferences between vibrational modes can also be destructive. With an increasing number of molecules in the volume elements ΔV_k, however, the negative contributions tend to average out.

As an example, Figure 2.4 compares the excitation density of a cubic SF_6 aerosol particle with the density of a spherical SF_6 particle. The contour plot illustrates that the vibrational excitation in a spherical particle is completely delocalized over the whole particle. The excitation density of the cube is highly structured in the surface region. Regions with high density (dark) and low density (light) alternate. The comparison with the IR spectrum reveals that it is this surface structure which determines the spectral band shape.

2.3   DESCRIPTION OF THE EXPERIMENTAL SETUP

We briefly summarize here the experimental setups that were used to record the aerosol IR spectra shown in Section 2.4. For the particle generation, we used collisional cooling [8,15,20,28–31] and jet expansions [18,20,32–35]. A sketch of the collisional cooling cell is depicted in Figure 2.5 (left). Aerosol particles are formed by injecting a warm sample gas into a precooled bath gas (helium or nitrogen) which leads to supersaturation and thus to particle formation. For the present study which focuses on ice particles, the temperature of the bath gas varied between 4 and 78 K. In addition to the temperature, the concentration and pressure of the sample gas and the pressure of the bath gas can be varied over a wide range. This allows us in a unique way to mimic the conditions in different planetary atmospheres. A crucial aspect in this context is the fact that in a collisional cooling cell the aerosol particles are in thermal equilibrium with the surrounding gas phase. For studying aerosols, this constitutes a major advantage over the otherwise extremely powerful jet expansion techniques which work far from the equilibrium. A high-pressure variant of a jet apparatus is sketched in Figure 2.5 (right). Here particles are formed by rapid expansion of a gas/supercritical fluid (pressures up to 500 bar) into vacuum, while the particle size can be tuned by varying the pressure of the gas/supercritical fluid. In collisional cooling cells as well as in jet expansions we characterize the

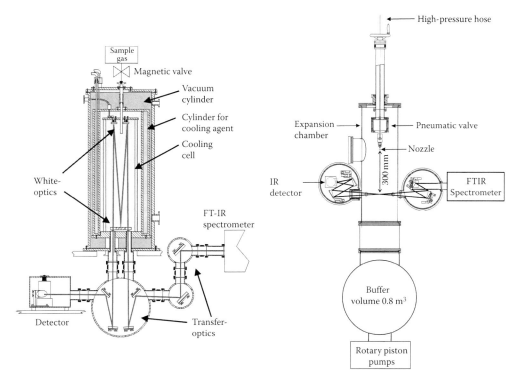

FIGURE 2.5 **(See color insert following page 206.)** Schematic representation of the experimental setup. Left: Collisional cooling cell. Right: Rapid expansion of supercritical solutions (RESS) setup.

aerosols and certain aspects of the processes by which they are formed *in situ* with rapid scan Fourier transform IR spectroscopy (see Figure 2.5). Our setup is especially well suited to investigate nanoparticles in the size range below 100 nm.

2.4 SOME EXAMPLES

2.4.1 The Influence of Particle Size and Internal Structure on IR Spectra

When performing a spectroscopic analysis of finite-size systems, a major goal is always the extraction of information on the size from the spectroscopic data. IR spectra of aerosol particles exhibit pronounced size effects, which we classify in the following as "direct" and "indirect" size effect. In the first case, the finite size directly influences the structure of the IR bands by modifying the exciton coupling. This effect shows up even without any changes in a particle's shape or its internal structure. Figure 2.6 shows as an example calculated IR spectra for carbon monoxide (CO) ice particles of different sizes. All particles have the same crystalline structure and a cubic shape. Figure 2.6 illustrates that each size has a unique IR spectrum. This is in particular true for particles with sizes below about 5–10 nm (traces a–d). A slight increase in size leads to a pronounced modification of the exciton structure. Between about 10 and 100 nm, the overall band shape remains almost unchanged (similar to trace d) except for a smoothing out of some fine structure, since variation in size no longer influences the exciton coupling strongly. The spectra are "size-converged" in this region. Above 100 nm, the size dependence of IR spectra is increasingly governed by elastic scattering of the IR light by the particles (trace e). Characteristic scattering patterns are the dispersion shape of the absorption band and the slanted baseline.

As opposed to direct size effects, indirect size effects are caused by size-dependent structural variations or size-dependent changes of the particles' shape. Frequently observed phenomena are

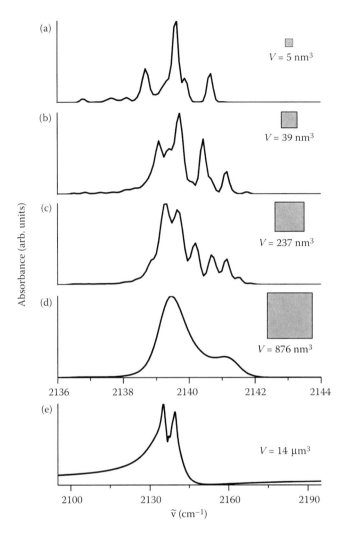

FIGURE 2.6 The effect of exciton coupling on the IR absorption spectrum of crystalline carbon monoxide particles with a cubic shape as a function of size. Pronounced changes (a–d) can be observed up to approximately 10 nm in size. A further increase in size has only a marginal effect on the band shapes until (e) scattering effects become significant. The particle's volume ranges from (a) 5 nm³ to (e) 14 μm³.

structural variations in the surface of the particles owing to the altered molecular environment compared to the core. Usually, the surface has a more amorphous structure than the core [12,17,21,29,36,37]. Figure 2.7 shows calculated IR spectra of spherical CO ice particles with a radius of 4 nm and a "crystalline core–amorphous shell" architecture. The increasing contribution of the amorphous shell from 0% to 100% by volume leads to a pronounced asymmetric broadening of the absorption band. The excitation density as a function of the radial coordinate, which is shown in Figure 2.7 for the particle with a 40% amorphous shell, confirms that the broad asymmetric contribution to the spectrum originates from the amorphous shell, which extends from 3.4 to 4 nm. Corresponding features in the excitation density are found for the particles with higher and lower amorphous content.

Figure 2.8 illustrates a similar structural surface effect observed for small ammonia aerosols, which have been found, for example, in Jupiter's atmosphere [38,39]. The experimental IR spectra (left) show a broadening of the absorption bands toward higher wave numbers with decreasing particle size. The analysis of these data with a combined exciton model/MD approach revealed that

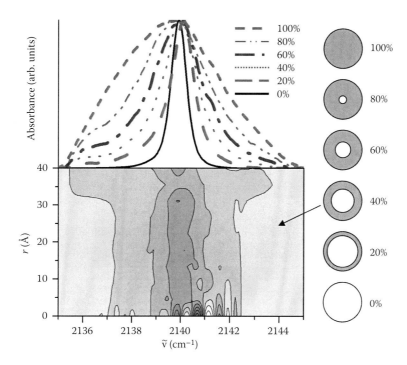

FIGURE 2.7 **(See color insert following page 206.)** Upper traces: Calculated absorption spectra for ensembles (10 elements) of crystalline core–amorphous shell carbon monoxide particles of spherical ($r = 4$ nm) shape. The amorphous shell contribution ranges from (thin full) 0 to (thick short dashed) 100 vol%. Lower panel: Excitation density for the ensemble with 40 vol% amorphous shell. The particles are partitioned into spherical shells 1 Å thick. The excitation density (Equation 2.24), increasing from light to dark, was integrated over each shell and normed by the number of molecules in the shell. The asymmetry and broadening of the IR band can be clearly traced back to the amorphous shell.

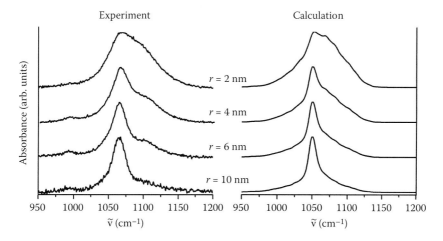

FIGURE 2.8 Experimental (left) and calculated (right) IR absorption spectra of NH_3 aerosols as a function of size in the region of the umbrella vibration. Particle radii are indicated for each spectrum.

this behavior is caused by a surface layer which is more disordered than the crystalline core [12]. The corresponding calculated spectra are depicted on the right. As a rather intriguing result, we found the thickness of the surface layer (around 1 nm) to be more or less independent of the size of the aerosol particles.

The internal structure of aerosol particles is not solely determined by the size of the particles. For example, the conditions under which they are formed can have a major influence or changing conditions can modify the internal structure of already existing particles. Among such processes, a particularly interesting case is the phase transition from a supercooled liquid droplet to a crystalline particle [22,37,40,41]. If the supercooled droplets are "long lived" species, they will able to play a more significant role in atmospheric processes. The crystallization kinetics of these phase transitions and the internal structures of the particles can be determined by time-dependent IR spectroscopy as illustrated in Figure 2.9 for ethane (left) and fluoroform (right). In both cases, supercooled liquid droplets are initially formed ($t = 0$ s), which over time convert into fully crystalline particles ($t = 847$ s and $t = 38$ s, respectively). Strong evidence for polar ethane clouds in the atmosphere of Saturn's moon Titan was recently obtained from the Cassini mission [42]. Our analysis of the crystallization kinetics of ethane reveals that supercooled ethane droplets are a "long-lived" species and thus might play a similar role in ethane clouds on Titan as supercooled liquid water droplets do in the Earth's atmosphere.

2.4.2 THE INFLUENCE OF THE PARTICLE SHAPE ON IR SPECTRA

Figure 2.10 a–d illustrate how for various ice aerosol particles the band structure of IR transitions varies if the shape of the aerosol particles changes [13,16–19,22]. The experimental spectra on the left are for particles with shapes that have similar axis ratios, such as cubes, cuboctahedra, or spheres. The spectra on the right show the same bands but for particles with an elongated shape. The qualitative change in the spectral features between shapes with similar axis ratios and elongated particles is the same for all particles types: The former exhibits one prominent band, while the latter shows in addition two pronounced shoulders (labeled with arrows), one on the low-frequency side and the other on the high-frequency side. The details of the spectral band shape depend on the type of substance (intermolecular interactions, crystal structure) and on the type of band (degeneracy). An obvious difference exists between the hydrogen-bonded ammonia particles and all the other particles. As a consequence of the hydrogen bonds, the ammonia band is broad and rather unstructured and the shape effects are thus not very pronounced in the ammonia spectra.

The differences in the spectra of carbon dioxide and fluoroform are mainly due to the differences in the crystal structure and in the degeneracy of the bands. The antisymmetric stretching vibration of CO_2 depicted in trace a is nondegenerate and the CO_2 particles have a cubic crystal structure. Both factors lead to a comparatively simple exciton structure. The exciton structure of fluoroform (trace c), by contrast, is more complicated since the crystal structure is monoclinic and two CF_3 stretching modes overlap in the region around 1140 cm^{-1}, one of which is twofold degenerate. SF_6 in trace d is a special case. Although the stretching band is threefold degenerate, the spectral features look similar to those observed for CO_2. The reason lies in the spherical symmetry of SF_6 with respect to the exciton Hamiltonian [13]. The resulting exciton predictions for the two different shapes of SF_6 particles are shown in trace e. The trends observed in the experiment are clearly reproduced by the exciton calculations. Note that the comparison is not perfect, because in the experiment we measure a spectrum for an ensemble of particles (some distribution in shape is possible) while the calculations are for single shapes.

2.4.3 SPECTRAL FEATURES OF MULTICOMPONENT PARTICLES

Multicomponent particles can have many different internal structures or architectures, which in turn can produce a wealth of differing spectroscopic features. Apart from homogeneous mixtures

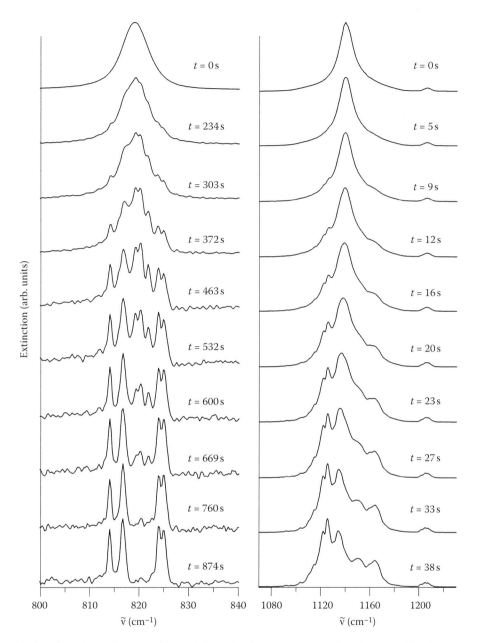

FIGURE 2.9 Time-dependent experimental IR extinction spectra capturing the crystallization process of supercooled liquid droplets of C_2H_6 (left) in the region of the ν_9 vibration and of CHF_3 (right) in the region of the ν_2/ν_5 vibrations.

(statistically mixed on a molecular level), they can form particles with a core–shell architecture or particles in which one compound forms inclusions in the matrix of another compound [13,16,19,43]. All these different structures/architectures modify the environment of individual molecules in an aerosol particle and thus the exciton coupling. The homogeneous mixing of different substances has a particularly strong influence as illustrated in Figure 2.1d, e. The pure elongated carbon dioxide particles in trace d show the characteristic band structure of elongated shapes discussed in the

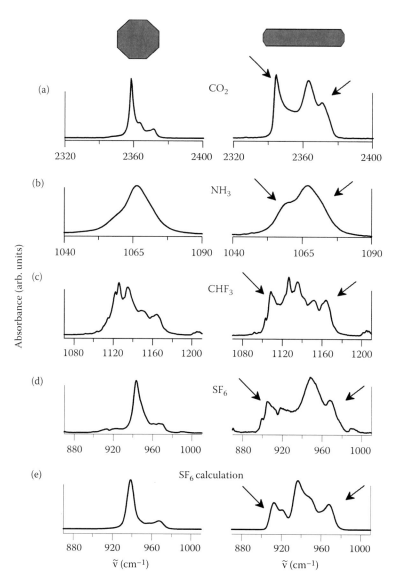

FIGURE 2.10 (a–d): Shape dependence of experimental IR absorption spectra for crystalline particles of "similar axis ratio" (left) and for elongated crystalline particles (right) composed of (a) CO_2 (v_3 band), (b) NH_3 (v_2 band), (c) CHF_3 (v_2/v_5 bands), and (d) SF_6 (v_3 band). "Shoulders" characteristic for elongated particles are indicated by arrows. (e) Corresponding calculated IR spectra for the SF_6 particles.

previous section. If carbon dioxide is homogeneously mixed with another substance (in this case, $^{13}CO_2$ instead of $^{12}CO_2$), this band structure is almost completely absent as illustrated by the experimental spectrum in trace e. The reason is the modified exciton coupling due to the presence of a second substance. This is confirmed by the corresponding exciton calculations shown in Figure 2.11 for the pure carbon dioxide particle on the left and the homogeneously mixed particle on the right. A general conclusion of these observations is the fact that the information on the particles' shape is lost in the IR spectra of homogeneously mixed aerosol particles.

Even though the information on the shape is lost for such homogeneously mixed particles, IR spectra can still allow us to extract information on the internal structure of the mixture. It is, for example, not clear whether an amorphous mixture or a mixed crystal phase is formed. Figure 2.12

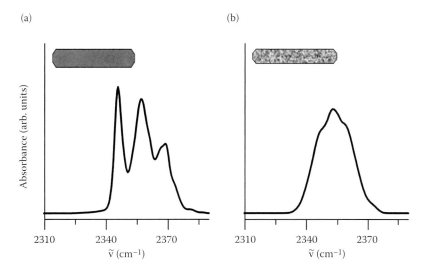

FIGURE 2.11 The effect of particle composition on IR absorption spectra illustrated by exciton calculations in the region of the antisymmetric stretch vibration of $^{12}CO_2$ for an elongated particle composed of (a) pure $^{12}CO_2$ and (b) a 1:1 statistical mixture of $^{12}CO_2$ and $^{13}CO_2$. Homogeneous mixing increases the distance between resonant oscillators, which causes the loss of information on the particle shape in the IR spectrum.

shows the spectrum of mixed acetylene/carbon dioxide particles with a mixing ratio of 1.5 (acetylene in excess) [28,41,44]. As was just discussed, the broad structureless band of $^{12}CO_2$ at 2350 cm^{-1} is an indication that the two substances form a homogenous mixture. Clear evidence that this is a mixed crystal and not an amorphous mixture comes from the antisymmetric stretching vibration of the isotopomer $^{13}CO_2$ (present because of its natural abundance of 1%) around 2280 cm^{-1} and from the CH-stretching vibration of C_2H_2 around 3240 cm^{-1}. Both show the sharp bands of a the crystalline phase with a well-defined shift in the mixed phase (dashed lines labeled "mixed" in Figure 2.12) relative to the pure phases (dotted lines labeled "pure" in Figure 2.12). In contrast, an amorphous

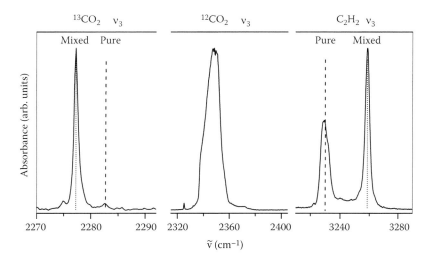

FIGURE 2.12 Experimental IR absorption spectrum of mixed CO_2/C_2H_2 particles in the regions of the v_3 bands of $^{13}CO_2$, $^{12}CO_2$, and C_2H_2. The C_2H_2:CO_2 ratio is 1.5:1; $^{13}CO_2$ is present in natural abundance. All bands are scaled to the same maximum absorbance.

mixture would necessarily have shown a broad distribution of shifts, which is clearly not observed. A more detailed analysis indicates that the two substances form a monoclinic mixed crystal with a 1:1 ratio of carbon dioxide and acetylene [28,41,44]. The fact that the spectrum in Figure 2.12 still shows features of pure acetylene is due to its excess relative to carbon dioxide (1.5:1 instead of 1:1) so that regions/particles of pure acetylene coexist with those consisting of the mixed phase.

Depending on the particle formation conditions (temperature, pressure) and on the properties of the different substances (phase transition data, intermolecular interactions), other architectures such as core–shell particles can be formed instead of homogeneous mixtures. Because of their possible relevance on Jupiter's moon Io, we have investigated the formation of mixed sulfur dioxide/carbon dioxide particles [43]. We found that cocondensation of SO_2 gas and CO_2 gas (premixed gas sample) does not lead to the formation of homogeneously mixed aerosol particles, but to the formation of core–shell particles with SO_2 in the core and CO_2 in the shell (IR spectrum shown in trace a in Figure 2.13a). The very different sublimation (195 K) and boiling (236 K) points of carbon dioxide and sulfur dioxide, respectively, as well as less favorable intermolecular interactions in the mixture compared to the pure substances are probably the major reasons why homogeneous mixtures do not form. The core–shell structure is reflected in the IR bands in Figure 2.13a. The SO_2 band structure is the same as that observed for pure SO_2 particles (not shown here) as is to be expected for an SO_2 core. The CO_2 band, however, shows a structure that is quite different from that observed for pure CO_2 particles (see Figure 2.10a). It is the characteristic structure of a CO_2 shell as confirmed by the simulated shell spectrum in trace c. The final proof is provided by the spectrum in trace b, where the core–shell architecture formation was enforced by forming SO_2 particles first and coating them subsequently with carbon dioxide.

In Figure 2.14, we finally present two examples that illustrate how IR spectroscopy combined with a microscopic model can be used to observe mixing and demixing processes within aerosol particles as a function of time. The left panel shows snapshots of the mixing of CO_2 and N_2O in

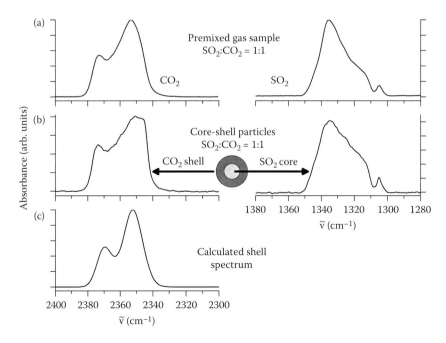

FIGURE 2.13   (a, b) Experimental IR spectra of SO_2/CO_2 aerosol with a 1:1 substance ratio in the region of the antisymmetric stretch vibration of CO_2 (left) and SO_2 (right). The particles are formed from a premixed gas sample (a) and by subsequent deposition of CO_2 on previously formed SO_2 particles (b). (c) Calculated IR spectrum of the CO_2 shell.

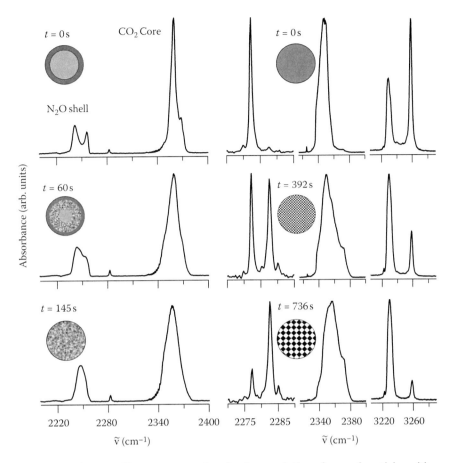

FIGURE 2.14 Top to bottom: IR spectra capturing the time evolution of aerosol particles with an unstable internal structure. Left: Mixing of core and shell molecules over time for a particle with an initial core–shell structure (CO_2 in the core and N_2O in the shell). The CO_2:N_2O ratio is 3:1. Right: Demixing over time of mixed CO_2/C_2H_2 aerosol particles that are initially formed in a metastable mixed crystal phase. Demixing can be observed via the ν_3 bands of $^{13}CO_2$, $^{12}CO_2$, and C_2H_2 (see also Figure 2.12).

aerosol particles starting from a core–shell architecture with pure carbon dioxide in the core (75% by volume) and pure nitrous oxide in the shell (25% by volume) (top trace). At the beginning ($t = 0$ s), the N_2O band shows the characteristic double structure of a shell. With increasing time ($t = 60$ s), the shell and the core start to mix which leads to the gradual disappearance of the double structure of the shell and also to a broadening of the core band. After mixing is complete (bottom trace), both bands are broad and unstructured, which is consistent with a homogenously mixed particle. The observation of the opposite process—demixing over time—is demonstrated in the right panel for the mixed crystalline phase of acetylene/carbon dioxide particles of Figure 2.12. In this case, the driving force is the energy difference between the metastable mixed crystalline phase and the (stable) separate pure crystalline phases. The progression of the phase separation with time is obvious from the increasing intensity of the "pure" $^{13}CO_2$ and "pure" C_2H_2 bands at the expense of the corresponding "mixed" bands (see also Figure 2.12). It is also reflected in the change of the band structure of the $^{12}CO_2$ band. A plausible scenario is that the demixing results in aerosol particles with a coarse-grained structure of regions with pure acetylene and regions with pure carbon dioxide as implied by the sketches. This will be subject of a future detailed analysis.

2.5 CONCLUSION

In this chapter, we have presented the theory behind the vibrational exciton model for the analysis of IR spectra of molecularly structured aerosols. The various examples demonstrate that IR absorption bands with strong molecular transition dipoles—and thus strong vibrational exciton coupling—are particularly sensitive to particle properties, such as size, shape, and architecture. They also illustrate how the exciton model can be used to unravel the information on particle properties hidden in experimental IR spectra.

In contrast to popular continuum approaches, such as Mie theory, the present model provides a molecular picture of the origin of the various observed spectroscopic patterns. Such molecular models represent a crucial step toward a better understanding of aerosol properties and lay the foundation for new developments in related fields of aerosol science.

ACKNOWLEDGMENTS

This work was supported by the Natural Sciences and Engineering Research Council of Canada (NSERC), the Canada Foundation for Innovation, and the A.P. Sloan Foundation (R.S.). T.P. acknowledges a graduate fellowship from the NSERC.

REFERENCES

1. van de Hulst, H. C. 1981. *Light Scattering by Small Particles*. New York, NY: Dover.
2. Bohren, C. F. and Huffman, D. R. 1998. *Absorption and Scattering of Light by Small Particles*. New York, NY: Wiley Interscience.
3. Mishchenko, M. I., Hovenier, J. W., and Travis, L. D. 2000. *Light Scattering by Nonspherical Particles: Theory, Measurements, and Applications*. San Diego, CA: Academic Press.
4. Berg, M. J., Sorensen, C. M., and Chakrabarti, A. 2005. Patterns in Mie scattering: Evolution when normalized by the Rayleigh cross section. *Appl. Opt.* 44:7487–7493.
5. Sigurbjörnsson, O. F., Firanescu, G., and Signorell, R. 2009. Intrinsic particle properties from vibrational spectra of aerosols. *Annu. Rev. Chem. Phys.* 60:127–146.
6. Fox, D. and Hexter, R. M. 1964. Crystal shape dependence of exciton states in molecular crystals. *J. Chem. Phys.* 41:1125–1139.
7. Cardini, G., Schettino, V., and Klein, M. L. 1989. Structure and dynamics of carbon dioxide clusters: A molecular dynamics study. *J. Chem. Phys.* 90:4441–4449.
8. Disselkamp, R. and Ewing, G. E. 1990. Infrared spectroscopy of large CO_2 clusters. *J. Chem. Soc. Faraday Trans.* 86:2369–2373.
9. Wales, D. J. and Ewing, G. E. 1992. Spectroscopic signature of fractal excitons. *J. Chem. Soc. Faraday Trans.* 88:1359–1367.
10. Disselkamp, R. and Ewing, G. E. 1993. Large CO_2 clusters studied by infrared spectroscopy and light scattering. *J. Chem. Phys.* 99:2439–2448.
11. Signorell, R. 2003. Verification of the vibrational exciton approach for CO_2 and N_2O nanoparticles. *J. Chem. Phys.* 118:2707–2715.
12. Firanescu, G., Luckhaus, D., and Signorell, R. 2006. Size effects in the infrared spectra of NH_3 ice nanoparticles studied by a combined molecular dynamics and vibrational exciton approach. *J. Chem. Phys.* 125:144501.
13. Firanescu, G., Luckhaus, D., and Signorell, R. 2008. Phase, shape, and architecture of SF_6 and SF_6/CO_2 aerosol particles: Infrared spectra and modeling of vibrational excitons. *J. Chem. Phys.* 128:184301.
14. Decius, J. C. and Hexter, R. M. 1977. *Molecular Vibrations in Crystals*. New York, NY: McGraw-Hill.
15. Barnes, J. A. and Gough, T. E. 1987. Fourier-transform infrared-spectroscopy of molecular clusters: The structure and internal mobility of clustered carbon dioxide. *J. Chem. Phys.* 86:6012–6017.
16. Signorell, R. and Kunzmann, M. K. 2003. Isotope effects on vibrational excitons in carbon dioxide particles. *Chem. Phys. Lett.* 371:260–266.
17. Jetzki, M., Bonnamy, A., and Signorell, R. 2004. Vibrational delocalization in ammonia aerosol particles. *J. Chem. Phys.* 120:11775–11784.

18. Bonnamy, A., Georges, R., Hugo, E., et al. 2005. IR signature of $(CO_2)_N$ clusters: Size, shape and structural effects. *Phys. Chem. Chem. Phys.* 7:963–969.

19. Signorell, R., Jetzki, M., Kunzmann, M., et al. 2006. Unraveling the origin of band shapes in infrared spectra of $N_2O-^{12}CO_2$ and $^{12}CO_2-^{13}CO_2$ ice particles. *J. Phys. Chem. A.* 110:2890–2897.

20. Firanescu, G., Hermsdorf, D., Ueberschaer, R., et al. 2006. Large molecular aggregates: From atmospheric aerosols to drug nanoparticles. *Phys. Chem. Chem. Phys.* 8:4149–4165.

21. Firanescu, G. and Signorell, R. 2009. Predicting the influence of shape, size and internal structure of CO aerosol particles on their infrared spectra. *J. Phys. Chem. B.* 113:6366–6377.

22. Sigurbjörnsson, O. F., Firanescu, G., and Signorell, R. 2009. Vibrational exciton coupling as a probe for phase transitions and shape changes of fluoroform aerosol particles. *Phys. Chem. Chem. Phys.* 11:187–194.

23. Snels, M. and Reuss, J. 1987. Induction effects on IR-predissociation spectra of $(SF_6)_2$, $(SiF_4)_2$ and $(SiH_4)_2$. *Chem. Phys. Lett.* 140:543–547.

24. Katsuki, H., Momose, T., and Shida, T. 2002. SF6 and its clusters in solid parahydrogen studied by infrared spectroscopy. *J. Chem. Phys.* 116:8411–8417.

25. Allen, M. P., Michael P., and Tildesley, D. J. 1988. *Computer Simulations of Liquids.* New York, NY: Oxford University Press.

26. Rapaport, D. C. 2004. *The Art of Molecular Dynamics Simulation*, 2nd Ed. Cambridge: Cambridge University Press.

27. Adcock, S. A. and McCammon, J. A. 2006. Molecular dynamics: Survey of methods for simulating the activity of proteins. *Chem. Rev.* 106:1589–1615.

28. Gough, T. E. and Wang, T. 1995. Vibrational spectroscopy of co-crystallized carbon dioxide and acetylene. *J. Chem. Phys.* 102:3932–3937.

29. Devlin, J. P., Joyce, C., and Buch, V. 2000. Infrared spectra and structures of large water clusters. *J. Phys. Chem. A* 104:1974–1977.

30. Bauerecker, S., Taraschewski, M., Weitkamp, C., et al. 2001. Liquid-helium temperature long-path infrared spectroscopy of molecular clusters and supercooled molecules. *Rev. Sci. Instrum.* 72:3946–3955.

31. Kunzmann, M. K., Signorell, R., Taraschewski, M., et al. 2001. The formation of N_2O nanoparticles in a collisional cooling cell between 4 and 110 K. *Phys. Chem. Chem. Phys.* 3:3742–3749.

32. Häber, T., Schmitt, U., and Suhm, M. A. 1999. FTIR-spectroscopy of molecular clusters in pulsed supersonic slit-jet expansion. *Phys. Chem. Chem. Phys.* 1:5573–5582.

33. Häber, T., Schmitt, U., Emmeluth, C., et al. 2001. Ragout-jet FTIR spectroscopy of cluster isomerism and cluster dynamics: From carboxylic acid dimers to N_2O nanoparticles. *Faraday Discuss.* 118:331–359.

34. Bonnamy, A., Georges, R., Benidar, A., et al. 2003. Infrared spectroscopy of $(CO_2)_N$ nanoparticles $(30 < N < 14,500)$ flowing in a uniform supersonic expansion. *J. Chem. Phys.* 118:3612–3621.

35. Bonnamy, A., Hermsdorf, D., Ueberschaer, R., et al. 2005. Characterization of the rapid expansion of supercritical solutions by Fourier transform infrared spectroscopy *in situ*. *Rev. Sci. Instrum.* 76:053904.

36. Buch, V., Bauerecker, S., Devlin, J. P., et al. 2004. Solid water clusters in the size range of tens-thousands of H_2O: A combined computational/spectroscopic outlook. *Int. Rev. Phys. Chem.* 23:375–433.

37. Signorell, R. and Jetzki, M. 2007. Phase behaviour of methane haze. *Phys. Rev. Lett.* 98:013401.

38. Atreya, S. K., Wong, A. S., Baines, K. H., et al. 2005. Jupiter's ammonia clouds—localized or ubiquitous? *Planet. Space Sci.* 53:498–507.

39. Reuter, D. C., Simon-Miller, A. A., Lunsford, A., et al. 2007. Jupiter cloud composition, stratification, convection, and wave motion: A view from new horizons. *Science* 318:223–225.

40. Sigurbjörnsson, O. F. and Signorell, R. 2008. Evidence for the existence of supercooled ethane droplets under conditions prevalent in Titan's atmosphere. *Phys. Chem. Chem. Phys.* 10:6211–6214.

41. Wang, C. C., Zielke, P., Sigurbjörnsson, O., Viteri, C. R., and Signorell, R. 2009. Vibrational spectroscopy of aerosols with low abundance in Titan's atmosphere: C_2H_6, C_2H_4, C_2H_2, and CO_2. *J. Phys. Chem. A* 113:11129–11137. DOI: 10.1021/jp904106e.

42. Griffith, C. A., Penteado, P., Rannou, P., et al. 2006. Evidence for a polar ethane cloud on Titan. *Science* 313:1620–1622.

43. Signorell, R. and Jetzki, M. 2008. Vibrational excitation coupling in pure and composite sulfur dioxide aerosols. *Faraday Discuss.*, 137:51–64.

44. Gough, T. E. and Rowat, T. E. 1998. Measurements of the infrared spectra and vapor pressure of the system carbon dioxide acetylene at cryogenic temperatures. *J. Chem. Phys.* 109:6809–6813.

3 Aerosol Nanocrystals of Water Ice

Structure, Proton Activity, Adsorbate Effects, and H-Bond Chemistry

J. Paul Devlin

CONTENTS

3.1 Introduction 50
3.2 Ice Aerosol Nanocrystal Formation and Particle Structure 51
    3.2.1 Formation and FTIR Sampling of Ice Aerosol Nanocrystals 52
    3.2.2 Evidence for Structural Layering: The Surface, Subsurface, and Core
           Spectra of D_2O and H_2O Nanocrystals 54
3.3 Structural Insights from Response of Particle Spectra to Weak Adsorbates: H_2
    and CO as Case Studies 59
    3.3.1 Vibrational Spectra of CO Adsorbed on ASW and Ice Nanocrystals 60
    3.3.2 Vibrational Spectra of H_2 Adsorbed on ASW and Ice Nanocrystals 61
3.4 Proton and Orientational L-Defect Activity in the Interior and at the Surface of
    Bare Ice Aerosol Nanocrystals 63
    3.4.1 Aerosol Particle-Core Proton and L-Defect Activity 65
    3.4.2 Enhanced Surface Proton Activity: Time Decay of Particle-Core Activity 66
3.5 Adsorbate Control of Point-Defect Activity of Ice Aerosol Nanocrystals 67
    3.5.1 Weak-Acid Control of Proton Activity in the Core of Ice
           Nanocrystals near 110 K 67
    3.5.2 Effect of the Base Dopants Ammonia and Mono-Methyl Amine on
           Ice-Defect Activity 69
    3.5.3 Surface-Based Dynamic Equilibrium of Ice Autoionization 69
    3.5.4 Managing Interior L-Defect Activity with SO_2 and H_2S Adsorbates 70
3.6 Quantitative Kinetics of Hydrate Formation and Transformations Near 120 K 71
    3.6.1 Kinetics of Ammonia-Hydrate Formation from Ice Nanocrystals 72
    3.6.2 Kinetics of Acid-Hydrate Formation from Ice Nanocrystals 72
    3.6.3 Kinetics of Rapid Clathrate-Hydrate Formation by H-Bonding Adsorbates
           near 120 K: A "Catalytic" Role for H-Bonding Guests 73
    3.6.4 Defect Structures/Dynamics that Promote Formation and Transformations
           of Clathrate Hydrates with H-Bonded Guests 73
3.7 Summary 74
References 75

3.1 INTRODUCTION

The ice particles of most interest here have diameters in the 4–40 nm range with the percentage of surface water molecules ranging from ~40% to 4% of the total. This highlights two reasons for interest in ice nanoparticles. The relative amount of bare ice surface is sufficient (1) to permit differentiation of the FTIR surface spectrum from that of the interior ice, and (2) to reveal clearly the response of the surface spectrum to an adsorbate, and the adsorbate spectrum to the interaction with the surface. We will see that the *ice surface vibrational spectrum is markedly different than that of the interior, to a degree that exceeds that for any other known molecular substance.* This uniqueness of the surface is accompanied by a second unusual property: the great difference between the surface and the ice interior is the source of *a subsurface transition region* with a unique structure and spectrum. Thus, as highlighted in Section 3.2, we can use spectroscopic and computational results to distinguish three parts of each ice nanocrystal: the core, the surface, and the subsurface. This three-part description applies to crystalline ice in general; that is, even the bulk forms of hexagonal and cubic ice, the two quite similar phases that are stable in terrestrial environments.[1]

Clear recognition and assignment of the surface bands of ice aerosol particles also means that the response of spectral features to adsorbates is uniquely definitive of the nature of surface–adsorbate interactions. This has enabled an operational categorization of ice adsorbates with three distinct classes: weak, intermediate, and strong.[2] Weak adsorbates, featured in Section 3.3 (H_2, N_2, Ar, CH_4, CO, CO_2, O_2, O_3, CF_4, etc.), produce observable but minor shifts of ice surface bands ranging from ~5 to 30 cm^{-1} but have no noticeable influence on the subsurface. Adsorbates of intermediate strength, such as H_2S, SO_2, HCN, NH_3, acetone, and small ethers, when present *at monolayer and submonolayer levels*, produce large shifts of the surface modes, modify the ice particle defect activity, and cause an observable rearrangement of the ice surface and subsurface, but do not penetrate to initiate H-bond chemistry within ice. However, *multilayer* coatings of some intermediate adsorbates, such as NH_3, H_2S, and small ether molecules do lead to the formation of molecular hydrates, even near 120 K.[3–5]

Strong adsorbates, such as HCl, HBr, and HNO_3 can attack the H-bond surface structure, at less than *monolayer levels* of dosing, generating new chemical entities with ion structures ranging from hydronium (H_3O^+) to Zundel ($H_5O_2^+$).[6–8] However, there is credible evidence that even strong acids remain largely molecular when present at <20% of a monolayer on nanocrystal[8] or amorphous ice surfaces[6–9] below ~70 K. This sensitivity to the dosing level has been related to the importance of self-solvation by the acids. A recent paper summarizes credible views of HCl–water interactions for surfaces ranging from small water clusters through crystalline ice.[10] Amorphous and crystalline acid hydrates, generated at low temperatures by strong acid adsorbates that are *present beyond monolayer levels*, will be discussed in Section 3.6.2. A general principle for adsorbates on solid water surfaces is that high-energy sites, in particular the dangling coordination sites of the three-coordinated water molecules, must first be occupied before adsorbate penetration beyond the surface is favored.

When ice aerosol nanocrystals are formed in an environment below 140 K, it is possible to isolate a few % D_2O molecules intact within an otherwise H_2O ice lattice. The isolated D_2O is an ideal probe of protonic and orientational[11] defect activity in ice, because the associated hop and turn steps embody the only known mechanism for transforming isolated D_2O units to isolated HDO molecules at thermal energies.[1,12] Since isolated D_2O and HDO each have a distinctly unique infrared spectrum, FTIR was used to characterize the point-defect activity within thick ice films in the 130–150 K range[12,13]; but such thick films lack a useful spectral probe of surface protonic activity. However, the ice-surface infrared spectrum of isolated HDO is shifted from that of isolated D_2O,[14] so, as is shown in Section 3.4, *the rate of defect-promoted isotopic exchange at nanocrystal surfaces can be measured and compared to that of the interior.*[15] Further, both the surface and interior protonic and orientational activities respond to the presence of weak acid and base adsorbates.[16] In Section 3.5, the response of isotopic-exchange rates to these adsorbates are used to further *reveal charge-transport characteristics of both the interior and surface of ice particles* (which may both be fundamental to terrestrial electrical storms; see, e.g., ref. [17]).

Many adsorbates on ice can also be categorized by the nature of the hydrate that forms when the ice surface is exposed to an abundance of the adsorbate at the proper temperature. We review, in Section 3.6, the *conversion of ice nanocrystals, near 120 K, to crystalline nanoparticles of ammonia, acid, and clathrate hydrates* on a subhour timescale when exposed, respectively, to NH_3, certain strong acids, and proton-acceptor guest molecules such as ethers, acetone, and formaldehyde. There is currently an extraordinary interest in the clathrate hydrates of methane, CO_2, and H_2 because of an abundance in nature,[18] a potential as a source of energy,[19] or the concern that the release of the guest molecules from natural hydrates, in particular methane, may have a capacity for a serious environmental impact.[18] Quantitative kinetic parameters for the low-temperature (~120 K) formation and transitions of clathrate hydrates, including those with CH_4 and CO_2 guests, have become available in recent years.[4,5] Using results from on-the-fly and empirical simulations, they are examined in Section 3.6 in the context of identification of a mechanism for the surprisingly facile low-temperature transformations.

3.2   ICE AEROSOL NANOCRYSTAL FORMATION AND PARTICLE STRUCTURE

Although we are interested in all components of an ice nanocrystal, that is, the surface, subsurface, and core, each of which constitutes a significant fraction of any nanocrystal, the focus will be on the particle surface. Initially, it is important to recognize what is known about the surfaces of related substances. Useful molecular-level information is available for (1) small cold water clusters, (2) water nanodroplets, (3) amorphous solid water (ASW) films, (4) thin ice films with oxygen-ordered surfaces, and (5) large single crystals of ice.

Experimental data from each of these water "phases" analyzed with the help of modern computational methods indicate that the degree of order at the surface increases through the list. For *small water clusters*, there are no interior water molecules as each participates in three or less hydrogen bonds (H-bonds)[20]; so, none of the local configurations resemble the interior of nanodroplets, ASW films or, in particular, the ice crystal phases for which four-coordination is an ice-rule (Bernal-Fowler: 2 donor and 2 acceptor H-bonds for each ice water molecule[1]). Structurally, water nanodroplets[21] and ASW[22] represent major steps from small water clusters towards crystalline ice. The compacted form of ASW, in particular, is considered to have a random tetrahedral network structure with an average water coordination number approaching 4. Still, spectra[23] and computations[24] indicate that a significant fraction of *porous-ASW molecules* manage to form only 2 or 3 hydrogen bonds and thus *resemble molecules in small clusters*; a structural and spectroscopic similarity, shared with the surface of water nanodroplets, that becomes less pervasive as the porosity of the ASW decreases.[25,26]

Strangely enough, the surface of crystalline ice also displays some similarity with small water clusters. This is because termination of an *oxygen-ordered ice structure* at a surface must create unsaturated coordination sites, with the simplest oxygen-ordered ice surface a molecular bilayer with three-coordinated molecules in the topmost layer and exclusively four-coordination in the second layer of the bilayer. Though oxygen ordered, such a surface has generally been viewed as "proton disordered"[27]; that is, an extinction of the proton disorder within the ice.[1] Clearly, these various phases do have common structural elements at the surface and like the smallest clusters, the outermost layer of either ASW or crystalline ice has uncoordinated sites with water molecules that either lack a donor hydrogen, at a lone-pair (acceptor) oxygen site, or have a protruding O−H group for which there is no acceptor. These two types of surface sites, which are intermixed "randomly" in a proton-disordered outermost layer, will be *denoted as d-O for dangling-oxygen and d-H for dangling hydrogen sites*. They are, spectroscopically, the most recognizable and, chemically, the most important aspects of various icy surfaces. They are the surface sites of highest energy which *respond most quickly and definitively when the ice surface is exposed to a new physical or chemical environment.*[2]

It appears that proton disorder is not necessarily rampant at the surface of smooth crystalline ice at low temperatures. Diffraction of an incident beam of helium atoms has confirmed that the usual *oxygen order does extend to the surface of a carefully prepared ice film* on a smooth metal substrate.[28] The diffraction data also raised speculation about *possible proton ordering of the surface.* Classical physics concepts, treating the d-O and d-H sites as opposite dipoles, suggest that the lowest energy state is one with a high level of proton ordering; that is, with the surface composed of parallel stripes of exclusively d-O and d-H sites.[29] Recent computational studies have affirmed the low energy of the striped-surface phase while pointing to the existence and likely importance of numerous outermost-layer configurations which deviate in minor ways from the perfectly striped surface.[30] In general, the energy is reduced for surfaces that limit the number of near-neighbor dangling-hydrogen sites.[31] It was also demonstrated that a striped-surface model best reproduces observed sum-frequency generation (SFG) spectra of a large ice single crystal near 120 K[32] as well as aspects of the helium-beam diffraction pattern.[28] It can be concluded that *a proton-ordered striped surface is energetically favored* but, because of barriers to molecular rotation, may never be fully achievable at the required low temperatures and is probably not a factor for the surface of the smaller nanocrystals.

3.2.1 Formation and FTIR Sampling of Ice Aerosol Nanocrystals

The question of particular interest is where, within this considerable range from complete disorder to full order, does the ice nanocrystal surface fit? Before directly addressing that question, the preparative methods and maturation levels of ice nanocrystals are considered. A cartoon view of the glass manifold and transfer lines are shown along with the sample cell in Figure 3.1.[33] The ice nanocrystal aerosols at the heart of this discussion were typically prepared from a single pulse, to a pressure of ~200 Torr, of ~1% mixtures of water in He(g) or N_2(g) into the thick-walled static cold-condensation cell held at 100 ± 30 K. The heat transfer to the walls by the carrier gas results in rapid cooling of the water molecules so that small aggregates form (~260 K), which quickly nucleate the nanodroplet phase (~250 K) that crystallizes at ~220 K.[34] With the limited water vapor nearly

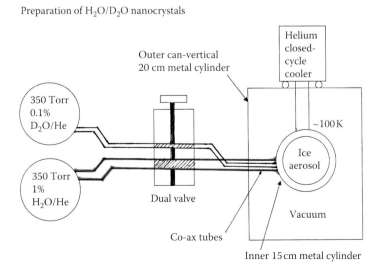

FIGURE 3.1 Illustration of the gas delivery manifold, the cold condensation cell, and the closed-cycle helium cooler for sampling of ice nanocrystals in the 40–160 K range.

depleted, the aerosol then cools to the chosen cell temperature. It is possible that, during the milliseconds required for this final cooling phase, the ice particles collect into loosely connected aggregates, but the transmission electron microscopy of ice nanocrystals, prepared in a similar manner, found the particles to be largely single crystals.[35] Larger *primary* nanocrystals are favored by (1) an increase in the % water in the initial mixture, (2) an increased cell formation pressure for a given mixture, and (3) a higher cell temperature.

Figure 3.1 includes two mixed water-gas storage bulbs, one for H_2O and the other for D_2O. A number of significant published aerosol spectra are of deuterated ice, which simplifies sampling through the avoidance of interfering ambient $H_2O(g)$ absorption bands. However, more pertinent to the current discussion is that the double storage bulbs can be used to generate ice nanocrystals with a low percentage of intact D_2O molecules incorporated substitutionally. A simultaneous release of the two mixtures into the coaxial tubing at the same total pressure allows both time and space overlap of the two pulses as they enter the cold cell. The resulting isolated D_2O is the probe for observation of point-defect activity, as described in Sections 3.3 and 3.4.

Once an ice aerosol is formed, the FTIR spectra are obtained by a direct transmission of the infrared beam along the axis of the cylindrical inner cell, while also passing through the ZnS inner and KBr outer windows. Reflective scatter is not a factor for such small particles (<40 nm) though the infrared spectra do reflect the nanocrystal shapes.[36] The main limitation of sampling the aerosol phase is derived from the relatively short settling half-lives. Despite the small particle size, the infrared absorbance typically decreases by ~50% over a 10-min period with spectral quality seriously degraded beyond ~30 min. Several studies emphasized in this chapter required sample observations extended over many hours. Then, an alternate sampling scheme is used in which the *aerosol particles are collected into 3D arrays* supported on the inner cell windows. This procedure is possible because ~5% of the fresh ice particles collide with and stick to the cell windows during the pulsing of a water-gas mixture into the cold cell. Multiple pump-load cycles can then be used to form an array of optimum optical thickness. Aerosols and arrays with effective ice thicknesses of 0.2–0.5 µm have been used in the studies described in Sections 3.3–3.6.

Information about average particle size is often required to analyze the spectroscopic data. This information is available for ice nanocrystals based on known infrared band intensities. The "linear" plot in Figure 3.2 from an earlier review[21] relates the average particle size to the ratio of the

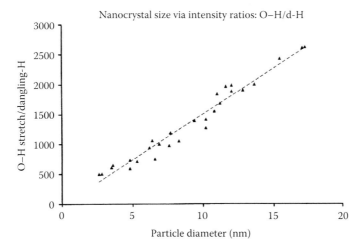

FIGURE 3.2 Plot of the experimentally determined ratio of the integrated intensities of the ice O−H stretch band and the 3692 cm^{-1} d-H surface band versus average particle diameter (particle diameters were determined separately from the ratio of the O−H stretch-band intensity to that of the asymmetric-stretch mode of a monolayer of adsorbed CF_4).

integrated intensity of the major ice band to that of the surface dangling-H mode near 3692 cm⁻¹. The plot linearity, which was established for particles with diameters ranging from 4 to 20 nm, implies that the density of d-H sites on the particle surface is "independent" of particle size. The plot was normalized to actual particle size using an alternative approach. The ratio of the known absolute absorbance intensities of the major ice band and of the band of the asymmetric-stretch mode of CF_4 allows the average particle size to be deduced from the ratio of the two intensities as measured with CF_4 present as an adsorbed monolayer on the ice particles.[21,37]

3.2.2 EVIDENCE FOR STRUCTURAL LAYERING: THE SURFACE, SUBSURFACE, AND CORE SPECTRA OF D_2O AND H_2O NANOCRYSTALS

In the limit of large nanocrystals, the infrared absorbance spectrum is dominated by interior crystalline ice. Careful measurements and the analysis of micron-sized particles have shown and made use of this dominance to obtain the optical constants of ice.[38] At the other extreme of size, the surface (and subsurface) absorbance has a major influence on infrared spectra. For example, at 4 nm *diameter*, the particle surface contains >40% of the water molecules,[39] as demonstrated in Figure 3.3. Although we are most interested in the nature of particles in the 20–40 nm diameter size range, for which both the surface and the interior contribute substantially to the observed absorbance, it is informative to review the stages of structural transitions from small water clusters through large single crystals.

Computational modeling of water clusters of 48 and 123 molecules indicates that the minimum energy structure is globular with an amorphous interior.[21] This is consistent with electron diffraction studies that found no evidence of a crystalline component for clusters smaller than 290 molecules.[40] Infrared spectra show that the component of the crystalline core is small (~10%) at 4 nm, that is, for $(H_2O)_{1000}$, and absent at 3 nm (Figures 3.3 and 3.4).[37] Nevertheless, with reduction of the particle size, the spectrum does not approach that of compacted ASW until ~2.8 nm, in close agreement with both diffraction and simulation results. This is evident that the particle interior is neither

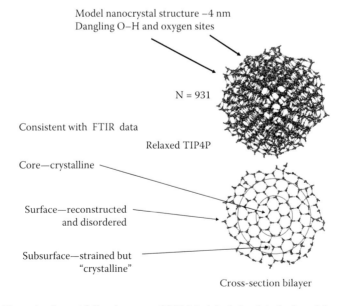

FIGURE 3.3 (See color insert following page 206.) Model of simulated relaxed 4-nm ice particle based on TIP4 potential. This simulated model, consistent with experimentation, shows the dangling surface groups and the reconstructed nature of the surface with reduced three-coordinated-water sites. The core, surface, and transitional subsurface are also indicated.

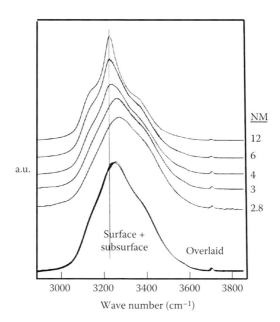

FIGURE 3.4 Comparison of FTIR spectra of H_2O ice nanoparticles at 100 K with decreasing size from the top (12, 6, 4, 3, and 2.8 nm). The four overlaid spectra (bottom) for 3, 4, 6, and 12 nm particles show that, after the removal of a core-component spectrum in the latter three cases, the remaining combined "surface and subsurface" components closely match each other as well as the 3 nm particle spectrum.

amorphous nor fully crystalline between ~3.5 and 3.0 nm. Notably, the spectra of 3.0-nm particles closely match the overlaid spectra obtained by removing the crystalline ice component from spectra of larger particles.[37,39] Removal of the crystalline core absorbance leaves that of the outer layers of ice particles, reasonably considered *a composite of the surface and subsurface*. Apparently, that composite structure is retained, even though the core is lost for particles slightly reduced in size. For this reason, the subsurface of nanocrystals is viewed as a distorted but nucleated phase.

The existence of both a surface and a subsurface region for ice, in general, is axiomatic. The surface is clearly unique with its top layer containing many three-coordinated molecules with d-H and d-O sites. With the surface greatly different than the interior, it follows that the second bilayer of molecules is also unique since the bilayer interacts with quite different water structures at its outer and inner interfaces. This is true regardless of the ordered or disordered nature of the surface layer; so, for all ice, a unique subsurface structure must exist. The difficult fundamental questions are *how many bilayers participate in the subsurface transition region*, between the surface and the ice core, and *what are the important features of the surface and subsurface structure and spectra?*

The nature of the subsurface, as characterized through computational simulations, can be viewed for a 4-nm particle in Figure 3.3. The simulation suggests, and experimental spectra affirm, that, for particles of 4 nm or greater, both *the surface and subsurface are ~2 bilayers* thick (though it would be surprising if this division would be fully independent of the experimental technique applied to the problem). Here we use FTIR spectra for D_2O particles varying from 4 to 16 nm in diameter (Figure 3.5) to demonstrate the separation of ice particle spectra and structure into the three components.[39] The core spectrum is available from much larger ice particles, where the core dominates. Also, there is little difficulty in determining the combined surface and subsurface spectrum, that is, the limiting spectrum as the average particle size is reduced until the core spectrum is nulled, that is, ~3.5 nm. This is shown in Figure 3.5, where the stability of this composite spectrum is confirmed by subtraction of the core component which gives the same residual spectrum for particles up to 10 nm in diameter (see also Figure 3.4).

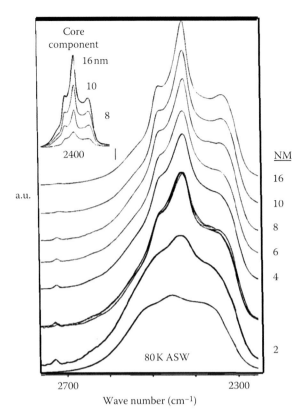

FIGURE 3.5 FTIR spectra at 100 K for *fully deuterated* ice nanoparticles of average size decreasing from 16 (top) to 2 nm (next to bottom). The overlaid spectra are for the 4, 6, 8, and 10 nm particles with the crystalline core components, shown in the inset, removed. The ASW spectrum at 80 K is shown for comparison.

The separation of this composite spectrum into its surface and subsurface components is not straightforward. One approach is based on the fact that warming of a particle array, formed at a low temperature (120 K) to a higher temperature (138 K), results in "Ostwald ripening"; that is, vapor from the more volatile smaller particles is condensed on larger particles. Significant relaxation of the new subsurface to core ice occurs, but relatively slowly. For this reason, vapor condensation thickens the subsurface of the growing particles, while the surface, capable of rapid relaxation, remains of constant width. The result, for a short annealing period, is a gain of core ice and loss of mostly surface H_2O, as the average particle size has increased. By contrast, a second longer 128 K anneal of the same particle array is accompanied by less vapor transport but more complete slow subsurface relaxation to core ice.[41] If the two 120 K spectra, obtained after separate periods of annealing at 138 K, are subtracted in turn and respectively from the original 120 K sample spectrum and the 120 K spectrum measured after the first anneal period, two difference spectra are obtained as presented for a D_2O array at the top of Figure 3.6.

The downward-going triplet of strong bands, in Figure 3.6 spectra (a) and (c), indicates the gain of core ice during each ripening period. When this gain is added out using the known core-ice spectrum one obtains the two spectra, (b) and (d), that reflect the combined loss of surface and subsurface ice during each anneal. The first (b) spectrum is dominated by surface and the second (d) spectrum by subsurface absorption. The result is two new spectra with differing amounts of the two components. Further, only the surface component has a band corresponding to the unique surface d-D mode at 2726 cm^{-1}. Thus, subtraction of (b) from (d), to null the d-D band, *gives the subsurface spectrum*. Then, removal of the subsurface spectrum from either spectrum (b) or (d) *reveals the*

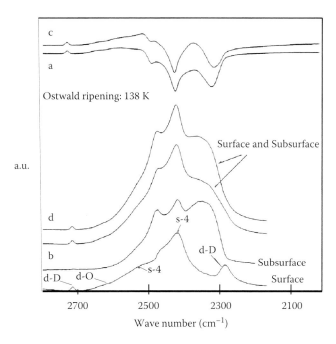

FIGURE 3.6 Sequence of difference spectra at 120 K showing the application of Ostwald ripening at 138 K to the determination of the surface and subsurface spectra of D_2O ice nanocrystals. Curves (a) and (c) compare differences from before and after ripening for 0.5 and 14.0 h, respectively. Curves (b) and (d) show the impact of removing the new ice core components from (a) and (c), leaving the spectra of the combined decrease in surface and subsurface from ripening. Resolution into the subsurface and surface components is based on the different relative contribution of the components to (b) and (d). The s-4 label refers to bands of surface four-coordinated molecules.

surface spectrum. The assignments of the individual surface features of bare ice nanocrystals, as labeled in the bottom spectrum of Figure 3.6, are from modeling of the simulated surface spectrum.[37] (Because of long-range dipole–dipole coupling among the s-4-molecule stretch modes, the features in the band-complex center are broad and the assignment is less definitive; here the s-4 label refers to surface four-coordinated water molecules.)

A nearly identical subsurface spectrum can also be obtained through a separate experimental approach. The adsorbates of intermediate strength SO_2 or HCN, following rapid adsorption onto the surface, cause a relatively rapid reorganization of the surface to a more ordered structure which, ultimately, causes a slow relaxation of much of the subsurface to core ice.[41] This sequence can be understood only by recognizing that the bare particle surface is reconstructed during formation, with loss of approximately half of the dangling three-coordination sites; a result of *high energy d-O and d-H sites reaching out to form strained H-bonds* as confirmed in particle simulations such as in Figure 3.3. This reconstruction is accompanied by a disordering of the ice particle surfaces. However, adsorbate molecules of intermediate strength can break open the strained H-bonds, a de-reconstruction that is observed spectroscopically as an increase in the dangling-bond band intensities. The resulting reduced surface strain favors a subsequent *slow relaxation to core ice* of a fraction of the subsurface, while the surface is unchanged. An FTIR difference spectrum for that change yields the (lost) subsurface spectrum since the known core spectrum can be used to add out the sizeable gain in absorbance from the new core ice. Finally, it is, of course, possible to use this subsurface spectrum to deduce the components of the combined surface and subsurface spectra of both Figures 3.5 and 3.6. The different approaches yield quite similar surface and subsurface spectra which have been reproduced many times, particularly for D_2O particles.

These methods for deducing surface and subsurface spectra do not translate well to nanoparticles of other molecular solids. It is uniquely possible for ice because of the well-known high sensitivity of water-molecule vibrational modes to their molecular configurations which, for the surface, range from molecules with dangling bonds, as in water vapor, to fully H-bonded water. It is this range of surface configurations and associated vibrational frequencies that provides the well-separated infrared bands that enable the resolution of the surface from the interior spectra. By contrast, attempts to deduce credible *surface* spectra of nanoparticles of solid methanol, ammonia, and HCl have been largely unsuccessful. Methanol favors ring configurations at the surface with strong H-bonds that differ little from those of the interior,[42,43] while $NH_3(s)$ intermolecular bonding is relatively weak so that in general the bands of the interior molecular vibrations are not well separated from those of the *surface*. However, the separation of the core spectrum of $NH_3(s)$ from a disordered outer *"transition" shell* has been possible because of broadening and a 19-cm^{-1} blue-shift of the shell ammonia symmetric bending mode.[44] The observation of surface-layer vibrational bands of CO_2 aerosol particles (the second most-studied of nanocrystalline molecular substances after water ice) has not been reported. Shifts, relative to the core, are expected to be weak and overridden by a strong dependence of the asymmetric-stretch-mode absorbance on particle shape.[45]

The ice *subsurface* bonding, spectra, and structure resemble that of the oxygen-ordered core ice much more than that of the surface. The rather modest increase in the range of bond strengths does not account directly for the observable change in the subsurface (vs. the core) spectrum but rather is a source of partial breakdown in the long-range vibrational dipole–dipole coupling that otherwise dictates the structure and breadth of the ice O—H stretch-band complex.[22,46] For both the subsurface and the core ice, *the band structure comes nearly exclusively from this long-range transition dipole coupling which extends over the entire particle core*; but the coupling range is reduced for distorted structures such as the ice subsurface.[47] That dipole coupling is the source of the O—D (or O—H) band structure is apparent from the fact that the O—D (or O—H) stretch-mode band of HDO in ice is a multiplet at high concentrations, but, when isolated, appears as a single smooth relatively narrow absorption band.[39,48] The reduced range of the dipole coupling for the subsurface modes causes a broadening of the subbands of the band complex relative to those of the core, as can be noted from Figures 3.5 and 3.6.

Despite presenting the case that the most stable form of the outermost layer of a smooth ice surface is very likely the Fletcher proton-ordered surface phase,[28,30] our discussion has presumed for the most part that the ice *nanocrystal surface is disordered*. The presumption is based on simulations that predict reconstruction to a badly disordered outermost layer with a reduced number of dangling coordination sites,[21,39] and the observations from FTIR that show adsorbed SO_2, in particular, de-reconstructs the disordered particle surfaces as signaled by an increase in the absorbance of the surface d-H and the crystalline core ice.[41] There is other evidence for the surface disorder that we have noted for crystals less than ~30 nm in diameter. This is based on the spectral observations of nanocrystals of a larger size when coated with the adsorbate CF_4. We have noted,[49] and simulations have confirmed,[27] that the dipole-intense triply degenerate asymmetric-stretch mode of CF_4, in a monolayer on smooth globular nanoparticles, produces a sharp transverse optical (TO) and longitudinal optical (LO) doublet with ~80-cm^{-1} splitting. However, the expected sharpness of the two bands, and to a lesser degree the magnitude of the splitting, decreases with roughening of the surface. From this, a sharpening of the CF_4 monolayer TO–LO bands, observed along with extensive Ostwald ripening during annealing at 140 K, suggests that the ice surface becomes smoother and perhaps moves from disorder towards order for larger (~45 nm diameter) nanocrystals. Facets on the larger particles may favor the Fletcher phase of alternating stripes of d-O and d-H sites as proposed for carefully prepared ice layers[30]; but there is no direct evidence that this is the case. A tentative conclusion is that the curvature of the surfaces of the smallest nanocrystals plays a part in their apparent surface disorder while particles with a diameter of 45 nm and larger likely have surface facets with oxygen order and a potential for some level of proton ordering; or, at least, structuring to reduce d-H–d-H interactions.[31]

Our focus has been on the ice stretch-mode infrared spectral region, but there is also much to be learned from the surface spectrum of the water bending mode.[37] This mode is very weak in the infrared spectra of the crystal phases of ice and also experiences serious interference (Fermi resonance) with the overtones of the librational modes. However, the intensity increases significantly with an increased asymmetry of binding of the water molecule.[37] For example, the bend-mode absorbance increases by a factor of ~3 as the coordination number decreases from 4 to 3. As a result, the bend-mode band complex for the outermost-layer of surface water molecules is readily observed for ice nanocrystals in the range 1600–1750 cm^{-1} and specific assignments have been made for d-H molecules at 1651 and d-O molecules at ~1710 cm^{-1}. The values for water vapor (1595), liquid water (1645), ASW (1711), and bulk ice (1735 cm^{-1})[50] bands show that the bending mode shifts up in frequency, that is, opposite to the O−H stretch mode, as the water H-bonding increases. Not surprisingly, the bands of three-coordinated surface molecules fall closest to the liquid water and ASW band positions.

3.3 STRUCTURAL INSIGHTS FROM RESPONSE OF PARTICLE SPECTRA TO WEAK ADSORBATES: H$_2$ AND CO AS CASE STUDIES

In a later section, the kinetics of the reactions of intermediate and strong adsorbates with ice nanocrystals will be examined. In the previous section, the role of intermediate adsorbates in modifying the overall structure of ice nanoparticles has been presented and used in de-convoluting the ice nanocrystal spectra into its three components. Here the interaction of weak adsorbates is examined. Perhaps the most interesting weak adsorbates are H$_2$ and CO, diatomic molecules of special importance to extraterrestrial studies of ice–adsorbate interactions.

As noted earlier, the influence of weak adsorbates on the structure of ice nanocrystals is trivial except for the surface, for which the infrared bands are of low intensity and the shifts induced by the adsorbates are small. As a result, the most useful approach is to compare a bare particle spectrum with that having a monolayer of adsorbate, with care to use precisely the same temperatures (as ice-band positions are notoriously temperature sensitive[51]). The FTIR difference spectra, *for coated-ice spectra subtracted from that of the bare ice*, which null out the unaffected subsurface and core absorption, are dominated by distinct positive features that mark the loss of the bare-ice surface bands, and negative ones reflecting gain in absorbance at the positions of the adsorbate-shifted surface water vibrations. For example, the three difference spectra near the bottom of Figure 3.7 for three weak adsorbates on D$_2$O nanocrystals, that is, H$_2$, N$_2$, and CO from top to bottom, are typical; and also remarkably similar to difference spectra for these weak adsorbates on D$_2$O ASW.[2] The major features, which have been assigned to the d-H, d-O, and s-4 surface sites based on simulated spectra, can be compared to the subbands of the ice surface spectrum in Figure 3.6. By contrast, the two difference spectra near the top of Figure 3.7, for the intermediate adsorbates H$_2$S and acetylene, are dominated by strong negative bands that reflect a large amount of core ice formed because of surface and subsurface ordering.[52] Several studies have been reported that transform the observed shifts/intensification of the d-D/d-H bands into a measure of the weak-adsorbate interaction strength with the surface sites (see, e.g., refs. [53,54]).

Because of the derivative nature of the spectral features of Figure 3.7, it is difficult to fix the precise positions of the shifted bands. This limitation can be overcome using a variation on the difference spectra. As we saw in Section 3.2.2, Ostwald ripening can be used to change ice particle size and thus reduce the amount of surface and subsurface of a particle array or aerosol. The difference between spectra measured before and after ripening, whether of bare or of coated ice, gives a particularly clear look at the d-D (or d-H) and the d-O bands at frequencies greater than that of the main ice band. This is shown in Figure 3.8 for bare ice and ice coated with CO and ethene. Comparison with the top bare-ice spectrum gives a clear view of the d-D and d-O shifts induced by a monolayer of the adsorbates despite the presence of broad subsurface absorption in each spectrum; for ethene, the shifts of both bands are double that for CO. Ethene is a borderline weak-to-intermediate adsorbate, but unlike acetylene shows no ability to reorder the subsurface so as to enlarge the ice core.

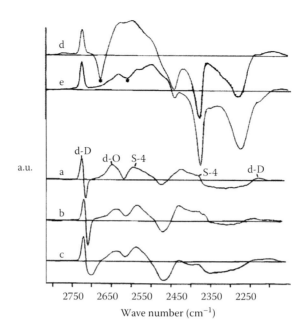

FIGURE 3.7   Difference spectra at 83 K (vs. the bare particle spectrum) showing the impact of weak (b, c: N_2, CO) and intermediate strength (d, e: C_2H_2, H_2S) adsorbates on the surface spectrum of ice nanocrystals (the spectrum for H_2 (a), is for 30 K). Note: the assignments of the features as given in (a) to the surface sites, the asterisks that mark the strongly shifted d-H mode in (d) and (e), and the intense negative-going features for new core ice generated by the intermediate adsorbates.

3.3.1   VIBRATIONAL SPECTRA OF CO ADSORBED ON ASW AND ICE NANOCRYSTALS

Information typically available from the FTIR spectra for weak adsorbates, besides the shift/ intensification of the ice surface vibrational bands, includes unique features that emerge in the frequency region of the adsorbate infrared-active modes. In the case of CO, the only molecular vibration is the stretch mode near 2140 cm^{-1}, the band of which has been studied extensively on

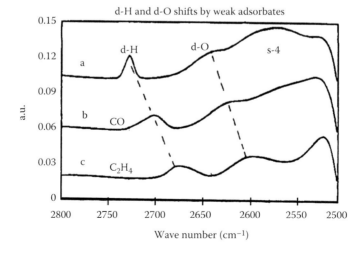

FIGURE 3.8   Difference spectra at 100 K, from un-annealed minus annealed (145 K) samples, which reveal the surface vibrational-mode band positions from the loss of surface through Ostwald ripening. Spectra a, b, and c are for bare, CO-coated, and ethane-coated ice nanocrystals.

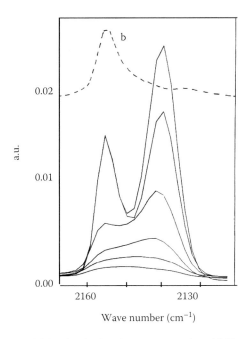

FIGURE 3.9   Infrared spectrum of CO adsorbed on ice nanocrystals at 40 K as % of surface d-H sites occupied. Coverage ranges from 20% to 100% from bottom up, and is for 20% CO and 80% N_2 coverage in (b).

ASW and nanocrystalline ice with parallel results for the two phases. Much of the interest in CO on ASW stems from the general view that interstellar ice is amorphous,[55,56] so that adsorbate CO bands on amorphous ice[57–59] might contribute to interstellar spectroscopy. Differences or similarities, between CO adsorbed on ASW compared to crystal particles, might offer a useful indication of the phase of the extraterrestrial ice in question.

Both ASW and nanocrystals of ice give CO adsorbate bands that are quite similar; that is, a doublet with the lower-frequency component near 2137 and the second near 2152 cm^{-1}, with the 2137 component more than double the integrated intensity of the other. However, both components are ~3 cm^{-1} higher in frequency for the nanocrystalline ice surface (Figure 3.9) than for ASW,[60] suggesting that differentiation of the ice phases through the adsorbate spectra is possible. Based largely on response to competition with different 2nd adsorbates,[2,58] the two bands are generally assigned to the interaction with the d-H (~2152) and the more numerous oxygen (~2137 cm^{-1}) surface sites. The 2137 cm^{-1} band is broader, likely reflecting a range of CO interactions with the oxygen of s-4 as well as d-O sites. Based on the peak maximum of ~2150 cm^{-1} for 20% CO cover, the bottom spectrum of Figure 3.9 reflects a preference of CO on ice nanocrystals for the 2152-cm^{-1} "sites." Further, the spectrum labeled "b" indicates that CO molecules compete only for the 2152 cm^{-1} "sites" when adsorbed from a dilute gaseous mixture with N_2. Not shown, however, is the reverse situation when the 2nd gas is the ether proton-acceptor ethylene oxide[57]; a result confirming that the 2152-cm^{-1} band is for CO occupying the d-H sites. The more general insights from this study are that *infrared bands of adsorbate molecules can be assigned in a meaningful way to the different surface sites that we have recognized in Section 3.2.2*, and that competition with a second well-chosen adsorbate can facilitate the assignments.

3.3.2   Vibrational Spectra of H_2 Adsorbed on ASW and Ice Nanocrystals

Like CO, there is considerable interest in H_2 absorbed on ice surfaces because of its abundance in extraterrestrial environments; most particularly in the cold regions of the dark clouds of interstellar space.[61,62] The stabilization/concentration of H_2 in icy media is also of great interest because of a

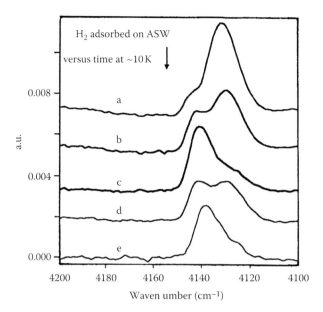

FIGURE 3.10 Infrared spectrum of H_2 adsorbed on ASW near 10 K. Time of adsorption increases from (a) to (c). Before (d) the H_2 was desorbed by warming and then re-adsorbed within the closed system. Shift of peak frequency of the induced H_2 infrared activity is, (a) to (c), from 4130 to 4140 cm^{-1}.

need to increase its ease of handling if a "hydrogen economy" is to develop and flourish.[19,63] Because of symmetry, the stretch mode of any isolated diatomic molecule, X_2, is forbidden in the infrared. This is true of H_2 (g), but adsorption on a surface can break that symmetry. It is, therefore, not surprising that ASW and ice nanocrystals, when exposed to H_2 (g), give rise to infrared bands in the H_2-stretch-mode region of the spectrum[64] near 4130 cm^{-1}. FTIR spectra of H_2 on porous ASW showed that a single band near 4132 cm^{-1} evolved into a doublet (4132 and ~4145 cm^{-1}), with the higher-frequency component dominating and shifting to 4140 cm^{-1} with time.

These puzzling spectra, shown in Figure 3.10, were identified with a behavior pattern related to the statistical probability of ortho-versus para-H_2 as a function of temperature and the relative interaction strength with the ASW surface. Computations showed that the ortho-H_2 $Q_1(1)$ state presents itself to the surface as more anisotropic than the para-H_2 $Q_1(0)$, which leads to a stronger association with the d-H and d-O surface sites. As a result *the freshly adsorbed hydrogen favors the ortho state* well beyond the 3 to 1 normal statistical distribution in the warm gas phase. With time, some paramagnetic impurity, likely O_2, catalyzes the ortho to para (ground state) conversion at the low sample temperature, resulting in primarily adsorbed para-H_2. The computations[64] also suggest why the para-hydrogen band shifts slowly from 4145 to 4140 cm^{-1} by identifying a large choice of surface configurations of similar but different energy. Molecules that ultimately manage an electrostatic bonding with the d-O molecules favor the lower frequency while also projecting a stronger vibrational dipole. Although the observation is apparently tentative,* a band for condensed hydrogen in interstellar clouds at 4141 cm^{-1} was reported and assigned to the Q_1 (1) state of H_2 in/on dirty amorphous ice.[61] However, the earlier results[64] suggest that Q_1 (0) (i.e., para-H_2) is a more appropriate assignment.[62]

The experimental and computational insights to hydrogen adsorbed on ASW[64] apparently apply, with slight modification, to adsorption on ice nanocrystals as well. The dashed bands of Figure 3.11, of (a) ortho- and (b) para-H_2 adsorbed *on ice nanocrystals* at ~16 K, are sharper and blue-shifted (+6 cm^{-1}) relative to the bands for an ASW sample. On the other hand, as for ASW, H_2 adsorbed on nanocrystals also showed strong preference for the ortho-H_2 in the initial adsorbed state, followed

---

* See a "weak" correction statement: *Science*, Vol. 287 (#5455), p. 976 (2000).

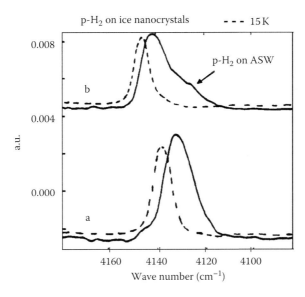

FIGURE 3.11 Comparison of the H$_2$ stretch-mode peak positions of ortho-H$_2$ in (a) and para-H$_2$ in (b) for adsorption on ice nanocrystals (- - - -) and ASW. The crystalline-to-amorphous ice shift is down by 6 cm^{-1}.

by a slow transition to the ground state of para-H$_2$. The evidence is that the ortho state is more strongly bound to the surface and that the ortho-to-para frequency shift is similar for the amorphous and crystalline ice. The para-H$_2$ band of ASW (4140) is closer to the "observed" H$_2$ band on interstellar ice (4141)[61] than that of the para-H$_2$ on ice nanocrystals (4146 cm^{-1}); consistent with both the amorphous nature of interstellar ice and the possible validity of the spectrum reported for the interstellar H$_2$. The H$_2$ frequency difference for the two ice phases confirms that *the nanocrystal surface is different than that of the porous ASW*, with the binding, at least of H$_2$ molecules that have significant vibrational-dipole strength, greater for the ASW.

There is also much to be learned about the surface binding of H$_2$ to both forms of ice by monitoring the H$_2$ impact on the surface spectra[2]; the most significant insights affirm those already noted from the observation and simulation of the H$_2$ spectra. For example, the d-D shift for the ortho-H$_2$ is ~5 cm^{-1} and is diminished for para-H$_2$ on both ASW and ice nanocrystal surfaces. Although the d-O shift is somewhat greater (~15 cm^{-1}), both *the d-H and the d-O shifts are the smallest observed for molecular adsorbates on ice* (CF$_4$ shifts are quite similar and certain adsorbates at submonolayer levels go nearly exclusively to one type of dangling site, leaving the other largely unshifted).

3.4 PROTON AND ORIENTATIONAL L-DEFECT ACTIVITY IN THE INTERIOR AND AT THE SURFACE OF BARE ICE AEROSOL NANOCRYSTALS

One might naively assume that the simple reliable manner of "measuring" proton (or hydroxide) mobility in ice would be observations on the conductivity of ice, particularly at temperatures above 200 K where the concentration of ionic defects is generally assumed to be significant. However there are severe problems with such measurements, perhaps best related to (1) uncontrollable space–charge build-up at electrical contacts with the surface, and (2) an unusually great amount of "interfering" proton transport along ice surfaces.[1] These measurement problems are so severe that, following a critical analysis by von Hipple et al. in 1971,[65] it was generally accepted that little valid information had been gained from numerous conductivity measurements to that date. The concept spread that, despite an elegantly descriptive theory developed by Onsager and Dupuis[66] and Jaccard,[67,68] *perhaps there is no such thing as proton hop steps within bulk ice* (i.e., steps that parallel the Grotthuss mechanism for conductivity in liquid water).

The problems noted above indicated that *a reliable probe of charge transport within ice must respond directly to changes in the ice interior* so as to avoid the dominant effects of the ice surface. This was reemphasized in results of Cowin et al. showing that hydronium ions soft-landed on an ice surface remain there over a large temperature range.[69] A direct interior probe was mentioned in the Introduction: namely a low percentage of D_2O molecules isolated intact and substitutionally in the ice interior. As described, if a D_2O molecule within ice converts to an HDO molecule, the evidence is strong, and generally accepted, that a proton or hydroxide ion has passed through the D_2O site. Regardless of which ion passes, its motion can be envisioned as a result of successive proton hops from one end of a sequence of hydrogen bonds to the other, with the charge center moving accordingly (Figure 3.12). Thus, the hop step that leads to isotopic exchange can be a source of ion conductivity. However, random hopping can only cause *H- or D-transfer back and forth within a hydrogen bond*. To actually fully separate the two D atoms, so as to form isolated HDO, a second class of defects must be invoked[12,13]; mobile orientational (L- or D-) defects capable of moving the deuterium atom to a neighbor hydrogen-bond position during defect passage through the $(HDO)_2$ "dimer" created by proton hopping (Figure 3.12). This is the same pair of defects invoked, in ice proton-conductivity theory, to reset the water orientations within proton-transfer chains so that sequential proton passage is possible.[1]

The conversion of probe D_2O molecules to $(HDO)_2$ and ultimately to HDO molecules isolated in ice, reflecting both the hop and turn steps, can be monitored by FTIR as each of the three D-isotopomers has a distinct infrared spectrum, as shown in Figure 3.13. This capability enabled the determination of defect activities in early studies of thick crystalline ice films in the 135–150 K range.[12,13] In both the early and more recent studies, the observed quantity is *labeled "activity" since it reflects a combination of defect mobility with defect population*. The results showed similar

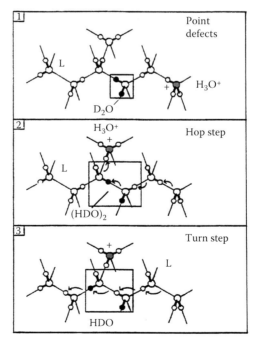

FIGURE 3.12 Top panel: representation of the orientational L-defect and protonic H_3O^+ point defects. Note the missing hydrogen of the L-defect while the classic D-defect would show 2 H atoms between the oxygens. Middle panel: passage of the protonic defect down an H_2O water chain and through a D_2O molecule (hop steps); bottom panel: a similar passage of the L-defect (turn steps). Hop steps produce $[HDO]_2$ and, combined with the turn steps, isolated HDO.

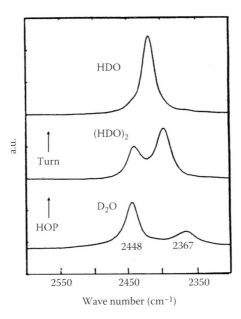

FIGURE 3.13   Spectra of the isolated isotopomer of D_2O (bottom) and the isotopomers [HDO]$_2$ and [HDO] that are produced by the passage of protons (hop) and L-defects (turn) through the D_2O crystal site. Note relationship to Figure 12.

interior proton and L-defect activities with roughly an hour required for passage of both a proton and an L-defect through 50% of the D_2O sites at 140 K, with defect-activation energies of ~9 and 12 kcal/mol, respectively. Moon et al.[70] have developed alternate methods of monitoring interior proton activity by isotopic exchange in layered films and have confirmed the low-temperature proton mobility; that is, *interior proton hopping occurs with low activation energy.*

By contrast, the hydroxide ion was shown not to readily engage in charge transport in thick ice films; lightly NH_3-doped samples displayed no isotopic exchange up to the maximum temperature sampled (~180 K).[71] Limited evidence that the D-defect is much less mobile than the L-defect was also reported for the thick crystalline ice films. Therefore, as we address observations on isotopic exchange in ice nanocrystals, *the impact of defect activities will generally be ascribed to the protonic ion defect and the orientational L-defect.*

3.4.1   AEROSOL PARTICLE-CORE PROTON AND L-DEFECT ACTIVITY

The formation of ice particles in aerosols with incorporated small amounts of intact D_2O, based on a coaxial $H_2O–D_2O$ delivery to the cold condensation cell, was described in Section 3.2 and Figure 3.1. An example of the spectra of components of the isotopomeric mixture that result from isotopic exchange is given in Figure 3.13 and as the subbands of Figure 3.14. The FTIR spectra that evolve as a function of time reflect the isotopic exchange[72,73] (Figure 3.14 is for SO_2-coated ice at 122 K). The change with time (not shown) is observed as the decreasing integrated intensity of the bands labeled D_2O and the increase in intensity of the HDO and (HDO)$_2$ bands, as identified in Figure 3.13. The resolved spectra are then used to determine the protonic and L-defect activities (after correction for the presence of HDO that forms during sample preparation).[12]

The bare-nanocrystal proton activity was judged to be quite similar to that for thick ice films for the same sample temperatures.[72] However, for the bare ice nanocrystal samples, there is little evidence of the intermediate (HDO)$_2$ isotopomer in the resolved spectrum. This implies that the L-defect step, that converts (HDO)$_2$ to isolated HDO molecules, is much faster than the proton hop

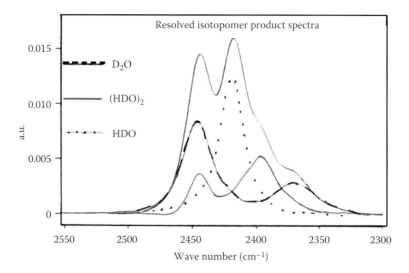

FIGURE 3.14   FTIR spectral band complex (top) from observations of *interior* isotopic exchange produced by protonic and L-defect activity. The sample, with SO_2 adsorbate, displays significant exchange during 20 min at 122 K. The magnitude of the subbands of the complex (D_2O, $[HDO]_2$, and HDO), determined using standardized spectra such as in Figure 3.3, indicate the exchange rates and defect activities.

step. This, in turn, signals an enhanced L-defect activity with respect to the case for thick ice films.[12] The L-defect turn step, which converts $(HDO)_2$ into isolated HDO, could only be observed by increasing the proton hop rate using a weak-acid adsorbate, in particular SO_2. Then, as in Figure 3.14, comparable exchange-product band intensities were obtained for HDO and $(HDO)_2$ following 20 min at 122 K.

The *enhanced nanocrystal L-defect activity* is not surprising, since the nearby surface of the nanocrystals can be viewed as a low-energy source of such defects. Rotation of a surface d-O molecule, to generate a d-H site plus an L-defect, can occur more easily, that is, with the breaking of fewer H-bonds, than the generation of an internal L-defect. Even more L-defects can be injected into an ice particle, if a proton-acceptor adsorbate, for example, an ether, is available to stabilize the d-H molecules. This enhancement of the L-defect activity by ether adsorbates has been reported.[72,73]

3.4.2   ENHANCED SURFACE PROTON ACTIVITY: TIME DECAY OF PARTICLE-CORE ACTIVITY

As noted above, even for bulk ice at relatively high temperatures, there is evidence that proton conductivity of the ice surface dominates over that of the interior. However, the isotopic exchange measurements for *thick ice films lack a probe to measure the surface versus the interior proton activity*; the FTIR signals of dilute isotopomeric surface species being too weak. This problem is overcome with ice nanocrystals because of the much greater surface-to-volume ratio. As shown in the left panel of Figure 3.15, the surface d-D of D_2O (2725 cm^{-1}) can be observed and differentiated from the surface d-D of HDO (2712 cm^{-1}).[15,33] Further, when the left panel of Figure 3.15, is compared with the right panel, which shows the interior isotopic exchange, it is clear that the *surface exchange rate is much greater than for the interior.* The relative rate of proton-induced isotopic exchange is a factor of ~20 greater for the surface than the interior. This factor increases with time as the interior rate decays; by a factor of ~40 as the % D_2O exchange approaches 80%.[73] This interior decay is thought to be a result of proton trapping at the ice surface where several levels of computation,[15,33] as well as related experiments,[69,74] have indicated significantly lower proton energy.

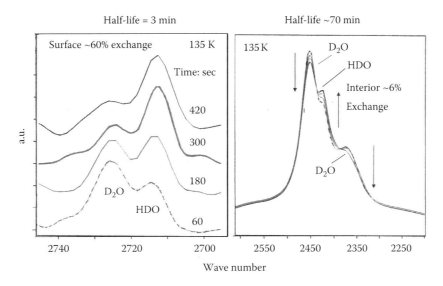

FIGURE 3.15 FTIR bands of surface D_2O and HDO with increasing time (left panel; bottom up) showing rapid exchange/surface proton activity. On the right, the minor simultaneous change in the bands of the interior-ice isotopomers indicates a proton activity reduced by a factor of ~23 relative to the surface activity.

3.5 ADSORBATE CONTROL OF POINT-DEFECT ACTIVITY OF ICE AEROSOL NANOCRYSTALS

Ice aerosol nanocrystals also have the desirable property that it is possible to easily monitor the impact of adsorbates on the activity of the protonic- and L-defect. The response of both surface and the interior defects can be observed to a degree. However there are serious limitations on following the response of *surface* defect activity to adsorbates, since the bands of the surface D_2O and HDO, always quite weak, shift and become hidden as the adsorbate dosing increases. For this reason, most published results for adsorbate impact on defect activity are for the interior of ice particles, where the adsorbates do not interfere with the probe spectra. Although the data examined here have been largely reproduced with ice nanocrystals in aerosols, the time stability of ice arrays favors their use for the quantitative longer-term measurement of adsorbate effects on particle defect activity.

3.5.1 Weak-Acid Control of Proton Activity in the Core of Ice Nanocrystals Near 130 K

A monolayer of H_2S has been observed to increase the interior proton activity by a factor of ~4 compared to that of bare ice particles. This refers to the early stages of isotopic exchange before the activity for bare ice decays extensively. The comparison for the two cases is presented in the bottom panel of Figure 3.16, where the projected zero-order half-life for exchange is plotted versus the cumulative exchange, with a compressed scale on the left for the slower exchange of the bare-ice sample. The rising curves are a semiquantitative measure of the decaying rate of exchange for both samples, with the *net decay* at a cumulative exchange of 80% nearly an order of magnitude greater for bare ice than the H_2S-doped ice particle array. If we apply the concept that the decay results from trapping of protons on the ice surface, then less decay in the proton activity for H_2S-doped particles can be attributed to the saturation of the deeper surface traps by protons released by the weak acid. Ultimately, the proton-activity ratio, for samples with and without H_2S dosing, approaches two orders-of-magnitude after 80% cumulative exchange.

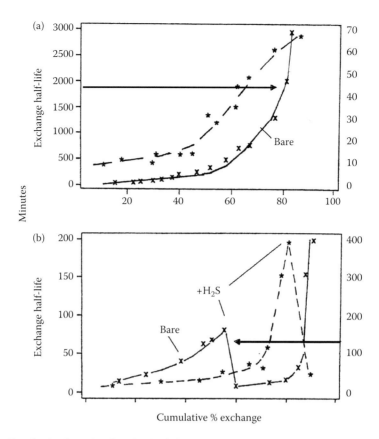

FIGURE 3.16 Panel (a): plots showing the half-lives (*t*/2) for *interior* isotopic exchange as observed for stages of the cumulative exchange. The plots show much longer (×4) *t*/2 values for bare ice particles versus ice with adsorbed H$_2$S. The decay in rates, shown by increasing *t*/2 with % exchange, is also much greater for the bare ice. Panel (b): discontinuities in the *t*/2 plots show the strong influence of adding or removing adsorbed H$_2$S from the ice nanocrystals at 136 K. In addition to the labeled additions, H$_2$S was removed at ~65% (dashed curve) and 80% (solid curve).

The influence of H$_2$S on the interior-ice proton activity represents proof that the weak acid undergoes a level of ionization as a semisolvated adsorbate on the ice surface. Perhaps more surprising is that adsorbed CO$_2$, which presumably forms surface-adsorbed (but undetected) H$_2$CO$_3$, also supplies protons to the ice interior at only a slightly lower rate. This behavior likely says more about the surface than the weak acid adsorbates; that is, the ionization is promoted by the surprising stability of the hydronium ion bonded at the ice surface[33]; a stability understood, from computations, in terms of the three strong H-bonds to surface-water d-O sites plus the compatibility, of the weakly negative backside of the hydronium ion, with a nonbonding configuration.

The adsorbate control of the interior proton activity is even more impressive when advantage is taken of the volatility of the adsorbed state of the two weak acids at 136 K. Then the acids can be used to alternately enhance and suppress the exchange rate by their addition and subsequent removal to vacuum. This behavior, which has been observed for both H$_2$S and CO$_2$, is displayed for H$_2$S in the top panel of Figure 3.16. The solid curve shows, on the right scale, that the bare-ice interior exchange rate decays significantly (half-life increases) up to 55% total exchange. Then, the addition of H$_2$S rapidly accelerates the exchange nearly an order of magnitude until evacuation, at 80% exchange, produces a dramatic rate decrease. On the other hand, the dashed plot shows, on the left scale, how the exchange rate for H$_2$S-coated ice changes slowly until the acid is removed near 65% cumulative exchange. Then the rate drops dramatically, as expected for bare ice, until the

H_2S adsorbate coat is reestablished near 80% total exchange. The added H_2S then dominates, causing a 10-fold acceleration of the exchange rate that reflects the *interior proton activity.*

3.5.2 EFFECT OF THE BASE DOPANTS AMMONIA AND MONO-METHYL AMINE ON ICE-DEFECT ACTIVITY

The weak acid results, besides showing that the population of the interior protons can be controlled from the surface, also suggest that ionized adsorbates are, indeed, the instruments of that control. This judgment is strengthened through observation of the impact of the base adsorbates, ammonia and mono-methyl amine, on proton activity within ice nanocrystals. Since the two bases behave qualitatively the same (with the amine base impact somewhat stronger) the focus here is on ammonia. Ammonia differs from the weak-acid adsorbates, for which the impact is minor for low doses, displaying a *major impact at <1% of a monolayer.* This level of NH_3 dosing *completely stops the isotopic exchange at 150 K*; both in the interior and on the surface of the nanocrystals. The effect has been observed for both aerosols[16] and particle arrays.[33] This result recalls the report that trace doping with ammonia or the organic base 7-azaindole stops proton activity in thick ice layers even at 180 K.[71] The nanocrystal result is both reassuring and informative. It is reassuring, because it affirms the relative immobility of the hydroxide ion while also indicating a high purity of the ice particles. *The low natural concentration of protons, expected for the pure state, is apparently further reduced by orders of magnitude* through the mass-action response of the natural dynamic proton—hydroxide autoionization equilibrium. A convincing effect is displayed of the shift from mobile protons to immobile hydroxide ions.

The result is further informative, because it shows that the *hydroxide immobility extends to the ice surface.* A recent high level on-the-fly simulation of the behavior of the hydroxide ion in ice found new deep off-the-lattice interior traps, while also showing strongly bound surface sites with the surface energies generally lower than in the interior.[16] However, a considerable range of trap depths is revealed in the simulations, consistent with the observation that increasing the ammonia dosing level beyond 10% of a monolayer results in a gradual revival of the interior isotopic exchange. In other words, the generation of more hydroxide ions by the surface NH_3 can saturate the deeper traps and ultimately, near a full monolayer, increase the interior exchange rate to the value for pure ice.[16] Apparently, hydroxide ion immobility originates from the presence of natural deep traps, with the hop step of protons from water molecules to the hydroxide ions otherwise possible. The lost proton activity, for the ice arrays lightly dosed with NH_3, was also recoverable through the addition of acid adsorbate molecules.

3.5.3 SURFACE-BASED DYNAMIC EQUILIBRIUM OF ICE AUTOIONIZATION

The most general conclusion derived from the observations of the enhanced proton activity at the surface of bare ice, and the response of the interior activity to surface adsorbates, is that *the controlling dynamic equilibrium is strictly surface based.*[73] The observed interior D_2O isotopic exchange rates imply an approximate time period required for a proton to pass through half of the sites of the ice lattice. At 140 K that is, respectively, an hour for the interior sites compared to a minute for the surface sites of bare ice. This implies that many hours would be required to fully establish equilibrium in the ice interior. However, observations show that *the interior proton activity achieves a steady state in a matter of minutes when a weak acid is adsorbed* on the particle surfaces. This can be deduced from Figure 3.16 from the sharp interior rate changes that are observed. It follows that the controlling autoionization equilibrium must develop on the surface where the much greater exchange rates imply that only a few minutes is required for relaxation at 140 K. The interior activity reflects this equilibrium and adjusts rapidly to it. That is, the interior proton population achieves a steady-state value via random access from the surface over finite energy barriers. In general, there are very few protons generated by autoionization in the interior. Otherwise, prior to being trapped, they would produce isotopic exchange, within the lightly ammonia-dosed particles, particularly at higher temperatures. That is not observed.

3.5.4 Managing Interior L-Defect Activity with SO_2 and H_2S Adsorbates

The discussion of the influence of adsorbates on defect activity has focused on protonic activity. We could extend that discussion to adsorbed SO_2 which has been observed to convert to the "weak" acid H_2SO_3 (Figure 3.17) that acts as a rich source of protons for the interior of the ice nanocrystals.[75] That behavior has been successfully modeled to (1) show that a large fraction of the adsorbed SO_2 prefers the H_2SO_3 form, and (2) identify likely steps, in the solvation near the ice surface, that ultimately lead to a separated pair of hydronium and bisulfite ions. However, SO_2 is also of interest because of a unique effect on the *suppression of the L-defect activity.*

It was remarked in Section 3.4.1 that bare nanocrystals reflect an enhanced L-defect activity relative to bulk ice, presumably because the disordered surface with missing H-bonds can act as a significant L-defect source through the rotational conversion of d-O to d-H sites. Adsorbed ether and H_2S molecules appear to add to the injection of L-defects from the surface to the interior through stabilization of the d-H sites. The contrast between H_2S as a source and SO_2 as a depressant of L-defect activity is apparent from the isotopic-exchange spectra. Since both adsorbates are a source of increased protonic activity, a build-up of the $(HDO)_2$ isotopomer could be expected. However, the spectra show that with H_2S the adsorbate, the D_2O proceeds quickly to primarily isolated HDO (Figure 3.18), indicating a significant increase in the L-defect activity. By contrast, most of the $(HDO)_2$ product of the hop state is retained over considerable time when the adsorbate is SO_2, as in Figure 3.14 (The HDO of Figure 3.14 is mostly "contaminant" from exchange during the sample preparation stage.)

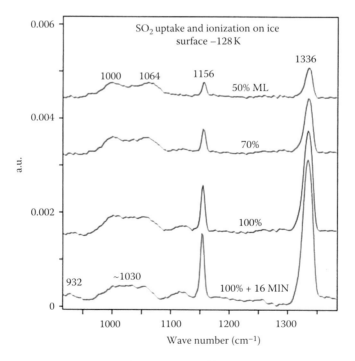

FIGURE 3.17 FTIR spectra of adsorbed SO_2 on an array of ice nanocrystals at 128 K as function of coverage/time from top. Bands, at 1000 and 1064 cm^{-1}, assigned to bisulfite ion grow and then saturate with SO_2 coverage, while SO_2 bands (1156 and 1336 cm^{-1}) continue to grow even after 100% of a monolayer.

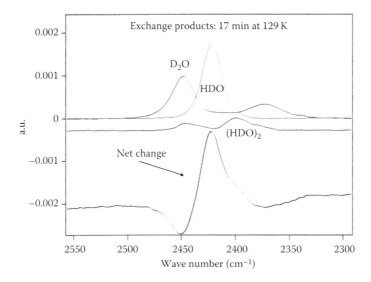

FIGURE 3.18 Difference spectrum (bottom) shows overall effect of isotopic exchange from D_2O, in the interior of ice nanocrystals, converting to [HDO]$_2$ and isolated HDO at 129 K after coating with H_2S adsorbate. The resolution of the bottom spectrum into components (top) reveals the amount of lost D_2O and gain of primarily isolated HDO. The large HDO gain implies a rapid turn step that depletes the [HDO]$_2$.

Speculation as to why SO_2 might depress the interior L-defect activity has focused on the possibility that the increased order induced in the subsurface region (described in some detail in Section 3.2.2) increases the energy for injection of L-defects to the ice interior. This is based on the known greater L-defect activity for ASW than for crystalline ice.[76] It is not so clear as for the protonic interior activity, but the ability to adjust the interior L-defect activity using adsorbates, and the evidence that the L-defect activity is greater near the surface,[72] raises the possibility that, as for protons, *the interior activity depends on a dynamic L and D-defect equilibrium in the surface region*, which serves as a source of the occasional L-defect able to surmount an energy barrier to the interior.

3.6 QUANTITATIVE KINETICS OF HYDRATE FORMATION AND TRANSFORMATIONS NEAR 120 K

The formation/transformation rates of solid hydrates of adsorbates are enhanced by the large surface-to-volume ratios of nanoparticles. This enables sampling on laboratory timescales in the 100–170 K range with quantitative kinetic data having been obtained for a range of hydrates.[3,4,5,77] The formation of the various hydrates; that is, ammonia, strong-acid, and clathrate, from adsorbates occurs by the nucleation of a particular phase at the surface, followed by reaction that forms an outer-crust of hydrate product. This establishes a *hydrate—ice-core interfacial reaction zone* which requires a supply of reactants for the product crust to further expand and take over the particle. Usually, the supply of reactant to the reaction interface is by *transport of adsorbed molecules through the outer product crust*, a transport based on molecular diffusion that often controls the rate of conversion of ice to hydrate.

Formation of a product, *from particle reaction with an adsorbate*, is often described by the "shrinking core model"[78] based on Fick's 2nd law of diffusion for migration down a concentration gradient. This model has been applied to convert kinetic data for the formation of various hydrates to diffusion coefficients and associated activation energies.[3–5,77] However, it is possible that in some instances the water mobility exceeds that of the adsorbate, in which case ice-core water molecules diffuse down a reversed gradient to the surface; which then becomes the reaction zone.[5,79] Here, the conversion of ice nanocrystals to acid and base hydrates is reviewed briefly. The low-temperature formation of nanocrystals of clathrate hydrates will be examined in more detail in Section 3.6.3.

3.6.1 Kinetics of Ammonia-Hydrate Formation from Ice Nanocrystals

Ammonia forms three different molecular crystalline hydrates, $NH_3 \cdot 2H_2O, NH_3 \cdot H_2O,$ and $2NH_3H_2O$ each of which has been characterized by spectroscopic/diffraction measurements on bulk samples (see, e.g., ref. [80]). Of these, the mono and hemihydrates form most readily when ice nanocrystals are exposed to ammonia near 120 K.[3] If the ammonia is limited, by admission through a leak valve, the product is the monohydrate, while the hemihydrate forms when an abundance of ammonia has access to the ice particles. To provide such access, it was necessary to form the ice array with incorporated alternating layers of ammonia and ice particles. This method is commonly used with ice nanocrystals to provide uniform availability of an adsorbate throughout the array during conversion to hydrate nanoparticles.

With an abundance of ammonia available, the formation of the hemihydrate was observed to follow the shrinking-core model with a half-life for formation of <5 min at 112 K. At lower temperatures, for example, 107 K, the product is initially the amorphous form of the hemihydrate which slowly crystallizes during the first hour of reaction. Once crystallized at 107 K, the reaction rate increased markedly despite the increasing diffusion length, suggesting that *ammonia diffuses through the crystalline product crust much more readily than through the amorphous phase*. The lattice structure of the crystalline hemihydrate includes sizeable channels that likely support rapid molecular diffusion,[80] providing an "open" path for ammonia to the reaction zone and enabling the rapid hydrate formation at such a low temperature. The crystalline diffusion coefficient was estimated as roughly 20 times greater than for the amorphous phase near 107 K. However, the possibility that the enhanced rate results from generation of grain boundaries during crystallization should be noted.

3.6.2 Kinetics of Acid-Hydrate Formation from Ice Nanocrystals

Like ammonia, strong acids such as HCl and HBr form a series of hydrates for which the water-to-acid ratio ranges from 1 to 6. The crystalline solids are ionic in nature with the cation a proton hydrated to different degrees.[81] The conversion of an array of ice nanocrystals to acid-hydrate nanoparticles occurs on a similar timescale and in the same temperature range as for the ammonia hydrates.[77] Further, as for NH_3, the original product at the lowest temperature (~110 K) is amorphous; in particular, the amorphous acid monohydrate which does not crystallize until warming to ~135 K. Crystalline mono-, di-, and trihydrates equilibrate with the ambient acid vapor, transitioning to the stable form in the 140–170 K range. Nanocrystals of higher hydrates form at higher temperatures when the available acid vapor is limited. The nanoparticles can be cycled between two phases in a few minutes by control of the temperature and ambient acid pressure.

Perhaps the most striking aspect of the kinetic results is the rapidity with which the molecular acid moves through the cold ionic nanoparticles; a requirement for formation and phase transitions. The mechanism by which the acid arrives at or leaves an interfacial reaction zone is not clear but could involve acids alternating between molecular and ionic states. One observation that may be related to HCl-molecule transport through the hydrates is the presence of an unassigned weak sharp band near 2100 cm^{-1} in the FTIR spectrum of each of the three hydrates, in both the crystalline and amorphous phases. This band has been related to (interstitial) *molecular HCl* bound to a halide anion.[81,82] This possibility is consistent with the relative ease with which the molecular acids are added to or removed from the three hydrates, and the decreasing intensity of the ~2100 cm^{-1} band with decreasing acid to water content. However, quite exhaustive on-the-fly computational studies have been unable to confirm that assignment.[81]

3.6.3 KINETICS OF RAPID CLATHRATE-HYDRATE FORMATION BY H-BONDING ADSORBATES NEAR 120 K: A "CATALYTIC" ROLE FOR H-BONDING GUESTS

Clathrate hydrates, C.H.s, are nonstoichiometric hydrate inclusion compounds in which small guest molecules occupy cages within a crystalline host lattice of "tetrahedral" H-bonding configurations resembling those of ordinary ice.[83] Two types of C.H.s have been studied extensively, the classic structure I, or s-I, and structure II, or s-II, hydrates. A 46-water-molecule cubic unit cell of type s-I includes eight cages of which two are small and six large. The small cages accommodate both atoms and small molecules ranging in size from argon to CO_2, while somewhat larger molecules, ranging from CO_2 through the size of trimethylene oxide, $(CH_2)_3O$, favor the large cages. The 16 small cages, of an s-II unit cell of 136 water molecules, are closely matched in size to those of s-I hydrates, while 8 larger cages can host molecules as large as CBr_2F_2, tetrahydrofuran, $(CH_2)_4O$, or even cyclohexanone.[84]

Since much interest has been stirred by the extensive presence of "gas" C.H.s in the natural environment,[18] most scientific studies of C.H.s have been at terrestrial temperatures (>220 K). Huge quantities of methane C.H.s are buried in the arctic tundra and under the sediment of the ocean shelves; along with much CO_2. There is presently a keen interest in (1) identifying mixed C.H.s, that is, with large guests in the large cages and gas molecules in the small cages, such that the smaller guests, such as methane or H_2, can be handled at more manageable pressures than required for the simple gas hydrates,[63,85] (2) developing an expanded base of information on the rates of formation and transformations of the gas and mixed C.H.s,[19] and (3) cataloging a broad base of information for use in advancing the control technology for gas hydrates, as required for mining and the management of potential environmental impacts.

A clathrate hydrate will commonly form at or above the melting temperature of ice, though a high pressure is required if "permanent" gases are included as guests. However, if the formation of the typical hydrate from ice is attempted at low temperatures, the rate is very slow such that, even for gas hydrates at high pressures, days are required to fully convert small ice particles to a clathrate hydrate.[86,87] However, it has been known since the mid-1980s that there is *something special about the formation rate of simple and mixed C.H.s of small ether molecules*; the formation was demonstrated at temperatures as low as 100 K in the direct codeposition with water.[88–90] Later, it was reported that formaldehyde also enables C.H. formation at such low temperatures.[91] Most recently, it was demonstrated that the formation, from ice nanocrystals, of ether C.H.s and mixed ether C.H.s with methane or CO_2, occurs at 120 K with guest diffusion rates comparable to that for simple methane C.H. formation at 240 K.[4,5] This established that *molecular transport through a C.H crust, the rate controlling factor, is enhanced by several orders of magnitude when one guest is a proton acceptor.*

3.6.4 DEFECT STRUCTURES/DYNAMICS THAT PROMOTE FORMATION AND TRANSFORMATIONS OF CLATHRATE HYDRATES WITH H-BONDED GUESTS

The extreme enhancement of the rate of C.H. formation by proton-acceptor guests has drawn attention to other properties of ether clathrate hydrates. Most particularly, it is known that the dielectric relaxation rates of the C.H.s of small ethers exceed those of non-H-bonding guests by orders of magnitude.[83] Since dielectric relaxation in ice-like substances is attributed to mobility associated with L-defects,[1] a conjecture was developed relating the exceptional relaxation rates to the ability of a proton-acceptor molecule to steal a proton from its cage wall on a transient basis.[92] This transfer, of a lattice O−H coordination to the guest, inserts an L-defect into the hydrate lattice reducing the energetics of relaxation. At the same time, a large hole that appears at the L-defect location in the cage wall was imagined to enable guest-molecule migration through the C.H.; accelerating formation and transformation reactions.[4,93]

However, recent computational results suggest that molecular transport is not greatly enhanced by transient L-defects in the cage walls.[5] Rather, the apparent role of a proton-acceptor guest is to

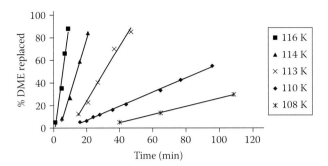

FIGURE 3.19 Plots showing the rate of replacement of DME by H_2S from the small cages of the s-II DME clathrate hydrate for temperatures ranging from 108 to 116 K. The straight lines reflect a constant replacement rate that is, unlike what is expected for a shrinking-core particle reaction and also indicate a surprisingly rapid exchange of the guests in the small cages.

stabilize and thus increase by orders of magnitude the population of vacancy defects in a C.H. lattice. The site of a missing water molecule in the C.H. lattice displays dangling high-energy d-H sites that the guest ether molecules can help stabilize. The missing water molecules create large passageways for guest molecules and were previously identified as the C.H. feature that enables the migration of CO_2 within its C.H.[94] (though at much higher temperatures). The implication is that equilibrium thermal populations of vacancy defects, comparable in magnitude to populations in most C.H.s at high temperatures (>200 K), occur at much lower temperatures within C.H.s with proton-acceptor guests.

Numerous examples of transformations of C.H.s with proton-acceptor guests, occurring on a minute timescale at ~120 K, have now been reported.[5] In addition to C.H. formations from ice nanocrystals, the transformations include the simple gain and loss of gas molecules from the small cages, the replacement of one small cage or one large cage molecule by a second, s-I transitions to s-II or the reverse, and phase transitions that accompany guest exchange. Perhaps most remarkable is the ability of H_2S to replace dimethyl ether (DME) molecules from the small cages of the s-II C.H. with a half-life of one minute at 120 K (Figure 3.19), and the loading of methane or CO_2 into empty small cages on a similar timescale.

One can question the term "catalyst" applied to the proton-acceptor guest molecules. They clearly lower the activation energy of formation compared to that of most clathrate-hydrates by facilitating a new transport mechanism, but they also participate as "reactants" in the process. However, *there are many examples where ether guest molecules do function in the full sense of a catalyst.*[5] Perhaps the simplest case is the rapid loading of CO_2/methane into the empty small cages of either the s-I or s-II clathrate of DME near 120 K. The guest ether molecules in the large cages do not change configuration or identity, while providing a new low-energy path for the loading of small-cage guests; *the H-bond chemistry that enables the small-guest transport is catalyzed.*

3.7 SUMMARY

Ice nanocrystals generated in aerosols have been shown to be a versatile agent in the study of the nature of the ice surface and subsurface regions, the influence of adsorbates on an ice surface and interior, the relative proton activity at the surface and in the interior of ice, and the quantitative kinetics of formation/transformations of numerous important naturally occurring hydrates. Knowledge of the nature of the surface is fundamental to the understanding of most of the physical and chemical properties examined in this review. In that respect, *a particularly basic point is that the ice surface, compared to other molecular solid surfaces, has sites with a truly exceptional range of bonding characteristics.* That range has been revealed through computational simulations and FTIR data, including spectral response to the great variety of interactions with weak, intermediate and strong adsorbates. An insight to the ice surface structure is required to explain the difference in surface bonding of *para-* and *ortho-*H_2. Changes in ice nanocrystals induced by stronger adsorbates would also be baffling without a valid concept of the surface structure. Cases in point include (a) the tendency to form as a reconstructed disordered surface, but to de-reconstruct in the presence of certain adsorbates, in particular SO_2, (b) the stability of isolated *molecules* of strong acids on the surface at <80 K that reflects the reduced ability of the surface to solvate an adsorbate following surface reconstruction (that decreases the number of active high-energy three-coordinated surface molecules), and (c) an unusual surface stability of the hydronium ion for neutral ice relative to the interior, and the influence of that stability on the ionization of weak acids and bases.

REFERENCES

1. Petrenko, V. F. and Whitworth, R. W. 1999. *Physics of Ice*. Oxford University Press, Oxford.
2. Devlin, J. P. and Buch, V. 1995. Surface of ice as viewed from combined spectroscopic and computer modeling studies. *J. Phys. Chem.* 99: 16534–16548.
3. Uras, N. and Devlin, J. P. 2000. Rate study of ice particle conversion to ammonia hemihydrate: Hydrate crust nucleation and NH_3 diffusion. *J. Phys. Chem. A* 104: 5770–5777.
4. Gulluru, D. B. and Devlin, J. P. 2006. Rates and mechanisms of conversion of ice nanocrystals to ether clathrate hydrates: Guest-molecule catalytic effects at ~120 K. *J. Phys., Chem. A.* 110: 1901–1906.
5. Buch, V., Devlin, J. P., Monreal, I. A., Jagoda-Cwiklik, B., Uras-Aytemiz, N., and Cwiklik, L. 2009. Clathrate hydrate with hydrogen-bonding guests. *Phys. Chem. Chem. Phys.* 11: 10245–10265, doi:10.1039/b911600c.
6. Delzeit, L., Rowland, B., and Devlin, J. P. 1993. Infrared spectra of HCl complexed/ionized in amorphous hydrates and at ice surfaces in the 15–90 K range. *J. Phys. Chem.* 97: 10312–10318.
7. Devlin, J. P., Uras, N., Sadlej, J., and Buch, V. 2002. Discrete stages in the solvation and ionization of hydrogen chloride adsorbed on ice particles. *Nature* 417: 269–271.
8. Buch, V., Sadlej, J., Uras, N. A., and Devlin, J. P. 2002. Solvation and ionization stages of HCl on ice nanocrystals. *J. Phys. Chem. A* 106: 9374–9389.
9. Kang, H., Shin, T. H., Park, S. C., Kim, I. K., and Han, S. J. 2000. Acidity of hydrogen chloride on ice. *J. Am. Chem. Soc.* 122: 9842.
10. Skvortsov, D., Lee, S. J., Choi, M. Y., and Vilesov, A. F. 2009. Hydrated HCl clusters, $HCl(H_2O)_{1\text{-}3}$, in helium nanodroplets: Studies of free OH vibrational stretching modes. *J. Phys. Chem. A* 113: 7360–7365.
11. Bjerrum, N. 1952. Structure and properties of ice. *Science* 115: 385.
12. Collier, W. B., Ritzhaupt, G., and Devlin, J. P. 1984. Spectroscopically evaluated rates and energies for proton transfer and bjerrum defect migration in cubic ice. *J. Phys. Chem.* 88: 363.
13. Wooldridge, P. J. and Devlin, J. P. 1988. Proton trapping and defect energetics in ice from FT-IR monitoring of photoinduced isotopic exchange of isolated D_2O. *J. Chem. Phys.* 92: 3086.
14. Devlin, J. P. 2000. Preferential deuterium bonding at the ice surface: A probe of surface water molecule mobility. *J. Chem. Phys.* 112: 5527–5529.
15. Buch, V., Milet, A., Vácha, R., Jungwirth, P., and Devlin, J. P. 2007. Water surface is acidic. *Proc. Natl. Acad. Sci. U.S.A.* 104: 7342–7347.
16. Cwiklik, L., Devlin, J. P., and Buch, V. 2009. Hydroxide impurity in ice. *J. Phys. Chem.* 113: 7482–7490.
17. Gaskell, W. and Illingworth, A. J. 1980. Charge transfer accompanying individual collisions between ice particles and its role in thunderstorm electrification. *Quart. J. R. Met. Soc.* 106: 841–854.
18. Kennett, J. P., Cannariato, K. G., Hendy, I. L., and Behl, R. J. 2000. Carbon isotopic evidence for methane hydrate instability during quaternary interstadials. *Science* 288: 128.
19. Sloan, E. D. 2003. Fundamental principles and applications of natural gas hydrates. *Nature* 426: 353.
20. Buck, U., Ettischer, M., Melzer, M., Buch, V., and Sadlej, J. 1998. Structure and spectra of three-dimensional $(H_2O)_n$ clusters, $n = 8, 9, 10$. *Phys. Rev. Lett.* 80: 2578.
21. Buch, V., Bauerecker, S., Devlin, J. P., Buck, U., and Kazimirski, J. K. 2009. Solid water clusters in the size range of tens-thousands of H_2O: A combined computational/spectroscopic outlook. *Int. Rev. Phys. Chem.* 23: 375.
22. Rice, S. A., Bergren, M. S., Belch, A. C., and Nielson, G. 1983. A theoretical analysis of the hydroxyl stretching spectra of ice I*h*, liquid water, and amorphous solid water. *J. Phys. Chem.* 87: 4295.

23. Rowland, B. and Devlin, J. P. 1991. Spectra of dangling OH groups at ice cluster surfaces and within pores of amorphous ice. *J. Chem. Phys.* 94: 812–813.

24. Buch, V. and Devlin, J. P. 1991. Spectra of dangling OH bonds in amorphous ice: Assignment to 2- and 3-coordinated surface molecules. *J. Chem. Phys.* 94: 4091–4092.

25. Rowland, B., Fisher, M., and Devlin, J. P. 1991. Probing icy surfaces with the dangling-OH-mode absorption: Large ice clusters and microporous amorphous ice. *J. Chem. Phys.* 95: 1378–1384.

26. Kimmel, G. A., Stevenson, K. P., Dohnalek, Z., Smith, R. S., and Kay, B. D. 2001. Control of amorphous solid water morphology using molecular beams. I. Experimental results. *J. Chem. Phys.* 114: 5284–5294.

27. Buch, V., Delzeit, L., Blackledge, C., and Devlin, J. P. 1996. Structure of the ice nanocrystal surface from simulated versus experimental spectra of adsorbed CF_4. *J. Phys. Chem.* 100: 3732–3744.

28. Glebov, A., Graham, A. P., Menzel, A., Toennies, J. P., and Senet, P. 2000. A helium atom scattering study of the structure and phonon dynamics of the ice surface. *J. Chem. Phys.* 112: 11011–11022.

29. Fletcher, N. H. 1992. Reconstruction of ice crystal surfaces at low temperatures. *Philos. Mag.* 66: 109.

30. Buch, V., Groenzin, H., Li, I., Shultz, M.J., and Tosatti, E. 2008. Proton order in the ice crystal surface. *Proc. Natl. Acad. Sci. U.S.A.* 105: 5969–5974.

31. McDonald, S., Ojamae, L., and Singer, S. J. 1998. Graph theoretical generation and analysis of hydrogen-bonded structures with application to the neutral and protonated water cube and dodecahedral clusters. *J. Phys. Chem. A.* 102: 2824–2832.

32. Groenzin, H., Li, I., Buch, V., and Shultz, M. J. 2007. The single-crystal, basal face of ice I*h* investigated with sum frequency generation. *J. Chem. Phys.* 127: 214502.

33. Vacha, R., Buch, V., Milet, A., Devlin, J. P., and Jungwirth, P. 2007. Autoionization at the surface of neat water: Is the top layer pH neutral, basic, or acidic? *Phys. Chem. Chem. Phys.* 9: 4736.

34. Huang, J. and Bartell, L. S. 1995. Kinetics of homogeneous nucleation in the freezing of water clusters. *J. Phys. Chem.* 99: 3924.

35. Delzeit, L. and Blake, D. 2001. A characterization of crystalline ice nanoclusters using transmission electron microscopy. *J. Geophys. Res.—Planets* 106 (E12): 33371.

36. Buch, V. and Devlin, J. P. 1999. A new interpretation of the OH-stretch spectrum of ice. *J. Chem. Phys.* 110: 3437–3443.

37. Devlin, J. P., Sadlej, J., and Buch, V. 2001. Infrared spectra of large H_2O clusters: New understanding of the elusive bending mode of ice. *J. Phys. Chem. A* 105: 974–983.

38. Clapp, M. L., Miller, R. E., and Worsnop, D. R. 1995. Frequency-dependent optical constants of water ice obtained directly from aerosol extinction spectra. *J. Phys. Chem.* 99: 6326.

39. Devlin, J. P., Joyce, C., and Buch, V. 2000. Infrared spectra and structures of large water clusters. *J. Phys. Chem. A.* 104: 1974–1977.

40. Torchet, G., Schwartz, P., Farges, J., de Feraudy, M. F., and Raoul, B. 1983. Structure of solid water clusters formed in a free jet expansion. *J. Chem. Phys.* 79: 6196.

41. Delzeit, L., Devlin, J. P., and Buch, V. 1997. Structural relaxation rates near the ice surface: Basis for separation of the surface and subsurface spectra. *J. Chem. Phys.* 107: 3726–3729.

42. Uras-Aytemiz, N., Devlin, J. P., Sadlej, J., and Buch, V. 2006. HCl solvation at the surface and within methanol clusters/nanoparticles II: Evidence for molecular wires. *J. Phys. Chem. B.* 110: 21751–21763.

43. Uras-Aytemiz, N., Devlin, J. P., Sadlej, J., and Buch, V. 2006. HCl solvation in methanol clusters and nanoparticles: Evidence for proton-wires. *Chem. Phys. Lett.* 422: 179–183.

44. Signorell, R. 2003. Verification of the vibrational exciton approach for CO_2 and N_2O nanoparticles. *J. Chem. Phys.* 118: 2707–2715.

45. Firanescu, G., Luckhaus, D., and Signorell, R. 2006. Size effects in the infrared spectra of NH_3 ice nanoparticles studied by a combined molecular dynamics and vibrational exciton approach. *J. Chem. Phys.* 125: 144501-1–144501-13.

46. Whalley, E. 1977. A detailed assignment of the O−H stretching bands of ice I. *Can. J. Chem.* 55: 3429.

47. Buch, V., Tarbuck, T., Richmond, G. L., Groenzin, H., Li, I., and Shultz, M. J. 2007. Sum frequency generation surface spectra of ice, water, and acid solution investigated by an exciton model. *J. Chem. Phys.* 127: 2047101–20471015.

48. Devlin, J. P., Wooldridge, P. J., and Ritzhaupt, G. 1986. Decoupled isotopomer vibrational frequencies in cubic ice: A simple unified view of the Fermi diads of decoupled H_2O, HOD, and D_2O. *J. Chem. Phys.* 84: 6095–6100.

49. Rowland, B., Kadagathur, N. S., and Devlin, J. P. 1995. Infrared spectra of CF_4 adsorbed on ice: Probing adsorbate dilution and phase separation with the υ_3 transverse-longitudinal splitting. *J. Chem. Phys.* 102: 13–19.

50. Bertie, J. E. and Devlin, J. P. 1984. Refined vibrational data for H₂O isolated in D₂O cubic ice. *J. Phys. Chem.* 88: 380–381.
51. Sivakumar, T. C., Rice, S. A., and Sceats, M. G. 1978. Raman spectroscopic studies of the OH stretching region of low density amorphous solid water and of polycrystalline ice I*h*. *J. Chem. Phys.* 69: 3468.
52. Delzeit, L., Devlin, M. S., Rowland, B., Devlin, J. P., and Buch, V. 1996, Adsorbate-induced partial ordering of the irregular surface and subsurface of crystalline ice. *J. Phys. Chem.* 100: 10076–10082.
53. Holmes, N. S. and Sodeau, J. R. 1999. A study of the interaction between halomethanes and water-ice. *J. Phys. Chem. A* 103: 4673.
54. Sadlej, J., Rowland, B., Devlin, J. P., and Buch, V. 1995. Vibrational spectra of water complexes with H₂, N₂, and CO. *J. Chem. Phys.* 102: 4804–4818.
55. Hagen, W., Tielens, A. G. G. M., and Greenberg, J. M. 1983. The three micron "ice" band in grain mantles. *A.&A.* 117: 132–140.
56. Palumbo, M. E. 2006. Formation of compact solid water after ion irradiation at 15 K. *A.&A.* 453: 903–909.
57. Devlin, J. P. 1992. Molecular interactions with icy surfaces: Infrared spectra of CO absorbed in microporous amorphous ice. *J. Chem. Phys.* 96: 6185–6188.
58. Palumbo, M. E. 1997. Infrared spectra and nature of the principal CO trapping sites in amorphous and crystalline H₂O ice. *J. Phys. Chem. A.* 101: 4298.
59. Al-Halabi, A., Fraser, H. J., Kroes, G. J., and van Dishoeck, E. F. 2004. Adsorption of CO on amorphous water-ice surfaces. *A.&A.* 422: 777.
60. Buch, V. and Devlin, J. P. 2002. Ice nanoparticles and ice adsorbate interactions: FTIR spectroscopy and computer simulations. In *Water in Confining Geometries*. V. Buch and J. P. Devlin (Ed). Springer, Berlin, pp. 425–462.
61. Sandford, S. A., Allamandola, L. J., and Geballe, T. R. 1993. Spectroscopic detection of molecular hydrogen frozen in interstellar ices. *Science* 262: 400.
62. Buch, V. and Devlin, J. P. 1994. Interpretation of the 4141 inverse centimeters (2.415 microns) interstellar infrared absorption feature. *Ap. J.* 431: L135–L138.
63. Uchida, T., Takeya, S., Wilson, L. D. et al. 2003. Measurements of physical properties of gas hydrates and *in situ* observations of formation and decomposition processes via Raman spectroscopy and x-ray diffraction. *Can. J. Phys.* 81: 351–357.
64. Hixson, H. G., Wojcik, M. J., Devlin, M. S., and Devlin, J. P. 1992. Experimental and simulated vibrational spectra of H₂ absorbed in amorphous ice: Surface structures, energetics, and relaxations. *J. Chem. Phys.* 97: 753–767.
65. von Hippel, A., Knoll, D. B., and Westphal, W. B. 1971. Transfer of protons through "pure" ice I*h* single crystals. I. Polarization spectra of ice. *J. Chem. Phys.* 54: 134–144.
66. Onsager, L. and Dupis, M. 1962. The electrical properties of ice. In *Electrolytes*. Ed. B. Pesce. Pergamon Press, Oxford, pp. 27–46.
67. Jaccard, C. 1959. Etude theorique et experimentale des proprietes de la glace. *Helvetica Phys. Acta* 32: 89–128.
68. Jaccard, C. 1964. Thermodynamics of irreversible processes applied to ice. *Phys. Kondensierten Mater.* 3: 99–118.
69. Cowin, J. P., Tsekouras, A. A., Iedema, M. J., Wu, K., and Ellison, B. B. 1999. Immobility of protons in ice from 30 to 190 K. *Nature* 398: 405.
70. Moon, E. S., Lee, C. W., and Kang, H. 2008. Proton mobility in thin ice films: A revisit. *Phys. Chem. Chem. Phys.* 10: 4814–4816.
71. Devlin, J. P. 1992. Defect activity in icy solids from isotopic exchange rates: Implications for conductance and phase transitions. In *Proton Transfer in Hydrogen-Bonded Systems*. Ed. T. Bountis. NATO Advanced Study Institute Series B: Physics. Plenum Press, New York, Vol. 291, pp. 249–260.
72. Uras-Aytemiz, N., Joyce, C., and Devlin, J. P. 2001. Protonic and bjerrum defect activity near the surface of ice at *T* < 145 K. *J. Chem. Phys.* 115: 9835–9842.
73. Devlin, J. P. and Buch, V. 2007. Evidence for the surface origin of point defects in ice: Control of interior proton activity by adsorbates. *J. Chem. Phys.* 127: 0911011–0911014.
74. Lee, C.W., Lee, P.R., and Kang, H. 2006. Protons at ice surfaces. *Angew. Chem.* 118: 5655–5659.
75. Jagoda-Cwiklik, B., Devlin, J. P., and Buch, V. 2008. Spectroscopic and computational evidence for SO₂ ionization on 128 K ice surface. *Phys. Chem. Chem. Phys.* 10: 4678–4684.
76. Fisher, M. and Devlin, J. P. 1995. Defect activity in amorphous ice from isotopic exchange data: Insight into the glass transition. *J. Phys. Chem.* 99: 11584–11590.
77. Devlin, J. P., Gulluru, D. B., and Buch, V. 2005. Rates and mechanisms of conversion of ice nanocrystals to hydrates of HCl and HBr: Acid diffusion in the ionic hydrates. *J. Phys. Chem. B* 109: 3392–3401.

78. Carter, R. E. 1961. Kinetic model for solid-state reactions. *J. Chem. Phys.* 34: 2010.
79. Staykova, D. K., Kuhs, W. F., Salamatin, A. N., and Hansen, T. 2003. Formation of porous gas hydrates from ice powders: Diffraction experiments and multistage model. *J. Phys. Chem.* 107: 10299.
80. Bertie, J. E. and Devlin, J. P. 1984. The infrared spectra and phase transitions of pure and isotopically impure $2ND_3.H_2O$, $2NH_3.D_2O$, $2NH_3.H_2O$, and $2ND_3.D_2O$ between 100 and 15 K. *J. Chem. Phys.* 81: 1559–1572.
81. Buch, V., Dubrovskiy, A., Mohamed, F. et al. 2008. HCl hydrates as model systems for protonated water. *J. Phys. Chem. A* 112: 2144–2161.
82. Haq, S., Harnett, J., and Hodgson, A. 2002. Adsorption and solvation of HCl into ice surfaces. *J. Phys. Chem. B.* 106: 3950–3959.
83. Davidson, D. W. 1973. Clathrate hydrates. In *Water: A Comprehensive Treatise*. F. Franks (Ed). Plenum Press, New York, Vol. 2, ch. 3, pp. 115–234.
84. Strobel, T. A., Hester, K. C., Sloan, E. D., and Koh, C. A. 2007. A hydrogen clathrate hydrate with cyclohexanone: Structure and stability. *J. Am. Chem. Soc.* 129: 9544–9545.
85. Strobel, T. A., Taylor, C. J., Hester, K. C. et al. 2006. Molecular hydrogen storage in binary $THF-H_2$ clathrate hydrates. *J. Phys. Chem. B* 110: 17121–17125.
86. Henning, R. W., Schultz, A. J., Thieu, V., and Halpern, Y. 2000. Neutron diffraction studies of CO_2 clathrate hydrate: Formation from deuterated ice. *J. Phys. Chem. A* 104: 5066.
87. Wang, X., Schultz, A. J., and Halpern, Y. 2002. Kinetics of methane hydrate formation from polycrystalline deuterated ice. *J. Phys. Chem. A* 106: 7304.
88. Bertie, J. E. and Devlin, J. P. 1983. Infrared spectroscopic proof of the formation of the structure I hydrate of oxirane from annealed low-temperature condensate. *J. Chem. Phys.* 78: 6340–6341.
89. Richardson, H. H., Wooldridge, P. J., and Devlin, J. P. 1985. FT-IR spectra of vacuum deposited clathrate hydrates of oxirane H_2S, THF, and ethane. *J. Chem. Phys.* 83: 4387–4394.
90. Fleyfel, F. and Devlin, J. P. 1988. FT–IR spectra of 90 K films of simple, mixed, and double clathrate hydrates of trimethylene oxide, methyl chloride, carbon dioxide, tetrahydrofuran, and ethylene oxide containing decoupled D_2O. *J. Phys. Chem.* 92: 631–635.
91. Ripmeester, J. A., Ding, L., and Klug, D. D. 1996. A clathrate hydrate of formaldehyde. *J. Phys. Chem.* 100: 13330.
92. Gough, S. R., Whalley, E., and Davidson, D. W. 1968. Dielectric properties of the hydrates of argon and nitrogen. *Can. J. Chem.* 46: 1673.
93. Wooldridge, P. J., Richardson, H. H., and Devlin, J. P. 1987. Mobile bjerrum defects: A criterion for ice-like crystal growth. *J. Chem. Phys.* 87: 4126–4131.
94. Demurov, A., Radhakrishnan, R., and Trout, B. L. 2002. Computations of diffusivities in ice and CO_2 clathrate hydrates via molecular dynamics and Monte Carlo simulations. *J. Chem. Phys.* 116: 702.

4 Infrared Extinction and Size Distribution Measurements of Mineral Dust Aerosol

Paula K. Hudson, Mark A. Young, Paul D. Kleiber, and Vicki H. Grassian

CONTENTS

4.1 Introduction 79
4.2 Instrument Design for Simultaneous Measurements of IR Extinction Spectra and Size Distributions 80
4.3 Analysis of the Data: Mie Theory 82
4.4 Comparison of Experimental IR Extinction Spectra of Components of Mineral Dust Aerosol to Mie-Simulated Spectra 83
    4.4.1 Clay Minerals 83
    4.4.2 Oxide and Carbonate Minerals..92
4.5 Conclusions and Future Work 98
Acknowledgments..98
References...98

4.1 INTRODUCTION

Mineral dust aerosol consists largely of wind-blown soil particles (Sheehy, 1992; Tegen and Fung, 1994; Henderson-Sellers, 1995; Buseck and Posfai, 1999; Prospero, 1999; Sokolik and Toon, 1999; Prospero and Lamb, 2003; Bauer et al., 2004). Aerosol flux into the troposphere varies with location and season, but the estimates of average atmospheric mineral dust loading total ~800–1500 Tg/year (Bauer et al., 2004). While the Sahara Desert is the largest global contributor, human activities (such as deleterious agricultural and grazing practices) have contributed to expanding desertification and increased dust loads (Sheehy, 1992; Tegen and Fung, 1994; Henderson-Sellers, 1995; Buseck and Posfai, 1999; Bauer et al., 2004). The frequency and intensity of dust events, and ultimately the atmospheric mineral aerosol loading, are expected to increase as long as improper land-use practices are driven by economic, social, and political circumstances.

Mineral dust aerosol has a significant effect on the physical and chemical equilibrium of the atmosphere. Like all particles, mineral dust in the atmosphere influences the global radiation balance through the direct scattering and absorption of light across the spectrum. Modeling the impact of mineral dust aerosol on atmospheric processes requires accurate knowledge of dust loading, as well as information about the dust composition, shape, and size distributions. Aerosol optical depths may be determined from remote sensing studies including high-resolution or narrowband infrared (IR) spectral data from satellites (Ackerman, 1997; Pierangelo et al., 2004). The IR region is of particular importance as satellite measurements determine the key atmospheric and oceanic properties such as the atmospheric temperature profile, water vapor and trace gas concentrations, and sea surface temperature through IR spectral measurements (Sokolik, 2002). Under high dust-loading conditions, the effect of atmospheric dust must be subtracted from the spectral measurements to accurately determine these atmospheric properties (DeSouza-Machado et al., 2006). When the effects of dust are not properly included in satellite retrievals, the error can influence the interpretation of climate trends (Ackerman, 1997). For example, a recent study found a change of 3 K in surface temperature determined from advanced very high resolution radiometer (AVHRR) data when aerosols were included in the model (Highwood et al., 2003).

Mie theory is commonly used for modeling the optical properties of aerosols in both radiative forcing and satellite retrieval algorithms (Wang et al., 2002; Conant et al., 2003; Moffet and Prather, 2005; DeSouza-Machado et al., 2006). Mie theory is easy to apply for a given particle size distribution and a set of optical constants. However, Mie theory is derived assuming uniform spherical particles (Bohren and Huffman, 1983). Atmospheric mineral dust aerosol, on the other hand, are often inhomogeneous, complex mixtures of particles and particle aggregates of varying composition (Claquin et al., 1999) and, moreover, are typically nonspherical (Dick et al., 1998), which may have a significant effect on the optical properties (Kalashnikova and Sokolik, 2002). This is especially true in the neighborhood of IR absorption resonances, where particle shape effects may cause significant variations in spectral line positions and band profiles, even for small particles (Bohren and Huffman, 1983). These effects can be important since field studies have shown that as much as 30% of the total submicron aerosols can be mineral dust during a dust event (Arimoto et al. 2006). This could result in errors in the dust retrievals depending on the overlap between the actual and calculated line profiles, and the specific narrowband IR sensor channels that are used to determine dust composition and aerosol optical depth (Ackerman, 1997; Pierangelo et al., 2004).

In this chapter, we show the experimental IR extinction spectra of components of mineral dust aerosol, which are measured simultaneously with particle size distributions. Mie theory is then used to generate a simulated spectrum from the measured size distribution and the available literature optical constants. This method allows an absolute comparison between the simulated and measured IR extinction spectra, with no adjustable parameters, thereby providing a quantitative evaluation of the errors associated with using Mie theory to model the optical properties of mineral dust aerosol.

4.2 INSTRUMENT DESIGN FOR SIMULTANEOUS MEASUREMENTS OF IR EXTINCTION SPECTRA AND SIZE DISTRIBUTIONS

An instrument designed to measure IR extinction spectra and size distributions is shown in Figure 4.1 (Hudson et al., 2007). The instrument consists of an atomizer (TSI, Inc. Model 3076) for aerosol generation, a Fourier-transform IR (FTIR) spectrometer (Thermo Nicolet Nexus Model 670) with a liquid nitrogen-cooled external MCT-A detector, and two particle sizing instruments: a scanning mobility particle sizer (SMPS) (TSI, Inc. Model 3936), consisting of a differential mobility analyzer (DMA) (TSI, Inc. Model 3080) and a condensation particle counter (CPC) (TSI, Inc. Model 3025A), and an aerodynamic particle sizer (APS) (TSI, Inc. Model 3321). The particles follow a flow stream from the aerosol generator through the path of the IR beam to the particle sizing instrumentation by a combination of conductive tubing and glass flow tubes. Size distributions and IR extinction spectra are then simultaneously measured for each atomized aerosol dust sample.

The extinction cell length from window to window is 100 cm. The cell also has window purge ports, 90 cm apart, used for blowing dry air over the windows, which additionally mixes with the particle stream. The distance between the particle stream entrance and exit ports in the observation tube is 75 cm. The observation tube is sealed at both ends with barium fluoride (BaF_2) windows with a transmission range from 800 to 4000 cm^{-1} in the IR spectral region. All IR spectra were measured by co-adding 256 scans at an instrument resolution of 8 cm^{-1}.

It should be noted that, due to the wide variability in the size of ambient aerosol particles, it is often necessary to combine measurements from multiple instruments to yield a full size

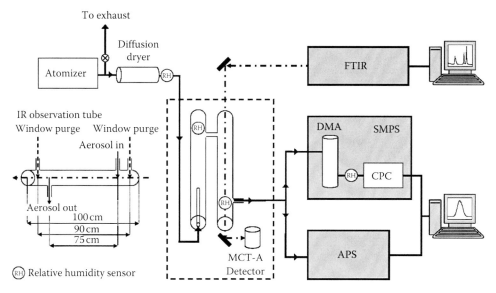

FIGURE 4.1   The main components of the Multi-Analysis Aerosol Reactor System (MAARS) designed to simultaneously measure aerosol IR extinction spectra and size distributions as a function of mobility and aerodynamic diameters with an SMPS and an APS, respectively, are shown. A detailed view of the IR observation tube is shown with physical lengths for the glass tube, distance between window purges and particle stream "in" and "out." See text for further details. (Reproduced from Hudson, P. K., Gibson, E. R., Young, M. A., Kleiber, P. D., and Grassian, V. H., 2007. *Aerosol Sci. Technol.* 41:701–710. With permission.)

distribution. After the extinction cell, the particle stream is divided by a 3/8″ stainless steel cross directing the particle stream to the SMPS, the APS, and an exhaust. The flow conditions are such that there is enough flow exiting the observation tube to supply the flow needs of both the SMPS (sheath flow = 2.0 lpm, aerosol flow = 0.2 lpm) and the APS (5 lpm). For high particle concentrations, one or two diluters (TSI, Inc. Model 3302A) are used with the APS to decrease the concentration to measurable levels. The SMPS measures the mobility diameter of the particle distribution from 20 to 900 nm and the APS measures the aerodynamic diameter of the particle distribution from 550 nm to 20 μm. The scan times for the IR, SMPS, and APS (210 s total) are synchronized for simultaneous measurement of the size distribution with the corresponding IR spectrum.

Because the two sizing instruments use different principles of operation, the measured mobility and aerodynamic diameters are converted to a common volume equivalent diameter. According to Hinds (1999), the volume equivalent diameter is related to the aerodynamic diameter by

$$D_{ve} = D_a \sqrt{\chi \frac{\rho_o}{\rho_p} \frac{C_s(D_a)}{C_s(D_{ve})}}, \tag{4.1}$$

where χ is the dynamic shape factor, ρ_o is the reference density (1 g cm^{-3}), ρ_p is the density of the particle, and $C_s(D_a)$ and $C_s(D_{ve})$ are the Cunningham slip factors for the aerodynamic diameter and the volume equivalent diameter, respectively. Similarly, the volume equivalent diameter can be related to the mobility diameter as follows (DeCarlo et al., 2004):

$$D_{ve} = D_m \frac{C_s(D_{ve})}{\chi C_s(D_m)} \tag{4.2}$$

where $C_s(D_m)$ is the Cunningham slip factor for the mobility diameter. Note that for a spherical particle, $\chi = 1$, the volume equivalent and mobility diameters are equal. In order to determine the relationship between aerodynamic and mobility diameters, Equations 1 and 2 can be combined:

$$D_m = D_a \chi^{3/2} \sqrt{\frac{\rho_o}{\rho_p}} \frac{C_s(D_m)\sqrt{C_s(D_a)}}{C_s(D_{ve})^{3/2}}. \tag{4.3}$$

Equation 4.3 relates the measured mobility and aerodynamic particle diameters. Since both D_m and D_a are measured experimentally, and ρ_p is known for our samples, Equation 4.3 can be used to empirically determine the dynamic shape factor χ. Once the SMPS and APS data have been combined on a common mobility diameter scale and χ is determined, the mobility diameter data can be converted to a volume equivalent diameter distribution for use in the Mie calculation using Equation 4.2.

It is important to note that our goal is not to determine precise aerodynamic shape factors, but rather to constrain the particle size distribution for use in Mie theory simulations of extinction spectra. Small errors in the shape factor or particle density have relatively little impact on the Mie simulations and will not significantly affect our results (*vide infra*). It is also worth pointing out that there is no direct correlation between aerodynamic shape and the optical properties of non-spherical particles.

4.3   ANALYSIS OF THE DATA: MIE THEORY

Mie theory can be used to predict the model optical properties of uniform spherical particles. Of particular importance are the angle-integrated extinction (C_{ext}), absorption (C_{abs}), and scatter (C_{sca}) cross-sections, given by

$$C_{ext} = \left(\frac{2\pi}{k^2}\right)\sum_{n=1}^{\infty}(2n+1)\mathrm{Re}(a_n(X,m)+b_n(X,m)) \tag{4.4}$$

$$C_{sca} = \left(\frac{2\pi}{k^2}\right)\sum_{n=1}^{\infty}(2n+1)\left(\left|a_n(X,m)\right|^2 + \left|b_n(X,m)\right|^2\right) \tag{4.5}$$

$$C_{abs} = C_{ext} - C_{sca}, \tag{4.6}$$

where the scattering coefficients $a_n(X, m)$ and $b_n(X, m)$ are expressed in terms of the Ricatti–Bessel functions that depend on X (where $X = \pi N D/\lambda$, D is the particle diameter, N is the refractive index of the surrounding medium with $N \sim 1$ for air, and λ is the wavelength of the incident light) and the full complex index of refraction $m = n + ik$.

The Mie simulation code used here (adapted from Hung and Martin, 2002) calculates an extinction spectrum for a given size distribution and set of optical constants. The code essentially computes the extinction spectrum for a given size bin, and then sums the spectra over the size bins weighted by the particle number density in each bin to obtain the final spectrum. Mie simulations and optical constants available in the literature are used here for comparison to the experimentally measured spectra as is typical for modeling the radiative effects of atmospheric dust in many applications (Sokolik and Toon, 1999).

Mie theory is commonly used for modeling the optical properties of aerosols in both radiative forcing calculations and satellite data retrieval algorithms (Wang et al., 2002; Conant et al., 2003; Moffet and Prather, 2005; DeSouza-Machado et al., 2006). However, Mie theory is strictly valid only for spheres. Atmospheric aerosol particles are generally highly irregular in shape, which can have a significant effect on the optical properties. It is known that Mie theory does a poor job in predicting the resonance line positions and band shapes for nonspherical particles, even for small particles that fall in the Rayleigh regime, $D \ll \lambda$ (Bohren and Huffman, 1983). This could lead to errors in aerosol retrievals depending on the spectral overlap between the actual and calculated line profiles, and the specific narrow-band IR sensor channels used to determine aerosol optical depth and dust composition.

Simple analytic relations have been derived in the small particle (Rayleigh) limit to model resonance absorption cross-sections for particles with characteristic shapes such as ellipsoids, disks, and needles (Bohren and Huffman, 1983). The analytic solutions for the absorption due to a continuous distribution of ellipsoidal (CDE), disk-shaped, or needle-shaped particles are given in Equations 4.7 through 4.9:

$$C_{abs}^{cde} = kv\,\text{Im}\left[\frac{2\varepsilon}{\varepsilon - 1}\log\varepsilon\right] \tag{4.7}$$

$$C_{abs}^{disk} = \frac{kv}{3}\left[\frac{1}{\varepsilon'^2 + \varepsilon''^2} + 2\right]\varepsilon'' \tag{4.8}$$

$$C_{abs}^{needle} = \frac{kv}{3}\left[\frac{8}{(\varepsilon' + 1)^2 + \varepsilon''^2} + 1\right]\varepsilon'' \tag{4.9}$$

where $\varepsilon = \varepsilon' + i\varepsilon''$, is the complex dielectric constant, $k = (2\pi/\lambda)$, and v is the volume of the particle, related to the diameter of an equivalent volume spherical particle by $v = (\pi D_{ve}^3/6)$.

4.4 COMPARISON OF EXPERIMENTAL IR EXTINCTION SPECTRA OF COMPONENTS OF MINERAL DUST AEROSOL TO MIE-SIMULATED SPECTRA

Mineral dust aerosol consists of a complex mixture of minerals. In these studies, we have focused on several important and abundant components of mineral dust: clays, oxides, and carbonates.

4.4.1 CLAY MINERALS

Figures 4.2 through 4.4 show experimental and simulated IR spectra for illite, kaolinite, and montmorillonite, respectively. Water has been reduced in the samples by passing the aerosol flow through two diffusion dryers prior to measuring the IR spectrum, but weak adsorbed water bands are still apparent in the spectra (*vide infra*). IR absorptions due to gas-phase water present in the spectrum due to slight changes in the conditioning tube have been subtracted. Gas-phase carbon dioxide (CO_2) is also kept to a minimum using a commercial dry air generator (Parker Balston, Model 75-62) but can be observed in the spectrum as a doublet centered at 2348 cm^{-1}. The volume equivalent size distributions determined with the sizing instruments are used as inputs in the Mie simulation in combination with optical constants drawn from the published literature (Querry, 1987; Roush et al., 1991).

Panel a of Figures 4.2 through 4.4 show the experimental spectrum and Mie simulation from 800 to 4000 cm^{-1} for illite, kaolinite, and montmorillonite, respectively. An expanded view of the prominent Si−O stretch resonance spectral region is shown in panel b of Figures 4.2 through 4.4. All three clay samples show the characteristic Si−O stretch (1036–1048 cm^{-1}) as well as structural O−H stretches near 3620 cm^{-1}.

Clear differences between the Mie theory simulation and the experimentally observed resonance line profiles are apparent for the three clay samples. Discrepancies are readily seen in the strong

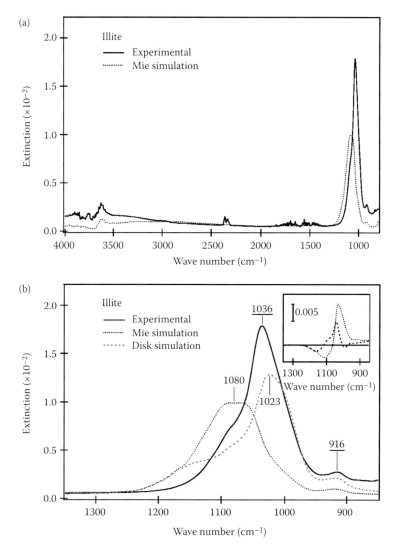

FIGURE 4.2 (a) The experimental IR spectrum (solid black line) for illite and the corresponding Mie simulation (dotted gray line). (b) An expansion of the resonance region of (a) from 850 to 1350 cm⁻¹. Included is the disk analytic simulation (dashed line). The underlined peak assignments correspond to the experimental spectrum. The inset shows the subtraction of both simulations from the experimental spectrum. The Mie simulation is blue-shifted 44 cm⁻¹ relative to the experimental spectrum for the largest resonance in this region, whereas the disk simulation is only red-shifted by 13 cm⁻¹ relative to the experimental spectrum. (Reproduced from Hudson, P. K., Gibson, E. R., Kleiber, P. D., Young, M. A., and Grassian, V. H. 2008a. *J. Geophys. Res. Atmos.* 113:D01201. With permission.)

Si–O stretch resonance region where there are significant differences in band position, band shape, and peak intensity. This spectral range is particularly important, because the silicate stretch region can be used to characterize mineral dust loading from narrow-band and high-resolution satellite spectral data (Ackerman, 1997; Sokolik, 2002; DeSouza-Machado et al., 2006). The experimental Si–O resonance peaks for illite, kaolinite, and montmorillonite are red-shifted ~27–44 cm⁻¹ relative to the Mie simulation. Peak assignments and mode descriptions for the experimental and Mie simulated spectra are given in Table 4.1 with a comparison to the literature values.

In addition to the discrepancy in the Si–O resonance region, the baseline slopes in the 2500–4000 cm⁻¹ range also differ somewhat. The poor simulation of the baseline slope may be due to

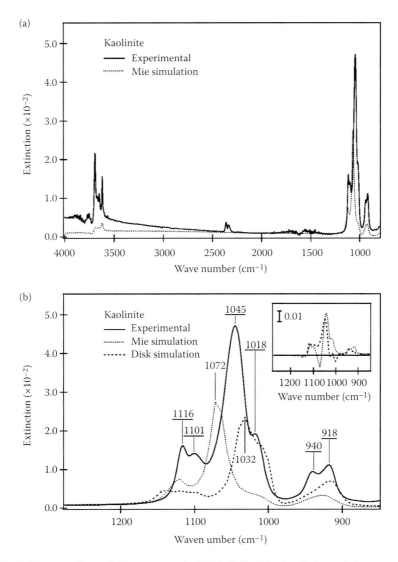

FIGURE 4.3 (a) The experimental IR spectrum (solid black line) for kaolinite and the corresponding Mie simulation (dotted gray line). (b) An expansion of the resonance region of (a) from 850 to 1300 cm^{-1}. Included is the disk analytic simulation (dashed line). The underlined peak assignments correspond to the experimental spectrum. The inset shows the subtraction of both simulations from the experimental spectrum. The Mie simulation is blue-shifted 27 cm^{-1} relative to the experimental spectrum for the largest resonance in this region, whereas the disk simulation is only red-shifted by 13 cm^{-1} relative to the experimental spectrum. (Reproduced from Hudson, P. K., Gibson, E. R., Kleiber, P. D., Young, M. A., and Grassian, V. H. 2008a. *J. Geophys. Res. Atmos.* 113:D01201. With permission.)

residual adsorbed water in the clay samples and is discussed in detail below. The adsorbed water is most evident in the montmorillonite sample shown in Figure 4.4a and b. In addition to the Si–O stretch observed near 1050 cm^{-1} and the inner O–H stretch at 3616 cm^{-1}, the O–H stretching of adsorbed water (3000–3400 cm^{-1}) and H_2O bending (1633 cm^{-1}) modes are present in the experimental spectrum. These modes are due to the adsorption of water into the inner layers of montmorillonite as it is swellable clay. The structure consists of two tetrahedral sheets sandwiching a central octahedral sheet (Si/Al ratio = 2:1) that allows water to penetrate into the interlayer molecular spaces. Because of their structures, illite and kaolinite are "less swellable" and the water signatures are correspondingly

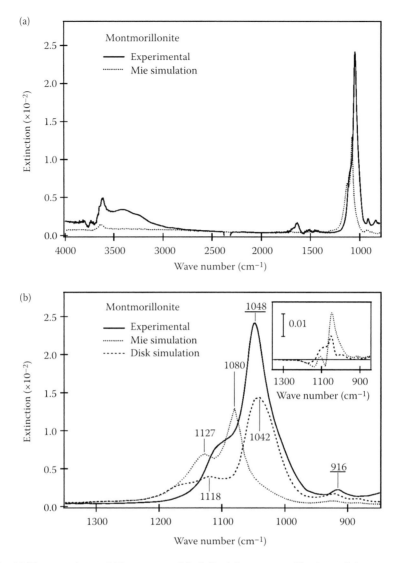

FIGURE 4.4 (a) The experimental IR spectrum (black line) for montmorillonite and the corresponding Mie simulation (dotted gray line). (b) An expansion of the resonance region of (a) from 850 to 1350 cm^{-1}. Included is the disk analytic simulation (dashed line). The underlined peak assignments correspond to the experimental spectrum. The inset shows the subtraction of both simulations from the experimental spectrum. The Mie simulation is blue-shifted 32 cm^{-1} relative to the experimental spectrum for the largest resonance in this region where as the disk simulation is only red-shifted by 6 cm^{-1} relative to the experimental spectrum. (Reproduced from Hudson, P. K., Gibson, E. R., Kleiber, P. D., Young, M. A., and Grassian, V. H. 2008a. *J. Geophys. Res. Atmos.* 113:D01201. With permission.)

weaker. There might also be some slight variations in resonance line shapes associated with water uptake in these clay samples. Recent work by the Tolbert group has shown that the presence of water does not lead to a major change in the silicate Si–O resonance peak structure (Frinak et al., 2005). For example, even in montmorillonite, the clay with the most water association, the shift in the Si–O resonance peak position under dry and wet conditions is less than 10 cm^{-1} (Frinak et al., 2005). Other studies in our laboratory involving attenuated total reflectance measurements on bulk clay powders have shown that there can be some slight differences in the line position and band shape among different mineral samples from different sources, although these variations are smaller

TABLE 4.1
Vibrational Assignments for Several Clay Compounds: Illite, Kaolinite, and Montmorillonite

Vibrational Assignment and Mode Description[a]	Literature[a]	Experimental	Mie Simulation	Disk Simulation
Illite (IMt-1)				
δ(Al—Al—OH)	920	916	920	920
ν(Si—O)	1035	1036	1080	1023
ν(O—H) structural hydroxyl groups	3615	3628	3616	N/A[b]
Kaolinite (KGa-1b)				
Δ(O—H) inner hydroxyl groups	915	918	928	918
δ(O—H) inner-surface hydroxyl groups	938	940	952 (sh)[d]	952 (sh)
ν(Si—O) in-plane	1011	1018	1016 (sh)	1012 (sh)
ν(Si—O) in-plane	1033	1045	1072	1032
ν(Si—O) perpendicular	1102	1101/1116	1121	1096–1134 (br)[e]
ν(O—H) inner hydroxyl	3620	3620	3624	N/A
ν(O—H) inner-surface hydroxyl	3653	3650	3656	N/A
ν(O—H) inner-surface hydroxyl	3669	3670	–	N/A
ν(O—H) inner-surface hydroxyl	3694	3691	3688	N/A
Montmorillonite (SAz-1)				
Δ(Al—Al—OH)	915	916	924	923
ν(Si—O)	1030	1048	1080	1042
ν(Si—O)	1109 (sh)	1105 (sh)	1127	1118/1176 (sh)
ν(O—H) structural hydroxyl groups	3620	3616	3632	N/A

Source: Reproduced from Hudson, P. K., Gibson, E. R., Kleiber, P. D., Young, M. A., and Grassian, V. H. 2008a. *J. Geophys. Res. Atmos.* 113:D01201. With permission.

[a] Van Olphen and Fripiat (1979).

[b] N/A means that the assignment is outside the simulated spectral region.

[c] Madejova and Komadel (2001) transmission data.

[d] sh = shoulder.

[e] br = broad.

than the line shifts observed here (Schuttlefield et al., 2007). The Mie simulations should not capture the adsorbed water features.

The Mie simulation is quantitatively compared to the experimental spectra in two ways. First, the integrated area of the Si—O resonance peaks can be compared. Specifically for kaolinite and montmorillonite, the experimental peak amplitudes are roughly a factor or two larger. Hudson et al. have discussed possible sources of error for the observed spectral differences in clays (Hudson et al., 2007). The largest systematic uncertainties are in the optical path length, and in the possibility that there may be some particle transmission losses between the extinction cell and the particle counting and sizing instrumentation. We estimated the total systematic error from these sources to be ~12% in the resonance peak amplitude, with minimal error (<2 cm^{-1}) in the resonance peak position

(Hudson et al., 2007). Second, the chi-square error is determined for the Mie simulation relative to the experimental spectrum over the spectral range from 900 to 1300 cm^{-1}. The chi-square error is defined as the sum of the square of the difference between the simulation and the experimental spectrum. The chi-square error between the experimental spectrum and Mie simulation, and the integrated area, normalized to the area of the experimental spectrum, are shown in Table 4.2. The chi-square error will be discussed with respect to the analytic shape simulations below.

Analytic shape simulations were also calculated according to Bohren and Huffman (1983) for the sphere, CDE, disk, and needle using the volume equivalent size distributions and the literature optical constants for comparison to the experimental spectra. These results are shown in Figure 4.5 for illite. The analytic results for spheres essentially reproduce the Mie theory results as expected, since our particle size distribution lies largely in the small particle range ($D \ll \lambda$). However, the calculated resonance line profiles for disks, needles, or for the "continuous distribution of ellipsoids" are all markedly different. The chi-square error was used to determine the best fit to the data. Although the more commonly used CDE model offers an improvement over the Mie results, the disk simulation gives the best fit quantitatively to the experimental data. As a result, the disk simulation has also been added to panel b of Figures 4.2 through 4.4. It can be seen that, with respect to the peak position and band shape of the Si–O resonance peak, the disk simulation results in a better match to the experimental data than the Mie simulation. As compared to the 27–44 cm^{-1} blue shift of the Si–O peak in the Mie simulation, this peak is now red-shifted to a much smaller magnitude (6–13 cm^{-1}). That small clay particles might be better fitted by a disk shape model is not at all surprising; Nadeau (1987) explored particle size and shape effects in the IR spectrum of kaolinite and suggested that these particles of kaolinite with $D < 2$ µm typically have a mean thickness of 0.03–0.15 µm.

Note that the analytic solutions are derived under the assumption that $X = (2\pi D/\lambda) \ll 1$ and $|m|X \ll 1$, where $|m|$ is the magnitude of the complex index of refraction (Bohren and Huffman, 1983). The value of $|m|$ varies dramatically across the resonance region, but peaks at ~2.5 in each case. Thus, the peak values of $|m|X$ range up to ~0.5. As in the case of X, the $|m|X$ values also cannot be considered to be $\ll 1$, yet the analytic shape theory results give a better overall agreement than Mie theory for modeling line positions, shapes, and integrated intensities. A subtraction of the Mie and disk simulation from the experimental results is shown as an inset in panel b of Figures 4.2 through 4.4 as well.

Systematic errors in the integrated absorbance for the major silicate resonance bands may affect radiative forcing calculations; in particular, a systematic underestimate in the integrated absorbance could lead to an underestimate of the positive forcing contributions of mineral dust aerosol, due to outgoing terrestrial IR absorbance. However, in this context, it is important to note that the aerosols

---

TABLE 4.2
Quantitative Chi-Squared Error and Relative Integrated Area of Resonance Region (900–1300 cm^{-1})

Compound	Chi-Squared ($\times 10^{-6}$)[a]		% Area Relative to Experimental Spectrum[a]	
	Mie	Disk	Mie	Disk
Illite	23.2 ± 8.9	6.2 ± 3.0	86 ± 9	102 ± 11
Kaolinite	112 ± 86.1	77.3 ± 52.8	47 ± 7	55 ± 9
Montmorillonite	36.3 ± 17.6	9.2 ± 5.7	59 ± 7	79 ± 9

Source: Reproduced from Hudson, P. K., Gibson, E. R., Kleiber, P. D., Young, M. A., and Grassian, V. H. 2008a. *J. Geophys. Res.* Atmos. 113:D01201. With permission.

[a] Statistical error from an average of 8–12 spectra for each compound.

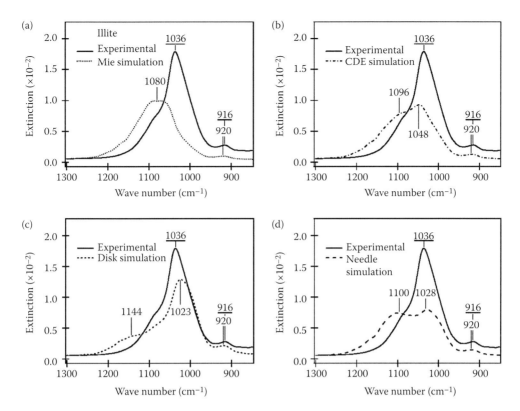

FIGURE 4.5 Comparison of the experimental extinction spectrum for illite with simulation results for several characteristic particle shapes: (a) spheres, (b) continuous distribution of ellipsoids, (c) disks, and (d) needles. The extinction cross-sections for these particle shapes are given in Bohren and Huffman (1983). (Reproduced from Hudson, P. K., Gibson, E. R., Kleiber, P. D., Young, M. A., and Grassian, V. H. 2008a. *J. Geophys. Res. Atmos.* 113:D01201. With permission.)

studied here ($D = 0.1–1$ μm) may give a relatively small contribution to overall aerosol optical depth in the IR owing to their small volume. However, it does raise the question whether similar errors in Mie theory simulations may also be present in estimating dust absorption for larger mineral aerosol particles, with $D > 1$ μm.

Sokolik (2002) has suggested that high-resolution studies in the silicate stretch spectral region, 1099–1200 cm^{-1}, may be particularly useful for determining dust composition because many of the abundant mineralogical constituents of dust aerosol (including the silicate clays and quartz) have characteristic vibrational resonance features in this region. However, errors in simulated peak position and line shape could adversely affect determinations of mineral composition based on high-resolution IR satellite data. For example, we have found that the clay Si−O stretch resonance peak is red-shifted by ca. 35 cm^{-1} from the Mie theory prediction. Indeed, the observed Si−O stretch modes in illite, kaolinite, and montmorillonite all overlie the strong O_3 resonance absorption line near 1042 cm^{-1}, which may make it problematic to quantify the clay silicate features in the spectrum. On the other hand, laboratory measurements of quartz dust spectra find that the Si−O stretch in quartz aerosol is also red-shifted from the Mie simulation and in fact lies near the silicate resonance band position for the clays predicted by Mie theory (Bohren and Huffman, 1983). Thus, due care must be taken when attempting to use Mie theory to derive compositional information from high-resolution IR spectral data in the Si−O stretch region. Spectral simulations based on the distributions of ellipsoid models might be preferable in modeling high-resolution spectra of mixed mineral dust aerosol obtained from field or satellite measurements.

In this regard, a recent modeling study of atmospheric dust loading based on AIRS data is note-worthy. DeSouza-Machado et al. (2006) show a series of clear-sky and dust-sky biases for retrieved optical depths from high-resolution spectral data from AIRS. The data show a characteristic "V"-shape dip, a signature that is commonly associated with silicate-based absorbers. DeSouza-Machado et al. (2006) have attempted to remove the effects of dust extinction from the biases to obtain more accurate temperature and water vapor measurements. This analysis uses a Mie-theory-based radiative transfer model to simulate the effects of Saharan dust, which is a largely silicate clay (Volz, 1973). The clear-sky bias for the retrieved optical depth provides evidence for a residual absorption band in the 1020–1060 cm^{-1} range that is attributed to errors in the ozone concentration. However, an alternative explanation may be that the residual dip in the brightness temperature (BT) spectrum is associated with the silicate clay absorption resonance centered near 1040 cm^{-1}, which was misplaced by the Mie simulation. The insets in panel b of Figures 4.2 through 4.4 show a dis-persion shape profile when the Mie simulation is subtracted from the experimental absorption spec-trum that is similar to the retrieved clear-sky bias data in DeSouza-Machado et al. (2006) in the 1020–1120 cm^{-1} range. A similar effect is observed in the disk simulation subtraction but to a lesser degree.

Although the overall agreement between the simulations and experimental spectra are poor, the way the simulations are utilized in satellite data retrievals may be more forgiving. Ackerman (1997) has suggested that comparing the difference measurements between specific narrowband IR sensor channels of a moderate resolution imaging spectroradiometer (MODIS) (BT_8–BT_{11} versus BT_{11}–BT_{12}) could be used to indicate the presence of dust. However, the simulated BT differences can also be affected by errors in peak placement and band shape. Sokolik (2002) has previously reported modeling results describing the effect that light and moderate dust loadings (where dust composition is varied as a function of a mixture of quartz and clays) has on BT retrievals for four narrowband sensors, namely the high-resolution infrared radiation Sounder (HIRS/2), AVHRR, MODIS, and geostationary operational environmental satellites (GOES-8) (Sokolik, 2002). Table 4.3 shows the narrowband channels for these satellites used for integration. For comparison, a top hat (rectangu-lar) integration over the narrowband range for each channel has been carried out for the experimen-tal and simulated spectra presented here. The ratios of integrated peak areas for the Mie and the analytic disk simulation, relative to the experimental extinction spectrum, are shown in Figure 4.6 for illite, kaolinite, and montmorillonite. Figure 4.6a–d presents the bands for each respective satellite, whereas Figure 4.6e shows a side-by-side comparison of all satellites divided according to

---

TABLE 4.3
Narrowband Satellite Sensors Operating in the IR Region

Sensor	BT_x Channel	Channel (μm)	Channel (cm^{-1})
HIRS/2	8	8.01–8.42	1248.4–1187.6
	11	10.89–11.32	918.3–883.4
AVHRR	11	10.3–11.3	970.9–885
	12	11.5–12.4	869.6–806.5
GOES-8	11	10.2–11.2	980.4–892.9
	12	11.5–12.5	869.6–800
MODIS	8	8.4–8.7	1190.5–1149.4
	11	10.78–11.28	927.6–886.5
	12	11.77–12.27	849.6–815

Source: Reproduced from Hudson, P. K., Gibson, E. R., Kleiber, P. D., Young, M. A., and Grassian, V. H. 2008a. *J. Geophys. Res.* Atmos. 113:D01201. With permission.

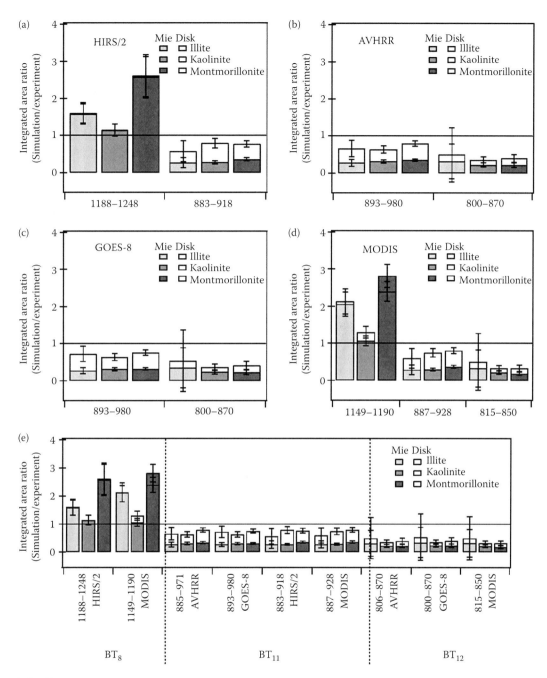

FIGURE 4.6 The relative ratio (in percent) of the integrated area for the Mie (shaded bars) and disk-shape simulations (open bars) to the experimental spectrum for illite, kaolinite, and montmorillonite for the (a) HIRS/2, (b) AVHRR, (c) GOES-8, and (d) MODIS narrowband sensors listed in Table 4.3. The solid line indicates where the area of the simulation is equal to the experimental spectrum. (e) A compilation of the data shown in (a–d) in order to compare satellites across the BT bands centered at 8, 11, and 12 μm, respectively. (Reproduced from Hudson, P. K., Gibson, E. R., Kleiber, P. D., Young, M. A., and Grassian, V. H. 2008a. *J. Geophys. Res. Atmos.* 113:D01201. With permission.)

the wavelength of integration. Note that Figure 4.6 does not present the results from simulated BT measurements, but the extinction results presented here are closely related to the BT spectra and give useful insights. Only slight differences are observed between satellite sensors for common wavelength integrations. For example, in the BT band centered at 8 μm (BT_8) using the HIRS 8.01–8.42 μm (1188–1248 cm^{-1}) and MODIS 8.4–8.7 μm (1149–1190 cm^{-1}) bands, the Mie and disk simulated integrated areas are larger than the experimental from 15% to 162% for, respectively, kaolinite and montmorillonite. However, in the HIRS and MODIS BT_8 bands, there is only a small difference between the Mie and disk simulations for all three clays. With the exception of the BT_8 bands, the disk simulation more closely approximates the integrated area relative to the experimental spectrum. There is no strong bias from satellite to satellite as the integrating range for each is not significantly different. The error bars represent the statistical error calculated from the standard deviation in the average of multiple measurements of experimental spectra with size distributions used in the Mie and disk simulations.

Note that Mie theory simulations could introduce an error in data retrieval from the BT_8–BT_{11} difference signals for dust clouds that are clay-rich, as Mie theory appears to systematically overestimate the extinction signal in the BT_8 range and underestimate the extinction in the BT_{11} and BT_{12} bands. The BT spectrum is directly related to the atmospheric transmission and, as such, appears as the inverse of the absorption spectrum. Thus, an overestimate in the BT_8–BT_{11} difference signal in the extinction (absorption) spectrum corresponds to an underestimate in the BT_8–BT_{11} difference signal in the BT spectrum. As a result, the BT_8–BT_{11} difference signals tend to be underpredicted in the Mie simulation for clay particles in this size range. The extinction signals in the BT_{11} and, especially, the BT_{12} band regions are weak and it is difficult to draw a firm conclusion about the BT_{11}–BT_{12} difference signal. However, it appears that the BT_{11}–BT_{12} differences calculated by Mie theory, as compared to the disk simulation, may be somewhat closer to the experimental result. That is, even though Mie theory may underestimate the total extinction for fine clay particles in the 11–12 μm spectral range, the BT_{11}–BT_{12} band differences, which are related to the slope of the extinction spectrum rather than its absolute value, may be more consistent with the Mie theory simulation. This latter point is important, because Sokolik (2002) has suggested using the slope in the 820–920 cm^{-1} (10.9–12.2 μm) spectral region as a determinant for dust. Thus, while the main resonance peak structure may not be well modeled in Mie theory, the measured extinction slope (and, by inference, the BT slope) in the 820–920 cm^{-1} range may be consistent with the Mie theory simulations for the fine silicate clay particles ($D = 0.1$–1 μm) studied here.

4.4.2 Oxide and Carbonate Minerals

Figure 4.7 shows typical results comparing the extinction spectrum for quartz aerosol with the results from model simulations for different characteristic particle shapes. Specifically, Figure 4.7 shows a comparison of the Si–O stretch resonance line spectrum for quartz aerosol with the results from four distinct simulations using either (a) Mie theory or the analytic theory results for shaped particles assuming the following: (b) a continuous distribution of ellipsoids (CDE)-shape model, (c) a disk-shape model, or (d) a needle-shape model for the particles. It is clear from the results in Figure 4.7 that the simple Rayleigh-based CDE model fits the data much better than Mie theory or the models for needle-like or disk-like particles. Quantitative comparisons of the chi-squared differences between the experimental spectra and the model simulations confirm this conclusion. Similar conclusions are obtained from an analysis of the calcite and dolomite spectra. For this reason, the results displayed in Figures 4.8 through 4.10, for quartz, calcite and dolomite, respectively, show only the comparisons between the experimental spectra and the simulations for Mie theory and the analytic CDE model.

As shown in the last section for silicate clay mineral dust aerosol samples in the fine particle size range ($D \sim 0.1$–1 μm), the analytic theory results for disk-shaped particles gave a better match to the

Quartz

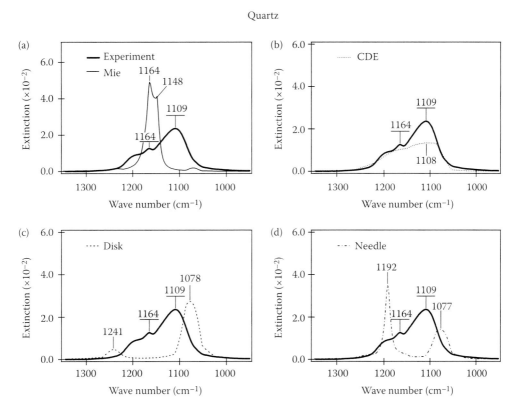

FIGURE 4.7 Experimental IR spectrum of quartz (bold line) compared to (a) Mie, (b) continuous distribution of ellipsoids (CDE), (c) disk, and (d) needle simulations in the Si–O resonance region from 950 to 1350 cm^{-1}. The CDE simulation best matches the peak position and band shape of the experimental spectrum. (Reproduced from Hudson, P. K., Young, M. A., Kleiber, P. D., and Grassian, V. H. 2008b. *Atmos. Environ.* 42:5991–5999. With permission.)

experimental spectra than Mie theory or the CDE model. This clear difference between the clays and non-clay mineral dust aerosol samples suggests important systematic differences associated with particle shape that are reflected in the measured extinction spectra.

Figures 4.8 through 4.10 show the comparison between the experimental spectrum, Mie simulations (left panels), and CDE simulations (right panels), for quartz, calcite, and dolomite, respectively. The Mie and CDE simulations for these birefringent compounds are calculated by averaging the output simulations using the *e*-ray and *o*-ray optical constants (spectral averaging [SA] method) or by averaging the *e*-ray and *o*-ray optical constants prior to the simulation (the optical constants averaging [OCA] method). Only the resonance region is presented for each compound, showing the Si–O stretching region from 950 to 1350 cm^{-1} for quartz (Figure 4.8), and the carbonate region from 1200 to 1725 cm^{-1} for calcite and dolomite (Figures 4.9 and 4.10).

Figure 4.8 shows the comparison between the experimental quartz spectrum and the model simulations for four different sets of optical constants (Table 4.4 presents a list of references for optical constants). The Mie-based simulation shows a consistently sharper peak that is blue-shifted by more than 30 cm^{-1}, relative to the experimental spectrum. Slight differences in the Mie calculated resonance profile fine structure are observed between the SA and OCA simulations when using the *o*-ray and *e*-ray optical constant data from oriented crystals. Steyer et al. (1974) pellet data (Figure 4.8d) provide the best Mie simulation match to the experimental spectrum. In general, the CDE simulation provides a much better match to the experimental spectrum than the Mie theory

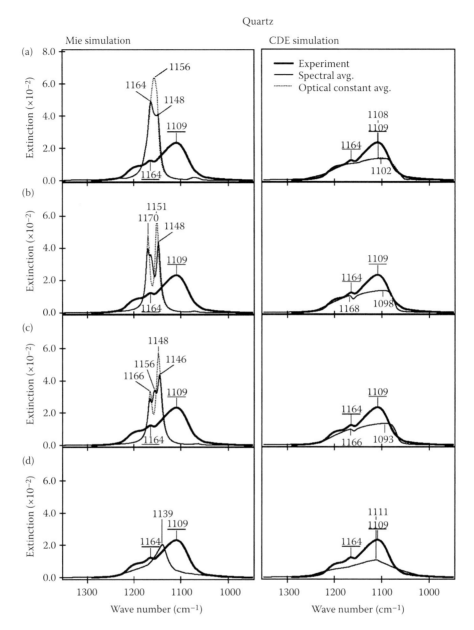

FIGURE 4.8 Mie (left panels) and continuous distribution of ellipsoid (CDE) (right panels) simulations compared to the experimental spectrum for quartz (bold line) using four different sets of optical constants from (a) Longtin et al., (b) Spitzer and Kleinman, (c) Wenrich and Christensen, and (d) Steyer et al., pellet data. (a)–(c) show Mie and CDE simulations calculated using spectral averages (solid line) or by averaging the optical constants (dotted line). See text for details. (Reproduced from Hudson, P. K., Young, M. A., Kleiber, P. D., and Grassian, V. H. 2008b. *Atmos. Environ.*42:5991–5999. With permission.)

simulation with respect to peak position and band shape for all four sets of optical constants. There are negligible differences between the spectrally averaged and optical constant averaged spectra with the CDE simulation (Figure 4.8a–d).

Figure 4.9 shows the comparison between the experimental spectrum, Mie simulation and CDE simulation for calcite. Figure 4.9a–b are from *e*-ray and *o*-ray optical constants, while Figure 4.9c, d

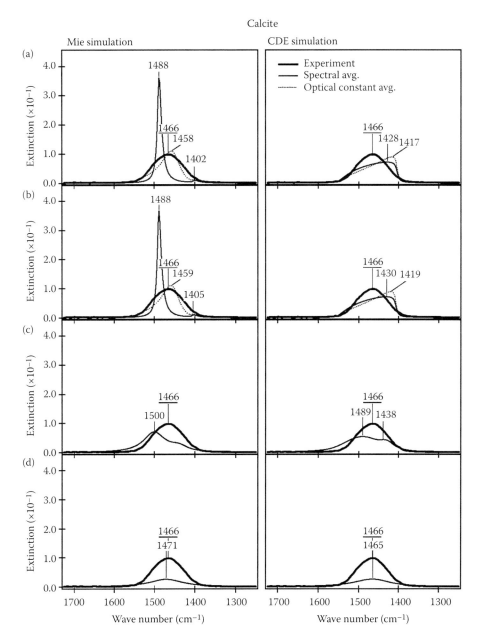

FIGURE 4.9 Mie (left panels) and continuous distribution of ellipsoid (CDE) (right panels) simulations compared to the experimental spectrum for calcite (bold line) using four different sets of optical constants from (a) Lane (1999), (b) Long et al. (1993), (c) Long et al. (1993) pellet data, and (d) Orofino et al. (2002) pellet data. (a) and (b) show Mie and CDE simulations calculated using spectral averages (solid line) or by averaging the optical constants (dotted line). See text for details. (Reproduced from Hudson, P. K., Young, M. A., Kleiber, P. D., and Grassian, V. H. 2008b. *Atmos. Environ.* 42:5991–5999. With permission.)

are from pellet data. Differences are observed between the two optical constants in both the v_3 and v_2 carbonate peaks at ca. 1470 and 880 cm^{-1} (not shown), respectively. Analogous to quartz, the v_3 and v_2 carbonate peaks are much sharper and greater in intensity in the Mie simulation than the experimental spectrum. Although the v_2 mode has the correct peak position (not shown) the v_3 mode is blue-shifted by 22 cm^{-1} relative to the experimental spectrum. Large differences are also observed

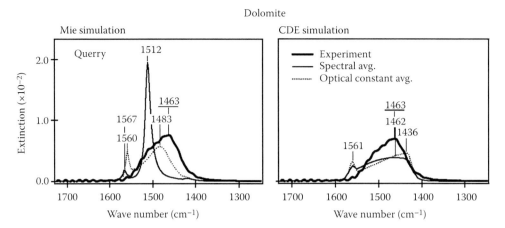

FIGURE 4.10 Mie (left panel) and continuous distribution of ellipsoid (CDE) (right panel) simulations compared to the experimental spectrum for dolomite (bold line) using Querry optical constants. The Mie and CDE simulations are calculated using spectral averages (solid line) or by averaging the optical constants (dotted line). See text for details. (Reproduced from Hudson, P. K., Young, M. A., Kleiber, P. D., and Grassian, V. H. 2008b. *Atmos. Environ.* 42:5991–5999. With permission.)

in the Mie simulation between the spectrally averaged and optical constant averaged results. Again, the CDE simulation more closely matches the experimental spectrum in band shape and position. Only small differences are apparent between the spectrally averaged and optical constant averaged CDE simulations. Differences are observed between the two sets of pellet data as well. Orofino et al. (2002) data greatly underpredict the resonance absorption strength. Although the Orofino et al.

TABLE 4.4
Optical Constant Data Sources for the Mineral Dust Samples Used in this Work

Compound	Reference	Method	Crystal or Pellet	Data Range (cm^{-1})
Quartz	Longtin et al.[a]	Table	E-ray and O-ray	30–50,000
Quartz	Spitzer and Kleinman[b]	Dispersion equations	E-ray and O-ray	270–2,000
Quartz	Wenrich and Christensen[c]	Dispersion equations	E-ray and O-ray	400–2,000
Quartz	Steyer et al.[d]	Table	Pellet	400–1,400
Calcite	Lane[e]	Dispersion equations	E-ray and O-ray	400–2,000
Calcite	Long et al.[f]	Dispersion equations	E-ray and O-ray	30–2,000
Calcite	Long et al.[f]	Dispersion equations	Pellet	30–4,000
Limestone	Orofino et al.[g]	Dispersion equations	Pellet	500–2,857
Dolomite	Querry et al.[h]	Table	E-ray and O-ray	250–4,000

Source: Reproduced from Hudson, P. K., Young, M. A., Kleiber, P. D., and Grassian, V. H. 2008b. *Atmos. Environ.* 42:5991–5999.

[a] Longtin et al. (1988).
[b] Spitzer and Kleinman (1961).
[c] Wenrich and Christensen (1996).
[d] Steyer et al. (1974).
[e] Lane (1999).
[f] Long et al. (1993).
[g] Orofino et al. (2002).
[h] Querry (1987).

(2002) data are specifically for limestone, the paper suggests that the optical properties should be valid in the submicron size regime for calcite grains that are randomly oriented.

Figure 4.10 shows the comparison between the experimental spectrum, Mie simulation and CDE simulation for dolomite. As dolomite is another carbonate compound, results similar to calcite are observed. The v_3 resonance peak in the Mie simulation is blue-shifted 48 cm^{-1} relative to the experimental simulation, but the position and band shape are dramatically improved when averaged optical constants are used. The chi-squared value improves by a factor of 10 between the spectrally averaged and optical constant averaged Mie simulation. Again, the CDE simulation washes out the fine structure seen in the Mie simulation, resulting in a broad peak centered near that of the experimental spectrum with slight improvement in using the optical constant averaged simulation.

The previous discussion of silicate clay mineral dust aerosol suggested that satellite retrievals could be adversely affected by using Mie theory for dust retrievals from narrowband satellite data, depending on the relative overlap between the actual and Mie predicted resonance positions, and the satellite band (Hudson et al., 2008a). For oxide minerals, there is an overlap between the quartz region and the BT band at 8 μm (BT$_8$) measured by the HIRS/2 and MODIS satellites. The HIRS/2 satellite integrates a band from 1188 to 1248 cm^{-1} and MODIS from 1149 to 1190 cm^{-1}. Figure 4.11 shows the area ratio between the simulated and experimental extinction signal for quartz integrated over the BT$_8$ detection band range for the HIRS/2 and MODIS satellites. There are dramatic differences between the experimental results and the Mie and CDE-shape simulations. A top-hat integration over these wavelength regions shows that the ratio of the Mie simulation to the experimental extinction underpredicts or overpredicts the absorption by approximately a factor of two for the HIRS/2 and MODIS satellites, respectively. The CDE simulation gives a much better result for both satellites. The closer the ratio of the integrated area of the CDE simulation relative to the experiment is to 1, the better is the simulation. The HIRS/2 satellite has a ratio of 0.80 compared to 0.41 with the Mie simulation. Similarly, the MODIS satellite has a ratio of 0.71 relative to 1.59. These results are independent of the particular choice of optical constant data sets. This further supports that Mie theory does not accurately model atmospheric mineral dust particles in the IR region.

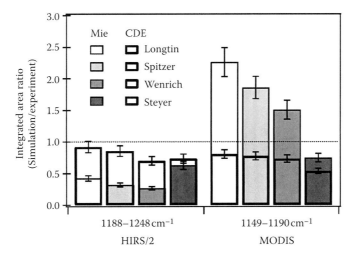

FIGURE 4.11 Ratio of the integrated areas of the Mie (light) or CDE (bold) simulation to the quartz experimental spectrum in the HIRS/2 and MODIS BT$_8$ bands for four sets of optical constants, Longtin et al. (open), Spitzer and Kleinman (light gray), Wenrich and Christensen (gray), and Steyer et al. (dark). The ratio is nearly one for all CDE experiments (darker-lined boxed area) where Mie theory either underpredicts (HIRS/2) or overpredicts (MODIS) the integrated area (lighter-line boxed area). (Reproduced from Hudson, P. K., Young, M. A., Kleiber, P. D., and Grassian, V. H. 2008b. *Atmos. Environ.* 42:5991–5999. With permission.)

Of course the Rayleigh models are very crude approximations. We expect that more sophisticated calculations based on T-matrix theory or using the distributed dipole approximation would likely give a more accurate model simulation, which could also be extended to much larger particle sizes (Draine and Flatau, 1994; Mishchenko et al., 2002). We intend to carry out such calculations in future work. However, for aerosols in the fine particle size mode ($D = 0.1–1$ μm) we were surprised at how well the simple Rayleigh-based theory can work for mineral dust aerosol despite the fact that the Rayleigh condition ($D \ll \lambda$) is not strictly satisfied. Apparently, for particles in this size range, the errors associated with applying Mie theory to nonspherical particles are substantially worse than the errors associated with using the Rayleigh results in a regime where D is not necessarily $\ll \lambda$.

4.5 CONCLUSIONS AND FUTURE WORK

These studies show the importance of shape effects in the IR resonance absorption line profiles of mineral dust aerosol. As most atmospheric particles are rarely found as individual components but in fact are complex mixtures of aggregates and coated particles. Future studies will focus on mixtures and coatings. Specifically, extinction measurements of multicomponent, complex mineral dust mixtures would allow rigorous evaluation of theoretical models over a broad range of particle characteristics. Furthermore, another important future effort is to investigate the impact of coatings and chemical processing effects on dust IR extinction spectra. Additionally, the analysis can be extended beyond Mie theory and analytic expressions to include other theoretical approaches (e.g., T-matrix theory) for calculating the extinction properties of mineral dust aerosol. The ultimate goal of this work to provide qualitative insight and quantitative data that can be directly incorporated into radiative transfer models for climate forcing calculations, remote sensing data retrievals, and atmospheric chemistry models.

ACKNOWLEDGMENTS

This chapter is based on work supported by the National Science Foundation under Grant No. ATM-0425989. Any opinions, findings, and conclusions or recommendations expressed in this chapter are those of the authors and do not necessarily reflect the views of the National Science Foundation. We thank Professor Scot Martin for providing us with the Mie simulation code and Professor Kuo-Ho Yang for his continued work on this code. We also thank Dr. Elizabeth R. Gibson for her contributions to the studies described in this chapter.

REFERENCES

Ackerman, S. A. 1997. Remote sensing aerosols using satellite infrared observations. *J. Geophys. Res.* 102:17069–17079.

Arimoto, R., Kim, Y. J., Kim, Y. P., Quinn, P. K., Bates, T. S., Anderson, T. L., Gong, S. et al. 2006. Characterization of Asian Dust during ACE-Asia. *Global Planet. Change* 52:23–56.

Bauer, S. E., Balkanski, Y., Schulz, M., Hauglustaine, D. A., and Dentener, F. 2004. Global modeling of heterogeneous chemistry on mineral aerosol surfaces: Influence on tropospheric ozone chemistry and comparison to observations. *J. Geophys. Res.* 109:D02304.

Bohren, C. F. and Huffman, D. 1983. *Absorption and Scattering of Light by Small Particles*. Wiley, New York, 530pp.

Buseck, P. R. and Posfai, M. 1999. Airborne minerals and related aerosol particles; Effects on climate and the environment. *Proc. Natl. Acad. Sci. U.S.A.* 96:3372–3379.

Claquin, T., Schulz, M., and Balkanski, Y. J. 1999. Modeling the mineralogy of atmospheric dust sources. *J. Geophys. Res.* 104:22243–22256.

Conant, W. C., Seinfeld, J. H., Wang, J., Carmichael, G. R., Tang, Y. H., Uno, I., Flatau, P. J., Markowicz, K. M., and Quinn, P. K. 2003. A model for the radiative forcing during ACE-Asia derived from CIRPAS Twin Otter and R/V Ronald H. Brown data and comparison with observations. *J. Geophys. Res.* 108:8661.

DeCarlo, P., Slowik, J., Worsnop, D., Davidovits, P., and Jimenez, J. 2004. Particle morphology and density characterization by combined mobility and aerodynamic diameter measurements. Part 1: Theory. *Aerosol Sci. Technol*. 38:1185–1205.

DeSouza-Machado, S. G., Strow, L. L., Hannon, S. E., and Motteler, H. E. 2006. Infrared dust spectral signatures from AIRS. *Geophys. Res. Lett*. 33:L03801.

Dick, W. D., Ziemann, P. J., Huang, P. F., and McMurry, P. H. 1998. Optical shape fraction measurements of submicrometre laboratory and atmospheric aerosols. *Meas. Sci. Technol*. 9:183–196.

Draine, B. T. and Flatau, P. J. 1994. Discrete-dipole approximation for scattering calculations. *J. Opt. Soc. Am. A—Opt. Image Sci. Vision* 11:1491–1499.

Frinak, E. K., Mashburn, C. D., Tolbert, M. A., and Toon, O. B. 2005. Infrared characterization of water uptake by low-temperature Na-montmorillonite: Implications for Earth and Mars. *J. Geophys. Res*. 110:D09308.

Henderson-Sellers, A. 1995. World Survey of Climatology—Future Climates of the World: A Modeling Perspective, A. Henderson-Sellers, Ed., Elsevier, Amsterdam, Vol. 16.

Highwood, E. J., Haywood, J. M., Silverstone, M. D., Newman, S. M., and Taylor, J. P. 2003. Radiative properties and direct effect of Saharan dust measured by the C-130 aircraft during Saharan Dust Experiment (SHADE): 2. Terrestrial spectrum. *J. Geophys. Res*. 108:8578.

Hinds, W. C. 1999. *Aerosol Technology: Properties, Behavior, and Measurement of Airborne Particles*. Wiley, New York, 483pp.

Hudson, P. K., Gibson, E. R., Kleiber, P.D., Young, M. A., and Grassian, V. H. 2008a. Coupled infrared extinction and size distribution measurements for several clay components of mineral dust aerosol. *J. Geophys. Res. Atmos*. 113:D01201.

Hudson, P. K., Young, M. A., Kleiber, P.D., and Grassian, V. H. 2008b. Coupled infrared extinction and size distribution measurements of several non-clay components of mineral dust aerosol (quartz, calcite, and dolomite). *Atmos. Environ*. 42:5991–5999.

Hudson, P. K., Gibson, E., Young, M. A., Kleiber, P. D., and Grassian, V. H., 2007. A newly designed and constructed instrument for coupled infrared extinction and size distribution measurements of aerosols. *Aerosol Sci. Technol*. 41:701–710.

Hung, H. M. and Martin, S. T. 2002. Infrared spectroscopic evidence for the ice formation mechanisms active in aerosol flow tubes. *Appl. Spectrosc*. 56:1067–1081.

Lane, M. D. 1999. Midinfrared optical constants of calcite and their relationship to particle size effects in thermal emission spectra of granular calcite. *J. Geophys. Res.—Planets* 104:14099–14108.

Long, L. L., Querry, M. R., Bell, R. J., and Alexander, R. W. 1993. Optical properties of calcite and gypsum in crystalline and powdered form in the infrared and far-infrared. *Infrared Phys*. 34:191–201.

Longtin, D. R., Shettle, E. P., Hummel, J. R., and Pryce, J. D. 1988. A wind dependent desert aerosol model: Radiative properties. AFGL-TR-88-0112, Air Force Geophysics Laboratory, Hanscom AFB, MA.

Kalashnikova, O. V. and Sokolik, I. N. 2002. Importance of shapes and compositions of wind-blown dust particles for remote sensing at solar wavelengths. *Geophys. Res. Lett*. 29:1398.

Madejova, J. and Komadel, P. 2001. Baseline studies of the Clay Minerals Society Source Clays: Infrared methods. *Clays Clay Miner*. 49:410–432.

Moffet, R. C. and Prather, K. A. 2005. Extending ATOFMS measurements to include refractive index and density. *Anal. Chem*. 77:6535–6541.

Mishchenko, M. I., Travis, L. D., and Lacis, A. A. 2002. *Scattering, Absorption, and Emission of Light by Small Particles*. Cambridge University Press, Cambridge, UK.

Nadeau, P. H. 1987. Relationships between the mean area, volume and thickness for dispersed particles of kaolinites and micaceous clays and their application to surface-area and ion-exchange properties. *Clay Miner*. 22:351–356.

Orofino, V., Blanco, A., Fonti, S., Marra, A. C., and Polimeno, N. 2002. The complex refractive index of limestone particles: An extension to the FIR range for Mars applications. *Planet. Space Sci*. 50:839–847.

Pierangelo, C., Chedin, A., Heilliette, S., Jacquinet-Husson, N., and Armante, R. 2004. Dust altitude and infrared optical depth from AIRS. *Atmos. Chem. Phys*. 4:1813–1822.

Prospero, J. M. 1999. Long-range transport of mineral dust in the global atmosphere: Impact of African dust on the environment of the southeastern United States. *Proc. Natl. Acad. Sci. U.S.A*. 96:3396–3403.

Prospero, J. M. and Lamb, P. J. 2003. African droughts and dust transport to the Caribbean: Climate change implications. *Science* 302:1024–1027.

Querry, M. R. 1987. *Optical Constants of Minerals and Other Materials from the Millimeter to the Ultraviolet*. U.S. Army, Aberdeen, MD, 331pp.

Roush, T., Pollack, J., and Orenberg, J. 1991. Derivation of midinfrared (5–25 μm) optical constants of some silicates and palagonite. *Icarus* 94:191–208.

Schuttlefield, J. D., Cox, D., and Grassian, V. H. 2007. An investigation of water uptake on clays minerals using ATR-FTIR spectroscopy coupled with quartz crystal microbalance measurements. *J. Geophys. Res.* 112:D21303.

Sheehy, D. P. 1992. A perspective on desertification of grazing-land ecosystems in North China. *Ambio* 21:303–307.

Sokolik, I. N. 2002. The spectral radiative signature of wind-blown mineral dust: Implications for remote sensing in the thermal IR region. *Geophys. Res. Lett.* 29:2154.

Sokolik, I. N. and Toon, O. B. 1999. Incorporation of mineralogical composition into models of the radiative properties of mineral aerosol from UV to IR wavelengths. *J. Geophys. Res.* 104:9423–9444.

Spitzer, W. G. and Kleinman, D. A. 1961. Infrared lattice bands of quartz. *Phys. Rev.* 121:1324–1335.

Steyer, T. R., Day, K. L., and Huffman, D. R. 1974. Infrared absorption by small amorphous quartz spheres. *Appl. Opt.* 13:1586–1590.

Tegen, I. and Fung, I. 1994. Modeling of mineral dust in the atmosphere: Sources, transport and optical thickness. *J. Geophys. Res.* 99:22897–22914.

Van Olphen, H. and Fripiat, J. 1979. *Data Handbook for Clay Minerals and Other Non-Metallic Materials.* Pergamon Press, Oxford, 346pp.

Volz, F. E. 1973. Infrared optical constants of ammonium sulfate, Sahara dust, volcanic pumice, and flyash. *Appl. Opt.* 12:564–568.

Wang, J., Flagan, R. C., Seinfeld, J. H., Jonsson, H. H., Collins, D. R., Russell, P. B., Schmid, B. et al. 2002. Clear-column radiative closure during ACE-Asia: Comparison of multiwavelength extinction derived from particle size and composition with results from Sun photometry. *J. Geophys. Res.* 107:4688.

Wenrich, M. L. and Christensen, P. R. 1996. Optical constants of minerals derived from emission spectroscopy: Application to quartz. *J. Geophys. Res.—Solid Earth* 101:15921–15931.

5 Infrared Spectroscopy of Dust Particles in Aerosols for Astronomical Application

Akemi Tamanai and Harald Mutschke

CONTENTS

5.1 Introduction 101
5.2 The Challenge of Identifying Dust Components from IR Spectra 102
5.3 An Aerosol Experiment for IR Spectroscopy of Free-Floating Dust Particles 105
    5.3.1 Experimental Setup 105
    5.3.2 Astrophysically Relevant Samples 107
    5.3.3 A Comparison of Aerosol and Pellet Spectra 109
5.4 Morphological Effects on IR Band Profiles 110
    5.4.1 Size Effect 110
    5.4.2 Shape Effect 112
    5.4.3 Agglomeration Effect 113
    5.4.4 Theoretical Simulations 114
5.5 Application to the Interpretation of Astronomical Spectra 116
5.6 Conclusion 119
Acknowledgments 120
References... 120

5.1 INTRODUCTION

Although the universe appears to be mostly empty space, a vast number of micron-sized dust grains together with gas are distributed widely throughout the space between the stars. These dust grains are formed from the heavier elements produced by nucleosynthesis in stellar interiors and set free when stars explode or lose their outer parts in the last stages of their "life." According to the abundance of these heavier elements, the main condensable compounds are silicates, oxides, carbon, iron or iron sulfide, and water ice, but under special conditions also more exotic solids can be formed. These dust grains produce detectable thermal emission in the wavelength region between submillimeter (cold dust) and infrared (IR) (warm dust), depending on the dust temperature. In interstellar space, this temperature is very low, but in stellar environments it is often high enough to allow the fundamental vibrations of the crystal lattice to be thermally excited and to dominate the emission spectrum. For instance, since many decades, the Si−O stretching band of amorphous silicates at wavelengths around 10 μm is known to be present in the spectra of giant stars (e.g., red giant branch (RGB) and asymptotic giant branch (AGB) stars).

Nowadays astronomical investigations concentrate on forming planetary systems (the so-called protoplanetary accretion disks) or on dust in the existing planetary systems produced by collisions of cometary or asteroidal bodies (the so-called debris disks). As IR telescope technology has advanced, the emission spectra obtained by ground-based telescopes such as the Very Large

Telescope (VLT) operated by the European Southern Observatory (ESO), the Subaru telescope operated by the National Astronomical Observatory of Japan (NAOJ), the Infrared Telescope facility (IRTF) operated by the National Aeronautics and Space Administration (NASA) and satellite telescopes such as the Spitzer Space Telescope operated by NASA, the Infrared Astronomical Satellite (IRAS) operated jointly by the United States, United Kingdom, and the Netherlands, the Akari operated by the Japan Aerospace Exploration Agency (JAXA) have become increasingly detailed. A great advantage of satellite telescopes is increased sensitivity and wavelength coverage because of the large amount of light from astronomical objects not absorbed by the atmosphere of Earth. Thus, highly sensitive emission spectra with less noise can be obtained by making use of space-based telescopes. Such emission spectra from circumstellar disks in principle provide a great deal of information on dust grain properties such as their crystallinity, shape, size, agglomeration state, temperature, and chemical composition. Grain growth which is a first step to form larger bodies such as planetesimals has been observed in protoplanetary disks.[1,2] Detailed mineralogical composition analysis of circumstellar and cometary dust grains has been performed by modeling.[3,4] In practice, the interpretation of the spectra is difficult considering these many influences modifying the band profiles. Apart from comparing peak positions and bandwidths with measured spectra from the laboratory or with theoretically calculated spectra, it is important to understand how band profiles vary with parameters such as temperature and the grain morphology.

In this chapter, we will demonstrate how aerosol spectroscopy can be applied for this purpose. The difficulty of the accurate analysis of the properties of dust grains in observed IR spectra is discussed in detail in Section 5.2, including problems regarding the comparison of experimentally measured and theoretically calculated spectra with observed spectra. The experimental apparatus and its setup for aerosol spectroscopy are described together with information about astrophysically important samples. Two experimental approaches (the aerosol and pellet techniques) are addressed and provide a new view on the identification of dust components (Section 5.3). The morphological effects on IR band profiles and the limitations of theoretical simulations on IR band profiles are explored precisely for a better understanding of both experimental and theoretical approaches in Section 5.4. The way to apply the aerosol spectra to observed spectra in order to identify each component of dust grains is shown in Section 5.5. We conclude with a brief summary in Section 5.6.

5.2 THE CHALLENGE OF IDENTIFYING DUST COMPONENTS FROM IR SPECTRA

IR spectra obtained from "dusty" cosmic places are usually thermal emission spectra of warm dust located in the vicinity of a star. The dust configurations can be optically thick at the IR wavelengths. In this case, radiative transfer modeling is necessary to understand the spectra. In many cases, however, the configuration can be taken to be optically thin. Even in the important case of protoplanetary accretion disks, the emission comes mainly from an optically thin surface layer heated by the stellar radiation. In these cases, the spectrum is simply the product of the emission cross-section of the dust with a Planck function given by the equilibrium temperature of the dust (in principle different cross-sections and different temperatures for each dust grain). While the multiplication with the Planck function may determine the relative intensities of bands seen in the spectra, the band profiles are simply given by the emission cross-section which, according to Kirchhoff's Law, is equivalent to the absorption cross-section.

Having calculated or measured the absorption spectra available for comparison, one can try to disentangle the contribution of different "dust species," that is, of different compositions, crystal structures, grain sizes to the spectra. This has been done over the last years to quite some extent, mainly with spectra obtained by the Infrared Space Observatory (ISO) and Spitzer satellites and also with ground-based observations, although their wavelength range is very limited. For comparison, simulated spectra have mostly been preferred, because they can be calculated in a very flexible way, varying parameters of the model such as composition of the grain, grain size, and to some extent the grain geometry. Simulations for simple geometrical models such as spherical[5] or ellipsoidal[6] grain shapes have been widely used to compare with observed spectra.

The most common analytical model for computing light scattering and absorption is the Mie theory,[5] which is a complete solution for the interaction between a polarized plane wave and a dielectric sphere of (in principle) arbitrary size, especially covering the case of particle dimensions comparable to the wavelength or even larger than the wavelength of the incident radiation. Dust particles in astrophysical environments are sometimes known to be up to several tens of microns in size, especially in planetary systems (debris dust), in molecular clouds, and partly in outflows of evolved stars. So considering even mid-IR wavelengths of thermal radiation, it may be necessary to simulate dust optical properties using this general approach (in terms of particle size). Input data are the wavelength-dependent complex refractive index $m = n + ik$ (n: real part; k: imaginary part) of the dielectric (or conducting) material of which the particle is composed and its size relative to the wavelength. The complex refractive index of astrophysically relevant materials can be procured from databases such as the one collected at AIU Jena (http://www.astro.uni-jena.de/Laboratory/OCDB). Mie theory is quite easy to be implemented into complex simulations, that is, the computational effort is small.

However, the restriction to spherical grain shapes may be a serious limitation, since dust particles in some environments may in fact be complex aggregates of larger and smaller subgrains composed of different materials. Even for single homogeneous grains, the shapes may be irregular, which has, for example, a strong impact on the profiles of resonance bands. Such complex geometries are often treated by the discrete dipole approximation (DDA) approach,[7,8] where a solid particle is replaced by an array of N point dipoles in a lattice, with the lattice spacing being smaller than the wavelength of the incident radiation. The response of the whole grain to the incident radiation is calculated by taking the interaction of all dipoles into account, and the positions of all dipoles retain the original grain shape. This approach allows modeling the absorption and scattering properties of any grain geometry including porosity.[8–10] However, the computational effort is large, which makes it difficult to take the statistical variation of geometrical properties in real particle ensembles into account.

There are some approaches to the modeling of statistical optical properties. Such approaches have to use simple but easily variable particle geometries. The most intuitive one is to average absorption cross-sections of ellipsoidal particles of widely varying axis ratios, for which an exact theory similar to Mie theory exists. Another approach is to use a distribution of hollow or multilayered spheres, which is also able to mimic optical properties of statistical particle ensembles.

The statistical models become especially easy to handle if the particles can be assumed to be small compared to the wavelength and, for instance, for thermal emission at mid-IR or far-IR wavelengths, this often may be possible to assume. Submicrometer-sized particles are in fact dominating the dust size distribution in most environments where dust is lately condensed and is not sustained at high densities for a sufficiently long time to allow aggregation to take place. Even if the size distribution extends to larger grains, the cross-sectional area may be dominated by the small grains. Moreover, resonance bands of larger grains tend to be strongly broadened, so that the sharper bands of the small grains dominate the spectra which are under investigation. If considerations are restricted to particles with dimensions d small compared to the wavelength λ (Rayleigh[11] limit: $|k*md| \ll 1$, $k = 2\pi/\lambda$, m being the complex refractive index), a quasi-static approach can be taken. Thus, the particles can be treated as single dipoles with simple analytical expressions for the polarizabilities $\alpha(m, L)$ in case of ellipsoids—L being the geometrical factor or form factor—and $\alpha(m, f)$ in case of the hollow spheres—f being the filling factor of the central void. The absorption/emission cross-section of a particle is simply the imaginary part of the dipole polarizability multiplied with its volume and the wave number k.

If the probability for the parameters L or f is taken to be equally distributed over their respective range of definition (in both cases 0 … 1), the polarizabilities can even be analytically averaged (in case of the ellipsoids also over the orientation of the particles) leading to still quite simple analytical formulae for the absorption cross-sections, known as the CDE (continuous distribution of ellipsoids) model[6] and the DHS (distribution of hollow spheres) model.[12] Compared to Mie calculations in the Rayleigh limit, these models produce broader absorption curves, which are much better approximations to laboratory measured spectra.[6] However, the probability functions can also be modified in order to achieve an even better agreement with experimental data. This has recently been done for the probability functions $P(L)$ in an attempt to fit the aerosol spectra (Section 5.4.4). This approach has been named the DFF (distribution of form factors) model.[13]

At that stage, simulated spectra may already give fine representations of experimental or observed spectra, although it is not yet possible to perfectly predict form factor distributions for real particle geometries. For instance, the influence of particle agglomeration on the spectra is not yet sufficiently understood. In addition, anisotropic materials pose a problem, because the independent calculation of the polarizabilities along the different crystal orientations is not physically correct, but their combined treatment requires a DDA calculation.[14]

Therefore, since cosmic dust grains could be anisotropic in crystallography, irregular in shape, inhomogeneous in size, and intricate in agglomeration state, a lot of difficult problems on the details of the morphological aspects of dust grains remain in simulations of dust emission spectra. On the other hand, it may be feasible to measure the absorption spectra of particles with complex geometries and compositions in the laboratory. The pellet technique has been commonly used for the IR spectroscopic analysis of solids.[15–19] In this technique, a small amount of a particulate sample is mixed with potassium bromide (KBr), cesium iodide (CsI), or polyethylene (PE) powder that have high transmission through certain IR wavelength ranges. The mixture is pressed at high load so as to make, for example, a 0.5 mm thick and 13 mm diameter pellet, containing about 0.5 mg sample distributed over 1.33 square centimeter area for spectral transmission measurements.

The advantages of the pellet technique are low cost, low sample consumption, longevity of the pellets in desiccators, and the exact column density of the particulate sample is known. However, a serious problem of this technique is that the spectra for particles embedded in medium (KBr, CsI, or PE) are affected by the electromagnetic polarizability of the embedding medium.[15,20–24] Hence, the band profiles of the KBr pellet spectra are not equivalent to those observed in space since the dielectric constant of KBr is higher than that of vacuum which gives rise to a change in the spectral features. The main effects appearing on the band profiles of pellet measurements are (1) a shift in the peak positions with regard to wavelength, (2) the modification of the band shape of spectral features.[15] Another disadvantage is that the morphology of the particles, especially the agglomeration, is influenced by the pressing of the pellet, for example, depending on the grain size of the KBr powder and cannot be analyzed easily.

The deficiencies of the pellet technique can be solved by adopting an aerosol technique[25] to the spectroscopic extinction measurements of dust grains in the laboratory.[26–28] In order to avoid any environmental effects, in this technique astrophysically relevant particulates are suspended and retained in nitrogen gas (N_2) at approximately normal pressure during the measurements. The refractive index of the gas is comparable to that of vacuum. Grains of about a few μm size and smaller can be kept suspended sufficiently long for IR spectroscopy at a high signal-to-noise ratio. A great advantage is that the particle morphology (size, shape, and agglomeration) is able to be analyzed after the measurements by electron microscopy.

Figure 5.1 shows the emission spectrum of the crystalline-silicate rich dust around the Vega-type star HD113766,[29] together with the band profiles obtained by both the aerosol and the KBr pellet measurements for irregular shaped forsterite particles (Mg_2SiO_4), as well as with calculated band profiles for the CDE,[6] the DHS,[12] and the DFF[14] models. Among the model spectra, the DFF model is the one reproducing best not only the aerosol spectrum, but also the observed spectrum. The CDE and DHS models predict the structure of the observed band correctly, but the peak positions are located at too long wavelengths. The DFF model spectrum still deviates from the experimental band profile in the 9.8 μm peak, which is because of the incorrect treatment of the crystal anisotropy. The peaks at 9.8 and 11 μm in the HD113766 spectrum are broad compared with both the laboratory and the calculated spectra. This may indicate the presence of larger grains in the disk of the Vega-type star.

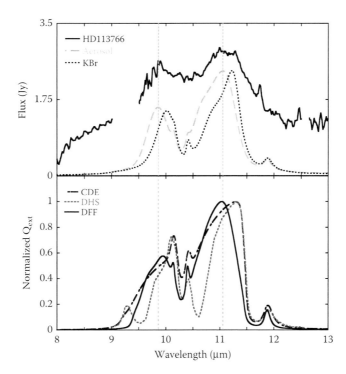

FIGURE 5.1 Top: Comparison of the emission spectrum of the dust around the main-sequence star HD113766 in the 10 μm wavelength range (solid line) with the experimental extinction band profiles obtained by aerosol (gray dashed line) and KBr (dotted line) pellet measurements on irregular shaped forsterite particles. Bottom: Calculated band profiles for forsterite particles using the CDE (dash-dotted line), DHS (gray dotted line), and DFF (solid line) models. The light dotted lines denote the peak positions of the aerosol spectrum. Note that the gap in the astronomical spectrum between 9.4 and 9.7 μm in wavelengths is because of uncorrectable atmospheric ozone features.

The details of this aerosol spectroscopy method will be discussed more thoroughly in the next section.

5.3 AN AEROSOL EXPERIMENT FOR IR SPECTROSCOPY OF FREE-FLOATING DUST PARTICLES

5.3.1 EXPERIMENTAL SETUP

A diagram of the aerosol spectroscopy apparatus is shown in Figure 5.2. A rotating-brush dust flow generator (Palas RBG1000) is set under a black anodized aluminum gas cell (MARS-8L/20V, Gemini Scientific Instr.) for dispersing a powdered sample in a N_2 gas stream. The dense aerosol along the gas flow passes through a two-stage impactor which controls for the desired gas flow and separates extremely large clumps from smaller dust particles ($d_{avg} \approx 2$–3 μm). At that time, the small-sized aerosol particles are concentrated by the impactor so that a concentration of 10^6 particles per cubic centimeter finally reaches the gas cell, which is 13.6 cm in diameter and 55.7 cm in total length.

The cell is composed of several important parts which are designed to provide high quality of optical performance. It is possible to adjust for an 18 m path length by taking advantage of multiple reflections between two gold mirrors (objective and field mirrors) mounted on both sides of the cell in order to increase the sensitivity. There are two inlets (not shown) for fresh N_2 gas to come into the cell constantly at a flow of 60 L/h. These inlets are located at both ends of the cell and provide

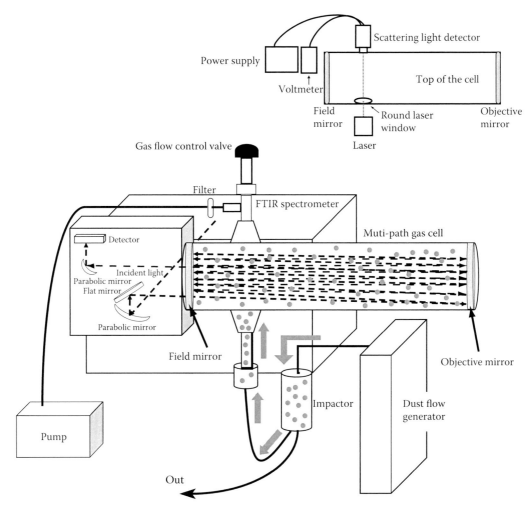

FIGURE 5.2 A schematic diagram of the experimental apparatus for aerosol spectroscopy. On the upper left diagram is a top view of the cell.

protection of the gold mirrors against deposition of particles from the aerosol, which is thereby concentrated in the middle of the cell.

A glass window 3.5 cm in diameter is installed in the central part of the cell and a diode laser is placed in front of the window (see top view of the cell in Figure 5.2). A lens with covered center mounted at the opposite side of the cell directs scattered laser light on a diode detector. Therefore, when the laser beam strikes the aerosol particles in the cell, forward scattered light is detected by the detector. This scattering light detector is connected to a voltmeter so as to determine whether a detectable amount of aerosol has reached the cell or not.

A Fourier Transformation Infrared Spectrometer (FTIR, Bruker 113v) with a DTGS detector and CsI windows is fixed to the cell and a detector box. The IR light from the spectrometer enters the detector and hits first a parabolic mirror, then a flat mirror. Thereupon, the light enters the cell and hits the objective mirror, then the flat mirror. After 16 passes back and forth the light finally returns to the upper part of the detector box, and hits a last parabolic mirror, then reaches the detector.

We describe the function of the two-stage impactor in detail because it plays a very important role by dividing the sample into two aerosol particle sizes (<1 and >1 μm). A cross-section of the originally manufactured impactor is represented in Figure 5.3. A large amount of particles produced by the dust flow generator are transported to the impactor by the N_2 gas. Once the aerosol enters the

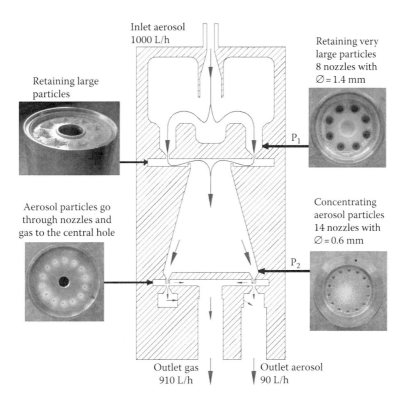

FIGURE 5.3 Cross-sectional diagram of a two stage impactor. Photos of each part of the impactor, a first impaction plate with eight nozzles (P_1), a plate under the P_1, a second impaction plate with 14 nozzles (P_2), and a second plate under the P_2 are included.

impactor, it passes through a set of eight small nozzles (diameter 1.4 mm) and goes at high velocity in the direction of a first flat impaction plate P_1. In front of this impaction plate P_1, the streamlines make a right-angled turn. If inertia of the aerosol particles exceeds a certain value, they are not able to remain in the streamline and directly hit the impaction plate P_1 and thus are removed from the aerosol. The smaller aerosol particles can remain in the gas-stream and reach to a second (in this case virtual) impactor stage. The purpose of this stage is to concentrate the aerosol by removing 91% of the gas. Here, the particles pass through 14 nozzles, which are 0.6 mm in diameter, in front of P_2 which contains holes of similar diameter. The aerosol particles are passing through these holes together with 9% of the gas (this flow of 90 L/h is controlled by a valve at the exit of the cell) whereas 91% of the total gas flow (1000 L/h) leaves to the exterior through the central hole in P_2. The concentrated aerosol is smoothly transported to the gas cell.

An ordinary filtration method is applied for capturing the aerosol particles on polyester capillary pore membrane filters, which have uniform cylindrical holes that are distributed randomly on them and are approximately perpendicular to the filter surface. We select two different pore diameter filters. The 13 mm filter diameter with 0.1 and 0.4 μm pore diameters are used depending on the particle size. The filter is mounted above the cell which is connected to a small pump in order to make use of sucking power to collect the particles in the cell. With their smooth surfaces the filters allow us to use SEM to analyze the physical properties of the particles due to their smooth surfaces.

5.3.2 ASTROPHYSICALLY RELEVANT SAMPLES

In fact, extraterrestrial dust grains have been collected in various places. Interplanetary dust particles (IDPs) are gathered in the stratosphere with high-altitude aircraft.[30,31] Cometary dust from comet 81P/Wild-2 was captured by the NASA Stardust spacecraft during the flyby.[32–34] After the 1960s, dust grains have been found in Greenland and Antarctic ice one after another.[35,36] Similarly, cosmic dust grains in deep-sea deposit have come under the spotlight after John Murray discovered them in 1876.[37–41] Those collected dust grains have been chemically, petrographically, and morphologically analyzed. From these investigations, if the collected grains exhibit extraordinary isotope abundances, they are evaluated as extraterrestrial dust grains.[42]

Another source to extract the information of dust properties is meteorites. Primitive meteorites such as chondrites contain micrometer-sized pre-solar grains. These grains are formed in main dust production sources (circumstellar shells and supernova ejecta), and are older than our solar system. The pre-solar grains do not undergo melting with the primal gas and other dust grains during the formation of the solar system. Hence, they hold the key to understand the origin of the chemical elements to the evolution of the solar system. The most abundant grain is diamond[43] which is the first pre-solar grain to be confirmed due to isotope anomalies. Silicon carbide was also discovered early.[44] Later, new types of pre-solar grains have been added, in the order of their discovery, graphite,[45] corundum,[46–48] spinel,[48] hibonite,[49,50] silicon nitride,[51] silicates,[52] and titanium oxide (TiO_2).[53] These grains are very small in size, which leads them to produce a strong absorption in the mid-IR range. Table 5.1 represents a summary of the investigations above. The dust grains are composed of heterogeneous substances. Most important of all is silicate dust which is the major dust component in most dusty media in space and is found in those collected samples.

Samples for experiments are selected based on the above information, especially focused on silicates and oxides (Table 5.2). In order to clarify the morphological effects on the mid-IR spectra, different shapes and sizes of samples such as Mg_2SiO_4, α- and β-TiO_2, α-Al_2O_3, and $MgAl_2O_4$ are used. The aerosol apparatus is highly effective only for particle sizes of less than approximately 1 μm. Thus, when the original size is much larger than 1 μm, a ball mill process (Si_3N_4 balls) is applied to grind the sample into smaller size. If the powder sample has a large size range, the sample is subjected to a sedimentation process in a solvent (acetone), which results in a size fraction less than 1 μm in diameter.

TABLE 5.1
Properties of Pre-solar (PS) Dust Grains in Meteorites and Extraterrestrial Dust Grains Discovered in Stratosphere, Antarctica, and Comet 81P/Wild 2

Mineralogical Name	Chemical Formula	Shape	Size (μm)	Presence in	Reference
Olivine	$(Mg, Fe)_2SiO_4$	Spherule, angular (fluffy, compact, porous agglomerates)	nm—100	Dust layers, cometary dust, meteorite (PS)	32–36, 43
Pyroxene	$(Mg, Fe)_2SiO_6$	Platelet, rod, ribbon (highly porous agglomerates)	0.1–1	IDPs, cometary dust, meteorite (PS)	30, 32–34, 43
Corundum	Al_2O_3		0.1–5	Meteorite (PS)	43
Spinel	$MgAl_2O_4$		0.1–5	Meteorite (PS)	43
Hibonite	$CaMg_{12}O_{19}$		0.1–1	Meteorite (PS)	43
Magnetite	Fe_3O_4	Spherule (compact agglomerates)	≤5–20	Dust layers	36
Silicon nitride	Si_3N_4		1	Meteorite (PS)	43
Silicon carbide	SiC		0.1–10	Meteorite (PS)	43
Graphite	C		0.1–10	Meteorite (PS)	43
Diamond	C		0.00026	Meteorite (PS)	43

TABLE 5.2
Properties of the Samples for Experiments

Classification	Chemical Formula	Particle Size (μm)	Particle Shape
Silicate	SiO_2, Mg_2SiO_4, $MgSiO_3$, $Mg_xFe_ySiO_4$, $Ca_xMg_ySi_zO_3$, $Mg_xFe_ySi_zO_3$, $CaSiO_3$, $Fe_xSi_yO_4$, $Al_2Si_2O_7 \cdot H_2O$, $Mg_xFe_ySi_4O_{10}$ $(OH)_2$	<1–5	Spherical (SiO_2), ellipsoid, irregular, platy, rodlike
Oxides	α-Al_2O_3, γ-Al_2O_3, χ–δ–κ-Al_2O_3, TiO, α-TiO_2, β-TiO_2, Ti_2O_3, Al_2TiO_5, $CaTiO_3$, $MgAl_2O_4$	10 nm–2	Irregular, thin, and rodlike, rounded, flaky

5.3.3   A Comparison of Aerosol and Pellet Spectra

As we mentioned in Section 5.2, the effect of an embedding medium on band profiles (the so-called matrix effect) has been a concern since the 1970s,[15] although the pellet technique has been widely used in IR spectroscopy. The effect is closely related to the complex dielectric functions of the particle ($\epsilon = \epsilon' + i\epsilon'' = m^2$) and of the surrounding medium (ϵ_m), because the polarizability of a particle essentially depends on the ratio ϵ/ϵ_m. The changes in the spectrum caused by an embedding of the particles into a solid matrix become especially pronounced if geometrical resonances (surface modes) occur. A condition for such resonances is that the real part ϵ' of the dielectric function becomes negative, which is often the case in the frequency range between the transverse (TO) and longitudinal (LO) optical phonon frequencies of crystalline lattices. For instance, for spherical particles small compared to the wavelength a resonance takes place at $\epsilon'/\epsilon_m = -2$. Hence, as the value of ϵ_m increases (N_2 1.0; KBr 2.3; CsI 3.0),[6] the peak positions are shifted to wavelengths where ϵ' is stronger negative, which is usually to longer wavelengths (toward the TO frequency).

For nonspherical particles, the effects are more complicated (see the grain shape effects—Section 5.4.2). However, also for such cases, which are of cause most abundant in reality, the matrix effects can be clarified by means of the aerosol spectroscopy and the comparison of aerosol and pellet spectra. Figure 5.4a shows the three normalized extinction spectra of TiO_2 (rutile) up to 22 μm in wavelength obtained by aerosol, KBr, and CsI pellet measurements.[28] The strongest peak of the aerosol spectrum (13.53 μm) shifts toward the red about 2 μm with the CsI pellet measurement. The shift is more substantial than in the case of the KBr spectrum.

In most cases, the matrix effect cannot be reduced to a simple peak shift. Particularly for strong bands, the spectra obtained by aerosol measurements occasionally show a band broadening toward longer wavelengths, a secondary peak, and a near-rectangular total profile. These effects are hardly ever seen in the spectra measured by the pellet technique. As shown in Figure 5.4b, the 12–29 μm band of TiO_2 (anatase) exhibits strongly pronounced differences between the aerosol and the CsI pellet spectra.[28] The CsI pellet spectrum produces two clear peaks at 17.78 and 29.19 μm, whereas a peak at 14.58 and smooth shoulder beyond the peak up to 19.89 μm is distinctly visible in the aerosol spectrum. The 29 μm peak of the CsI spectrum is quite sharp, while the 28 μm peak of the aerosol spectrum is flattened and forms a rectangular band profile. The disparity between the aerosol and pellet band profiles might additionally be related to the particle morphology, which may be caused by (a) using different dispersion methods; (b) transformation during the grinding procedure; or (c) powdered sample structure deformation caused by the high pressurization required for the pellet technique.

In order to compare the laboratory spectra with the observed spectra, it is best if the laboratory spectra are obtained under conditions as similar to those found in the astrophysical environment as possible, for instance, vacuum, suspended dust, unsettled dust agglomerates. With aerosol spectroscopy, it is possible to measure the particulates with negligible environmental effects. We will compare the results obtained from aerosol spectroscopy and observed spectra in Section 5.5.

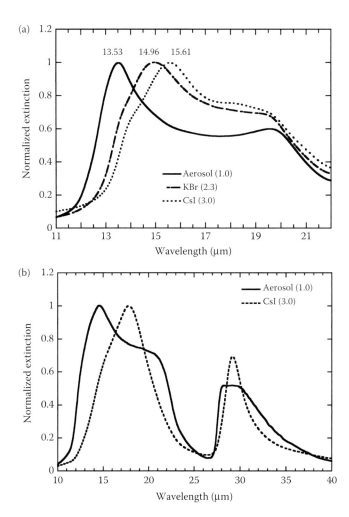

FIGURE 5.4 (a) Normalized extinction versus wavelength of TiO_2 (rutile) spectra obtained by aerosol (solid line), CsI (dotted line), and KBr pellet measurements (dash-dotted line). The values in the legend are the dielectric constant of each medium. (b) Normalized extinction spectra of TiO_2 (anatase: CA2) obtained by aerosol (solid line) and CsI pellet (dotted line) measurements.

5.4 MORPHOLOGICAL EFFECTS ON IR BAND PROFILES

In the preceding section, we discussed the experimental setup for the aerosol spectroscopy. The measurements of the extinction spectra for various powder samples in the IR region, 10–50 μm, paying special attention to the morphological effects are possible by making use of the apparatus. We verify a variety of morphological effects one by one comparing with extinction calculations by sphere, CDE, and DFF in this section.

5.4.1 Size Effect

The normalized extinction spectra of the four different sizes ($d = 0.5$, 1.0, 1.5, and ~5 μm) for amorphous SiO_2 monosphere particles are measured by the aerosol spectroscopy as shown in Figure 5.5. The peak position slightly shifts to longer wavelengths as the particle size increases among the 0.5, 1.0, and 1.5 μm particles. The conspicuous band broadening and the red-shift of the peak position are seen in the extinction spectrum of the ~5 μm particles. As the particle size increases, higher-order

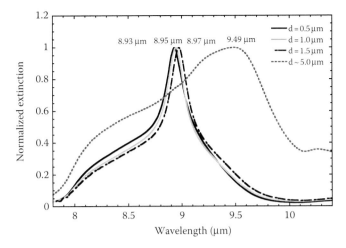

FIGURE 5.5   Normalized extinction spectra of amorphous SiO_2 monosphere particles 0.5 (black solid line), 1.0 (gray solid line), 1.5 (dash-dotted line), and 5 μm (dotted line) in diameter measured by the aerosol spectroscopy.

mode appears and lowers resonance energies as soon as the particle becomes comparable in size with the wavelength of the incident radiation. In consequence, the absorption bands from the surface- and bulk-modes are broadened, and the red shift of the peak position can be observed. Moreover, the higher-order mode causes the scattering which may possibly come into view much stronger than the absorption with large particles.[6,54,55] Hence, the band profile of the ~5 μm particles is greatly affected by the scattering.

The mass absorption coefficient (κ) of forsterite (Mg_2SiO_4) samples with two different size ranges (small: <1 μm and large: 3 ~ 50 μm in size) is represented in Figure 5.6 together with the SEM images. Not only the band shift and broadening appear clearly in these spectra, but also they show the extensive difference in the absorption strengths. The spectrum obtained by the small forsterite

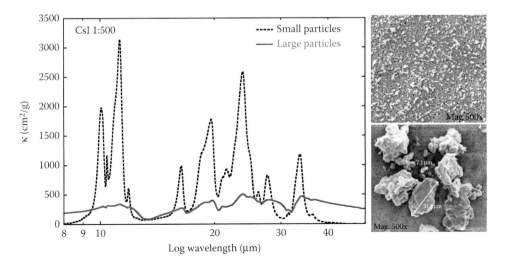

FIGURE 5.6   The mass absorption coefficient (κ) of forsterite particles in two different size ranges (small: <1 μm (dashed line) and large: 3–50 μm (solid line)) obtained by the CsI pellet measurements with regard to the wavelength in logarithmic scale. The SEM images of the two different size ranges: the small (upper) and the large particles (bottom) with magnification 500×.

particles exhibits a factor of approximately 9.3 higher absorption strengths than the spectrum measured for the large particles. Both spectra were measured with the CsI pellet technique, since precise concentrations are required for comparing the absorption strength of these spectra. The aerosol measurements are hardly quantitative in this respect.

5.4.2 SHAPE EFFECT

The SEM images of two forsterite (Mg_2SiO_4) samples in different shapes and sizes are shown in Figure 5.7. The forsterite 1 (left image) particles are of irregular with sharp edges, whereas those of the forsterite 2 (right image) are rather rounded. The extinction spectra measured by aerosol spectroscopy for these samples plainly differ in the sense that the peaks of the IR bands are shifted by up to 0.22 μm toward shorter wavelengths for the rounded forsterite particles (cf. Figure 5.8). A closer look at the spectra reveals that this peak shift results from a change in the band profile. Each band always covers approximately the same wavelength interval between the corresponding LO and TO phonon frequencies[6]; however, within that interval, absorption at different wavelengths is enhanced depending on the grain shape. The enhancement usually takes place at shorter wavelengths for rounded grain shapes and at longer wavelengths for irregular grain shapes.

It cannot be straightforwardly proven that the difference in the band profiles between the two samples (Forsterite 1 and 2) is a pure consequence of the particle shape. Nevertheless, it is reasonable because the geometrical resonances of a sphere take place at such shorter wavelengths (cf. the peak predicted by the CDE model, see Section 5.4.4), and the shape of the forsterite 1 particles is closer to a spherical shape compared to the forsterite 2 particles. Strongly nonspherical, for example, needle-like particles, which can be described as ellipsoids with $L = 0, 0, 1$ for fields along their three axes, cause resonances mainly at the TO frequency (resonance of $L = 0$, max. of ϵ''), which is at the long-wavelength edge of that interval. This may provide a qualitative understanding of the fact that, for example, the CDE model predicts a peak there.

Moreover, very similar differences between the IR band profiles of rounded versus irregular sharp-edged particles are found in the spectra of other particulate materials. The differences are especially pronounced for the strong bands of crystalline mineral powders such as corundum, spinel, rutile, and anatase powders,[28] which are strongly influenced by surface modes. Further, similar results have been found in attempts to modify the grain shapes of a single material by milling.[56]

In these comparisons, the agglomeration state may take part in the variation of the spectral features as well. Hence, we will consider how the agglomeration state of particles influences on them in Section 5.4.3. A better understanding of the grain shape effects may also be reached by means of

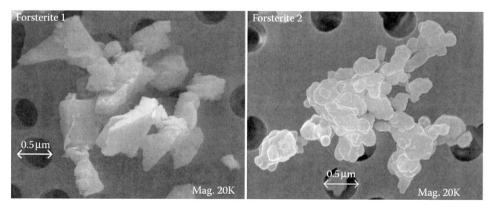

FIGURE 5.7 The SEM images (aerosol particles) of two forsterite samples. Left: irregular grain shapes with sharp edges. Right: nearly spheroidal and ellipsoidal shapes (rounded). The black dots are the holes of the polyester filter having a diameter of ~0.4 μm.

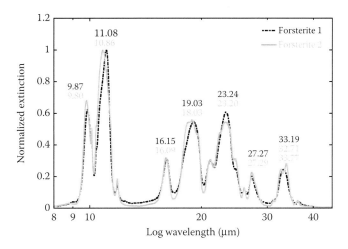

FIGURE 5.8   Normalized extinction versus wavelength in logarithmic scale of the forsterite in the shapes of irregular (Forsterite 1: dash-dotted line) and rounded (Forsterite 2: gray solid line) particles.

a theoretical modeling either using the DDA or a statistical approach. Our attempts to apply the latter for a simulation of grain-shape-dependent spectra will be described in Section 5.4.4.

5.4.3   AGGLOMERATION EFFECT

Commonly, dust grains present in nature do not exist as a single particle. They are composed of many heterogeneous particulates with intricate shapes. As the particle size becomes smaller, it gets more difficult to detach particles from surfaces when the relationship between adhesive and separating forces are given consideration.[25] The TEM and SEM images of the four different rutile (TiO_2) samples are shown in Figure 5.9, and the normalized extinction spectra obtained by the aerosol spectroscopy are represented in Figure 5.10. Most of the spectra are characterized by a strong, often double-peaked, absorption band between 12 and 22 µm, a second absorption band between 22 and 27 µm, and a third very broad one between 27 and 50 µm. Comparing the Rutile 1 (R1) and Rutile 2

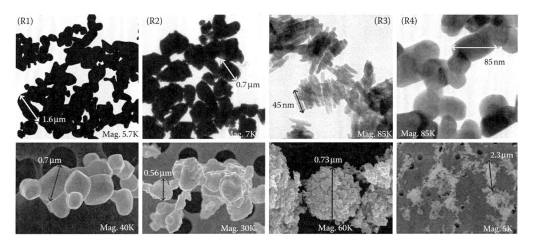

FIGURE 5.9   The TEM (upper) and SEM (bottom) images of four different rutile (TiO_2) samples. (a) R1: irregular shape with rounded edges (0.1–0.5 µm in size); (b) R2: irregular shape with sharp and rounded edges (0.1–0.5 µm in size); (c) R3: thin and long shapes (0.01–0.08 µm); and (d) R4: spheroid and ellipsoid shapes (rounded) (0.01–0.1 µm in size).

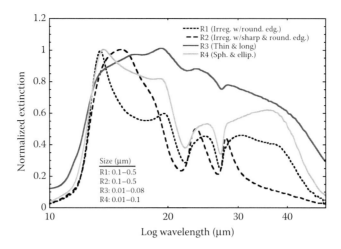

FIGURE 5.10 Normalized extinction spectra (measured in aerosol) of the four rutile samples shown in Figure 5.9 (R1 (dotted line), R2 (dashed line), R3 (dark gray solid line), and R4 (light gray solid line) with regard to wavelength in logarithmic scale.

(R2) spectra, the individual particle sizes are nearly the same, but not the shape. Particles of irregular shape with sharp edges have tendency to produce a relatively distinctive single peak at a considerable longer wavelength than that of rounded particles. In case of R1 and R2, the difference in peak position is about 1.8 μm. In case of the forsterite, the differences are smaller, but the tendency is the same (cf. Figure 5.8). Although the spectrum of R2 exhibits a weak shoulder at 18 μm, the clear secondary peak at 19 μm is not seen in this case. The spectrum of R1 shows a similar band profile to that of Rutile 4 (R4). Although the particle sizes of these samples are different from each other, there are two major similarities. The individual particle shape is irregular with rounded edges. The agglomeration state is composed of many elongated and porous agglomerates. It is a characteristic of aerosol particles that chainlike agglomerates are formed by charged particles.[25] Since the individual particle size of the R4 sample is smaller than that of R1, the agglomerate size of R4 is much larger than that of R1. Since small particles (<10 μm) are greatly influenced by the adhesive force on a particle more than any other forces such as gravity and air current, they easily coagulate and constitute larger agglomerates.[25] Increasing agglomerate sizes might cause additional broadening of the bands such as the sharp 13 μm band, which appears broader in the spectrum of R4 compared to that of R1. Unlike these three rutile samples, Rutile 3 (R3) spectrum does not show any clear peaks between 10 and 50 μm. The three major characteristic bands are merged into a very broad complex. If the agglomeration state is close-packed, a broader band profile can be produced and hide the peaks from view. It is possible to confirm this trend via theoretical calculations.

In the same way as with increasing particle size, the geometrical resonance shifts to longer wavelengths as the agglomeration size grows. This agglomeration size strongly depends on the individual particle size distribution. Moreover, the agglomeration state (e.g., fluffy, close-packed, elongated) is also an important factor to influence the band profile in the mid-IR region. So far, it is virtually impossible to quantify the relative importance of each morphological effect.

5.4.4 THEORETICAL SIMULATIONS

We referred to the controversial point of theoretical calculations for IR band profiles of particulate samples in Section 5.2. As we mentioned there, statistical approaches for grain geometries can approximate the characteristics of the IR spectra quite well, as long as the particles are small compared to the wavelength, that is, are submicron grains. We prefer the DFF model[13] for this purpose because of (1) its intuitive relation to the surface modes of ellipsoidal particles and (2) the possibility to calculate form factor distributions from a spatial discretization of a particle shape. As introduced already in Section 5.2, the form factor distribution $P(L)$ is the probability function of dipole polarizabilities $\alpha = ((\epsilon - 1)^{-1} + L)^{-1}$ depending on form factors L ($0 \leq L \leq 1$), which contribute to the extinction cross-section via their imaginary parts:

$$\langle C_{ext} \rangle / V = (2\pi/\lambda) * \int P(L) * \text{Im } \alpha (\epsilon, L) \, dL \tag{5.1}$$

Strong contributions to the polarizability occur for $1 - \epsilon \approx 1/L$, which links the form factor distribution to certain spectral positions where the ϵ value allows this condition to be fulfilled (the surface modes).

Min et al.[13] have calculated form factor distributions for aggregates of spherical particles with different fractal dimensions and for single particles having an irregular shape represented by the Gaussian random sphere (GRS) model. Using aerosol-measured IR spectra of spinel, corundum, and forsterite powders, Mutschke et al.[14] have demonstrated that the experimental spectra of rounded particles could be well reproduced by the DFF model for the spherical aggregates (with fractal dimensions around 2.0), whereas the DFF model of the GRS particles with a surface modulation standard deviation of $\sigma = 0.3$ reproduced well the characteristics of spectra of irregular grains.

We demonstrate, in Figure 5.11, that this is also true for spectra of rutile particles. We show the rutile normalized extinction spectra (R2 and R4) obtained by the aerosol spectroscopy together with the Mie (single spherical particle),[5] CDE (all ellipsoidal shapes with equal probability),[6] and the mentioned DFF[13,14] models. The form factor distributions for the latter are given in Figure 5.12. The CDE corresponds to a DFF model with $P(L) = 2(1 - L)$. As mentioned, all models assume the Rayleigh limit, where spectra do not depend on particle size apart from a wavelength-independent factor. The spectra are normalized for comparison.

As can be seen in the plot, the simulated spectrum of a spherical particle exhibits sharp-pointed peaks with the narrowest bandwidth occurring at short wavelengths within the band profiles. In contrast to this, the measured band profiles are broad and correspond better to a distribution of ellipsoidal shapes (CDE). The peaks of the CDE spectrum generally occur at considerably longer wavelengths than those of the Mie spectrum. In case of the 12–20 μm band, the CDE peak

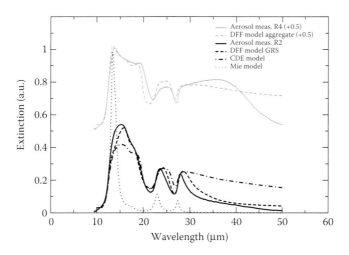

FIGURE 5.11 Normalized extinction spectra of two different rutile samples (R2: lower curves and R4: upper curves, extinction shifted by +0.5) measured by the aerosol spectroscopy compared with simulated spectra obtained with the DFF model. The form factor distributions for the two cases aggregate of spherical particles and Gaussian random sphere (GRS) are displayed in Figure 5.12. Spectra resulting from the CDE and Mie models are shown for comparison. (Note: The optical constants of rutile derived by Gervais and Periou[64] are applied for the calculations.)

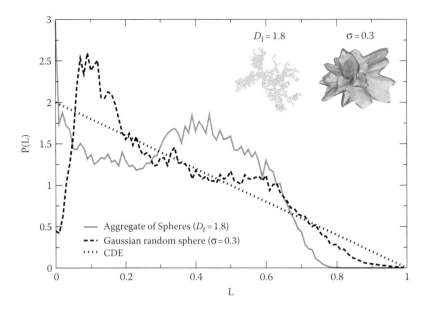

FIGURE 5.12 Form factor distributions for aggregates of spherical grains with a fractal dimension $D_f = 1.8$ and for Gaussian Random Spheres with a standard deviation $\sigma = 0.3$ as used for the spectra simulations shown in Figure 5.11. Examples of the synthetic particle shapes from which the distributions have been calculated are shown in the inset.

corresponds well with the spectrum of the irregular R2 particles, whereas the peak of the R4 spectrum (rounded particles; see Figure 5.10) at approximately 13–14 μm coincides quite well with that of the spherical grains. However, a look at the other bands (22–27 and 27–50 μm) shows that the CDE spectrum is far too much broadened to reproduce the experimental spectrum of the R2 sample satisfactorily, in these ranges it actually resembles the spectrum of the R4 sample.

A much better agreement with the R2 spectrum is reached with the DFF model originating from a GRS particle geometry. The reason is, that in this model very small form factors are less probable (see Figure 5.12), diminishing the extinction (surface modes) at the long-wavelength side of the bands. Surprisingly, such a contribution of very small form factors is needed to reproduce the spectrum of the near-spherical R4 particles (see Figure 5.9). Together with the typical short-wavelength peak of spherical particles, this long-wavelength contribution leads to nearly trapezoidal band profiles. As already referred to in Section 5.4.3, this behavior may be caused by the agglomeration, although this is not proven yet. The DFF models of the aggregates of spherical grains, especially at a low fractal dimension, reproduce quite well this type of spectrum which has also been found for corundum particles.[14] The 27–50 μm band, however, is again not as much broadened as the model predicts, that is, at wavelengths longer than 40 μm the extinction falls steeply instead of extending beyond 50 μm. The reason for this is not clear; the effect could be possibly related to the independent treatment of the crystallographic axes in the models, which is not correct for complicated particle shapes. It could also point to some deviation in the material data for our small particles compared to the bulk data of rutile.

5.5   APPLICATION TO THE INTERPRETATION OF ASTRONOMICAL SPECTRA

In this section, we demonstrate how laboratory-measured extinction spectra can be used for the interpretation of astronomical IR spectra dominated by emission bands of dust particles. We will show two examples, the first one for an evolved giant star with a molecular outflow where dust particles condense, the second for a main sequence star with a planetary system which has an asteroidal belt producing debris dust particles (see Section 5.1).

The emission spectrum of the semiregular variable AGB star g Her (Hercules) is shown in Figure 5.13 in comparison with the aerosol-measured extinction spectra of two contrastive morphological types of anatase (TiO_2) and spinel ($MgAl_2O_4$) submicron-sized particulate samples. As explained in Section 5.2, the extinction coefficient in this case is equal to the emission efficiency and the band positions and widths can be directly compared with emission spectra apart from their relative strength, which is determined by the temperature of the emitting dust. The observed emission spectrum of g Her exhibits sharp bands at 13, 18, 19.5, 28, and 32 μm. It can be directly seen that anatase and spinel particles produce peaks in the wavelength ranges 13–15 μm, 17–22 μm, and 28–32 μm. The precise peak positions, however, depend strongly on the grain morphology (see Section 5.4). In particular, the 13 μm band of anatase (CA1) particles, which are of irregular shape with rounded edges and in the size of 0.1–0.6 μm, corresponds well with the 13 μm peak position of the observed spectrum. Nevertheless, the different shape and size of anatase (CA2; long-narrow shape with rounded edges and size in ~0.05 μm) particles produce a peak at 14.58 μm which does not correspond to an observed band. In this case, also the agglomerate effect strongly influences the peak shift, because the CA2 particles are small enough in size to form large agglomerates easily (see Section 5.4.3). The same might be true for the 18 and 28 μm bands, which could possibly also be explained by CA1-like dust.

Since the 13 μm feature in the g Her spectrum is rather a prominent peak, it is conceivable that spinel particles with similar properties as the CSp1 sample (spherical shape and ~0.05 μm in size) enhance the prominence of the 13 μm peak. In the same way as anatase mentioned above, the 13 μm peak cannot be intensified by the irregular shaped spinel (CSp2; size in 0.04–0.5 μm), since the peak position of this sample is at 14.08 μm. The CSp1 spectrum also matches even better the observed 18 μm feature and also possesses a band at 32 μm, which altogether makes the presence of spherical spinel grains in the outflow of this star quite likely. In addition, an amorphous-silicate feature clearly appears in the observed spectrum at about 10 μm wavelength. The irregular shaped amorphous Mg_2SiO_4 spectrum obtained by aerosol spectroscopy is well consistent with the 10 μm features for the g Her emission spectrum, but in the case of amorphous silicates the grain shape effect on the IR bands is less pronounced, so that grain shapes cannot be well constrained.

However, for the potential stellar oxide dust particles in the outflows of stars, this contrast clearly demonstrates that the particle morphology plays a decisive role in comparisons of spectral features. Spherical particle shapes seem to be preferred when particles condense in outflows, but there are also counterexamples.

In Figure 5.14, we show a fit of the emission spectrum (Spitzer IRS data[57]) of the main-sequence star HD69830 (K0 V star)[58] by making use of the silicate extinction spectra obtained by aerosol spectroscopy. HD69830 is a very interesting object, because three extrasolar planets around it have been discovered.[59] The contribution of the star has been subtracted from the emission spectrum in order to display only the excess emission of the dust particles ("debris dust") which are located in an asteroid belt 0.3 to 0.5 AU away from the star.[60] The spectrum has already been analyzed by Lisse et al.[61] using laboratory spectra measured in KBr pellets for comparison. These authors claimed to have detected six different species of crystalline silicate dust (three olivines and three pyroxenes), of which some like ferrosilite ($FeSiO_3$) have not been seen before in asteroidal or cometary dust.

The fit we have obtained using aerosol-measured spectra is much simpler. Here, the details of the emission bands are almost perfectly fitted by a combination of irregularly shaped crystalline olivine ($Mg_{1.60}Fe_{0.40}SiO_4$) particles and amorphous silicate particles (also irregularly shaped) of olivine stoichiometry, where both are in the size range of less than 1 μm. This size range is consistent with the constraints for debris dust around a K star. For the latter species, we have used a spectrum of amorphous Mg_2SiO_4 particles produced by a sol–gel process at the Jena Astrophysical Laboratory. However, the iron content in such an amorphous silicate is not very tightly constrained by the spectroscopic properties, and so an amorphous silicate containing iron would certainly also provide a good fit.

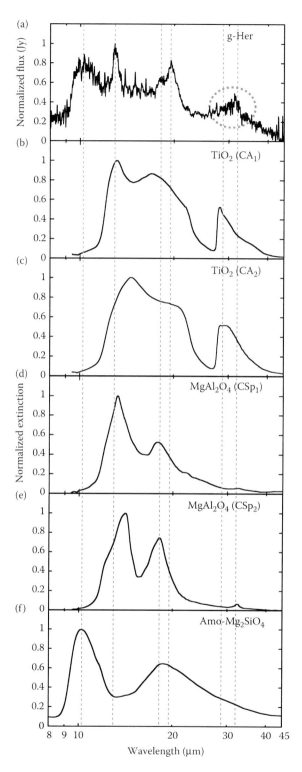

FIGURE 5.13 The normalized emission spectrum of g Her compared with the spectra of two anatase (CA1, CA2), and spinel (CSp1, CSp2) samples, and the spectrum of amorphous Mg_2SiO_4 obtained by aerosol spectroscopy. The gray dotted lines denote band positions at approximately 11, 13, 18, 19.5, 28, and 32 μm wavelengths.

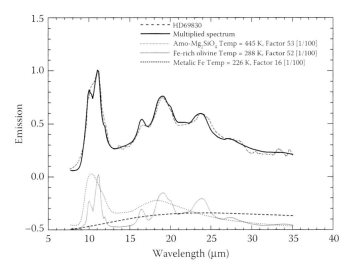

FIGURE 5.14 A fit of the observed emission spectrum of the main-sequence star HD69830 (dashed line on upper plot) and a simulated spectrum (solid line on upper plot) obtained by using laboratory measured aerosol spectra of amorphous Mg_2SiO_4 (dotted line), crystalline Fe-rich olivine (gray solid line), and metallic Fe particles (dash-dotted line) as emission efficiencies. The spectra are multiplied with Planck functions at the indicated temperatures and summed up with the indicated weight factors (see text).

On the other hand, the crystalline olivine behaves completely different from the case of amorphous silicate. Here, especially the bands at 16.4 and 23.9 µm shift strongly with the replacement of magnesium by iron ions[62] constraining the iron content for the majority of the crystalline silicate particles in HD69830 to values around 20% relative to the total number of metal ions (Mg + Fe). For olivine particles having only half of this iron content such as the standard olivine from San Carlos, Arizona ($Mg_{1.85}Fe_{0.15}SiO_4$), the bands are located at 16.2 and 23.5 µm, which would be in clear mismatch with the observed spectrum. Using KBr pellet spectra, one comes to different results because of the peak shift to longer wavelengths.

In addition to the silicates, a dust component producing a featureless spectrum is required. We have used aerosol data of small iron particles for this purpose. Metallic iron is found in cometary grains as an inclusion into glassy silicate particles.[63] The same featureless background contribution could be achieved by adding carbon (another conducting material) or a population of large grains, for which the bands are sufficiently broadened to actually disappear (sizes >10 µm).

The three aerosol spectra have been multiplied with Planck functions corresponding to different temperatures, which have been automatically adjusted together with the relative proportions of the species in order to match the observed spectrum as closely as possible. Even this simple fit is not unique, especially for the proportions and temperatures of the compounds providing only broad structures to the spectrum (i.e., iron and amorphous silicate). For the olivine, this is different because of the pronounced narrowbands present over the whole spectrum. Anyway, the proportions of the species are arbitrary values, because the aerosol spectra used are not quantitative. This can be largely overcome by normalizing the aerosol extinction spectra through a combination of simulation calculations and pellet measurements, given that the complex refractive index of the material is known. However, in case of conducting particles the simulations are too uncertain, and a quantitative analysis was beyond the scope of this demonstration.

5.6 CONCLUSION

Extinction spectra of astrophysically relevant dust grains in the mid-IR region (2–50 µm) can be obtained by means of aerosol spectroscopy. Electromagnetic interaction with media is negligible by dint of suspending dust particles in a nitrogen gas. As a result, the band profiles of the aerosol spectra are equivalent to those of observed ones in space (vacuum). The spectra obtained by aerosol spectroscopy tend to shift toward shorter wavelengths because of geometrical resonances of the grains compared to the spectra measured by the pellet technique, specifically for strong bands of crystalline materials.

Since dust particle size, shape, and agglomeration state are crucial factors to produce an effect on the band profiles, we have used electron microscope investigation of the same particulates to associate these morphological influences with the experimentally measured profiles. As particles increase in size, the band broadening and the red-shift of the peak positions are detected clearly because of an appearance of the higher-order modes. Individual particle shapes are also a decisive factor to determine the band profile. The irregular shaped grains appear to produce geometrical resonances at longer wavelengths in general. This is in the vicinity of the transverse optical lattice frequencies. Rounded particulates develop a tendency to produce double-peaked (rectangular) band profiles in contrast to irregular-shaped ones. Agglomerate state affects effectively the strength of the strong bands. Close-packed agglomerates give rise to a broadening of bands more than elongated ones.

We have given two examples for the application of aerosol-measured IR spectra of particulate samples to the interpretation of astronomical spectra. On the one hand, we demonstrated that morphological effects are important to understand the bands of oxide particles in the emission spectra of giant-star outflows. On the other hand, we have applied the laboratory-measured spectra for fitting the observed IR emission spectrum of dust grains in a massive asteroid belt around a main-sequence star. This fit revealed a quite simple mineralogical composition of the dust, which is dominated by small crystalline and amorphous olivine particles. The irregular shape and the small sizes of these particles are consistent with dust produced by collisions of larger bodies.

Astrophysical dust grain analysis based on IR spectra is not straightforward. The band profiles are strongly influenced by various physical and chemical properties of the dust grains. However, these spectra are indubitably an important source of information about the conditions under which dust grains exist and evolve in astronomical environments from the places of their formation to the places where planets are formed out of them. Thus, our task is to analyze the observed spectra as precise as possible by making use of our knowledge obtained from laboratory measurements. Aerosol spectroscopy is an important tool to make progress in this field.

ACKNOWLEDGMENTS

We are grateful to J. Blum, Th. Henning, Th. Posch, C. Koike, M. Min, D. R. Alexander, and S. Nietzsche for their assistance in improving our experimental investigations. We express our thanks to G. Born and W. Teuschel for their technical support. Our project has been funded by Deutsche Forschungsgemeinschaft (DFG) under Grant MU 1164/6.

REFERENCES

1. van Boekel, R., Waters, L. B. F. M., Dominik, C., Bouwman, J., de Koter, A., Dullemond, C. P., and Paresce, F. 2003. Grain growth in the inner regions of Herbig Ae/Be star disks. *A&A* 400: L21–L24.
2. van Boekel, R., Min, M., Leinert, Ch., Waters, L. B. F. M., Richichi, A. et al. 2004. The building blocks of planets within the "terrestrial" region of protoplanetary disks. *Nature* 432: 479–482.
3. Lisse, C. M., VanCleve, J., Adams, A. C., A'Hearn, M. F., Femández, Y. R. et al. 2006. Spitzer spectral observations of the deep impact Ejecta. *Science* 313: 635–640.
4. Lisse, C. M., Kraemer, K. E., Nuth, J. A. III, Li, A., and Joswiak, D. 2007. Comparison of the composition of the Temple 1 ejectra to the dust in Comet/Hale-Bopp 1995 O1 and YSO HD 100546. *Icarus* 187: 69–86.
5. Mie, G. 1908. Beiträge zur Optik trüber Medien speziell kolloidaler Metallösungen. *Ann. Phys.* 330: 377–445.
6. Bohren, C. F. and Huffman, D. R. 1983. *Absorption and Scattering of Light by Small Particles*. John Wiley & Sons Inc., New York.

7. Purcell, E. M. and Pennypacker, C. R. 1973. Scattering and absorption of light by nonspherical dielectric grains. *ApJ* 186: 705–714.
8. Draine, B. T. 1988. The discrete dipole approximation and its application to interstellar graphite grains. *ApJ* 333: 848–872.
9. Draine, B. T. and Goodman, J. 1993. Beyond Clausius–Mossotti: Wave propagation on a polarizable point lattice and the discrete dipole approximation. *ApJ* 405: 685–697.
10. Draine, B. T. and Flatau, P. J. 1994. Discrete-dipole approximation for scattering calculations. *J. Opt. Soc. Am. A* 11: 1491–1499.
11. Rayleigh, R. 1910. The incidence of light upon a transparent sphere of dimensions comparable with the wave-length. *Proc. Roy. Soc. London A* 84: 25–46.
12. van Boekel, R., Min, M., Waters, L. B. F. M., de Koter, A., Dominik, C., van den Ancker, M. E., and Bouwman, J. 2005. A 10 mm spectroscopic survey of Herbig Ae star disks: Grain growth and crystallization. *A&A* 437: 189–208.
13. Min, M., Hovenier, J. W., Dominik, C., de Koter, A., and Yurkin, M. A. 2006. Absorption and scattering properties of arbitrarily shaped particles in the Rayleigh domain: A rapid computational method and theoretical foundation for the statistical approach. *J. Quant. Spectrosc. Radiat. Transfer* 97: 161–180.
14. Mutschke, H., Min, M., and Tamanai, A. 2009. Laboratory-based grain-shape models for simulating dust infrared spectra. *A&A* 504: 875–882.
15. Dorschner, J., Friedemann, C., and Gürtler, J. 1978. Laboratory spectra of phyllosilicates and the interstellar 10-micrometer absorption band. *Astron. Nachr.* 299: 269–282.
16. Koike, C., Hasegawa, H., Asada, N., and Hattori, T. 1981. The extinction coefficients in mid- and far-infrared of silicate and iron-oxide minerals of interest for astronomical observations. *Ap&SS* 79: 77–85.
17. Oronfino, V., Mennella, V., Blanco, A., Bussoletti, E., Colangeli, L., and Fonti, S. 1991. Experimental extinction properties of granular mixtures of silicon carbide and amorphous carbon. *A&A* 252: 315–319.
18. Posch, T., Kerschbaum, F., Mutschke, H., Fabian, D., Dorschner, J., and Hron, J. 1999. On the origin of 13 μm feature: A study of ISO-SWS spectra of oxygen-rich AGB stars. *A&A* 352: 609–618.
19. Chihara, H., Koike, C., Tsuchiyama, A., Tachibana, S., and Sakamoto, D. 2002. Compositional dependence of infrared absorption spectra of crystalline silicates I. Mg–Fe pyroxenes. *A&A* 391: 267–273.
20. Fabian, D., Posch, Th., Mutschke, H., Kerschbaum, F., and Dorschner, J. 2001. Infrared optical properties of spinels: A study of the carrier of the 13, 17, 32 μm emission features observed in ISO-SWS spectra of oxygen-rich AGB stars. *A&A* 373: 1125–1138.
21. Papoular, R., Cauchetier, M., Begin, S., and LeCaer, G. 1998. Silicon carbide and the 11.3 μm feature. *A&A* 329: 1035–1044.
22. Henning, Th. and Mutschke, H. 2000. Optical properties of cosmic dust analogs. In *Thermal Emission Spectroscopy and Analysis of Dust, Disks, and Regoliths*. Eds. M. L. Sitko, A. L. Sprague and D. K. Lynch. Sheridan Books, Inc., MI, Vol. 196, pp. 253–271.
23. Speck, A. K., Hofmeister, A. M., and Barlow, M. J. 2000. Silicon carbide: The problem with laboratory spectra. In *Thermal Emission Spectroscopy and Analysis of Dust, Disks, and Regoliths*. Eds. M. L. Sitko, A. L. Sprague, and D. K. Lynch. Sheridan Books, Inc., MI, Vol. 196, pp. 281–290.
24. Clément, D., Mutschke, H., Klein, R., and Henning, Th. 2003. New laboratory spectra of isolated b-SiC nanoparticles: Comparison with spectra taken by the Infrared Space Observatory. *ApJ* 594: 642–650.
25. Hinds, W. C. 1999. *Aerosol Technology: Properties, Behavior, and Measurement of Airborne Particles*. John Wiley & Sons Inc., New York.
26. Tamanai, A., Mutschke, H., Blum, J., and Neuhäuser, R. 2006. Experimental infrared spectroscopic measurement of light extinction for agglomerate dust grains. *J. Quant. Spectrosc. Radiat. Transfer* 100: 373–381.
27. Tamanai, A., Mutschke, H., Blum, J. and Meeus, G. 2006. The 10 μm infrared band of silicate dust: A laboratory study comparing the aerosol and KBr pellet techniques. *ApJ* 648: L147–L150.
28. Tamanai, A., Mutschke, H., Blum, J., Posch, Th., Koike, C., and Ferguson, J. W. 2009. Morphological effects on IR band profiles: Experimental spectroscopic analysis with application to observed spectra of oxygen-rich AGB stars. *A&A* 501: 251–267.
29. Schütz, O., Meeus, G., and Sterzik, F. 2005. Mid-IR observations of circumstellar disks II. Vega-type stars and a post-main sequence object. *A&A* 431: 175–182.
30. Bradley, J. P., Brownlee, D. E., and Veblen, D. R. 1983. Pyroxene whiskers and platelets in interplanetary dust: Evidence of vapour phase growth. *Nature* 301: 473–477.
31. Jessberger, E. K., Bohsung, J., Chakaveh, S., and Traxel, K. 1992. The volatile element enrichment of chondritic interplanetary dust particles. *Earth Planet Sci. Lett.* 112: 91–99.

32. Zolensky, M. E., Zega, T. J., Yano, H., Wirick, S., Westphal, A. J., Weisberg, M. K. et al. 2006. Mineralogy and petrology of Comet 81P/Wild 2 nucleus samples. *Science* 314: 1735–1739.

33. Keller, L. P., Bajt, S., Baratta, G. A., Borg, J., Brandley, J. P. et al. 2006. Infrared spectroscopy of Comet 81P/Wild 2 samples returned by stardust. *Science* 314: 1728–1731.

34. Kearsley, A. T., Borg, J., Graham, G. A., Burchell, M. J., Cole, M. J. et al. 2008. Dust from comet Wild 2: Interpreting particle size, shape, structure, and composition from impact features on the Stardust aluminum foils. *Meteorit. Planet. Sci.* 43: 41–73.

35. Engrand, C., Narcisi, B., Petit, J.-R., Cobrica, C., and Duprat, J. 2008. Cosmic dust layers in EPICA-DOME C deep ice core. In *39th Lunar and Planetary Science. Lunar and Planetary Science XXXIX: LPI Contribution* No. 1391: 1154.

36. Misawa, K., Tomiyama, T., Kohno, M., Noguchi, T., Nagano, K. et al. 2008. Extraterrestrial dust layers in dome Fuji ice core, East Antarctica. In *39th Lunar and Planetary Science. Lunar and Planetary Science XXXIX: LPI Contribution* No. 1690: 1391.

37. Murray, J. 1884. Volcanic ashes and cosmic dust. *Nature* 29: 585–590.

38. Brownlee, E. D., Bates, B. A., and Wheelock, M. M. 1984. Extraterrestrial platinum group nuggets in deep-see sediments. *Nature* 309: 693–695.

39. Brownlee, D. E., Bates, B., and Schramm, L. 1997. The elemental composition of stony cosmic spherules. *Meteorit. Planet. Sci.* 32: 157–175.

40. Bonte, Ph., Jehanno, C., Maurette, M., and Brawnlee, D. E. 1987. Platinum metals and microstructure in magnetic deep sea cosmic spherules. *J. Geophys. Res.* 92: E641–E648.

41. Parashar, K. and Shyam Prasad, M. 2008. Preliminary investigation on cosmic dust collected by "MACDUC" experiment from Central Indian Ocean. In *39th Lunar and Planetary Science. Lunar and Planetary Science XXXIX: LPI Contribution* No. 1391: 1045.

42. Heide, F. and Wlotzka, F. 1995. *Meteorites: Messenger from Space*. Springer-Verlag, Berlin.

43. Ott, U. and Hoppe, P. 2006. Pre-solar grains in meteorites and interplanetary dust: An overview. In *Highlights of Astronomy. IAU XXVI General Assembly*. Ed. Karel A. van der Hucht, Cambridge University Press, Cambridge, U.K., Vol. 14: 341–344.

44. Lewis, R. S., Ming, T., Wacker, J. F., Anders, E., and Stell, E. 1987. Interstellar diamonds in meteorites. *Nature* 326: 160–162.

45. Bernatowics, T., Fraundorf, G., Ming, T., Anders, E. et al. 1987. Evidence for interstellar SiC in the Murray carbonaceous meteorite. *Nature* 330: 728–730.

46. Amari, S., Anders, A., Virag, A. and Zinner, E. 1990. Interstellar graphite in meteorites. *Nature* 345: 238–240.

47. Huss, G. R., Hutcheon, I. D., Fahey, A., and Wasserburg, G. J. 1993. Oxygen isotope anomalies in Orgueil corundum: Confirmation of presolar origin. *Meteoritics* 28: 369.

48. Hutcheon, I. D., Huss, H. G., Fahey, A. J., and Wasserburg, G. J. 1994. Extreme ^{26}Mg and ^{17}O enrichments in an Orgueil Corundum: Identification of a presolar oxide grain. *ApJ* 425: L97–L100.

49. Nittler, L. R., O'D Alexander, C. M., Gao, X., Walker, R. M., and Zinner, E. K. 1994. Interstellar oxide grains from the Tieschitz ordinary chondrite. *Nature* 370: 443–446.

50. Nittler, L. R., Hoppe, P., O'D Alexander, C. M., Amari, S., Eberhardt, P. et al. 1995. Silicon nitride from supernovae. *ApJ* 453: L25–L28.

51. Choi, B. G., Wasserburg, G. J., and Huss, G. R. 1999. Circumstellar Hibonite and Corundum and nucleosynthesis in asymptotic giant branch stars. *ApJ* 522: L133–L136.

52. Nguyen, A. N. and Zinner, E. 2004. Discovery of ancient silicate stardust in a meteorite. *Science* 303: 1496–1499.

53. Nittler, L. R., O'D Alexander, C. M., Gallino, R., Hoppe, P., Nguyen, A. N. et al. 2008. Aluminum-, calcium- and titanium-rich oxide stardust in ordinary chondrite meteorites. *ApJ* 682: 1450–1478.

54. Tolstoy, V. P., Chernyshova, I. V., and Skryshevsky, V. A. 2003. *Handbook of Infrared Spectroscopy of Ultrathin Films*. John Wiley & Sons Inc., NJ.

55. Akimov, Y. A., Koh, W. S., and Ostrikov, K. 2009. Enhancement of optical absorption in thin-film solar cells through the excitation of higher-order nanoparticle plasmon modes. *Opt. Express* 17: 10195–10205.

56. Imai, Y., Koike, C., Chihara, H., Murata, K., Aoki, T., and Tsuchiyama, A. 2009. Shape and lattice distortion effects on infrared absorption spectra of olivine particles. *A&A* 507: 277–281.

57. Beichman, C. A., Bryden, G., Gautier, T. N., Stapelfeldt, K. R., Werner, M. W., Misselt, K., Rieke, G., Stansberry, J., and Trilling. D. 2005. An excess due to small grains around the nearby K0 V star HD 69830: Asteroid or cometary debris? *ApJ* 626: 1061–1069.

58. Song, I., Caillault, J.-P., Barrado y Navascués, D., Stauffer, J. R., and Randich, S. 2000. Ages of late spectral type Vega-like stars. *ApJ* 533: L41–L44.
59. Lovis, C., Mayor, M., Pepe, F., Albert, Y., Benz, W. et al. 2006. An extrasolar planetary system with three Neptune-mass planets. *Nature* 441: 305–309.
60. Charbonneau, D. 2006. A Neptunian triplet. *Nature* 441: 292–293.
61. Lisse, C. M., Beichman, C. A., Bryden, G., and Wyatt, M. C. 2007. On the nature of the dust in the debris disk around HD69830. *ApJ* 658: 584–592.
62. Koike, C., Chihara, H., Tsuchiyama, A., Suto, H., Sogawa, H., and Okuda, H. 2003. Compositional dependence of infrared absorption spectra of crystalline silicate II. Natural and synthetic olivines. *A&A* 399: 1101–1107.
63. Bradley, J. P. 1994. Chemically anomalous, preaccretionally irradiated grains in interplanetary dust from comets. *Science* 265: 925–928.
64. Gervais, F. and Piriou, B. 1974. Temperature dependence of transverse- and longitudinal-optic modes in TiO_2 (rutile). *Phys. Rev. B* 10: 1642–1654.

Section II

Raman Spectroscopy

6 Linear and Nonlinear Raman Spectroscopy of Single Aerosol Particles

N.-O. A. Kwamena and Jonathan P. Reid

CONTENTS

6.1 Introduction 127
6.2 Raman Spectroscopy 128
    6.2.1 Spontaneous Raman Spectroscopy.. 128
    6.2.2 Stimulated Raman Spectroscopy.. 134
6.3 Instrumentation... 138
    6.3.1 Laser Sources.. 139
    6.3.2 Microdroplet Generation and Sampling 140
    6.3.3 Spectrometers and Detectors 140
    6.3.4 Spectral Interferences 141
6.4 Recent Applications of Raman Spectroscopy in Studies of Single
    Liquid Microdroplets.. 141
    6.4.1 Applications of Spontaneous Raman Scattering 142
        6.4.1.1 Electrodynamic Balance.. 142
        6.4.1.2 Optical Levitation 142
        6.4.1.3 Optical Tweezers... 143
    6.4.2 Applications of SRS.. 145
        6.4.2.1 Droplet Train 145
        6.4.2.2 Optical Levitation 146
        6.4.2.3 Optical Tweezers... 147
6.5 Conclusions.. 150
Acknowledgments.. 150
References.. 150

6.1 INTRODUCTION

Single-particle techniques coupled with Raman spectroscopy have found wide applicability in the fields of colloid, pharmaceutical, aerosol, and environmental sciences. To this end, being able to directly monitor the size, phase, composition, and refractive index of an aerosol in real-time *in situ* can provide a route to better understand their physical and chemical properties.[1–3] Studies on a single-particle basis have a number of advantages over ensemble measurements: the inherent averaging over particle size and compositional distributions, and the short time-scales often associated with ensemble studies, can lead to some ambiguity in examining aerosol processes. In addition, wall effects can be minimized through studies on single particles. One disadvantage of single aerosol particle measurements is that they are currently limited to particles larger than 1 μm in diameter.[4] The dominant size range of aerosol in the atmosphere is smaller than 1 μm and, thus, care must be taken when considering the relevance of these measurements to atmospheric aerosol. However, single particle studies do provide a link between ensemble studies and bulk/surface investigations.[5]

In the first section of this chapter, we will review the theoretical basis of spontaneous and stimulated Raman spectroscopy. This will be followed by a discussion of the instrumental requirements for performing measurements on single aerosol particles. Finally, we will describe a range of applications of Raman spectroscopy in studies of aerosol properties and dynamics.

6.2   RAMAN SPECTROSCOPY

Both linear and nonlinear (spontaneous and stimulated) Raman spectroscopy have been applied to studies of aerosol dynamics. We shall first describe the basis of linear spontaneous Raman scattering before considering the specific application of Raman scattering in the characterization of aerosols. We will conclude this section by extending our discussion to the nonlinear, stimulated Raman scattering (SRS) variant of the technique.

6.2.1   Spontaneous Raman Spectroscopy

The first Raman scattering measurement was made by Chandrasekhara Ventaka Raman in 1928 using sunlight as the illumination source and $CHCl_3$ as a liquid sample. He rapidly received recognition for this work, receiving the Nobel Prize for this pioneering discovery in 1930. The routine application of Raman spectroscopy, however, was not possible until the advent of the laser, which provided a source of monochromatic radiation of sufficient intensity that the extremely weak inelastic scattering could be resolved, even from gas-phase samples. It should be noted that one inelastically scattered photon occurs for every 10^6–10^8 photons undergoing elastic scattering.

A molecule irradiated with light of frequency ω_e, experiences a time-dependent electric field of amplitude

$$E(t) = E_0 \sin(2\pi\omega_e t). \tag{6.1}$$

The presence of the electromagnetic field leads to a perturbation of the electron density within the molecule and induces a time-dependent dipole given by

$$\mu(t) = \alpha(t)E(t) \tag{6.2}$$

where $\mu(t)$ is the induced dipole at time t and $\alpha(t)$ is the time-varying polarizability of the species. The polarizability of the molecule may be described using a harmonic oscillator approximation for the molecule, with an oscillation at the molecular vibrational frequency ν_k:

$$\alpha(t) = \alpha_0 + A\sin(2\pi\nu_k t) \tag{6.3}$$

Thus, the dipole induced by the applied field may be expressed as

$$\mu(t) = (\alpha_0 + A\sin(2\pi\nu_k t))(E_0 \sin(2\pi\omega_e t)). \tag{6.4}$$

The induced dipole can then be shown to have frequency components at the frequency of the incident light and at frequencies higher and lower than the incident light by an amount equivalent to the frequency of the molecular vibration, referred to as anti-Stokes and Stokes scattering, respectively.

$$\mu(t) = \alpha_0 E_0 \sin(2\pi\omega_e t) + \frac{1}{2}AE_0(\cos(2\pi(\omega_e - \nu_k)t) - \cos(2\pi(\omega_e + \nu_k)t)) \tag{6.5}$$

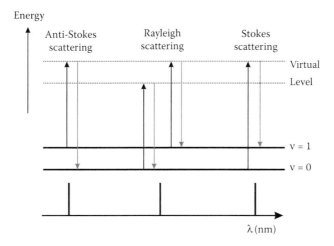

FIGURE 6.1 Schematic of the energy levels and transitions involved in a Raman scattering process. Stoke's scattering leads to vibrational excitation of the molecule. Anti-Stoke's scattering leads to a loss of vibrational energy from the molecule. The laser energy is indicated by a solid black line and the scattered energy by a gray line.

A schematic of the inelastic and elastic scattering processes in shown in Figure 6.1.

In the linear classical scattering limit, the number of Raman photons scattered per second, N_R (s^{-1}), over a solid angle $\Delta\Omega$ (sr) depends on the incident laser irradiance, I_0 (W cm^{-2}), the frequency of the incident light, ν (s^{-1}), the number density of molecules in the sample volume, N (molecule cm^{-3}), the beam cross-section, S (cm^2), the Raman scattering cross-section ($\partial\sigma/\partial\Omega$) (cm^2 sr^{-1}), and the interaction length, L (cm).

$$N_R = \left(\frac{\partial\sigma}{\partial\Omega}\right)(\Delta\Omega) \times \frac{I_0}{h\nu} \times N \times S \times L \qquad (6.6)$$

Illustrative values of Raman cross sections are given in Table 6.1.

The incident field can be resolved into polarization components along the x, y, and z axes. Similarly, the polarizability of the molecule can be resolved into components, α_{ij}, where the index i refers to the polarization of the molecule along one of the three orthogonal Cartesian coordinates x, y, and z, and j refers to the polarization axis of the external electric field. Thus, the molecular dipole moment induced parallel to the x-axis through interaction with the field, μ_x, can be written as

$$\mu_x = \alpha_{xx}E_x + \alpha_{xy}E_y + \alpha_{xz}E_z \qquad (6.7)$$

TABLE 6.1
Some Illustrative Values of Raman Cross-Sections Used Throughout this Work

Species	Pump Wavelength (nm)	Cross-Section (cm^2 molecule^{-1} sr^{-1})
$H_2O_{(l)}$	532	4×10^{-30}
$H_2O_{(l)}$	355	5×10^{-29}
$NO_{3(aq)}^-$	514.5	9.1×10^{-29}
$SO_{4(aq)}^{2-}$	514.5	9.9×10^{-29}

Analogous expressions can be written for the induced dipole components along the y and z directions. Thus, the molecular polarizability can be written as a tensor

$$\begin{bmatrix} \mu_x \\ \mu_y \\ \mu_z \end{bmatrix} = \begin{bmatrix} \alpha_{xx} & \alpha_{xy} & \alpha_{xz} \\ \alpha_{yx} & \alpha_{yy} & \alpha_{yz} \\ \alpha_{zx} & \alpha_{zy} & \alpha_{zz} \end{bmatrix} \begin{bmatrix} E_x \\ E_y \\ E_z \end{bmatrix}. \tag{6.8}$$

The polarizability tensor relates the induced dipoles along the three orthogonal Cartesian axes to the polarizability components and the electric field.

From Equation 6.5 it can be seen that for the molecular vibration to lead to inelastic scattering, at least one component of the polarizability tensor must vary during the period of the vibrational oscillation. This condition may be written as

$$\left(\frac{\partial \alpha_{ij}}{\partial Q_k} \right)_e \neq 0 \tag{6.9}$$

where Q_k is the normal coordinate associated with a vibration of frequency ν_k, and the derivative is evaluated at the equilibrium position.

Incident light is usually plane polarized along one axis, either x or y, and propagates in one direction, usually specified as the z direction. A schematic of a typical Raman arrangement is shown in Figure 6.2. Commonly, laser radiation is incident upon a sample and scattered light is collected at

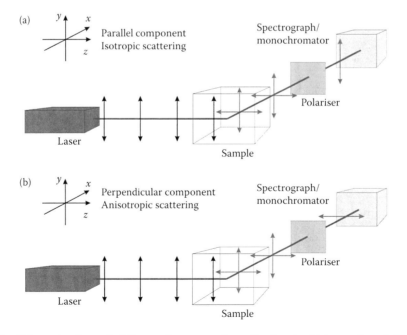

FIGURE 6.2 Schematic of a typical Raman instrument for collecting the (a) isotropic and (b) anisotropic scattering components. The relative configuration of the laser source, sample, and spectrograph/monochromator are shown for a typical collection angle of 90°. (Reprinted from Reid, J. P.; Meresman, H.; Mitchem, L.; Symes, R. *International Reviews in Physical Chemistry* 2007, *26*, 139. Copyright 2007, Taylor and Francis. With permission.)

90° to the incidence axis, often within the horizontal plane. Thus, if the light is vertically polarized along the y-axis, the polarizability tensor can be reduced to

$$\begin{bmatrix} \mu_x \\ \mu_y \\ \mu_z \end{bmatrix} = \begin{bmatrix} \alpha_{xy}E_y \\ \alpha_{yy}E_y \\ \alpha_{zy}E_y \end{bmatrix}.$$ (6.10)

The second component of the induced dipole ($\alpha_{yy}E_y$) leads to scattering with a polarization parallel to the incident polarization. This is described as the parallel component of scattering. The first and third components led to scattered light of polarization x or z, described as the perpendicular scattering components.

The polarizations of the incident and scattered light are usually selected by analyzers allowing the different components of the polarizability tensor to be measured independently. Isotropic scattering is recorded when the parallel component of the scatter is measured, anisotropic scattering is recorded when the perpendicular component is measured. The ratio of the intensity of anisotropic to isotropic scattering is referred to as the depolarization ratio. Symmetric vibrations are characterized by low depolarization ratios (close to zero) and the scattering is dominated by the isotropic component. Vibrations with asymmetric character led to depolarized scatter and a strong anisotropic scattering component. Measurements of the depolarization ratio can provide a confirmation of the assignment of vibrational modes in a Raman spectrum.

It is convenient to introduce the Porto notation, which permits an unambiguous description of the geometry of a scattering experiment, and allows an easy identification of the polarizability tensor components to which a particular experiment is sensitive. The Porto notation [A(B, C)D] describes an experiment in which incident radiation propagates parallel to the direction A with an electric vector parallel to B. Light is collected propagating in direction D, along which direction an analyzer transmits light of polarization parallel to C. The polarizability tensor element that is responsible for the detectable scatter is then α_{BC}. If the incident beam is vertically polarized and scatter is collected irrespective of polarization along a perpendicular axis in the horizontal plane, this can be written as [Z(Y, Y + Z)X]. Thus, the experiment is sensitive to the polarizability tensor elements α_{YY} and α_{YZ}, both isotropic and anisotropic scattering components will be recorded, and the spectrum may show features arising from symmetric and nonsymmetric vibrations. If a horizontally polarized incident beam is used with the same collection geometry, [Z(X, Y + Z)X], the possible polarizability tensor components that contribute to the detected scatter are α_{XY} and α_{XZ} leading to the observation of anisotropic scattering, that is, scatter due to totally symmetric vibrations will not be observed.

Raman spectra arising from Stoke's scattering from bulk samples and droplets of aqueous sodium chloride (NaCl) solution are compared in Figure 6.3 for unpolarized collection and for selected polarizations equivalent to the individual parallel and perpendicularly scattered components. The Raman features can be assigned to a bending vibration at 1645 cm^{-1}, a very weak combination band at 2125 cm^{-1} of the 1645 cm^{-1} bending and the 686 cm^{-1} librational mode, and the broad OH stretching band, which has a maximum at 3480 cm^{-1} with a strong shoulder visible at 3290 cm^{-1}.[6–8] It is apparent that the Raman band contours from the droplet measurements are identical to those recorded from the bulk-phase sample. This confirms that recording spectra from droplets is equivalent to recording spectra from bulk samples and is not accompanied by changes in the polarization of the scattered light induced by the sample.[9–11]

The dependence of the band shape of the OH stretching vibration on the polarization of the scattered light can be rationalized as follows.[6,9,12] The low Stoke's frequency side of the band transforms as a totally symmetric vibration, appearing in the Raman spectrum recorded when the incident and scattered light are chosen to have parallel polarizations. The high-frequency side of the Stoke's band transforms as an asymmetric vibration giving rise to depolarized Raman scattering that can be recorded either when the scattered polarization is selected to be parallel or perpendicular to the

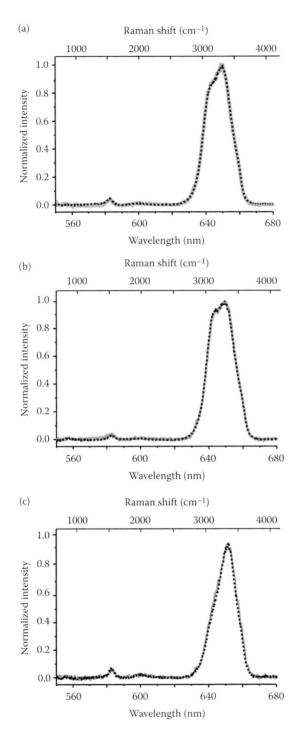

FIGURE 6.3 Comparison of the spontaneous Raman spectra from water droplets (radius ~38–42 μm, gray thick line) and a bulk-phase sample (black dotted line). Illumination with vertically polarized light and (a) unpolarized collection of scatter [Z(Y, Y + Z)X]; (b) vertically polarized scatter [Z(Y, Y)X]; (c) horizontally polarized scatter [Z(Y, Z)X]. Spectra were averaged over many droplets over a time period of ~30 s. (Reprinted from Reid, J. P.; Meresman, H.; Mitchem, L.; Symes, R. *International Reviews in Physical Chemistry* 2007, *26*, 139. Copyright 2007, Taylor and Francis. With permission.)

incident polarization. Although the high-frequency side of the band is expected to arise from the higher frequency asymmetric stretching normal mode and the low-frequency side from the lower frequency symmetric stretching normal mode, consistent with the observed degree of polarization in the scattered light, care must be taken in interpreting the spectra as the OH vibrational line shape.

The band contour is highly sensitive to the distribution of hydrogen bonding environments experienced by the water molecules giving rise to the Raman signature.[8,13,14] Indeed, the presence of structure breaking ions such as chloride has been shown to lead to systematic changes in the shape of the OH band as the presence of ions increasingly disrupts the intermolecular hydrogen bonds that form in the aqueous phase with increasing solute concentration.[8,13,14] While more strongly hydrogen-bonded water molecules are expected to give rise to lower frequency O—H stretching vibrations, less strongly hydrogen-bonded water molecules (i.e., stronger intramolecular O—H bond) are anticipated to give rise to higher frequency scatter. Mixture and continuum models of the water structure have been suggested to account for the shape of the band contour.[13,15–18]

Raman spectra from aqueous droplets containing nitrate and sulfate anions recorded from experiments with the geometry [Z(Y, Y + Z)X] are shown in Figure 6.4. Raman band assignments and

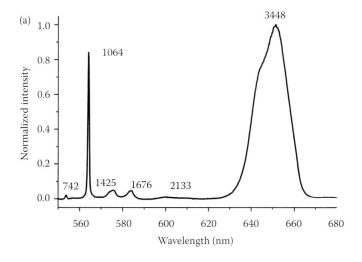

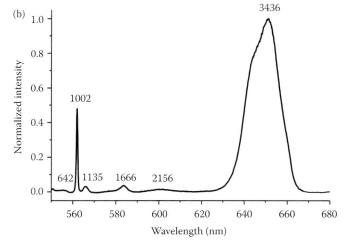

FIGURE 6.4 Normalized spontaneous Raman spectra [Z(Y, Y + Z)X] from solutions of (a) 1 M $NaNO_{3(aq)}$ and (b) 1 M $Na_2SO_{4(aq)}$. Numbers refer to the Stoke's shift of each band (in cm^{-1}) and the spectra are recorded with an excitation wavelength of 532 nm. (Reprinted from Reid, J. P.; Meresman, H.; Mitchem, L.; Symes, R. *International Reviews in Physical Chemistry* 2007, *26*, 139. Copyright 2007, Taylor and Francis. With permission.)

frequencies are reported in Table 6.2. These measurements once again confirm the consistency of bulk and droplet measurements. They also illustrate that Raman spectroscopy provides an unambiguous method for identifying molecular constituents of the aerosol phase. In addition, Equation 6.6 clearly indicates that quantitative compositional measurements can be made provided that the volume of the sample droplet is known. Although a determination of size is difficult, the signal is often normalized by the intensity of the Raman signature from the solvent, usually water, accounting for the variation in sample volume.

6.2.2 STIMULATED RAMAN SPECTROSCOPY

A sequence of Raman spectra recorded from an optically trapped aqueous aerosol droplet over 30 min is shown in Figure 6.5. The spontaneous Raman scattering can be assigned to the excitation of the OH stretching vibration, as discussed above. In addition, the Raman scattering intensity is amplified at discrete wavelengths, which are observed to vary with time. At wavelengths commensurate with Mie resonances, also commonly referred to as whispering gallery modes (WGMs) or morphology-dependent resonances, the spherical droplet behaves as a low-loss optical cavity.[19–21] Raman scattered light, at wavelengths commensurate with the Mie resonances, can circulate within the droplet and surpass a threshold intensity above which Raman scattering is amplified through the stimulation of further scattering.

WGMs are identified in Mie scattering calculations as the ripple structure, apparent in the extinction efficiency.[20] A single resonance arises from the dominant contribution of a single term to the Mie expansion of the internal or scattered field and, thus, the light has a well-defined angular and radial dependence within the droplet. From a geometric optics viewpoint, light propagating within the droplet approaches the surface at an angle beyond the critical angle, undergoing total internal reflection each time it encounters the surface. The resonant mode can also be considered as a standing wave formed from an integer number of wavelengths around the droplet circumference, referred to as the mode number, n. The internal light intensity calculated for a particular resonant mode is shown in Figure 6.6a. The light penetrates to a depth of approximately a/m from the droplet surface,

---

TABLE 6.2
Raman Shifted Wavelengths and Vibrational Frequencies for the Sulfate and Nitrate Anions and the Water Molecule

Species	Vibration	ν (cm^{-1})	λ (nm) (532 nm pump)
$SO_{4(aq)}^{2-}$	ν_2	451.5	545.1
$SO_{4(aq)}^{2-}$	ν_4	619	550.1
$SO_{4(aq)}^{2-}$	ν_1	980.4	561.3
$SO_{4(aq)}^{2-}$	ν_3	1107	565.3
H_2O	ν_2	1650	583.2
H_2O	ν_1	3280	644.5
H_2O	ν_3	3490	653.3
$NO_{3(aq)}^{-}$	ν_1	1049	563.4
$NO_{3(aq)}^{-}$	ν_2	830	556.6
$NO_{3(aq)}^{-}$	ν_3	1395	574.6
$NO_{3(aq)}^{-}$	ν_4	~723	553.3

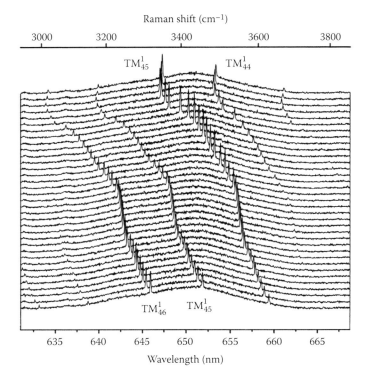

FIGURE 6.5 An example of the temporal evolution in droplet spectrum during a period of ~30 min (time increases from bottom to top). The droplet decreases from a size of 3.986 μm to a size of 3.913 μm over the time frame shown. The evaporation of the droplet is apparent from the tracking of the resonant modes to lower wavelength. The mode assignment is shown. This is recorded with an excitation wavelength of 532 nm for TE and TM modes. The electric field vector is indicated. (Reprinted from Mitchem, L.; et al. *Journal of Physical Chemistry A* 2006, *110*, 8116. Copyright 2006, American Chemical Society. With permission.)

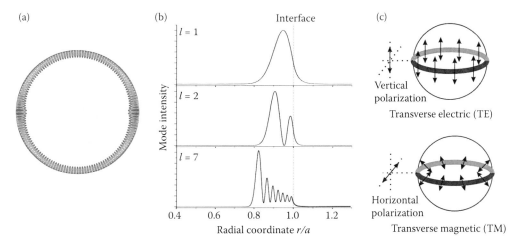

FIGURE 6.6 (a) Internal distribution of light intensity within a Mie resonance (WGM) of mode number 71 and first order. (b) Examples of the calculated variation in light intensity for $l = 1, 2,$ and 7 modes. (c) Schematic indicating the relative polarizations of light for TE and TM modes. (Reprinted from Reid, J. P.; Meresman, H.; Mitchem, L.; Symes, R. *International Reviews in Physical Chemistry* 2007, *26*, 139. Copyright 2007, Taylor and Francis. With permission.)

where a is the droplet radius and m is the real part of the refractive index of the droplet medium.[20,22] For a wavelength λ, the value of n can be estimated from the relationship

$$n\frac{\lambda}{m} = 2\pi a \qquad (6.11)$$

Approximate wavelength spacing, $\Delta\lambda$, of Mie resonances can be estimated from the asymptotic expression[20]

$$\Delta\lambda = \frac{\lambda^2 \tan^{-1}(m^2-1)^{1/2}}{2\pi a (m^2-1)^{1/2}} \qquad (6.12)$$

for Mie resonances around the wavelength λ. If the wavelengths of resonant modes can be measured, the size of a droplet can be estimated with nanometer accuracy by iteratively comparing the observed wavelengths with those predicted by Mie scattering theory.[4,9,23]

A Mie resonance may exhibit more than one maximum in the radial distribution of the light intensity within the droplet.[3] The mode order, l, specifies the number of maxima in the light intensity, as shown in Figure 6.6b. Modes of higher-order penetrate further into the droplet from the surface. The polarization of the light circulating within a mode must also be defined; modes with no radial dependence in the electric field are designated as transverse electric (TE) modes and modes with no radial dependence in the magnetic field as transverse magnetic (TM) modes. These are shown schematically in Figure 6.6c. The Mie resonances leading to the enhanced Raman scattering intensity are indicated in Figure 6.5 with the subscript denoting the mode number and the superscript the mode order.

The existence of Mie resonances leads to considerably lower incident threshold intensities for nonlinear processes in spherical droplets than in bulk-phase measurements. For example, SRS,[24–26] coherent and stimulated anti-Stokes Raman scattering,[27] lasing,[28,29] stimulated Brillouin scattering,[30] and third-order sum-frequency generation[31] have all been observed. Unlike SRS from bulk samples, SRS from droplets is only observed at the discrete wavelengths commensurate with WGMs, shows a temporal delay from the incident field, and has an isotropic scattering distribution.[21]

The intensity of SRS at a Stoke's frequency ω, $I_{SRS}(\omega)$, depends nonlinearly on the intensity of the spontaneous scattering, I_R, and can be expressed as

$$I_{SRS}(\omega) = I_R(\omega)\exp\left[\left(B(\omega)I_R - L\right)a\Phi\right] \qquad (6.13)$$

where $B(\omega)$ depends on the spatial overlap of the pump light and the Raman scattered light at the Stoke's shifted frequency at which the scattering undergoes stimulation, the Purcell factor, which describes the quantum electrodynamic enhancement of the Raman scattering due to a modified density of final states limited by spatial confinement of the cavity modes, and the Raman gain, which depends on species concentration. L describes losses due to radiative decay, absorption, and scattering from the cavity. The path length is expressed in terms of the particle radius, a, multiplied by the angular change on propagation of the light within the cavity, Φ. For the SRS intensity to exceed the spontaneous scattering intensity, the exponent must take a positive value, that is, the gain per unit length must exceed the cavity loss.

The cavity-enhanced Raman scattering (CERS) signal follows the spontaneous band contour at low gain without any exponential scaling, as illustrated in Figure 6.7a. This is the case when droplets are illuminated with low power continuous wave (CW) laser beams and the light intensities at the probe and Raman scattered intensities remain low. However, at the much high irradiances achievable over short durations with pulsed lasers, the SRS can be significantly above threshold and the Raman band follows an exponential scaling of the spontaneous band contour.[21] At Raman

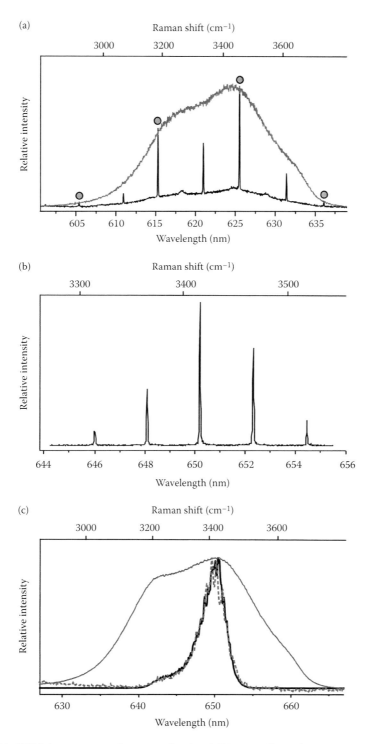

FIGURE 6.7 (a) CERS from a tweezed aqueous droplet, 4.681 μm radius, with an excitation wavelength of 514.8 nm (black line), illustrating that modes of the same order and polarization (indicated by the filled circles) follow the band contour of spontaneous Raman scattering from a bulk measurement (gray line). (b) Stimulated Raman spectrum recorded by pulsed laser illumination from a droplet ~24.5 μm in radius. (c) Composite SRS spectrum recorded from ~1000 water droplets ~27 μm radius (black line) compared with an exponential scaling (dotted gray line) of the spontaneous Raman band (gray line).

shifted wavelengths commensurate with Mie resonances toward the most intense central part of a band, the intensity of SRS grows more significantly than at the edges of the band due to gain narrowing and in accord with the exponential scaling of the spontaneous line shape. An example of this is shown in Figure 6.7b. It should also be noted that the illumination geometry may have a significant impact on the WGMs that lead to SRS. Edge illumination with a laser pulse leads to evanescent coupling into any WGM that is close to resonance in wavelength with the incident light.[21] This leads to an enhancement in the pump light intensity within the input mode volume of a particular mode number and order and the selective growth of SRS in output modes that are optimal in spatial overlap. Thus, although the droplet may be of a size sufficient to support high-quality modes of a range of mode orders, one mode order may dominate in the wavelength-resolved emission, arising from the increased gain associated with satisfying the double-resonance condition.[21]

An example of a composite spectrum formed from the addition of SRS spectra from many aqueous droplets of variable size is shown in Figure 6.7c. The variation in size leads to fluctuation in Mie resonance wavelengths from droplet to droplet and, hence, variation in the exact wavelengths at which SRS occurs. By forming a composite spectrum from many droplets, a smooth SRS band contour is observed. When compared with exponential scaling of the shape of a spontaneous band, a gain factor can be derived. Following from Equation 6.13, the normalized SRS spectral intensity at wavelength λ, $I_{SRS}(\lambda)$ (maximum intensity normalized to 1) can be expressed as scaling of the normalized spontaneous spectrum, $I_0(\lambda)$, and a gain factor G:

$$I_{SRS}(\lambda) = I_0(\lambda) \exp\left[GI_0(\lambda) - 1 \right] \tag{6.14}$$

This expression can allow the consistency of the SRS and spontaneous Raman spectra to be assessed. Although $G \ll 1$ for Raman spectra recorded from optically trapped aerosol droplets illuminated with a CW laser beam of ~10 mW power and irradiance <1 MW cm^{-2} values[23] of G between 12 and 15 have been measured for aqueous droplets illuminated with a laser pulse of irradiance 0.3 GW cm^{-2} and 10 ns duration.[21] The consequences of stimulated enhancement on the Raman band contour are clearly evident in Figure 6.7c.

The WGMs apparent in the Raman fingerprint can allow the size of the droplet to be estimated with high accuracy. For droplets smaller than ~12 μm diameter and illuminated by visible light, only the lowest few mode orders are observed in the spectral signature and the droplet size can be determined with nanometer accuracy.[4,9,23] For droplets larger than this, many mode orders can support SRS and the difficulty of assigning the correct mode order to an observed mode leads to ambiguity in the size determination. Thus, Equation 6.12 is used to estimate the droplet size from the mode spacing, introducing an uncertainty of the order of ±200 nm.[32] Indeed, not only can the droplet size be determined, but the relative intensities of SRS from different chemical components within a droplet can allow an accurate determination of the composition near the droplet surface. For example, the concentration of ethanol and water within mixed component droplets can be determined with an accuracy of ±0.2% by volume of the alcohol in the outer 3% of the droplet radius.[33,34] An example of the dependence of the SRS on composition averaged over many droplets is shown in Figure 6.8a along with a calibration plot in Figure 6.8b, which shows the expected nonlinear growth in alcohol signal with an increase in alcohol concentration.

6.3 INSTRUMENTATION

In general, there are three main components of an experimental setup for performing Raman spectroscopy on single-liquid microdroplets: a laser source, a means of droplet generation or sampling, and a dispersive grating spectrometer for resolving a spectrum coupled to a sensitive detector. We have introduced in Figure 6.2 a general schematic of the spectroscopic components and configuration for these types of studies. However, the experimental set up may vary from laboratory to laboratory depending on the individual requirements of the particular experiment.

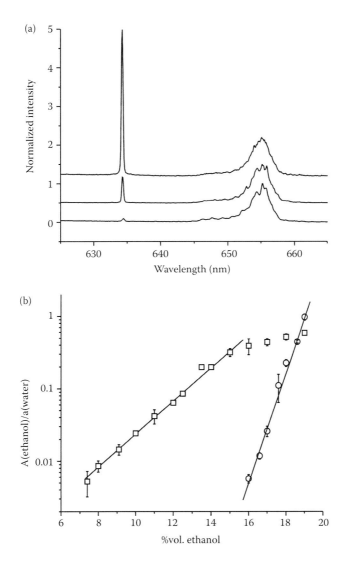

FIGURE 6.8 (a) SRS spectrum accumulated from 1000 ethanol/water droplets with compositions of 15%, 12%, 8% vol. ethanol (top, middle, and bottom spectra, respectively). The CH stretching vibration at 635 nm and the OH vibration at ~655 nm are apparent in the spectrum, acquired with an excitation wavelength of 532 nm. (b) Calibration curves for the integrated intensity of the CH signature relative to the water OH band for ethanol/water droplets recorded at excitation wavelengths of 532 and 355 nm (squares and circles, respectively). (Reprinted from Hopkins, R. J.; Symes, R.; Sayer, R. M.; Reid, J. P. *Chemical Physics Letters* 2003, *380*, 665. Copyright 2003, Elsevier. With permission.)

6.3.1 LASER SOURCES

Raman scattering transitions are weak and based on small frequency shifts, therefore, lasers are an ideal light source because they provide an intense, monochromatic source of light radiation. A wide range of lasers is commercially available for probing single particles by Raman scattering. The most commonly used lasers are listed in Table 6.3. The best suited lasers are those that fall in the blue and green regions of the spectrum since Raman scattering cross-sections vary as the fourth power of frequency. In addition, most optical detectors have optimal sensitivity to radiation in this region.

Continuous wave (CW) lasers are commonly used for spontaneous Raman scattering studies, whereas pulsed lasers have been used for stimulated Raman measurements. The most frequently

TABLE 6.3
Common Laser Sources Used in Raman Spectroscopy of Liquid Microdroplets. The Most Intense Laser Lines are Listed for the Ar^+ and Kr^+ Laser Sources

Laser Source	Wavelength (nm)	Type of Output
Ar^+	487.9 or 514.5	CW
Kr^+	530.2 or 647.0	CW
HeNe	632.8	CW
Diode	782 or 830	CW and pulsed
Nd:YAG	1064	Pulsed

used CW laser for spontaneous Raman is the Ar ion laser, a source that has high-power laser lines that fall in the blue and green regions of the visible part of the electromagnetic spectrum, as well as lines that fall in the near-UV region. For stimulated Raman studies, the solid-state pulsed Nd:YAG laser is often used because its second harmonic laser line (532 nm) provides a stable high-power output. Lasers are also advantageous for investigating the Raman scattering of microparticles, because they provide a small diameter collimated light beam that can be easily focused. Further, if the laser is linearly polarized then depolarization measurements can also be performed.

6.3.2 MICRODROPLET GENERATION AND SAMPLING

Measurements on a single suspended particle or with a monodisperse particle stream (or droplet train) are the two most common methodologies used in droplet Raman scattering studies. Ultrasonic nebulizers, droplet-on-demand generators and vibrating orifice aerosol generators (VOAG) can be used to produce liquid droplets. Solid particles can then be formed by drying.

An ultrasonic nebulizer uses a piezoelectric crystal to generate ultrasonic waves that are focused onto a liquid surface. Capillary waves form at the surface and then break to form a dense aerosol cloud. Ultrasonic nebulizers generate polydisperse particle populations with diameters less than 10 µm at concentrations greater than 10^7 particles cm^{-3}. Droplet-on-demand generators allow selective generation of droplets of a predefined volume, driving the constriction of a micro-capillary of chosen diameter using a piezoelectric transducer. Droplet-on-demand generators and ultrasonic nebulizers are frequently coupled with electrostatic or optical methods for isolating and trapping individual particles.[4] Such an approach can allow time-resolved measurements of evolving particle size, phase, and composition over indefinite time-periods, providing an opportunity to interrogate the details of aerosol equilibrium, chemical, and optical properties.

A monodisperse stream of liquid microdroplets can be obtained using a VOAG.[35] The VOAG operates by forcing a liquid through a small orifice, forming a liquid jet. The liquid jet is unstable to a mechanical disturbance and by applying a regular mechanical vibration to the jet, monodisperse droplets can be generated. The frequency of the modulating vibration, provided by a piezoelectric crystal, and the size of the orifice dictate the size of the droplets generated. Droplet diameters of 5–200 µm can be readily achieved with concentrations of 10–500 particles cm^{-3}.

6.3.3 SPECTROMETERS AND DETECTORS

In order to limit the radiation that enters the spectrometer to wavelengths longer than the source, holographic interference filters (or notch filters) are used, enabling rejection of the high intensity Rayleigh scattering line and allowing only the Stokes portion of the Raman spectrum to be retained. Spectral resolution of the Raman signature is achieved with a dispersive grating spectrometer. The greater the number of groves on the diffraction grating and/or the longer the focal length of the instrument, the greater is the resolution of the spectrum. The entrance slit width is also important in defining the spectral resolution achieved.

The wavelength dispersed light is then directed to a detector. A very sensitive photodetection system that is able to detect the weakest lines in the spectrum is required. In the past photomultiplier tubes (PMT) have been used as detectors in Raman spectroscopy. As single-channel detectors, these only allow the measurement of the scattered light intensity at a single wavelength at any one instance, with the spectrally resolved light first passing through an exit slit. Such a system is optimal for performing high-resolution measurements with low scattered light intensities. However, the diffraction grating must be rotated to obtain a frequency resolved spectrum.

Currently, the most efficient detectors used in Raman spectroscopy are intensified photodiode arrays (PDAs) and charge-coupled devices (CCDs). A PDA is a linear array of photosensitive elements that convert the energy of photons incident on the photodiode into an electrical charge. The charge pattern is related to the intensity of radiation impinging on the photodiode and the pattern is then read as a Raman spectrum. PDAs are similar to PMTs, but they are limited in their sensitivity and signal-to-noise. An intensifier, which is mounted in front of the PDA, can be used to increase the photodiode's performance. The intensifier consists of three parts: a photocathode, a multichannel plate (MCP) and a phosphor screen. The intensifier can increase the sensitivity of the detector by a factor of up to 10^3–10^4. However, PDAs are subject to dark counts that limit the exposure time of the spectrometer. This can be reduced by synchronously gating the intensifier on a pulsed laser to collect light over a narrow time window.

The CCD is a two-dimensional optical array detector and is currently the most popular detector for Raman spectroscopy. The CCD is a type of charge transfer device that integrates signal information when radiation is incident upon it. In comparison to the PDA, the CCD has lower detection limits and a smaller dynamic range. In addition, CCDs are subject to cosmic ray interference and dark counts. These dark counts can be reduced by cryogenic or thermoelectric cooling of the CCD. Intensifiers can also be used to increase the performance of CCDs.

6.3.4 SPECTRAL INTERFERENCES

Fluorescence can lead to significant degradation in the performance of a Raman instrument, as broad fluorescence bands can often obscure Raman signals. Fluorescence may arise from the absorption of laser light by the dominant components present in a droplet or even by impurities that are present. Fluorescence due to impurities can be removed by purifying the solution used for particle generation. However, if fluorescence is due to the droplet constituents, other steps must be taken, such as shifting the exciting wavelength to longer wavelengths (i.e., near-infrared lasers), using repetitive scanning with background subtraction or using a pulsed laser and gating the detection on the short duration laser pulse. Near IR lasers can also be used for samples that are sensitive to photodecomposition.

6.4 RECENT APPLICATIONS OF RAMAN SPECTROSCOPY IN STUDIES OF SINGLE LIQUID MICRODROPLETS

The following discussion introduces numerous applications of single-particle Raman spectroscopy combined with electrostatic trapping, optical levitation, and optical tweezing techniques for isolating single particles. Examples of spontaneous and stimulated Raman measurements are discussed separately for each.

An electrodynamic balance (EDB), for which the Millkian oil droplet experiment was an early forerunner, requires that the gravitational force on a charged particle be balanced by an electrostatic repulsion from a charged electrode with an applied DC field.[4,36] Confinement in three-dimensions is achieved through the strong restoring force exerted on a particle within a time-varying AC field superimposed on the DC component. The second and third techniques for isolating particles are both optical in origin. The gravitational force exerted on a particle is balanced by an upward radiation pressure (scattering) force exerted by a loosely focused laser beam leading to optical levitation.[4,36] By contrast, the tight focusing of a laser beam with a high-numerical aperture microscope objective leads to a single-beam gradient force optical trap (optical tweezers), in which the high gradient in light intensity draws the particle into the region of highest light intensity, dominating the weaker scattering force.[2,4,37]

6.4.1 APPLICATIONS OF SPONTANEOUS RAMAN SCATTERING

Raman active species give rise to characteristic Raman shifts. As such, spontaneous Raman scattering can provide information on particle composition. The examples presented below illustrate how monitoring the evolution of the spontaneous Raman scattering of aerosol yields information on the chemical composition of homogeneous and mixed phase droplets, phase transitions, particle concentration, and temperature changes. The progression of heterogeneous reactions can also be probed using the spontaneous Raman scattering of aerosol droplets using different experimental techniques.

6.4.1.1 Electrodynamic Balance

Stoke's shifted frequencies are very sensitive to the chemical composition, phase, and molecular interactions in a sample. Therefore, combining Raman spectroscopy with an EDB provides a unique tool that allows continuous *in situ* characterization of micron-sized droplets in the subsaturated relative humidity (RH) regimes.[38,39] Chapter 7 by Lee and Chan provides an in-depth presentation of the applications of EDB/Raman spectroscopy and only a brief summary of some of these applications is provided below.

One of the most common applications of the EDB has been the investigation of the thermodynamic properties of inorganic and mixed organic/inorganic aerosol.[39–41] The addition of Raman spectroscopy extends the operation of the EDB to real-time measurements of the chemical composition of trapped aerosol particles. For example, the full-width half-height (FWHH) of the spontaneous peaks in the Raman spectrum of ammonium sulfate droplets containing dicarboxylic acids was used to monitor the phase of the individual components of the droplet. Solid components could be distinguished from liquid components as the Raman signatures from solid components generally exhibit sharper peaks.[42] Additionally, the hygroscopicity of coated inorganic seed particles can be studied, because the observed changes in particle mass can be related to water partitioning to the aerosol as the RH in the surrounding environment is varied. Chan et al.[43] were able to show that solid particles coated with a soluble organic like glutaric acid can form internally mixed particles once deliquesced, therefore, changing the thermodynamic properties and the morphology of the final particle. Raman spectroscopy was used to measure the chemical composition of the organic coating.

The heterogeneous oxidation of unsaturated fatty acids (i.e., oleic, linoleic, and linolenic acids) was studied by Chan and coworkers using their EDB/Raman system.[44–46] The loss of the fatty acid due to oxidative aging was determined by the disappearance of the C=C Raman feature. The formation of the oxidation products during the course of a reaction was observed from the evolution of the Raman signature peaks for O–O, C=O and OH groups. Once oxidized, these particles were shown to be more hygroscopic than prior to ozonolysis.

6.4.1.2 Optical Levitation

The spontaneous Raman spectra obtained from optically levitated droplets have been applied in studies examining the effect of changes in the environmental conditions on the equilibrium state of an aerosol and the change in the composition of droplets from heterogeneous reactions. The influence of temperature on phase transitions and the size of optically levitated sulfate aerosol has been studied by Mund and Zellner.[47] They developed a thermostated cell to perform experiments on optically levitated droplets at temperatures as low as 180 K in gas flows up to 2 cm s^{-1}.[48] The spontaneous Raman scattering from water/glycerol/ammonium sulfate droplets was used to investigate the freezing nucleation of sulfate aerosol by monitoring the composition of droplets as they were either cooled or supercooled.[48,49] Trunk and coworkers also looked at the phase transitions of water/glycerol/ ammonium sulfate droplets. Crystallization of these droplets was observed from changes in the frequency (from ~981 to 976 cm^{-1}) and bandwidth of the symmetric stretch of the sulfate ion.[50]

Raman spectroscopy has been used to monitor the polymerization and copolymerization of optically levitated droplets by examining changes in the spontaneous Raman spectrum. As the polymerization reaction progressed, Musick and coworkers[51] observed the disappearance of the C=C stretching mode at 1629 cm^{-1}, indicating the consumption of the styrene monomer and the appearance of a peak at 1581 cm^{-1} due to aliphatic carbons, indicative of the presence of polystyrene. Esen and coworkers also monitored photo polymerization in droplets and observed changes in the spontaneous Raman spectra, including a decrease in the C=C and the =CH$_2$ band intensities and a corresponding increase in the CH$_3$ band intensity as the polymerization reaction inside the droplet proceeded.[52]

6.4.1.3 Optical Tweezers

Optical tweezers have recently become a frequently used tool for capturing and manipulating single particles or arrays of aerosol.[2,4,53] The restoring gradient force exerted on a tweezed particle provided by a tightly focused laser beam is of orders of magnitude larger than the gravitational force acting on a particle. This should be contrasted with a conventional optical levitation trap in which the gravitational force exerted on a particle is always exactly balanced by an upward radiation pressure force. Not only can the strong gradient force allow a particle to be retained within a gas flow or as the surrounding pressure is reduced below atmospheric, but also particles are effectively immobilized exhibiting only limited Brownian diffusion within the harmonic potential created by the large gradient in light intensity. Further, the positions of particles can be actively controlled by beam steering: many of the established techniques used to manipulate particles in liquids, such as holographic optical tweezers (HOT), have been extended to aerosols.[2,4] The close proximity of a high numerical aperture objective, which is used to form the optical trap, also provides a highly efficient method for collecting Raman scattered light. Raman spectra can be routinely collected with 1 s time resolution, and even with a time resolution of 10's of milliseconds. We will now consider specific examples of the coupling of Raman spectroscopy with optical tweezers.

6.4.1.3.1 *Concentration and Temperature-Dependent Raman Measurements*

The spontaneous Raman scatter of an optically tweezed aerosol provides useful chemical information with regard to the droplet's solute concentration. To illustrate this, the spontaneous Raman scattering from a trapped droplet has been shown to exhibit the same polarization-dependent behavior as scatter from bulk samples at the same solute concentration.[9,23] Therefore, the scatter from droplets can be calibrated to bulk samples and the concentration can be determined. In addition, information about the intermolecular interactions occurring between molecules within the droplet can be obtained from the shape and location of the Raman peaks.

The OH stretching band of an aqueous NaCl droplet appears at a Stokes shift of 2900–3700 cm^{-1} from the incident laser beam, with the maximum occurring at 3480 cm^{-1} and a shoulder at 3290 cm^{-1}. An increase in the ionic concentration of an aqueous NaCl droplet results in greater disruption to the hydrogen-bonding network, as observed by a shift in the maximum peak position of the OH band to higher frequencies and a decrease in the intensity of the band shoulder.[9,23] Monitoring changes in the shape of the OH band can provide information regarding temperature changes occurring within the droplet. A decrease in the intensity of the shoulder at 3290 cm^{-1} is observed if there is a temperature increase within the droplet. Again, the change in the spontaneous Raman spectrum is a result of disruptions to the hydrogen bonding network in the bulk of the droplet.

Information on the phase partitioning within a droplet can be obtained by examining the influence of hydrophobic organics on the hydrogen bonding structure of an aqueous droplet. In

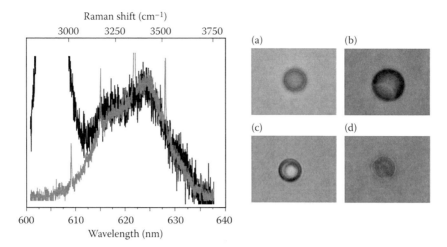

FIGURE 6.9 A sample of images of coagulated decane/water droplets (a) to (d). The OH band shape in the spectrum (shown in black) from a multiphase droplet (c) is compared with that for a pure water droplet (gray). The similarity in shapes of the OH stretching bands confirms that the hydrogen-bonding network within the aqueous component is not altered by the presence of the decane component, that is, the two components are phase-separated. (Reprinted from Mitchem, L.; Buajarern, J.; Ward, A. D.; Reid, J. P. *Journal of Physical Chemistry B* 2006, *110*, 13700. Copyright 2006, American Chemical Society. With permission.)

particular, the invariance of the OH band shape, when a water droplet is mixed with a decane droplet, has been used to confirm that the organic component is phase segregated and does not perturb the hydrogen bonding within the aqueous subphase.[54,55] An example of this is shown in Figure 6.9.

6.4.1.3.2 Chemical Composition of Mixed Organic/Inorganic Aqueous Droplets

Using the intensity of the OH band as an internal standard, the concentration of a second Raman active species can also be obtained from the spontaneous Raman spectrum.[9] The chemical composition of aqueous droplets dosed with an organic compound either by gas-phase adsorption, controlled coalescence of nebulized particles, or the coalescence of submicron aerosol produced by homogeneous nucleation can be probed using optical tweezers/Raman spectroscopy. Prior to the introduction of organics into the aerosol optical tweezers system, the Raman scatter of an aqueous/NaCl droplet consists of only the broad OH band (~3200–3600 cm^{-1}). The incorporation of an organic, such as ethanol, into the optically trapped droplet results in the appearance of a CH band in the region of 2852–3016 cm^{-1}.[56] The intensity of the CH band increases as the amount of organic (i.e., ethanol) within the droplet increases. Similar observations have been made for aqueous NaCl droplets dosed with decane and oleic acid.[55,57]

Buajarern and coworkers[58] used a dual trap optical tweezers technique to probe phase transitions of aqueous droplets. Initially, an aqueous NaCl droplet dosed with a high sodium dodecyl sulfate (SDS) concentration was trapped. A comparison of the Raman signature recorded from the droplet with that from a bulk-phase sample of the same concentration is shown in Figure 6.10. Water subsequently evaporated from the droplet resulting in the formation of a metastable gel. As the composite droplet formed a microgel, the CH band intensity increased and dominated the OH band intensity. Further, the shape of the OH band contour was observed to change, indicating a significant change in the hydrogen-bonding network within the droplet.

An inverse correlation in the temporally varying intensities of the CH and OH bands was observed for aqueous droplets containing a solid, insoluble palmitic acid inclusion.[59] Estimates for the time of diffusion of the inclusion within the droplet were determined and it was concluded from the autocorrelation of the CH and OH intensities that the organic inclusion was able to freely diffuse within

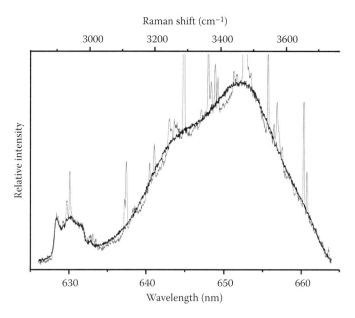

FIGURE 6.10 Comparison of the Raman spectra observed from a trapped aqueous droplet (structured gray spectrum) and a bulk-phase solution (black), both composed of 0.2 M SDS and 0.2 M NaCl. (Reprinted from Buajarern, J.; Mitchem, L.; Reid, J. P. *Journal of Physical Chemistry A* 2007, *111*, 13038. Copyright 2007, American Chemical Society. With permission.)

the aqueous host droplet. Further, these experiments demonstrated that optical tweezers can be used to retain nonspherical, solid particles.

6.4.1.3.3 Heterogeneous Reactions

The progression of heterogeneous reactions of single particles has been observed by monitoring the appearance and/or disappearance of product and reactant peaks, respectively, as a function of reaction time. The oxidation of oleic acid in synthetic seawater droplets by ozone was investigated by monitoring the spontaneous Raman scattering signal characteristic for oleic acid and its oxidation products until no signal was observed.[60] Similar studies have also been performed for fumarate and benzoate ions in aqueous droplets as well as α-pinene on dodecane and pentadecane aerosols.[61]

6.4.2 APPLICATIONS OF SRS

As described earlier, the CERS fingerprint of a droplet consists of a spontaneous Raman band with resonances from SRS superimposed. The spontaneous Raman band can be used for chemical composition determination, whereas a comparison of the SRS peaks with Mie theory permits droplet size to be monitored. Further, the relative intensities of SRS from different chemical components can be used to investigate the evolution of the near-surface composition of rapidly evaporating droplets containing volatile components. The accuracy with which the size of the droplet can be determined is dependent on the spectral dispersion of the spectrograph used to record the spectra and the accuracy with which the refractive index of the droplet is known.[3]

6.4.2.1 Droplet Train

The nonlinear dependence of stimulated Raman intensity on composition provides an extremely sensitive method for probing small changes in composition. In addition, the signal arises from the near-surface region and is ideally positioned to provide a signature of composition during evaporation. Lin and Campillo were the first to use the bulk and nonlinear responses to spatially resolve the changes in composition during the evaporation of methanol–water droplets generated by a VOAG.[62] Measurements identified a depletion of the organic component relative to water in the outer shell corresponding to a penetration depth from the surface of 10% of the droplet radius, or ~0.7 μm. They monitored the depletion of alcohol occurring over the first 150 μs following droplet generation. They concluded that SRS is ideally suited for radially profiling the composition of droplets in the diameter range of 5–30 μm.

Reid and coworkers have extended this approach to study the evaporation of alcohol–water droplets, in which the alcohol (methanol, ethanol or 1-propanol) was present between 15% and 23% by premixed volume.[34,63] Droplet radii studied were in the range 20–60 μm. Although larger than those studied by Lin and Campillo,[62] these measurements were not dependent on assigning the specific orders of the Mie resonances appearing in the Raman fingerprint. Instead, the droplet size generated by the VOAG was swept over a narrow range of sizes, ensuring that the spectrum acquired from each droplet was different; the stimulated Raman occurred at different sets of Mie resonant wavelengths for each single droplet/single laser pulse measurement. Spectra from more than 1000 droplets were summed to provide a composite/average fingerprint. By comparison of the ratio of CH and OH intensities with a calibration performed on droplets of known composition (Figure 6.8), the change in composition due to evaporation was investigated, with the mass flux of the more volatile alcohol component dominating the change in composition at an early time. Seeding of the droplets with Rhodamine B also allowed the evolving temperature of the droplet bulk to be monitored by laser-induced fluorescence: the bandwidth of the fluorescence spectrum is dependent on temperature and can be used to infer temperature changes.[64]

Evaporation measurements of evolving composition were performed for droplets initially of 20–60 μm in radius, at surrounding dry nitrogen gas pressure of between 10 and 100 kPa, and at evaporation times between 0.2 ms and 2.5 ms. An example of the change in Raman fingerprint with the exposure time is shown in Figure 6.11. At early exposure times, the consequences of an increased rate of gas diffusion away from the droplet with decreasing pressure were observed in an increasing level of depletion of the alcohol from the near-surface region. It was concluded that the results were consistent with quasi-steady models of evaporation if the signal was assumed to arise from the outer 3% of the droplet radius, between 0.6 and 1.8 μm for the droplet sizes considered.

Measurements at longer times up to 2.5 ms have highlighted the deficiencies in using such a quasi-steady model to interpret the evolving composition.[63] For small droplets at low pressure and at the longest evaporation times, a slight recovery of near-surface composition was measured. This observation was compared to a more rigorous model of unsteady heat and mass transfer. When the spatially resolved simulations were convoluted with possible radial distributions for the expected signal, it was concluded that consistency between the model predictions and experiments could only be achieved if the stimulated Raman signal was once again assumed to originate from the outer shell to a depth of (3 ± 1)% of the radius from the droplet surface. This is consistent with the more highly confined probe region expected when the stimulated Raman signal arises from double-resonance excitation. The recovery in ethanol concentration near the droplet surface at long times was shown to be consistent with the bulk diffusion of the alcohol from the droplet core to the surface, driven by the initial rapid depletion at the surface. This provided the first example of a direct measurement of the internal diffusion/mixing of components within an aerosol particle during heat and mass transfer by Raman spectroscopy.

6.4.2.2 Optical Levitation

Mie resonances were first observed in Raman scattering in measurements from optically levitated droplets. Thurn and Kiefer[65] observed SRS at wavelengths commensurate with Mie resonances superimposed on the broad underlying envelope of spontaneous scattering in their studies of optically levitated water/glycerol droplets. Comparison of the wavelengths of the resonances observed with predictions from Mie theory allowed the size of the droplet to be determined.[65] This approach was adopted by Lettieri and Preston[66] for their investigation of the condensational growth of

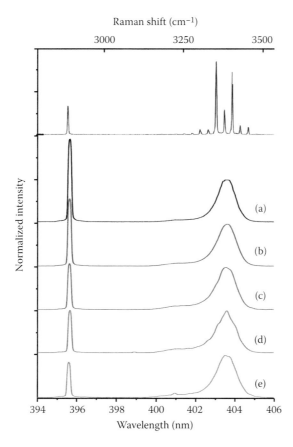

FIGURE 6.11   CERS composite spectra for 37.8 μm radius 18% v/v ethanol/water droplets at 0.2, 0.8, 1.3, 1.7, and 2.0 ms after generation (a–e, respectively), evaporating in a dry nitrogen atmosphere at 13.2 kPa. The depletion of the alcohol can be seen from the decrease in the CH signature at ~395.5 nm. A CERS fingerprint from a single ethanol/water droplet is shown for comparison. (Reprinted from Hopkins, R. J.; Reid, J. P. *Journal of Physical Chemistry A* 2005, *109*, 7923. Copyright 2005, American Chemical Society. With permission.)

dioctyl phthalate droplets. They observed frequency shifts of the resonances as the size of the particle changed.

6.4.2.3   Optical Tweezers

Raman spectra recorded from optically tweezed aerosol particles are similar to those described for optically levitated droplets, consisting of both spontaneous and stimulated components. We will describe how this technique can be used to investigate the equilibrium state of aqueous aerosol and the changes occurring on aerosol coalescence by monitoring the evolution of the stimulated-scattering signature.

6.4.2.3.1   Comparative Growth and Evaporation Experiments

The amplification of Raman scattering at wavelengths commensurate with Mie resonances has been used to monitor the evolving size of aqueous droplets exposed to water and organic vapors.[23,56,67] The equilibrium size of an aqueous droplet responds to changes in the RH surrounding the droplet as the vapor pressure of the droplet adjusts to the partial pressure of water in the surrounding environment. The response of an optically trapped droplet to changes in RH over 50 h was determined by sizing the droplet over the course of an experiment.[23] The size information was obtained by comparing the stimulated Raman wavelengths to Mie theory. An experimental Köhler curve (i.e., wet particle size as function of RH) was compared to a theoretically determined Köhler curve for a system with comparable solute concentrations. In the work by Mitchem et al.[23] the RH range investigated was between 80 and 92%. Hargreaves et al.[67] used a similar technique to investigate how an aqueous NaCl droplet changes with variation in RH in the supersaturated regime (i.e., RH 40–70%). An example of the variation in size recorded with change in RH is shown in Figure 6.12. In both the works by Mitchem et al.[23] and Hargreaves et al.[67] the RH was measured using capacitance probes located outside the optical tweezers cell, a conventional method that is routinely used in many applications. These macroscopic capacitance RH probes have slow response times (~1 min) and have accuracies of approximately ±2% below 90%. In addition, they do not capture the local RH surrounding the microscopic droplet, nor do they capture any RH gradients that may be present.

Spectroscopic probes of gas-phase composition are routinely used, for example, in absorption spectroscopy. However, it is also possible to use the Raman signature of a trapped droplet as a probe of gas-phase composition. The optical trapping of a second control droplet in a second optical trap, in addition to the droplet of interest, can vastly improve the accuracy of RH measurements within the optical tweezers cell. The control droplet, often chosen to be an aqueous NaCl droplet, acts as a highly accurate, localized and responsive probe of RH, describing the RH conditions in close proximity to the droplet of interest with high time-resolution.[2] Relative humidity determination using the control droplet can be achieved provided the dry particle size is known and the solute of the probe droplet is one that has been well studied. If these parameters are known, then the dependence of wet particle size with RH can be easily determined. The accuracy of RH determined within the cell using an aqueous NaCl droplet as a probe of RH is better than ±0.09%.[68]

6.4.2.3.2 Comparative Thermodynamic Measurements

The use of a control droplet as a RH probe in dual-trap optical tweezers has been used to perform thermodynamic measurements on aqueous droplets containing a soluble organic compound, glutaric acid, at high RH.[69] This benchmark experiment demonstrated that comparative dual trap experiments permit investigations of the hygroscopic properties of mixed aqueous organic/ inorganic droplets. For these comparative measurements, a pure aqueous NaCl droplet was trapped along with an aqueous NaCl or ammonium sulfate droplet doped with glutaric acid. The stimulated

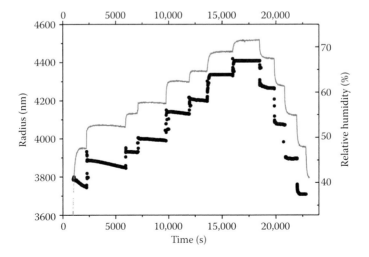

FIGURE 6.12 Time dependence of the evolving wet droplet size with change in relative humidity for an aqueous NaCl droplet. The RH trend is shown by the gray line and the wet droplet size by the black filled circles. (Reprinted from Hargreaves, G.; et al. *Journal of Physical Chemistry A* 2010, In press. Copyright 2010, American Chemical Society. With permission.)

Raman peaks were used to monitor the change in the size of both droplets with RH, as the RH decreased from 99%. The change in RH was determined from the control droplet and therefore permitted an experimental Köhler curve to be inferred. Similar comparative experiments have also been performed on aqueous NaCl droplets containing other dicarboxylic acids at RHs close to saturation and in the solute supersaturated regime. In all cases, experimental Köhler curves were in good agreement with theory and measured equilibrium sizes routinely lie within ±0.5% of those predicted, providing a stringent test for models of the aerosol equilibrium state.

Butler and coworkers[53] have extended the comparative thermodynamic measurements even further by performing such experiments on arrays of optically tweezed aqueous NaCl droplets using HOT. With this technique, one of the droplets of the array is used as an RH probe for the remaining droplets. Further, the same spectroscopic techniques used for the dual trap experiments have been applied to arrays of droplets manipulated by HOT. One trapping position within the aerosol array, often a carousel arrangement, was aligned to the spectroscopic detection system, as shown in Figure 6.13. By sequential rotation of the carousel, the Raman spectra of all droplets were recorded on a timescale of less than a minute. Experimental Köhler curves were in good agreement with theory for all the droplets of the array. This approach has recently been extended to allow a comparison of the equilibrium state of chemically distinct droplets held in different optical traps.

6.4.2.3.3 Coagulation of Miscible and Immiscible Droplets

The outcome of the controlled coagulation of two aqueous droplets has been explored by examining the CERS fingerprint for each droplet and the combined droplet postcoagulation.[37] The volume of each of the droplets was determined once they had equilibrated with the surrounding environment and stabilized in size, as determined by continuously monitoring the Mie resonances of each droplet. Upon coagulation, the volume of the combined droplet was compared to the sum of the volumes of the two individual droplets. The calculated and expected volumes were in excellent agreement with an error of $\pm 3 \times 19$ m^{-3} or an error less than 0.1%.[37]

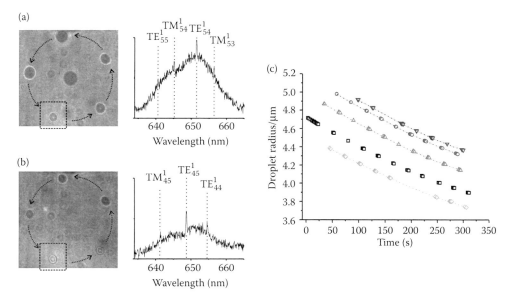

FIGURE 6.13 Holographic optical tweezing can be used to capture and manipulate the arrays of aerosol droplets. (a) The image and spectrum of a droplet (inside the dotted box) immediately after capture. The dotted arrows illustrate the direction of rotation of the carousel. (b) The same droplet after 5 min in the carousel. During this time, the RH in the cell has dropped by 2% and the droplets have all decreased in size through evaporation. (c) The evolution in size of the five droplets as a function of time. (Reprinted from Butler, J. R.; et al. *Lab on a Chip* 2009, *9*, 521. With permission.)

The coagulation of droplets of differing composition may result in the formation of a multiphase droplet. Further, by monitoring the CERS of the composite mixed phase droplet its mixing state can be determined. For example, the coagulation of aqueous NaCl droplets with either ethanol and decane droplets has been studied to look at the mixing state and phase segregation of organics within an aqueous host droplet.[54,55,57] The coagulation of an ethanol droplet with an aqueous NaCl droplet resulted in the observation of WGM not only on the OH band, but also on the CH band.[54,70] Although a homogeneous droplet is to be expected, given the miscibility of ethanol and water, the possible formation of a core–shell structure was also considered with an aqueous core surrounded by an ethanol shell. In order to distinguish between the two configurations, Mie calculations for both a homogeneous droplet and a core–shell droplet were performed. For such a layered droplet, a spectral shift between the relative wavelengths of TE and TM modes, referred to as the mode offset, is expected and compared with that predicted for a homogeneous droplet. In the specific case of a mixed ethanol aqueous NaCl droplet, the results of the Mie calculations were consistent with the expected configuration of a homogeneous droplet.

The mode offset that is observed for core–shell droplets is a result of differences in the refractive indices of the two phases.[57] The coagulation of an aqueous NaCl droplet with a decane droplet was studied to examine the influence of an organic layer thickness on mode offset. At thin layer thicknesses (<10 nm), the mode offset is similar to that of a homogeneous droplet, but increases with increasing layer thickness. For core–shell structures of very thick shells (>500 nm), the WGM circulate entirely within the shell and the spectrum appears as one from a homogeneous droplet composed of the shell material. The thickness of a decane layer on aqueous NaCl droplets was estimated by performing core–shell calculations on the Raman spectra of optically trapped mixed droplets. It was assumed that the refractive indices of the core and shell were that of water and decane, respectively. The calculated Mie resonances were compared with the experimental resonances and best estimates of core size and layer thickness were obtained. Based on these calculations, layer thicknesses can be estimated with an accuracy of better than ±8 nm.[57]

6.5   CONCLUSIONS

We have reviewed the basis of Raman spectroscopy and its application to the characterization of aerosol dynamics. Not only can information about the composition and phase of an aerosol particle be gained, but also the fundamental intermolecular interactions between chemical components can be interrogated. These measurements can be performed on a single-particle basis and the time-evolving properties interrogated over an indefinite time period, examining the response of the particle to changes in environmental conditions. In addition, SRS can provide an extremely sensitive measure of the near-surface composition of liquid droplets and the size of a droplet can be determined with nanometer accuracy. The details of aerosol processes can be investigated on a single-particle basis with unprecedented detail.

ACKNOWLEDGMENTS

The authors gratefully acknowledge support from the Engineering and Physical Sciences Council (EPSRC) through the award of a Leadership Fellowship (JPR). NOAK acknowledges the Natural Sciences and Engineering Council (NSERC) of Canada for the award of a postdoctoral research fellowship and a Canada–United Kingdom Millennium Research Award.

REFERENCES

1. Schweiger, G. Raman-scattering on single aerosol-particles and on flowing aerosols—a review. *Journal of Aerosol Science* 1990, *21*(4), 483–509.
2. Mitchem, L.; Reid, J. P. Optical manipulation and characterisation of aerosol particles using a single-beam gradient force optical trap. *Chemical Society Reviews* 2008, *37*(4), 756–769.

3. Reid, J. P.; Mitchem, L. Laser probing of single aerosol droplet dynamics. *Annual Review in Physical Chemistry* 2006, *57*, 245–271.

4. Reid, J. P. Particle levitation and laboratory scattering. *Journal of Quantitative Spectroscopy & Radiative Transfer* 2009, *110*(14–16), 1293–1306.

5. Wills, J. B.; Knox, K. J.; Reid, J. P. Optical control and characterisation of aerosol. *Chemical Physics Letters* 2009, *481*(4–6), 153–165.

6. Scherer, J. R.; Go, M. K.; Kint, S. Raman-spectra and structure of water from –10 to 90 degrees. *Journal of Physical Chemistry* 1974, *78*(13), 1304–1313.

7. Walrafen, G. E. Raman and infrared spectral investigations of water structure. In *Water: A Comprehensive Treatise*, Franks, F. (Ed.), Plenum Press: New York, NY, 1972; Vol. 1; p. 151.

8. Lilley, T. H. Raman spectroscopy of aqueous electrolyte solutions. In *Water: A Comprehensive Treatise*, Franks, F. (Ed.), Plenum Press: New York, NY, 1973; Vol. 3; p. 265.

9. Reid, J. P.; Meresman, H.; Mitchem, L.; Symes, R. Spectroscopic studies of the size and composition of single aerosol droplets. *International Reviews in Physical Chemistry* 2007, *26*(1), 139–192.

10. Fung, K. H.; Tang, I. N. Polarization measurements on Raman-scattering from spherical droplets. *Applied Spectroscopy* 1992, *46*(7), 1189–1193.

11. Vehring, R.; Schweiger, G. Optical determination of the temperature of transparent microparticles. *Applied Spectroscopy* 1992, *46*(1), 25.

12. Cunningham, K.; Lyons, P. A. Depolarization ratio studies on liquid water. *Journal of Chemical Physics* 1973, *59*(4), 2132–2139.

13. Li, R. H.; Jiang, Z. P.; Chen, F. G.; Yang, H. W.; Guan, Y. T. Hydrogen bonded structure of water and aqueous solutions of sodium halides: A Raman spectroscopic study. *Journal of Molecular Structure* 2004, *707*(1–3), 83–88.

14. Terpstra, P.; Combes, D.; Zwick, A. Effects of salts on dynamics of water: A Raman spectroscopy study. *Journal of Chemical Physics* 1990, *92*, 65.

15. Wang, Z. H.; Pakoulev, A.; Pang, Y.; Dlott, D. D. Vibrational substructure in the OH stretching band of water. *Chemical Physics Letters* 2003, *378*(3–4), 281–288.

16. Wernet, P.; Nordlund, D.; Bergmann, U.; Cavalleri, M.; Odelius, M.; Ogasawara, H.; Naslund, L. A.; et al. The structure of the first coordination shell in liquid water. *Science* 2004, *304*(5673), 995–999.

17. Stenger, J.; Madsen, D.; Hamm, P.; Nibbering, E. T. J.; Elsaesser, T., Ultrafast vibrational dephasing of liquid water. *Physical Review Letters* 2001, *8702*(2), 027401.

18. Brubach, J. B.; Mermet, A.; Filabozzi, A.; Gerschel, A.; Roy, P. Signatures of the hydrogen bonding in the infrared bands of water. *Journal of Chemical Physics* 2005, *122*(18), 184509.

19. Campillo, A. J.; Eversole, J. D.; Lin, H. B. Cavity QED modified stimulated and spontaneous processes in microdroplets. In *Optical Processes in Microcavities*, Chang, R. K., Campillo, A. J. (Eds.); World Scientific: Singapore, 1996.

20. Hill, S. C.; Benner, R. E., Morphology-dependent resonances. In *Optical Effects Associated with Small Particles*, Barber, P. W., Chang, R. K. (Eds.); World Scientific: 1988; Vol. 1, p. 3.

21. Symes, R.; Sayer, R. M.; Reid, J. P. Cavity enhanced droplet spectroscopy: Principles, perspectives and prospects. *Physical Chemistry Chemical Physics* 2004, *6*, 474.

22. Eversole, J. D.; Lin, H. B.; Campillo, A. J. Input-output resonance correlation in laser-induced emission from microdroplets. *Journal of the Optical Society of America B-Optical Physics* 1995, *12*(2), 287–296.

23. Mitchem, L.; Buajarern, J.; Hopkins, R. J.; Ward, A. D.; Gilham, R. J. J.; Johnston, R. L.; Reid, J. P. Spectroscopy of growing and evaporating water droplets: Exploring the variation in equilibrium droplet size with relative humidity. *Journal of Physical Chemistry A* 2006, *110*(26), 8116–8125.

24. Lin, H. B.; Eversole, J. D.; Campillo, A. J. Continuous-wave stimulated Raman-scattering in microdroplets. *Optics Letters* 1992, *17* (11), 828–830.

25. Snow, J. B.; Qian, S. X.; Chang, R. K. Stimulated Raman-scattering from individual water and ethanol droplets at morphology-dependent resonances. *Optics Letters* 1985, *10*(1), 37–39.

26. Biswas, A.; Latifi, H.; Armstrong, R. L.; Pinnick, R. G. Double-resonance stimulated Raman-scattering from optically levitated glycerol droplets. *Physical Review A* 1989, *40*(12), 7413–7416.

27. Qian, S. X.; Snow, J. B.; Chang, R. K. Coherent Raman mixing and coherent anti-stokes Raman-scattering from individual micrometer-size droplets. *Optics Letters* 1985, *10*(10), 499–501.

28. Qian, S. X.; Snow, J. B.; Tzeng, H. M.; Chang, R. K. Lasing droplets—highlighting the liquid-air interface by laser-emission. *Science* 1986, *231*(4737), 486–488.

29. Tzeng, H. M.; Wall, K. F.; Long, M. B.; Chang, R. K. Laser-emission from individual droplets at wavelengths corresponding to morphology-dependent resonances. *Optics Letters* 1984, *9*(11), 499–501.

30. Wirth, F. H.; Juvan, K. A.; Leach, D. H.; Swindal, J. C.; Chang, R. K.; Leung, P. T. Phonon-retention effects on stimulated Brillouin-scattering from micrometer-sized droplets illuminated with multiple short laser-pulses. *Optics Letters* 1992, *17*(19), 1334–1336.

31. Leach, D. H.; Chang, R. K.; Acker, W. P.; Hill, S. C. 3rd-order sum-frequency generation in droplets—experimental results. *Journal of the Optical Society of America B-Optical Physics* 1993, *10*(1), 34–45.

32. Sayer, R. M.; Gatherer, R. D. B.; Gilham, R. J. J.; Reid, J. P. Determination and validation of water droplet size distributions probed by cavity enhanced Raman scattering. *Physical Chemistry Chemical Physics* 2003, *5*(17), 3732–3739.

33. Hopkins, R. J.; Symes, R.; Sayer, R. M.; Reid, J. P. Determination of the size and composition of multi-component ethanol/water droplets by cavity-enhanced Raman scattering. *Chemical Physics Letters* 2003, *380*(5–6), 665–672.

34. Hopkins, R. J.; Reid, J. P. A comparative study of the mass and heat transfer dynamics of evaporating ethanol/water, methanol/water, and 1-propanol/water aerosol droplets. *Journal of Physical Chemistry B* 2006, *110*(7), 3239–3249.

35. Berglund, R. N.; Liu, B. Y. H. Generation of monodisperse aerosol standards. *Environmental Science & Technology* 1973, *7*, 147.

36. Davis, E. J. A history of single aerosol particle levitation. *Aerosol Science and Technology* 1997, *26*(3), 212–254.

37. Hopkins, R. J.; Mitchem, L.; Ward, A. D.; Reid, J. P. Control and characterisation of a single aerosol droplet in a single-beam gradient-force optical trap. *Physical Chemistry Chemical Physics* 2004, *6*(21), 4924–4927.

38. Sjogren, S.; Gysel, M.; Weingartner, E.; Baltensperger, U.; Cubison, M. J.; Coe, H.; Zardini, A. A.; Marcolli, C.; Krieger, U. K.; Peter, T. Hygroscopic growth and water uptake kinetics of two-phase aerosol particles consisting of ammonium sulfate, adipic and humic acid mixtures. *Journal of Aerosol Science* 2007, *38*(2), 157–171.

39. Chan, C. K.; Kwok, C. S.; Chow, A. H. L. Study of hygroscopic properties of aqueous mixtures of disodium fluorescein and sodium chloride using an electrodynamic balance. *Pharmaceutical Research* 1997, *14*(9), 1171–1175.

40. Ha, Z. Y.; Choy, L.; Chan, C. K. Study of water activities of supersaturated aerosols of sodium and ammonium salts. *Journal of Geophysical Research-Atmospheres* 2000, *105*(D9), 11699–11709.

41. Tang, I. N.; Munkelwitz, H. R. Water activities, densities, and refractive-indexes of aqueous sulfates and sodium-nitrate droplets of atmospheric importance. *Journal of Geophysical Research-Atmospheres* 1994, *99*(D9), 18801–18808.

42. Ling, T. Y.; Chan, C.K. Partial crystallization and deliquescence of particles containing ammonium sulfate and dicarboxylic acids. *Journal of Geophysical Research-Atmospheres* 2008, *113*(D14), D14205.

43. Chan, M. N.; Lee, A. K. Y.; Chan, C. K. Responses of ammonium sulfate particles coated with glutaric acid to cyclic changes in relative humidity: Hygroscopicity and Raman characterization. *Environmental Science & Technology* 2006, *40*(22), 6983–6989.

44. Lee, A. K. Y.; Chan, C. K. Single particle Raman spectroscopy for investigating atmospheric heterogeneous reactions of organic aerosols. *Atmospheric Environment* 2007, *41* (22), 4611–4621.

45. Lee, A. K. Y.; Chan, C. K. Heterogeneous reactions of linoleic acid and linolenic acid particles with ozone: Reaction pathways and changes in particle mass, hygroscopicity, and morphology. *Journal of Physical Chemistry A* 2007, *111*(28), 6285–6295.

46. Lee, A. K. Y.; Ling, T. Y.; Chan, C. K. Understanding hygroscopic growth and phase transformation of aerosols using single particle Raman spectroscopy in an electrodynamic balance. *Faraday Discussions* 2008, *137*, 245–263.

47. Mund, C.; Zellner, R. Optical levitation of single microdroplets at temperatures down to 180 K. *A European Journal of Chemical Physics and Physical Chemistry* 2003, *4*(6), 630–638.

48. Mund, C.; Zellner, R. Freezing nucleation of levitated single sulfuric acid/H$_2$O micro-droplets. A combined Raman- and Mie spectroscopic study. *Journal of Molecular Structure* 2003, *661*, 491–500.

49. Mund, C.; Zellner, R. Raman- and Mie-spectroscopic studies of the cooling behaviour of levitated, single sulfuric acid/H$_2$O microdroplets. *A European Journal of Chemical Physics and Physical Chemistry* 2003, *4*(6), 638–645.

50. Trunk, M.; Lubben, J. F.; Popp, J.; Schrader, B.; Kiefer, W. Investigation of a phase transition in a single optically levitated microdroplet by Raman-Mie scattering. *Applied Optics* 1997, *36*(15), 3305–3309.

51. Musick, J.; Popp, J.; Trunk, M.; Kiefer, W. Investigations of radical polymerization and copolymerization reactions in optically levitated microdroplets by simultaneous Raman spectroscopy, Mie scattering, and radiation pressure measurements. *Applied Spectroscopy* 1998, *52*(5), 692–701.

52. Esen, C.; Kaiser, T.; Schweiger, G. Raman investigation of photopolymerization reactions of single optically levitated microparticles. *Applied Spectroscopy* 1996, *50*(7), 823–828.

53. Butler, J. R.; Wills, J. B.; Mitchem, L.; Burnham, D. R.; McGloin, D.; Reid, J. P. Spectroscopic characterisation and manipulation of arrays of sub-picolitre aerosol droplets. *Lab on a Chip* 2009, *9*(4), 521–528.

54. Mitchem, L.; Buajarern, J.; Ward, A. D.; Reid, J. P. A strategy for characterizing the mixing state of immiscible aerosol components and the formation of multiphase aerosol particles through coagulation. *Journal of Physical Chemistry B* 2006, *110*(28), 13700–13703.

55. Buajarern, J.; Mitchem, L.; Reid, J. P. Characterizing multiphase organic/inorganic/aqueous aerosol droplets. *Journal of Physical Chemistry A* 2007, *111*(37), 9054–9061.

56. Mitchem, L.; Hopkins, R. J.; Buajarern, J.; Ward, A. D.; Reid, J. P. Comparative measurements of aerosol droplet growth. *Chemical Physics Letters* 2006, *432*(1–3), 362–366.

57. Buajarern, J.; Mitchem, L.; Reid, J. P. Characterizing the formation of organic layers on the surface of inorganic/aqueous aerosols by Raman spectroscopy. *Journal of Physical Chemistry A* 2007, *111*(46), 11852–11859.

58. Buajarern, J.; Mitchem, L.; Reid, J. P. Manipulation and characterization of aqueous sodium dodecyl sulfate/sodium chloride aerosol particles. *Journal of Physical Chemistry A* 2007, *111*(50), 13038–13045.

59. Laurain, A. M. C.; Reid, J. P. Characterizing internally mixed insoluble organic inclusions in aqueous aerosol droplets and their influence on light absorption. *Journal of Physical Chemistry A* 2009, *113*(25), 7039–7047.

60. King, M. D.; Thompson, K. C.; Ward, A. D. Laser tweezers Raman study of optically trapped aerosol droplets of Seawater and oleic acid reacting with ozone: Implications for cloud-droplet properties. *Journal of the American Chemical Society* 2004, *126*(51), 16710–16711.

61. King, M. D.; Thompson, K. C.; Ward, A. D.; Pfrang, C.; Hughes, B. R. Oxidation of biogenic and water-soluble compounds in aqueous and organic aerosol droplets by ozone: A kinetic and product analysis approach using laser Raman tweezers. *Faraday Discussions* 2008, *137*, 173–192.

62. Lin, H. B.; Campillo, A. J. Radial profiling of microdroplets using cavity-enhanced Raman spectroscopy. *Optics Letters* 1995, *20*(15), 1589.

63. Homer, C. J.; Jiang, X.; Ward, T. L.; Brinker, C. J.; Reid, J. P. Measurements and simulations of the near-surface composition of evaporating ethanol–water droplets. *Physical Chemistry Chemical Physics* 2009, *11*, 7780–7791.

64. Hopkins, R. J.; Howle, C. R.; Reid, J. P. Measuring temperature gradients in evaporating multicomponent alcohol/water droplets. *Physical Chemistry Chemical Physics* 2006, *8*(24), 2879–2888.

65. Thurn, R.; Kiefer, W. Structural resonances observed in the Raman-spectra of optically levitated liquid droplets. *Applied Optics* 1985, *24*(10), 1515–1519.

66. Lettieri, T. R.; Preston, R. E. Observation of sharp resonances in the spontaneous Raman-spectrum of a single optically levitated microdroplet. *Optics Communications* 1985, *54*(6), 349–352.

67. Hargreaves, G.; Kwamena, N.-O. A.; Zhang, Y.-H.; Butler, J. R.; Rushworth, S.; Clegg, S. L.; Reid, J. P. Measurements of the equilibrium size of supersaturated aqueous sodium chloride droplets at low relative humidity using aerosol optical tweezers and an electrodynamic balance. *Journal of Physical Chemistry A* 2010, *114*, 1806–1815.

68. Butler, J. R.; Mitchem, L.; Hanford, K. L.; Treuel, L.; Reid, J. P. *In situ* comparative measurements of the properties of aerosol droplets of different chemical composition. *Faraday Discussions* 2008, *137*, 351–366.

69. Hanford, K. L.; Mitchem, L.; Reid, J. P.; Clegg, S. L.; Topping, D. O.; McFiggans, G. B. Comparative thermodynamic studies of aqueous glutaric acid, ammonium sulfate and sodium chloride aerosol at high humidity. *Journal of Physical Chemistry A* 2008, *112*(39), 9413–9422.

70. Buajarern, J.; Mitchem, L.; Ward, A. D.; Nahler, N. H.; McGloin, D.; Reid, J. P. Controlling and characterizing the coagulation of liquid aerosol droplets. *Journal of Chemical Physics* 2006, *125*(11), 114506.

7 Raman Spectroscopy of Single Particles Levitated by an Electrodynamic Balance for Atmospheric Studies

Alex K. Y. Lee and Chak K. Chan

CONTENTS

7.1   Introduction 156
7.2   Experiment and Theory 159
    7.2.1   Generation of Solution Droplets 159
    7.2.2   The Design of the Electrodynamic Balance 160
    7.2.3   Hygroscopic Measurements 160
    7.2.4   Particle Mass Yield Measurements due to Reactions 161
    7.2.5   Raman Spectroscopy of Single Levitated Particles 161
    7.2.6   Morphology and Light Scattering Pattern Measurements 163
7.3   Single-Particle Raman Spectroscopy Applications 163
    7.3.1   Partial Crystallization and Deliquescence of $(NH_4)_2SO_4$–Dicarboxylic
           Acid Particles 163
        7.3.1.1   Hygroscopic Measurements of the $(NH_4)_2SO_4$–Malonic Acid
                (AS/MA) System 164
        7.3.1.2   Raman Characterization and fwhh Analysis 166
        7.3.1.3   Comparison to E-AIM Predictions 166
    7.3.2   Hygroscopicity of Organically Coated Inorganic Seed Particles 167
        7.3.2.1   Water-Insoluble Organic Coating: Octanoic Acid 169
        7.3.2.2   Water-Soluble Organic Coating: Glutaric Acid 171
    7.3.3   Formation of Metastable Salts from Crystallization of Solution Droplets 174
        7.3.3.1   Formation of Metastable Double Salts: $(NH_4)_2SO_4$–NH_4NO_3
                Mixed Particles 175
        7.3.3.2   Transformation of $3AN \cdot AS$ to $2AN \cdot AS$ and its
                Dependence on RH 177
    7.3.4   Heterogeneous Reactions of Organic Aerosol Particles 179
        7.3.4.1   Raman Characterization of Ozone-Processed Unsaturated
                Fatty Acid Particles 180
        7.3.4.2   Autoxidation and Ozonolysis of the Autoxidation Products 180
        7.3.4.3   Changes in Particle Mass Yield Caused by Heterogeneous Reactions 184
        7.3.4.4   Effects of Ozone Concentration on Production Formation 185
7.4   Summary and Prospective 185
Acknowledgments 187
References 187

7.1 INTRODUCTION

Aerosol particles are ubiquitous in the Earth's atmosphere. They arise from natural sources, such as windborne dust, sea spray, and volcanic action; from anthropogenic activities, such as fossil fuel and biomass combustion; and from reactions between gaseous emissions [1]. Atmospheric aerosols have been implicated in numerous regional problems, including adverse human health effects, poor air quality, visibility degradation, and acid deposition [1–5]. On the global scale, they not only have a direct effect on modifying the radiative balance of the atmosphere by scattering and/or absorbing solar radiation, but also act as cloud condensation nuclei (CCN), which modify the physics and chemistry and precipitation rates of clouds, consequently influencing the global climate [6,7]. A good understanding of the composition and properties of atmospheric aerosols as well as their related atmospheric processes is therefore particularly important for air quality and global climate predictions.

The Earth's atmosphere is like a huge chemical reactor in which atmospheric aerosols can physically and chemically interact continuously with any gaseous species in the surrounding air throughout their atmospheric lifetimes. The properties (e.g., size, shape, phase, composition, hygroscopicity, reactivity, etc.) of atmospheric aerosols are therefore always changing once they are suspended in the ambient environment [8–10]. Over the last few decades, many experimental strategies have been developed to investigate gas–particle interactions of atmospheric interest in laboratory studies. Common approaches include using a coated wall flow tube, an aerosol flow tube, and an aerosol chamber coupled with different types of state-of-the-art instruments for characterizing the average properties of aerosol ensembles, such as their size distribution, hygroscopicity, and CCN activity. To monitor the composition of aerosols, offline chromatographic techniques are often employed to characterize filter-based aerosol ensembles, whereas state-of-the-art aerosol mass spectrometers (AMS) and optical spectroscopy allows the analysis of the composition of individual particles with high time resolution.

While the aerosol ensemble analysis is useful for gaining the whole picture and for gathering overall kinetic and mechanistic information of various atmospheric processes and phenomena, single-particle levitation techniques with *in situ* nonintrusive optical spectroscopy have played a very unique and vital role in investigating the microphysics and chemistry of atmospherically relevant aerosol particles over the past few decades. Single particles, in liquid, solid, or mixed phases, can be stably trapped by optical or electrostatic levitation methods. The details of the working principles and designs of these methods have been explicitly documented by Davis and Schweiger [11]. In particular, the electrostatic approach allows accurate measurements of relative particle mass changes in response to the surrounding air by simply adjusting the strength of the electric fields required to hold the levitated particles stationary. This feature makes the levitation system act as a very sensitive picobalance or a so-called electrodynamic balance (EDB), allowing the investigation of any physical uptake processes and gas–particle interactions that would cause particle mass modifications under well-controlled environments [12–14]. Another distinct advantage of the EDB is that a single particle can be levitated inside the trap for an extended period of time (days), facilitating the investigation of various atmospheric processes and phenomena involving kinetic effects (e.g., mass transfer limitation and low reaction rate) and repeated experimental cycles [13,15–19].

Hygroscopicity is one of the properties of aerosols, closely related to the light-scattering efficiency of atmospheric aerosols as their size, phase, and composition changes when they are equilibrated at different relative humidities (RH). By using the levitation system alone to record particle mass, the EDB has been demonstrated to be a very reliable tool for measuring the phase transformations (i.e., deliquescence and crystallization) and hygroscopic growth of aerosols, including inorganic salts and organic compounds, as well as their mixtures of atmospheric interest. Table 7.1 summarizes the measurements of the phase transformations and hygroscopic growth of numerous single organic component and mixed organic aerosol particles measured by our group. Since the levitated liquid droplet is not in contact with any foreign surface, heterogeneous nucleation of

TABLE 7.1

Phase Transformation and Hygroscopicity of Different Single Component and Mixed Organic/Inorganic Particles of Atmospheric Interest Measured by our EDB

Chemical Species	Study (Ref.)	Instrument	Remarks
Adipic acid	Chan et al. [25]	EDB	
	Yeung et al. [26]	EDB/Raman	In mixture with $(NH_4)_2SO_4$ only
Azelaic acid	Chan et al. [25]	EDB	
Citric acid	Peng et al. [13]	EDB	UNIFAC comparison and parameterization
	Choi and Chan [12]	SEDB	In mixture with $(NH_4)_2SO_4$ and NaCl only
Fulvic acid	Chan and Chan [27]	EDB	Nordic Aquatic References and Suwannee River, also in mixtures with $(NH_4)_2SO_4$ and NaCl
Glutaric acid	Peng et al. [13]	EDB	UNIFAC comparison and parameterization
	Choi and Chan [12]	SEDB	In mixture with $(NH_4)_2SO_4$ and NaCl only
	Choi and Chan [20]	SEDB	Also in mixture with MA
	Chan et al. [17]	SEDB, EDB/Raman	Coating on $(NH_4)_2SO_4$ solid particles
	Ling and Chan [28]	EDB/Raman	In mixture with $(NH_4)_2SO_4$ only
Glyerol	Choi and Chan [12]	SEDB	In mixture with $(NH_4)_2SO_4$ and NaCl only
	Choi and Chan [20]	SEDB	
Glyoxylic acid	Chan et al. [25]	EDB	UNIFAC comparison and Kappa prediction
Maleic acid	Choi and Chan [20]	SEDB	Also in mixture with malic acid
Malic acid	Peng et al. [13]	EDB	UNIFAC comparison and parameterization
	Choi and Chan [20]	SEDB	Also in mixture with maleic acid
MA	Peng et al. [13]	EDB	UNIFAC comparison and parameterization
	Choi and Chan [12]	SEDB	In mixture with $(NH_4)_2SO_4$ and NaCl only
	Choi and Chan [20]	SEDB	Also in mixture of glutaric acid
	Ling and Chan [28]	EDB/Raman	In mixture with $(NH_4)_2SO_4$ only
4-Methylphthalic acid	Chan et al. [25]	EDB	UNIFAC comparison and Kappa prediction
Octanoic acid	Chan and Chan [18]	SEDB	Coat on $(NH_4)_2SO_4$ solid particles
Oleic acid	Lee and Chan [15]	EDB/Raman	Also ozone-processed oleic acid
Oxalic acid	Peng et al. [13]	EDB	UNIFAC comparison and parameterization
Pinonic acid	Chan et al. [25]	EDB	Kappa prediction
Pimelic acid	Chan et al. [25]	EDB	UNIFAC comparison and Kappa prediction
Succinic acid	Peng et al. [13]	EDB	UNIFAC comparison and parameterization
	Choi and Chan [12]	SEDB	In mixture with $(NH_4)_2SO_4$ and NaCl only
	Ling and Chan [28]	EDB/Raman	In mixture with $(NH_4)_2SO_4$ only
Tartaric acid	Peng et al. [13]	EDB	UNIFAC comparison and parameterization
Organic salts	Peng and Chan [29]	EDB	Include sodium: formate, acetate, succinate, pyruvate, malonate, maleate, methanesulfonate, oxalate, and ammonium oxalate
Amino acid	Chan and Chan [30]	EDB, SEDB	Include glycine, alanine, serine, glutamine, threonine, arginine, and asparagines, UNIFAC comparison and parameterization
Biomass burning derived organics	Chan and Chan [30]	EDB	Include levoglucosan, mannosan, and galactosan
Monosaccharide	Chan et al. [25]	EDB	Include fructose, mannose, and glucose, UNIFAC comparison, and Kappa prediction
Disaccharide	Chan et al. [25]	EDB	Include maltose and lactose, UNIFAC comparison, and Kappa prediction
Poly-unsaturated fatty acids	Lee and Chan [16]	EDB/Raman	Include linoleic and linolenic acid, also ozone-processed linoleic and linolenic acid

supersaturated solutions can be suppressed, allowing the levitated droplet to reach a very high degree of supersaturation before crystallization takes place. In addition to the static approach, the recently developed scanning EDB (SEDB) technique allows quick water activity measurements of levitated droplets within an hour, which facilitates hygroscopic measurements of semivolatile organic species [12,20] and studies of the influence of organic coatings on the kinetics and mass transfer accompanying hygroscopic growth of inorganic salts [17,18]. The hygroscopic properties of aerosols measured by the EDB have been applied to modify the interaction parameters of the universal functional activity coefficient (UNIFAC) method, to provide data for aerosol thermodynamic models (e.g., the Extended Aerosol Inorganic Model (E-AIM)) for improving the accuracy of aerosol property, air quality, and global climate predictions [21–23] and to calculate Kappa values for predicting CCN activities [24,25].

The typical hygroscopic growth and evaporation curves of a pure ammonium sulfate (AS) particle that can be generated from EDB measurements are shown in Figure 7.1 as an illustration. In growth measurements, a $(NH_4)_2SO_4$ particle is in a solid state at very low RH (e.g., 5% RH). As the RH increases, the particle remains solid (i.e., $m/m_0 \approx 1$) until the RH reaches 80% RH at which point the solid particle spontaneously absorbs water, indicated by the sudden increase in m/m_0. The RH at which the solid-to-liquid phase transition occurs is known as the deliquescence relative humidity (DRH). A further increase of the RH leads to additional water condensation onto the particle to maintain thermodynamic equilibrium. In evaporation measurements, the m/m_0 of the $(NH_4)_2SO_4$ particle decreases continuously as the RH decreases owing to evaporation of the water. However, the particle does not crystallize at the DRH but exists as a supersaturated solution droplet. The particle subsequently crystallizes when the RH further decreases to about 35% RH. The RH at which the liquid-to-solid phase transition occurs is known as the crystallization relative humidity (CRH). It is important to note that some particles do not exhibit phase transitions as the RH changes (e.g., H_2SO_4). These noncrystallizing particles exist in liquid state and absorb and desorb water reversibly as the RH increases and decreases, respectively.

Although both static EDB and SEDB approaches are excellent for hygroscopic measurements and investigation of the phase transformations of levitated aerosol particles, some observed hygroscopic phenomena are somewhat difficult to explain if only the particle mass data are available. For example, Chan et al. [31] observed a significant reduction in the evaporation rate of highly concentrated magnesium sulfate droplets. They attributed this reduction to gel formation inside the droplets, but they did not provide any molecular-level information to support their hypothesis. Choi and Chan [12] investigated the effects of organic components on the hygroscopic growth of inorganic

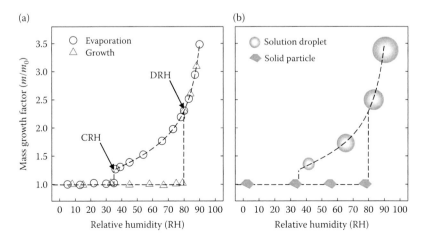

FIGURE 7.1 (a) Hygroscopic growth and evaporation curves of pure AS particles. (b) The phase state of AS particles as a function of RH (as a percentage) during the hygroscopic cycle.

aerosols. They reported that both NaCl– and $(NH_4)_2SO_4$–malonic acid (MA) mixtures absorb a significant amount of water before their respective DRH, because MA absorbs water reversibly without crystallization. Even though they provided a very reasonable explanation, the evolution of the phases of MA and the inorganic components in response to the ambient RH were not experimentally verified.

The combination of EDB and *in situ* nonintrusive single-particle optical spectroscopy improves the application of EDB to hygroscopic measurements and other potential research areas, because the spectroscopic information allows researchers to gain a better understanding of the physics and the chemistry of aerosol particles from a microscopic point of view. Rapid acquisition of Mie-scattering data together with the interpretation of resonance Raman spectra have made it possible to study numerous evaporation and condensational processes as well as determine the thermodynamic and transport properties of the evaporating species using particle levitation devices [32–35]. Furthermore, many researchers have used spontaneous Raman spectroscopy to study the morphology-dependent resonances, phase transformations, chemical reactions, and molecular interactions, especially under supersaturated conditions, of levitated particles [36–44]. Stimulated Raman spectroscopy can be used to characterize the chemical composition of the near-surface region and to determine the changes in droplet size with nanoscale accuracy, allowing high-quality investigations of mass transfer processes (e.g., condensation or evaporation of water vapor) [11,45]. In addition to Raman characterization, laser-induced fluorescence spectroscopy has been recently used to probe pH values [46] and to examine solute–water interactions, such as the relative abundance of solvated water and free water and efflorescence of levitated solution droplets [47,48].

While most of the previous studies employed single-particle Raman spectroscopic techniques to study transport processes and physiochemical properties of the levitated particles, in this chapter, we present some typical examples to demonstrate how the combination of the EDB with single-particle Raman spectroscopy (EDB/Raman) can be used as a chemical tool to improve our understanding of the microphysics and chemistry of aerosol particles, including their phase transformations, hygroscopic properties, molecular interactions, and chemical reactions, which are related to atmospheric applications. Four specific topics are discussed. They are (1) partial crystallization and deliquescence of $(NH_4)_2SO_4$–dicarboxylic acid particles; (2) hygroscopicity of organically coated inorganic seed particles; (3) formation of metastable salts from the crystallization of solution droplets and; (4) heterogeneous reactions of organic aerosol particles. More examples of the use of the EDB/Raman system for atmospheric applications can be found in some of our previous publications [43,49,50].

7.2 EXPERIMENT AND THEORY

In this section, the method used to generate solution droplets, the design of the EDB and its working principle for conducting hygroscopic measurements as well as mass yield determination owing to reactions, the design of the EDB/Raman system, and the method to examine the morphology of a levitated particle, are presented. The motivation for and relevant experimental details of each example will be presented separately in their corresponding sections.

7.2.1 Generation of Solution Droplets

To generate solution droplets, a stock solution with the desired chemical composition is first prepared by dissolving the solute into either water or organic solvent, depending on its solubility. A small amount of this solution is introduced into a piezoelectric droplet generator. By applying electric pulses to the piezoelectric droplet generator, solution droplets with a diameter of a few tens of micrometers are generated and subsequently charged by passing them through a metal induction plate before they enter into the EDB. A charged solution droplet can be levitated and kept stationary with proper adjustment of the AC and DC electric fields inside the EDB as described below.

7.2.2 THE DESIGN OF THE ELECTRODYNAMIC BALANCE

The development, applications, and some common configurations of the EDB used by various research groups have been reviewed by Davis and Schweiger [11]. A schematic diagram of the EDB used in this study is shown in Figure 7.2. The EDB comprises three electrodes insulated with each other. The middle part of the EDB is an AC ring electrode, which provides a time-varying force to force a charged particle to move toward the center of the balance. The two end caps made of metal are positioned symmetrically and are connected to the positive and negative DC electrodes, respectively. A charged particle can be balanced against gravity by a DC potential established between the two end-cap electrodes. Hence, by adjusting the AC and DC potentials applied to the electrodes, a charged particle can be levitated and held stationary at the null point (the geometric center) of the EDB.

When a levitated particle is held stationary at the geometric center of the EDB and there is no drag force acting on the balanced particle, the electrostatic force acting on the particle imposed by the DC potential is equal to the weight of the particle. As a result, the particle mass can be determined as

$$m = \frac{C_0 q}{g Z_0} V, \tag{7.1}$$

where m is the particle mass, C_0 is the balance calibration constant, q is the charge on the particle, g is the gravitational constant, Z_0 is the distance between the top and the bottom electrodes, and V is the DC voltage required for balancing the particle at the geometric center of the balance. Assuming that there is no loss or gain of charge, the mass of the particle is directly proportional to the applied DC voltage. The direct proportionality of the particle mass to the DC voltage forms the basis for the use of the EDB as an analytical balance, in addition to being a levitation tool.

7.2.3 HYGROSCOPIC MEASUREMENTS

Static hygroscopic measurements involve the equilibrium mass measurements of a particle under stepwise changes of the RH. In evaporation measurements, a particle is initially equilibrated at a high RH (e.g., 85% RH). After recording the balancing DC voltage, the particle is re-equilibrated at a lower RH. The RH is decreased in discrete steps for consecutive runs and the balancing DC voltage at each RH is recorded. The same procedure is applied to growth measurements in which case a particle is initially equilibrated at a low RH (e.g., 5% RH) and the RH is increased in discrete steps. When deliquescence or crystallization occurs, a particle absorbs or loses water abruptly, resulting in a relatively sharp change in the particle mass between the two RH values.

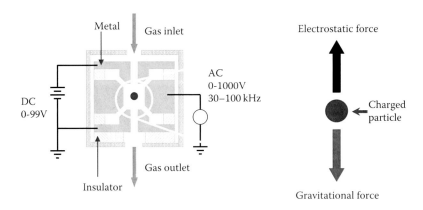

FIGURE 7.2 A schematic diagram of EDB.

At equilibrium, the mass of the particle, $m(RH)$, at a given RH can be determined by measuring the balancing DC voltage, $V(RH)$, using Equation 7.1. It is noted that the values of C_0, q, Z_0, and g are constant and are needed if the absolute particle mass is to be determined. In hygroscopic measurements, the particle mass change is attributed only to water absorption or evaporation in response to the RH changes in the EDB by assuming that the evaporative loss of the solutes is negligible. Hence, the relative mass change in particles due to evaporation or condensation of water is proportional to the balancing DC voltage before and after the RH change and can be determined from

$$\frac{m(RH_1)}{m(RH_2)} = \frac{V(RH_1)}{V(RH_2)}, \tag{7.2}$$

where $m(RH_1)$, $m(RH_2)$, $V(RH_1)$, and $V(RH_2)$ are the particle masses and balancing DC voltages at RH_1 and RH_2, respectively.

When a droplet is equilibrated with the surrounding environment inside the EDB, the water activity (a_w) of the droplet is related to the RH as follows: $a_w = P_w/P_w^{Sat} = RH/100$, where P_w is the partial pressure of water in the ambient environment and P_w^{Sat} is the saturation water vapor pressure at the ambient temperature. The Kelvin effect to correct for vapor pressure dependence from the curvature of a droplet can be ignored, since the particles are larger than 10 μm in diameter. The molar water-to-solute ratio (WSR) or the mass fraction of the solute (*mfs*) of the droplets is changed by altering the ambient RH, and both are determined by the DC balancing voltage measurements [13]. The experimental results can also be presented in the form of the mass ratio (m/m_0), which is defined as the ratio of the particle mass, m, at a given RH to the particle mass, m_0, at the reference RH ($RH_0 < 5\%$), when the composition of the particles is highly uncertain and/or no bulk data can be used for identification of the reference state [14,16,27]. In some cases, the *mfs* and mass ratio data are converted to a growth factor (G_f), which is defined as the diameter of particles at a given RH to the diameter of particles at the reference RH [18,30], for comparison with data obtained from measurements made with a Hygroscopic Tandem Differential Mobility Analyzer (H-TDMA).

The RH inside the EDB is adjusted by mixing a stream of saturated air and dry air at controlled flow rates. The RH of an airstream is determined by measuring its dew point with a dew point hygrometer (EdgeTech Dew Prime I) and the ambient temperature. The overall experimental error of the mass ratios obtained from the hygroscopic measurements is within 1% for droplets, and the error in the determination of RH is estimated to be ±1% at RH = 40–80%.

7.2.4 PARTICLE MASS YIELD MEASUREMENTS DUE TO REACTIONS

In addition to performing hygroscopic measurements, the EDB is an ideal tool to directly measure relative mass changes in the levitated particles due to the addition of gas-phase reactants or evaporative loss of particle-phase products via chemical reactions under a constant RH environment [14–16]. In our earlier work, a levitated particle was first equilibrated at a specific RH and the balancing DC voltage was recorded. Afterward, the gas-phase reactant was fed into the EDB while the RH inside the EDB was kept constant. The balancing DC voltage was recorded at several time intervals (e.g., 1–2 h) over the whole period of the gas-phase reactant exposure (e.g., 20 h). The experimental results are presented in the form of mass ratios (m_t/m_0), which is the ratio of the particle mass at a given gas-phase reactant exposure time, m_t, to the initial particle mass, m_0. Hence, the time-series profile of the relative particle mass changes due to chemical reactions can be obtained.

7.2.5 RAMAN SPECTROSCOPY OF SINGLE LEVITATED PARTICLES

Taking advantage of the distinct ability to levitate a single particle stationary in the EDB, Raman scattering of the levitated particle can be measured for aerosol phase and composition analyses. In our works, a Raman spectroscopy system consisting of a 5 W argon ion laser (Coherent I90–5) and a 0.5 m monochromator (Acton SpectraPro 500) attached to a charge-coupled detector (CCD) (Andor Technology DV420-OE) was integrated with the EDB system. The 514.5 nm line of an argon ion laser with output power between 25 and 50 mW was used as the source of excitation. A schematic diagram of the EDB/Raman system used in our work is shown in Figure 7.3a. A pair of lenses matching the f/7 optics of the monochromator was used to focus the 90° scattering of the levitated droplet in the EDB onto the slit of the monochromator. A 514.5 nm Raman notch filter was placed between the two lenses to remove the strong Rayleigh scattering. A 300 g/mm grating of the monochromator was selected. The resolution of the spectra obtained was about 6 cm^{-1}. All measurements were made at ambient temperatures of 22–24°C.

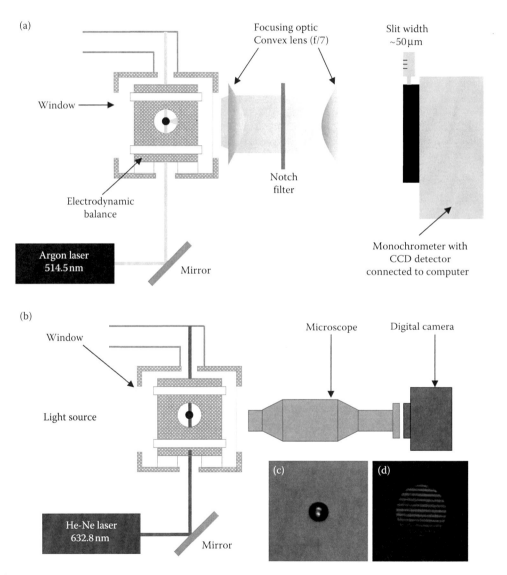

FIGURE 7.3 (a) A schematic diagram of the EDB/Raman system. (b) The experimental setup used for capturing morphology and light-scattering patterns of levitated particles. (c) An image of a $(NH_4)_2SO_4$ droplet. (d) A light-scattering pattern from a $(NH_4)_2SO_4$ homogeneous droplet.

7.2.6 Morphology and Light Scattering Pattern Measurements

The morphology and laser-illuminated light-scattering pattern of the levitated particles can be monitored and captured through the window of the EDB using a microscope (5× objective and 20× eyepiece) coupled with a digital camera (Nikon Cooplix 990). To capture the images of the levitated particle, we placed the light source at the opposite side or at an angle of 90° with respect to the microscope so that the particle's shape could be recorded by the digital camera via the microscope (Figure 7.3b, c). The phase of each particle was examined by analyzing the laser-illuminated light-scattering pattern from the levitated particle [51,52]. The laser beam (632.8 nm) illuminated the levitated particle from the bottom of the EDB and the microscope was positioned at an angle of 90° with respect to the laser beam. The light-scattering image was first focused so that two clear light spots could be observed via the eyepiece of the microscope. After that, the eyepiece was removed to obtain the laser-illuminated light-scattering pattern (Figure 7.3d). If the levitated particle was a homogeneous liquid droplet, then the light-scattering pattern consisted of a parallel fringe structure (Mie scattering). In contrast, a solid levitated particle resulted in highly irregular and fluctuating patterns because of the rotation of the particle. If the levitated particles contained solid inclusions, the fringe became noisy because of the combination of the Mie spectrum and the random scattering from the solid inclusions. Parsons et al. [52] and Olsen et al. [53] used a similar experimental approach to study the nucleation of levitated droplets.

7.3 SINGLE-PARTICLE RAMAN SPECTROSCOPY APPLICATIONS

7.3.1 Partial Crystallization and Deliquescence of $(NH_4)_2SO_4$–Dicarboxylic Acid Particles

Dicarboxylic acids are the most abundant water-soluble organics present in atmospheric aerosols and they have been shown to have significant impacts on the phase transformation, hygroscopicity, and optical properties of inorganic aerosols such as NaCl and $(NH_4)_2SO_4$ [12,52,54–60]. Because of their atmospheric importance, some thermodynamic models have been developed to understand the mixtures of inorganics with dicarboxylic acids [22,23,61]. In the evaporation branch of the hygroscopic measurements, their water content can be reasonably estimated by a number of models if homogeneous supersaturated droplets are assumed. In contrast, the water content of partially deliquesced particles with increasing RH is not easy to predict because of deficiencies in our understanding of the physical properties of multicomponent systems and of experimental data for fitting the correction parameters. In some cases, we must assume the sequence of dissolution and the initial phase for each species to fit the experimental data [22]. A better understanding of the phase of each species in a multicomponent system as a function of RH is therefore essential for the accurate parameterization of partial deliquescence in thermodynamic models.

The measurement of particle mass alone can accurately measure the amount of water present in levitated particles during the partial crystallization and deliquescence processes, whereas a laser-illuminated light-scattering pattern allows us to study solid inclusions in aqueous droplets and the morphology of complex particles. However, these measurements still do not provide any information about the phase (aqueous vs. solid) of individual components inside the mixed particles. To supplement the information gained from the EDB and light-scattering pattern measurements of the partially crystallized/deliquesced particles, the phase of individual species can be determined from the full-width-at-half-height (fwhh) of the Raman peaks for each species. Because the Raman peaks of solid phases are sharper than that of the same species in aqueous form, relative changes in fwhh served as a very useful indicator of the phase transition. Lee et al. [43] recently demonstrated the use of fwhh analysis of the Raman peaks to characterize the partial deliquescence of NH_4NO_3–$(NH_4)_2SO_4$ mixed particles (see Figure 7.4 for details). Ling and Chan [28] attempted to extend the Raman investigation of the phase transition and hygroscopicity of multicomponent particles to

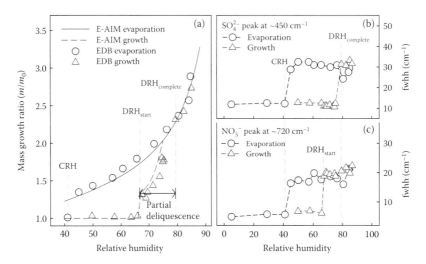

FIGURE 7.4 (a) Hygroscopicity of NH_4NO_3-$(NH_4)_2SO_4$ (1:1 molar ratio) mixed particles measured by EDB. (b) Dependence of fwhh of a SO_4^{2-} peak at 450 cm^{-1} on RH; (c) Dependence of fwhh of a NO_3^- peak at 720 cm^{-1} on RH. The particles crystallize at ~40% RH and start to absorb water (DRH_{start}) at ~67% and completely deliquesce ($DRH_{complete}$) at ~79%. During partial deliquescence, the fwhh analysis shows that NO_3^- first dissolved, but most of SO_4^{2-} remained in solid phase at 67% RH and then gradually dissolved when RH reached 79%. (Adapted from Lee, A. K. Y., Ling, T. Y., and Chan, C. K. 2008. *Faraday Discussions* 137: 245–263. With permission.)

systems with both inorganics and organics. They applied a similar Raman spectroscopic analysis to examining the partial crystallization and deliquescence behavior of a few $(NH_4)_2SO_4$-dicarboxylic acid mixed particles. In this section, the results from the $(NH_4)_2SO_4$-malonic acid (mole ratio = 1:1) system measured by Ling and Chan [28] are presented in detail as an illustration of the method. More examples and detailed discussions of partial crystallization and deliquescence of $(NH_4)_2SO_4$-dicarboxylic acid mixtures can be found in Ling and Chan [28] and Yeung et al. [26].

7.3.1.1 Hygroscopic Measurements of the $(NH_4)_2SO_4$–Malonic Acid (AS/MA) System

Malonic acid (MA) is a noncrystallizing, water-soluble dicarboxylic acid [13] and its influence on the phase transition and hygroscopicity of NaCl and $(NH_4)_2SO_4$ have been investigated using the SEDB technique [12]. Choi and Chan [12] reported that MA in AS/MA particles did not crystallize and the mixture became anhydrous ($m/m_0 = 1$) at RH < 20%. Because of the noncrystallizing MA, the AS/MA particles could absorb a significant amount of water reversibly prior to the DRH, which is defined as the point at which the particles are completely deliquesced, of pure AS and MA. It is important to note that this observation was not experimentally verified at the molecular level as the evolution of the phase of each species in response to the ambient RH could not be identified based on the SEDB data alone.

Figure 7.5 shows the relative mass changes of AS/MA particles undergoing evaporation and growth experiments using the static EDB approach. The measurements show a gradual loss of water with decreasing RH, without any sudden decrease in mass. The results are in general comparable with those from Choi and Chan [12], except that abrupt crystallization was not observed during evaporation. Although the gradual reduction of the particle mass did not confirm if crystallization had taken place, the elastic light-scattering pattern did change as the RH decreased from 20% to 16% (Figure 7.6), indicating the occurrence of phase changes between the two RHs. The regular pattern at 20% RH suggests that the AS/MA particle remained as a homogeneous supersaturated aqueous droplet. On the contrary, the distorted fringe pattern observed at 16% RH indicates the

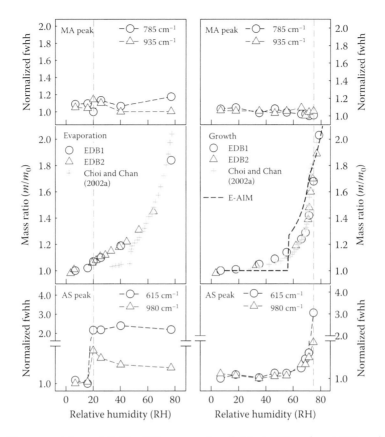

FIGURE 7.5 Hygroscopicity and fwhh of Raman peaks of AS at 615 cm^{-1} and 980 cm^{-1} and of MA at 785 cm^{-1} and 935 cm^{-1} for AS/MA particles undergoing evaporation (left panel) and growth (right panel). Errors in fwhh were estimated to be about 5–15%. (Adapted from Ling, T. Y. and Chan, C. K. 2008. *Journal of Geophysical Research-Atmospheres* 113.)

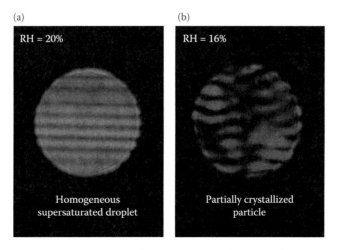

FIGURE 7.6 Elastic light-scattering pattern of an AS/MA particle undergoing evaporation at (a) 20% RH and (b) 16% RH. (Adapted from Ling, T. Y. and Chan, C. K. 2008. *Journal of Geophysical Research-Atmospheres* 113.)

presence of solid inclusions within the aqueous solution. Had the particle been completely crystal-lized, an irregular light pattern without any fringes, instead of the distorted fringe pattern, would have been observed. The results clearly show that partial crystallization of one (or more) species took place between 20% and 16% RH. The observed partial crystallization was also consistent with the mass measurements at 16% RH, where m/m_0 (=1.02) did not reach unity. Further reduction of RH continued to decrease the particle mass.

Upon increasing RH, partially crystallized AS/MA particles absorbed water gradually and con-tinuously, without any abrupt increase in mass (Figure 7.5). Distortion in the fringe pattern became less significant as the RH increased. Complete deliquescence was achieved at 79% RH, as con-firmed from the overlapping of the growth and evaporation m/m_0 data in the subsaturated region, as well as the restoration of regular fringe in the light-scattering pattern. The DRH reported here is slightly higher than the values of $DRH_{complete}$ reported by Choi and Chan [12] and Parsons et al. [52] (73.4% and 72.5%, respectively with similar particle compositions). $DRH_{complete}$ is the RH at which complete deliquescence occurs during hygroscopic growth. It should be noted that the RH was changed in steps in our experiments, so that the $DRH_{complete}$ actually fell somewhere between 75% (the last data point measured before complete deliquescence) and 79% (the first data point measured after complete deliquescence).

7.3.1.2   Raman Characterization and fwhh Analysis

Figure 7.7 shows a series of Raman spectra for an AS/MA particle undergoing a hygroscopic cycle as described in the previous section. The homogeneous AS/MA droplet evaporated down to 16% RH, followed by growth up to 79% RH. The fwhh analysis was performed on the AS peaks at 615 and 980 cm^{-1} and the MA peaks at 785 and 935 cm^{-1} (Figure 7.5). During evaporation, at 16% RH, there was a sudden drop in the fwhh of the AS peaks, and this suggested that crystallization of the AS had taken place. On the other hand, negligible changes in the fwhh of the MA peaks suggest that most MA remained in the aqueous solution. However, a detailed comparison of the single-particle spectrum taken at 16% RH and the bulk solid MA spectrum reveals some possible solid MA char-acteristics in the AS/MA particle at 16% RH: a shoulder at 1697 cm^{-1} and a double peak at ≈3000 cm^{-1} correspond to MA solids (Figure 7.8). Together with the absence of phase changes as suggested by the constant fwhh, it appears that the majority of the MA stayed in aqueous form, which had a larger value of fwhh than the corresponding solid peaks. The crystallized AS may have induced heteroge-neous crystallization of part of the supersaturated MA, although it has been reported that AS does not promote the nucleation of MA [59]. Further reduction of the RH did not lead to observable changes in the Raman characteristics.

In the growth experiment, while it is plausible that the residual aqueous MA contributed to the water uptake at low RH, deliquescence of the AS indeed gradually took place starting from 69% RH. The fwhh of AS increased slowly between 69% and 71% RH, indicating the presence of aqueous phase AS. A sharp increase in fwhh occurred at 75% RH, in accordance with the mass measurements when significant water uptake took place. The DRH of AS was lowered in the presence of MA, consistent with the findings reported by Braban and Abbatt [59] and Parsons et al. [52]. On the other hand, the minor double-peak MA feature persisted at 71% RH, and the feature disappeared when the particle completely deliquesced. This indicated that once MA crystallizes, it deliquesces at an RH higher than 71%, comparable to the DRH of 72.5% derived from thermodynamic predictions [22]. Furthermore, Clegg and Seinfeld [22] suggested that MA was not completely dissolved as they predicted more absorbed water than the measured growth data reported by Choi and Chan [12] for RH < 65%.

7.3.1.3   Comparison to E-AIM Predictions

Figure 7.5 depicts the water uptake of AS/MA particles as predicted by E-AIM during partial deli-quescence. The predictions of the RH at which the particles start taking up water (DRH_{start}) and completely deliquescence ($DRH_{complete}$) as well as the solid species present are compared with the experimental results as shown in Table 7.2. Apparently, E-AIM overpredicted the water uptake of

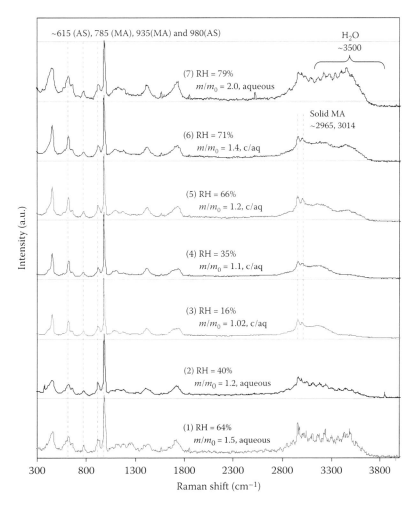

FIGURE 7.7   Raman spectra for an AS/MA particle undergoing evaporation (1–3) and growth (3–7). Spectra labeled "c/aq" refer to aqueous droplets containing solid crystals. (Adapted from Ling, T. Y. and Chan, C. K. 2008. *Journal of Geophysical Research-Atmospheres* 113.)

AS/MA particles. The difference in the identity of the solid phases present during partial deliquescence can account for the discrepancies in the amount of water uptake between model predictions and experimental results. In the AS/MA system, it was predicted that the organic acid completely dissolves at the DRH$_{start}$. This resulted in a larger amount of dissolved MA, which contributed to more water uptake prior to the complete deliquescence, compared to the experimental results with both solid MA and AS present during partial deliquescence. Table 7.2 also summarizes the partial deliquescence behaviors of $(NH_4)_2SO_4$–glutaric acid (AS/GA), $(NH_4)_2SO_4$–succinic acid (AS/SA), and $(NH_4)_2SO_4$–adipic acid (AS/AA) systems based on the fwhh analysis and the E-AIM predictions. A detailed discussion on the phase transformation and Raman characterization of these mixtures can be found in Ling and Chan [28] and Yeung et al. [26].

7.3.2   HYGROSCOPICITY OF ORGANICALLY COATED INORGANIC SEED PARTICLES

Organic coatings have been commonly found in atmospheric aerosols and the effects of organic coatings on the hygroscopicity of aerosol particles are usually assessed by comparing the

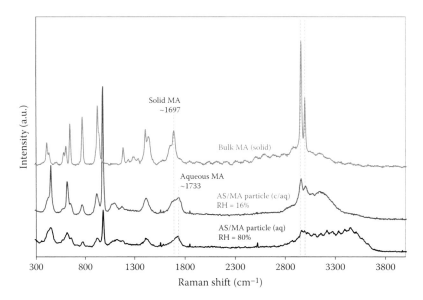

FIGURE 7.8 Comparison of the Raman spectra of a single AS/MA particle at 16% RH and 80% RH with bulk MA crystals. (Adapted from Ling, T. Y. and Chan, C. K. 2008. *Journal of Geophysical Research-Atmospheres* 113.)

hygroscopicity of the coated particles, which typically consist of a solid inorganic salt core (e.g., $(NH_4)_2SO_4$ and NaCl) and an organic coating, with that of uncoated inorganic salt particles [62–67]. In general, inorganic salts coated with hydrophobic (or water-insoluble) organics can still grow and deliquesce. Compared to inorganic salts, a lower hygroscopic growth of the coated particles per unit dry particle mass was observed at RH above the deliquescence point because of the negligible growth and association of water with the hydrophobic organic coating. However, although hydrophobic organic coatings have been found to retard the transport rate of water molecules from aerosol particles [8,68], the mass transfer effects of such organic coatings on the equilibrium time of laboratory-generated or atmospheric particles are not well characterized. On the other hand, water-soluble organic coatings on inorganic salt particles may dissolve into droplets once the inorganic

TABLE 7.2
Comparison of the DRH$_{Start}$, DRH$_{complete}$, and the Identity of the Solid Phases Observed from the Experiments and the Predictions from the E-AIM

	Raman Characterization			E-AIM		
	DRH$_{Start}$	DRH$_{Complete}$	Solid Phase	DRH$_{Start}$	DRH$_{Complete}$	Solid Phase
AS/MA[1]	NA[c]	79	AS, MA	57	72	AS
AS/GA[a]	70	80	AS, GA	71	75	AS
AS/SA[a]	80	>90	SA	79	97	SA
AS/AA[b]	70	80	AS, AA	NA[#]	80	AA

[a] Ling and Chan [28].

[b] Yeung et al. [26].

[c] DRH$_{Start}$ is not defined as the AS/MA particles remained partially crystallized at low RH, [#]DRH$_{Start}$ is not defined as partial deliquescence of AS/AA (3 and 10 wt% AA) particles cannot be predicted by E-AIM.

core deliquesces and hence the effects of water-soluble organic coatings on aerosol hygroscopicity can be more complicated than those of hydrophobic coatings.

Our group has successfully demonstrated the capability of the SEDB method for examining the mass transfer effects of organic coatings on the water uptake or evaporation rates of $(NH_4)_2SO_4$ particles. The details of the experimental procedures, such as the generation of organic coatings and RH–time calibration curves, can be found in Chan et al. [17] and Chan and Chan [18]. In the SEDB method, upon a step change in the RH in the feed stream to the EDB, the balancing voltage of the particles is continuously monitored and recorded as a function of time. The fractional change in the balancing voltage, $\delta = (V_i - V(t))/(V_i - V_f)$, as a function of time is analyzed. V_i and V_f are the initial and the final balancing voltages of the experiment, respectively, and $V(t)$ is the balancing voltage at time t. The effects of the organic coatings on the water transport rates can be investigated by examining the differences between the δ values of the solution droplet and the δ values of the respective calibration curves utilizing $CaCl_2$ droplets, $\Delta = \delta_{exp} - \delta_{cal}$. The SEDB measurements assume that the evaporation from and the condensation of water onto the trapped droplets are much faster than the change in RH in the EDB. If the Δ values significantly deviate from zero, the assumption of quasi-equilibrium is no longer held and a new rate-limiting step exists. This analysis only applies when the aerosol particles are solution droplets. In this section, the hygroscopic measurements of water-insoluble octanoic acid and highly water-soluble glutaric acid-coated $(NH_4)_2SO_4$ particles measured by Chan et al. [17] and Chan and Chan [18] are presented as illustrations.

7.3.2.1 Water-Insoluble Organic Coating: Octanoic Acid

$(NH_4)_2SO_4$ particles coated with different amounts of octanoic acid were subjected to RH cycling to effect two cycles of deliquescence and crystallization. Figure 7.9a shows that lightly coated particles (9 wt%) exhibited hygroscopicity and had DRH and CRH similar to those of pure $(NH_4)_2SO_4$ particles in the two cycles. The abrupt changes in the particle sizes were presumably due to deliquescence of the solid $(NH_4)_2SO_4$ cores and crystallization of $(NH_4)_2SO_4$ solution droplets. As shown in Figure 7.9d, upon increasing RH, the heavily coated particles (34 wt%) did not absorb any water until 83.2% RH, which was higher than the DRH of the $(NH_4)_2SO_4$ particles (80.1% RH). During evaporation, the particles decreased in size and remained as solution droplets and subsequently crystallized at 41.6% RH, which was lower than the CRH of $(NH_4)_2SO_4$ particles (45.6% RH) measured in this study. In both cases, the hygroscopic data were almost identical in the two cycles, indicating that most octanoic acid remained on the surface of $(NH_4)_2SO_4$ throughout the experiments.

Figure 7.9b, c shows that the δ values of the lightly coated $(NH_4)_2SO_4$ particles had trends similar to those of uncoated $(NH_4)_2SO_4$ particles. As depicted in Figure 7.9b, the sharp increase in δ at about 25 min signaled deliquescence and the Δ values were close to zero after the deliquescence point. In the crystallization experiments, the Δ values were within the experimental uncertainty prior to the crystallization point. These observations suggest that there was no retardation in the 9 wt% coated particles as compared to the rates of $(NH_4)_2SO_4$ particles in the two cycles. The SEDB method provided equilibrium measurements for the lightly coated particles. Nevertheless, the possibility of an incomplete coating on the solid cores and solution droplets cannot be ruled out.

Figure 7.9e shows that there was a significant delay of about 25 min between the abrupt increases in δ values of the 34 wt% coated $(NH_4)_2SO_4$ particles and that of the uncoated $(NH_4)_2SO_4$ particles. Since the deliquescence of uncoated particles is almost instantaneous, the delay for the coated particles was attributed to the mass transfer effects of water molecules from the gas phase to and through the coating to reach the solid $(NH_4)_2SO_4$ core. The characteristic time, τ, for the water molecules to diffuse through the octanoic acid coating was estimated to be 6.3×10^{-4} s by $\tau = R^2/(\pi^2 D)$, where R is the thickness of coating and D is the diffusivity of water in octanoic acid. However, this short diffusion time cannot explain the observed delay in deliquescence. After deliquescence, the δ values followed the trend of $(NH_4)_2SO_4$ particles (i.e., $\Delta \approx 0$), suggesting that the octanoic acid coating did not retard the growth rates of the $(NH_4)_2SO_4$ droplets in the timescale of our experiments.

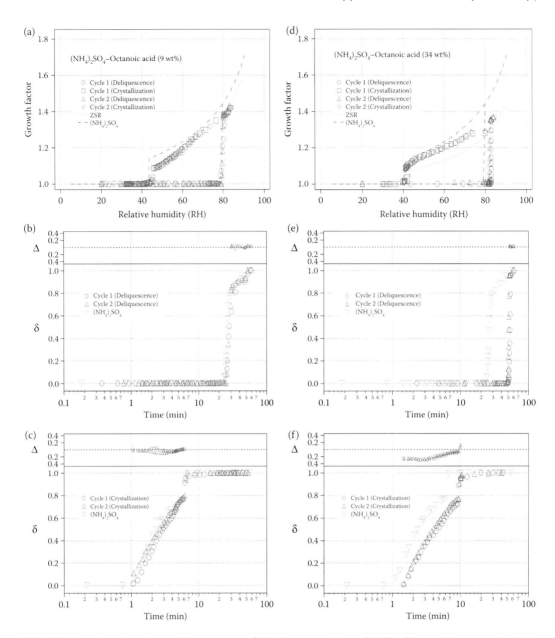

FIGURE 7.9 **(See color insert following page 206.)** Hygroscopicity of $(NH_4)_2SO_4$ particles coated with (a) 9 wt% and (d) 34 wt% octanoic acid in two deliquescence and crystallization cycles. The δ and Δ values of $(NH_4)_2SO_4$ particles coated with (b) 9 wt% and (e) 34 wt% octanoic acid as a function of time are shown from two deliquescence cycles. The δ and Δ values of $(NH_4)_2SO_4$ particles coated with (c) 9 wt% and (f) 34 wt% octanoic acid are shown as a function of time in two crystallization cycles. (Adapted from Chan, M. N. and Chan, C. K. 2007. *Atmospheric Environment* 41: 4423–4433.)

This was possibly because the coating became much thinner after deliquescence as the droplet size significantly increased and the amount of water transferred into the particle during droplet growth was much less than that during deliquescence. In the crystallization experiments (Figure 7.9f), the δ values of the 34 wt% coated particles were always smaller than their corresponding values of $(NH_4)_2SO_4$ particles (i.e., $\Delta < 0$), suggesting that the relatively thick octanoic acid coating impeded the evaporation rate of the water molecules compared to that of the uncoated particles.

On the basis of the δ and Δ analyses, 34 wt% octanoic acid coating retarded the transport rate of the water molecules and altered the deliquescence and crystallization points of the $(NH_4)_2SO_4$ particles. Chan and Chan [18] performed additional static EDB measurements with 18 and 30 wt% coated particles to confirm that the octanoic acid coating did not alter the DRH and CRH of the $(NH_4)_2SO_4$ particles, even with high organic mass, when sufficient time was allowed. Hence, the alteration in the observed DRH and CRH values was due to the kinetic limitation that occurred in the timescale of the SEDB method instead of the chemical interactions between octanoic acid and $(NH_4)_2SO_4$. In other words, the mass transfer of water molecules to/from and through the thick coating was slower than the change in RH in the EDB during the SEDB measurements. Furthermore, Chan and Chan [18] reported that the octanoic acid-coated particles had a water accommodation coefficient on the order of 10^{-3} and concluded that the presence of water-insoluble organic coatings can lower the mass accommodation coefficient of water and significantly affect the measured hygroscopicity of atmospheric particles.

7.3.2.2 WATER-SOLUBLE ORGANIC COATING: GLUTARIC ACID

Figure 7.10a, d illustrates the hygroscopicity of $(NH_4)_2SO_4$ particles coated with 13 and 49 wt% glutaric acid during two deliquescence and crystallization cycles, respectively using the SEDB method. In general, $(NH_4)_2SO_4$ particles with a thin coating had very similar hygroscopicity in the two cycles. The hygroscopic growth of the particles was comparable with that of $(NH_4)_2SO_4$ particles. For the heavily coated particles, the main difference in the water uptake behavior between the two cycles was that while freshly coated particles started to deliquesce at 80.9% RH, which was close to the DRH of $(NH_4)_2SO_4$ particles, the particles formed after the first measurement cycle absorbed a small amount of water at about 50% RH before the deliquescence and started to deliquesce at about 74% RH. The different deliquescence characteristics in the two cycles can be explained by the formation of homogeneously mixed $(NH_4)_2SO_4$–glutaric acid particles after the deliquescence in the first cycle, which was confirmed by Raman characterization and visual observations of the particle morphology in the static EDB experiment as shown below. The detailed description of the hygroscopicity of $(NH_4)_2SO_4$ particles both lightly and heavily coated with glutaric acid obtained by the SEDB method can be found in Chan et al. [17].

In the static EDB measurement, a pure solid $(NH_4)_2SO_4$ particle (Figure 7.11a) was first coated with 47 wt% glutaric acid. The coated particle was shaped like a snowflake (Figure 7.11b, c), suggesting that the coating was formed via condensation/coagulation of small glutaric acid particles. The Raman measurements clearly showed that the core region of the freshly coated particle mainly consisted of solid $(NH_4)_2SO_4$ with small glutaric acid signals (Figure 7.11b). When the position of the laser beam was adjusted to irradiate the coating, many strong Raman hydrocarbon signatures (2900–3100 cm^{-1}) were obtained, suggesting that the major component of the coating was glutaric acid (Figure 7.11c). It is interesting to note that the Raman spectrum of the coating differed from that of a solid glutaric acid particle (Figure 7.12c). The polymorphic forms of glutaric acid have been reported in the literature. The α- and β-form of glutaric acid have two and three strong hydrocarbon signals between 2900 and 3100 cm^{-1}, respectively. Therefore, the coating was in the α-form of glutaric acid, whereas the glutaric acid particles formed from crystallization of glutaric acid solution droplets was in the β-form. Investigation of the polymorphism of glutaric acid is underway in our laboratory.

Since the glutaric acid coating may have been highly porous or it may have incompletely covered the surface of the $(NH_4)_2SO_4$ particle, water molecules could penetrate the organic coating and reach the $(NH_4)_2SO_4$ directly, making the freshly coated particle deliquesce close to the DRH of $(NH_4)_2SO_4$ particles instead of the DRH of glutaric acid. Once the freshly coated particle deliquesced, the organic coating dissolved into the aqueous phase to form a mixed $(NH_4)_2SO_4$–glutaric acid solution droplet with a spherical shape as shown in Figure 7.11d. Afterward, the mixed $(NH_4)_2SO_4$–glutaric acid solid particle was formed after crystallization of the solution droplet in the

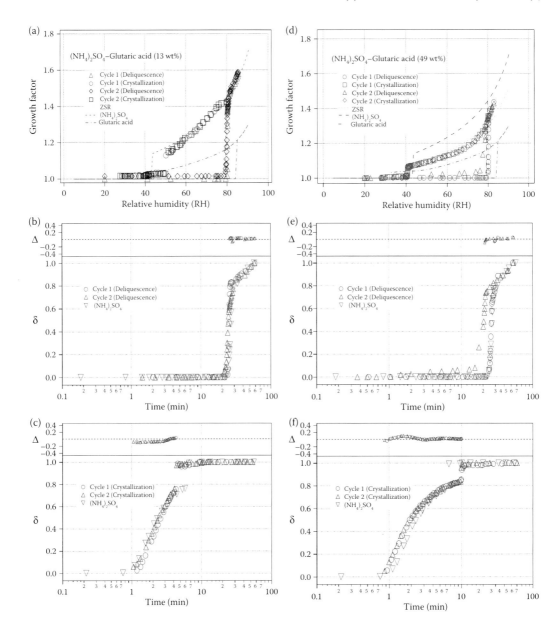

FIGURE 7.10   (See color insert following page 206.) Hygroscopicity of $(NH_4)_2SO_4$ particles coated with (a) 13 wt% and (d) 49 wt% glutaric acid in two deliquescence and crystallization cycles. The δ and Δ values of $(NH_4)_2SO_4$ particles coated with (b) 13 wt% and (e) 49 wt% glutaric acid are shown as a function of time in two deliquescence cycles. The δ and Δ values of $(NH_4)_2SO_4$ particles coated with (c) 13 wt% and (f) 49 wt% glutaric acid are shown as a function of time in two crystallization cycles. (Adapted from Chan, M. N., Lee, A. K. Y., and Chan, C. K. 2006. *Environmental Science and Technology* 40: 6983–6989. With permission.)

first cycle (Figure 7.11e). The morphologies of the aqueous droplet and the solid particle formed in the second cycle were similar to those formed in the first cycle. The Raman spectra of the aqueous droplets (Figure 7.11d, f) and solid particles (Figure 7.11e, g) formed after deliquescence of the glutaric acid-coated solid $(NH_4)_2SO_4$ particles in the two cycles are almost identical to those of mixed $(NH_4)_2SO_4$–glutaric acid aqueous droplets and solid particles (Figure 7.12a, b) generated by premixed

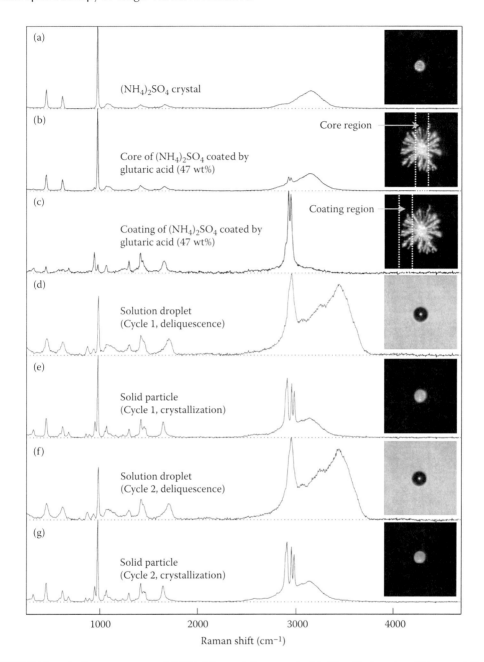

FIGURE 7.11 The Raman spectra of $(NH_4)_2SO_4$ particles coated with 47 wt% glutaric acid in two deliquescence and crystallization cycles: (a) $(NH_4)_2SO_4$ particles; (b) solid $(NH_4)_2SO_4$ particles coated with glutaric acid (center); (c) solid $(NH_4)_2SO_4$ particles coated with glutaric acid (branch/coating); (d) solution droplets in the first deliquescence cycle; (e) solid particles in the first crystallization cycle; (f) solution droplets in the second deliquescence cycle; and (g) solid particles in the second crystallization cycle. The light source was positioned at the opposite side of the microscope and at an angle of $90°$ with respect to the microscope to capture images of the solution droplets and the solid particle, respectively. (Adapted from Lee, A. K. Y., Ling, T. Y., and Chan, C. K. 2008. *Faraday Discussions* 137: 245–263. With permission.)

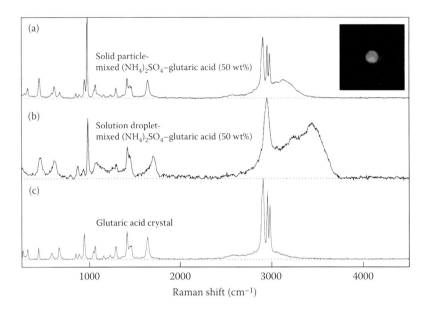

FIGURE 7.12   The Raman spectra of preprepared mixed $(NH_4)_2SO_4$–glutaric acid particles (50 wt% glutaric acid): (a) solid particles; (b) solution droplets; and (c) the Raman spectra of solid glutaric acid particles (Adapted from Lee, A. K. Y., Ling, T. Y., and Chan, C. K. 2008. *Faraday Discussions* 137: 245–263. With permission.)

solutions of a similar amount of glutaric acid (50 wt%). Our observations clearly show that mixing of $(NH_4)_2SO_4$ and glutaric acid lowers the DRH value in the second cycle, which is consistent with previous studies that used mixed $(NH_4)_2SO_4$–glutaric acid particles generated from the premixed solution of known $(NH_4)_2SO_4$ and glutaric acid concentrations [12,57,60].

Similar to the case of the octanoic acid coating experiments, δ and Δ analyses were also performed to determine the kinetic effects of glutaric acid coating on aerosol hygroscopicity. For the 13 wt% coated particles, the Δ values of the particles were close to the zero line within acceptable uncertainty in the two cycles (Figure 7.10b, c). This provided evidence that the assumption of quasi-equilibrium between the particle and its surroundings was valid. Even a thick coating of glutaric acid on the solid particle did not significantly impede the evaporation and condensation rates of water molecules from the particles in the two deliquescence and crystallization cycles compared to the rates of $(NH_4)_2SO_4$ particles (Figure 7.10e, f) [17]. The shifts of CRH in the two cycles and DRH in the second cycle observed in the case of thick coatings were attributed to chemical effects, but not to kinetic effects. Overall, glutaric acid coating does not impede the evaporation and condensation rates of water molecules compared to the rates of $(NH_4)_2SO_4$ particles in the two cycles. Nevertheless, Peng et al. [13] observed that pure glutaric acid exhibits mass transfer delays in the deliquescence experiments. The glutaric acid coating may not only change the DRH of the coated particles, but this suggests that it may also cause mass transfer delays during deliquescence if the coating completely covers the $(NH_4)_2SO_4$ particles.

7.3.3   FORMATION OF METASTABLE SALTS FROM CRYSTALLIZATION OF SOLUTION DROPLETS

Atmospheric particles can be in solid or in aqueous states, depending on the RH history of their immediate environment. While equilibrium thermodynamic models are available, there have been laboratory aerosol studies reporting the formation of metastable salts, not accounted for by the thermodynamic predictions. For example, inorganic salts such as Na_2SO_4, $LiClO_4$ [40], and $NaClO_4$ [69] and water-soluble organic salts such as sodium formate and sodium acetate [29] form metastable anhydrates and hydrates that are not predicted by the bulk-phase thermodynamics. Because of the high supersaturation at which a droplet crystallizes, metastable crystals with different chemical identities may form when a solution droplet contains more than one solute. Colberg et al. [70] found that $H_2SO_4/NH_3/H_2O$ droplets with ammonium-to-sulfate ratios equal to 1 (corresponding to the composition of ammonium bisulfate) did not crystallize into ammonium bisulfate, which is the product predicted by the AIM. Braban and Abbatt [59] and Badger et al. [71] used an aerosol flow tube-Fourier Transform Infrared Spectroscopy (FTIR) to study the phase transition of mixtures of $(NH_4)_2SO_4$ with MA/humic acid. They proposed the formation of an $(NH_4)_2SO_4$–humic acid complex upon crystallization, which converts into pure AS before deliquescence.

In a series of aerosol flow tube-FTIR experimental studies, Martin and coworkers [72–74] examined the crystallization properties of the $SO_4^{2-}/NO_3^-/NH_4^+/H^+$ system. Schlenker et al. [73] described the dependence of the types of crystals formed on the overall aqueous composition. Most of their results are consistent with the predictions of the AIM. However, they also observed the formation of metastable crystals. In some cases, part of the metastable salts is chemically transformed into the thermodynamically stable forms at elevated RH below the DRH. These results further support the possibility of metastable crystal formation and the proposition that RH affects the transformation of metastable components. In this section, the application of the EDB/Raman system to the characterization of crystals formed from supersaturated solution droplets and to monitoring the transformation of the metastable particles *in situ* is described. The effects of RH on the transformation are also examined. The formation of metastable salts produced by the crystallization of $(NH_4)_2SO_4$–NH_4NO_3 (mole ratio = 1:1) supersaturated droplets measured by Ling and Chan [75] are presented here in detail for illustration. Furthermore, the formation of metastable polymorphs formed by the crystallization of some $(NH_4)_2SO_4$–dicarboxylic acid mixed droplets and their transformation were also observed in other single particle studies and the details can be found in Ling and Chan [28].

7.3.3.1 Formation of Metastable Double Salts: $(NH_4)_2SO_4$–NH_4NO_3 Mixed Particles

Ammonium nitrate (AN), ammonium sulfate (AS), and protons (H^+) are the predominant inorganic components in atmospheric aerosols. In the $SO_4^{2-}/NO_3^-/NH_4^+/H^+$ system, complex salt formation is possible upon crystallization from solutions. Martin et al. [72] described the crystals formed in the $SO_4^{2-}/NO_3^-/NH_4^+/H^+$ system as predicted by the AIM and presented the results in the form of an isothermal phase diagram. In the subsystem containing mixtures of AN and AS only, two pure salts (AN and AS) and two double salts $3(NH_4NO_3) \cdot (NH_4)_2SO_4 (3AN \cdot AS)$ and $2(NH_4NO_3) \cdot (NH_4)_2SO_4$ ($2AN \cdot AS$) are predicted to form, depending on the AN/AS mixing ratios. To identify the multiple salts formed in single-particle crystallization experiments, reference Raman spectra were first acquired for the four possible salts (i.e., AN, AS, $3AN \cdot AS$, and $2AN \cdot AS$) in bulk samples as shown in the bottom panel of Figure 7.13. Table 7.3 summarizes the spectral characteristics of the bulk salts.

The AIM model predicts the formation of $2AN \cdot AS$ and pure AS in equal amounts during crystallization from an equimolar AN/AS solution with no $3AN \cdot AS$ formed. However, in the single-particle crystallization experiments, Raman features of both $3AN \cdot AS$ and $2AN \cdot AS$, as well as pure AS, were observed in the freshly crystallized particles. The amounts of $3AN \cdot AS$ and $2AN \cdot AS$ varied among numerous crystallization experiments based on their Raman signatures. Three examples of single-particle Raman spectra are shown in the upper panel of Figure 7.13, corresponding to the three particles containing (1) the weakest $3AN \cdot AS$ signal (relative to $2AN \cdot AS$), (2) a moderate $3AN \cdot AS$ signal, and (3) the strongest $3AN \cdot AS$ signal observed upon crystallization. These Raman spectra were taken within 15 min after crystallization of the supersaturated droplets.

The strong signal at ~1050 cm^{-1} is an indicator for the chemical nature of nitrates. It was assumed that whenever a particle contained more than one salt, their Raman signals would superimpose linearly. In case 3 of Figure 7.13, a single slightly asymmetric peak at 1053 cm^{-1} and the two shoulders around the 717 cm^{-1} peak suggests that the solid particle probably consisted mainly of $3AN \cdot AS$. In contrast, cases 1 and 2 of Figure 7.13 show obvious two resolvable peaks (at ~1046 cm^{-1}

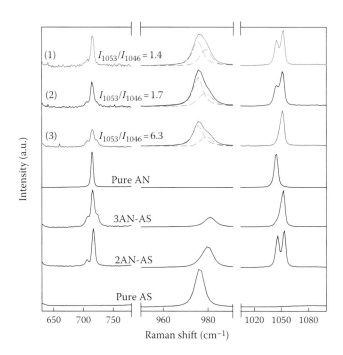

FIGURE 7.13 Raman spectra of three freshly crystallized particles. Contributions of AS_{pure} (peak at 975 cm^{-1}) and AS_{double} (peak at 980 cm^{-1}) were estimated and shown by dotted lines. The corresponding I_{1053}/I_{1046} ratios are also indicated. Reference Raman spectra of four bulk salts are included for comparison. The intensity of the Raman peaks below 800 cm^{-1} were rescaled to enlarge the spectral features. (Adapted from Ling, T. Y. and Chan, C. K. 2007. *Environmental Science and Technology* 41: 8077–8083. With permission.)

and ~1053 cm^{-1}) with different relative intensities. Although the minor peak at 1046 cm^{-1} might represent either 2AN·AS or pure AN, the dominance of the shoulder on the lower wave number side compared to the other side around the 717 cm^{-1} peak suggests that the peak at 1046 cm^{-1} indeed originated from 2AN·AS. Therefore, the single peak of 3AN·AS at 1053 cm^{-1} was apparently superimposed on the two resolvable peaks (with $I_{1053}/I_{1046} = 1.1$) of 2AN·AS. Depending on the relative amount of 3AN·AS and 2AN·AS, the I_{1053}/I_{1046} ratios were different, but always larger than 1.1 for all freshly crystallized particles. A large I_{1053}/I_{1046} ratio implies that the particle was relatively rich in 3AN·AS, while a particle with a ratio close to 1.1 was relatively rich in 2AN·AS.

The above analysis of the relative abundance of the double salts based on the nitrate peaks can be further supported by the sulfate peaks at ~980 cm^{-1}. As the overall mole ratio of AN to AS remained 1:1, AS should have been present both as pure AS (AS_{pure}) and as double salts (AS_{double}). The AS_{pure}/AS_{double} ratio is equal to 1 and 2 when all nitrate was present as 2AN·AS and 3AN·AS, respectively. Therefore, the AS_{pure}/AS_{double} ratio should range from 1 to 2 for particles containing both double salts. A 3AN·AS-rich particle had more pure AS than one containing less 3AN·AS (or 2AN·AS-rich particle). The reference spectra show sulfate signals at 975 and 980 cm^{-1} for AS_{pure} and AS_{double}, respectively. In Figure 7.13, the freshly crystallized particles show an asymmetric Raman sulfate peak at ~980 cm^{-1} with a shoulder on the high wave number side. This indicates that both AS_{pure} and AS_{double} were present, and their signals were superimposed. The growth of the shoulder (shown from the deconvolution of the sulfate peak) on the higher wave number side from a 3AN·AS-rich particle to a 2AN·AS-rich particle is in agreement with the evidence from the nitrate signals. Table 7.3 summarizes the peak characteristics for 3AN·AS-rich and 2AN·AS-rich particles, which can be indicators for following the transformation of the metastable component into the stable form.

TABLE 7.3
Summary of Raman Spectral Features of (1) Bulk Samples and (2) 3AN·AS-Rich and 2AN·AS-Rich Single Particles

Peak Location	Bulk Pure AN	Bulk 3AN·AS	Bulk 2AN·AS	Bulk Pure AS
717 cm^{-1} nitrate signal	No shoulders around the main peak	Two shoulders around the main peak	A single shoulder on the lower wave number side of the main peak	N/A
~980 cm^{-1} sulfate signal	N/A	Symmetric peak at 980 cm^{-1}	Symmetric peak at 980 cm^{-1}	Symmetric peak at 975 cm^{-1}
~1050 cm^{-1} nitrate signal	Symmetric peak at 1046 cm^{-1}	Asymmetric peak at 1053 cm^{-1}	Double peaks at 1046 and 1053 cm^{-1}, with peak intensity ratio (I_{1053}/I_{1046}) equals to 1.1	N/A

Peak Location	3AN·AS-Rich Particles	2AN·AS-Rich Particles
717 cm^{-1} nitrate signal	Two shoulders around the main peak	A single shoulder on the lower wave number side of the main peak
~980 cm^{-1} sulfate signal	A smaller shoulder on the higher wave number side	A dominant shoulder on the higher wave number side
~1050 cm^{-1} nitrate signal	Single peak at 1053 cm^{-1}, large I_{1053}/I_{1046} ratio	Increased dominance of 1046 cm^{-1} peak for double peaks, small I_{1053}/I_{1046} ratio (close to 1.1)

Source: Adapted from Ling, T. Y. and Chan, C. K. 2007. *Environmental Science and Technology* 41: 8077–8083. With permission.

Note: $I_{1053}/I_{1046} = 1.1$ is regarded as a feature of 2AN·AS. A deviation in this value means the Raman signal originates from samples other than 2AN·AS or mixtures of 2AN·AS with other salts.

Ling and Chan [75] also estimated the relative amount of 3AN·AS, 2AN·AS, and pure AS present in 15 freshly crystallized particles by assuming that the residual signal after subtracting the 2AN·AS signal (i.e., $I_{1053}-1.1 \times I_{1046}$) arose from 3AN·AS. The results suggest that freshly crystallized particles contained both 2AN·AS and 3AN·AS in various amounts. Since the same procedures were applied in crystallizing the particles, the differences in the composition of the freshly crystallized particles could not have been due to experimental variations. The differences were possibly due to the random nature of the nucleation events. The AIM predicts that 2AN·AS and pure AS are thermodynamically stable crystals formed from equimolar AN/AS solutions. 3AN·AS should therefore be a metastable component. The observations thus provide evidence that freshly crystallized particles can have various degrees of metastability.

7.3.3.2 Transformation of 3AN·AS to 2AN·AS and Its Dependence on RH

Although the freshly crystallized particles differed in their initial compositions, there was a consistent tendency for metastable 3AN·AS-rich particles to transform into thermodynamically more stable 2AN·AS-rich particles. A typical series of spectra obtained at 55% RH as a function of time is shown in Figure 7.14. The freshly crystallized solid particle showed a strong peak at 1053 cm^{-1}, a large initial I_{1053}/I_{1046} ratio, and shoulders on both sides of the peak at 717 cm^{-1}, indicating that it was a 3AN·AS-rich particle. Under such RH conditions, the 3AN·AS-rich particle starts to transform into a 2AN·AS-rich one, as revealed by the changes in the three spectral features as listed in Table 7.3.

As shown in Figure 7.14, a distinctive signal for 2AN·AS at 1046 cm^{-1} gradually developed (i.e., the I_{1053}/I_{1046} ratio decreased) with time, indicating the occurrence of the following chemical transformation: 4AS + 2(3AN·AS) → 3(2AN·AS) + 3AS. This conclusion is also supported by comparing the

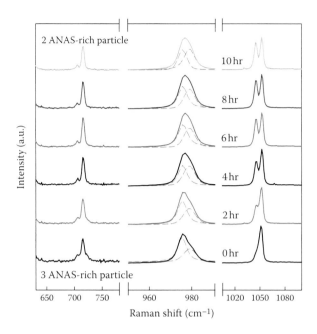

FIGURE 7.14   A series of Raman spectra of a crystallized particle obtained at 55% RH showing the chemical transformation as a function of time. The peaks are normalized by the intensity of 420 cm⁻¹ sulfate peak, which showed little spectral change during the transformation. Contributions of AS_{pure} and AS_{double} were estimated and are shown as dotted lines. (Adapted from Ling, T. Y. and Chan, C. K. 2007. *Environmental Science and Technology* 41: 8077–8083. With permission.)

signal changes at ~720 cm⁻¹ and 980 cm⁻¹. The degeneration of the shoulder on the higher wave number side of the nitrate peak at ~720 cm⁻¹ indicates the disappearance of $3AN \cdot AS$ while the shoulder on the lower wave number side at the end of the transformation suggests the formation of $2AN \cdot AS$. The growth of the shoulder on the higher wave number side of the sulfate signal at ~980 cm⁻¹ also suggests the proposed transformation, as more AS_{double} is formed from AS_{pure} during the transformation. Even though AN might also have been contributing to the peak at 1046 cm⁻¹, $3AN \cdot AS$ would not be expected to transform into AN in these experiments, because the shoulder features at ~720 cm⁻¹ suggested the dominance of $2AN \cdot AS$. Besides, the transformation to AN would require the formation of more AS_{pure} to maintain the overall material balance (i.e., $4AS + 2(3AN \cdot AS) \rightarrow 6AN + 6AS$) and this is not consistent with the observed sulfate peak changes.

The following discussions on the dependence of RH on the chemical transformation focus on the changes in the I_{1053}/I_{1046} ratio, which were obvious and readily quantified for representing the relative extent of the transformation. Figure 7.15 shows the I_{1053}/I_{1046} ratio as a function of time at several RH conditions. In the experiment, the same particle was subjected to several crystallization–deliquescence cycles to eliminate any particle size effects from the transformation. Generally, a higher RH accelerated the transformation of $3AN \cdot AS$ into $2AN \cdot AS$. It has been proposed that higher RH promotes the adsorption of water vapor onto the crystal surface, which enhances ion mobility [73]. The observed RH dependency of the transformation rate supports this hypothesis. It should be noted that the initial rates were not compared, since the difference in the degree of metastability may also affect the transformation rate. Hence, the effects of RH on the overall transformation were evaluated by using the same "initial state" of the fixed I_{1053}/I_{1046} ratio, instead of the initial rate.

The observations show that it takes from several hours to more than a day, depending on RH, for a freshly crystallized metastable particle to reach its stable form, suggesting a high possibility that metastable solids persist in the atmosphere when the ambient RH is sufficiently low. Furthermore,

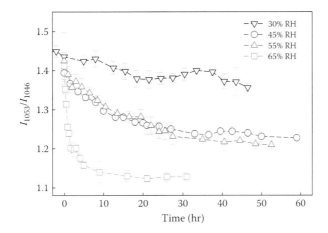

FIGURE 7.15 Plot of I_{1053}/I_{1046} ratio as a function of time. As there can be differences in the initial I_{1053}/I_{1046} ratios among several crystallization processes, to facilitate the comparison of the data, times with a comparable ratios were artificially designated as time = 0. The period between the actual initial I_{1053}/I_{1046} and the recalibrated zero time was plotted in the negative time domain. The chemical transformation was monitored at RH = 30%, 45%, 55%, and 65%. (Adapted from Ling, T. Y. and Chan, C. K. 2007. *Environmental Science and Technology* 41: 8077–8083. With permission.)

the formation of metastable polymorphs formed by the crystallization of $(NH_4)_2SO_4$–dicarboxylic acid mixed droplets has been recently observed by Ling and Chan [28]. Since the identity of the salts, which can differ from thermodynamic predictions, plays an important role in affecting the deliquescence properties of the particles, it is particularly important to improve our understanding of the formation of metastable salts or polymorphs as well as their transformation.

7.3.4 HETEROGENEOUS REACTIONS OF ORGANIC AEROSOL PARTICLES

Atmospheric aerosols can physically and chemically interact with various atmospheric components throughout their lifetimes, resulting in great uncertainty in global climate predictions. In atmospheric chemical aging processes, aerosol particles may react with atmospheric gas-phase oxidants such as ozone, hydroxyl and nitrate radicals, and some reactive organic vapors (referred as heterogeneous reactions), consequently altering the physical and chemical properties of atmospheric aerosols. In particular, heterogeneous reactions of organic aerosols with atmospheric oxidants are potentially important in enhancing the hygroscopicity and possibly the CCN activity of atmospheric organic aerosols [6,7]. The oxidation products that remain in the particle phase are generally more oxygenated and hydrophilic than are their parent molecules [8,76,77]. However, our understanding of the heterogeneous reactions of organic aerosols and their atmospheric implications is rather limited.

As the chemical composition (or functional groups) of reacting organic aerosols can be easily probed by Raman spectroscopy, the EDB/Raman system is particularly suitable to study heterogeneous organic reactions. More importantly, the EDB/Raman system allows long duration particle levitation (days) and thus long exposure of particles to gas-phase oxidants or organics at concentrations relevant to atmospheric applications is possible. The EDB approach also facilitates the direct mass yield measurements of the levitated particles during the exposure of gas-phase reactants and allows the identification of any changes in the aerosol hygroscopicity caused by the heterogeneous reactions. In this section, we describe the application of the EDB/Raman system to the investigation of heterogeneous oxidation of unsaturated fatty acid particles, including oleic acid (18:1), linoleic acid (C18:2), and linolenic acid (C18:3), with ozone (200–280 ppb) under ambient temperatures (22–24°C) and dry conditions (RH < 5%) during 20 h of exposure. The chemical structures of these fatty acids are shown in Figure 7.16. The details of the experimental procedures were described by

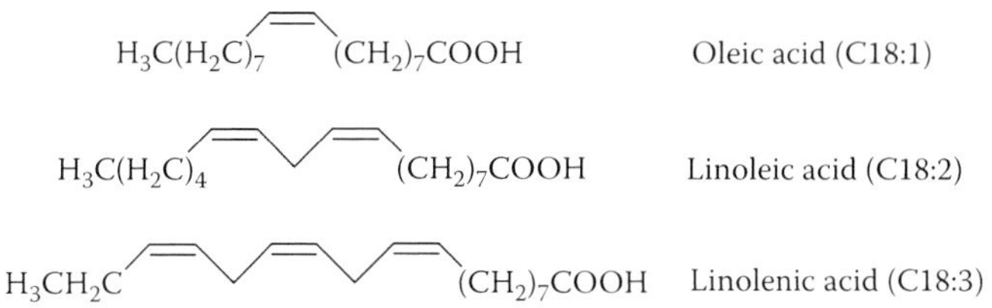

H₃C(H₂C)₇ ⎓ (CH₂)₇COOH　　　　Oleic acid (C18:1)

H₃C(H₂C)₄ ⎓⎓ (CH₂)₇COOH　　　Linoleic acid (C18:2)

H₃CH₂C ⎓⎓⎓ (CH₂)₇COOH　　Linolenic acid (C18:3)

FIGURE 7.16　Chemical structures of oleic acid, linoleic acid, and linolenic acid.

Lee and Chan [15,16]. The possible reaction pathways are proposed based on the Raman analysis and the effects of ozone concentrations on the reaction pathways and particle morphology are also briefly discussed.

7.3.4.1　Raman Characterization of Ozone-Processed Unsaturated Fatty Acid Particles

The Raman spectra of pure and ozone-processed oleic acid particles are shown in Figure 7.17a. The spectral changes illustrate that oleic acid ozonolysis results in the consumption of carbon–carbon double bonds (C=C, peaks at ~973, ~1269, ~1655, and ~3008 cm^{-1}) of oleic acid molecules and leads to the formation of peroxidic products (O—O, peak at ~850 cm^{-1}), carbonyl groups (C=O, peak at ~1740 cm^{-1}) and hydroxyl groups (O—H, peak at ~3450 cm^{-1}) [15]. These changes in functional groups are in accordance with the predictions of the Criegee mechanisms as well as the results reported in the recent literature. In brief, the primary reaction of oleic acid ozonolysis proceeds via the addition of an ozone molecule across the double bond of an oleic acid molecule to form a primary ozonide. The primary ozonide is unstable and thus decomposes to generate carbonyl compounds and Criegee intermediates. Peroxidic products are subsequently formed via the reactions between stabilized Criegee intermediates and other existing organic compounds. Some possible reaction pathways involving stabilized Criegee intermediates proposed in the recent literature have been summarized by Zahardis and Petrucci [78].

In the cases of linoleic acid and linolenic acid, the consumption of C=C bonds (peaks at ~973, ~1269, and ~3008 cm^{-1}) and similar product peak formations (O—O, C=O, and O—H groups) were also observed as shown in Figures 7.17b, c, respectively. The close similarities of the functional group characteristics of the reaction products generated in ozone-processed oleic acid, linoleic acid, and linolenic acid particles indicate that both linoleic acid and linolenic acid can also undergo direct ozonolysis within the ozone exposure period. In addition, the new product peaks formed at ~1590 cm^{-1} and ~1640 cm^{-1} are assigned to the symmetric and asymmetric C=C stretching vibrations of the conjugated dienes (C=C—C=C), respectively (see Figures 7.17b, c). However, it is important to note that conjugated dienes cannot be produced via the direct ozonolysis of linoleic acid and linolenic acid based on the Criegee mechanism. Instead, these spectral changes can be contributed by the autoxidation (or peroxidation) process [79]. A new band that peaks at ~1170 cm^{-1} is additional evidence of the autoxidation mechanisms as this peak has been observed in reported studies of lipid or oil autoxidation [80,81]. This peak, as well as the peaks at ~1590 and ~1640 cm^{-1}, is appreciably more intense for linolenic acid particles, which is consistent with the fact that linolenic acid is more likely to undergo autoxidation. Overall, our Raman measurements show that ozone-induced autoxidation, in addition to direct ozonolysis, is a plausible pathway in the reactions between ozone and linoleic acid and linolenic acid particles.

7.3.4.2　Autoxidation and Ozonolysis of the Autoxidation Products

Autoxidation is a typical free radical reaction consisting of chain initiation, propagation, and termination steps. The reaction mechanisms of the ozone-induced autoxdiation of linoleic acid and

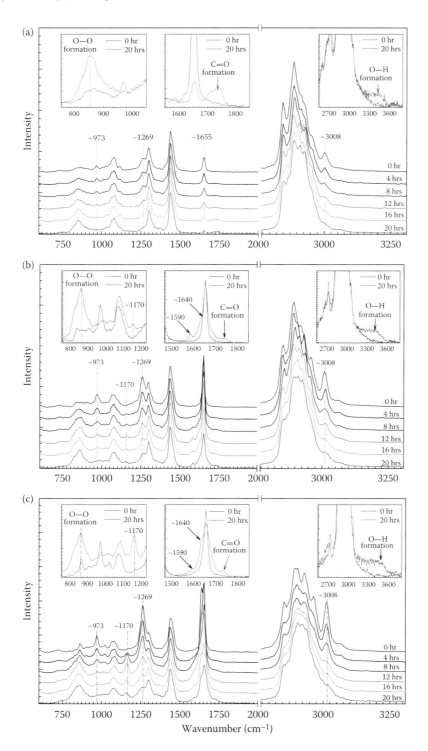

FIGURE 7.17   (See color insert following page 206.) The Raman spectra of (a) oleic acid (b) linoleic acid, and (c) linolenic acid particles at different ozone exposures. The Raman spectra are normalized to the intensity of the peak located at 1443 cm⁻¹. The inserts highlight the difference of spectral features between pure and ozone-processed particles. (Adapted from Lee, A. K. Y., Ling, T. Y., and Chan, C. K. 2008. *Faraday Discussions* 137: 245–263. With permission.)

linolenic acid are shown in Figure 7.18a, b, respectively, as illustrations. The details of the reaction mechanisms are available in Lee and Chan [16]. It is important to note that autoxidation does not eliminate the number of C=C bonds in the hydrocarbon skeletons of fatty acid molecules and that ozone can also react with the autoxidation products by attacking their C=C bonds. The ozonolysis of the autoxidation products can be evaluated by monitoring the intensity changes of the distinct Raman signals of conjugated dienes ranging from 1560 to 1640 cm⁻¹ (Figure 7.18a). In the case of linolenic acid, the intensity of these two peaks increased rapidly and started to decline after 10–12 h of ozone exposure (left axis of Figure 7.19a). The two-step kinetic behavior indicates the complex chemical mechanism involving accumulation of the intermediate products followed by their loss.

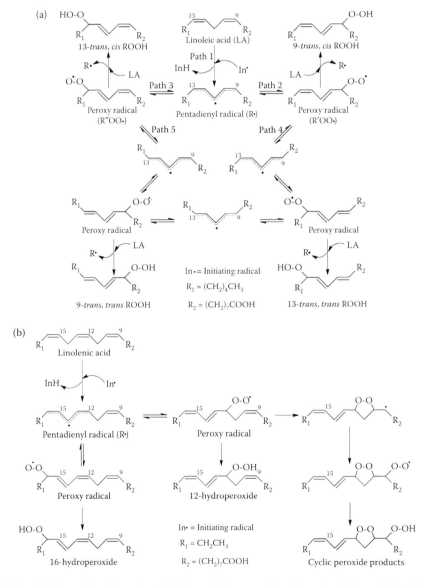

FIGURE 7.18 (a) Formation of conjugated diene hydroperoxides via autoxidation of linoleic acid. (b) Formation of conjugated diene hydroperoxides and cyclic peroxide products via autoxidation of linolenic acid. (Adapted from Lee, A. K. Y. and Chan, C. K. 2007. *Journal of Physical Chemistry A* 111: 6285–6295. With permission.)

The initial rapid increase in Raman intensities suggests that ozone-induced autoxidation is the prevailing reaction pathway and it is much faster than the decomposition of the conjugated diene hydroperoxides via ozonolysis. During the latter period of exposure, the rate of autoxidation slows down owing to depletion of the linolenic acid, and the ozonolysis of the autoxidation products becomes more dominant, leading to a decrease in the Raman signals. This analysis can be also applied to the case of linoleic acid (right axis of Figure 7.19a). In contrast, no Raman signal corresponding to the conjugated diene structures was observed in the ozone-processed oleic acid particle.

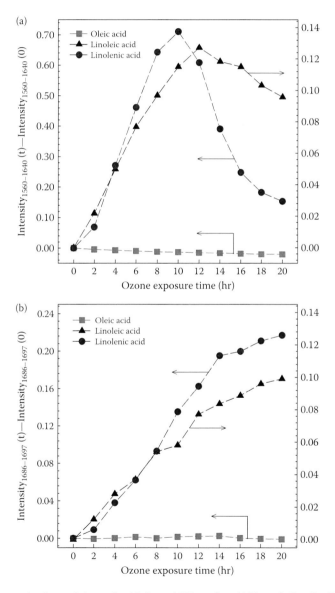

FIGURE 7.19 Changes in the peak intensity (a) from 1560 cm^{-1} to 1640 cm^{-1} (C=C—C=C structures) and (b) from 1686 cm^{-1} to 1697 cm^{-1} (C=C—C=O structures) of the polyunsaturated fatty acid particles (squares: oleic acid, triangles: linoleic acid, circles: linolenic acid) after different ozone exposure periods. The Raman intensities of these peaks are normalized to the intensity of the peak located at 1443 cm^{-1}. The experimental errors in intensity are within 10%. (Adapted from Lee, A. K. Y. and Chan, C. K. 2007. *Journal of Physical Chemistry A* 111: 6285–6295. With permission.)

This is consistent with the fact that conjugated diene would not be produced even if autoxidation of oleic acid takes place inside the particles.

The ozonolysis of the conjugated diene can generate products consisting of C=C—C=O conjugated systems in their hydrocarbon skeletons based on the predictions of the Criegee mechanism. In Figure 7.17b, c, the product peak at ~1690 cm^{-1} is likely to be caused by the C=O stretching vibrations of C=C—C=O conjugated systems. The continuous increase in the peak intensity ranging from 1686 to 1697 cm^{-1}, as shown in Figure 7.19b, for both linoleic acid and linolenic acid particles is an indication of the ozonolysis of the intermediate products produced by ozone-induced autoxidation. The result of oleic acid ozonolysis is also shown in Figure 7.19b, which indicates that there was no formation of the C=C—C=O conjugated system. As linolenic acid is more likely than linoleic acid to undergo autoxidation, ozone-processed linolenic acid particles produce more C=C—C=O conjugated structures than the ozone-processed linoleic acid particles.

7.3.4.3 Changes in Particle Mass Yield Caused by Heterogeneous Reactions

Figure 7.20 shows the mass ratios of the ozone-processed unsaturated fatty acid particles as a function of the ozone exposure time. Our direct mass measurements show that heterogeneous oxidation of organic aerosols can either increase or decrease the particle mass yields to different extents depending on the molecular structures of the chemical species. It is worth noting that the mass of the ozone-processed linoleic acid and linolenic acid particles initially increased rapidly and then slightly declined or remained at constant levels. This two-step behavior indicates that less volatile organics were produced by autoxidation at the beginning followed by the generation of more volatile species via ozonolysis in the later exposure period. According to Figure 7.18a, b, autoxidation pathways did not break down the skeleton of these two fatty acids to generate volatile species. Instead, the autoxidation pathways involved the addition of oxygen molecules to increase the

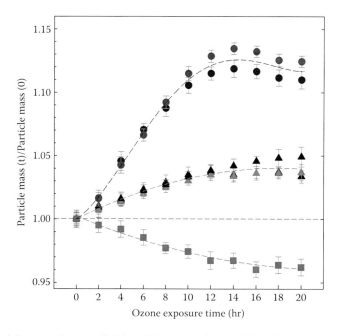

FIGURE 7.20 Particle mass changes of oleic acid (squares, first particle), linoleic acid (triangles, three particles), and linolenic acid (circles, two particles) particles during ozone exposure periods. The lines are to guide the eye and have no physical meaning. (Adapted from Lee, A. K. Y. and Chan, C. K. 2007. *Journal of Physical Chemistry A* 111: 6285–6295. With permission.)

molecular weight of the fatty acids. The mass gain in the linolenic acid particles was the most significant, because linolenic acid is the species that most favors autoxidation among the fatty acids studied here.

It should be noted that oleic acid cannot undergo autoxidation except under elevated temperature conditions. Therefore, the mass loss in the ozone-processed oleic acid particles was mainly caused by the evaporative loss of volatile organic products formed via ozonolysis, which was larger than the mass gain caused by the addition of ozone molecules. Lee and Chan [15] reported that about 5% of the total particle mass evaporated from a completely reacted oleic acid particle. Another possible reason for the observed particle mass gain is that direct ozonolysis of linoleic acid and linolenic acid only generates volatile species in low yields. It is possible that the poly-unsaturated fatty acids and existing intermediate products provide a significant number of active sites to react with Criegee intermediates. Since Criegee intermediates can also undergo isomerization to form short chain (<C9) carboxylic acids and some intermediate products are volatile or semivolatile, the rapid reactions of Criegee intermediates with the poly-unsaturated fatty acids and intermediate products would probably produce low volatility organics that would not escape to the gas phase.

7.3.4.4 Effects of Ozone Concentration on Production Formation

Lee and Chan [16] also measured the Raman spectra of linoleic acid and linolenic acid particles after exposure to high ozone concentrations (~10 ppm), much higher than the preceding discussion of measurements at concentrations lower than 300 ppb. The Raman spectra show that exposure to high ozone concentrations leads to the formation of O—O, C=O, and O–H functional groups but that there is less formation of conjugated diene structures (C=C—C=C characteristic peaks at ~1590 and ~1640 cm^{-1}). The intensity of the peak at ~1170 cm^{-1} is also more intense in the low ozone concentration experiments. These spectral differences are most obvious in the case of linolenic acid. These observations indicate that low ozone levels (~200–250 ppb) favor the autoxidation pathways compared with high ozone levels (~10 ppm). A possible explanation for this is that most poly-unsaturated fatty acids are forced to react rapidly with ozone molecules under extremely high ozone concentrations instead of undergoing autoxidation.

Another interesting observation is that the ozone concentration at which oxidation proceeds can affect the particle morphology based on the captured particle images and laser-illuminated light-scattering patterns. In Figure 7.21a–d, the particles remained in spherical shapes before and after the reaction if the ozone concentration was about 200–250 ppb. In addition, the laser-illuminated light-scattering patterns of the pure and ozone-processed particles consisted of a few sharp horizontal lines, indicating that the particles had spherical shapes and were completely in the liquid phase. In contrast, at very high ozone concentrations (~10 ppm), the morphology and light-scattering patterns were appreciably modified as shown in Figure 7.21e–f. The particle shape changed from spherical to highly irregular under high ozone concentrations. The laser-illuminated light-scattering patterns of the ozone-processed particles became irregular, which was evidence of the irregular particle shapes and/or the formation of solid materials. It is particularly important to note that the solid-coated particles absorbed a small amount of water but that the irregular particle shape and laser-illuminated light-scattering pattern still could be observed when the RH inside the EDB was increased to 85–90% for about 3 h. This observation suggests that part of the solid materials coated on the surface of the ozone-processed particles may have relatively high DRH and possibly form a transport barrier for water uptake [18]. Furthermore, the solid coating may also reduce the reactive uptake ability of atmospheric oxidants of organic aerosols [78].

7.4   SUMMARY AND PROSPECTIVE

The four examples discussed in this chapter have clearly demonstrated the capability of the EDB/Raman system in studying numerous physical and chemical processes or phenomena of atmospheric

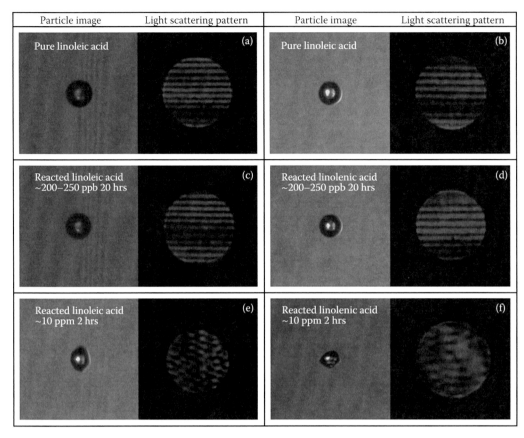

FIGURE 7.21   Particle images and laser-illuminated light-scattering patterns of the (a) pure linoeic acid, (b) pure linolenic acid, (c and e) ozone-processed linoleic acid (d and f), and ozone-processed linolenic acid particles. (Adapted from Lee, A. K. Y. and Chan, C. K. 2007. *Journal of Physical Chemistry A* 111: 6285–6295. With permission.)

interest. The special features of the EDB/Raman system for studies of atmospheric particles are summarized as follows:

1. Changes in the fwhh of Raman peaks are useful indicators for monitoring phase transformations, especially partial deliquescence and crystallization processes, of single levitated particles. In particular, the phase transformation behavior of individual components in a binary mixture can be monitored based on the fwhh analysis. A similar approach can be also applied to investigate the phase transformations of particles deposited on a hydrophobic substrate using a custom-made flow cell coupled with micro-Raman spectroscopic system [26]. Furthermore, our previous single-particle studies have shown that changes in Raman peak positions can be used to examine the molecular interactions between ions/molecules at high concentrations as well as the phase transformation [49,50].

2. The SEDB method allows the investigation of hygroscopic properties of semivolatile species as the measurements can be completed within an hour. As demonstrated in the organic coating study, the SEDB approach also allows evaluation of the importance of kinetic effects caused by some chemical species in repeated hygroscopic measurements and Raman spectroscopy can provide extra information to explain the observed hygroscopic properties. Choi and Chan [20] observed the slow growth rate of $MgSO_4$ solution droplets using the SEDB method and this was attributed to the formation of contact ion pairs which were characterized by a single-particle Raman study [49].

3. The solid particle obtained from the crystallization of supersaturated droplets can be in metastable form. *In situ* single-particle Raman characterization allows the formation of metastable salts to be probed as demonstrated in this chapter. The long levitation time and the controlled RH system can help us to understand the transformation of metastable salts in the atmospheric environment. The formations of metastable polymorphs in $(NH_4)_2SO_4$–dicarboxylic acid mixed particles as well as their transformation have been observed by Ling and Chan [28]. A similar approach is also applicable to the custom-made flow cell coupled with micro-Raman spectroscopic system.

4. Using the EDB/Raman approach, a single particle can be exposed to gas-phase reactants at an atmospherically relevant level for a long period of time, allowing the evaluation of the significance and details of certain heterogeneous reactions in the real atmosphere. In addition to the oxidative aging of organic aerosol particles, the static EDB approach has been used to investigate secondary organic aerosol formation by acid-catalyzed reactions of gas-phase organic precursors [14]. The mass yield measurement provides direct evidence to evaluate the occurrence of the acid-catalyzed reaction and its atmospheric importance.

The prolonged particle levitation allows the study of more than one atmospheric process in a single experiment (i.e., multiphysical and/or chemical aging), which helps improve our understanding on the complex evolution of atmospheric aerosols [10]. For example, an oxidative aged organic particle can be subjected to hygroscopic measurements to determine whether it experiences any changes in hygroscopic growth and phase transformations [15,16]. The repeated cycle of hygroscopic measurements of coated particles and the transformation of metastable salts mentioned in this chapter also demonstrated the importance of this experimental merit in atmospheric studies. Furthermore, although the EDB has been extensively used for water uptake and phase transformation of aerosol particles in the past, it is entirely possible to use it for studying the physical and/or reactive uptake of volatile or semivolatile organics onto the levitated particles [14,82], which is a very important process in the formation and evolution of atmospheric organic aerosols. The EDB/Raman system can possibly play a very unique role in studying the gas-particle partitioning of organics and formation of organic aerosols in the future.

ACKNOWLEDGMENTS

C. K. Chan dedicate this paper to Prof. E. James Davis of the University of Washington. The pioneering work of Prof. Davis in EDB studies has inspired C. K. Chan on his research on a number of spectroscopic applications of the EDB, as presented in this chapter. This work was funded by an Earmarked Grant (600208) from the Research Grants Council of the Hong Kong Special Administrative Region, China.

REFERENCES

1. Seinfeld, J. H. and Pandis, S. N. 2006. *Atmospheric Chemistry and Physics: From Air Pollution to Climate Change*, John Wiley, New York.
2. Watson, J. G. 2002. Visibility: Science and regulation. *Journal of the Air and Waste Management Association* 52: 628–713.
3. Dockery, D. W., Pope, C. A., Xu, X. P. et al. 1993. An association between air-pollution and mortality in 6 United-States cities. *New England Journal of Medicine* 329: 1753–1759.
4. Sloane, C. S. and White, W. H. 1986. Visibility—An evolving issue. *Environmental Science and Technology* 20: 760–766.
5. Davidson, C. I., Phalen, R. F., and Solomon, P. A. 2005. Airborne particulate matter and human health: A review. *Aerosol Science and Technology* 39: 737–749.
6. IPCC. 2007. *Climate Change 2007: The Physical Science Basis.* Contribution of working group I to the fourth assessment report of the IPCC, Solomon, S., Qin, D., Manning, M., Chen, Z., Marquis, M., Averyt, K. B., Tignor, M., and Miller H. L (Eds), Cambridge University Press, Cambridge.

7. Kanakidou, M., Seinfeld, J. H., Pandis, S. N. et al. 2005. Organic aerosol and global climate modelling: A review. *Atmospheric Chemistry and Physics* 5: 1053–1123.
8. Rudich, Y. 2003. Laboratory perspectives on the chemical transformations of organic matter in atmospheric particles. *Chemical Reviews* 103: 5097–5124.
9. Rudich, Y., Donahue, N. M., and Mentel, T. F. 2007. Aging of organic aerosol: Bridging the gap between laboratory and field studies. *Annual Review of Physical Chemistry* 58: 321–352.
10. Kroll, J. H. and Seinfeld, J. H. 2008. Chemistry of secondary organic aerosol: Formation and evolution of low-volatility organics in the atmosphere. *Atmospheric Environment* 42: 3593–3624.
11. Davis, E. J. and Schweiger, G. 2002. *The Airborne Microparticle: Its Physics, Chemistry, Optics and Transport Phenomena*, Springer Verlag, Heidelberg.
12. Choi, M. Y. and Chan, C. K. 2002. The effects of organic species on the hygroscopic behaviors of inorganic aerosols. *Environmental Science and Technology* 36: 2422–2428.
13. Peng, C., Chan, M. N., and Chan, C. K. 2001. The hygroscopic properties of dicarboxylic and multifunctional acids: Measurements and UNIFAC predictions. *Environmental Science and Technology* 35: 4495–4501.
14. Lee, A. K. Y., Li, Y. J., Lau, A. P. S., and Chan, C. K. 2009. A re-evaluation on the atmospheric significance of octanal vapor uptake by acidic particles: Roles of particle acidity and gas-phase octanal concentration. *Aerosol Science and Technology* 42: 992–1000.
15. Lee, A. K. Y. and Chan, C. K. 2007. Single particle Raman spectroscopy for investigating atmospheric heterogeneous reactions of organic aerosols. *Atmospheric Environment* 41: 4611–4621.
16. Lee, A. K. Y. and Chan, C. K. 2007. Heterogeneous reactions of linoleic acid and linolenic acid particles with ozone: Reaction pathways and changes in particle mass, hygroscopicity, and morphology. *Journal of Physical Chemistry A* 111: 6285–6295.
17. Chan, M. N., Lee, A. K. Y., and Chan, C. K. 2006. Responses of ammonium sulfate particles coated with glutaric acid to cyclic changes in relative humidity: Hygroscopicity and Raman characterization. *Environmental Science and Technology* 40: 6983–6989.
18. Chan, M. N. and Chan, C. K. 2007. Mass transfer effects on the hygroscopic growth of ammonium sulfate particles with a water-insoluble coating. *Atmospheric Environment* 41: 4423–4433.
19. Chan, M. N. and Chan, C. K. 2005. Mass transfer effects in hygroscopic measurements of aerosol particles. *Atmospheric Chemistry and Physics* 5: 2703–2712.
20. Choi, M. Y. and Chan, C. K. 2002. Continuous measurements of the water activities of aqueous droplets of water-soluble organic compounds. *Journal of Physical Chemistry A* 106: 4566–4572.
21. Clegg, S. L., Seinfeld, J. H., and Brimblecombe, P. 2001. Thermodynamic modelling of aqueous aerosols containing electrolytes and dissolved organic compounds. *Journal of Aerosol Science* 32: 713–738.
22. Clegg, S. L. and Seinfeld, J. H. 2006. Thermodynamic models of aqueous solutions containing inorganic electrolytes and dicarboxylic acids at 298.15 K. 1. The acids as nondissociating components. *Journal of Physical Chemistry A* 110: 5692–5717.
23. Clegg, S. L. and Seinfeld, J. H. 2006. Thermodynamic models of aqueous solutions containing inorganic electrolytes and dicarboxylic acids at 298.15 K. 2. Systems including dissociation equilibria. *Journal of Physical Chemistry A* 110: 5718–5734.
24. Petters, M. D. and Kreidenweis, S. M. 2007. A single parameter representation of hygroscopic growth and cloud condensation nucleus activity. *Atmospheric Chemistry and Physics* 7: 1961–1971.
25. Chan, M. N., Kreidenweis, S. M., and Chan, C. K. 2008. Measurements of the hygroscopic and deliquescence properties of organic compounds of different solubilities in water and their relationship with cloud condensation nuclei activities. *Environmental Science and Technology* 42: 3602–3608.
26. Yeung, M. C., Lee, A. K. Y., and Chan, C. K. 2009. Phase transition and hygroscopic properties of internally mixed ammonium sulfate and adipic acid (AS–AA) particles by optical microscopic imaging and Raman spectroscopy. *Aerosol Science and Technology* 43: 387–399.
27. Chan, M. N. and Chan, C. K. 2003. Hygroscopic properties of two model humic-like substances and their mixtures with inorganics of atmospheric importance. *Environmental Science and Technology* 37: 5109–5115.
28. Ling, T. Y. and Chan, C. K. 2008. Partial crystallization and deliquescence of particles containing ammonium sulfate and dicarboxylic acids. *Journal of Geophysical Research-Atmospheres* 113: D14205, doi:10.1029/2008JD009779.
29. Peng, C. G. and Chan, C. K. 2001. The water cycles of water-soluble organic salts of atmospheric importance. *Atmospheric Environment* 35: 1183–1192.

30. Chan, M. N., Choi, M. Y., Ng, N. L., and Chan, C. K. 2005. Hygroscopicity of water-soluble organic compounds in atmospheric aerosols: Amino acids and biomass burning derived organic species. *Environmental Science and Technology* 39: 1555–1562.

31. Chan, C. K., Flagan, R. C., and Seinfeld, J. H. 1998. *In situ* study of single aqueous droplet solidification of ceramic precursors used for spray pyrolysis. *Journal of the American Ceramic Society* 81: 646–648.

32. Davis, E. J. and Ray, A. K. 1977. Determination of diffusion-coefficients by submicron droplet evaporation. *Journal of Chemical Physics* 67: 414–419.

33. Ray, A. K., Davis, E. J., and Ravindran, P. 1979. Determination of ultralow vapor-pressures by submicron droplet evaporation. *Journal of Chemical Physics* 71: 582–587.

34. Davis, E. J., Ravindran, P., and Ray, A. K. 1980. A review of theory and experiments on diffusion from sub-microscopic particles. *Chemical Engineering Communications* 5: 251–268.

35. Davis, E. J. 1983. Transport phenomena with single aerosol-particles. *Aerosol Science and Technology* 2: 121–144.

36. Preston, R. E., Lettieri, T. R., and Semerjian, H. G. 1985. Characterization of single levitated droplets by Raman spectroscopy. *Langmuir* 1: 365–367.

37. Thurn, R. and Kiefer, W. 1985. Structural resonances observed in the Raman spectra of optically levitated liquid droplets. *Applied Optics* 24: 1515–1519.

38. Fung, K. H. and Tang, I. N. 1988. Raman-spectra of singly suspended supersaturated ammonium bisulfate droplets. *Chemical Physics Letters* 147: 509–513.

39. Chan, C. K., Flagan, R. C., and Seinfeld, J. H. 1991. Resonance structures in elastic and Raman-scattering from microspheres. *Applied Optics* 30: 459–467.

40. Tang, I. N., Fung, K. H., Imre, D. G., and Munkelwitz, H. R. 1995. Phase transformation and metastability of hygroscopic microparticles. *Aerosol Science and Technology* 23: 443–453.

41. Aardahl, C. L., Foss, W. R., and Davis, E. J. 1996. The effects of optical resonances on Raman analysis of liquid aerosols. *Journal of Aerosol Science* 27: 1015–1033.

42. Esen, C., Kaiser, T., and Schweiger, G. 1996. Raman investigation of photopolymerization reactions of single optically levitated microparticles. *Applied Spectroscopy* 50: 823–828.

43. Lee, A. K. Y., Ling, T. Y., and Chan, C. K. 2008. Understanding hygroscopic growth and phase transformation of aerosols using single particle Raman spectroscopy in an electrodynamic balance. *Faraday Discussions* 137: 245–263.

44. King, M. D., Thompson, K. C., Ward, A. D., Pfrang, C., and Hughes, B. R. 2008. Oxidation of biogenic and water-soluble compounds in aqueous and organic aerosol droplets by ozone: A kinetic and product analysis approach using laser Raman tweezers. *Faraday Discussions* 137: 173–192.

45. Reid, J. P., Meresman, H., Mitchem, L., and Symes, R. 2007. Spectroscopic studies of the size and composition of single aerosol droplets. *International Reviews in Physical Chemistry* 26: 139–192.

46. Sayer, R. M., Gatherer, R. D. B., and Reid, J. P. 2003. A laser induced fluorescence technique for determining the pH of water droplets and probing uptake dynamics. *Physical Chemistry Chemical Physics* 5: 3740–3747.

47. Choi, M. Y. and Chan, C. K. 2005. Investigation of efflorescence of inorganic aerosols using fluorescence spectroscopy. *Journal of Physical Chemistry A* 109: 1042–1048.

48. Choi, M. Y., Chan, C. K., and Zhang, Y. H. 2004. Application of fluorescence spectroscopy to study the state of water in aerosols. *Journal of Physical Chemistry A* 108: 1133–1138.

49. Zhang, Y. H. and Chan, C. K. 2000. Study of contact ion pairs of supersaturated magnesium sulfate solutions using Raman scattering of levitated single droplets. *Journal of Physical Chemistry A* 104: 9191–9196.

50. Zhang, Y. H., Choi, M. Y., and Chan, C. K. 2004. Relating hygroscopic properties of magnesium nitrate to the formation of contact ion pairs. *Journal of Physical Chemistry A* 108: 1712–1718.

51. Braun, C. and Krieger, U. K. 2001. Two dimensional angular light-scattering in aqueous NaCl single aerosol particles during deliquescence and efflorescence. *Optics Express* 8: 314–321.

52. Parsons, M. T., Riffell, J. L., and Bertram, A. K. 2006. Crystallization of aqueous inorganic-malonic acid particles: Nucleation rates, dependence on size, and dependence on the ammonium-to-sulfate. *Journal of Physical Chemistry A* 110: 8108–8115.

53. Olsen, A. P., Flagan, R. C., and Kornfield, J. A. 2006. Single-particle levitation system for automated study of homogenous solution nucleation. *Review of Scientific Instruments* 77: 073901.

54. Cruz, C. N. and Pandis, S. N. 2000. Deliquescence and hygroscopic growth of mixed inorganic–organic atmospheric aerosol. *Environmental Science and Technology* 34: 4313–4319.

55. Lightstone, J. M., Onasch, T. B., Imre, D., and Oatis, S. 2000. Deliquescence, efflorescence, and water activity in ammonium nitrate and mixed ammonium nitrate/succinic acid microparticles. *Journal of Physical Chemistry A* 104: 9337–9346.

56. Brooks, S. D., Wise, M. E., Cushing, M., and Tolbert, M. A. 2002. Deliquescence behavior of organic/ammonium sulfate aerosol. *Geophysical Research Letters* 29: 1917, doi:10.1029/2002GL014733.

57. Prenni, A. J., De Mott, P. J., and Kreidenweis, S. M. 2003. Water uptake of internally mixed particles containing ammonium sulfate and dicarboxylic acids. *Atmospheric Environment* 37: 4243–4251.

58. Wise, M. E., Surratt, J. D., Curtis, D. B., Shilling, J. E., and Tolbert, M. A. 2003. Hygroscopic growth of ammonium sulfate/dicarboxylic acids. *Journal of Geophysical Research-Atmospheres* 108: 4638, doi:10.1029/2003JD003775.

59. Braban, C. F. and Abbatt, J. P. D. 2004. A study of the phase transition behavior of internally mixed ammonium sulfate-malonic acid aerosols. *Atmospheric Chemistry and Physics* 4: 1451–1459.

60. Pant, A., Fok, A., Parsons, M. T., Mak, J., and Bertram, A. K. 2004. Deliquescence and crystallization of ammonium sulfate-glutaric acid and sodium chloride-glutaric acid particles. *Geophysical Research Letters* 31: L12111, doi:10.1029/2004GL020025.

61. Topping, D. O., McFiggans, G. B., and Coe, H. 2005. A curved multi-component aerosol hygroscopicity model framework: Part 2—Including organic compounds. *Atmospheric Chemistry and Physics* 5: 1223–1242.

62. Hansson, H. C., Wiedensohler, A., Rood, M. J., and Covert, D. S. 1990. Experimental-determination of the hygroscopic properties of organically coated aerosol-particles. *Journal of Aerosol Science* 21S: S241–S244.

63. Hansson, H. C., Rood, M. J., Koloutsou-Vakakis, S. et al. 1998. NaCl aerosol particle hygroscopicity dependence on mixing with organic compounds. *Journal of Atmospheric Chemistry* 31: 321–346.

64. Andrews, E. and Larson, S. M. 1993. Effect of surfactant layers on the size changes of aerosol-particles as a function of relative-humidity. *Environmental Science and Technology* 27: 857–865.

65. Wagner, J., Andrews, E., and Larson, S. M. 1996. Sorption of vapor phase octanoic acid onto deliquescent salt particles. *Journal of Geophysical Research-Atmospheres* 101: 19533–19540.

66. Xiong, J. Q., Zhong, M. H., Fang, C. P., Chen, L. C., and Lippmann, M. 1998. Influence of organic films on the hygroscopicity of ultrafine sulfuric acid aerosol. *Environmental Science and Technology* 32: 3536–3541.

67. Garland, R. M., Wise, M. E., Beaver, M. R. et al. 2005. Impact of palmitic acid coating on the water uptake and loss of ammonium sulfate particles. *Atmospheric Chemistry and Physics* 5: 1951–1961.

68. Gill, P. S., Graedel, T. E., and Weschler, C. J. 1983. Organic films on atmospheric aerosol-particles, fog droplets, cloud droplets, raindrops, and snowflakes. *Reviews of Geophysics* 21: 903–920.

69. Zhang, Y. H. and Chan, C. K. 2003. Observations of water monomers in supersaturated $NaClO_4$, $LiClO_4$, and $Mg(ClO_4)(2)$ droplets using Raman spectroscopy. *Journal of Physical Chemistry A* 107: 5956–5962.

70. Colberg, C. A., Krieger, U. K., and Peter, T. 2004. Morphological investigations of single levitated H_2SO_4/NH_3/H_2O aerosol particles during deliquescence/efflorescence experiments. *Journal of Physical Chemistry A* 108: 2700–2709.

71. Badger, C. L., George, I., Griffiths, P. T. et al. 2006. Phase transitions and hygroscopic growth of aerosol particles containing humic acid and mixtures of humic acid and ammonium sulphate. *Atmospheric Chemistry and Physics* 6: 755–768.

72. Martin, S. T., Schlenker, J. C., Malinowski, A., Hung, H. M., and Rudich, Y. 2003. Crystallization of atmospheric sulfate-nitrate-ammonium particles. *Geophysical Research Letters* 30, 2012, doi: 10.1029/2003/GLO17930.

73. Schlenker, J. C., Malinowski, A., Martin, S. T., Hung, H. M., and Rudich, Y. 2004. Crystals formed at 293 K by aqueous sulfate-nitrate-ammonium-proton aerosol particles. *Journal of Physical Chemistry A* 108: 9375–9383.

74. Schlenker, J. C. and Martin, S. T. 2005. Crystallization pathways of sulfate-nitrate-ammonium aerosol particles. *Journal of Physical Chemistry A* 109: 9980–9985.

75. Ling, T. Y. and Chan, C. K. 2007. Formation and transformation of metastable double salts from the crystallization of mixed ammonium nitrate and ammonium sulfate particles. *Environmental Science and Technology* 41: 8077–8083.

76. Decesari, S., Facchini, M. C., Matta, E. et al. 2002. Water soluble organic compounds formed by oxidation of soot. *Atmospheric Environment* 36: 1827–1832.

77. Asad, A., Mmereki, B. T., and Donaldson, D. J. 2004. Enhanced uptake of water by oxidatively processed oleic acid. *Atmospheric Chemistry and Physics* 4: 2083–2089.

78. Zahardis, J. and Petrucci, G. A. 2007. The oleic acid-ozone heterogeneous reaction system: products, kinetics, secondary chemistry, and atmospheric implications of a model system—a review. *Atmospheric Chemistry and Physics* 7: 1237–1274.
79. Porter, N. A., Caldwell, S. E., and Mills, K. A. 1995. Mechanisms of free-radical oxidation of unsaturated lipids. *Lipids* 30: 277–290.
80. Tachikawa, H., Polasek, M., Huang, J. Q. et al. 1998. Initiation reactions of lipid peroxidation: A Raman spectroscopic and quantum-mechanical study. *Applied Spectroscopy* 52: 1479–1482.
81. Muik, B., Lendl, B., Molina-Diaz, A., and Ayora-Canada, M. J. 2005. Direct monitoring of lipid oxidation in edible oils by Fourier transform Raman spectroscopy. *Chemistry and Physics of Lipids* 134: 173–182.
82. Chan, L. P., Lee, A. K. Y., and Chan, C. K. 2009. 2010. Gas-particle partitioning of alcohol vapors on organic aerosol particles. *Environmental Science and Technology*, 44: 257–262.

8 Micro-Raman Spectroscopy for the Analysis of Environmental Particles

Sanja Potgieter-Vermaak, Anna Worobiec, Larysa Darchuk, and Rene Van Grieken

CONTENTS

8.1 Introduction 193
8.2 From the Sample to the Result.. 195
    8.2.1 Sampling.. 195
        8.2.1.1 Samplers.. 195
        8.2.1.2 Substrates 199
    8.2.2 Analysis and Interpretation Strategy 200
        8.2.2.1 Experimental Conditions 200
        8.2.2.2 Point Analysis versus Mapping..................................... 201
        8.2.2.3 Spectral Library 202
8.3 MRS in Parallel with Elemental Analyses 202
    8.3.1 Stand-Alone Approach 202
        8.3.1.1 XRF for Bulk Elemental Concentrations...................... 202
        8.3.1.2 Sem/eds with the Possibility of Computer-Controlled Analysis for Single Particles 202
    8.3.2 Interfaced Approach 203
        8.3.2.1 Advantages of MRS Interfaced with SEM/EDS 203
        8.3.2.2 Challenges... 203
8.4 Conclusions and Future 205
References.. 205

8.1 INTRODUCTION

The analysis of environmental particles of various size ranges by means of Raman spectroscopy is a frequent topic in open literature and ranges from the analysis of soils (Jehlička et al., 2005), sediments (Moody et al., 2005; Stefaniak et al., 2006), atmospheric aerosols (Schweiger, 1999; Sengupta et al., 2005; Sobanska et al., 2006; Godoi et al., 2006; Lee and Chan, 2007), minerals (Hope et al., 2001; Stefaniak et al., 2006; Potgieter-Vermaak, 2007), and biocompounds (Baena and Lendl, 2004; Guedes et al., 2009), to many more. It certainly is therefore not a topic to be exploited in a single chapter and the authors will limit the focus to the application of micro-Raman spectroscopy (MRS) on particulate matter (PM) of atmospheric origin. The advantages and challenges of utilizing this technique on a stand-alone basis, as well as in parallel/interfaced with scanning electron microscopy (SEM) coupled with energy-dispersive x-ray spectrometric detection (SEM/EDS, also computer-controlled SEM/EDS—CCSEM/EDS), electron probe microanalysis with x-ray detection (EPXMA), and x-ray fluorescence spectroscopy (XRF), will be discussed.

CCSEM/EDS allows the characterization of large numbers of individual particles in a fast, automated way and provides reliable statistical results. A reverse Monte Carlo quantification procedure is used to calculate elemental concentrations within 15% relative accuracy, even for low-Z elements, if a thin-window detector is used (thin-window SEM/EDS) (Ro et al., 1991; Szalóki et al., 2000, 2001). The determination of beam-sensitive particles such as ammonium sulfate and nitrate are facilitated by using a liquid-nitrogen cooled sample stage (Szalóki et al., 2001; Worobiec et al., 2003). The results obtained by following this strategy of single particles analysis (SPA) are then used in conjunction with bulk elemental analysis obtained by XRF to provide information on the chemical composition and properties of the aerosol particles (Godoi et al., 2006; Worobiec et al., 2007). This information is of prime importance when interpreting air pollution trends and its subsequent influence on environmental monitoring and preventative conservation. Krueger et al. (2003) used this methodology to investigate the heterogeneous reactions incurred when silica, calcite, sodium chloride, and sea salt particles were exposed to nitric acid vapors. Further applications of this methodology are quite varied; the determination of atmospheric marine salt depositions (Delalieux et al., 2006), indoor air quality in schools and museums (Worobiec et al., 2006; Avigo et al., 2008), and the impact of industrial activities on human health (Godoi et al., 2008a), to name but a few.

The sensitivity of Raman spectroscopy (which enables one to investigate the composition, phase, crystallinity, crystal orientation, and, in some cases, doping of materials on a micrometer scale) makes it an ideal tool to characterize individual heterogeneous particles in the fine particle size range. Available literature on the application of MRS for the analysis of single atmospheric particles is sparse, and even more so when it is used in combination with or coupled to SEM/EDS. Some of the very first publications on Raman spectroscopy and aerosol particles were those of Fung and Tang (Fung and Tang, 1989, 1992; Tang and Fung, 1989), who reported on an extended study of artificial inorganic salt particles. A sophisticated setup was built, in which levitated single particles of known composition and in solid and liquid states could be analyzed by Raman. A valuable contribution has been made by this study in that the result of mixed salts, typically found in aerosols, were characterized for their respective Raman spectra. Since the early 1990s, various studies on levitated particles exposed to different conditions and their consequent heterogeneous reactions were studied, some with the use of Raman spectroscopy (e.g., Aardahl et al., 1996; Trunk et al., 1998; Musick and Popp, 1999). Wu et al. (2008) used MRS to observe the effect of potassium nitrate on the phase transitions of ammonium nitrate and came to the conclusion that ammonium nitrate may exist in phase form III at lower temperatures in the presence of potassium nitrate and therefore this phase transition should be considered possible during modeling.

The publications discussed in the previous paragraph revolved around carefully controlled laboratory setups utilizing synthetically generated particles, of which the theory and further application have been elegantly discussed in a review article by Schweiger (1999). The group of articles that dealt with the carbonaceous content of PM also deserves attention. Sze et al. (2001) published an article where the authors discuss the possibility to use the characteristic graphitic (G) and disorder (D) band ratios of carbonaceous molecules to fingerprint the carbonaceous matter found in PM. In this way, they speculate that the characteristic morphologies thus determined could be used to identify their respective sources. In a recent review on diesel soot by Maricq (2007), it was indicated that PM with diesel as source had still only been investigated for its C-containing component, but that preliminary studies showed extended possibilities. Escribano et al. (2001) also investigated PM among other C-containing compounds and concluded that the Raman spectra could give information on the physical character of the samples, that is, size, degree of disorder of the carbon microcrystalline domains, and so on. These authors also investigated the PM by means of SEM to characterize the morphology. Finally, a preliminary investigation on the use of MRS on PM collected on various substrates was reported (Potgieter-Vermaak and Van Grieken, 2006). The authors were able to identify various compounds, apart from soot, and discussed the practical challenges of analyzing PM collected by impaction. In addition, the ambiguous characterization of conglomerates and the necessity of complementing the molecular characterization with an elemental characterization of the same particle were mentioned.

A group at the University of Sciences and Technologies of Lille (Batonneau et al., 2001; Dupuy and Batonneau, 2003; Falgayrac et al., 2006; Choël et al., 2006; Batonneau et al., 2006; Sobanska et al., 2006; Uzu et al., 2009) used Raman mapping, mostly in conjunction with SEM/EDS and environmental SEM (ESEM/EDS) to characterize size-segregated atmospheric particles. Various sophisticated multivariate analysis techniques were used to extract intensities of pure species from a matrix of up to 2400 spectra. Individual particle analysis and/or CCSEM/EDS was discussed only in two of these papers (Choël et al., 2006; Uzu et al., 2009). This group has sampled dust emissions at a battery recycling plant during three campaigns, as well as PM some distance from the smelter during two campaigns in the late 1990s early 2000s. Although the focus was different in all their publications, the analytical methodology with regard to the Raman characterization remained the same. In one of their recent publications (Choël et al., 2006), for example, dust collected at the smelter, size segregated after resuspension, and PM collected up- and down-wind were analyzed for their bulk elemental composition (inductively coupled plasma spectrometry—ICP, XRF, and atomic absorption spectroscopy—AAS), water-soluble ionic composition (ion chromatography—IC), morphology (ESEM), SP composition (CCESEM) and, finally, for their molecular composition with MRS mapping. The atmospheric particles could not be analyzed for the low-Z elements due to constraints imposed by the instrumentation. Electron beam-sensitive particles could therefore not be analyzed by SEM or SPA and conclusions on the presence and quantities of these particles were based on total ionic concentrations. Although all these techniques provided data that could be used to answer the various aims in their publications, it is nowhere evident that the exact particle was analyzed by SEM/EDS and MRS.

A group at the University of Antwerp started with the application of MRS and SEM/EDS (computer controlled and manual) in 2005 and published an article on nano-manipulated particles of atmospheric origin in 2006 (Godoi et al., 2006). Particles collected in an indoor environment (Rubens House museum in Antwerp, Belgium) were transferred from the collection substrate to a TEM grid with the aid of tweezers. The exact same particles were manually analyzed with SEM/EDS and MRS. Various articles followed, whereby analysis on the same particles were performed by both instruments (Worobiec et al., 2006, 2007; Darchuk et al., 2008; Godoi et al., 2008b).

It is evident from the selected publications discussed in the previous paragraphs that the application of Raman spectroscopy to atmospheric particles has a long and interesting history. Numerous publications can be found on synthetic particle investigations. Although its application to carbon-containing atmospheric particles is fairly extensive, the number of publications that could be found on bulk analysis has been limited, even more so if the technique has been used together with an elemental characterization technique, such as SEM/EDS.

8.2 FROM THE SAMPLE TO THE RESULT

8.2.1 SAMPLING

8.2.1.1 Samplers

In environmental research, the correct sampling is crucial to obtain objective, representative, and correct information about the physical and chemical properties of the environmental particles. Under normal circumstances, the sampling technique and methodology depend heavily on the type of analytical techniques that will follow, particularly if it involves research on atmospheric PM. The sampling strategy should be adapted in such way as to protect the original chemical and physical properties of the PM under investigation. In addition, sample preparation should be avoided. PM sampling techniques (samplers) can generally be divided into two groups, depending on the information that is expected to be obtained:

- Sampling to perform bulk elemental analysis.
- Sampling to perform the analysis of individual particles for its chemical composition.

A detailed overview of the various samplers is given by Cahill (1990), Akselsson (1984), Annegarn et al. (1988), and Van Grieken and Markowicz (1993).

8.2.1.1.1 Sampling to Perform Bulk Analysis

Collection of the aerosol is done by high- or low-volume filtration and may be size segregated as is the case with the stacked filter unit (Otten, 1991) and the dichotomous sampler or virtual impactor (Van Grieken and Markowicz, 1993). Currently, numerous different kinds of active air suspended PM samplers are being used and operate at differing flow rates. Mostly, such devices can be operated online, with little restriction on time intervals, and could therefore provide information about the measured parameter in a time-dependence manner.

In cases where more detailed or sophisticated data are required, the preferred choice of sampling is "offline," even if the analysis can be done *in situ* and under "field conditions." As mentioned elsewhere, the sampling methodology will depend on the analysis technique employed. For bulk elemental, mass, and ionic concentration analyses, the most convenient method of sample collection is via impaction or filtration. For the collection of total air suspended PM, a stacked filter unit (SFU) was used in order to collect coarse (2.5 µm) and fine (<2.5 µm) aerosol particles (Otten, 1991). The SFU collects particles on 47-mm-diameter Nuclepore filters. Coarse particles were collected on an 8.0 µm pore-size filter while fine particle collection was performed on a 0.4 µm pore-size high-density filter. Filters were protected from outside elements by a Plexiglas cylinder which did not influence the size and the mass distribution of the aerosol samples. Deposition of particles forms a homogeneous layer and could then be analyzed by, for example, energy dispersive x-ray fluorescence (EDXRF).

Air quality guidelines throughout the world used to focus on PM10 (PM with an aerodynamic diameter below 10 µm), but of late PM2.5 guideline values have also been proposed. Impactors that allow collection in the required particle range are the Harvard-type PM1, PM2.5, and PM10 impactors (MS and T Area Samplers, Air Diagnostics and Engineering, Inc., Harrison, ME, USA), illustrated in Figure 8.1. Harvard-type impactors are equipped with quiet oil-free pumping units (Air Diagnostics and Engineering, air sampling pump, model SP-280E) (Stranger, 2005).

As depicted in Figure 8.1, inside the MS and T Area Sampler the air stream is drawn through an inlet and a nozzle. After the nozzle, particles larger than 1 or 2.5 or 10 µm (depending on the impactor type) impact on a sintered metal impactor plate and particles smaller than 1, 2.5, or 10 µm are consequently collected on a Teflon membrane filter (37 mm, 2 µm pore size, Anderson Teflon Membrane filters), placed onto a support ring. The integrity of the filters is maintained by being supported by a drain disc (Polyolefin drain discs, Whatman Inc., Maidstone, England). The pumping

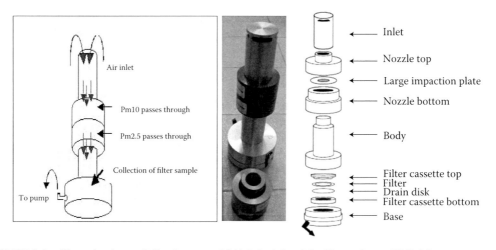

FIGURE 8.1 **(See color insert following page 206.)** Principle of the Harvard-type PM2.5 impactor.

unit is operated at a flow rate of 10 L min^{-1} for the fractions PM2.5 and PM10, and 23 L min^{-1} for the PM1 fraction. A custom built-in timer allows the pump to operate at specified time intervals. A homogeneous layer of PM allows XRF analysis for bulk elemental analysis, mass concentration determination by gravimetry, water-soluble ionic content after desorption by means of IC, and molecular analysis by means of MRS. The collection time depends on the research aims and the analytical limits and requirements. In general, a 24-h sampling period is applied to investigate the daily changes in the composition of the air suspended PM. Sufficient loading (thin layer) then enables the detection of most common elements by means of EDXRF. SPA could also be performed on these filters by means of EPXMA (or CCSEM/EDS) or MRS provided that particle collection takes place in such a manner that PM is collected as spatially separated particles. It should be noted that low-Z elemental analyses are compromised due to the chemical composition of the filter.

8.2.1.1.2 Samplers and Collection for SPA

Although the alteration of PM during collection cannot be totally avoided, it can be minimized by using the appropriate filtering device and substrate. For analysis by means of MRS and SEM/EDS (or EPXMA), it is crucial that the particle in its initial state be preserved to enable the correct determination of its chemical composition, morphology, shape, and diameter. In addition, minimum or no sample preparation should be required.

Cascade impactors, a multistage sampling device, are widely used for the collection of individual particles. By the impingement of the aerosol on a sequence of solid discs (referred to as stages) arranged in a vertical manner, as illustrated in Figure 8.2, particles are collected in a size-segregated way with the largest particles impacting first and the smallest last. By controlling the sampling time, a thin layer of separated particles can be obtained. The construction of the impactors allows the use of different collection substrates, which are discussed in the next section.

Due to its versatility, the most suitable impactor for the collection of individual particles on different substrates is the Berner-type cascade impactor, as illustrated in Figure 8.3. This 8-stage cascade impactor has been widely used in recent aerosol studies in Europe and in the United States and collects PM with cut-off diameters of 8/4/2/1/0.5/0.25/0.125/0.0625 μm. The jets are arranged in a circle. The collection plate has a large hole in the centre for the air to pass through to the

FIGURE 8.2 Scheme of one cascade impactor.

FIGURE 8.3   Assembled cascade impactor.

subsequent stage. This flow arrangement gives a symmetrical flow pattern, and thus deposits under each jet. A critical orifice, downstream of the final collection stage, controls the final flow rate through the impactor (see Figure 8.4). In general, the sampling time for atmospheric aerosol collection varies between 1 min (for stage 3) and 240 min (for stage 8), but it strongly depends on the particle concentration in the airstream. Various substrates could be used simultaneously without compromising the collection efficiency (see further discussion in Section 8.2.1.2).

Another type of cascade impactor is the 1 L/min single-orifice Battelle-type cascade impactor, as modified and commercialized by PIXE International Corporation (RPCI—PIXE International Corporation, Tallahassee, FL) (Baumann et al., 1981; Van Grieken and Markowicz, 1993). The schematic view and the technical details of the impactor are shown in Figure 8.5. The impactor differentiates the aerosol in up to 10 size fractions (>16, 16–8, 8–4, 4–2, 2–1, 1.0–0.5, 0.5–0.25, 0.25–0.12, 0.12–0.06,

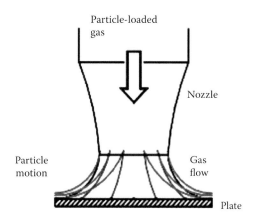

FIGURE 8.4   Particle separation in the impactor.

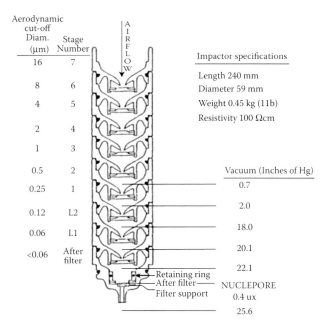

FIGURE 8.5   Cross section view of the PIXE cascade.

and <0.06 μm), provides good size resolution in the submicrometer range, and can be operated from small battery-powered pumps. Samples can be impacted onto Nuclepore (PCTE) membranes, TEM grids with an ultrathin carbon layer or other impaction substrate backings, and they are suitable for analysis using ion-beam techniques such as PIXE and Proton Elastic-Scattering Analysis (PESA) or SEM/EDS or MRS. These small impactors are ideally suited for use in confined areas such as show-cases, often used in museums to protect fragile or very valuable artefacts.

8.2.1.2   Substrates

For the analysis of single particles of atmospheric origin by means of MRS, a suitable substrate is of prime importance. The publication by Potgieter–Vermaak and Van Grieken (2006) clearly illustrates the contribution of the various substrates currently in use, namely; Si wafers, Nuclepore filters, Ag foil, Teflon membranes, and TEM grids. From the spectra it is obvious that all these substrates would cause ambiguous characterization if the particle is smaller than the laser spot or transparent to the laser, allowing penetration of the laser through the particle. This limits the application of MRS to PM of size ≥500 nm. In addition, substrates such as Ag foil, which could undergo aging owing to oxidation, should be treated with caution. The main disadvantages of the various substrates investigated are Nuclepore and Teflon filters—characteristic Raman spectrum, resulting in masking of the analyte spectrum; Si-wafer—characteristic Raman shift at 520.5 nm; Ag foil—characteristic spectrum of oxidation products of Ag, resulting in the masking of amorphous C, and TEM grids—characteristic spectrum for amorphous C from the C overlay, resulting in the masking of amorphous C.

In a follow-up study, a rather fundamental discussion around a suitable collection substrate for analysis by MRS and CCSEM/EDS was given (Godoi et al., 2006). In automated particle analysis, the backscattered electron image/signal (BEI) is used for particle recognition and is based on the contrast in mean atomic number (Z) between the substrate and particles. The BEI line profile of the various substrates investigated, that is, polycarbonate filters (an etched thin polymer film referred to as Nuclepore), Ag foil, Al foil, Si wafers, Be discs, C tapes, and TEM grids, supporting a pure calcium carbonate particle, indicated that the TEM grid with an ultrathin C layer outperformed all other substrates in terms of the sharpness of analytical signal and background noise size. The contribution of the substrate to the x-ray signal was also investigated and the only substrate that had a minimal contribution was the TEM grid. Finally, particle size dependence was investigated and as expected the contribution of the substrate reduced with an increase in size. This chapter further illustrates that the use of a TEM grid to collect size-segregated PM would be most successful if it has to be analyzed with both techniques. The additional feature of the unique midpoint of the TEM grid, enabling easy relocation of particles, makes it the preferred choice. The use of TEM grids is restricted to sampling by the so-called PIXE impactor. Application of this technique of sampling in large open spaces could yield ambiguous results, because it is statistically not representative of the bulk sample.

From past experience, a high-quality Ag foil, suitably cleaned, handled, and stored, could also be used as a substrate, and is the preferred choice in the Berner impactor, if large volumes of air need to be sampled. Ag foil has also been identified as the preferred choice if both techniques [electron probe microanalysis (EPMA) and MRS as stand-alone and combined instruments] are employed (Worobiec et al., 2009). Not only does Ag foil have more application possibilities in various types of impactors (in contrast to TEM grids), but has surface enhancement Raman spectroscopy (SERS) properties. The Raman effect, which is rather weak by nature, can greatly be enhanced if the molecules are attached to nanometre-sized metal structures with appropriate "roughness." Ag is, in practice, most widely used as a SERS material. However, the literature related to this aspect describes in most cases an etched Ag surface or Ag colloidal solution coating a smooth substrate. Unfortunately, a modified Ag foil with a significantly rough surface becomes useless for the analyses of small particles by MRS or SEM/EDS. Recent experiments showed, however, that even a nontreated Ag foil displayed enhancement of signal intensity, as shown in Figure 8.6 (Worobiec et al., 2009).

8.2.2 ANALYSIS AND INTERPRETATION STRATEGY

8.2.2.1 Experimental Conditions

The general and specific experimental conditions for the analysis of individual atmospheric particles by means of CCEM/EDS or EPXMA can be found in Van Grieken and Markowicz (1993) and the discussion in this section will be directed toward MRS. In MRS, a continuous laser beam is focused on a sample via microscope objectives. The photons interact with the molecules of the sample by the phenomenon of Raman scattering, in which largely the vibration modes of the molecules are detected

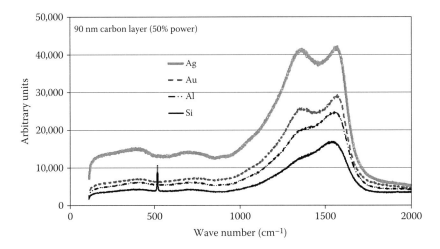

FIGURE 8.6 Raman spectrum of amorphous C analyzed under the same experimental conditions on various substrates.

by characteristic energy transfers to and from the photon. The optical spectrum of the scattered photons, called the Raman spectrum, is highly diagnostic of the molecular composition of compounds present in the sample analyzed. The analytical sample can be a bulk solid or a single particle of micrometer size. Specific problems of SPA are related to the small sizes of the particles. The main problem is the difficulty to perform quantitative analysis, because of the uncertainty in the determination of the interaction volume and the absence of suitable particle standards.

Some publications (Potgieter–Vermaak and Van Grieken, 2006) report the results from the analysis of samples of aerosol particles, which were analyzed with an InVia micro Raman spectrometer from Renishaw (Wotton-under-Edge, the United Kingdom). The general procedure followed is described in the following paragraph. For vibrational excitation, two lasers were applied: an argon laser (514.5 nm, Spectra Physica) with a maximum laser power of 50 mW and a diode laser (785 nm, Renishaw) with a maximum laser power of 300 mW. The laser beam power was attenuated by using density filters to ensure that beam damage is avoided without compromising on spectral resolution and acceptable S/N ratio. The Raman scattering detector is a highly sensitive Peltier CCD detector. Raman spectra were obtained using a synchroscan mode in the spectral range 100–3200 cm^{-1} or by applying a static mode. Calibration of the laser beam was done using the 520.5 cm^{-1} line of a Si wafer.

Measurement of times between 10 and 30 s have been used to collect the Raman spectra with a signal-to-noise ratio of better than 100/1 at a resolution of 2 cm^{-1}. The spectra were acquired by coaddition of several scans (between 2 and 50, depending on the scanning mode). Spectra of single particles were obtained using $100 \times$ (NA = 0.95; theoretical spot size is of 0.36 μm and 0.50 μm for 514.5 nm and for 785 nm laser, respectively) and $50 \times$ (NA = 0.70; theoretical spot size is of 0.45 μm and 0.68 μm for 514.5 nm and for 785 nm laser, respectively) magnification objectives. The laser beam spot size depends on the objectives and therefore the volume of sample analyzed could to a certain extent be regulated. A substrate with the impacted particles was put on the microscope stage without any prior sample preparation. Analyses of the aerosol particles (minimum 100 particles per sample) from different areas of the spot have been performed to get representative results. Particles were located and analyzed manually. Data acquisition was carried out with the Wire™ and Spectracalc software packages from Renishaw (Wotton-under-Edge, the United Kingdom).

8.2.2.2 Point Analysis versus Mapping

The use of computer-controlled micro-Raman mapping on collected atmospheric particles has been shown to be beneficial in bulk molecular characterization (Sobanska et al., 2006). The strategy followed by this group (Windig and Stephenson, 1992; Batonneau et al., 2001; Dupuy and Batonneau, 2003; Choël et al., 2006; Falgayrac et al., 2006; Batonneau et al., 2006; Sobanska et al., 2006; Uzu et al., 2009) is briefly described below. With the use of an XY-motorized stage and computer-controlled point-by-point scanning with 0.1 μm resolution, 1 μm reproducibility, and 90 mm × 60 mm maximum spatial range, atmospheric particles were analyzed for molecular composition. The spot size of the laser focused by the objective at the sample was of 1 μm^2. Spectra were indexed on the two-dimensional grid defined by the user. A systematic Raman mapping is carried out with a fixed step size shift (e.g., 1 μm) corresponding to the lateral resolution of the instrument, over a large sample area defined by a width (X) and a length (Y) expressed in ($m \times n$) pixels or μm^2. A three-dimensional table ($m \times n \times N$) was obtained and consisted of $m \times n$ spectra with the $N = 3770$ spectral points (for the spectral region from 100 to 3200 cm^{-1}). A narrow spectral range corresponding to characteristic Raman bands of a compound of interest could be selected from the whole spectral range. By integration, an estimation of the relative concentration of the compound of interest could be determined and provided a ($m \times n$) Raman image of the compound.

Unfortunately, this conventional procedure has many drawbacks for heterogeneous samples. The large heterogeneity at the level of the spatial resolution generates severe overlaps of the characteristic Raman peaks of the several compounds of interest in the pixel spectrum. Application of a multivariate resolution to the ($m \times n \times N$) spectral set yields Raman information about the nature of the compounds present in the sample and the amount of each compound present in each pixel. The method used for self-modeling mixture analysis is the SIMPLe-to-use Interactive Self-modeling Mixture Analysis (SIMPLISMA) approach. The mathematical principle of SIMPLISMA is based on the presence of pure variables (a wave number for Raman scattering spectra), where a pure variable is a variable which has intensity contribution from only one of the components of the mixture. When the pure variables of all components are known, it is possible to resolve the spectra of the pure components from the mixture spectra. The pure variables can be determined by mathematical means without prior knowledge of the pure components.

SIMPLISMA is then a powerful tool to determine the pure variables followed by the calculation of the pure spectra and associated "contributions." Some applications of SIMPLISMA have already been reported in the literature on Fourier transformed Raman spectra of airborne particles collected in an industrial zone (Sobanska et al., 2006) and tropospheric Pb- and Zn-rich particles (Choël et al., 2006).

8.2.2.3   Spectral Library

For the identification of Raman spectra, the experimental spectra (band wavenumber and relative intensities) are compared to reference spectra using commercially available Spectral Libraries, such as Search ID 301 software (Spectracalc software package GRAMS, Galactic Industries, Salem, NH, USA). For an unambiguous identification of the spectra, an in-house library is rather recommended. Recent research shows that the different instrumental functions and the experimental conditions such as temperature, atmosphere, laser power, laser wavelength, and so on can significantly influence the spectra and some small deviations in band positions are observed (Worobiec et al., 2009). This is of particular importance in case of band identification when samples of general unknown composition are analyzed.

8.3   MRS IN PARALLEL WITH ELEMENTAL ANALYSES

8.3.1   STAND-ALONE APPROACH

8.3.1.1   XRF for Bulk Elemental Concentrations

As already mentioned elsewhere, intimate knowledge of the chemical composition of air suspended PM is crucial when atmospheric pollution and global climate changes are studied. XRF, and especially EDXRF, allow bulk analysis at the sub-ppm level. Unifying, for example, for atmospheric aerosol studies (and for environmental particles in general), bulk XRF with single-particle EPXMA allows the identification of the sources of such particles and predict their impact on pollution, with relative ease. The chemical composition and size of single particles are important for assessing the potential of these particles to penetrate into the alveoli of the lungs and to be deposited on works of art, but bulk analysis with much better sensitivity allows information to be obtained about, for example, heavy metals (Van Grieken et al., 2009).

In the analytical procedure, the bulk analysis by EDXRF (also possible with other atomic spectrometric techniques, that is, inductively coupled plasma emission or mass spectrometry) gives the initial information about the elemental composition and concentration of the samples, apart from the mass concentration that is obtained gravimetrically. The information on the elemental composition is used to develop the calculation model for the elemental weight concentrations required to process the data from the individual particle analysis by EPXMA. Elemental composition is also helpful during spectral identification of Raman spectra in both semibulk and individual particle MRS analyses.

8.3.1.2   SEM/EDS with the Possibility of Computer-Controlled Analysis for Single Particles

SEM combined with x-ray emission for visualization and elemental analysis has been widely used in the study of atmospheric particles, because it enables visualization with excellent spatial resolution and a large depth of field. Although it reveals detailed information about the sample morphology and size at a submicrometer scale, high-quality elemental analysis and the lack of molecular information is evident, making the analysis of complex heterogeneous compounds particularly challenging. MRS, on the other hand, provides chemical, physical, and structural information about the materials analyzed, but the visual resolution is much poorer than in SEM/EDS. The combination of these two different techniques has proven to assist greatly in the unambiguous chemical and structural characterization of atmospheric particles.

This analytical approach, as has been shown by Worobiec et al. (2009), still needs optimization in various fundamental and analytical aspects. Worobiec et al. (2009) discussed the following practical considerations; sample preparation for sequential SEM/EDS/MRS analysis, particle recognition techniques, the compatibility of electron and laser spots, beam damage, and microchemical reactions; vacuum sample chamber versus ambient analysis.

Relocation of the same particle in a sample when analyzing it with more than one instrument poses various challenges and the best solution was found by using a TEM grid as a coordination system, of which the application is elegantly illustrated in Godoi et al. (2006). Concerning the compatibility of the two beam spots, the difficulty in exposing the particle to the same excitation volume is highlighted. The diameter of the electron beam has to be correlated with the laser spot size, in order to cover precisely the same area of interest. Second, the probing depth is discussed. In both cases, the nature of the sample plays a role. For SEM/EDS analysis this is controlled by the energy of the electron beam (accelerating voltage). For MRS, however, the collected volume of scattered light can be only be minimized by using a high-magnification objective, working in confocal mode (to reduce the volume of scattered light) and, in the case of some instruments, introducing a pinhole that reduces the beam volume before it excites the sample. These measures of control, however, are always at the expense of the quality of the optical image and consequently the intensity of the signal. Objectives of a large magnification are characterized by a small depth of field, which can cause loss of detail in the particle morphology and consequently lead to the incorrect characterization of the object (Stefaniak et al., 2006). Beam damage can be mitigated by using a cold stage for SEM/EDS and the appropriate power density for the laser.

8.3.2   Interfaced Approach

8.3.2.1   Advantages of MRS Interfaced with SEM/EDS

The advantages of stand-alone SEM/EDS and MRS analyses for the determination of elemental and molecular profiles of atmospheric particles have been illustrated in previous paragraphs, despite the challenges faced. It is also evident from the discussion that a hybrid instrument, combining these two techniques, would transform the SEM into a powerful characterization tool. Such a combination allows morphological, elemental, chemical, and physical analysis without moving the sample between instruments. The hybrid approach is, however, not without its own limitations and challenges as elegantly described in Worobiec et al. (2009) and briefly outlined in the following paragraphs.

Some of the challenges faced with the sequential analyses are solved, such as object relocation, since the superior imaging properties of the SEM provides the best spatial resolution and derives contrast either from surface topography or mean atomic number. Features of interest can be easily located using the SEM, and can then be rapidly identified with MRS. Consequently, the elemental and molecular spectra, the SEI, and the white light images, can be acquired from the same sample position. The data which can be acquired include (1) morphology and mean atomic number from SEM (SEI and BEI); (2) elemental composition from EDS analysis; (3) chemical composition and identification from MRS; and (4) physical structure (crystallographic and mechanical data) from MRS.

8.3.2.2   Challenges

Some challenges remain, however, and new ones appear. In Worobiec et al. (2009), the most important aspects are highlighted: (a) vacuum versus ambient sample chamber for MRS; (b) beam damage

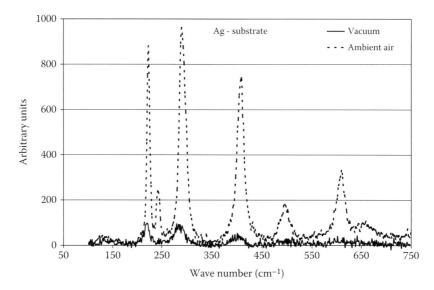

FIGURE 8.7   Raman spectrum of hematite: effect of the atmosphere (vacuum/ambient air) on the intensities; laser power: 0.5 mW (effective on sample); substrate: Ag-foil; acquisition tile of 1 scan: 60 s.

and molecular changes; (c) the deposition of amorphous carbon in SEM/EDS sample chamber; and (d) the relocation of fine particles. One of the disadvantages of the hybrid instrument is that the Raman intensities are in general 10× lower than those obtained in ambient conditions with the same laser power density, as illustrated in Figure 8.7.

This inevitably results in the loss of analytical signal and the spectrum of particles with chemical components that have small Raman cross-sections could be masked by the noise. During investigations of Worobiec et al. (2009), it became apparent that beam damage in the hybrid system occurs in a totally different manner and the physical and chemical phenomena underlying the damage remain unclear. The presence and continuous observation of amorphous carbon due to apparent deposition has been a problem with the hybrid instrument. A typical example is shown in Figure 8.8.

The build-up of carbon on sample surfaces in the SEM is well known and frequently observed, but remains relatively poorly understood. It is thought that carbon absorption on the surface

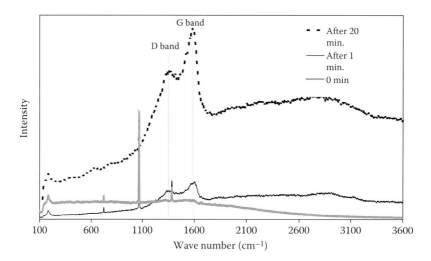

FIGURE 8.8   Raman intensities of amorphous C detected during the analysis of artificially generated sodium nitrate particles, deposited on Ag-foil.

(from ambient exposure or contamination of the vacuum chamber) is cross-linked as the electron beam impinges on the CH_x species, resulting in the polymerization of the species and the consequent stabilization of carbon species. This effect results in a "growing" carbon layer on the sample surface. This can be frequently observed as what is often (and erroneously) referred to as a "charge square" on the sample. In Figure 8.8, it is clearly observed that longer exposure times result in higher intensities of the characteristic D and G bands. This phenomenon could then result in the masking of the Raman signal of particles of interest. This is especially a concern when x-ray mapping is done, where the high probe currents and pixel dwell times make thick carbon overlayer growth inevitable. Consequently, x-ray mapping should only be carried out *after* SEM observation, x-ray point analysis and SEM-MRS analysis has been carried out.

Although the issue of particle relocation is essentially solved by the use of a hybrid instrument, it is not necessarily without challenges when one considers atmospheric particles of submicrometer size. Excitation of precisely the same area is strongly dependent on the precision of the reconfiguration of the laser or electron beam that needs to take place during switching between the two modes of analysis. When the appearances of particles are different due to the difference of the depth of field, it becomes nearly impossible to ensure that exactly the same micrometer particle and particle part are being analyzed. The current geometry and position of the detectors and probes in the sample chamber of the commercially available hybrid instrument makes it impossible to collect the x-ray spectra and MRS spectra simultaneously. There is little or no control on the exact position of the particle of interest, other than to rely on the precision of the mechanical insertion and retraction mechanism. Geometrical shifts on micrometer scale attributable to the SEM may occur and therefore the analysis of particles smaller than 3 µm might be rendered unreliable.

8.4 CONCLUSIONS AND FUTURE

Raman analysis applied to atmospheric particles has undoubted advantages and, on its own, it provides useful information on the molecular character of the aerosol under investigation. This can be either accomplished with manual point analyses or by mapping, each with its own constraints and challenges. The additional information gained if it is used parallel or sequential to SEM/EDS (stand-alone or hybrid) adds significant value to the interpretation of air pollution trends and its subsequent influence on the environment. This approach of characterization, however, is also not without challenges and the research showed that the influences of the sample presentation, preparation, and analysis strategy need to be very clearly understood and planned.

One of the most important criteria is that such an approach must guarantee that the environmental particle under question is analyzed in the original state of collection, and particular care must be taken to avoid changes of the particles owing to the nature of the analysis.

The analytical procedure (i.e., the determination of mass concentration, bulk elemental concentration, bulk ionic composition, single particle elemental concentration, and clusterification followed by molecular characterization of the individual particle) supplies the valuable information required to interpret the influence of PM on pollution phenomena. The challenges faced in some of the techniques involved in this analytical procedure have been illustrated in the previous paragraph. If these challenges can be addressed to a satisfactory degree, the end result could prove to be of great significance for various applications where the analysis of heterogeneous particles is beneficial, and specifically for the analysis of atmospheric particles. MRS provides a unique, elegant, nondestructive way to characterize the molecular make-up of PM.

REFERENCES

Aardahl, C. L., Foss, W. R., and Davis, E. J. 1996. The effects of optical resonances on Raman analysis of liquid aerosols. *J. Aerosol. Sci.* 27:1015–1033.

Akselsson, K. R. 1984. Aerosol sampling and samplers matched to PIXE analysis. *Nucl. Instr. Meth.* B3:425–430.

Annegarn, H. J., Cahill, T. A., Sellschop, J. P. F., and Zucchiatti, A. 1988. Time sequence particulate sampling and nuclear analysis. *Phys. Scripta* 37:282–291.

Avigo, Jr., D., Godoi, A. F. L., Janissek, P. R., Makarovska, Y., Krata, A., Potgieter-Vermaak, S., Alföldy, B., Van Grieken, R., and Godoi, R. H. M. 2008. Particulate matter analysis at elementary schools in Curitiba, Brazil. *Anal. Bioanal. Chem.* 391:1459–1468.

Baena, J. R. and Lendl, B. 2004. Raman spectroscopy in chemical bioanalysis. *Curr. Opin. Chem. Biol.* 8:534–539.

Batonneau, Y., Laureyns, J., Merlin, J.-C., and Brémard, C. 2001. Self-modelling mixture analysis of Raman micro spectrometric investigations of dust emitted by lead and zinc smelters. *Anal. Chim. Acta* 446:23–37.

Batonneau, Y., Sobanska, S., Lauryens, J., and Bremard, C. 2006. Confocal microprobe Raman imaging of urban tropospheric aerosol particles. *Eniron. Sci. Technol.* 40:1300–1306.

Baumann, S., Houmere, P.D., and Nelson, J.W. 1981. Cascade impactor samples for PIXE analysis. *Nucl. Instr. Meth.* 181:499–502.

Cahill, T. A. 1990. Analysis of air pollutants by PIXE: The second decade. *Nucl. Instr. Meth.* B49:345–350.

Choël, M., Deboudt, K., Flament, P., Lecornet, G., Perdrix, E., and Sobanska, S. 2006. Fast evolution of tropospheric Pb- and Zn-rich particles in the vicinity of a lead smelter. *Atmos. Environ.* 40:4439–4449.

Darchuk, L., Stefaniak, E. A., Worobiec, A., Kiro, S., Oprya, M., Ennan A., Bekshaev, A., Kontush, S., Ospitali, F., and Van Grieken, R. 2008. Influence of experimental conditions for MRS investigation of Fe-rich particles formed at manual metal arc welding. In R. Withnall and B. Z. Chowdhry, Eds, *Proceedings of the XXIst International Conference on Raman Spectroscopy*, IM Publications LLP, Chichester, UK, 1140–1141.

Delalieux, F., Van Grieken, R., and Potgieter, J. H. 2006. Distribution of atmospheric marine salt depositions over Continental Western Europe. *Mar. Pollut. Bull.* 52:606–611.

Dupuy, N. Y. and Batonneau, Y. 2003. Reliability of the contribution profiles obtained through the SIMPLISMA approach and used as reference in a calibration process. Application to Raman micro-analysis of dust particles. *Anal. Chim. Acta* 495:205–215.

Escribano, R., Sloan, J. J., Sidique, N., Sze, S.-K., and Dudev, T. 2001. Raman spectroscopy of carbon containing particles. *Vib. Spectros.* 26:179–186.

Falgayrac, G., Sobanska, S., Laureyns, J., and Brémard, C. 2006. Heterogeneous chemistry between $PbSO_4$ and calcite microparticles using Raman microimaging. *Spectrochim. Acta A* 64:1095–1101.

Fung, K. H. and Tang, I. N. 1989. Composition analysis of suspended aerosol particles by Raman spectroscopy: Sulphates and nitrates. *J. Colloid Inter. Sci.* 130:219–224.

Fung, K. H. and Tang, I. N. 1992. Aerosol particle analysis by resonance Raman spectroscopy. *J. Aerosol Sci.* 23:309–307.

Godoi, R. H. M., Braga, D. M., Makarovska, Y., Alföldy, B., Carvalho Filho, M. A. S., Van Grieken, R., and Godoi, A. F. L. 2008a. Inhalable particulate matter from lime industries: Chemical composition and deposition in human respiratory tract. *Atmos. Environ.* 42:7027–7033.

Godoi, R. H. M., Potgieter-Vermaak, S., Godoi, A. F. L., Stranger, M., and Van Grieken, R. 2008b. Assessment of aerosol particles within the Rubens' House Museum in Antwerp, Belgium. *X-Ray Spectrom.* 37:298–303.

Godoi, R. H. M., Potgieter-Vermaak, S., De Hoog, J., Kaegi, R., and Van Grieken, R. 2006. Substrate selection for optimum qualitative and quantitative single atmospheric particles analysis using nano-manipulation, sequential thin-window electron probe x-ray microanalysis and micro-Raman spectroscopy. *Spectrochim. Acta B* 61:375–388.

Guedes, A., Ribeiro, N., Ribeiro, H., Oliveira, M., Noronha, F., and Abreu, I. 2009. Comparison between urban and rural pollen of *Chenopodium alba* and characterization of adhered pollutant aerosol particles. *Aerosol Sci.* 40:81–86.

Hope, G. A., Woods, R., and Munce, C. G. 2001. Raman microprobe mineral identification. *Min. Eng.* 14:1565–1577.

Jehlička, J., Edwards, H. G. M., Villar, S. E. J., and Pokorny, J. 2005. Raman spectroscopic study of amorphous and crystalline hydrocarbons from soils, peats and lignite. *Spectrochim. Acta A* 61:2390–2398.

Krueger, B. J., Grassian, V. H., Iedema, M. J., Cowin, J. P., and Laskin, A. 2003. Probing heterogeneous chemistry of individual atmospheric particles using scanning electron microscopy and energy-dispersive x-ray analysis. *Anal. Chem.* 75:5170–5179.

Lee, A. K. Y. and Chan C. K. 2007. Single particle Raman spectroscopy for investigating atmospheric heterogeneous reactions of organic aerosols. *Atmos. Environ.* 41:4611–4621.

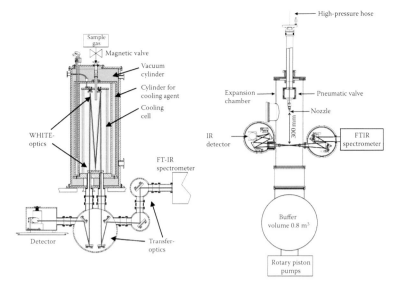

FIGURE 2.5 Schematic representation of the experimental setup. Left: Collisional cooling cell. Right: Rapid expansion of supercritical solutions (RESS) setup.

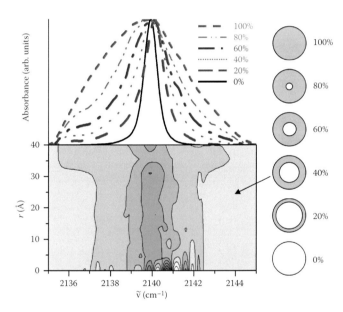

FIGURE 2.7 Upper traces: Calculated absorption spectra for ensembles (10 elements) of crystalline core–amorphous shell carbon monoxide particles of spherical ($r = 4$ nm) shape. The amorphous shell contribution ranges from (thin full) 0 to (thick short dashed) 100 vol%. Lower panel: Excitation density for the ensemble with 40 vol% amorphous shell. The particles are partitioned into spherical shells 1 Å thick. The excitation density (Equation 2.24), increasing from light to dark, was integrated over each shell and normed by the number of molecules in the shell. The asymmetry and broadening of the IR band can be clearly traced back to the amorphous shell.

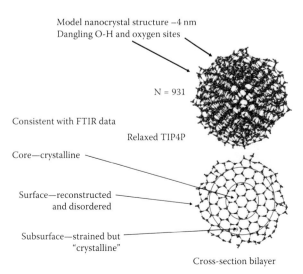

FIGURE 3.3 Model of simulated relaxed 4-nm ice particle based on TIP4 potential. This simulated model, consistent with experimentation, shows the dangling surface groups and the reconstructed nature of the surface with reduced three-coordinated-water sites. The core, surface, and transitional subsurface are also indicated.

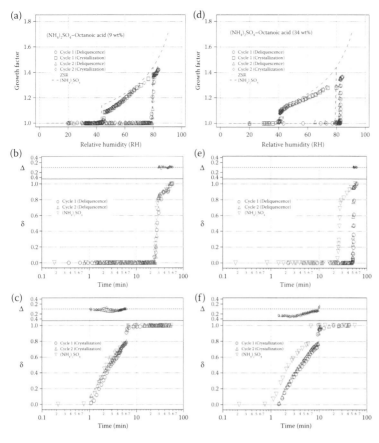

FIGURE 7.9 Hygroscopicity of (NH₄)₂SO₄ particles coated with (a) 9 wt% and (d) 34 wt% octanoic acid in two deliquescence and crystallization cycles. The δ and Δ values of (NH₄)₂SO₄ particles coated with (b) 9 wt% and (e) 34 wt% octanoic acid as a function of time are shown from two deliquescence cycles. The δ and Δ values of (NH₄)₂SO₄ particles coated with (c) 9 wt% and (f) 34 wt% octanoic acid are shown as a function of time in two crystallization cycles. (Adapted from Chan, M. N. and Chan, C. K. 2007. *Atmospheric Environment* 41: 4423–4433.)

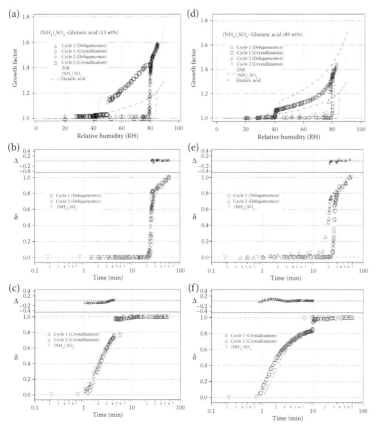

FIGURE 7.10 Hygroscopicity of (NH₄)₂SO₄ particles coated with (a) 13 wt% and (d) 49 wt% glutaric acid in two deliquescence and crystallization cycles. The δ and Δ values of (NH₄)₂SO₄ particles coated with (b) 13 wt% and (e) 49 wt% glutaric acid are shown as a function of time in two deliquescence cycles. The δ and Δ values of (NH₄)₂SO₄ particles coated with (c) 13 wt% and (f) 49 wt% glutaric acid are shown as a function of time in two crystallization cycles. (Adapted from Chan, M. N., Lee, A. K. Y., and Chan, C. K. 2006. *Environmental Science and Technology* 40: 6983–6989. With permission.)

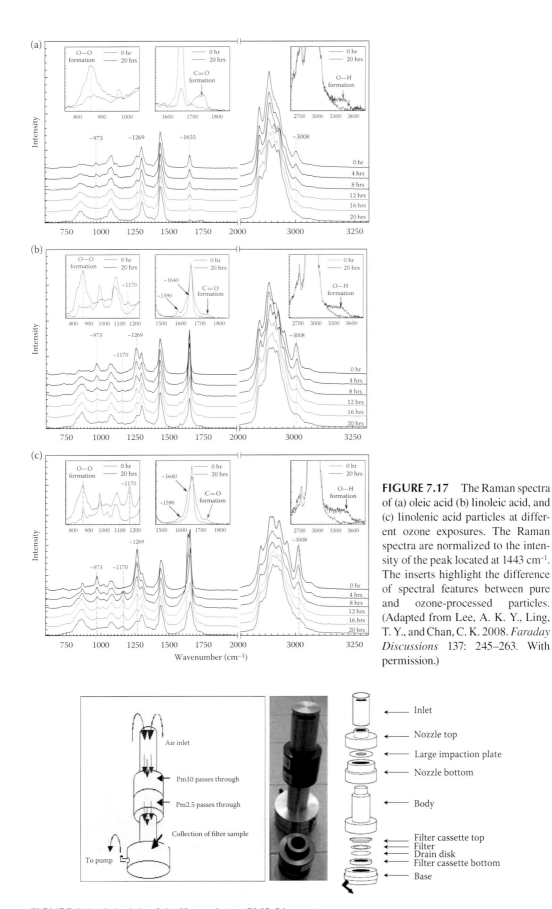

FIGURE 7.17 The Raman spectra of (a) oleic acid (b) linoleic acid, and (c) linolenic acid particles at different ozone exposures. The Raman spectra are normalized to the intensity of the peak located at 1443 cm⁻¹. The inserts highlight the difference of spectral features between pure and ozone-processed particles. (Adapted from Lee, A. K. Y., Ling, T. Y., and Chan, C. K. 2008. *Faraday Discussions* 137: 245–263. With permission.)

FIGURE 8.1 Principle of the Harvard-type PM2.5 impactor.

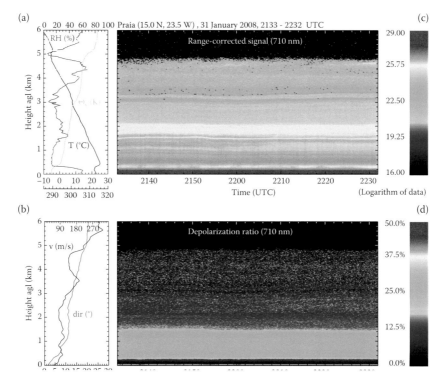

FIGURE 9.6 Dust and smoke observed with BERTHA over Praia, Cape Verde, on 31 January 2008. (b) Range-corrected 710-nm signal (arbitrary units) and (d) volume depolarization ratio at 710 nm are shown with 15-m vertical and 10-s temporal resolution. Radiosonde profiles of (a) relative humidity RH, temperature T, and virtual potential temperature θ_v (evening sonde, Vaisala RS80, launch at 2123 UTC) and of (c) horizontal wind speed v and direction dir (morning sonde, Vaisala RS92, launch at 1110 UTC) are shown in addition. The lower of the two x-axes in (a) refers to the virtual potential temperature. The upper x-axis in (c) refers to the wind direction. (Plot adapted from Figure 1 in Tesche, M., Ansmann, A., Müller, D., Althausen, D., Engelmann, R., Freudenthaler, V., and Groß, S. 2009. *J. Geophys. Res.* 114:D13202.)

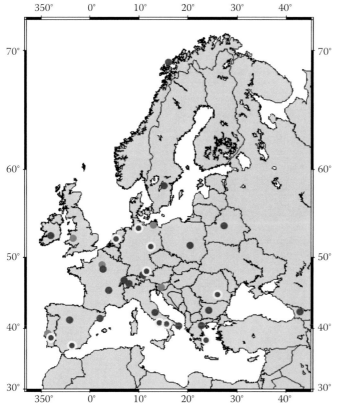

FIGURE 9.9 Distribution of all the EARLINET lidar stations in Europe. Blue dots denote stations that are no longer active. Red dots denote active stations. All stations except for those in Belarus, Bulgaria, Georgia, Spain (Madrid) operate at least one Raman channel for extinction profiling. Yellow circles denote stations that operate multiwavelength Raman lidars. Status as of 2010.

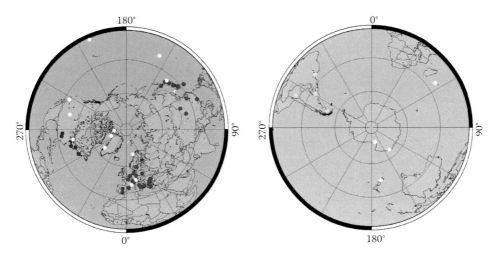

FIGURE 9.10 Distribution of stations as available through the cooperation between existing networks. The different networks are indicated by the dot color: AD-NET violet, ALINE yellow, CISLiNet green, EARLINET red, MPLNET brown, NDACC white, CREST blue. See the GALION report[62] for technical details of the lidar stations. (Plot adapted from GALION Technical Report. 2008. Plan for the Implementation of the GAW Aerosol Lidar Observations Network GALION, Report No. 178. World Meteorological Organization, Global Atmospheric Watch. http://www.wmo.ch/pages/prog/arep/gaw/gaw-reports.html.)

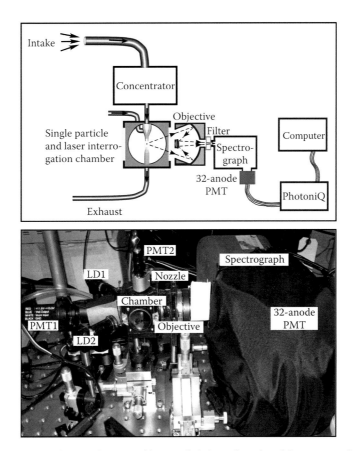

FIGURE 12.1 Schematic of the PFS. Aerosol is sampled through a virtual impactor and a focusing nozzle inlet, which concentrates super-micron particles into a flowing laminar jet. As single particles within the aerosol jet are drawn through the sampling volume, they are probed with a pulsed UV laser, which excites fluorors within the particles. The fluorescence emitted is collected by a reflective objective, focused onto a spectrograph slit, and detected with a 32-anode PMT detector. This arrangement permits rapid measurement of single-particle fluorescence spectra, with aerosol nominal sample rates of about 100 L/min for supermicron particles. (From Pan, Y.L. et al. 2007. *J. Geophys. Res. Atmos.*, 112: doi:1029/2007JD008741. With permission.)

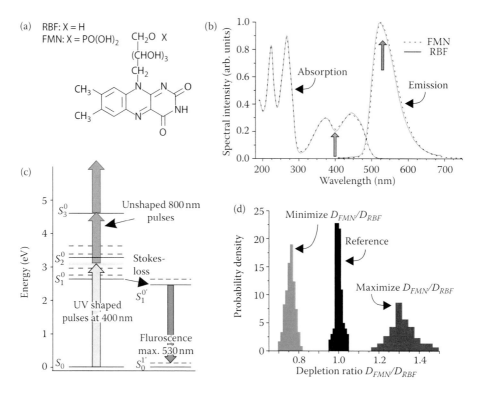

FIGURE 13.13 RBF and FMN are structurally (a) and spectroscopically (b) quite similar. A shaped UV pulse coordinated with a time-delayed IR pulse enables selectivity (c) and is able to both maximize and minimize the fluorescence depletion ratio by ±28% (d). Arrows at 400 nm and 530 nm in (b) locate regions of UV control and collected fluorescence, respectively. (Reprinted from Roth, M., Guyon, L., Roslund, J. et al. 2009. *Physical Review Letters* 102: 253001. Copyright 2009, American Physical Society. With permission.)

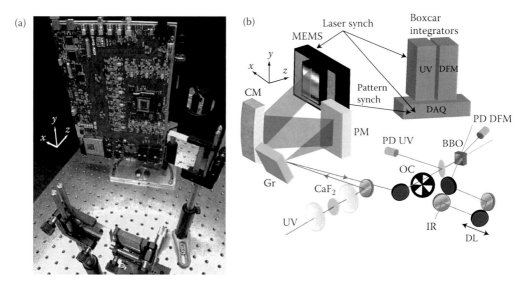

FIGURE 13.14 (a) Picture of the shaper setup. We recognize the Fraunhofer MEMS chip, placed in the middle right of the driving board. (b) Experimental scheme. Gr: grating; PM: plane mirror; CM: cylindrical mirror; DL: temporal delay line; OC: optical chopper. PD UV/DFM: detection photodiodes, DAQ: acquisition card. The OC is synchronized to half the repetition rate of the laser to reject every second UV pulse. (From Rondi, A., Extermann, J., Bonacina, L. et al. 2009. *Applied Physics B—Lasers and Optics* 96. With kind permission from Springer + Science Media.)

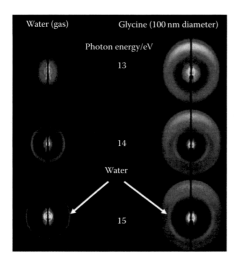

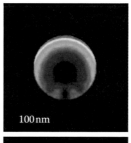

FIGURE 15.12 Velocity map images of 100 nm glycine particles recorded at photon energies of 13, 14, and 15 eV (right). Images of the gas-phase components (water) that travel with the aerosol beam are shown at the left.

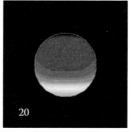

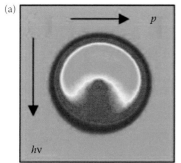

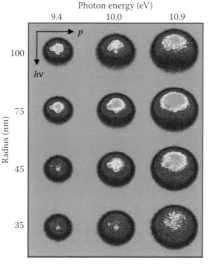

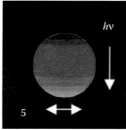

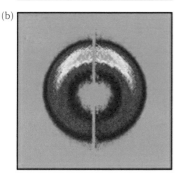

FIGURE 15.14 (a) Raw and (b) reconstructed photoelectron images of NaCl (polydisperse size distribution) obtained at a photon energy of 12 eV. Arrows show the photon (hv) and particle (p) beam propagation directions. (Reproduced from Wilson, K. R.; Zou, S. L.; Shu, J. N.; Rühl, E.; Leone, S. R.; Schatz, G. C.; Ahmed, M. 2007. *Nano Lett.* 7:2014–2019. Copyright 2007 American Chemical Society. With permission.)

FIGURE 15.16 Raw velocity map images of size-selected NaCl aerosol as function of photon energy. The arrows at the top of the figure indicate the photon (hv) and particle (p) beam propagation directions. (Reproduced from Wilson, K. R.; Zou, S. L.; Shu, J. N.; Rühl, E.; Leone, S. R.; Schatz, G. C.; Ahmed, M. 2007. *Nano Lett.* 7:2014–2019. Copyright 2007 American Chemical Society. With permission.)

FIGURE 15.18 Mie theory calculations of the internal electric field amplitude as a function of particle radius. (Reproduced from Wilson, K. R.; Zou, S. L.; Shu, J. N.; Rühl, E.; Leone, S. R.; Schatz, G. C.; Ahmed, M. 2007. *Nano Lett.* 7:2014–2019. Copyright 2007 American Chemical Society. With permission.)

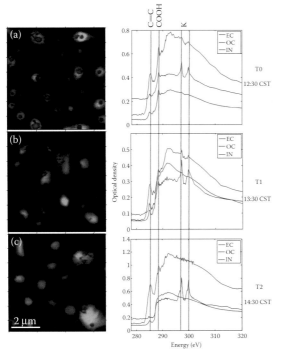

FIGURE 17.17 (Left) SVD maps of particles from the three sampling sites showing soot (red), inorganic regions (blue), and organic regions (green). Panels a, b, and c refer to sampling sites T0, T1, and T2 located progressively farther from the urban center of Mexico City. Soot was defined as regions with $\geq 35\%$ sp^2 hybridized carbon, inorganic regions were defined as having a pre-edge (278–280 eV) to postedge (320 eV) ratio of 0.6 and organic regions contained absorption at the COOH peak. (Right) Spectra used to produce SVD maps shown on the left.

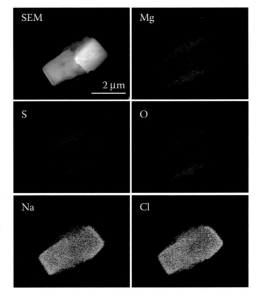

FIGURE 18.8 SEM image and EDX elemental maps of the sea-salt particle studied in the ESEM experiment shown in Figure 18.7. Segregation of Na- and Mg-containing salts is observed. As a wet sea-salt particle dries out, highly soluble Mg salts crystallize last, forming outer structures on the surface of less soluble NaCl, which crystallizes first.

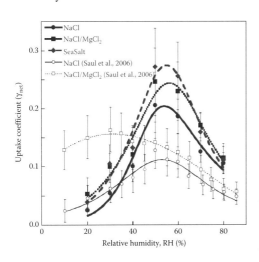

FIGURE 18.12 Values of initial uptake coefficient γ_{net} as a function of RH for NaCl, mixture of NaCl/MgCl$_2$ ($X_{Mg/Na} = 0.114$), and sea-salt particles. Solid symbols are experimental data from microscopy study with deliquesced particles of dry size $D_p \sim 0.9\ \mu m$. The open symbols represent data from single-particle mass spectrometry study of Saul et al. (2006) for deliquesced particles of dry size $D_p \sim 0.1\ \mu m$. (From Liu et al. 2007. *Journal of Physical Chemistry A* 111:10026–10043. Copyright 2007, American Chemical Society. With permission.)

Maricq, M. M. 2007. Review chemical characterization of particulate emissions from diesel engines: A review. *Aerosol Sci.* 38:1079–1118.

Moody, C. D., Jorge Villar, S. E., Edwards, H. G. M., Hodgsonc, D. A., Doran, P. T., and Bishop, J. L. 2005. Biogeological Raman spectroscopic studies of Antarctic lacustrine sediments. *Spectrochim. Acta A* 61:2413–2417.

Musick, J. and Popp, J. 1999. Investigations of chemical reactions between single levitated magnesium chloride microdroplets with SO_2 and NO_x by means of Raman spectroscopy and elastic light scattering. *Phys. Chem. Chem. Phys.* 1:5497–5502.

Otten, Ph. 1991. Transformation, concentrations and deposition of North Sea aerosols. PhD Thesis, Antwerp, Belgium.

Potgieter-Vermaak, S. 2007. Surface characterization of a heavy mineral sand with micro-Raman spectroscopy. In *The Proceedings of the 6th International Heavy Minerals Conference 'Back to Basics,'* Nyala Game Lodge, Natal, South Africa, The Southern African Institute of Mining and Metallurgy.

Potgieter-Vermaak, S. S. and Van Grieken, R. 2006. Preliminary evaluation of micro-Raman spectroscopy for the characterization of individual aerosol particles. *Appl. Spectroc.* 60:39–47.

Ro, C.-U., Osán, J., and Van Grieken, R. 1991. Determination of low-Z elements in individual environmental particles using windowless EPMA. *Anal. Chem.* 71:1521–1528.

Schweiger, G. 1999. Raman scattering on single aerosol particles and on flowing aerosols: A review. *J. Aerosol Sci.* 21:483–509.

Sengupta, A., Laucks, M. L., Dildine, N., Drapala, E., and Davisb, E. J. 2005. Bioaerosol characterization by surface-enhanced Raman spectroscopy (SERS). *Aerosol Sci.* 36:651–664.

Sobanska, S., Falgayrac, G., Laureyns, J., and Br'em, C. 2006. Chemistry at level of individual aerosol particle using multivariate curve resolution of confocal Raman image. *Spectrochim. Acta A* 64: 1102–1109.

Stefaniak, E. A., Worobiec, A., Potgieter-Vermaak, S., Alsecz, A., Török, S., and Van Grieken, R. 2006. Molecular and elemental characterisation of mineral particles by means of parallel micro-Raman spectroscopy and scanning electron microscopy/energy dispersive x-ray analysis. *Spectrochim. Acta B* 61:824–830.

Stranger, M. 2005. Characterisation of health related particulate and gas-phase compounds in multiple indoor and outdoor sites in Flanders. PhD thesis. University of Antwerp, Belgium.

Szalóki, I., Osán, J., Ro, C.-U., and Van Grieken, R. 2000. Quantitative characterisation of individual aerosols particles by thin window electron probe microanalysis combined with iterative simulation. *Spectrochim. Acta B* 55:1017–1030.

Szalóki, I., Osán, J., Worobiec, A., de Hoog, J., and Van Grieken, R. 2001. Optimisation of experimental conditions of thin-window EPMA for light-element analysis of individual environmental particles. *X-ray Spectrom.* 30:143–155.

Sze, S.-K., Siddique, N., Sloan, J. J., and Escribano, R. 2001. Raman spectroscopic characterization of carbonaceous aerosols. *Atmos. Environ.* 35:561–568.

Tang, I. N. and Fung, K. H. 1989. Characterization of inorganic salt particles by Raman spectroscopy. *J. Aerosol Sci.* 20:609–617.

Trunk, M., Popp, J., and Kiefer, W. 1998. Investigations of the composition changes of an evaporating, single binary-mixture microdroplet by inelastic and elastic light scattering. *Chem Phys. Lett.* 284:377–381.

Uzu, G., Sobanska, S., Aliouane, Y., Pradere, P., and Dumat, C. 2009. Study of lead phytoavailability for atmospheric industrial micronic and sub-micronic particles in relation with lead speciation. *Environ. Pollut.* 157:1178–1185.

Van Grieken, R. and Markowicz, A. 1993. *Handbook of X-ray Spectrometry, Methods and Techniques, 1993, in Practical Spectroscopy Series*, Vol. 14, Marcel Dekker, New York, ISBN 0-8247-0600-5.

Van Grieken, R., Potgieter-Vermaak, S., Darchuk, L., and Worobiec, A. December 2009. Integration of analysis techniques of different scales using X-ray induced and electron induced X-ray spectrometry for application s in preventative conservation and environmental monitoring. *IAEA XRF Newsletter* 18:9–13.

Windig, W. and Stephenson, D. A. 1992. Self-modeling mixture analysis of second-derivative near-infrared spectral data using the SIMPLISMA approach. *Anal. Chem.* 64:2735–2742.

Worobiec, A., de Hoog, J., Osán, J., Szalóki, I., and Van Grieken, R. 2003. The study of the behaviour of beam sensitive atmospheric aerosol particles on different collection substrates using the thin window EPMA and the cold stage technique. *Spectrochim. Acta B* 58:479–495.

Worobiec, A., Osán, J., Szalóki, I., Maenhout, W., Stefaniak E. A., and Van Grieken, R. 2007a. Characterization of sources and process of individual atmospheric aerosol particles in Amazon Basin. *Atmos. Environ.* 41:9217–9230.

Worobiec, A., Stefaniak, E. A., Kiro, S., Oprya, M., Bekshaev, A., Spolnik, Z., Potgieter-Vermaak, S. S., Ennan, A., and Van Grieken, R. 2007b. Comprehensive microanalytical study of welding aerosols with X-ray and Raman based methods. *X-Ray Spectrom.* 36:328–335.

Worobiec, A., Potgieter-Vermaak, S., Brooker, A., Darchuk, L., Stefaniak, E., and Van Grieken, R. 2010. Interfaced SEM/EDS and micro Raman spectroscopy for the characterisation of heterogeneous environmental particles—Fundamental and practical challenges, *Microchem. J.* 94:65–72.

Worobiec, A., Stefaniak, E. A., Kontozova, V., Samek, L., Karaszkiewicz, P., Van Meel, K., and Van Grieken, R. 2006. Characterisation of individual atmospheric particles within the Royal Museum of the Wawel Castle in Cracow, Poland. *e-Preservation Science* 3:63–69.

Wu, H. B. and Chan, C. K. 2008. Effects of potassium nitrate on the solid phase transitions of ammonium nitrate particles. *Atmos. Environ.* 42:313–322.

9 Raman Lidar for the Characterization of Atmospheric Particulate Pollution

Detlef Müller

CONTENTS

9.1 Introduction 210
    9.1.1 Importance of Aerosols in the Climate System 210
    9.1.2 Observation Techniques of Aerosols 210
9.2 Lidar as a Tool for the Vertically Resolved Sounding of Aerosols 211
    9.2.1 History of Aerosol Characterization with Raman Lidar 211
    9.2.2 Basic Instrument Setup 212
    9.2.3 The Lidar Equation 214
    9.2.4 Simple Backscatter Lidar 214
9.3 Raman Lidar 215
    9.3.1 Raman Lidar Equation 215
    9.3.2 Raman Effect 216
    9.3.3 Multiwavelength Raman Lidar: Example of an Instrument 218
    9.3.4 Optical Parameters of Advanced Multiwavelength Aerosol Raman Lidars 220
    9.3.5 Microphysical Particle Parameters from Multiwavelength Raman Lidar 221
9.4 Current Status 223
    9.4.1 Example: A Mixed Dust/Smoke Plume from West Africa 223
    9.4.2 EARLINET: The First Aerosol Raman Lidar Network 226
9.5 Outlook 228
    9.5.1 Instruments 228
        9.5.1.1 Identification of Chemical Signatures of Aerosol Particles 228
        9.5.1.2 New Measurement Wavelengths: Ultraviolet Region 228
        9.5.1.3 New Measurement Wavelengths: Infrared Region 229
        9.5.1.4 Extending Measurements into Daytime 230
        9.5.1.5 New Data Products 231
    9.5.2 Inversion Algorithms 231
    9.5.3 GALION: The Potential Future Global Lidar Network 232
References 233

9.1 INTRODUCTION

9.1.1 Importance of Aerosols in the Climate System

Coordinated, three-dimensional monitoring of the global aerosol distribution and its temporal variability is a basic requirement of climate research. Such four-dimensional observations are needed for an adequate consideration of particles in atmospheric models that are used to simulate the regional and intercontinental aerosol transport and the influence of aerosol particles on short-term weather and the Earth's radiation budget (direct climate effect), and on cloud processes (indirect climate effect).[*] However, such a four-dimensional monitoring system is not available. Thus, the uncertainty in estimating the aerosol impact on future climate is unacceptably large for constraining climate simulations and climate change projections.[1–3†] Quantifying the aerosol impact would require a multi-instrument, global observing system and a capacity for integrating diverse data.[4–7] A program for coordinating and integrating these observations, dubbed the Progressive Aerosol Retrieval and Assimilation Global Observing Network (PARAGON) initiative, has been proposed by Diner et al.[8]

A particularly large gap remains in the vertically resolved description of optical and physical (commonly referred to as "microphysical") properties, and of geometrical features of free-tropospheric particles and their impact on climate and air quality.[9–13] Information on the altitude at which aerosol particle layers are present is important for calculating aerosol radiative forcing. The atmosphere's albedo has significant impact on radiative forcing, and it makes a significant difference in the radiative impact whether the aerosol particles are above a cloudy marine boundary layer or between the clouds within the boundary layer.[14] The uppermost aerosol layer in the lower troposphere has the largest effect on the radiation budget.[15]

Optical and microphysical properties of boundary layer aerosols that originate from local and regional emissions of particles and gases are usually very different from free-tropospheric particles, which are often advected over large distances from other continents. Aerosol layers in the free troposphere often are optically very thin. Complex aerosol-cloud interaction processes result, if cirrus clouds or alto-cumulus clouds are on top of these plumes or the plumes are embedded in these clouds. Thus, even faint particle layers may become important,[16] in particular if they are transported over long distances. Changes in aerosol composition and concentration lead to changes in the number and size of cloud droplets, and thus influence the lifetime and reflectivity of clouds. The phase state of cloud droplets may be changed.[17] The changes of cloud properties may alter the hydrological cycle, for example, Ramanathan et al.[3] Correlation studies between cloud and aerosol properties derived from satellite observations imply that some aerosols suppress cloud formation and precipitation, for example, Rosenfeld[18] and Koren et al.[19] Other aerosol types, in contrast, can enhance precipitation, for example, Feingold et al.[20] and Rudich et al.[21]

9.1.2 Observation Techniques of Aerosols

The main reason for the lack of documenting and characterizing of particles in the free troposphere is the limit of observational techniques. *In situ* surface observations provide detailed optical, chemical, and microphysical characterization of particles. Measurement platforms at the surface, however, only deliver a point-like description of the particle conditions in space. These platforms may be subject to strong influence by local aerosol sources, and make it nearly impossible to accurately monitor particles imported from aerosol sources in the far-field of the observational area. In particular, any observation of pollution aloft is impossible. Airborne *in situ* platforms extend this one-dimensional view. They give insight into particle properties in greater heights of the atmospheric column, for example, Masonis et al.[22] The disadvantage of such instrumentation is that only observations during intensive field phases rather than long-term monitoring of aerosol conditions is possible.

---

[*] Intergovernmental Panel on Climate Change (IPCC), 2007 (http://www.ipcc.ch/).
[†] See also IPCC, 2007 (http://www.ipcc.ch/).

During the past decade, sun photometers have become a very successful device for particle monitoring. The Aerosol Robotic Network (AERONET)[23] provides a wealth of optical and microphysical particle properties[24,25] under ambient atmospheric conditions. The derived quantities describe column-averaged conditions, that is, there is no separation between locally influenced boundary layer aerosols and free-tropospheric particle layers. The same holds for passive satellite sensors such as MODIS (http://modis.gsfc.nasa.gov/), MISR (http://www-misr.jpl.nasa.gov/), TOMS (http://toms.gsfc.nasa.gov/), OMI (http://aura.gsfc.nasa.gov/instruments/omi.html), POLDER (http://smsc.cnes.fr/POLDER/), and PARASOL (http://www.cnes.fr/web/CNES-en/1474-parasol.php) which were or still are used for aerosol monitoring.

In summary, it can be stated that only two-dimensional images of aerosol conditions can be acquired with either one or a combination of several different aforementioned remote sensing platforms. Vertically resolved lidar observations therefore are a basic requirement to complement this two-dimensional, column-integrated view on which the global aerosol distribution is presently based.[26] Since the mid 1990s, Raman lidar, and in particular multiwavelength Raman lidar has contributed significantly to the task of characterizing atmospheric particles.

The following text summarizes the current status of Raman lidar used for the characterization of atmospheric aerosols. Additional information on that topic can be found in the following four lidar text books that were published since 1976: see Refs.[27–30] The reader will also find a thorough overview on many other lidar techniques in these textbooks. The basic principle of Raman lidar will be explained. A brief overview describes how important optical and microphysical properties of particulate pollution are derived from the measured signals. A measurement example shows the achievable potential of modern Raman lidars. An outlook on future work is given at the end of this chapter.

9.2   LIDAR AS A TOOL FOR THE VERTICALLY RESOLVED SOUNDING OF AEROSOLS

9.2.1   History of Aerosol Characterization with Raman Lidar

Raman lidars have proven to be most useful for quantitative studies of optical properties of tropospheric aerosol particles. Raman lidar measures the volume extinction and backscatter coefficients of particles simultaneously and independently of each other, and this was shown in the early 1990s.[31,32] The potential of Raman lidar was first shown after the eruption of Mount Pinatubo in 1991.[33] This eruption was the most intense volcanic event in the last century. The evolution of the volcanic plume that entered deep into the stratosphere was monitored with Raman lidar. The data showed the benefit of measuring directly profiles of particle extinction coefficients.

The most important feature of Raman lidar is that no critical assumptions on atmospheric input parameters are needed in data analysis. This is in contrast to data analysis for the widely applied standard backscatter lidar technique.[34,35] Raman lidar also allows us to measure the particle extinction-to-backscatter ratio or lidar ratio with high accuracy. This is in contrast to standard backscatter lidar for which we need to assume this lidar ratio in order to derive optical quantities. A wrong assumption of the lidar ratio results in wrong results for the inferred optical quantities.

The lidar ratio provides qualitative information on the particle type under investigation. This parameter depends on particle size, shape, and chemical composition. Examples can be found in Refs.[36–44]

Despite the impressive examples of vertical profiling of the particle extinction coefficient, the full potential of the Raman lidar technique was hardly exploited toward a quantitative particle characterization until around 1995. Quantitative particle characterization means that other important parameters such as mean particle size, complex refractive index, and the single-scattering albedo should be inferred too.

This dissatisfying situation changed with two new Raman lidars which came into operation at the Leibniz Institute for Tropospheric Research (IfT) in the mid-1990s.[45,46] These systems emit laser radiation at several wavelengths, in contrast to the single-wavelength Raman lidars that were used for the observations of the Mount Pinatubo plume. The most important new feature of these systems was that they provided particle extinction information at two wavelengths on the basis of Raman signals, and that particle backscatter coefficients of high accuracy could be inferred at a minimum of three wavelengths.

The two systems at IfT at that time by far surpassed any other existing lidar dedicated to investigations on aerosols. After the systems became operational,[47,48] it was shown rather quickly that the qualitative particle characterization could be put on a quantitative base with multiwavelength Raman lidar.[49] Profiles of extinction and backscatter coefficients simultaneously measured at several wavelengths between 355 and 1064 nm wavelength are an essential requirement to retrieve, for example, mean (effective) particle radius and single-scattering albedo,[42,43,50–54] the latter parameter is one of the most important parameters controlling radiative forcing by aerosols.

Multiwavelength Raman lidar thus presented a milestone regarding particle characterization by lidar. The task of identifying the investigated aerosol type is strongly constrained by lidar ratios simultaneously measured at two wavelengths, for example, Refs.[41–43,54–56]. A minimum number of two wavelengths is required for calculating the Ångström exponent. This aerosol index[57] is most commonly used to characterize anthropogenic particles because it is sensitive to changes of particle size in the accumulation mode (particle diameters <1 μm) of the aerosol size distribution. Examples are given by, for example, Refs.[36,58,59]

Since the beginning of multiwavelength Raman lidar measurements at IfT in 1996, these systems were involved in long-term observations of particulate pollution at the institute in the framework of the German Lidar Network from 1997 to 2000[60] and the European Aerosol Research Lidar Network (EARLINET) since 2000,[61] and in large, integrated field campaigns such as the Second Aerosol Characterization Experiment (Tellus, 52B, No. 2, 2000), the Lindenberg Aerosol Characterization Experiment 1998 (*J. Geophys. Res.*, 107, No. D21, 2002), the Indian Ocean Experiments in 1999/2000 (*J. Geophys. Res.*, 106, No. D22, 2001 and 107, No. D19, 2002), and the Saharan Mineral Dust Experiments (SAMUM) in 2006 (*Tellus*, 61, 2009) and 2008.

The obvious advantage of Raman lidar over conventional elastic backscatter lidar triggered other lidar groups in the world, such as, for instance, lidar groups that participate in EARLINET, to upgrade their systems to at least single-wavelength Raman lidars.[62] Several groups in EARLINET have progressed one step further and recently (status as of 2010) installed multiwavelength Raman lidars.[63–67]

9.2.2 Basic Instrument Setup

Figure 9.1 shows the basic setup of a lidar. The laser transmits short-duration light pulses. The beam usually is expanded in order to reduce the divergence of the outgoing laser radiation. The laser transmitter and the optical receiver unit can be operated in two modes. If the laser beam travels along the line of sight of the receiver telescope we call it parallel configuration (monoaxial configu-

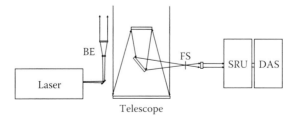

FIGURE 9.1   Block diagram of a typical lidar system. List of abbreviations: BE—beam expander, FS—field stop, SRU—signal receiver unit, DAS—data acquisition system. (Plot adapted from Wandinger, U. 2005. *Lidar, Range-Resolved Optical Remote Sensing of the Atmosphere*. Weitkamp, C. (Ed.). Springer, Berlin. pp. 241–271. With permission.)

ration). In collinear arrangement (biaxial configuration), the laser beam is coupled from the side into the receiver field of view of the telescope.

The backscattered light from molecules and particles is collected with a telescope. The intensity of the light elastically backscattered by atmospheric molecules and particles is measured versus time by passing it through the telescope receiver, collimating optics, a bandpass filter for daylight suppression, an interference filter for suppression of return signals at other wavelengths to an appropriate detector. Photomultiplier tubes are usually used. The signal profile is recorded by an analog-to-digital converter or by a photon-counting device and subsequently stored on a computer.

Lidar signals are accumulated for a selected integration period. This period, that is, the number of laser shots, depends on the signal-to-noise ratio we wish to obtain, the data analysis methodology, and the specific scientific questions that are tackled. The signal integration time may thus cover time intervals from less than a second to many minutes.

The main parameters that define the instrument characteristics are the wavelengths at which the laser operates, laser power or energy per laser pulse, collecting area of the receiver telescope, transmission characteristics of the optical system, out-of-band suppression of the return signals, and detector efficiency, linearity, and dynamic range of the detector array. Other important parameters that may be considered under circumstances are eye-safety, system reliability, and long-term stability of adjustment.

Operating wavelengths for an aerosol Raman lidar are chosen on the basis of several factors. The lidar measurements can be performed in any spectral region where the atmosphere is reasonably transparent. For this reason, the wavelength of the laser radiation should be chosen such that absorption of laser radiation by trace gases is kept to a minimum level. We also need to keep in mind that molecular scattering cross-sections are proportional to λ^{-4}. For that reason Raman lidar is practically limited to the visible and ultraviolet and near infrared part of the electromagnetic spectrum. Aerosol particles possess a scattering cross-section that is typically proportional to λ^{-1}. For large particles, this dependence is weaker, and for smaller particles the dependence may be stronger. Particle diameters accessible to the measurement technology described in this chapter are approximately between 100 nm to 5 µm. As we explain in Section 9.3, we can make use of the wavelength dependence of the backscatter and extinction signals to estimate size distribution parameters.

The choice of the laser wavelength is also determined by the way lidar signal profiles are calibrated. We chose atmospheric regions where Rayleigh scattering dominates over scattering from particles. In this way, we make use of the fact that the Rayleigh scattering cross-sections can be calculated from theory. This calibration procedure requires us to choose laser wavelengths for which the Rayleigh signal is well above the sensitivity threshold of the detectors used. We need detectors with a sufficiently high sensitivity to the lidar return signals. In principle, laser sources are available in the spectral range from about 300 nm to 10 µm. In state-of-the-art aerosol Raman lidars, we use highly developed industrial lasers such as the Nd:YAG (neodymium:yttrium–aluminum–garnet). This type of laser allows an easy way of generating additional laser light by using the so-called frequency doubling and tripling crystals.

We distinguish between the so-called backscatter lidar systems and Raman lidars.[68] The first type of instruments detects only radiation that is elastically scattered from both aerosol particles and molecules. The second instrument type also detects molecular scattering separately from particle scattering. This molecular scattering can result either from vibrational Raman scattering from nitrogen or oxygen molecules in the atmosphere, or from pure rotational Raman scattering from nitrogen and oxygen molecules. Backscatter lidar and Raman lidar can be operated at multiple wavelengths simultaneously. In Section 9.3 the Raman lidar type is presented in more detail.

It should be mentioned that if we want to distinguish between spherical particles and particles with irregular shape we may add measurement channels for detecting the depolarized light. These additional detectors thus may be very useful in distinguishing between the presence of industrial/urban pollution which usually consists of rather spherical particles and dust particles, such as, for instance, from desert regions, which usually have nonspherical shape. However, this instrument type is not further covered in this chapter. Another instrument type is High Spectral Resolution Lidar (HSRL) in which Rayleigh scattering is the molecular signal. This system type is also not further discussed in this chapter. Details on lidars used for the detection of depolarized light and HSRL technology may be found in Ref.[30]

9.2.3 The Lidar Equation

The lidar equation for return signals due to elastical backscatter by air molecules and aerosol particles can, in its simplest form, be written as[69]

$$P(R) = \frac{E_0 \eta}{R^2} O(R) \beta(R) \exp\left[-2\int_0^R \alpha(r)\,dr\right].$$

(9.1)

The term $P(R)$ is the signal owing to Rayleigh and particle scattering received from distance R, E_0 is the transmitted laser pulse energy, η contains lidar parameters describing the efficiencies of the optics and detectors, and $O(R)$ describes the overlap between the outgoing laser beam and the receiver field of view.

The lidar equation links the observable (the energy returned as a function of time) to two unknown quantities (the backscatter coefficient of the aerosol and the two-way transmission through the atmosphere which comes from the integration of the extinction with height). The parameters $\beta(R)$ (often written in units of $km^{-1}\,sr^{-1}$) and $\alpha(R)$ (written in units of km^{-1}) are the backscatter coefficients and extinction coefficients caused by particles (index par) and molecules (index mol), respectively:

$$\begin{aligned}
\beta(R) &= \beta_{par}(R) + \beta_{mol}(R), \\
\alpha(R) &= \alpha_{par}(R) + \alpha_{mol}(R).
\end{aligned}$$

(9.2)

If the laser wavelengths are chosen appropriately, the absorption of light by molecules can be ignored. In any case, absorption effects have to be corrected in the measured signals before applying the data analysis methods.

9.2.4 Simple Backscatter Lidar

The simplest form of an aerosol lidar is a backscatter lidar which can be operated either at one or more wavelengths. For such an "elastic" lidar[29] the wavelength of the return signal is the same as the wavelength of the emitted light pulse. From the technological point of view, these instruments require least technological sophistication to implement. The retrieval of quantitative optical information of the aerosol particles requires rather sophisticated data analysis, and the uncertainties imposed by the assumption on lidar ratio remain.

Comparably strong assumptions regarding particle properties have to be made in the data analysis process. The reason for that is seen in the lidar equation 9.1 which links the return signal to two unknown quantities. From a mathematical point of view, it is impossible to find a unique solution of the lidar equation unless we make assumptions in the solution. We must make assumptions on the backscatter properties of the aerosol particles which are present along the lidar beam in order to derive the other missing quantity, that is, extinction by aerosol particles. We can alternatively make assumptions on particle extinction in order to derive particle backscatter.

This problem can be put in other words: we must assume the particle extinction-to-backscatter or lidar ratio. What makes the situation so complicated is the fact that the lidar ratio depends on the microphysical (particle size), chemical (complex refractive index), and morphological (particle shape) properties of the particles. All these properties, in turn, depend on relative humidity. The lidar ratio can vary strongly with height, as for example, different aerosol types are present in layers above each other, for example, Ansmann et al.[36,70] Lidar ratios may vary from 20 to 100 sr, depending on the aerosol type and measurement wavelength. Examples can be found in Ref.[71] Many techniques are discussed in the literature to work around this difficulty, for example, Refs.[32,34,35,69,72–80] Despite the success of the various techniques, the variations of the lidar ratio make it practically impossible to estimate trustworthy extinction profiles after Equation 9.1.

9.3 RAMAN LIDAR

The unsatisfactory situation of a simple elastic backscatter lidar is improved significantly with Raman lidar. It permits an accurate vertical profiling of the particle extinction coefficient. First attempts to infer particle extinction properties from Raman signal profiles were reported by Gerry and Leonard.[81] First accurate horizontal transmission measurements with Raman lidar were done by Leonard and Caputo.[82] First results with a lidar pointing in vertical direction were presented by Ansmann et al.[31,32] The Raman lidar observations of the volcanic plume generated by the Pinatubo eruption in 1991 showed that profiles of the volume scattering coefficient can be obtained with ground-based Raman lidars also up to stratospheric heights.[33,83]

9.3.1 RAMAN LIDAR EQUATION

Raman lidar measures lidar return signals elastically backscattered by air molecules and particles and inelastically (Raman) backscattered by nitrogen and/or oxygen molecules.[68] The determination of the particle extinction coefficient from molecular backscatter signals is rather straightforward. The lidar-ratio does not need to be assumed in solving the lidar equation. Other critical assumptions, that is, the reference value and Ångström exponents can be reduced to acceptable levels. The advantage of the Raman lidar over conventional elastic backscatter lidar can be seen from the respective lidar equation for the molecular backscatter signal:

$$P(R, \lambda_{Ra}) = \frac{E_0 \eta \lambda_{Ra}}{R^2} O(R,\, \lambda_{Ra}) \beta_{Ra}(R,\, \lambda_0) \exp\left\{ -\int_0^R \left[\alpha(r,\, \lambda_0) + \alpha(r,\, \lambda_{Ra})\right] dr \right\}. \tag{9.3}$$

The coefficient β_{Ra} denotes Raman backscattering from molecules. Particle backscattering does not appear in Equation 9.3. The only particle-scattering effect on the signal strength is attenuation. The term $\alpha(R, \lambda_0)$ describes the extinction on the way up to the backscatter region. The term $\alpha(R, \lambda_{Ra})$ is the extinction on the way back to the lidar. For the rotational Raman case, $\alpha(R, \lambda_0) = \alpha(R, \lambda_{Ra})$ can be used. However, in the case of a vibration–rotational Raman signal the shift of the wavelength from $\alpha(R, \lambda_0)$ to $\alpha(R, \lambda_{Ra})$ after the scattering process must be considered. If, for example, a Nd:YAG laser wavelength of 532 nm is transmitted, the first Stokes vibration–rotation Q branch of nitrogen is centered at $\alpha(R, \lambda_{Ra}) = 607$ nm.

The molecular backscatter coefficient is calculated from (a) the molecular number density N_{Ra}, which is the nitrogen or oxygen molecule number density for the Raman case and the air-molecule number density for the Rayleigh case, and (b) the molecular (differential) cross-section $d\sigma_{Ra}/d\Omega(\pi, \lambda_0)$ for the Raman scattering process at the laser wavelength λ_0 and the scattering angle π:

$$\beta_{Ra}(R,\, \lambda_0) = N_{Ra}(R) \frac{d\sigma_{Ra}}{d\Omega}(\pi,\, \lambda_0). \tag{9.4}$$

The molecular number density profile is calculated from actual radiosonde observations or standard atmospheric temperature and pressure profiles.

Given these expressions, we can solve the particle extinction coefficient in a rather straightforward manner. The exact solution procedure is described in Ansmann and Müller.[84] A reference value for particle backscattering at R_0 must be estimated. Furthermore, we must estimate the wavelength dependence between the return signal at the emission wavelength λ_0 and the Raman-shifted wavelength λ_{Ra}. To reduce the effect of the uncertainty in this estimate on the solution, one should choose the reference height in the upper troposphere where particle scattering is typically negligible compared to Rayleigh scattering. Then only the air density, molecular backscattering, and atmospheric extinction properties must be estimated to solve Equation 9.3. Meteorological profiles or standard atmosphere data are used to calculate air density and molecular backscatter terms.

The particle transmission ratio for the height range between R_0 and R is estimated from the measured particle extinction profile with the assumption that α is proportional to λ^k. In the case of the rotational Raman signals, spectral transmission corrections are not necessary. Given our knowledge of the particle extinction coefficient, and thus the transmission term, we can solve Equation 9.1 for the profile of the particle backscatter coefficient $\beta_{par}(R, \lambda_0)$. Details of the solution procedure are given by Ansmann and Müller.[84]

The height profile of the particle lidar ratio

$$S_{par}(R,\lambda_0) = \frac{\alpha_{par}(R,\lambda_0)}{\beta_{par}(R,\lambda_0)}, \tag{9.5}$$

can be obtained from the profiles of $\alpha_{par}(R, \lambda_0)$ and $\beta_{par}(R, \lambda_0)$.

9.3.2 RAMAN EFFECT

A photon that is absorbed by a molecule excites an electron into a virtual energy state, that is, the electron cannot remain in this excited level. It drops back into a stable state of energy and emits a photon. However, there are several possibilities for the energy of this emitted photon. If the molecule has the same state of energy as before the absorption process, we speak of Rayleigh scattering, and the energy of the emitted photon is the same as the energy of the photon that was absorbed by the molecule. If the energy of the emitted photon is less than the energy of the absorbed photon we speak of Stokes Raman scattering. If the energy of the emitted photon is higher than that of the absorbed photon we speak of anti-Stokes scattering.

Figure 9.2 shows in more detail the complexity of the Raman spectrum that arises if a molecule is excited into a virtual energy state from which it then relaxes. This relaxation may either be to its initial state with no change of rotation state (Rayleigh scattering), a change of the rotation state in its ground state (the vibration level is zero ($v = 0$) and we obtain rotational Raman lines around the Rayleigh line), or a change of the vibration–rotation state (we obtain vibration–rotational Raman lines).

Figure 9.3 shows, for example, the position of the rotational and vibrational–rotational Raman lines in relation to the Rayleigh line for oxygen, nitrogen, and water-vapor molecules, if the laser excitation wavelength is at 355 nm.

The technologically easiest way is to use vibrational Raman scattering from nitrogen (or oxygen). The Raman shift of 2331 cm^{-1} (in case of nitrogen) is high. The signal can therefore be separated reliably from the elastic particle scattering with standard optical beam splitters and interference filters. The technical implementation of separating the signals is rather easy; the scattering cross-section, and hence the received signal, is very low. For this reason, a high-power laser and narrow (for daytime applications) bandwidth filters are needed. In general, the filters should have high transmission at the detection wavelength and a good suppression of signals at all other wavelengths.

The background noise and operation mode need to be considered, when selecting the optical components. At nighttime, the background radiation is comparably low. In that case, we may use optical filters with a relatively broad bandwidth, that is, 1–5 nm full-width at half-maximum (FWHM). At daytime, the background radiation by the sun is high, and we need filters with an FWHM of 0.3 nm. The laser pulse energy should be at least 150–200 mJ at 30 Hz to obtain signal profiles with reasonable signal-to-noise ratio.

Pure rotational Raman scattering by nitrogen and oxygen offers a scattering cross-section that is about a factor of 30 higher than vibrational Raman scattering. The downside is that the Raman shift

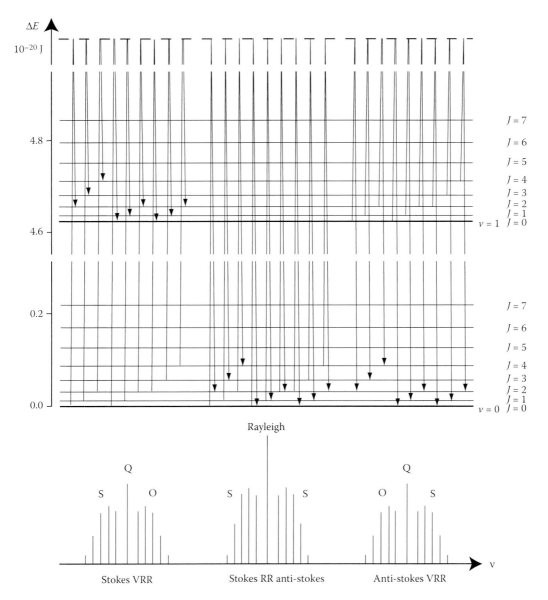

FIGURE 9.2 Energy levels of the vibration–rotation states of the nitrogen molecule and the Raman transitions. The resulting spectrum is shown. (Plot adapted from Wandinger, U. 2005. Raman lidar, in *Lidar, Range-Resolved Optical Remote Sensing of the Atmosphere*. Weitkamp, C. (Ed.). Springer, Berlin, pp. 241–271. With permission.)

is quite small, about 30 cm^{-1} only, so that the separation from the elastic particle backscatter is more challenging, keeping in mind that out-of-band blocking has to be on the order of 10^{-8}. Both filter techniques and double-grating polychromators have been demonstrated for this approach. In particular, the combination with a Fabry–Perot comb filter can suppress daylight sufficiently to allow for daytime operation.[44] A more sophisticated setup also allows one to retrieve the temperature profile simultaneously. With this technique, the price to be paid for better system efficiency is higher system complexity with corresponding sensitivity to misalignment.

A further increase in system efficiency for the signal from molecular scattering can be achieved by separate the Rayleigh scattering from the particle scattering. The Rayleigh scattering

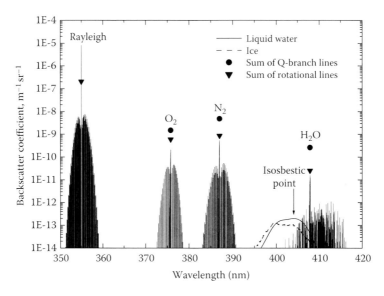

FIGURE 9.3 Schematic illustration of the position of the rotational and vibrational–rotational Raman lines of oxygen (O_2), nitrogen (N_2), and water vapor (H_2O). The solid and dashed traces denote the Raman-continuum of liquid and solid water.

cross-section of air is more than three orders of magnitude greater than that for vibrational Raman scattering. The spectral separation is based on the Doppler broadening of the Rayleigh line, leading to a line width of about 0.01 cm^{-1}. This line surrounds the much narrower peak from particle scattering. This separation is done in the HSRL technique; for details see Ref.[30] The advantage of this technique is that it is suitable for daytime operation. The disadvantage is the high complexity of such a system.

9.3.3 MULTIWAVELENGTH RAMAN LIDAR: EXAMPLE OF AN INSTRUMENT

Figure 9.4 shows the sketch of a multiwavelength Raman lidar. Table 9.1 summarizes the most important system parameters of the instrument.

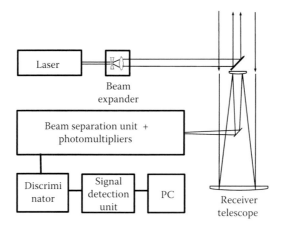

FIGURE 9.4 Aerosol/temperature/relative-humidity Raman lidar (MARTHA) operated at the Leibniz Institute for Tropospheric Research, Leipzig, Germany.

TABLE 9.1
Instrument Characteristics of the Multiwavelength Raman Lidar System MARTHA; Status as of 2009

Laser Type	Nd:YAG
Emitted wavelengths	355, 532, 1064 nm
Pulse energies	450, 450, 450 mJ
Beam expansion	15-fold
Beam divergence	0.1 mrad
Repetition rate	30 Hz
Type of receiver telescope	Cassegrain
Diameter of main mirror	80 cm
Wavelength separation	Polychromator made of interference filters and gratings
Signal detection	Photon counting
Signal processing	300-MHz Counter
Receiver Channels	
Elastic scattering	355, 532, 1064 nm
Raman scattering: nitrogen	387, 607 nm
Raman scattering: water vapor	408 nm
Depolarization	532 nm
Temperature	Around 532 m

The stationary Multiwavelength–Aerosol–Raman–Temperature–Humidity Apparatus (MARTHA) is described by Mattis[46] and Mattis et al.[85] One Nd:YAG laser is used for generating laser pulses at 355, 532, and 1064-nm wavelength with a repetition rate of 30 Hz. The laser beam is expanded 15-fold and vertically transmitted into the atmosphere. The backscattered radiation is collected by a Cassegrain telescope and transmitted to the signal detection unit.

Polychromators are used for separating the signals at different wavelengths. Interference filters in front of the detectors transmit the backscatter signals at selected wavelengths. Narrowband interference filters are used for the detectors that collect the Raman signals. A suppression of the elastic signals by 8–10 orders of magnitude is achieved. Details of the characteristics of the optical components are described by Althausen et al.[45] and Mattis.[46]

Figure 9.5 shows a sketch of the signal receiver box. The return beam is separated according to wavelength. Preferably the signals at the shorter wavelengths are separated from the signals at the longer wavelengths by transmitting the latter, and reflecting the light pulses at the shorter wavelengths. Another point to keep in mind is that Raman signals are much weaker than the elastic return signals. For this reason as little optical components as possible should be in the way of the Raman signals. In this way we obtain an optical setup that needs to be optimized as much as possible for the various signal types. The signals are detected with photomultipliers, which are operated in the current mode. Standard photomultiplier tubes are used for signal detection. The signals are collected on computer cards that operate with a sampling rate of 300 MHz.

Temperature profiles are obtained from signals of the Stokes and anti-Stokes branches of the pure rotational Raman spectrum of nitrogen excited by radiation of the second harmonic of the Nd:YAG laser. The technique is described in Refs.[85,86] The discrimination of the pure rotational Raman spectrum against the sky background is done with a Fabry–Perot interferometer according to the technique described in Ref.[87] The temperature channels can be used to determine profiles of the particle extinction coefficient at daytime conditions.[44] For that purpose one takes the sum of the signals of the Stokes and the anti-Stokes branches of the pure rotational Raman signals. The particle extinction profiles are then calculated with the Raman method.[32]

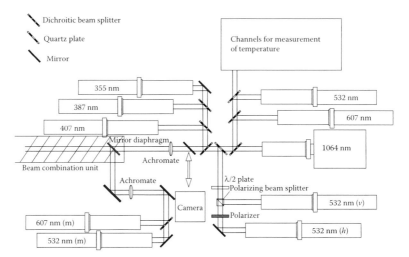

FIGURE 9.5 Signal receiver box of MARTHA. Numbers denote the wavelength at which signals are detected. Regarding work principle of the temperature channels see Mattis et al.[85] The index h denotes light that is polarized like the state of polarization of the outgoing laser beam. The index v denotes light polarized perpendicular to the state of polarization of the outgoing laser beam. The index me denotes channels that are used for the detection of multiple scattered light.

9.3.4 Optical Parameters of Advanced Multiwavelength Aerosol Raman Lidars

Table 9.2 lists the optical parameters that can be determined with multiwavelength Raman lidars. Particle volume backscatter and extinction coefficients are so-called extensive parameters. This means that these parameters change in value with changing particle number concentration. In contrast, particle lidar ratio, extinction- and backscatter-related particle Ångström exponents, and the linear particle depolarization ratio are intensive particle parameters, that is, they do not change with changing number concentration.

The parameters in Table 9.2 can be used for a qualitative separation of the observed aerosol particles into the so-called aerosol types. Particle lidar ratios depend on particle size and shape and their complex refractive index (real and imaginary part). The extinction-related particle Ångström exponents mainly depend on particle size. The backscatter-related particle Ångström exponents depend on particle size, complex refractive index, and particle shape. Finally, the linear particle depolarization ratio strongly depends on particle shape. Recent results from the SAMUM field experiments have shown that depolarization ratios slightly depend on the mean particle size.[88]

TABLE 9.2
Optical Parameters That are Determined with MARTHA

Parameter	Symbol	Wavelength, Respectively Wavelength Range (Indicated as Wavelength Pair)
Particle volume backscatter coefficient	β_{par}	355, 532, 1064
Particle volume extinction coefficient	α_{par}	355, 532
Particle lidar ratio	S_{par}	355, 532
Extinction-related particle Ångström exponent	$\mathring{a}_{\alpha,par}$	355/532
Backscatter-related particle Ångström exponent	$\mathring{a}_{\beta,par}$	355/532, 532/1064
Linear particle depolarization ratio	δ_{par}	532

A two-wavelength Raman lidar (355 and 532 nm) with polarization channels already allows us to identify several aerosol types.[71] Maritime particles show low lidar ratios at both wavelengths (20–30 sr) and a low depolarization ratio. Dust, urban haze, and smoke show similar lidar ratios (40–70 sr) at 532 nm. The lidar ratio of urban haze can be slightly larger at 355 than at 532 nm. Lidar ratios of dust are similar. In contrast, the lidar ratio of forest-fire smoke often is clearly lower at 355 nm compared to 532 nm.[71] To distinguish urban haze from dust, the depolarization ratio is needed. The backscatter-related Ångström exponents also can be used for a distinction. This parameters is around 0 for large dust particles (wavelength pair 355/532 nm), whereas it may easily exceed the value 1 in the case of urban pollution.

9.3.5 Microphysical Particle Parameters from Multiwavelength Raman Lidar

During the past decade, sophisticated computational procedures have been developed and successfully tested that permit the retrieval of microphysical properties of particles such as volume and surface-area concentration, effective radius, refractive index characteristics, and single-scattering albedo from multiwavelength Raman lidar observations.

It has been shown that the method of inversion with regularization is a practical method.[49,89–93] A summary of the theoretical model is given by Ansmann and Müller.[84] A minimum number of three measurement wavelengths[52,91] as well as a combination of particle backscatter and particle extinction coefficients[49,89,91,94] are needed for a successful retrieval of microphysical particle properties.

Most common is the use of vibrational Raman scattering at 355 nm and 532 nm, that is, the second and third harmonics of the Nd:YAG laser. In combination with the backscatter at 355, 532, and 1064 nm, such a system allows one to estimate microphysical properties using just a single laser source. A substantial number of systems using this technique are presently operated.[62] A similar approach using pure rotational Raman scattering at the same wavelengths is under test, the necessary filter techniques are available.

Inversion (in its correct mathematic definition) means that one explicitly solves the mathematical integral equations that link the measured aerosol optical properties, here of multiwavelength lidar, to the underlying mathematical quantities, that is, particle size distribution, particle shape, and complex refractive index. The methods applied make use of the spectral information contained in the backscatter coefficients and extinction coefficients at multiple wavelengths and its change with particle size.[95,96]

A historic overview on what has led to currently used algorithms is found in Ansmann and Müller.[84] The method of inversion with regularization with constraints[97] has become the standard method for the retrieval of microphysical parameters of tropospheric particles from multiwavelength lidar observations.

Profiles of the physical particle properties follow from the numerical inversion of the vertically and spectrally resolved particle backscatter and particle extinction coefficients. The optical data are related to the physical quantities through Fredholm integral equations of the first kind:

$$g_i(\lambda_k) = \int_{r_{\min}}^{r_{\max}} K_i(r, m, \lambda_k, s) v(r) \, dr + \varepsilon_i^{\exp}(\lambda_k), \quad i = \beta_{\text{par}}, \alpha_{\text{par}}, k = 1, \ldots, n. \qquad (9.6)$$

The term $g_i(\lambda_k)$ denotes the optical data at wavelengths λ_k in a specific height R (or range from the lidar). For easier reading, reference to height R is omitted in the following discussion. Subscript i denotes the kind of information, that is, whether it is the particle backscatter β_{par} or particle extinction α_{par} coefficient. The optical data have an error $\varepsilon_i^{\exp}(\lambda_k)$. The expression $K_i(r, m, \lambda_k, s)$ describes the kernel efficiencies of backscatter and extinction, respectively. They depend on the radius r of the particles, their complex refractive index m, the wavelength λ_k of the interacting light, as well as the shape s of the particles.

For spherical particle geometry, the kernel functions $K_i(r, m, \lambda_k, s)$ are calculated from the respective extinction and backscatter efficiencies $Q_i(r, m, \lambda_k, s)$ for individual particles[98] weighted with their geometrical cross-section πr^2:

$$K_i(r, m, \lambda_k) = (3/4r)Q_i(r, m, \lambda_k). \tag{9.7}$$

The term $v(r)$ in Equation 9.6 describes the volume concentration of particles per radius interval dr. The lower integration limit is defined by r_{min}, the radius down to which particles are optically efficient. For measurement wavelengths larger than 355 nm, the minimum particle size is around 50 nm in radius. The upper limit, r_{max} is the radius at which concentrations are so low that particles no longer contribute significantly to the signal. For typical particle size distributions in the troposphere r_{max} is below 10 µm.

In the inversion of Equation 9.6, the volume or surface-area concentrations are in general preferred over the number concentration, because they increase the sensitivity of the kernel efficiencies in the optically active range of the investigated particle size distribution. On average this shift leads to a stabilization of the inverse problem. A detailed description of the solution of the set of Equations 9.6 is given by Ansmann and Müller.[84]

The low number of measured optical particle properties requires introducing physical and mathematical constraints in the inversion algorithms in order to come up with sensible microphysical particle parameters. These algorithms cannot derive the exact shape of particle size distributions, which might not be achievable even in the near future owing to the low number of measured optical information of multiwavelength Raman lidar and the lack of appropriate mathematical tools. Nevertheless, in recent years, there have been first attempts to derive the approximate shape of the particle size distributions.[53]

Table 9.3 lists the parameters that can be derived with these inversion algorithms. Mean parameters such as the effective radius (cross-section weighted mean radius) of the particle size distribution are derived with comparably high accuracy. As for the specific inversion algorithm presented in Refs.[47–49,89], the accuracy of that parameter is on the order of ±25% in the range of effective radii from around 0.1–1.5 µm, and on the basis of the available measurement wavelengths. At present it does not seem possible to fully retrieve particles in the so-called coarse mode of particle size distributions[66,92] which is largely determined by particles from natural sources such as mineral dust. However, particles from anthropogenic activities are mainly present in the fine mode fraction which is accessible to the inversion algorithms.

Other size parameters such as volume concentration and surface-area concentrations can be derived to accuracies better than ±50%, if measurement errors of the optical parameters are less than 20%. Such measurement accuracies can be achieved with Raman lidars. There are first exploratory attempts to derive particle number concentration. Errors often are still higher than

---

TABLE 9.3
Parameters That are Inferred with the Inversion Algorithm

Parameters

Mean (effective) radius

Number concentration

Surface-area concentration

Volume concentration

Real part of complex refractive index

Imaginary part of complex refractive index

Single-scattering albedo

100%, but recent work shows that uncertainties can be reduced to less than 50% for particles with radii above 50 nm.

The complex refractive index can be derived to approximately ±0.05 in real part and ±50% in imaginary part, which allows one to derive the single-scattering albedo to an accuracy as good as 0.05 (at 532 nm) under favorable measurement conditions. Single-scattering albedo does not follow directly from the inversion algorithm. Particle size distribution and complex refractive index are used to calculate this parameter with a Mie-scattering algorithm.[98]

The complex refractive index is derived as wavelength- and size-independent quantity. There is the question regarding the representativeness of this wavelength- and size-independency, in view of the fact that complex refractive indices of atmospheric aerosol particles are generally wavelength- and size-dependent. We still lack in detailed simulation studies that may help answer this question of representativeness. Preliminary studies show that this mean, wavelength- and size-independent quantity is located inside the range of variation of the true values of the complex refractive index (for the measurement wavelength ranges from 355 to 1064 nm). In view of the overall uncertainties that are connected to data inversion, this accuracy can be considered sufficient at the present state of algorithm development.

We compute single-scattering albedo on the basis of the wavelength- and size-independent complex refractive index. In general, we prefer to compute this parameter at 532 nm wavelength, as it is in the center of the measurement wavelength range of nowadays used aerosol Raman lidars (355, 532, and 1064 nm).

In conclusion, we state that large progress has been made since the mid 1990s regarding the development of inversion methodology. This progress is largely owing to the continuous effort of combining theoretical work, simulations with synthetic data sets, and application to experimental data. Table 9.4 summarizes this work effort. The publications are categorized according to their main topic, that is, theoretical work, simulations with synthetic data sets, and application to experimental data.

9.4 CURRENT STATUS

9.4.1 Example: A Mixed Dust/Smoke Plume from West Africa

The following measurement is an example that shows the potential of modern multiwavelength Raman lidar. Measurements of multiwavelength aerosol Raman lidar in combination with polarization lidar were carried out at Praia (14.9°N, 23.5°W), Republic of Cape Verde, off the west coast of West Africa. The observations were performed during the Saharan Mineral Dust Experiment (SAMUM) in January and February 2008.[114] A detailed description of the measurement is given by Tesche et al.[109]

TABLE 9.4
Categorization of Publications that Deal with Inversion with Regularization Applied to Data Collected with Multiwavelength Lidar

	Reference
Theory	52, 89–94, 99, 100
Simulation	49, 90–93, 99, 100
Case study	48, 49, 52, 56, 91–93, 99, 101
Application	42, 43, 48–52, 54–56, 91–93, 99, 101–112
Instrument combination	53, 66, 99, 101, 113

The data were taken with the Backscatter Extinction lidar-Ratio Temperature Humidity lidar Apparatus (BERTHA).[45,109,114] Additional data needed for data analysis were taken with the three-wavelength Multiwavelength Lidar System (MULIS).[88]

BERTHA delivers profiles of the particle backscatter coefficients at 355, 400, 532, 710, 800, and 1064 nm, the depolarization ratio is measured at 710 nm. Nitrogen Raman signals at 387 and 607 nm are used to infer the particle extinction coefficients at 355 and 532 nm. The relative error of the volume backscatter coefficients ranges from 5% to 15%. The relative uncertainty in the extinction coefficients is about 10% and up to 50% for extinction values above and below 100 Mm^{-1} (M = 10^{-6}), respectively. Details on the data analysis procedure are given by Tesche et al.[109,114] Figure 9.6 provides an overview of the atmospheric situation on January 31, 2008. Shown are the aerosol lidar observations at 710 nm and radiosonde profiles of meteorological parameters. The height–time display of the 710-nm volume depolarization ratio (Figure 9.6d) shows a two-layer system consisting

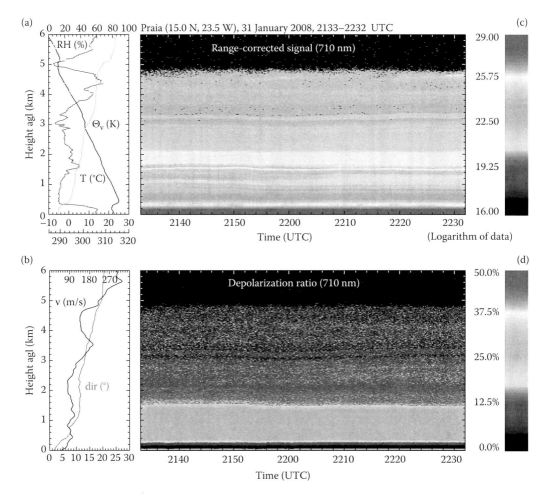

FIGURE 9.6 **(See color insert following page 206.)** Dust and smoke observed with BERTHA over Praia, Cape Verde, on 31 January 2008. (b) Range-corrected 710-nm signal (arbitrary units) and (d) volume depolarization ratio at 710 nm are shown with 15-m vertical and 10-s temporal resolution. Radiosonde profiles of (a) relative humidity RH, temperature T, and virtual potential temperature θ_v (evening sonde, Vaisala RS80, launch at 2123 UTC) and of (c) horizontal wind speed v and direction dir (morning sonde, Vaisala RS92, launch at 1110 UTC) are shown in addition. The lower of the two x-axes in (a) refers to the virtual potential temperature. The upper x-axis in (c) refers to the wind direction. (Plot adapted from Figure 1 in Tesche, M., Ansmann, A., Müller, D., Althausen, D., Engelmann, R., Freudenthaler, V., and Groß, S. 2009. *J. Geophys. Res.* 114:D13202.)

of a pure dust layer below 1500 m above ground level and a lofted aerosol layer consisting of a mixture of dust and smoke. The volume depolarization ratio indicates the presence of nonspherical dust particles. The African aerosol plume extends to about 5 km height. Nighttime lidar measurements (for the time period shown in Figure 9.6) show particle optical depths of 0.6 at 532 nm. Sun photometer measurements carried out at daytime show similar values of optical depth.

The radiosonde humidity and temperature profiles (Figure 9.6a) indicate a shallow marine boundary layer up to 350 m. A dry, almost well-mixed Saharan dust layer extends from 400 to 1500 m. A lofted aerosol layer with enhanced and varying relative humidity is present above 1.5 km height. The lofted layer can be separated into two layers according to the profiles of the relative humidity and the potential temperature. One layer extends from 1.5 to 3 and the second layer reaches from 3 to 5 km height. Thin cumulus clouds developed above 4.6 km after 2230 UTC (Universal Time Coordinated). Thus, swollen aerosol particles are present in the moist layer above 3.8 km height in Figure 9.6.

The combination of signals from BERTHA and MULIS and auxiliary information on pure Saharan mineral dust collected during SAMUM 2006 in Morocco (special issue in *Tellus* B, Vol. 61, 2009) permit us to separate the optical properties of desert dust and biomass burning particles as a function of height. Figure 9.6 shows a typical situation of aerosol particle layering in the tropical outflow regime of western Africa during the winter season. Above a dense desert dust layer (with an optical depth of about 0.25 at 532 nm) which reached 1500 m, a lofted layer consisting of desert dust (0.08 optical depth) and biomass burning smoke (0.24 optical depth) extended from 1500 to 5000 m height. Extinction values are 20 ± 10 Mm^{-1} (desert dust) and 20–80 Mm^{-1} (smoke) in the lofted plume. The smoke extinction-to-backscatter ratios are rather high, with values up to more than 100 sr.

Microphysical properties of the smoke aerosol are estimated using the inversion algorithm described in Section 9.3. The algorithm is applied to the spectrally resolved backscatter and extinction coefficients shown in Figure 9.7. Figure 9.8 shows the inversion results for the two most important parameters. The height profiles of the effective radius (surface-area weighted mean radius) and

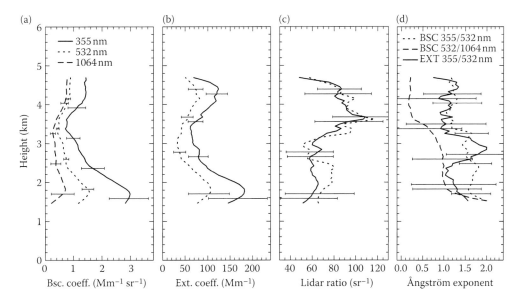

FIGURE 9.7 Smoke (a) backscatter coefficients, (b) extinction coefficients, (c) lidar ratios, and (d) backscatter- and extinction-related Ångström exponents retrieved from the BERTHA observations on January 31, 2008. The 1-h mean signal profiles are smoothed with 660-m window length to reduce the statistical uncertainty. Error bars (one standard deviation) indicate the total retrieval uncertainty. (Plot has been adapted from Figure 5 in Tesche, M., Ansmann, A., Müller, D., Althausen, D., Engelmann, R., Freudenthaler, V., and Groß, S. 2009. *J. Geophys. Res.* 114:D13202.)

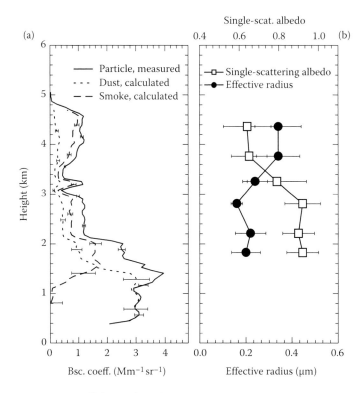

FIGURE 9.8   (a) Backscatter coefficients of particles, dust, and smoke at 532 nm and (b) effective radius and single scattering albedo (SSA, 532 nm) retrieved with the inversion algorithm[49,89] described in Section 3 and applied to the profiles of backscatter and extinction coefficients of smoke down shown in Figure 9.7. (Plot adapted from Figure 6 in Tesche, M., Ansmann, A., Müller, D., Althausen, D., Engelmann, R., Freudenthaler, V., and Groß, S. 2009. *J. Geophys. Res.* 114:D13202.)

the single-scattering albedo (scattering-to-extinction ratio) of the smoke particles are shown in Figure 9.8 together with the backscatter coefficients (total, dust, smoke). The uncertainty of the inversion products considers errors introduced by the inversion procedure itself and 20% uncertainty in each of the five optical input parameters (three backscatter and two extinction coefficients).

The effective radius shows values around 0.2 μm in the lower part of the lofted plume (dry aerosol particles) and values up to 0.35 μm in the upper part (swollen aerosol particles, some activated particles). The single-scattering albedo ranges from 0.9 (less absorbing particles) in the lower part of the lofted plume to rather low values of 0.60–0.65 in the upper part, describing highly absorbing smoke particles. The latter finding is in agreement with the rather high lidar ratios of 80–100 sr which indicate highly light-absorbing particles.

9.4.2   EARLINET: The First Aerosol Raman Lidar Network

The European Aerosol Research Lidar Network (EARLINET) is the first aerosol lidar network on a continental scale. The network was established in 2000.[61] One of the main goals is to provide a comprehensive, quantitative, and statistically significant database for the aerosol distribution on a continental scale. The use of these data can contribute significantly to the quantification of aerosol concentrations, radiative properties, long-range transport and budget, and prediction of future trends in aerosol pollution levels. It can also contribute to improved transport model treatment on a wide range of scales, and it offers a rich data set serving to allow for a better exploitation of present and future data from satellite remote sensing for a variety of parameters.

Figure 9.9 shows the location of the 27 stations that are presently operated in Europe (status as of 2010). This network currently consists of 7 single backscatter lidar stations, 10 Raman lidar stations with the ultraviolet Raman channel for independent measurements of aerosol extinction and back-scatter, and 10 multiwavelength Raman lidar stations (elastic channel at 355, 532, and 1064 nm, Raman channels at 387 and 607 nm, plus in some cases a depolarization channel that is operated either at 355 or at 532 nm) for the retrieval of aerosol microphysical properties. The network activity is based on scheduled measurements, a rigorous quality assurance program addressing both instruments and evaluation algorithms, and a standardized data exchange format. All network stations participated in intercomparisons both at instrument and algorithm levels with standardized procedures.[115-116] EARLINET data have already been used for a first statistical analysis of the aerosol optical properties over Europe,[117] climatological studies,[117-119] studies on Saharan dust events,[44,63,65,120,121] volcanic eruptions,[122] biomass burning,[54,123] long-range transport,[53,54,124] and solar aerosol radiative forcing.[125] The microphysical retrieval algorithms[89,91,93] that were developed have been applied to data collected using the multiwavelength Raman lidar systems.

EARLINET observations data have been used for first comparisons with models.[126,127] Another example of the benefit of these Raman lidars is the important role they play in validating and exploiting data of the CALIPSO mission. This first satellite with a backscatter lidar (CALIOP) onboard was launched in April 2006.[128,129] EARLINET started correlative measurements for CALIPSO in June 2006.[130,131] EARLINET Raman lidar stations are an optimal tool to validate CALIPSO lidar data, as EARLINET provides measured profiles of aerosol extinction and lidar ratio for the aerosol retrievals from the backscatter lidar.

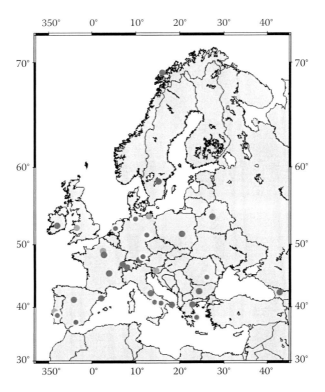

FIGURE 9.9   (See color insert following page 206.) Distribution of all the EARLINET lidar stations in Europe. Blue dots denote stations that are no longer active. Red dots denote active stations. All stations except for those in Belarus, Bulgaria, Georgia, Spain (Madrid) operate at least one Raman channel for extinction profiling. Yellow circles denote stations that operate multiwavelength Raman lidars. Status as of 2010.

9.5 OUTLOOK

9.5.1 INSTRUMENTS

Many technical parameters need to be improved in future generations of multiwavelength Raman lidars in order to fill gaps in particle characterization with aerosol lidar. Reducing the uncertainty of the data products will help reduce the uncertainties of climate impact studies. The following discussion presents an overview on the complex issue of necessary technical improvements.

9.5.1.1 Identification of Chemical Signatures of Aerosol Particles

The inversion algorithm described in Section 9.3 delivers only average values for particle size and complex refractive index. These values then may be attributed to the likely presence of certain particle components such as small anthropogenic particles. If the complex refractive index is at the lower end of reasonable numbers we may infer that sulfates or nitrates are contained in the detected particles. But it is a fact that current multiwavelength Raman lidar observations do not let us directly infer the presence and/or concentration of *water-soluble* and *water-insoluble* components. We cannot distinguish between, for instance, sulfate-containing particles and nitrate-containing particles. This means that a direct characterization of the particles in terms of chemical components is impossible.

Recent advances in measurement techniques may offer a solution, although. Tatarov and Sugimoto[132] determined for the case of a mixed mineral-dust/urban-pollution plume over East Asia the quartz concentration in that pollution plume. This concentration measurement was done by detecting Raman signals that directly originated from the silicon-dioxide in the dust plume. The excitation energy was at 532 nm and the return signal from quartz was detected at around 545 nm.

It is not clear yet how much concentration of mineral quartz is needed in order to obtain reasonable signal-to-noise ratios. It also remains unknown whether this Raman return signal is always located at the same wavelength. We know that mineral dust has different chemical compositions (and thus different wavelength- and size-dependent complex refractive index) in various desert regions of the world. The internal structure of the quartz crystals may vary with location on the globe. The literature offers only little information on the scattering-cross section of quartz particles useful for atmospheric applications. For this reason, the question on the general applicability of this new methodology remains open.

In early 2009, another test on the feasibility of this methodology was launched. A multiwavelength Raman lidar was upgraded with such a novel Raman quartz channel.[133–135] However, the excitation wavelength of the laser light was set at 355 nm leading to a Raman quartz signal at around 361 nm. First tests in fact show a signal from quartz Raman scattering, which adds an important corner stone to advancing Raman lidar into the stage of measuring certain chemical compounds of atmospheric aerosols.

9.5.1.2 New Measurement Wavelengths: Ultraviolet Region

We need to expand the range of measurable optical particle parameters. Such work implicitly would reduce the uncertainties of the retrieved microphysical parameters. As for the technical limitation, we need to expand the range of measurements into the ultraviolet and infrared range of the electromagnetic spectrum. We have to keep in mind that present-day multiwavelength Raman lidars operate at wavelengths of 355, 532, and 1064 nm.

It is commonly known from Mie-scattering theory that measurements at wavelengths on the order of the size of the investigated particles are needed in order to obtain size resolved information of the scattering particles. The literature shows that size-resolved observations are limited to approximately 50–80 nm particle radius at the lower end of the particle size spectrum. For smaller particle radii, the applied inversion algorithms generate unacceptably large errors. In fact, if we apply strict quality constraints on the inversion results, we have to state that errors already become critical at particle radii less than 100 nm. In this way, we do not have a possibility to investigate aerosol particles in the Aitken mode, for example Ref.[136], of the particle size spectrum.

Feasibility tests on the basis of simulations with synthetic data were carried out for various wavelengths as low as 266 nm. Results suggest that the size resolution of the investigated particle size distribution can be significantly improved for Aitken mode particles. In these studies, we neglected the fact that the quality of the optical data, that is, measurement errors, may be severely impaired by light absorption by, for example, ozone, which strongly influences the signals that we obtain from the aerosol particles. The height range of the detected signals may also be severely limited.

Given these limitations, extinction profiles in the ultraviolet wavelength range would greatly help us. We need a good estimate of ozone absorption in the ultraviolet wavelength range. For instance, if we measure Raman scattering signals of oxygen molecules and nitrogen molecules simultaneously, we may obtain a good estimate of the Ångström exponent which is needed for deriving particle extinction coefficients. However, theoretically we can go one step further. If we know the absorption cross-section of ozone accurately enough, we may use this information in combination with an estimated aerosol Ångström exponent to infer the extinction coefficient at 266 nm. Yet we have to emphasize that his concept is a highly exploratory starting point. In combination with the backscatter coefficient at 266 nm, we would obtain the lidar ratio for particles in the Aitken mode and we expect significant improvements of particle characterization.

9.5.1.3   New Measurement Wavelengths: Infrared Region

Work was done in recent years to explore the limit of particle size characterization at the upper range of the particle size spectrum. In this context, it should be mentioned that sun photometer observations are carried out in the frame of AERONET.[23] Particle size distributions are inferred to particle radii up to 15 μm.[137] According to results discussed in the literature, the coarse mode fraction of the particle size distribution, which is defined as particles of radius above approximately 1 μm, is retrievable.

In view of our present knowledge from simulation studies we must state that the coarse mode fraction cannot be retrieved from data acquired with present-day aerosol lidar. This restriction in part is caused by the maximum available measurement wavelength of 1064 nm employed in aerosol Raman lidar systems. Great effort has been put in determining an upper limit of the retrieval of particle size distributions from multiwavelength Raman lidar observations. Veselovskii et al.[94] indicate that this upper retrievable size range is at particle effective radius of approximately 2 μm at maximum. This number converts into a size-resolved characterization of the investigated particle size distribution of less than 5 μm in particle radius, if we consider a monomodal particle size distribution as the basis of this estimate. This limit for the effective radius unfortunately has to be seen even more critically as it can be only derived if additional information, as for instance the complex refractive index, is known. In summary, we must accept the fact that particle size distributions with effective radii larger than 1 μm may not be derived, which basically restricts our particle microphysical characterization to the fine mode fraction of the particle size distribution, for example, Refs.[42,43,52–55,91–93,101]

How can we resolve the problem of this limitation? We need measurement wavelengths larger than 1064 nm, and preferably we also need particle extinction measurements at least at 1064 nm. Such larger wavelength for particle extinction will once more improve particle microphysical characterization. But we have to keep in mind that a trustworthy retrieval of optical properties, particularly of the extinction coefficients at infrared wavelengths, will become increasingly difficult with increasing wavelength. The reason for this difficulty is the fact that we need to separate the molecular signal component from the particle signal component. We need to calibrate the profile in height ranges where molecular scattering is dominant. The identification of such areas is difficult, because the signal strength from molecular scattering decreases with wavelength (λ) as λ^4. The signal strength is so low that we lack in sufficiently strong signals that can be calibrated.

Even if we may be able to resolve the problem of signal calibration in the molecular atmosphere, we still cannot directly measure the particle backscatter coefficient. We also need to know the particle extinction-to-backscatter ratio, which is the second critical input parameter for the measurement of the backscatter coefficient. The literature shows the large variation of this parameter at measurement wavelengths of 355 and 532 nm.[71]

We may have ways of estimating the lidar ratio at 1064 nm, but we certainly have no concept of the variation of this parameter at wavelengths larger than 1064 nm. Estimates on the basis of simulation studies with synthetic data are inappropriate. This problem may be overcome if we can measure the particle extinction coefficient at least at near-infrared wavelengths, as this parameter intrinsically provides us with the particle lidar ratio. Measuring the extinction coefficient, however, requires detection of the extremely weak Raman signals.

The Raman signals of nitrogen are at around 1415 nm for an excitation wavelength at 1064 nm. Photomultipliers with high quantum efficiency till date are not available for that wavelength range. We may somewhat go around this problem simply by using the Stokes rather than the anti-Stokes Raman-shifted signals. In that case, we end up at a wavelength of around 851 nm, which is just about in the range of detectors that still have reasonable sensitivity. Unfortunately, the signals from Stokes Raman scattering are significantly weaker than the signals from anti-Stokes Raman scattering, which reduces the advantage that we gain with the better detector efficiency.

In summary, this work is highly exploratory and we do not yet have any experience regarding additional technical limitations that may arise from implementing such channels, nor the problems in data analysis that may occur from low signal-to-noise ratios. We must state that inferring extinction profiles in the near-infrared wavelength range is highly important but still out of reach with currently available lidar technology. Yet the concept appears attractive enough to tackle it in the coming years.

9.5.1.4 Extending Measurements into Daytime

Another technical challenge that particularly impairs the usefulness of Raman lidar is connected to the problem of measuring particle extinction coefficients under daylight conditions. Lidar instruments need to be upgraded to the point that they can deliver useful signal profiles at the Raman-shifted wavelengths at daylight conditions. First feasibility studies and prototype channels on the basis of the Fabry–Perot filter technique have shown that useable measurements under daylight conditions are possible.[44] A technically less demanding solution would be narrow interference filters.

There are several reasons why daytime measurements with Raman lidar are needed. The wealth of information on aerosol optical and microphysical particle parameters currently comes from observations at night time. Such data sets have obvious disadvantages. Detailed studies on the change of aerosol properties during the course of a day are impossible. Observations on the effects of mixing of air in the planetary boundary layer during sunrise and sunset are restricted. We have no real opportunity for detailed characterization of particle properties during overpasses of satellites carrying passive aerosol sensors. The A-Train (http://www.nasa.gov/mission_pages/cloudsat/multimedia/a-train.html) currently is the most prominent example of an array of such aerosol sensors. The daylight limitation severely constrains the benefit of aerosol Raman lidar measurements for satellite-based platforms, which can give us a global view of aerosol pollution. We have to keep in mind that satellite sensors need ground-based validation points. AERONET sun photometers in that regard have evolved into an indispensable tool. We have to state that at present (multiwavelength) Raman lidar is not yet a method of choice for satellite validation studies at daylight conditions.

This limitation also restricts the usefulness of observations with a combination of lidar with sun photometer. Such measurement geometry however is needed for a true integrative approach of lidar–satellite–sun photometer observations of aerosols. Sun photometers not only measure optical depth, but also can be operated in the sky-brightness scanning mode (Almucantar measurements). From this geometry, we obtain the particle phase function at different measurement wavelengths. This information in turn provides an extensive set of particle microphysical properties such as particle size distributions and complex refractive index,[137] from which we may infer, for example, the single-scattering albedo. Measurements of the polarization state of the detected radiation are important in the case of scattering of electromagnetic radiation by particles of nonspherical shape. Furthermore, sun photometer–satellite observations are important to correct the surface–albedo effect that needs to be considered in the analysis of aerosol observations with satellites. For all the points listed here, multiwavelength Raman lidar could add valuable data.

9.5.1.5 New Data Products

We have to think of additional data products that can be measured with future generations of aerosol lidars. These new data products do not necessarily follow from the explicit use of the measured Raman signals. But we have to keep in mind that the Raman detectors are a vital part of complex lidar systems. These systems deliver optical data which in part only have reasonable quality (low uncertainty), because Raman signals are available. On the basis of the data inversion from optical into microphysical particle properties, which in fact also only works because high-quality extinction coefficients from Raman signal measurements are available, we may derive additional useful parameters.

The most attractive new candidate of data products is the particle phase function. This parameter can be obtained from so-called bi-static lidar systems. The transmitted laser beam is observed with an external optical receiver system under sideward scattering geometry. One scans the laser beam at various height levels, which in turn delivers scattering signals other than at 180°. Such instrument geometry does not give us the complete phase function between 0° and 180°. However, if laser transmitter and receiver unit are both at the same ground level, we can theoretically obtain the phase function from a minimum angle of 90° up to a maximum angle of 180°.

A disadvantage of this instrument setup is given by the fact that we can observe the scanned laser beam only at one specific angle in each height bin of the optical profiles, respectively. Thus, we have different information on the phase function for different height levels. This constraint may be relaxed again, if we observe well-mixed particle plumes that extend over several kilometers in the atmosphere. In that case, we may combine the scattering information retrieved at different angles and different height intervals into one scattering phase function. In any case, the combination of a bistatic lidar system that possesses one or more channels for particle extinction measurements would greatly expand the range of application of Raman lidar.

Particle depolarization measurements at several wavelengths would even further increase the usefulness of multiwavelength Raman lidar. At present, lidar systems usually only employ one measurement wavelength to infer the so-called linear particle depolarization ratio. Multiwavelength dust depolarization-ratio observations were carried out with several ground-based and airborne lidars in the frame of the SAMUM. Results presented by Freudenthaler et al.[88] suggest that an improved characterization of nonspherical mineral dust particles is possible, if we use at least two measurement wavelengths for particle depolarization ratio measurements.

9.5.2 Inversion Algorithms

As described in Section 9.3.5, particle effective radius, volume, and surface-area concentration can be derived with inversion algorithms. The complex refractive index can be estimated to an accuracy that allows a sensible retrieval of the single-scattering albedo. Given these data products and the first useable results on the fine mode fraction of the particle size distribution, it seems feasible to roughly estimate the light-absorption profile in future versions of the methodology. However, a final proof of that would require considerable work into algorithm development, and feasibility studies with synthetic and experimental data. One key factor that would decide on the improvement of the data products again would be the quality of the optical profiles, and particularly the quality of the Raman signal profiles that are used for the particle extinction measurements.

There are several problems in data inversion, which have to be tackled in order to further improve the quality of the derived parameters. The identification of particles in the coarse mode of the particle size distribution (particle radii >1 μm) is not possible in a satisfactory manner. First explor-atory studies on that specific problem have begun.[66,94,99] An extension of the measurement wavelength range of lidar, which includes the detection of Raman signals at wavelengths above 607 nm, might be a way of improving the inversion data products regarding coarse mode particles.

9.5.3 GALION: THE POTENTIAL FUTURE GLOBAL LIDAR NETWORK

Twenty-seven lidar stations participate in EARLINET (status as of August 2010). Until around 2004, only IfT operated multiwavelength aerosol lidar instruments in Europe. The work at IfT, however, triggered instrument development in the aerosol lidar community to the point that now (status as of August 2010) nine more stations have upgraded their systems to multiwavelength Raman lidar capability. EARLINET thus is the first lidar network capable not only of measuring optical particle properties, but also of providing important microphysical properties at several key sites in Central and South Europe.

The Asian Dust lidar network, AD-Net (http://www-lidar.nies.go.jp/) and the National Institute for Environmental Studies (NIES, Japan) lidar network are the East Asian counterpart to EARLINET. Work has begun to upgrade one instrument to multiwavelength capability, which will include measurements of particle extinction coefficients at two wavelengths. However, in that case, the HSRL[135,138,139] instrument type rather than the Raman lidar type will be used.

Lidar systems operating at Spitsbergen near the polar circle,[140] South Korea,[141] and Japan[103] show the potential of observing aerosols vertically resolved with multiwavelength Raman lidar in aerosol hot spots on the globe. Raman lidars, if distributed over the northern hemisphere (about 10 lidars per continent) would significantly enhance our knowledge of vertical layering and long-range trans-port of natural aerosols and anthropogenic haze, and of the interaction of aerosols with water and ice clouds. Such systems could be incorporated in large lidar networks. EARLINET has achieved this goal in Europe. The AD-Net and the NIES lidar network are successfully operated in East Asia. The Cooperative Remote Sensing Science and Technology (CREST) Lidar Network (CLN) covers the northeastern part of North America.

In summary, these networks could be linked together to form a global lidar network in which Raman lidar technology can play a key role. The first steps regarding such a global network have just recently been undertaken. The GAW Aerosol Lidar Observations Network (GALION) has been inaugurated in 2008.[62] Figure 9.10 provides an overview on the number and distribution of the lidar stations that are linked in GALION. Given the logistic problems, such a global network in the begin-ning certainly can only act as an "idea." We need to develop concepts regarding exchange of exper-tise, and we need to establish common software and instrument standards. An attractive idea in that respect certainly is the establishment of several multiwavelength Raman lidar stations on each continent.

However, we can go one step further. This next step will make multiwavelength Raman lidar a key player in validation studies of a new generation of multiwavelength HSRL. The HSRL tech-nique has not been covered in this chapter, but it will be the future prime candidate for extinction profiling aboard aircraft and space-borne platforms. This technology is considerably more complex than Raman lidar, but only this instrument type will allow us to obtain extinction profiles on a laser shot-by-shot basis. Fast flying research aircraft and satellites simply do not allow signal averaging over several laser shots, as in that way we would lose most of the information on the aerosol particle distribution along the horizontal dimension.

It must be stated that the research work with multiwavelength Raman lidar since the mid 1990s certainly has pushed the aerosol lidar community into developing multiwavelength HSRL (MW-HSRL) systems. A first airborne MW-HSRL will soon become operational.[138] Multiwavelength Raman lidars will play an outstanding role in forthcoming missions on aerosol sounding with lidar aboard satellites. New lidar systems such as the Atmospheric Lidar (ATLID) on ESA's (European Space Agency) Earth Clouds, Aerosols and Radiation Explorer (EarthCARE) platform, and the

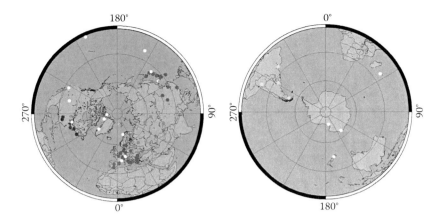

FIGURE 9.10 **(See color insert following page 206.)** Distribution of stations as available through the cooperation between existing networks. The different networks are indicated by the dot color: AD-NET violet, ALINE yellow, CISLiNet green, EARLINET red, MPLNET brown, NDACC white, CREST blue. See the GALION report[62] for technical details of the lidar stations. (Plot adapted from GALION Technical Report. 2008. Plan for the Implementation of the GAW Aerosol Lidar Observations Network GALION, Report No. 178. World Meteorological Organization, Global Atmospheric Watch. http://www.wmo.ch/pages/prog/arep/gaw/gaw-reports.html.)

Atmospheric Laser Doppler Instrument (ALADIN) aboard the ADM—Aeolus satellite will be only the first revolutionary steps toward future space-borne MW-HSRL.

The first generation of comparably simple backscatter lidar aboard the CALIPSO satellite was launched in 2006. EarthCARE and ADM—Aeolus are in the design stage. These satellite–lidars scan the atmosphere in a curtain-like track. The distance between tracks and the extremely short time a satellite needs to cross any spot on Earth call for point-source-like observations of sophisticated ground-based lidars organized in well-coordinated networks such as EARLINET, in which (multiwavelength) Raman lidars will play an outstanding role.

REFERENCES

1. Houghton, J. T., Ding, Y., Griggs, D. J., Noguer, M., van der Linden, P. J., and Xiaosu D. (Eds). 2001. *Climate Change 2001: The Scientific Basis*. Cambridge University Press, New York, 944pp.
2. Penner, J. E., Zhang, S. Y., and Chuang, C. C. 2003. Soot and smoke aerosol may not warm climate. *J. Geophys. Res.* 108:4657.
3. Ramanathan, V., Crutzen, P. J., Lelieveld, J., Mitra, A. P., Althausen, D., Anderson, J., Andreae, M. O., et al. 2001. The Indian Ocean experiment: An integrated analysis of the climate forcing and effects of the great Indo-Asian haze. *J. Geophys. Res.* 106:28371–28398.
4. Anderson, T. L., Charlson, R. J., Bellouin, N., Boucher, O., Chin, M., Christopher, S. A., Haywood, J. et al. 2005. An "A-Train" strategy for quantifying direct climate forcing by anthropogenic aerosols. *Bull. Am. Meteor. Soc.* 86:1795–1809.
5. Charlson, R. J. 2001. Extending atmospheric aerosol measurements to the global scale. *IGAC Newslett.* 25: 11–14.
6. Heintzenberg, J., Graf, H.-F., Charlson, R. J., and Warneck, P. 1996. Climate forcing and the physico-chemical life cycle of the atmospheric aerosol—Why do we need an integrated inter global research programme. *Contr. Atmos. Phys.* 69:261–271.
7. Seinfeld, J. H., Charlson, R., Durkee, P. A., Hegg, D., Huebert, B. J., Kiehl, J., McCormick, M. P., et al. (Eds). 1996. *A Plan for a Research Program on Aerosol Radiative Forcing and Climate Change*. National Academy Press, Washington, DC, 180pp.
8. Diner, D. J., Menzies, R. T., Kahn, R. A., Anderson, T. L., Bösenberg, J., Charlson, R. J., Holben, B. N., et al. 2004. PARAGON: An integrated approach for characterizing aerosol climatic and environmental interactions. *Bull. Am. Meteor. Soc.* 85:1491–1501.

9. Collins, W. J., Stevenson, D. S., Johnson, C. E., and Derwent, R. G. 2000. The European regional ozone distribution and its links with the global scale for the years 1992 and 2015. *Atmos. Environ.* 24:255–276.

10. Creilson, J. K., Fishman, J., and Wozniak, A. E. 2003. Intercontinental transport of tropospheric ozone: A study of its seasonal variability across the North Atlantic utilizing tropospheric ozone residuals and its relationship to the North Atlantic Oscillation. *Atmos. Chem. Phys.* 3:2053–2066.

11. Jacob, D. J., Logan, J. A., and Murti, P. P. 1999. Effect of rising Asian emissions on surface ozone in the United States. *Geophys. Res. Lett.* 26:2175–2178.

12. McKendry, I. G., Hacker, J. P., Stull, R., Sakiyama, S., Mignacca, D., and Reid, K. 2001. Long-range transport of Asian dust to the Lower Fraser Valley, British Columbia, Canada. *J. Geophys. Res.* 106:18,361–18,370.

13. Prather, M., Gauss, M., Berntsen, T., Isaksen, I., Sundet, J., Bey, I., Brasseur, G., et al. 2003. Fresh air in the 21 century? *Geophys. Res. Lett.* 30:1100.

14. Haywood, J. M. and Shine, K. P. 1997. Multispectral calculations of the direct radiative forcing of the tropospheric sulphate and soot aerosols using a column model, *Q. J. R. Meteorol. Soc.* 123:1907–1930.

15. Quijano, A. L., Sokolik, I. N., and Toon, O. B. 2000. Radiative heating rates and direct radiative forcing by mineral dust in cloudy atmospheric conditions. *J. Geophys. Res.* 105:12,207–12,219.

16. Ansmann, A., Mattis, I., Müller, D., Wandinger, U., Radlach, M., and Althausen, D. 2005. Ice formation in Saharan dust over central Europe observed with temperature/humidity/aerosol Raman lidar. *J. Geophys. Res.* 110:D18S12.

17. Lohmann, U. and Feichter, J. 2005. Global indirect aerosol effects: A review. *Atmos. Chem. Phys.* 5:715–737.

18. Rosenfeld, D. 2000. Suppression of rain and snow by urban and industrial air pollution. *Science* 287:1793–1796.

19. Koren, I., Kaufman, Y. J., Remer, L. A., and Martins, J. V. 2004. Measurement of the effect of Amazon smoke on inhibition of cloud formation. *Science* 303:1342–1345.

20. Feingold, G., Cotton, W. R., Kreidenweis, S. M., and Davis, J. T. 1999. Impact of giant cloud condensation nuclei on drizzle formation in marine stratocumulus: Implications for cloud radiative properties. *J. Atmos. Sci.* 56:4100–4117.

21. Rudich, Y., Khersonsky, O., and Rosenfeld, D. 2002. Treating clouds with a grain of salt. *Geophys. Res. Lett.* 29:2064.

22. Masonis, S. J., Franke, K., Ansmann, A., Müller, D., Althausen, D., Ogren, J. A., Jefferson, A., and Sheridan, P. J. 2002. An intercomparison of aerosol light extinction and 180°-backscatter as derived using *in situ* instruments and Raman lidar during the INDOEX field campaign. *J. Geophys. Res.* 107:8014.

23. Holben, B. N., Eck, T. F., Slutsker, I., Tanré, D., Buis, J. P., Setzer, A., Vermote, E., et al. 1998. AERONET—A federated instrument network and data archive for aerosol characterization. *Remote Sens. Environ.* 66:1–16.

24. Dubovik, O., Holben, B. N., Eck, T. F., Smirnov, A., Kaufman, Y. J., King, M. D., Tanré, D., and Slutsker, I. 2002. Variability of absorption and optical properties of key aerosol types observed in worldwide locations. *J. Atmos. Sci.* 59:590–608.

25. Holben, B. N., Tanré, D., Smirnov, A., Eck, T. F., Slutsker, I., Abuhassan, N., Newcomb, W. W., et al. 2001. An emerging ground-based aerosol climatology: Aerosol optical depth from AERONET. *J. Geophys. Res.* 106:12,067–12,097.

26. Kinne, S., Lohmann, U., Feichter, J., Schulz, M., Timmreck, C., Ghan, S., Easter, R., et al. 2003. Monthly averages of aerosol properties: A global comparison among models, satellite data, and AERONET ground data. *J. Geophys. Res.* 108:4634.

27. Hinkley, E. D. (Ed.). 1976. *Laser Monitoring of the Atmosphere*. Springer-Verlag, Berlin, 380pp.

28. Measures, R. M. (Ed.). 1992. *Laser Remote Sensing: Fundamentals and Applications*. Krieger Publishing Company, Malabar, FL, 524pp.

29. Kovalev, V. A. and Eichinger, W. E. (Eds). 2004. *Elastic Lidar: Theory, Practice, and Analysis Methods*. Wiley-VCH, Weinheim, 615pp.

30. Weitkamp, C. (Ed.). 2005. *Lidar–Range-Resolved Remote Sensing of the Atmosphere*, Springer-Verlag, Singapore, 460pp.

31. Ansmann, A., Riebesell, M., and Weitkamp, C. 1990. Measurements of atmospheric aerosol extinction profiles with a Raman lidar. *Opt. Lett.* 15:746–748.

32. Ansmann, A., Wandinger, U., Riebesell, M., Weitkamp, C., and Michaelis, W. 1992. Independent measurement of extinction and backscatter profiles in cirrus clouds by using a combined Raman elastic-backscatter lidar. *Appl. Opt.* 31:7113–7131.

33. Wandinger, U., Ansmann, A., Reichardt, J., and Deshler, T. 1995. Determination of stratospheric aerosol microphysical properties from independent extinction and backscattering measurements with a Raman lidar. *Appl. Opt.* 34:8315–8329.

34. Bissonnette, L. R. 1986. Sensitivity analysis of lidar inversion algorithm. *Appl. Opt.*, 25:2112–2125.

35. Fernald, F. G. 1984. Analysis of atmospheric lidar observations: Some comments. *Appl. Opt.* 23:652–653.

36. Ansmann, A., Wagner, F., Müller, D., Althausen, D., Herber, A., von Hoyningen-Huene, W., and Wandinger, U. 2002. European pollution outbreaks during ACE 2: Optical particle properties inferred from multiwavelength lidar and star/sun photometry. *J. Geophys. Res.* 107:4259.

37. Ferrare, R. A., Turner, D. D., Heilman-Brasseur, L., Feltz, W. F., Dubovik, O., and Tooman, T. P. 2001. Raman lidar measurements of the aerosol extinction-to-backscatter ratio over the Southern Great Plains. *J. Geophys. Res.* 106:20,333–20,347.

38. Franke, K., Ansmann, A., Müller, D., Althausen, A., Wagner, F., and Scheele, R. 2001. One-year observations of particle lidar ratio over the tropical Indian Ocean with Raman lidar. *Geophys. Res. Lett.* 28:4559–4562.

39. Franke, K., Ansmann, A., Müller, D., Althausen, D., Venkataraman, C., Reddy, M. S., Wagner, F., and Scheele, R. 2003. Optical properties of the Indo-Asian haze layer over the tropical Indian Ocean. *J. Geophys. Res.* 108:4059.

40. Mattis, I., Ansmann, A., Müller, D., Wandinger, U., and Althausen, D. 2002. Dual-wavelength Raman lidar observations of the extinction-to-backscatter ratio of Saharan dust. *Geophys. Res. Lett.* 29:9.

41. Mattis, I., Ansmann, A., Wandinger, U., and Müller, D. 2003. Unexpectedly high aerosol load in the free troposphere over central Europe in spring/summer 2003. *Geophys. Res. Lett.* 30:2178.

42. Müller, D., Ansmann, A., Wagner, F., and Althausen D. 2002. European pollution outbreaks during ACE 2: Microphysical particle properties and single-scattering albedo inferred from multiwavelength lidar observations. *J. Geophys. Res.* 107:4248.

43. Müller, D., Franke, K., Ansmann, A., Althausen, D., and Wagner, F. 2003. Indo-Asian pollution during INDOEX: Microphysical particle properties and single-scattering albedo inferred from multiwavelength lidar observations. *J. Geophys. Res.* 108:4600.

44. Müller, D., Mattis, I., Wandinger, U., Althausen, D., Ansmann, A., Dubovik, O., Eckhardt, S., and Stohl, A. 2003. Saharan dust over a central European EARLINET–AERONET site: Combined observations with Raman lidar and Sun photometer. *J. Geophys. Res.* 108:4345.

45. Althausen, D., Müller, D., Ansmann, A., Wandinger, U., Hube, H., Clauder, E., and Zörner, S. 2000. Scanning 6-wavelength 11-channel aerosol lidar. *J. Atmos. Oceanic Technol.* 17:1469–1482.

46. Mattis, I. 2002. Aufbau eines Feuchte–Temperatur–Aerosol-Ramanlidars und Methodenentwicklung zur kombinierten Analyse von Trajektorien und Aerosolprofilen (Construction of a humidity–temperature– aerosol Raman lidar and development of methods for the combined analysis of trajectories and aerosol profiles). PhD thesis, Universität Leipzig, Germany.

47. Müller, D. 1997. Inversionsalgorithmus zur Bestimmung physikalischer Partikeleigenschaften aus Mehrwellenlängen-Lidarmessungen (Inversion algorithm for the determination of particle properties from multiwavelength lidar measurements). PhD thesis, Universität Leipzig, Germany.

48. Müller, D., Wandinger, U., Althausen, D., Mattis, I., and Ansmann, A. 1998. Retrieval of physical particle parameters from lidar observations of extinction and backscattering at multiple wavelengths. *Appl. Opt.* 37:2260–2263.

49. Müller, D., Wandinger, U., and Ansmann, A. 1999. Microphysical particle parameters from extinction and backscatter lidar data by inversion with regularization: Simulation. *Appl. Opt.* 38:2358–2368.

50. Müller, D., Wagner, F., Althausen, D., Wandinger, U., and Ansmann, A. 2000. Physical properties of the Indian aerosol plume derived from six-wavelength lidar observations on 25 March 1999 of the Indian Ocean Experiment. *Geophys. Res. Lett.* 27:1403–1406.

51. Müller, D., Wagner, F., Wandinger, U., Ansmann, A., Wendisch, M., Althausen, D., and von Hoyningen– Huene, W., 2000. Microphysical particle parameters from extinction and backscatter lidar data by inversion with regularization: Experiment. *Appl. Opt.* 39:1879–1892.

52. Müller, D., Wandinger, U., Althausen, D., and Fiebig, M. 2001. Comprehensive particle characterization from three-wavelength Raman-lidar observations: Case study. *Appl. Opt.* 40:4863–4869.

53. Müller, D., Mattis, I., Wehner, B., Althausen, D., Wandinger, U., Ansmann, A., and Dubovik, O. 2004. Closure study on optical and microphysical properties of a mixed urban and Arctic haze air mass observed with Raman lidar and Sun photometer. *J. Geophys. Res.* 109:D13206.

54. Müller, D., Mattis, I., Wandinger, U., Ansmann, A., Althausen, D., and Stohl, A. 2005. Raman Lidar observations of aged Siberian and Canadian forest-fire smoke in the free troposphere over Germany in 2003: Microphysical particle characterization. *J. Geophys. Res.* 110:D17201.

55. Müller, D., Franke, K., Wagner, F., Althausen, D., Ansmann, A., Heintzenberg, J., and Verver, G. 2001. Vertical profiling of optical and physical particle properties over the tropical Indian Ocean with six-wavelength lidar, 2. Case studies. *J. Geophys. Res.* 106:28,577–28,595.

56. Wandinger, U., Müller, D., Böckmann, C., Althausen, D., Matthias, V., Bösenberg, J., Weiß, V., et al. 2002. Optical and microphysical characterization of biomass-burning and industrial-pollution aerosols from multiwavelength lidar and aircraft measurements. *J. Geophys. Res.* 107:8125.

57. Ångström, A. 1964. The parameters of atmospheric turbidity. *Tellus*, 16:64–75.

58. Eck, T. F., Holben, B. N., Reid, J. S., Dubovik, O., Smirnov, A., O'Neill, N. T., Slutsker, I., and Kinne, S. 1999. Wavelength dependence of the optical depth of biomass burning, urban, and desert dust aerosols. *J. Geophys. Res.* 104:31,333–31,350.

59. Reid, J. S., Eck, T. F., Christopher, S. A., Hobbs, P. V., and Holben, B. 1999. Use of the Ångström exponent to estimate the variability of optical and physical properties of aging smoke particles in Brazil. *J. Geophys. Res.* 103:27,473–27,489.

60. Bösenberg, J., Alpers, M., Althausen, D., Ansmann, A., Böckmann, C., Eixmann, R., Franke, A., et al. 2001. *The German Aerosol Lidar Network: Methodology, Data, Analysis.* Report No. 317, Max Planck Institute for Meteorology, Hamburg, Germany.

61. Bösenberg, J., Matthias, V., Amodeo, A., Amiridis, V., Ansmann, A., Baldasano, J. M., Balin, I., et al. 2003. *EARLINET: A European Aerosol Research Lidar Network to Establish an Aerosol Climatology*, Report No. 348, Max Planck Institute for Meteorology, Hamburg, Germany.

62. GALION Technical Report. 2008. Plan for the Implementation of the GAW Aerosol Lidar Observations Network GALION, Report No. 178. World Meteorological Organization, Global Atmospheric Watch. http://www.wmo.ch/pages/prog/arep/gaw/gaw-reports.html

63. Papayannis, A., Amiridis, V., Mona, L., Tsaknakis, G., Balis, D., Bösenberg, J., Chaikovski, A., et al. 2008. Systematic lidar observations of Saharan dust over Europe in the frame of EARLINET (2000–2002). *J. Geophys. Res.* 113:D10204.

64. Pisani, G., Armenante, M., Boselli, A., Frontoso, M.G., Spinelli, N., and Wang, X. 2007. Atmospheric aerosol characterization during Saharan dust outbreaks at Naples EARLINET station, in *Remote Sensing of Clouds and the Atmosphere XII*, Comerón, A., Picard, R. H., Schäfer, K. Slusser, J. R., and Amodeo A. (Eds.). SPIE, Vol. 6745, www.spiedl.org.

65. Mona, L., Amodeo, A., Pandolfi, M., and Pappalardo, G. 2006. Saharan dust intrusions in the Mediterranean area: Three years of Raman lidar measurements. *J. Geophys. Res.* 111:D16203.

66. Balis, D., Giannakaki, E., Müller, D., Amiridis, V., Kelektsoglou, K., Rapsomanikis, S., and Bais, A. 2009. Estimation of the microphysical aerosol properties over Thessaloniki, Greece, during the SCOUT-O₃ campaign with the synergy of Raman lidar and Sun photometer data. *J. Geophys. Res.* 115:D08202.

67. Alados Arboledas, L., Müller, D., Guerrero-Rascado, J. L., Navas-Guzmán, F., Pérez-Ramírez, D., and Olmo, F. J. 2010. Optical and microphysical properties of fresh biomass burning retrieved by Raman lidar, star- and sun-photometry. *Geophys. Res. Lett.* submitted.

68. Wandinger, U. 2005. Raman lidar, in *Lidar, Range-Resolved Optical Remote Sensing of the Atmosphere.* Weitkamp, C. (Ed.). Springer, Singapore, pp. 241–271.

69. Fernald, F. G., Herman, B. M., and Reagan, J. A. 1972. Determination of aerosol height distributions by lidar. *J. Appl. Meteorol.* 11:482–489.

70. Ansmann, A., Wagner, F., Althausen, D., Müller, D., Herber, A., and Wandinger, U. 2001. European pollution outbreaks during ACE 2: Lofted aerosol plumes observed with Raman lidar at the Portuguese coast. *J. Geophys. Res.* 106:20,725–20,733.

71. Müller, D., Ansmann, A., Mattis, I., Tesche, M., Wandinger, U., Althausen, D., and Pisani, G. 2007. Aerosol-type-dependent lidar-ratios observed with Raman lidar. *J. Geophys. Res.* 112:D16202.

72. Ackermann, J. 1998. The extinction-to-backscatter ratio of tropospheric aerosol: A numerical study. *J. Atmos. Oceanic Technol.* 15:1043–1050.

73. Gonzales, R. 1988. Recursive technique for inverting the lidar equation. *Appl. Opt.* 27:2741–2745.

74. Klett, F. G. 1981. Stable analytical solution for processing lidar returns. *Appl. Opt.* 20:211–220.

75. Klett, F. G. 1985. Lidar inversion with variable backscatter/extinction ratios. *Appl. Opt.* 24:1638–1643.

76. Kovalev, V. A. 1995. Sensitivity of the lidar solution to errors of the aerosol backscatter-to-extinction ratio: Influence of a monotonic change in the aerosol extinction coefficient. *Appl. Opt.* 34:3457–3462.

77. Kunz, G. J. 1996. Transmission as an input boundary value for an analytical solution of a single-scatter lidar equation. *Appl. Opt.* 35:3255–3260.

78. Sasano, Y., Browell, E. V., and Ismail, S. 1985. Error caused by using a constant extinction/backscatter ratio in the lidar solution. *Appl. Opt.* 24:3929–3932.

79. Collis, R. T. H. and Russell, P. B. 1976. Lidar measurement of particles and gases by elastic backscattering and differential absorption, Laser monitoring of the atmosphere. In *Topics in Applied Physics*. Hinkley, E. D. (Ed.). Vol. 14, Springer, Berlin, pp. 71–151.

80. Welton, E. J., Voss, K. J., Quinn, P. K., Flatau, P. J., Markowicz, K., Campbell, J. R., Spinhirne, J. D., et al. 2002. Measurements of aerosol vertical profiles and optical properties during INDOEX 1999 using micropulse lidars. *J. Geophys. Res.* 107:8019.

81. Gerry, E. T. and Leonard, D. A. 1967. Airport glide slope visual range indicator using laser Raman scattering, in *First International Conference on Laser Applications*, Paris, France.

82. Leonard, D. A. and Caputo, B. 1974. A single-ended atmospheric transmissometer. *Opt. Eng.* 13:10–14.

83. Donovan, D. P. and Carswell, A. I. 1997. Principle component analysis applied to multiwavelength lidar aerosol backscatter and extinction measurements. *Appl. Opt.* 36:9406–9424.

84. Ansmann, A. and Müller, D. 2005. Lidar and atmospheric aerosol particles, in *Lidar. Range-Resolved Optical Remote Sensing of the Atmosphere*. Weitkamp, C. (Ed.). Springer, Singapore, pp. 105–114.

85. Mattis, I., Ansmann, A., Althausen, D., Jaenisch, V., Wandinger, U., Müller, D., Arshinov, Y. F., et al. 2002. Relative-humidity profiling in the troposphere with a Raman lidar. *Appl. Opt.* 41:6451–6462.

86. Arshinov, Y., Bobrovnikov, S., Serikov, I., Ansmann, A., Wandinger, U., Althausen, D., Mattis, I., and Müller, D. 2005. Daytime operation of a pure rotational Raman lidar by use of a Fabry–Perot interferometer. *Appl. Opt.* 44:3593–3603.

87. Arshinov, Y. F. and Bobrovnikov, S. M. 1999. Use of a Fabry–Perot interferometer to isolate pure rotational Raman spectra of diatomic molecules. *Appl. Opt.* 38:4635–4638.

88. Freudenthaler, V., Esselborn, M., Wiegner, M., Heese, B., Tesche, M., Ansmann, A., Müller, D., et al. 2009. Depolarization-ratio profiling at several wavelengths in pure Saharan dust during SAMUM 2006. *Tellus.* 61B:165–179.

89. Müller, D., Wandinger, U., and Ansmann, A. 1999. Microphysical particle parameters from extinction and backscatter lidar data by inversion with regularization: Theory. *Appl. Opt.* 38:2346–2357.

90. Böckmann, C. 2001. Hybrid regularization method for the ill-posed inversion of multiwavelength lidar data in the retrieval of aerosol size distributions. *Appl. Opt.* 40:1329–1342.

91. Veselovskii, I., Kolgotin, A., Griaznov, V., Müller, D., Wandinger, U., and Whiteman, D. N. 2002. Inversion with regularization for the retrieval of tropospheric aerosol parameters from multiwavelength lidar sounding. *Appl. Opt.* 41:3685–3699.

92. Veselovskii, I., Kolgotin, A., Griaznov, V., Müller, D., Franke, K., and Whiteman, D. N. 2004. Inversion of multiwavelength Raman lidar data for retrieval of bimodal aerosol size distribution. *Appl. Opt.* 43:1180–1195.

93. Böckmann, C., Miranova, I., Müller, D., Scheidenbach, L., and Nessler, R. 2005. Microphysical aerosol parameters from multiwavelength lidar. *J. Opt. Soc. Am. A* 22:518–528.

94. Veselovskii, I., Kolgotin, A., Griaznov, V., Müller, D., and Whiteman, D. N. 2005. Information content of multiwavelength lidar data with respect to microphysical particle properties derived from eigenvalue analysis. *Appl. Opt.* 44:5292–5303.

95. Uthe, E. E. 1982. Particle size evaluations using multiwavelength extinction measurements. *Appl. Opt.* 21:454–459.

96. Uthe, E. E. 1982. Airborne lidar measurements of smoke plume distribution, vertical transmission, and particle size. *Appl. Opt.* 21:460–463.

97. Tikhonov, A. N. and Arsenin V. Y. (Eds). 1977. *Solutions of Ill-posed Problems*. John Wiley, New York. 258pp.

98. Bohren, C. F. and Huffman, D. R. (Eds). 1983. *Absorption and Scattering of Light by Small Particles*. John Wiley, Hoboken, NJ. 530pp.

99. Pahlow, M., Müller, D., Tesche, M., Eichler, H., Feingold, G., Eberhard, W. E., and Cheng, Y. F. 2006. Retrieval of aerosol properties from combined multiwavelength lidar and sun photometer measurements: Simulations. *Appl. Opt.* 45:7429–7442.

100. Kolgotin, A. and Müller, D. 2008. Theory of inversion with two-dimensional regularization: Profiles of microphysical particle properties derived from multiwavelength lidar measurements. *Appl. Opt.* 47:4472–4490.

101. Müller, D., Tesche, M., Eichler, H., Engelmann, R., Althausen, D., Ansmann, A., Cheng, Y. F., et al. 2006. Strong particle light-absorption over the Pearl River Delta (South China) and Beijing (North China) determined from combined Raman lidar and sun photometer observations. *Geophys. Res. Lett.* 33:L20811.

102. Eixmann, R., Böckmann, C., Fay, B., Matthias, V., Mattis, I., Müller, D., Kreipl, S., et al. 2002. Tropospheric aerosol layers after a cold front passage in January 2000 as observed at several stations of the German Lidar Network. *Atmos. Res.* 63:39–58.

103. Murayama, T., Müller, D., Wada, K., Shimizu, A., Sekigushi, M., and Tsukamato, T. 2004. Characterization of Asian dust and Siberian smoke with multiwavelength Raman lidar over Tokyo, Japan in spring 2003. *Geophys. Res. Lett.* 31:L23103.

104. Müller, D., Mattis, I., Ansmann, A., Wandinger, U., and Althausen, D. 2007. Raman lidar for monitoring of aerosol pollution in the free troposphere, in *Advanced Environmental Monitoring.* Kim, Y. J. and Platt, U., Springer, New York, pp. 155–166.

105. Müller, D., Mattis, I., Ansmann, A., Wandinger, U., Ritter, C., and Kaiser, D. 2007. Multiwavelength Raman lidar observations of particle growth during long-range transport of forest-fire smoke in the free troposphere. *Geophys. Res. Lett.* 34:L05803.

106. Engelmann, R., Wandinger, U., Ansmann, A., Müller, D., Žeromskis, E., Althausen, D., and Wehner, B. 2008. Lidar observations of the vertical aerosol flux in the planetary boundary layer. *J. Atmos. Oceanic Tech.* 25:1296–1306.

107. Ansmann, A., Baars, H., Tesche, M., Müller, D., Althausen, D., Engelmann, R., Pauliquevis, T., and Artaxo, P. 2009. Dust and smoke transport from Africa to South America: Lidar profiling over Cape Verde and the Amazon rainforest. *Geophys. Res. Lett.* 36:L11802.

108. Noh, Y. M., Müller, D., Shin, D. H., Lee, H., Jung, J. S., Lee, K. H., Cribb, M., et al. 2009. Optical and microphysical properties of severe haze and smoke aerosol measured by integrated remote sensing techniques in Gwangju, Korea. *Atmos. Environ.* 43:879–888.

109. Tesche, M., Ansmann, A., Müller, D., Althausen, D., Engelmann, R., Freudenthaler, V., and Groß, S. 2009. Vertically resolved separation of dust and smoke over Cape Verde using multiwavelength Raman and polarization lidars during Saharan Mineral Dust Experiment 2008. *J. Geophys. Res.* 114:D13202.

110. Noh, Y. M., Müller, D., Mattis, I., and Kim, Y. J. 2010. Vertically resolved light-absorption characteristics and the influence of relative humidity on particle properties: Multiwavelength Raman lidar observations of East Asian aerosol types over Korea. *J. Geophys. Res.*, submitted.

111. Mattis, I., Seifert, P., Müller, D., Tesche, M., Hiebsch, A., Kanitz, T., Schmidt, J., et al. 2009. Volcanic aerosol layers observed with multiwavelength Raman lidar over central Europe in 2008–2009. *Geophys. Res. Lett.*, 115:D00L04.

112. Althausen, D., Engelmann, R., Baars, H., Heese, B., Ansmann, A., Müller, D., and Komppula, M. 2009. Portable Raman lidar Polly^{XT} for automated profiling of aerosol backscatter, extinction, and depolarization. *J. Atmos. Oceanic Technol.*, 26:2366–2378.

113. Tesche, M., Müller, D., Ansmann, A., Hu, M., and Zhang, Y. 2008. Retrieval of microphysical properties of aerosol particles from one-wavelength Raman lidar and multiwavelength sun photometer observations. *Atmos. Environ.* 42:6398–6404.

114. Tesche, M., Ansmann, A., Müller, D., Althausen, D., Mattis, I., Heese, B., Freudenthaler, V., et al. 2009. Vertical profiling of Saharan dust with Raman lidars and airborne HSRL during SAMUM. *Tellus* B. 61:144–164.

115. Matthias V., Freudenthaler, V., Amodeo, A., Balin, I., Balis, D., Bösenberg, J., Chaikovsky, A., et al. 2004. Aerosol lidar intercomparison in the framework of the EARLINET project. 1. Instruments: Erratum. *Appl. Opt.* 43:2578–2579.

116. Böckmann, C., Wandinger, U., Ansmann, A., Bösenberg, J., Amiridis, V., Boselli, A., Dalaval, A., et al. 2004. Aerosol Lidar Intercomparison in the Framework of the EARLINET Project. 2. Aerosol Backscatter Algorithms. *Appl. Opt.* 43:977–989.

117. Matthias, V. and Bösenberg, J. 2002. Aerosol climatology for the planetary boundary layer derived from regular lidar measurements. *Atmos. Res.* 63:221–245.

118. Mattis, I., Ansmann, A., Müller, D., Wandinger, U., and Althausen, D. 2004. Multiyear aerosol observations with dual-wavelength Raman lidar in the framework of EARLINET. *J. Geophys. Res.* 109: D13203.

119. Amiridis, V., Balis, D. S., Kazidzis, S., Bais, A., and Giannakaki, E. 2005. Four-year aerosol observation with a Raman lidar at Thessaloniki, Greece, in the framework of the European Aerosol Research Lidar Network (EARLINET). *J. Geophys. Res.* 110:D21203.

120. Ansmann, A., Bösenberg, J., Chaikovsky, A., Comerón, A., Eckhardt, S., Eixmann, R., Freudenthaler, V., et al. 2003. Long-range transport of Saharan dust to northern Europe: The 11–16 October 2001 outbreak observed with EARLINET. *J. Geophys. Res.* 108:4783.

121. De Tomasi, F., Blanco, A., and Perrone, M. R. 2003. Raman lidar monitoring of extinction and backscattering of African dust layers and dust characterization. *Appl. Opt.* 42:1699–1709.

122. Pappalardo, G., Amodeo, A., Mona, L., Pandolci, M., Pergola, N., and Cuomo, V. 2004. Raman lidar observations of aerosol emitted during the 2002 Etna eruption. *Geophys. Res. Lett.* 31:L05120.
123. Balis, D. S., Amiridis, V., Zerefos, C., Gerasopoulos, E., Andreae, M., Zanis, P., Kazantzidis, A., Kazadzis, S., and Papayannis, A. 2003. Raman lidar and sunphotometric measurements of aerosol optical properties over Thessaloniki, Greece, during a biomass burning episode. *Atmos. Environ.* 37:4529–4538.
124. Wandinger, U., Mattis, I., Tesche, M., Ansmann, A., Bösenberg, J., Chaikovski, A., Freudenthaler, V., et al. 2004. Air mass modification over Europe: EARLINET aerosol observations from Wales to Belarus. *J. Geophys. Res.* 109:D24205.
125. Wendisch, M., Müller, D., Mattis, I., and Ansmann, A. 2006. Potential of lidar backscatter data to estimate solar aerosol radiative forcing. *Appl. Opt.* 45:770–783.
126. Guibert, S., Matthias, V., Schulz, M., Bösenberg, J., Eixmann, R., Mattis, I., Pappalardo, G., Perrone, M. R., Spinelli, N., and Vaughan, G. 2005. The vertical distribution of aerosol over Europe—Synthesis of one year of EARLINET aerosol lidar measurements and aerosol transport modeling with LMDzT-INCA. *Atmos. Environ.* 39:2933–2943.
127. Pérez, C., Nickovic, S., Baldasano, J. M., Sicard, M. Rocadenbosch, F., and Cachorro, V. E. 2006. A long Saharan dust event over the Western Mediterranean: Lidar, sun photometer observations and regional dust modeling. *J. Geophys. Res.* 111:D15214.
128. Winker, D. M., Hunt, W. H., and McGill, M. J. 2007. Initial performance assessment of CALIOP. *Geophys. Res. Lett.* 34:L19803.
129. Hunt, W. H., Winker, D. M., Vaughan, M. A., Powell, K. A., Lucker, P. L., and Weimer, C. 2009. CALIPSO lidar description and performance assessment. *J. Atmos. Oceanic Technol.* 26:1214–1228.
130. Pappalardo, G., Wandinger, U., Mona, L., Hiebsch, A., Mattis, I., Amodeo, A., Ansmann, A., et al. 2010. EARLINET correlative measurements for CALIPSO: First intercomparison results. *J. Geophys. Res.* 115:D00H19.
131. Mattis, I., Mona, L., Müller, D., Pappalardo, G., Alados Arboledas, L., D'Amico, G., Amodeo, A., et al. 2007. EARLINET correlative measurements for CALIPSO, in *Lidar Technologies, Techniques, and Measurements for Atmospheric Remote Sensing III.* Singh, U. and Pappalardo, G. (Eds). *Proceedings of SPIE*, SPIE, Vol. 6750, Bellingham, WA, USA, www.spiedl.org.
132. Tatarov, B. and Sugimoto, N. 2005. Estimation of quartz concentration in the tropospheric mineral aerosols using combined Raman and high-spectral-resolution lidars. *Opt. Lett.* 30:3407–3409.
133. Mattis, I., Müller, D., Shin, D. H. Noh, Y., Tatarov, B., Choi, T., and Chae, N. 2009. Determination of quartz concentration in mineral dust from measurements of quartz Raman scattering with lidar at two wavelength, *International Tropospheric Profiling (ISTP)*, Utrecht, The Netherlands, October 2009.
134. Müller, D., Noh, Y. M., Shin, D. H., Shin, S. K., Lee, K. H., Kim, Y. J., Mattis, I., et al. 2010. Mineral quartz concentration measurements in mixed mineral dust/urban haze pollution plumes over Korea with multiwavelength/aerosol/Raman-quartz lidar. *Geophys. Res. Lett.*, In press.
135. Tatarov, B., Sugimoto, N., Matsui, I., Shin, D. H., and Müller, D. 2010. Raman spectra of chemical components of atmospheric aerosols obtained by multi-channel lidar spectrometer, *25th International Laser Radar Conference (ILRC)*, St. Petersburg, Russia, 5–9 July.
136. Hinds, W. C. 1982. *Aerosol Technology: Principles, Techniques, and Applications*. John Wiley, New York, 1160pp.
137. Dubovik, O. and King, M. D. 2000. A flexible inversion algorithm for retrieval of aerosol optical properties from sun and sky radiance measurements. *J. Geophys. Res.* 105:20,673–20,696.
138. Hair, J. W., Hostetler, C. A., Cook, A. L., Harper, D. B., Ferrare, R. A., Mack, T. L., Welch, W., Izquierdo, L. R., and Hovis, F., E. 2008. Airborne high spectral resolution lidar for profiling aerosol optical properties, *Appl. Opt.* 47:6734–6752.
139. Eloranta, E. W. 2005. High spectral resolution lidar, in *Lidar. Range-Resolved Optical Remote Sensing of the Atmosphere.* Weitkamp, C. (Ed.). Springer, Berlin, pp. 143–163.
140. Ritter, C., Kirsche, A., and Neuber, R. 2004. Tropospheric aerosol characterized by a Raman lidar over Spitsbergen. Pappalardo, G., Amodeo, A., and Warmbein, B. (Eds). *Proceedings of the 22nd International Laser Radar Conference (ILRC 2004)*. ESA Publications Division, Matera, Italy, pp. 459–462.
141. Hong, C. S., Lee, K. H., Kim, Y. J., and Iwasaka, Y. 2004. Lidar measurements of the vertical aerosol profile and optical depth during the ACE-Asia 2001 IOP at Gosan, Jeju Island, Korea. *Environ. Monitor. Assessment* 92:43–57.

Section III

VIS/UV Spectroscopy, Fluorescence, and Scattering

10 UV and Visible Light Scattering and Absorption Measurements on Aerosols in the Laboratory

Zbigniew Ulanowski and Martin Schnaiter

CONTENTS

10.1 Introduction 243
10.2 Measurement of UV–VIS Spectral Absorption and Extinction 244
    10.2.1 Experimental Methods 246
        10.2.1.1 Extinction Spectroscopy 246
        10.2.1.2 Absorption Spectroscopy...248
    10.2.2 Laboratory and Aerosol Chamber Results 250
        10.2.2.1 Carbonaceous Aerosol 251
        10.2.2.2 Mineral Dust Aerosol..253
10.3 Measurement of Angular Dependence of Scattering 254
    10.3.1 Multiple Particles...256
    10.3.2 Single Particles 256
        10.3.2.1 Phase Function Measurement 256
        10.3.2.2 Azimuthal Scattering...258
        10.3.2.3 Two-Dimensional Scattering 260
References..263

10.1 INTRODUCTION

This chapter describes the techniques and recent results pertaining to the measurement of elastic scattering and absorption of aerosols. The information on aerosol properties that can be retrieved from measurements of spectral and angular dependence is not dissimilar. Most notably, both the wavelength dependence of scattering and extinction and the angle dependence of scattering are functions of the size parameter, the ratio of the particle circumference to the wavelength of the light. Therefore, spectroscopy and angular measurement of intensity are closely related, and both can be used to retrieve the size, or size distribution, of aerosol particles. However, we observe that additional properties can be obtained from the angular dependence of scattering, by virtue of stronger sensitivity of scattering to particle shape, symmetry, and orientation and, for two-dimensional (2D) scattering patterns, the greater information content of the data. Moreover, for single particles, scattering is much easier to measure than extinction. In contrast, the opposite can be said to be true of measurements on multiple particles. Therefore, extinction and angular measurements are in some ways complementary to each other.

10.2 MEASUREMENT OF UV–VIS SPECTRAL ABSORPTION AND EXTINCTION

Extinction and absorption spectroscopy in the UV–VIS spectral range is an interdisciplinary diagnostic tool that is applied in the laboratory to characterize particulate matter in the context of astrophysical, atmospheric, and combustion research.[1–3] The optical extinction by aerosols, that is, the removal of light from a beam penetrating a particulate medium, is the sum of light scattering, where the light is redistributed in different directions, and light absorption, where the light is converted into thermal energy that heats the particles and their surroundings.

A thorough presentation of the fundamental equations and definitions related to the theory of light scattering and absorption can be found in the chapter by Leisner and Wagner in this book. Here, we give only a brief summary of those definitions and equations that are necessary to understand the content of Section 10.2.1.

When light penetrates an ensemble of identical particles with a particle number density of n and a geometrical thickness of l, the extinction cross section C_{ext} is related to the optical depth τ of the particulate medium by

$$\tau = -\ln\left(\frac{I}{I_0}\right) = n \cdot C_{\text{ext}} \cdot l, \tag{10.1}$$

where I and I_0 are the transmitted and incident intensities, respectively. The extinction coefficient b_{ext} is the ratio between the optical and geometrical thicknesses and is therefore a measure of light attenuation per standard distance, given, for instance, in meters:

$$b_{\text{ext}} = n \cdot C_{\text{ext}}. \tag{10.2}$$

Relating the extinction cross section to the geometrical cross section A_p or the mass M_p of the particles gives the extinction efficiency

$$Q_{\text{ext}} = \frac{C_{\text{ext}}}{A_p} \tag{10.3}$$

and the mass-specific cross section

$$\sigma_{\text{ext}} = \frac{C_{\text{ext}}}{M_p}, \tag{10.4}$$

respectively.

As already mentioned above, the extinction cross section C_{ext} is the sum of the scattering and absorption cross sections C_{sca} and C_{abs} and thus can be expressed as

$$C_{\text{ext}} = C_{\text{sca}} + C_{\text{abs}}. \tag{10.5}$$

The optical cross sections given in Equation 10.5 are dependent on the size distribution of the particle ensemble and the particle material. The optical behavior of the particle material is represented by the complex refractive index

$$m(\lambda) = n(\lambda) + ik(\lambda). \tag{10.6}$$

The real part n of the complex refractive index of atmospheric aerosols in the visible spectrum is not very sensitive to the chemical composition. The index is variable within the rather limited range of $n = 1.3$–1.65 for most aerosol types and relative humidity values. The imaginary part of the index

k is much more variable, ranging from 1×10^{-9} (e.g., water droplets) to more than 0.1 in very sooty or iron-rich dusty aerosols.[4]

Extinction spectroscopy is widely used in conjunction with inversion algorithms to retrieve the size distribution of nonabsorbing or weakly absorbing aerosols in the submicrometer size range. Here, the problem is that aerosol particles are in general nonspherical, and scattering by nonspherical particles is generally different from that of spherical particles. Thus, an assumption of spherical, homogeneous particles in the inversion process may result in errors. However, when the particle radius a is less than about $\lambda/2$, that is, $x < 0.3$, where $x = 2\pi a/\lambda$ is the size parameter, the extinction curves for both nonspherical and spherical particles are very similar.[5] The same holds if the particle radius is larger than about 4λ ($x > 25$, Figure 10.1) which is true for particle radii larger than about 1–3 µm depending on the wavelength used in the UV–VIS spectral range. It is clear from Figure 10.1 that the extinction curve is highly sensitive to the actual particle shape in the intermediate particle size range $\lambda/2 < a < 4\lambda$, where more detailed microphysical properties of the particles have to be known in order to get a more reliable retrieval of the particle size distribution from a measured UV–VIS extinction spectrum.

Absorption spectroscopy is of particular importance for the climate impact of the atmospheric aerosol, since absorbing particles modify the radiation fluxes in two different ways: directly, by the absorption of shortwave solar radiation and semidirectly, by modifying the temperature distribution of the atmosphere which initiates a number of feedback mechanisms.[6] The contribution of absorption by refractory carbonaceous particles (black carbon, BC) to global warming is still poorly understood and induces a significant uncertainty.[7] Moreover, the determination of the spectrally absorption properties of aerosols becomes increasingly important especially for weakly absorbing aerosols, such as organic aerosols, such as brown carbon (BrC) or humic-like substances (HULIS), and mineral dust (MD) aerosol. For these aerosols, the fraction of absorption in light extinction, that is, the single scattering albedo

$$\omega = \frac{C_{\text{sca}}}{C_{\text{sca}} + C_{\text{abs}}}, \tag{10.7}$$

is the most important factor that determines the shortwave radiative forcing.

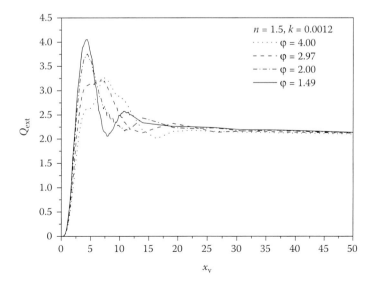

FIGURE 10.1 Extinction efficiency Q_{ext} as a function of the size parameter $x_v = 2\pi a/\lambda$ for oblate spheroidal particles with different aspect ratios φ. The spectra were calculated using the T-Matrix method by Mishchenko.[36] Note that the extinction curve is highly sensitive to the actual particle shape in the intermediate particle size range $\lambda/2 < a < 4\lambda$. (Diagram courtesy of Robert Wagner.)

In climate modeling, BC is often assumed to be the only light-absorbing aerosol with specific optical properties throughout the atmosphere. This picture is now changing, since it becomes evident that (1) the absorption properties of BC are linked to the combustion, chemical, and mixing processes during emission and atmospheric transport,[8,9] (2) there is a high continuum of organic carbonaceous particles with varying spectral absorption properties but with almost unknown refractive indices,[10] and (3) MD can contribute to light absorption depending on the mineralogical composition of the aerosol. The quantification of the UV–VIS spectral dependence of aerosol absorption, usually expressed by a power law

$$\frac{b_{abs}(\lambda_1)}{b_{abs}(\lambda_1)} = \left(\frac{\lambda_1}{\lambda_2}\right)^{-\alpha_{abs}},$$ (10.8)

where α_{abs} is the absorption Ångström exponent, is an important step to characterize the variability in the electronic structure of carbonaceous and mineral aerosol particles.

10.2.1 EXPERIMENTAL METHODS

10.2.1.1 Extinction Spectroscopy

By combining Equations 10.1 and 10.2 the most fundamental method to determine the extinction coefficient b_{ext} experimentally is to measure the light intensities I_{free} and $I_{aerosol}$ after passing through an aerosol-free and aerosol-filled optical extinction cell (OEC):

$$b_{ext} = -\frac{1}{l_{OEC}} \ln\left[\frac{(I/I_r)_{aerosol}}{(I/I_r)_{free}}\right],$$ (10.9)

where l_{OEC} is the optical path length and I_r is the reference intensity to account for potential drifts of the light source.

This method has been used in many laboratory investigations to calibrate filter-based absorption measurement techniques by measuring independently the extinction and scattering coefficients b_{ext}, b_{sca} and calculating the difference of both to get the absorption coefficient b_{abs}.[11–14] Henceforth, we refer to this method as *difference method* (DM). While the first OECs measured the extinction coefficient at a single wavelength,[15] more recent instruments measure the extinction coefficient at several discrete wavelengths[16] or even spectrally resolved.[17] The OEC designed by Virkkula et al.[16] uses three light-emitting device (LED) sources with weighted average wavelengths of 467, 530, and 660 nm. The cell is a tube of 366 cm length extended by purge airflow sections to prevent particles from depositing on the lenses at both ends of the tube. A flat mirror at the far end of the tube is used to generate a single-folded optical path with an effective optical path length of 659 ± 1 cm. The lenses at both ends of the cell generate slightly convergent light beams back and forth through the tube and focus the light beams to the flat mirror on one end and to the photodetector on the other end of the cell. The instrument was calibrated and tested during the Reno Aerosol Optics Study (RAOS[18]) in June 2002 by comparing it with different scattering and absorption measurement methods and for different scattering (white) and absorbing (black) aerosols. The accuracy and detection limits of the OEC for white and black aerosols were deduced by comparing b_{ext} measured by the OEC with b_{sca} measured with a commercial three color integrating nephelometer (TSI, type 3563). The accuracy was calculated from the relative difference $|b_{ext} - b_{sca}|/b_{sca}$ in case of white aerosol and was found to vary from experiment to experiment within 1% and 10%. Detection limits were deduced from the standard deviations of $|b_{ext} - b_{sca}|$ and $|(b_{ext} - b_{sca}) - b_{abs}|$ for white (ammonium sulfate and polystyrene latex) and black (soot) aerosol, respectively. The absorption coefficient b_{abs} was measured independently with a photoacoustic instrument (see Section 10.2.1.2 for a description of this method). Maximum detection limits at the blue wavelength around 30 Mm^{-1} and 48 Mm^{-1} were deduced for the extinction and absorption measurements, respectively. A validation of spectral data gathered by the DM requires an analysis of the error in the spectral characteristic of the deduced absorption, that is, the absorption Ångström exponent α_{abs} (Equation 10.8), in dependence on b_{abs}. Flame soot aerosol is well suited for this purpose, since it has an α_{abs} value of ~1.0 throughout the visible spectral range.[19] Owing to the above noise in the b_{abs} deduction, especially for the blue wavelength, the OEC designed by Virkkula et al.[16] clearly deviates from this α_{abs} value for b_{abs} values less than about 100 Mm^{-1}.

A similar single-folded path OEC was designed by[17] for spectroscopic measurements of b_{ext} in the 200–1015 nm wavelength range (Figure 10.2). Instead of using a set of discretely emitting light sources, the long path extinction spectrometer (LOPES) employs a broadband deuterium/halogen lamp combination in conjunction with a diode array spectrometer to acquire extinction spectra in the above wavelength range with a spectral resolution of 2.5 nm. The lamp combination and the spectrometer are connected to the OEC via a special fiber system. Light beam shaping and path folding are realized by a 90° off-axis parabolic mirror on the emitting and receiving side and a corner cube mirror on the far end of the cell (Figure 10.2). Two LOPES instruments with cell lengths of 3.5 and 5 m are operated at the aerosol chambers of the aerosol and cloud simulation chamber facility aerosol interactions and dynamics in the atmosphere (AIDA) of KIT.[3] The accuracy of the LOPES instruments in terms of extinction coefficients is defined by the instrument noise which is 20 and 40 Mm^{-1} for the 10 and 7 m version, respectively. These OEC measurements use Equation 10.9 to determine the extinction coefficient b_{ext} from subsequent measurements of the transmitted intensities $I_{aerosol}$ and I_{free}. Since, in practice, any measurement of $I_{aerosol}$ and I_{free} covers a finite acceptance angle θ_{acc}, the measured intensity in the forward direction ($\theta = 0$) is systematically enhanced by the fraction of light scattered in the near-forward direction ($0 < \theta \leq \theta_{acc}$). For OECs, this enhances the measured intensity I and consequently reduces the measured extinction (see Equation 10.1). The magnitude of this bias increases with θ_{acc} and the size parameter x. Using scalar diffraction theory[20,21] and assuming nonabsorbing aerosol ($\omega = 1$), the error in b_{ext} can be estimated for given θ_{acc} and x. The LOPES systems have acceptance angles θ_{acc} of about 4 mrad, which limits the useful particle radius to $a < 2.75$ μm, if the acceptance error is limited to be less than 1% in b_{ext} over the whole spectral region covered by the LOPES instruments.[17]

The LOPES instruments were also used in a series of laboratory studies on the optical properties of soot,[13,19,22] biomass burning aerosol,[17] and MD.[23] The main focus of these studies was to deduce

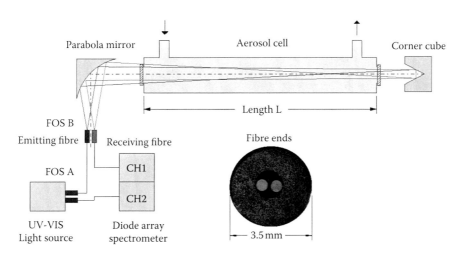

FIGURE 10.2 Schematics of the LOPES system of the aerosol chamber AIDA.[17] The magnified image of the fiber connector jacket shows the side-by-side alignment of the emitting and receiving fibers.

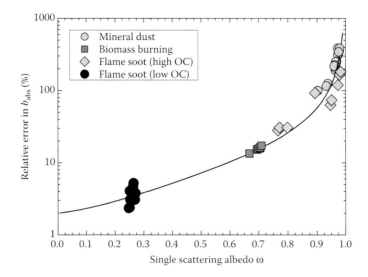

FIGURE 10.3 Dependence of the measurement error of the absorption coefficient b_{abs}, as deduced by the DM, on the single scattering albedo ω of different aerosol types. The theoretical error (black line) was calculated based on Equation 10.10.

the wavelength-dependent mass-specific absorption cross section of the different aerosol species by applying the DM, that is, by measuring independently the scattering coefficient with the TSI 3563 integrating nephelometer. Owing to systematic uncertainties resulting from angular truncation and a noncosine light source, the nephelometer measurements have to be corrected carefully. This is especially an issue for weakly absorbing aerosols, such as MD or biomass burning aerosol, since the DM is based on the difference of two numbers and the relative error of the deduced absorption coefficient $\Delta b_{abs}/b_{abs}$ is increasing with increasing single scattering albedo ω[17]

$$\frac{\Delta b_{abs}}{b_{abs}} = \frac{1}{1-\omega} \sqrt{\left(\frac{\Delta b_{ext}}{b_{ext}}\right)^2 + \left(\frac{\Delta b_{sca}}{b_{sca}}\omega\right)^2}. \tag{10.10}$$

Figure 10.3 shows the increase of the measurement error b_{abs} with ω, which is mainly induced by the uncertainties in determining the nephelometer correction factors. Therefore, the OEC/integrating nephelometer device combination is a useful set of instruments for laboratory studies on the optical properties of aerosols and yields wavelength-dependent absorption coefficients with a reasonable accuracy of about 20–5% for aerosol species with a single-scattering albedo up to about 0.7–0.8.

10.2.1.2 Absorption Spectroscopy

As we have seen in the previous section, the quantitative determination of the absorption coefficient of weakly absorbing aerosols is difficult and cannot be achieved by using the DM. Therefore, sensitive techniques are needed that measure the absorption coefficient directly on airborne particles. Photoacoustic (PA) spectroscopy is a highly promising technique in this respect. An excellent literature review and a methodical description of the photoacoustic technique are given in.[24] The basic principle of the photoacoustic effect which is utilized to measure the aerosol absorption coefficient is illustrated in Figure 10.4. Energy that is deposited in an aerosol particle by light absorption results in a fast heating up of the particle and in a subsequent heat transfer to the surrounding air. This results in an expansion of the surrounding air and, if the incident laser light is modulated, in a

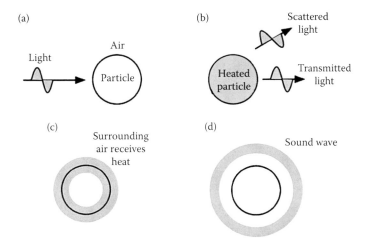

FIGURE 10.4 Physical concept of the photoacoustic method: modulated light incident on a particle (a) is partially absorbed (b) and transferred to a periodical heating of the surrounding (c) which results in a pressure modulation and finally a sound wave (d). (Adapted from Moosmuller, H., Chakrabarty, R. K., and Arnott, W. P. 2009. *J. Quant. Spectros. Radiat. Transfer* 110:844–878. With permission.)

periodical pressure disturbance or sound wave. Therefore, the photoacoustic effect is the conversion of light to sound. In a typical photoacoustic setup, the temperature increase of the particle is typically less than 1 K; hence, the photoacoustic technique should not be confused with the laser-induced incandescence technique where strongly absorbing particles are heated to very high temperatures of several thousand K and their visible thermal radiation is analyzed.[25] By drawing the aerosol through an acoustic resonator and modulating the laser power at the resonant frequency of the resonator, the photoacoustic sound is amplified and can be easily detected by a calibrated microphone attached to the cavity. The most frequently applied PA cavity geometry is a cylindrical tube or pipe for the generation of radial, azimuthal, and longitudinal modes.[26] Two types of longitudinal plane wave resonator designs are currently used for aerosol applications with a resonator length of half wavelength[23,27] or full wavelength.[28,29] According to Rosencwaig,[30] the measured pressure amplitude P_m of the acoustic wave is directly related to the absorption coefficient b_{abs} by

$$b_{abs} = P_m \frac{\pi^2 A_{res} f_0}{P_L Q(\gamma - 1)}, \tag{10.11}$$

where A_{res}, f_0, and Q are the cross-sectional area, the acoustic resonance frequency, and the quality factor of the resonator. P_L is the modulated average laser power and γ is the ratio of the isobaric and isochoric specific heats of the carrier gas ($\gamma_{air} = 1.4$). The accuracy of a photoacoustic system depends on the uncertainties in P_m, A_{res}, f_0, P_L, and Q, and hence to a large degree on the stability and noise of the microphone and the laser. Typical detection limits of recent PA instruments in terms of b_{abs} are in the range of 0.1–1 Mm^{-1}.[27,31] While single-wavelength PA instruments have been used extensively for aerosol absorption measurements,[17,19,31,32] novel multiwavelength instruments have emerged during the last few years,[23,33] which is mainly a consequence of progress in laser technology. The instrument by Lewis et al.[33] employs two power-modulated laser diodes emitting at 405 and 870 nm merged by a dichroic beam splitter to form a collimated light beam which is directed through the cavity of a single full wavelength PA resonator. The PA cavity of this instrument comprises a reciprocal nephelometer which facilitates the simultaneous measurement of the scattering coefficient and hence the single scattering albedo ω of the aerosol. Figure 10.5 displays the optical scheme of

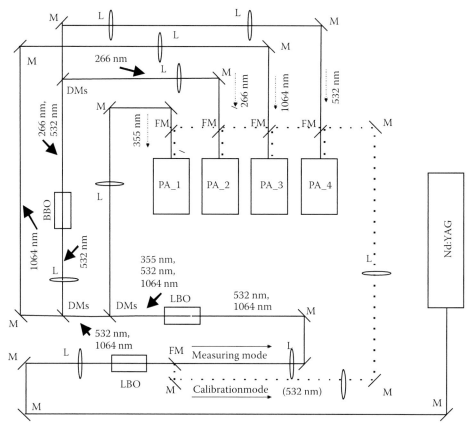

FIGURE 10.5 Schematic view of a multiwavelength photoacoustic system based on the second-, third-, and fourth-harmonics generation of a Nd:YAG laser using a set of nonlinear crystals. (Diagram courtesy of Tibor Ajtai.)

another multiwavelength instrument that utilizes the fundamental (1064 nm), doubled (532 nm), tripled (355 nm), and quadrupled (266 nm) emission of a pulsed Q-switched Nd:YAG laser. The repetition rate of the laser is set to 4 kHz to match the resonance frequency of the four PA cells. Instrument calibration is done by operating the four PA resonators at 532 nm and drawing a sufficient and known amount of absorbing NO_2 gas through the cells. The gas-based calibrations of this instrument have been checked with high concentrations of absorbing particulate matter and by comparing the measured absorption coefficients with results deduced from simultaneous measurements with the DM (see Figure 10.6). Uncertainties introduced by the calibration procedure are typically in the range of a few percent.

10.2.2 Laboratory and Aerosol Chamber Results

In the following two sections, we highlight recent results from laboratory studies on the UV–VIS spectroscopic properties of carbonaceous and MD aerosols. The experiments have been conducted in the aerosol chambers of the AIDA facility which has been established during the past decade as a unique experimental facility to study the optical properties of complex aerosol particles and to investigate the formation of cloud particles under realistic atmospheric conditions. Readers interested in the technical details of the chamber and the chamber instrumentation are referred to Wagner et al.[3]

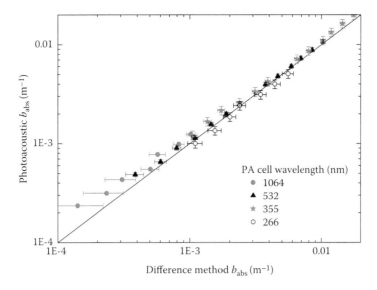

FIGURE 10.6 Comparison of the absorption coefficients measured by multiwavelength photoacoustic spectroscopy and deduced by the DM for flame soot aerosol.

10.2.2.1 Carbonaceous Aerosol

In an experiment series conducted in 2003, in the smaller 4 m³ chamber of the AIDA facility, the spectral absorption properties of combustion aerosol emitted from a propane diffusion flame operated at different combustion conditions, that is, at different C/O ratios, were investigated by utilizing the DM described in Section 10.2.1.[19] The fuel-to-air ratio of the burner was varied over a wide range from 0.24 to 0.98 in terms of the C/O atomic ratio which resulted in vastly different contents of organic carbon (OC) in the emitted particulate matter. The variation in the ratio of organic to refractory (black) carbon had in turn a strong influence on the spectral optical properties of the emitted particles which is clearly reflected by the compilation of spectral results given in Figure 10.7. The mass-specific absorption cross section σ_{abs} decreases with increasing OC content and the relative strength of this decrease is wavelength-dependent and is higher for longer wavelengths resulting in a steeper wavelength-dependence of σ_{abs} and, thus, a higher absorption Ångström exponent α_{abs} for aerosol particles with a higher OC content. A flat wavelength dependence of the absorption coefficient with α_{abs} values around 1 was found for BC particles with a low OC content below about 20%. Such a spectral behavior is theoretically expected from Mie theory in case of small particles compared to the wavelength, that is, for Rayleigh particles[20]

$$\sigma_{abs} = \frac{6\pi}{\lambda\rho} \mathrm{Im}\left\{\frac{m^2 - 1}{m^2 + 2}\right\}, \tag{10.12}$$

where ρ is the particle density and m is the complex refractive index of the particle material. Note that Equation 10.12 is applicable also for BC particle aggregates, given that the individual particles of the aggregate are Rayleigh particles, the aggregate has an open fractal-like structure, and multiple scattering and self-interactions within the aggregate can be neglected.[13] From Equation 10.12 it is also clear that α_{abs} values around 1 imply a weak wavelength dependence of the term $(m^2 - 1)/(m^2 + 2)$ and, thus, of the complex refractive index of the particle material. A large Ångström exponent of 3.5 or even above, as measured for flame particles with OC contents above 50%, in turn means that the refractive index of the particle material is strongly wavelength dependent. The strong wavelength-dependence was assigned to a separate particle nucleation mode which was observed in

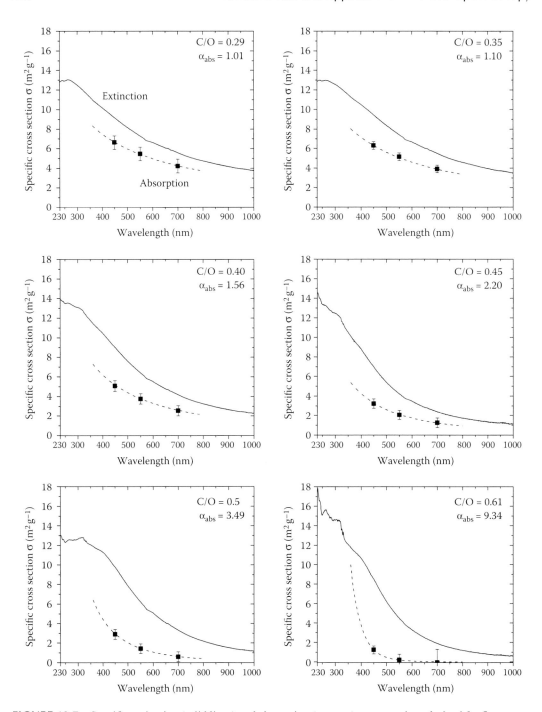

FIGURE 10.7 Specific extinction (solid lines) and absorption (squares) cross sections derived for flame soot aerosols produced at different burning conditions, that is, fuel-to-oxygen ratios C/O. The absorption cross sections were deduced from extinction (LOPES) and total scattering (nephelometer) measurements (DM). Dashed lines represent power law fits to deduce the absorption Ångström exponent α_{abs} given in the plots. (Adapted from Schnaiter, M., Gimmler, M., Llamas, I., Linke, C., Jager, C., and Mutschke, H. 2006. *Atmos. Chem. Phys.* 6:2981–2990. With permission.)

the size distribution of the flame emission and which obviously consists of condensed organic species (most likely polycyclic aromatic hydrocarbons) with a steep absorption edge toward the near-UV. This interpretation is consistent with results from aerosol mass spectrometric investigations on premixed propane/O_2 flames.[34] In the discussion of these results in the context of other laboratory or atmospheric measurements, one has to keep in mind that (1) b_{abs} was not measured directly but was deduced from the DM which might introduce a strong uncertainty for aerosols with a single scattering albedo above 0.8 (see Figure 10.3), and (2) the power law definition of the Ångström exponent (Equation 10.8) is certainly not unique over the whole UV–VIS spectral range and is limited to the spectral range covered by the experimental method. So, future experiments should rely on multiwavelength photoacoustic spectroscopy with as many wavelength positions as possible especially in the near-UV and visible spectral region to overcome these limitations.

10.2.2.2 Mineral Dust Aerosol

MD aerosols mostly consist of a mixture of different mineral phases, such as quartz, carbonates, sulfates, and clay minerals. Depending on their mineralogical composition, the particles can be strong or weak light absorbers. In particular, the contribution of iron oxide phases such as hematite and goethite significantly increases the absorption cross sections of the MD aerosols at visible and near-UV wavelengths.

In a series of measurement campaigns at the AIDA facility, the wavelength dependence of the specific extinction and absorption cross sections of a variety of Saharan MD samples have been measured. The absorption cross sections were directly measured at the wavelength positions 1064, 532, 355, and 266 nm with the Nd:YAG laser-based, multiwavelength photoacoustic spectrometer described in Section 10.2.1. Extinction spectra were recorded in parallel from 230 to 1000 nm with the OEC LOPES described in Section 10.2.1. A first study conducted in 2004 with soil samples from different locations in the Sahara desert has indicated that mineral aerosols feature a strong and variable spectral dependence of the absorption coefficient with absorption Ångström exponents as high as 5 in the visible to mid-UV spectral range. Moreover, the results point to the fact that the presence of iron oxide phases in the dust significantly increases the absorption cross section of the MD aerosol in the visible and especially in the near- and mid-UV.[23] These spectral absorption characteristics were affirmed in a more recent study conducted in 2007 with the Saharan soil samples BURKINA FASO and SAMUM B3 which were collected during the field campaigns AMMA and SAMUM, respectively.[3] Figure 10.8 shows the mass-specific absorption cross sections (left panel) and the single scattering albedo (SSA) (right panel) for both MD aerosols (symbols), illustrating the strong wavelength dependence of MD absorption in the visible and UV spectral regimes. The increase of the mass-specific absorption cross section and the decrease in SSA toward UV wavelengths is more pronounced for the BURKINA FASO dust sample. This can only be attributed to differences in the mineralogical composition, given that the size distribution parameters of both dust samples were almost identical (compare the lognormal parameters given in Figure 10.8). Therefore, the mineral composition of both dust samples was analyzed by x-ray-powder diffractometry (XRD). The XRD analyses indicate that the finest-sieved fraction of the BURKINA FASO sample, as it was used in the experiments, indeed contains a higher amount of both goethite and hematite as strongly absorbing iron oxide phases compared to the SAMUM B3 sample. This is also evidenced by the modeling results for the wavelength dependence of the mass-specific absorption cross section and the single scattering albedo, as shown in Figure 10.8 (lines). In these calculations, the dust samples were modeled as an internal mixture of illite (component 1, dielectric function ε_1) and hematite (component 2, ε_2), using the Bruggeman approximation[35] to compute the effective complex dielectric function ε_{mix} of the two-component mixture

$$(1-x_2)\frac{\varepsilon_1 - \varepsilon_{mix}}{\varepsilon_1 + 2\varepsilon_{mix}} + x_2\frac{\varepsilon_2 - \varepsilon_{mix}}{\varepsilon_2 + 2\varepsilon_{mix}} = 0, \qquad (10.13)$$

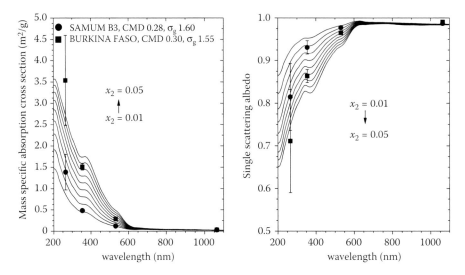

FIGURE 10.8 Wavelength-dependent mass-specific absorption cross section (left) and single scattering albedo (right) for the dust samples BURKINA FASO (squares) and SAMUM B3 (dots); comparison between measurements (symbols) and a set of Mie calculations (lines) for different hematite volume fractions x_2; see text for details. Note that the size distribution parameters of the aerosols, that is, the count median diameter (CMD) and the geometric standard deviation σ_g, are almost identical. So, the differences in the spectral behavior can be attributed to differences in the mineralogical composition. (Adapted from Wagner, R., Linke, C., Naumann, K. H., Schnaiter, M., Vragel, M., Gangl, M., and Horvath, H. 2009. *J. Quant. Spectros. Radiat. Transfer* 110:930–949. With permission.)

where x_2 is the volume fraction of hematite. Using the measured size distribution parameters as input in Mie calculations, a series of spectra was calculated for x_2 ranging from 0.01 to 0.05 in increments of 0.005. The best agreement between computations and measurements is obtained with $x_2 \sim 0.04$ for the BURKINA FASO and $x_2 \sim 0.02$ for the SAMUM B3 sample, illustrating that already minor changes in the hematite content strongly alter the absorption properties of the dust samples toward near-UV and mid-UV wavelengths. This example nicely illustrates the potential of multiwavelength photoacoustic spectroscopy in terms of analyzing the composition of weakly absorbing aerosols, given that a good spectral coverage in the blue to mid-UV regime is realized.

Currently, a retrieval algorithm is being developed to deduce the wavelength-dependent real and imaginary parts of the complex refractive index for the various dust probes from the measured extinction and absorption spectra.

10.3 MEASUREMENT OF ANGULAR DEPENDENCE OF SCATTERING

The measurement of scattering as a function of angle, often referred to as multiangle photometry, has a long and rich history in the field of particle characterization.[37,38] In a recent review of multi-angle scattering on single particles, Kaye et al.[39] divided this area into measurements in one (scattering) plane, azimuthal measurements at one polar (scattering) angle, and detection over both azimuthal and polar angles, that is, 2D scattering. The latter area can be further subdivided into measurements at a small, discrete number of angles or high-resolution measurements using 2D imaging detectors. We adopt this classification here except that in the 2D scattering category, owing to the dominance of detectors such as charge-coupled devices (CCDs), less attention will be given to discrete angle measurements. We also describe measurements on multiple particles as a separate category, as they almost exclusively consist of measurements in a single, polar (scattering) plane.

Polar nephelometry, sometimes known as laser diffractometry, is concerned with the measurement of light as a function of scattering angle, whether for a single small particle, an ensemble of particles or a macroscopic specimen. The technique can give information about properties such as the size, refractive index, or coarse internal structure of small particles, including living cells. The polar dependence of scattering is variously referred to as the scattering diagram, differential light-scattering pattern, angular scattering coefficient (if the absolute magnitude is not known) or the phase function. The last term is often used in this general sense; however, in its strict meaning, and in the context of atmospheric scattering, the phase function is usually normalized so that its average value over all directions is equal to 1.[40] However, it should be noted that another definition which is in use states that the *integral* of the function over all directions should be 1. The latter definition leads to values smaller by a factor of 4π and represents the probability that the scattering will occur in a particular solid angle (expressed in steradians, sr).

The phase functions of aerosol particles have traditionally been measured using a detector, typically a photomultiplier, rotating around a central area containing a stream of particles or in some cases a single immobilized (trapped) particle.[41–47] Such an arrangement can be called a "laser diffractometer," by analogy with x-ray diffractometry. Discrete detector arrays have also been employed[48,49] (see also Section 10.3.2.2). If a narrow range of scattering angles is needed, linear or 2D photodetector arrays can be used.[50,51]

A different design, originally developed by Gucker et al.[52] uses an annular section of an ellipsoidal reflector to collect light scattered in one plane and focus it on a single detector via a scanning aperture providing angular dependence.[53,54] Like the optical fiber design described above, such systems are capable of fast scanning but their alignment is more difficult, good-quality reflectors are difficult to make, and the detection optics reduces open access to the scattering volume. A modification of this approach relies on combining a reflector with an imaging detector.[55]

An alternative diffractometer design uses a wide-angle, semicircular fiber-optics array coupled to a scanning system allowing the use of a single, stationary photomultiplier[56]—see Figure 10.9. This brings several advantages over "classical" multiangle photometer designs using rotating detectors. First, faster scanning (up to 100 scattering patterns/s) is possible than in systems employing a moving detector. Second, good access to the scattering volume is available making it possible to augment the system in various ways, for example, by including a particle trap and/or observation microscopes. Third, there is little possibility of misalignment, as timing and scanning functions are

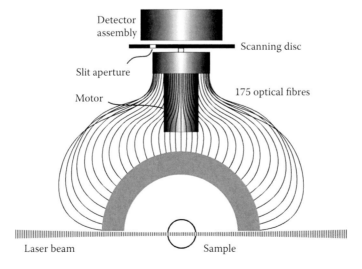

FIGURE 10.9 Fiber optics laser diffractometer. The input ends of the fibers point toward the sample, the output ends, scanned by a spinning disc with a small aperture, toward a photomultiplier.[56]

closely integrated. The disadvantages are fixed angular resolution and a difficulty with measuring different polarization states of scattered light.

Scattering from atmospheric ice crystals has recently grown in significance because of the impact that cold clouds have on climate.[57] As a consequence of this growing interest, many measurements have been carried out both *in situ* and in the laboratory, as described in the following sections. Of particular relevance from the standpoint of remote sensing and the radiative impact of clouds are phase functions and the asymmetry parameter; the latter because it is often used to represent the properties of ice particles in climate models.

Correct normalization of phase functions and the estimation of the asymmetry parameter g require the knowledge of the full angular range of scattering. Furthermore, the asymmetry parameter strongly depends on forward scattering, putting special emphasis on low scattering angles, where measurement is especially difficult.[40] In a system described by Ulanowski et al. the lower scattering angle limit was reduced to 0.5° using a linear photodiode as an additional detector. The narrow forward scattering range below 0.5° could then be accurately extrapolated using Mie theory, or in the case of large, highly nonspherical particles, using Fraunhofer diffraction on an ensemble of equal average cross-sectional area ellipses.[58]

10.3.1 MULTIPLE PARTICLES

Complete scattering matrices of a large number of samples containing multiple particles, including minerals, volcanic ash, cosmic dust analogues and various single-cell organisms in water have been determined at the University of Amsterdam.[41,42] The measurements continue using improved apparatus at the Institute of Astrophysics in Granada.[47] A database of these results is freely accessible through the internet and includes size distributions and scanning electron microscope images and other ancillary data.[59] Other recent examples include complex refractive index and size distribution of aerosol,[60] refractive index of aerosol,[61] sulfate aerosols,[55,62] quartz dust,[55] marine aerosols,[45] agglomerates,[46] phytoplankton and silt.[63] Measurements on ice crystals were carried out in cloud chambers,[49,64] on the ground,[65] and in clouds.[48,66]

10.3.2 SINGLE PARTICLES

10.3.2.1 Phase Function Measurement

Section 10.3 outlined general techniques used for the measurement of angular dependence of scattering. Many of these techniques have been applied to single particles. A common application is the determination of spherical particle size; a somewhat different one is refractive index retrieval by observing the rainbow peaks from trapped droplets.[50]

Measurement of the phase function in the true sense requires that correct normalization is carried out, implying that near-full angular range has to be covered, as discussed in Section 10.3. Unfortunately, few measuring systems have met this requirement in the context of single particles, as the lower scattering angle limit has typically been 5° or greater. Smaller limits include 3°[56,67] and 0.5°,[54,58] but care must be taken that the finite angular widths of the detector as well as the incident beam are allowed for the latter because the scattering diagram is convolved with the profile of the beam.[68,69]

There is much commonality between measurement systems used for aerosols and for particles in suspended liquids, examples of the latter including both flow[56,70,71] and trapped particle[54,72,73] instruments. Much of this work concerns living cells.[54,70–74] In the case of cells, the refractive index retrieved from the scattering measurements can be converted into solid or water content of the cell using simple relationships. If an assumption is made that the structure of the cells can be approximated by an inhomogeneous model, such as the coated sphere, the distribution of water can be determined in a noninvasive way.[75]

In the case of single particles in liquids, the scattering intensity from the suspending medium can be much higher than that from the particle, even in the absence of contamination, owing to density fluctuations. Providing some means of subtracting the background component from the light-scattering patterns is therefore advantageous. In a system described by Ulanowski et al.[56] subtraction was carried out at the earliest possible, namely at the photomultiplier output, by subtracting from the anode current a previously recorded background signal using a 16-bit digital to analogue converter. This feature allows extending the dynamic range of the measured intensity, because the only part of the system subjected to the whole intensity is the photomultiplier—a device characterized by excellent linearity over broad range of inputs. Consequently, less severe demands are put on the electronic circuits, bringing a significant advantage over systems based on multielement detector arrays.

Particle trapping is a valuable addition to laser diffractometry, as it allows prolonged measurements on particles in isolation from any solid supports, as well as measurements on particles in controlled orientations. Single beam optical gradient traps, or laser tweezers, have long been applied in liquid media,[54,72,73,76] and more recently in gaseous media.[77–79] For airborne particles, electrodynamic trapping is frequently employed.[50,58,67,79–81] Traps with open geometry are especially useful.[82–86] Acoustic levitation can also be employed for large particles.[79,87] Additional opportunities are provided by microwave analogue measurements. Not only is the orientation of the particle tightly controlled, but the particle is larger due to the scaling provided by the increased wavelength; so manufacturing elaborate particle geometries and structures "to order" becomes possible.[88]

For single nonspherical particles, an additional difficulty associated with obtaining correctly normalized phase functions is the need to average over all particle orientations. Such randomization can be achieved by taking advantage of angular oscillations in the electrodynamic trap. Angular instability of particles in such traps was investigated and modeled by Hesse et al.[84] When instability is produced, nonspherical objects first undergo angular oscillations and eventually, when entering deeper into instability conditions, rotate or show apparently chaotic angular motions. This approach has been used to measure the phase functions and the asymmetry parameter of ice analogue particles, representing atmospheric ice crystals, as already described in Section 10.3.[58]

It is possible to circumvent the need for orientational randomization if a uniform population of particles is available—an unusual situation. A further possibility is to record the identity of each measured particle, for example, using high-resolution imaging, so that measured "partial" scattering diagrams can be assigned to specific particle shapes and sizes, and a set of phase functions associated with each particle class can be built up by summation. This approach is necessary for *in situ* measurements and has been adopted for ice particles by Shcherbakov et al.[65,66] and the Particle Habit Imaging and Polar Scattering (PHIPS) probe described below. However, the difficulty with this method is that highly anisotropic scattering produced by ice crystals in fixed orientation[89,58] demands that a very large number of exemplar patterns is gathered for adequate averaging.

The PHIPS probe is the first single-particle instrument that simultaneously measures the particle geometry, the particle orientation, and the light-scattering diagram of the particle. The instrument is routinely operated in ice cloud characterization studies at the cloud chamber AIDA of KIT and is currently redesigned for *in situ* aircraft applications. The imaging and scattering parts of PHIPS are linked by particle detection and triggering system that ensures that only those particles are recorded which are in the scattering center and in the field of view and depth of field of the stereo imaging system.

Figure 10.10 shows the arrangement of the angular light scattering measurement by PHIPS. The instrument records the polar scattering diagram at 30 discrete angles with an angular resolution of 1° in the 1°–10° angular range and 8° in the 18–170° range. The apertures of 1 mm acrylic glass fibers and the scattering distance of 200 mm confine the solid angles to 2×10^{-5} sr of the first 10 scattering detectors. An additional fiber located a 0° is used to align the laser. The side-scattering and back-scattering fibers are located at a shorter distance of 60 mm from the scattering center and are equipped with convex lenses with apertures of 5 mm resulting in a solid angle of 5.5×10^{-3} sr for these detection angles. The fibers guide the scattered light to 30 individual photoamplifiers that use Si-PIN photo diodes to convert the light pulses to electrical signals. These signals are processed by

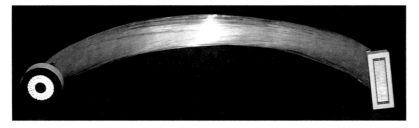

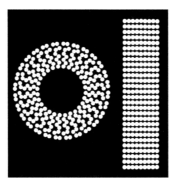

FIGURE 10.10   Fiber optics array used in the SID, version 2 to collect azimuthal scattering patterns (top). The array is divided into 28 segments of nine fibers each, arranged in azimuthal sectors at the input end (bottom left) and stacked parallel bars at the output end (bottom right), the latter coupled to a multichannel photomultiplier. (Images courtesy of Paul Kaye.)

a high-speed data acquisition card consisting of 32 individual ADC channels for simultaneous sampling across all scattering channels. The collimated emission ($\lambda = 532$ nm) of a diode-pumped, solid-state continuous wave laser with 300 mW power is used for particle detection and light scattering measurement. The incident laser beam is polarized parallel to the scattering plane, that is, in the drawing plane of Figure 10.10. A sharp-edged movable mirror is used to reflect the laser beam into a beam dump after passing the center of the cell which results in an already low background signal in the 1° scattering channel. The intersection between the laser beam and the field of view of the trigger detector (not shown in Figure 10.10 but facing the 90° scattering detector) defines the sensitive volume of the instrument. A pin hole is used in the trigger detector to adapt its field of view to the 100 μm depth of field of the stereo imaging system. The 30 scattering channels were calibrated based on glass beads and water droplets.

In case of pristine ice crystals, such as columns and plates, it is possible to reconstruct the 3D habit and particle orientation with respect to the scattering plane using the information extracted from the two corresponding images from the stereo imaging system of PHIPS. Figure 10.11 shows an example of a 3D model of a hexagonal ice plate that was reconstructed from the two images obtained by PHIPS in a synthetic ice cloud generated in the AIDA cloud chamber. So, the instrument provides the angular scattering function of ice crystals (Figure 10.11) whose 3D geometry and orientation is well characterized. This is an important step to validate optical particle models, such as the improved geometric optics model by Hesse et al.[90]

As already mentioned, scattering from atmospheric ice crystals is important in the context of atmospheric modeling, especially climate. Measurements on electrodynamically levitatated ice crystals have been carried out by Bacon et al.[81] and on ice analogues by Ulanowski et al.[58,67] The Polar Nephelometer has been used for measurements both on the ground and from aircraft.[48,65,66]

10.3.2.2   Azimuthal Scattering

The knowledge of purely azimuthal distribution of scattered intensity can provide some particles characteristics, including orientation, sphericity, and certain other shape information. Of these, the

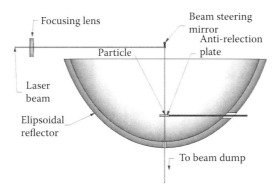

FIGURE 10.11 Experimental setup used to obtain 2D scattering patterns from a single particle deposited on an antireflection-coated glass plate. The particle is in the primary focus of the diffuse ellipsoidal reflector, the camera at the secondary focus. The 3 mm mirror and 5 mm plate are mounted on thin rods, to reduce obscuring the pattern. The plate can be tilted to permit measurements at different particle orientations.[101] Diagram courtesy of Chris Stopford.

sphericity is the easiest one to assess, because homogeneous spheres illuminated by unpolarized or circularly polarized light produce azimuthally uniform scattering. To quantify particle departure from spherical in the context of azimuthal scattering, Kaye et al.[91–94] defined the *asphericity factor Af*, which is a ratio of the standard deviation to the mean response of discrete detector elements, normalized so that it varies between 0 for perfectly uniform response to 100 for most nonuniform response (i.e., when only one element gives nonzero output). The *Af* is proportional to the coefficient of variation of detector response, CoV:

$$Af = 100(n-1)^{1/2}\frac{\sigma}{m} = 100\,\mathrm{CoV}/(n-1), \tag{10.14}$$

where n is the number of elements, σ and m represent the standard deviation and mean, respectively, of individual element responses. The *Af* is also related to the *sphericity index*,[95,96] *SPX*:

$$SPX = 1 - (n)^{1/2}\,Af/100. \tag{10.15}$$

One area where particle sphericity is extremely important is atmospheric science. The radiative properties of clouds and their entire behavior in the context of weather and climate depend crucially on the thermodynamic phase of cloud particles. The phase cannot be determined solely from the knowledge of temperature, as supercooled water droplets are frequently present. *In situ* measurement of cloud properties therefore demands that the phase is determined: this can be accomplished in most cases by assessing whether the particles are spherical (droplets) or not (ice crystals). As discussed in the next section, direct imaging of cloud particles is not the best method of accomplishing this task. For these reasons, several cloud probes have been developed over the last decade at the University of Hertfordshire, collectively known as Small Ice Detectors (SIDs). Successive models

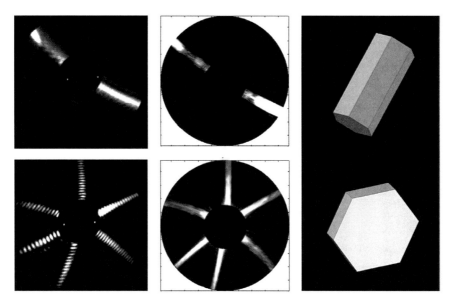

FIGURE 10.12 Experimental scattering patterns from a SID-3 type instrument (left) for an ice column and plate,[127] in comparison with ray tracing with diffraction on facets (RTDF) theory (center)[89] and the 3D shapes used for the computations, in the actual orientations (right). (Diagram courtesy of Ahmed Abdelmonem.)

obtain scattering patterns with progressively increasing angular resolution. The latest ones use intensified CCD cameras to capture 2D scattering patterns, and are described in the next section. The earlier models rely on multielement detectors measuring mainly the azimuthal scattering. The first of these, SID-1, was mainly intended to discriminate between water droplets and ice crystals by determining their sphericity. SID-1 contained six photomultipliers arranged symmetrically around the azimuth centered at the polar angle of 30°, in addition to a particle "trigger" detector defining the sensing volume, and a forward scattering detector for particle sizing.[94] This instrument was superseded by SID-2, containing between 24 and 28 detector elements arranged azimuthally in an annulus. In an early embodiment utilizing an intensified hybrid photodiode array,[93] three additional sensing elements were present inside the main annulus.[97] Later designs of SID-2 combine a fiber optic bundle with a multichannel photomultiplier—see Figure 10.12. SID instruments have been employed in both aircraft campaigns and in laboratory experiments, most notably at the AIDA cloud chamber.[97–101]

The relative simplicity of azimuthal scattering patterns in combination with their sensitivity to particle shape offers opportunities for fast particle classification.[92,102] Azimuthal patterns can also provide the size and aspect ratio of particles such as fibers[103] and prismatic ice crystals.[101] Some particle information is retained even in the azimuthal frequency spectrum, as demonstrated by the recovery of the size and aspect ratio of prismatic ice crystals.[104] However, as for azimuthal scattering in general, such frequency analysis is limited to cases when some *a priori* information on particle shape exists, and even then some ambiguity may be present.[101,105]

Finally, a natural outcome from azimuthal scattering is information on particle orientation.[58,89,106,107] For fibrous particles, not only the azimuthal orientation angle can be recovered, but also the tilt with respect to the incident direction.[106,108]

10.3.2.3 Two-Dimensional Scattering

The 2D pattern of light scattered by a particle, sometimes referred to as the spatial distribution of scattering, or two-dimensional angular optical scattering (TAOS), contains azimuthal in addition to polar (scattering) angle dependence. The pattern is a function of the properties of the size, shape, dielectric structure, surface properties, and orientation of the particle. Therefore, the analysis of 2D patterns can be expected to provide ways to determine these characteristics in great detail. Early research at University of Hertfordshire explored the potential of scattering pattern analysis for particle shape classification and demonstrated how such techniques can be implemented in real-time airborne particle measurement systems designed for aerosol characterization and fiber detection.[109–111] Broad reviews of this area were presented by Kaye et al.[38,39]

A simple way to capture a 2D scattering pattern is to use an imaging array detector directly.[73] However, interference patterns can sometimes be seen in this configuration, for example, due to windows,[112] and the angular range of the measurement is narrow. Somewhat wider coverage can be obtained by projecting the 2D scattering pattern onto an opaque or transparent, flat screen, for recording by a camera. This arrangement yields useful information, for example, for theory verification, but intensity correction may be required.[58,89,86,73,113] A cylindrical screen has also been used.[69] Similar angular coverage is obtainable, in principle, through the use of a Fourier lens to project the distribution of scattering onto a detector array.[87,83,114–116] Ellipsoidal reflectors, first introduced into this field by Kaye et al.[109,110] provide even wider angular coverage.[117–119]

However, the ellipsoidal reflector must be of high optical quality so as not to introduce significant wave front distortion, otherwise artefacts may appear. Moreover, aberrations increase quickly with increasing distance of the particle from the focal point. This may result in angle and intensity errors, including spurious patterns.[117] A useful compromise is to employ a similar optical configuration containing an ellipsoidal reflector with diffuse instead of reflecting surface—see Figure 10.13. The demands on surface quality become relaxed, as subwavelength accuracy is no longer required, and very precise alignment is not needed. The arrangement is appropriate for stationary, immobilized particles.[101,105]

In any of these techniques, the particle can be immobilized to control its orientation. Thin glass or carbon fibers,[58,89,118] antireflection coated windows,[58,101,105] and optical[73] and electrodynamic traps[51,112,115,116] have been used for this purpose.

The range of particle types and processes that can be studied using 2D scattering is wide. Examples include particle clusters,[73,87,114,120–122] fibers,[110,111] deliquescence and efflorescence,[112] freezing (nucleation),[51,83,116] droplet distortion and inclusions,[123] tire dust,[124] ambient aerosol,[117] polyhedral microparticles (bars, cubes, flakes),[111] and ice and ice analogues (see below).[51,58,67,89,115,125] Another area where 2D scattering is increasingly used, sometimes in connection with other techniques such as fluorescence, is the detection and classification of biological aerosols.[39] Recent work in this area includes measurements on pollen[118–119] and bacterial spores.[121,126]

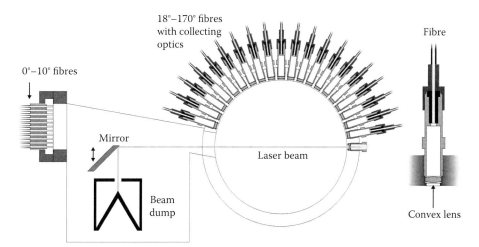

FIGURE 10.13 Plane section through the measuring cell of PHIPS that represents the scattering plane. A magnified illustration of a fiber equipped with collection optics is shown on the right. (Images courtesy of Paul Kaye.)

Shape characterization is of special importance for cloud particles, mainly because the radiative properties of clouds strongly depend on particle shape.[57] Unfortunately, a direct imaging of cloud particles is difficult because of conflicting demands of high optical resolution, large sample volume, and the need for "noninvasive" measurement avoiding particle breakup artefacts; consequently, smaller ice crystals *in situ* cannot be imaged in detail.[127,128] The SID instruments are designed to address this issue. The earlier ones measure the azimuthal dependence of scattering; hence they are described in the previous section. The most recent design, SID-3, relies on a CCD camera with an image intensifier to captures 2D patterns. It exists in various laboratories as well as aircraft versions.[125,127,129] A tangible demonstration of the power of 2D scattering patterns to encode particle shape and orientation is provided by the examples shown in Figure 10.14 where patterns from ice crystals, obtained using a laboratory version of SID3, are compared with theory.

Particle size, shape, and internal or surface structure are all "encoded" in the 2D pattern. Holler et al.[114,122] established in experiments with sphere clusters that the number of intensity "islands" (peaks) in the patterns was proportional to the overall size of the clusters. Similar, qualitative relationship can be seen in patterns from pollen grains.[118–119] In contrast, Auger et al. found the island number to be proportional to the individual element size (i.e., sphere diameter)[130]; however, the relationship between these properties, or between the number of islands and the periodicity of undulations on the surface of theoretically modeled Chebyshev particles, was not monotonic.[131] Therefore, caution is needed when interpreting such data: "cross-talk" between different properties is likely, and furthermore some conclusions may be affected by the image processing routines used for feature extraction. Multivariate statistical analysis of frequency spectra of 2D patterns, such as principal component and discriminant function analysis, offers ways of isolating relationships between particle and pattern properties, and allows some degree of classification of particle aggregates.[122] Other statistical parameters, such as features of the gray-level co-occurrence matrix, show correlation with surface roughness of particles.[125]

Particle roughness may turn out to have high significance in scattering from atmospheric particles, such as MD aerosols or ice crystals. While theoretical models of scattering from particles with rough surfaces are still being developed and tested, there is growing evidence that smooth spheroids or smooth polyhedral shapes do not adequately represent real atmospheric particles.[57,66,132–134]

Experimental 2D patterns from smooth and rough ice analogue crystals showed distinct differences: while the former showed sharp, well-defined bright arcs and spots, the latter had more

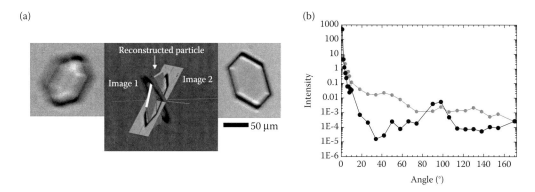

FIGURE 10.14   (a) The 3D model reconstruction principle of PHIPS which gives the particle geometry and orientation (laser direction is indicated by the thin horizontal line). The hexagonal ice plate was grown in the AIDA chamber at a temperature around 11°C. Note that the particle was out of the depth of field of the microscopic unit 1 (left image). (b) Angular scattering measurements with PHIPS for the ice plate shown in (a) (black dots) and for a second, similarly sized plate with a different orientation (gray dots). (Diagram courtesy of Ahmed Abdelmonem.)

random, "speckly" appearance, but with greater azimuthal symmetry. Significantly, the smooth and rough ice analogues were characterized by very different asymmetry parameters (see Section 10.3.2.1).[58] The distinguishing features of the 2D patterns may provide information on surface roughness of atmospheric particles. Indeed, recent data obtained using the SID-3 probe indicate that rough ice crystals may be prevalent in some types of clouds,[125] corroborating indirect evidence obtained with the Polar Nephelometer.[66]

REFERENCES

1. Henning, T. and Schnaiter, M. 1998. Carbon—From space to laboratory. *Earth Moon Planets* 80:179–207.
2. Sorensen, C. M. 2001. Light scattering by fractal aggregates: A review. *Aerosol Sci. Technol.* 35:648–687.
3. Wagner, R., Linke, C., Naumann, K. H., Schnaiter, M., Vragel, M., Gangl, M., and Horvath, H. 2009. A review of optical measurements at the aerosol and cloud chamber AIDA. *J. Quant. Spectrosc. Radiat. Transfer* 110:930–949.
4. Sokolik, I. N. and Toon, O. B. 1999. Incorporation of mineralogical composition into models of the radiative properties of mineral aerosol from UV to IR wavelengths. *J. Geophys. Res. Atmos.* 104:9423–9444.
5. Kocifaj, M. and Horvath, H. 2005. Inversion of extinction data for irregularly shaped particles. *Atmos. Environ.* 39:1481–1495.
6. Vogel, B., Vogel, H., Baumer, D., Bangert, M., Lundgren, K., Rinke, R. and Stanelle, T. 2009. The comprehensive model system COSMO-ART— Radiative impact of aerosol on the state of the atmosphere on the regional scale. *Atmos. Chem. Phys.* 9:8661–8680.
7. Solomon, S., Qin, D., Manning, M., Marquis, M., Averyt, K., Tignor, M. M. B., and Miller, H. L. *Climate Change 2007—The Physical Science Basis*. Cambridge University Press, Cambridge, 2007.
8. Schnaiter, M., Gimmler, M., Llamas, I., Linke, C., Jäger, C., and Mutschke, H. 2006. Strong spectral dependence of light absorption by organic carbon particles formed by propane combustion. *Atmos. Chem. Phys.* 6:2981–2990.
9. Schnaiter, M., Linke, C., Möhler, O., Naumann, K. H., Saathoff, H., Wagner, R., Schurath, U., and Wehner, B. 2005. Absorption amplification of black carbon internally mixed with secondary organic aerosol. *J. Geophys. Res. Atmos.* 110:D19204.
10. Andreae, M. O. and Gelencsér, A. 2006. Black or brown carbon? The nature of light-absorbing carbonaceous aerosols. *Atmos. Chem. Phys.* 6:3131–3148.
11. Bond, T. C., Anderson, T. L., and Campbell, D. 1999. Calibration and intercomparison of filter-based measurements of visible light absorption by aerosols. *Aerosol Sci. Technol.* 30:582–600.
12. Reid, J. S., Hobbs, P. V., Liousse, C., Martins, J. V., Weiss, R. E., and Eck, T. F. 1998. Comparisons of techniques for measuring shortwave absorption and black carbon content of aerosols from biomass burning in Brazil. *J. Geophys. Res. Atmos.* 103:32031–32040.
13. Schnaiter, M., Horvath, H., Mohler, O., Naumann, K. H., Saathoff, H., and Schock, O. W. 2003. UV–VIS–NIR spectral optical properties of soot and soot-containing aerosols. *J. Aerosol Sci.* 34:1421–1444.
14. Weingartner, E., Saathoff, H., Schnaiter, M., Streit, N., Bitnar, B., and Baltensperger, U. 2003. Absorption of light by soot particles: Determination of the absorption coefficient by means of aethalometers. *J. Aerosol Sci.* 34:1445–1463.
15. Weiss, R. E., Kapustin, V. N., and Hobbs, P. V. 1992. Chain-aggregate aerosols in smoke from the Kuwait Oil Fires. *J. Geophys. Res. Atmos.* 97:14527–14531.
16. Virkkula, A., Ahlquist, N. C., Covert, D. S., Sheridan, P. J., Arnott, W. P., and Ogren, J. A. 2005. A three-wavelength optical extinction cell for measuring aerosol light extinction and its application to determining light absorption coefficient. *Aerosol Sci. Technol.* 39:52–67.
17. Schnaiter, M., Schmid, O., Petzold, A., Fritzsche, L., Klein, K. F., Andreae, M. O., Helas, G. et al. 2005. Measurement of wavelength-resolved light absorption by aerosols utilizing a UV–VIS extinction cell. *Aerosol Sci. Technol.* 39:249–260.
18. Sheridan, P. J., Arnott, W. P., Ogren, J. A., Andrews, E., Atkinson, D. B., Covert, D. S., Moosmuller, H. et al. 2005. The Reno Aerosol Optics Study: An evaluation of aerosol absorption measurement methods. *Aerosol Sci. Technol.* 39:1–16.

19. Schnaiter, M., Gimmler, M., Llamas, I., Linke, C., Jager, C., and Mutschke, H. 2006. Strong spectral dependence of light absorption by organic carbon particles formed by propane combustion. *Atmos. Chem. Phys.* 6:2981–2990.

20. Bohren, C. F. and Huffman, D. R. *Absorption and Scattering of Light by Small Particles*. Wiley, New York, 1983.

21. Moosmüller, H. and Arnott, W. P. 2003. Angular truncation errors in integrating nephelometry. *Rev. Sci. Instrum.* 74:3492–3501.

22. Schnaiter, M., Linke, C., Mohler, O., Naumann, K. H., Saathoff, H., Wagner, R., Schurath, U., and Wehner, B. 2005. Absorption amplification of black carbon internally mixed with secondary organic aerosol. *J. Geophys. Res. Atmos.* 110:D19204.

23. Linke, C., Mohler, O., Veres, A., Mohacsi, A., Bozoki, Z., Szabo, G., and Schnaiter, M. 2006. Optical properties and mineralogical composition of different Saharan mineral dust samples: A laboratory study. *Atmos. Chem. Phys.* 6:3315–3323.

24. Moosmuller, H., Chakrabarty, R. K., and Arnott, W. P. 2009. Aerosol light absorption and its measurement: A review. *J. Quant. Spectrosc. Radiat. Transfer* 110:844–878.

25. Schwarz, J. P., Gao, R. S., Fahey, D. W., Thomson, D. S., Watts, L. A., Wilson, J. C., Reeves, J. M. et al. 2006. Single-particle measurements of midlatitude black carbon and light-scattering aerosols from the boundary layer to the lower stratosphere. *J. Geophys. Res. Atmos.* 111:D16207.

26. Miklos, A., Hess, P., and Bozoki, Z. 2001. Application of acoustic resonators in photoacoustic trace gas analysis and metrology. *Rev. Sci. Instrum.* 72:1937–1955.

27. Lack, D. A., Lovejoy, E. R., Baynard, T., Pettersson, A., and Ravishankara, A. R. 2006. Aerosol absorption measurement using photoacoustic spectroscopy: Sensitivity, calibration, and uncertainty developments. *Aerosol Sci. Technol.* 40:697–708.

28. Schmid, O., Artaxo, P., Arnott, W. P., Chand, D., Gatti, L. V., Frank, G. P., Hoffer, A., Schnaiter, M., and Andreae, M. O. 2006. Spectral light absorption by ambient aerosols influenced by biomass burning in the Amazon Basin. I: Comparison and field calibration of absorption measurement techniques. *Atmos. Chem. Phys.* 6:3443–3462.

29. Tian, G. X., Moosmuller, H., and Arnott, W. P. 2009. Simultaneous photoacoustic spectroscopy of aerosol and oxygen A-Band absorption for the calibration of aerosol light absorption measurements. *Aerosol Sci. Technol.* 43:1084–1090.

30. Rosencwaig, A. 1980. Photoacoustic-spectroscopy. *Annu. Rev. Biophys. Bio.* 9:31–54.

31. Arnott, W. P., Moosmuller, H., Rogers, C. F., Jin, T. F., and Bruch, R. 1999. Photoacoustic spectrometer for measuring light absorption by aerosol: Instrument description. *Atmos. Environ.* 33:2845–2852.

32. Arnott, W. P., Moosmuller, H., Sheridan, P. J., Ogren, J. A., Raspet, R., Slaton, W. V., Hand, J. L., Kreidenweis, S. M., and Collett, J. L. 2003. Photoacoustic and filter-based ambient aerosol light absorption measurements: Instrument comparisons and the role of relative humidity. *J. Geophys. Res. Atmos.* 108:D16203.

33. Lewis, K., Arnott, W. P., Moosmuller, H., and Wold, C. E. 2008. Strong spectral variation of biomass smoke light absorption and single scattering albedo observed with a novel dual-wavelength photoacoustic instrument. *J. Geophys. Res. Atmos.* 113:D16203.

34. Slowik, J. G., Stainken, K., Davidovits, P., Williams, L. R., Jayne, J. T., Kolb, C. E., Worsnop, D. R., Rudich, Y., DeCarlo, P. F., and Jimenez, J. L. 2004. Particle morphology and density characterization by combined mobility and aerodynamic diameter measurements. Part 2: Application to combustion-generated soot aerosols as a function of fuel equivalence ratio. *Aerosol Sci. Technol.* 38:1206–1222.

35. Ossenkopf, V. 1991. Effective-medium theories for cosmic dust grains. *Astron. Astrophys.* 251:210–219.

36. Mishchenko, M. I. and Travis, L. D. 1998. Capabilities and limitations of a current FORTRAN implementation of the T-matrix method for randomly oriented, rotationally symmetric scatterers. *J. Quant. Spectrosc. Radiat. Transfer* 60:309–324.

37. Kerker, M. 1997. Light scattering instrumentation for aerosol studies: An historical overview. *Aerosol Sci. Technol.* 27:522–540.

38. Kaye, P. H. 1998. Spatial light scattering as a means of characterising and classifying non-spherical particles. *Meas. Sci. Technol.* 9:141–149.

39. Kaye, P. H., Aptowicz, K., Chang, R. K., Foot, V., and Videen, G. 2007. Angularly resolved elastic scattering from airborne particles. In *Optics of Biological Particles*. Hoekstra, A., Maltsev, V., and Videen, G., (Eds). Springer, Dordrecht, pp. 31–61.

40. Mishchenko, M. I., Travis, L. D., and Lacis, A. A. 2002. *Scattering, Absorption, and Emission of Light by Small Particles*, Cambridge University Press, Cambridge.

41. Hovenier, J. W. 2000. Measuring scattering matrices of small particles at optical wavelengths. In *Light Scattering by Nonspherical Particles*. Eds. Mishchenko, M. I., Hovenier, J. W., and Travis, L. D., Academic Press, San Diego, pp. 355–365.

42. Volten, H., Muñoz, O., Rol, E., de Haan, J. F., Vassen, W., Hovenier, J. W., Muinonen, K., and Nousiainen, T. 2001. Scattering matrices of mineral aerosol particles at 441.6 nm and 632.8 nm. *J. Geophys. Res. Atmos.* 106:17375–17401.

43. Kuik, F., Stammes, P., and Hovenier, J. W. 1991. Experimental determination of scattering matrices of water droplets and quartz particles. *Appl. Opt.* 30:4872–4881.

44. Verhaege, C., Shcherbakov, V., and Personne, P. 2008. Limitations on retrieval of complex refractive index of spherical particles from scattering measurements. *J. Quant. Spectrosc. Radiat. Transfer* 109:2338–2348.

45. Quinby-Hunt M. S., Erskine L. L., and Hunt, A. J. 1997. Polarized light scattering by aerosols in the marine atmospheric boundary layer. *Appl. Opt.* 36:5168–5184.

46. Hadamcik, E., Renard, J.-B., Levasseur-Regourd, A.C., Lasue, J., Alcouffe, G., and Francis, M. 2009. Light scattering by agglomerates: Interconnecting size and absorption effects (PROGRA2 experiment) *J. Quant. Spectrosc. Radiat. Transfer* 110:1755–1770.

47. Muñoz, O., Moreno, F., Guirado, D., Ramos, J. L., López, A., Girela, F., Jerónimo, J. M., Costillo, L. P., and Bustamante, I. 2010. Experimental determination of scattering matrices of dust particles at visible wavelengths: The IAA light scattering apparatus. *J. Quant. Spectrosc. Radiat. Transfer* 111:187–196.

48. Oshchepkov, S., Harumi, I., Gayet, J. F., Sinyuk, A., Auriol, F., and Havemann, S. 2000. Microphysical properties of mixed-phase and ice clouds retrieved from *in situ* airborne "polar nephelometer" measurements. *Geophys. Res. Lett.* 27:209–212.

49. Barkey, B., Bailey, M., Liou, K. N., and Hallett, J. 2002. Light-scattering properties of plate and column ice crystals generated in a laboratory cold chamber. *Appl. Opt.* 41:5792–5796.

50. Duft, D. and Leisner, T. 2004. The index of refraction of supercooled solutions determined by the analysis of optical rainbow scattering from levitated droplets. *Int. J. Mass Spectrom.* 233:61–65.

51. Krämer, B., Hübner, O., Vortisch, H., Wöste, L., Leisner, T., Schwell, M., Rühl, E., and Baumgärtel, H. 1999. Homogeneous nucleation rates of supercooled water measured in single levitated microdroplets. *J. Chem. Phys.* 111:6521–6527.

52. Gucker, F. T., Tuma, J., Lin, H. M., Huang, C. M., Ems, S. C., and Marshall T. R. 1973. Rapid measurement of light-scattering diagrams from single particles in an aerosol stream and determination of latex particle size. *J. Aerosol Sci.* 4:389–404.

53. Kaller, W. A. 2004. A new polar nephelometer for measurement of atmospheric aerosols. *J. Quant. Spectrosc. Radiat. Transfer* 87:107–117.

54. Watson, D., Hagen, N., Diver, J., Marchand, P., and Chachisvilis, M. 2004. Elastic light scattering from single cells: Orientational dynamics in optical trap. *Biophys. J.* 87:1298–1306.

55. Curtis, D. B., Aycibin, M., Young, M. A., Grassian, V. H., and Kleiber, P. D. 2007. Simultaneous measurement of light-scattering properties and particle size distribution for aerosols. *Atmos. Environ.* 41:4748–4758.

56. Ulanowski, Z., Greenaway, R. S., Kaye, P. H., and Ludlow, I. K. 2002. Laser diffractometer for single particle scattering measurements. *Meas. Sci. Technol.* 13:292–296.

57. Baran, A. J. 2009. A review of the light scattering properties of cirrus. *J. Quant. Spectrosc. Radiat. Transfer* 110:1239–1260.

58. Ulanowski, Z., Hesse, E., Kaye, P. H., and Baran, A. J. 2006. Light scattering by complex ice-analogue crystals. *J. Quant. Spectrosc. Radiat. Transfer* 100:382–392.

59. Amsterdam Scattering Database. http://www.iaa.es/scattering/

60. Zhao, F. 1999. Determination of the complex index of refraction and size distribution of aerosols from polar nephelometer measurements. *Appl. Opt.* 38:2331–2336.

61. Dick, W. D., Ziemann, P. J., and McMurray, P. H. 1998. Shape and refractive index of submicron atmospheric aerosols from multiangle light scattering measurements. *J. Aerosol Sci.* 29:S103–S104.

62. Barkey, B., Paulson S., and Chung A. 2007. Genetic algorithm inversion of dual polarization polar nephelometer data to determine aerosol refractive index. *Aerosol Sci. Technol.* 41:751–760.

63. Volten, H., de Haan, J. F., Hovenier, J. W., Schreurs, R., Vassen, W., Dekker, A. G., Hoogenboom, H. J., Charlton, F., and Wouts, R. 1998. Laboratory measurements of angular distributions of light scattered by phytoplankton and silt. *Limnol. Ocean.* 43:1180–1197.

64. Sasaki, Y., Nishiyama, N., and Furukawa, Y. 1998. Experimental study on light scattering from an artificial ice cloud. *Polar Meteor. Glaciol.* 12:130–139.

65. Shcherbakov, V., Gayet, J. F., Baker, B., and Lawson, P. 2006. Light scattering by single natural ice crystals. *J. Atmos. Sci.* 63:1513–1525.

66. Shcherbakov, V., Gayet, J. F., Jourdan, O., Ström, J., and Minikin, A. 2006. Light scattering by single ice crystals of cirrus clouds. *Geophys. Res. Lett.* 33:L15809.

67. Ulanowski, Z., Hesse, E., Kaye, P. H., Baran, A. J., and Chandrasekhar, R. 2003. Scattering of light from atmospheric ice analogues. *J. Quant. Spectrosc. Radiat. Transfer* 79–80C:1091–1102.

68. Lock, J. A. 1997. Scattering of a diagonally incident focused Gaussian beam by an infinitely long homogeneous circular cylinder. *J. Opt. Soc. Am. A* 14:640–652.

69. Ulanowski, Z. and Hesse, E. 2003. Scattering of non-uniform incident fields by long cylinders. In *Proc. 7th Int. Conf. Electromagnetic Light Scatt. Nonsph. Part, Bremen,* 362–365.

70. Maltsev, V. P. 2000. Scanning flow cytometry for individual particle analysis. *Rev. Sci. Instrum.* 71:243–255.

71. Maltsev, V. P. and Semyanov, K. A. 2004. *Characterisation of Bio-Particles from Light Scattering,* VSP, Utrecht.

72. Doornbos, R. M. P., Schaeffer, M., Hoekstra, A. G., Sloot, P. M. A., de Grooth, B. G., and Greve, J. 1996. Elastic light-scattering measurements of single biological cells in an optical trap. *Appl. Opt.* 35:729–734.

73. Neukammer, J., Gohlke, C., Höpe, A., Wessel, T., and Rinneberg, H. 2003. Angular distribution of light scattered by single biological cells and oriented particle agglomerates. *Appl. Opt.* 42:6388–6397.

74. Berdnik, V. V., Gilev, K., Shvalov, A., Maltsev, V., and Loiko, V. A. 2006. Characterization of spherical particles using high-order neural networks and scanning flow cytometry. *J. Quant. Spectrosc. Radiat. Transfer* 102:62–72.

75. Ulanowski, Z. and Ludlow, I. K. 1989. Water distribution, size and wall thickness in *Lycoperdon-pyriforme* spores. *Mycol. Res.* 93:28–32.

76. Ulanowski, Z. and Ludlow, I. K. 2000. Compact optical trapping microscope using a diode laser. *Meas. Sci. Technol.* 11:1778–1785.

77. Taji, K., Tachikawa, M., and Nagashima, K. 2006. Laser trapping of ice crystals. *Appl. Phys. Lett.* 88:141111.

78. Mitchem, L. and Reid, J. P. 2008. Optical manipulation and characterisation of aerosol particles using a single-beam gradient force optical trap. *Chem. Soc. Rev.* 37:756–769.

79. Reid, J. P. 2009. Particle levitation and laboratory scattering. *J. Quant. Spectrosc. Radiat. Transfer* 110:1293–1306.

80. Steiner, B., Berge, B., Gausmann, R., Rohmann, J., and Rühl, E. 1999. Fast *in situ* sizing technique for single levitated liquid aerosols. *Appl. Opt.* 38:1523–1529.

81. Bacon, N. J. and Swanson, B. D. 2000. Laboratory measurements of light scattering by single levitated ice crystals. *J. Atmos. Sci.* 57:2094–2104.

82. Davis, E. J. 1997. A history of single aerosol particle levitation. *Aerosol Sci. Technol.* 26:212–254.

83. Shaw, R. A., Lamb, D., and Moyle, A. M. 2000. An electrodynamic levitation system for studying individual cloud particles under upper-tropospheric conditions. *J. Atmos. Oceanic Technol.* 17:940–948.

84. Hesse, E., Ulanowski, Z., and Kaye, P. H. 2002. Stability characteristics of cylindrical fibres in an electrodynamic balance designed for single particle investigation. *J. Aerosol Sci.* 33:149–163.

85. Zhu J.H., Zheng F., Laucks M. L., and Davis, E. J. 2002. Mass transfer from an oscillating microsphere. *J. Colloid Interface Sci.* 249:351–358.

86. Kolomenskii, A. A., Jerebtsov, S. N., Stoker, J. A., Scully, M. O., and Schuessler, H. A. 2007. Storage and light scattering of microparticles in a ring-type electrodynamic trap. *J. Appl. Phys.* 102:094902.

87. Holler, S., Surbek, M., Chang, R. K., and Pan, Y. L. 1999. Two-dimensional angular optical scattering patterns as droplets evolve into clusters. *Opt. Lett.* 24:1185–1187.

88. Gustafson, B. A. S. 1996. Microwave analog to light-scattering measurements: A modern implementation of a proven method to achieve precise control. *J. Quant. Spectrosc. Radiat. Transfer* 55:663–672.

89. Clarke, A. J. M., Hesse, E., Ulanowski, Z., and Kaye, P. H. 2006. A 3D implementation of ray tracing combined with diffraction on facets: Verification and a potential application. *J. Quant. Spectrosc. Radiat. Transfer* 100:103–114.

90. Hesse, E., Mc Call, D. S., Ulanowski, Z., Stopford, C., and Kaye, P. H. 2009. Application of RTDF to particles with curved surfaces, *J. Quant. Spectrosc. Radiat. Transfer,* 110:1599–1603.

91. Kaye, P. H., Eyles, N. A., Ludlow, I. K., and Clark, J. M. 1991. An instrument for the classification of airborne particles on the basis of size, shape and count frequency. *Atmos. Environ.* 25A:645–654.

92. Kaye, P. H., Alexander-Buckley, K., Hirst, E., and Saunders, S. 1996. A real-time monitoring system for airborne particle shape and size analysis. *J. Geophys. Res.* 101:19215–19221.

93. Kaye, P. H., Barton, J. E., Hirst, E., and Clark, J. M. 2000. Simultaneous light scattering and intrinsic fluorescence measurement for the classification of airborne particles. *Appl. Opt.* 39:3738–3745.

94. Hirst, E., Kaye, P. H., Greenaway, R. S., Field, P., and Johnson, D. W. 2001. Discrimination of micrometre-sized ice and super-cooled droplets in mixed-phase cloud. *Atmos. Environ.* 35:33–47.

95. Dick, W. D., McMurry, P. H., and Bottiger, J. R. 1994. Size- and composition-dependent response of the DAWN—A multiangle optical detector. *Aerosol Sci. Technol.* 20:345–362.

96. Dick, W. D., Ziemann, P. J., Huang, P.-F., and McMurry, P. H. 1998. Optical shape fraction measurements of submicrometre laboratory and atmospheric aerosols. *Meas. Sci. Technol.* 9:183–196.

97. Cotton, R., Osborne, S., Ulanowski, Z., Hirst, E., Kaye, P. H., and Greenaway, R. S. 2009. The ability of the Small Ice Detector (SID2) to characterise cloud particle and aerosol morphologies obtained during flights of the FAAM BAe146 research aircraft. *J. Atmos. Oceanic Technol.* 27:290–303.

98. Field, P. R., Cotton, R. J., Noone, K., Glantz, P., Kaye, P. H., Hirst, E., Greenaway, R. S. et al. 2001. Ice nucleation in orographic wave clouds: Measurements made during INTACC. *Quart. J. Royal Met. Soc.* 127:1493–1512.

99. Moehler, O., Field, P. R., Connolly, P., Benz, S., Saathoff, H., Schnaiter, M., Wagner, R. et al. 2006. Efficiency of the deposition mode ice nucleation on mineral dust particles. *Atmos. Chem. Phys.* 6:3007–3021.

100. Field, P. R., Möhler, O., Connolly, P., Krämer, M., Cotton, R., Heymsfield, A. J., Saathoff, H., and Schnaiter, M. 2006. Some ice nucleation characteristics of Asian and Saharan desert dust. *Atmos. Chem. Phys.* 6:2991–3006.

101. Stopford, C. 2010. Ice crystal classification using two-dimensional light scattering patterns. PhD thesis, University of Hertfordshire, Hertfordshire.

102. Kaye, P. H., Hirst, E., and Wang-Thomas, Z. 1997. A neural network based spatial light scattering instrument for hazardous airborne fiber detection. *Appl. Opt.* 36:6149–6156.

103. Barthel, H., Sachweh, B., and Ebert, F. 1998. Measurement of airborne mineral fibres using a new differential light scattering device. *Meas. Sci. Technol.* 9:210–220.

104. Ulanowski, Z., Stopford, C., Hesse, E., Kaye, P. H., Hirst, E., and Schnaiter, M. 2007. Characterization of small ice crystals using frequency analysis of azimuthal scattering patterns. In *Proc. 10th Int. Conf. on Electromagnetic and Light Scatt.*, Bodrum, 225–228.

105. Stopford, C., Ulanowski, Z., Hesse, E., Kaye, P. H., Hirst, E., Schnaiter, M., and McCall, D. 2008. Initial investigation into using Fourier spectra as a means of classifying ice crystal shapes. In *Proc. 11th Int. Conf. on Electromagnetic and Light Scatt.*, Hatfield, 247–250.

106. Hirst, E., Kaye, P. H., Buckley, K. M., and Saunders, S. J. 1995. A method of investigating the orientational behaviour of fibrous particles in gaseous flow. *Part. Part. Syst. Char.* 12:3–9.

107. Ulanowski, Z., Kaye, P. H., and Hirst, E. 1998. Respirable asbestos detection using light scattering and magnetic alignment. *J. Aerosol Sci.* 29:S13–S14.

108. Morriss, C. D. 2005. Fibre alignment in aerosol delivery systems. PhD thesis, University of Hertfordshire, Hertfordshire.

109. Kaye, P. H., Hirst, E., Clark, J. M., and Micheli, F. 1992. Airborne particle shape and size classification from spatial light scattering profiles. *J. Aerosol Sci.* 23:597–611.

110. Hirst, E., Kaye P. H., and Guppy, J. R. 1994. Light scattering from nonspherical airborne particles: theoretical and experimental comparisons. *Appl. Opt.* 33:7180–7187.

111. Hirst, E. and Kaye P. H. 1996. Experimental and theoretical light scattering profiles from spherical and non-spherical particles. *J. Geophys. Res.* 101:19231–19235.

112. Braun, C. and Krieger, U. K. 2001. Two-dimensional angular light-scattering in aqueous NaCl single aerosol particles during deliquescence and efflorescence. *Opt. Expr.* 8:314–321.

113. Ulanowski, Z. 2005. Ice analog halos. *Appl. Opt.* 44:5754–5758.

114. Holler, S., Pan, Y., Chang, R. K., Bottiger, J. R., Hill, S. C., and Hillis, D. B. 1998. Two-dimensional angular optical scattering for the characterization of airborne microparticles. *Opt. Lett.* 23:1489–1491.

115. Kraemer, B., Schwell, M., Huebner, O., Vortisch, H., Leisner, T., Ruehl, E., Baumgaertel, H., and Woeste, L. 1996. Homogeneous ice nucleation observed in single levitated micro droplets, *Ber. Bunsenges. Phys. Chem.* 100:1911–1914.

116. Weidinger, I., Klein, J., Stöckel, P., Baumgärtel, H., and Leisner, T. 2003. Nucleation behavior of n-alkane microdroplets in an electrodynamic balance. *J. Phys. Chem. B* 107:3636–3643.

117. Aptowicz, K. B., Pinnick, R. G., Hill, S. C., Pan, Y. L., and Chang, R. K. 2006. Optical scattering patterns from single urban aerosol particles at Adelphi, Maryland, USA: A classification relating to particle morphologies. *J. Geophys. Res.* 111:D12212.

118. Surbek, M., Esen, C., and Schweiger, G. 2009. Elastic light scattering on single pollen: Scattering in a large space angle. *Aerosol Sci. Technol.* 43:679–684.
119. Surbek, M., Esen, C., Schweiger, G., and Ostendorf, A. 2010. Pollen characterization and identification by elastically scattered light. *J. Biophot.* doi:10.1002/jbio.200900088.
120. Holler, S., Auger, J.-C., Stout, B., Pan, Y., Bottiger, J. R., Chang, R. K., and Videen, G. 2000. Observations and calculations of light scattering from clusters of spheres. *Appl. Opt.* 39:6873–6887.
121. Fernandes, G. E., Pan, Y.-L., Chang, R. K., Aptowicz, K., and Pinnick, R. G. 2006. Simultaneous forward- and backward-hemisphere elastic-light-scattering patterns of respirable-size aerosols. *Opt. Lett.* 31:3034–3036.
122. Holler, S., Zomer, S., Crosta, G. F., Pan, Y.-L., Chang, R. K., and Bottiger, J. R. 2004. Multivariate analysis and classification of two-dimensional angular optical scattering patterns from aggregates. *Appl. Opt.* 43:6198–6206.
123. Secker, D. R., Kaye, P. H., Greenaway, R. S., Hirst, E., Bartley, D. L., and Videen, G. 2000. Light scattering from deformed droplets and droplets with inclusions. I. Experimental results. *Appl. Opt.* 39:5023–5030.
124. Crosta, G. F., Zomer, S., Pan, Y.-L., and Holler, S. 2003. Classification of single-particle two-dimensional angular optical scattering patterns and heuristic scatterer reconstruction. *Opt. Eng.* 42:2689–2701.
125. Ulanowski, Z., Kaye, P. H., Hirst, E., and Greenaway, R. S. 2010. Light scattering by ice particles in the Earth's atmosphere and related laboratory measurements. In *Proc. 12th Int. Conf. on Electromagnetic Light Scatt.*, Helsinki, 294–297.
126. Auger, J.-C., Aptowicz, K. B., Pinnick, R. G., Pan, Y.-L., and Chang, R. K. 2007. Angularly resolved light scattering from aerosolized spores: Observations and calculations. *Opt. Lett.* 32:3358–3360.
127. Kaye, P. H., Hirst, E., Greenaway, R. S., Ulanowski, Z., Hesse, E., DeMott, P. J., Saunders, C., and Connolly, P. 2008. Classifying atmospheric ice crystals by spatial light scattering. *Opt. Lett.* 33:1545–1547.
128. Connolly, P. J., Flynn, M. J., Ulanowski, Z., Choularton, T. W., Gallagher, M. W., and Bower, K. N. 2007. Calibration of 2 D imaging probes using calibration beads and ice crystal analogues: The depth of field. *J. Atmos. Oceanic Technol.* 24:1860–1879.
129. Kaye, P. H., Hirst, E., Ulanowski, Z., Hesse, E., Greenaway, R. S., and DeMott, P. J. 2008. A light scattering instrument for investigating cloud ice microcrystal morphology. In *Proc. 11th Int. Conf. on Electromagnetic and Light Scatt.*, Hatfield, 235–238.
130. Auger, J.-C., Fernandes, G. E., Pan, Y.-L., Aptowicz, K. B., and Chang, R. K. 2007. Influence of microparticle surface roughness on TAOS patterns: Experimental and theoretical studies. *PIERS Online* 3:897–899.
131. Auger, J.-C., Fernandes, G. E., Aptowicz, K. B., Pan, Y.-L., and Chang, R. K. 2010. Influence of surface roughness on the elastic-light scattering patterns of micron-sized aerosol particles. *Appl. Phys. B* 99:229–234.
132. Nousiainen, T. and Muinonen, K. 2007. Surface-roughness effects on single-scattering properties of wavelength-scale particles. *J. Quant. Spectrosc. Radiat. Transfer* 106:389–397.
133. Zubko, E., Muinonen, K., Shkuratov, Y., Videen, G., and Nousiainen, T. 2007. Scattering of light by roughened Gaussian random particles. *J. Quant. Spectrosc. Radiat. Transfer* 106:604–615.
134. Yang, P., Kattawar, G. W., Hong, G., Minnis, P., and Hu, Y. 2008. Uncertainties associated with the surface texture of ice particles in satellite-based retrieval of cirrus. *IEEE Trans. Geosci. Remote Sens.* 46:1940–1957.

11 Progress in the Investigation of Aerosols' Optical Properties Using Cavity Ring-Down Spectroscopy
Theory and Methodology

Ali Abo Riziq and Yinon Rudich

Contents

11.1 Introduction 269
11.2 Interaction of Aerosols with Solar Radiation and Mie Theory 271
11.3 CRD-S: Methodology and Applications..274
    11.3.1 Pulsed-Laser CRD...275
    11.3.2 Continuous Wave Cavity Ring-Down (cw-CRD) Spectroscopy 277
    11.3.3 Sensitivity and Detection Limit..278
11.4 Application of CRD-AS for Measuring the Extinction Coefficient (α_{ext}), Extinction
    Cross Section (σ_{ext}) and Extinction Efficiency (Q_{ext}) of Aerosols........................279
    11.4.1 Measuring the Complex Refractive Index (*m*) of Aerosols by CRD-AS.................281
    11.4.2 Homogenously Mixed Aerosols..282
    11.4.3 Coated Aerosols...286
    1.4.4 Recent Advances in CRD-AS...288
11.5 Summary 291
Acknowledgments...291
References...291

11.1 INTRODUCTION

Aerosols are defined as small solid or liquid particles suspended in the atmosphere for a period of a few hours to a few days, depending on their size. They have a complex chemical composition that mainly depends on their source, which may be natural and/or anthropogenic. Natural aerosols may originate from forest fires, sea spray, dust storms, and volcanoes, whereas anthropogenic aerosols are emitted due to human activities, such as biomass burning, industry, and transportation (i.e., car, truck, ship, and airplane exhaust). The coupling between the two sources can further complicate the composition, and thus the processing, of aerosols in the atmosphere.

It is well established that aerosols play a pivotal role in Earth's radiation balance, but large uncertainties exist regarding the magnitude and overall effect of their contribution (IPCC, 2007). Aerosols have a direct effect on the radiation budget (RB) of the atmosphere by absorbing and scattering solar radiation (Lohmann and Feichter, 2001; Ramanathan et al., 2001; Kaufman, 2002; Bellouin et al.,

2005; Bates et al., 2006). In addition, aerosols indirectly affect the RB in several ways: their microphysical properties and ability to act as cloud condensation nuclei (CCN) affects cloud formation, and their concentration can change the reflectivity and lifetime of clouds (Koren et al., 2004; Kaufman et al., 2005; Ramanathan et al., 2005; Koren et al., 2007). Specifically, the optical properties, such as the complex refractive index and phase function, of aerosols affect the RB as they determine an aerosol's interaction with incoming and outgoing solar radiation (Kim and Ramanathan, 2008).

Aerosols undergo various processes that may alter their characteristics (Kesselmeier and Staudt, 1999; Kulmala et al., 2004; Poschl, 2005; Rudich et al., 2007). These are known as secondary processes, in which primary aerosols react with radicals or oxidants that can change the physical and chemical properties of their surfaces as well as their bulk composition (Finlayson-Pitts, 1999; Rudich, 2003; Seinfeld and Pandis, 2006; Rudich et al., 2007). Aerosols also serve as condensation nuclei for semivolatile organic compounds and acids forming coated or mixed aerosols, which can significantly alter their optical properties (Schwarz et al., 2008; Lack et al., 2009), such as their absorption or water content. For example, it has been shown recently that the absorption of light by soot aerosols can be significantly enhanced, due to the presence of an outer layer of nonabsorbing compounds (i.e., coating) (Saathoff et al., 2003; Bond and Bergstrom, 2006; Schwarz et al., 2008; Adler et al., 2010; Lack et al., 2009). The coating of soot by organics, for example, occurs primarily during its release into the atmosphere and condensation of semivolatile and nonvolatile organic compounds when released from car exhaust (Robinson et al., 2007).

Aerosols can both absorb and scatter solar radiation. Aerosols that only scatter light contain mainly inorganic species: inorganic salts, sulfate aerosols, and saturated carboxylic hydrocarbons (e.g., glutaric acid). These aerosols exert a net negative effect on the radiative forcing, often referred to as the "cooling effect," Aerosols that scatter and absorb light are classified into two different types: weakly absorbing and strongly absorbing aerosols. The first type includes organic carbon (OC) aerosols emitted from fossil fuel and biofuel burning, also known as "brown carbon" (Jacobson et al., 2000; Gelencser, 2004; Lund Myhre and Nielsen, 2004; Maria et al., 2004; Andreae and Gelencser, 2006; Decesari et al., 2006; Hoffer et al., 2006). OC aerosols can also be formed by natural biogenic emissions of either primary aerosols (aerosols directly emitted into the atmosphere) or by formation of secondary aerosols from the condensation of semivolatile organic matter onto preexisting aerosols leading to the formation of new aerosols with different physical and chemical properties. Therefore, these aerosols can have complex chemical composition. Dust aerosols can also absorb solar radiation, mostly at shorter wavelengths, which can cause them to undergo chemical transformation in the atmosphere that affect their optical properties (Sokolik and Toon, 1996; Sokolik et al., 1998; Laskin et al., 2005). The second type, strongly absorbing aerosol, consists mainly of black carbon (BC), a product of incomplete combustion, particularly of coal, diesel fuel, and biomass burning generated by industrial activities, transportation, fires, and household burning. This is the strongest absorber of visible and near ultraviolet light among aerosols (Schnaiter et al., 2003; Bond and Bergstrom, 2006; Decesari, 2006; Ramanathan and Carmichael, 2008; Chow et al., 2009).

A major gap in quantifying aerosols' direct climatic effect is an insufficient knowledge of their optical properties. Improved aerosol characterization is required, which can be partially achieved by improving *in situ* and laboratory measurements of their optical properties. This requires fast and accurate measurements that can address the uncertainty associated with fast changes in the chemical and physical properties of aerosols in the atmosphere. In addition, it is necessary to study how optical properties change by various atmospheric processes (reaction, water adsorption, etc.) and how to calculate the optical properties of more complex aerosols. Such studies are usually carried out in laboratory conditions. Recently, cavity ring-down spectroscopy (CRD-S) has proved to be a robust technique for measuring the optical properties of aerosols. This technique was extensively used in different setups for both field and laboratory measurements and showed excellent agreement, within a few percent, with theoretical values of the optical properties of different known aerosols.

In this chapter, we introduce the CRD aerosol spectrometry (CRD-AS) technique, discuss recent experimental progress, and establish the theoretical framework needed for understanding the optical parameters studied in the CRD technique. The basis of this theoretical framework is Mie theory, which is the analytical solution of Maxwell's equations for scattering and the absorption of electromagnetic radiation by spherical particles of known refractive index and size at a specific wavelength.

11.2 INTERACTION OF AEROSOLS WITH SOLAR RADIATION AND MIE THEORY

Solar radiation reaching the Earth's atmosphere interacts with gases and aerosols. These interactions lead to attenuation of the incident light. The attenuation (extinction) is expressed as $\alpha_{ext} = \alpha_{abs} + \alpha_{scat}$, in which α_{ext} is the extinction coefficient, α_{abs} and α_{scat} are the absorption and scattering coefficients, respectively (usually they are expressed by common units, cm^{-1}, m^{-1} or Mm^{-1} (10^{-6} m^{-1})). The key to understanding the aerosols' direct effect on climate is through the determination of their optical properties, such as the single scattering albedo (SSA) (ω), which is the ratio of the scattered light to the total attenuated (extinction) light and expressed as $\omega = \alpha_{scat}/(\alpha_{abs} + \alpha_{scat})$, mass absorption and scattering efficiencies (E_{abs} and E_{scat}), the scattering phase function, $P(\theta)$, which is the angular distribution of light scattered by a particle at a given wavelength and the complex refractive index (m) (Bellouin et al., 2005; Bates et al., 2006; Chen et al., 2008). The complex refractive index is given by the equation: $m = n + ik$, where (n) is the real part and (k) is the imaginary part (normally called the optical constants) of the refractive index. The real part (n) indicates refraction and is defined as the ratio between the speed of light in vacuum (c) to the material (v), and is given by (c/v), while the imaginary part (the imaginary term) is related to the absorption. The absorption coefficient of the material is related to the imaginary part of the complex refractive index via the relation (Bohren and Huffman, 1998):

$$\alpha_{abs} = \frac{4\pi k}{\lambda} \tag{11.1}$$

where λ is the wavelength of the incident light. The refractive index of aerosols is defined as the ratio between the complex refractive indices of the aerosols and the surrounding medium (the surrounding medium is usually considered to be $m = 1$ for aerosols, since the surrounding medium is air). These parameters depend on the wavelength of the incident light, the chemical composition, and physical properties (such as shape and layers of materials) of aerosols.

In the process of absorption, incident light is converted into thermal energy, whereas scattering is the process in which light is diffracted, refracted, and reflected. The attenuation of light due to these two processes can be determined by the Beer–Lambert law, which describes the exponential decrease of the light traveling through a medium. This attenuation can be quantitatively expressed by

$$T(l) = \frac{I(l)}{I_0} = e^{-\alpha_{ext}l} \tag{11.2}$$

where $T(l)$ is the transmitted light passing through a distance l; I_0 and $I(l)$ are the intensities of the incident and transmitted light, respectively. The rate at which the intensity of the light decreases is proportional to α_{ext}. The total extinction in the atmosphere is defined by the sum of the total extinction due to gases and the total extinction due to aerosols ($\alpha_{ext(total)} = \alpha_{ext(g)} + \alpha_{ext(aerosols)}$). Therefore, the total extinction due to the gases and aerosols can be expressed as

$$\alpha_{ext(total)} = \alpha_{scat(g)} + \alpha_{abs(g)} + \alpha_{scat(aerosol)} + \alpha_{abs(aerosol)} \tag{11.3}$$

Here, $\alpha_{scat(g)}$ and $\alpha_{abs(g)}$ are the scattering and absorption coefficients of the gases and $\alpha_{scat(aerosols)}$ and $\alpha_{abs(aerosols)}$ are the scattering and absorption coefficients of the aerosols. The amount of scattering and absorption of light by gases in the atmosphere is well defined, due to the extensive measurements performed on gases in the atmosphere. Furthermore, the magnitude of the scattering of incident light due to gases is negligible compared to that of aerosols. Therefore, in this chapter we will only discuss the attenuation caused by aerosols, theoretically and experimentally.

The extinction coefficient is a function of the wavelength of the incident light and the particles' concentration and is given by

$$\alpha_{ext} = \sigma_{ext}(\lambda)N \tag{11.4}$$

where σ_{ext} is the extinction cross section, and N is the number concentration of the aerosols. The common units used for σ_{ext} are cm^2 or m^2.

The extinction cross section (σ_{ext}) is a more useful parameter, since it is the extinction normalized by the particle concentration (α_{ext}/N). Similar to the extinction coefficient, the extinction cross section is the sum of the absorption cross section (σ_{abs}) and the scattering cross section (σ_{scat}):

$$\sigma_{ext} = \sigma_{abs} + \sigma_{scat} \tag{11.5}$$

The extinction, scattering, and absorption cross sections, σ_{ext}, σ_{scat}, and σ_{abs}, can be normalized by their geometrical cross-sectional areas to yield dimensionless quantities, termed efficiency of extinction, scattering, and absorption:

$$Q_{ext} = \frac{\sigma_{ext}}{G}, \quad Q_{scat} = \frac{\sigma_{scat}}{G}, \quad Q_{abs} = \frac{\sigma_{abs}}{G} \tag{11.6}$$

where G is the geometrical cross section (equal to πr^2 for spherical particles in which r is the particle radius). Similar to the extinction coefficient (α_{ext}) and extinction cross section (σ_{ext}), the extinction efficiency is given as the sum of the absorption and scattering efficiency factors ($Q_{ext} = Q_{abs} + Q_{scat}$). The extinction efficiency (Q_{ext}) is usually a more convenient way for the presentation of the optical properties of aerosols than the extinction cross section. At a given wavelength, Q_{ext} is a function of only the size parameter (x), which is the ratio between the size of a particle and the wavelength of the incident light and can be simply given by

$$x = \frac{2\pi r}{\lambda} \tag{11.7}$$

where (r) is the particle radius and λ is the wavelength of the incident light. The size parameter (x) is significant, because the value of this parameter can immediately determine the scattering regime. Light scattering by particles falls into one of three different regimes, depending on the value of the size parameter. The first regime is called the *Rayleigh scattering* regime. In this regime, the size of the particles is small compared to the wavelength of the incident light and therefore, the size parameter values are $x \ll 1$. The extinction and scattering efficiencies (Q_{ext} and Q_{scat}) for monodispersed particles are given by (Bohren and Huffman, 1998; Seinfeld and Pandis, 2006)

$$Q_{ext} = 4x \, \mathrm{Im}\left\{\frac{m^2-1}{m^2+2}\left[1+\frac{x^2}{15}\left(\frac{m^2-1}{m^2+2}\right)\frac{m^4+27m^2+38}{2m^2+3}\right]\right\} + \frac{8}{3}x^4 \, \mathrm{Re}\left\{\left(\frac{m^2-1}{m^2+2}\right)^2\right\} \tag{11.8}$$

$$Q_{scat} = \frac{8}{3}x^4 \left| \frac{m^2 - 1}{m^2 + 2} \right|^2 \tag{11.9}$$

where m is the complex refractive index, and Im and Re denote the imaginary and real parts of the terms in the brackets. From the relation of the efficiency factors ($Q_{ext} = Q_{abs} + Q_{scat}$), the absorption efficiency can be simplified to

$$Q_{abs} = 4x \, \mathrm{Im} \left| \frac{m^2 - 1}{m^2 + 2} \right|^2 \tag{11.10}$$

If the term $[(m^2 - 1)/(m^2 + 2)]$ is a weak function of the wavelength, then Q_{scat} is proportional to x^4 or (λ^{-4}) and Q_{abs} is proportional to x or (λ^{-1}).

The second regime of scattering, called the geometric regime, is the case when particles are much larger than the incident wavelength ($x \gg 1$). The scattering of such large particles ($r > 4\ \mu m$) in the visible region can be calculated according to geometrical optics. For compact particles such as spheres, the extinction efficiency approaches a limiting value of 2 over the entire range of wavelengths: $\lim_{r \to \infty} Q_{ext} = 2$. This value is twice the value calculated by geometrical optics; this is referred to as the "extinction paradox" (Bohren and Huffman, 1998; Zakowicz, 2002). The extinction paradox is explained by the fact that for large particles, two different phenomena take place: diffraction and the geometrical optics effects of reflection, refraction, and absorption. The third regime, called the Mie scattering regime, is where the particle diameter is approximately the same size as the incident wavelength.

The Rayleigh and geometrical regimes (very small size parameters and very large size parameters) are not enough to describe any given particle size. Therefore, for isotropic and homogenous spherical particles of any given size, the extinction and scattering efficiencies can be obtained from more complicated calculations, developed by Mie. The Q_{ext} and Q_{scat} are given by (Bohren and Huffman, 1998, Seinfeld, 2006)

$$Q_{ext} = \frac{2}{x^2} \sum_{n=1}^{\infty} (2n+1)\mathrm{Re}(a_n + b_b)$$

$$Q_{scat} = \frac{2}{x^2} \sum_{n=1}^{\infty} (2n+1)\left(|a_n|^2 + |b_n|^2 \right) \tag{11.11}$$

where a_n and b_n are the Mie scattering coefficients or amplitude functions, which are a function of the size parameter and complex refractive index (m) and can be calculated with complex Bessel–Riccatti function (Bohren and Huffman, 1998; Scholz et al., 1998; Cox et al., 2002).

The Q_{ext}, Q_{scat}, and Q_{abs} are usually calculated as a function of the size parameter at a fixed complex refractive index (m). To understand the behavior of Q_{ext} as a function of size parameter and complex refractive index, we will discuss two different types of aerosols: the first, nonabsorbing aerosols ($k = 0$), in which the real part (n) of the refractive index is different. Alternatively, for the second type, n is fixed and k varies for different aerosols. To demonstrate the behavior of Q_{ext} as a function of the size parameter for spherical particles of the first type, we simulate two different curves with relative refractive indices of $m = 1.3 + i0.0$ and $m = 1.6 + i0.0$ at wavelength 532 nm that are presented in Figure 11.1a. The imaginary part of the refractive indices is equal to zero. Therefore, $Q_{ext} = Q_{scat}$ and Q_{abs} equals zero. It can be seen that Q_{ext} rapidly vanishes for small-size parameters, while Q_{ext} approaches the asymptotic value 2 at large-size parameter values. In the intermediate region the Q_{ext} value can exceed the geometrical-optics value and it increases as n increases: for

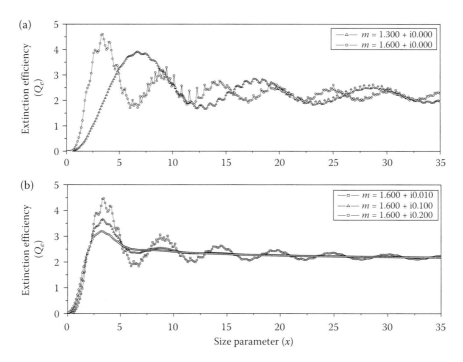

FIGURE 11.1 The extinction efficiency Q_{ext} factor versus the size parameter (x) for spherical particles at 532 nm. (a) Nonabsorbing particles with relative refractive indices $m = 1.300 + i0.000$ and $1.600 + i0.000$. (b) Absorbing particles with constant real part of the refractive index and variable imaginary part (0.010, 0.100 and 0.200).

example, in Figure 11.1a for $n = 1.3$, Q_{ext} reaches a maximum value of 3.919 while for $n = 1.6$ it reaches a maximum value of 4.615. The Q_{ext} as a function of size parameter shows several oscillations of maxima and minima and it approaches asymptotically the limiting value of 2 due to the "extinction paradox," described previously. These oscillations are a result of the interference between near-forward transmission and diffraction. The first peak is always higher than the following peaks and the subsequent peaks become lower and lower. In addition to the large oscillations, a ripple structure (sharp and small) is observed in each peak. These ripples are caused by the resonance behavior of a_n and b_n Mie scattering coefficients. It is important to note that it is not possible to observe these ripples experimentally due to the size distribution of the spherical particles, which leads to a distribution in the size parameter.

The dependence of Q_{ext} on the imaginary part (k) of the refractive index is presented in Figure 11.1b. Three different values of k are taken (0.01, 0.1, and 0.2) to examine the dependence of the curve on the imaginary part. The ripples structure in each peak is rapidly diminished when the values of k slightly increase; however, the shape of the peaks remains unchanged. When the values of k increase significantly, the resonance peaks are also eliminated.

Recently, CRD-AS was extensively used to study optical properties of aerosols and more specifically to measure Q_{ext} as a function of size parameters (x) of different aerosols to determine the complex refractive index (m). Other applications of the CRD-AS will be detailed later. In the following sections, we will introduce this technique and scrutinize recent work which has given rise to the CRD-AS technique as one of the most widely used for studying the optical properties of aerosols.

11.3 CRD-S: METHODOLOGY AND APPLICATIONS

CRD-S is a highly sensitive technique for the direct measurement of absorption (extinction) of gases and particles suspended in gas (e.g., aerosols). It was first developed over two decades ago by O'Keefe and Deacon, who performed highly sensitive gas-phase spectroscopy (O'Keefe and Deacon, 1988). The early applications of CRDS were mainly for spectroscopy, photochemistry, trace gas detection, environmental analysis, weak absorption transitions measurements, and kinetic studies (Romanini and Lehmann, 1995; Scherer et al., 1995; Paul et al., 1997; Scherer and Rakestraw, 1997; Cheskis et al., 1998; Campargue et al., 1999; Campargue et al., 2006) (Jongma et al., 1995; Wheeler et al., 1997; Newman et al., 1998; Howie et al., 2000; Kraus et al., 2002) (Fawcett, 2002) (Yu and Lin, 1994; Atkinson and Hudgens, 1997; Atkinson and Hudgens, 1999; Atkinson et al., 1999; Atkinson and Hudgens, 2000; Brown, 2002; Brown et al., 2002; Mazurenka et al., 2003; Parkes et al., 2003; Crunaire et al., 2006; Pradhan et al., 2008). There are a number of comprehensive reviews describing the technique (Scherer, 1997; Busch and Busch, 1999; Wheeler et al., 1998; Berden et al., 2000; Berden et al., 2000; Atkinson, 2003; Brown, 2003; Vallance, 2005). Recently, it has been widely adapted for studying atmospheric aerosols. The CRD system typically consists of two highly reflective plano-concave mirrors set opposite to one another to form a stable optical resonator. The absorption (extinction) is determined by measuring the exponential decay of the light leaking out of the resonator. By measuring the decay rate rather than the change in light intensity, which is normally used in conventional absorption spectroscopy techniques, the problem of the intensity fluctuations of the light source is eliminated. In addition, the CRD technique takes advantage of the very long effective absorption path length, which can reach up to several kilometers for a cavity 1 m in length, making this method one of the most sensitive absorption spectroscopy techniques available. Within the last few years, CRD was modified, improved, and adapted for a wide range of studies. Herein, we will focus only on two main methods, the pulsed-CRD and the continuous wave (cw) CRD, both of which are widely used in studies of the optical properties of aerosols.

11.3.1 Pulsed-Laser CRD

Pulsed-CRD is the simplest configuration of CRD. A typical experimental setup of pulsed-CRD is shown in Figure 11.2. Normally, a 10 ns pulsed laser is coupled to a stable optical resonator composed of two highly reflective plano-concave mirrors with a radius of curvature (r) (the typical radius of curvature ranges from -0.25 m to -1 m, however, -6 m mirrors also exist). These two mirrors are separated by a distance L to form a stable cavity. For a stable cavity, L should be $0 < L < r$ or $r < L < 2r$. A small portion of the laser intensity enters the cavity from the front side. This amount depends on the reflectivity of the mirrors (R). The light injected into the cavity travels back and forth between the two mirrors. Each time the light performs a round trip, it loses intensity due to leaking from both sides of the cavity, generating an exponential decay. The number of round-trip passes (1) necessary for the light inside the cavity to equal $1/e$ of its initial intensity is given by

$$a = -\frac{1}{2\ln(R)} \tag{11.12}$$

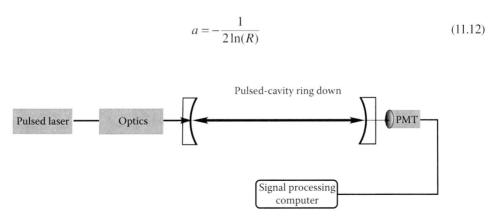

FIGURE 11.2 Schematic diagram showing the basic principles and the essential characteristics and components for simple pulsed-CRD.

where a is the number of round trip passes and R is the reflectivity of the mirrors. It is obvious that the decay time is dependent on the reflectivity of the mirrors and the distance between the mirrors. For example, if light is trapped inside a cavity with mirrors that have a reflectivity of 0.99995, the laser light experiences 10,000 round trip passes. The decay time of the laser pulse can be monitored by measuring the leaking light using a photomultiplier on the exit side of the cavity. In the absence of absorbers inside the cavity (empty cavity), the decay is mainly due to losses inside the cavity and through the mirrors.

The time-dependent light intensity transmitted through the mirrors is described by

$$I = I_o e^{(-t/\tau)} \tag{11.13}$$

The time constant for an exponential curve (τ) is defined as the time it takes to reduce the initial value by $1/e$. For an empty cavity, the decay time (τ_0) is described by

$$\tau_0 = \frac{L}{c(1-R)} \tag{11.14}$$

where L is the length of the cavity, c is the speed of light, and R is the reflectivity of the mirrors. This equation depicts the dependence of the ring-down time on the cavity length and the mirror reflectivity. Hence, with a larger cavity or higher reflective mirrors, the time constant becomes longer. Figure 11.3a represents a typical exponential decay curve. This decay was determined for an empty cavity using pulsed-CRD. The logarithmic scale of the decay (b) clearly shows a single exponential decay. The decay time can be extracted from the slope of this line by using simple linear fitting (108.27 μs

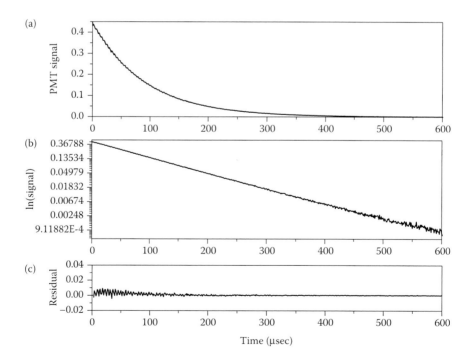

FIGURE 11.3 (a) Typical CRD decay obtained for an empty cavity (trace a). (b) The natural logarithm of the decay signal exhibiting a linear behavior as expected. The decay time measured for this cavity was 108.27 μs. (c) Represents the residual obtained from subtracting the decay signal from its fitting curve. Note the different scale values for the residual.

for this cavity). The difference between the decay and its fitting is shown in Figure 11.3c (note the scale of the residual). The decay time is directly related to the natural absorption coefficient of the optical cavity and a large number of reflections mean that the effective path length is now quite long. The other major advantage of CRDS concerns the elimination of light loss through fluctuations in the light source. A single pulse of light is the minimum requirement to form a ring down curve, which is the resulting exponential decay curve of the measured light pulse. Therefore, even if the source fluctuates from pulse to pulse, the decay time is not sensitive to these fluctuations.

These are good fundamentals for measuring the optical losses of mirrors and hence, the reflectivity of the mirrors, but what about the spectroscopy and optical properties of species? If the cavity is filled with an absorbing (or scattering) medium, an interesting effect is observed. In addition to light being absorbed (and transmitted) by the mirrors, the molecules or particles also reduce the intensity of the light on each pass through light absorption (extinction) at specific wavelengths. This would appear as a ring-down trace with a shorter time constant, due to additional terms in the ring-down expression, and the time constant will be described by

$$\tau = \frac{L}{c[(1-R)+\alpha_{ext}d]} \tag{11.15}$$

where α_{ext} is the extinction coefficient of the molecules or particles inside the cavity, and d is the actual length of the cavity filled with absorbing molecules or particles. The extinction coefficient (α_{ext}) is determined by subtracting the time constant of the empty cavity and the filled cavity. The subtracted equations can be written as

$$\alpha_{ext} = \frac{L}{cd}\left(\frac{1}{\tau} - \frac{1}{\tau_0}\right) \tag{11.16}$$

11.3.2 Continuous Wave Cavity Ring-Down (cw-CRD) Spectroscopy

Despite the simplicity of the pulsed-CRD system that was described above, in which the pulsed laser oscillates inside an optical resonator, this method suffers from several limitations. First, we have to consider the mode structure of the laser pulse. The intrinsic bandwidth of the laser for pulsed laser complicates the data analysis, due to mode beating and nonexponential decay. In addition, pulsed-CRD has a limited repetition rate. CW-CRD spectroscopy was introduced as a technique that overcomes these limitations by fast data acquisition that can reach 75 kHz as well as higher spectral resolution (Spence et al., 2000). The choice of using pulsed-CRD or cw-CRD depends on the application of interest. Pulsed-CRD can be applied to a wide range of wavelengths using dye lasers or optical parametric oscillators (OPOs) lasers. CW-CRD, however, offers higher spectral resolution and repetition rate; however, it has a limited wavelength range especially in the shorter wavelength (blue) region. Furthermore, cw-CRD adds an additional complexity when compared to the simple system of the pulsed-CRD, due to mode-matching limitations since the spectral overlap between the laser frequency and one of the cavities modes is not guaranteed anymore. This is due the narrow line width of the cw laser and the high finesse of the cavity. A power buildup in the selective mode of the cavity can be obtained by mode matching the cw laser with cavity modes. This increases the intensity of the laser power leaking out of the cavity and ease the detection of the decay time. To obtain mode matching, several approaches are used, but only one, introduced by Romanini et al. and used by many others for different applications including cw-CRD for investigating optical properties of aerosols, is described here. Figure 11.4 represents a typical cw-CRD spectrometer utilizing Romanini's method (Romanini et al., 1997). This cavity consists of two mirrors one of which is fixed on a mirror holder and the second moves along the cavity axis by a piezo-electric

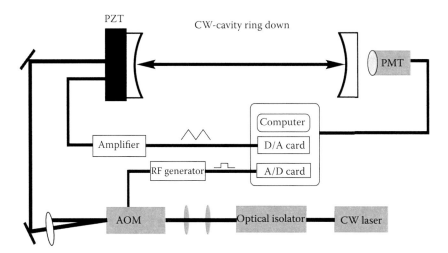

FIGURE 11.4   Schematic diagram showing the basic principles of one of the suggested cw-CRD setups.

transducer (PZT) which is driven by a 1 kHz signal with amplitude ~10 Vpp while maintaining stable alignment. A continuous laser beam is directed through an optical isolator and a telescope into an acousto-optic modulator (AOM) or a Pockels cell. The AOM is modulated by the buildup of light intensity in the cavity. The mirror mounted on the PZT travels back and forth at some frequency to scan the cavity length over a distance that is greater than $\lambda/2$ (λ is the laser wavelength) to modulate the cavity mode in order to achieve mode matching with the laser mode. When the length of the cavity matches the laser mode, light is efficiently trapped and the rapid intensity buildup is used to directly trigger the AOM that effectively blocks the laser beam. The exponential decay of the intensity in the cavity is digitized and processed for each "pulse." It is possible to achieve mode matching by utilizing only the changes in the cavity length brought by external vibrations. However, such mode matching relies only on accidental resonances in the cavity length, which are not controlled and therefore the previous approach using a PZT is more efficient.

Single-mode continuous sources are more difficult to utilize than pulsed sources. As the continuous light is directed into the cavity, intensity build up (constructive interference) occurs until a predetermined threshold is reached. At this point, the light must be interrupted and allowed to ring down. Mode matching is critical because of the appearance of transverse modes. In order to improve resolution (through exciting only longitudinal modes), mode matching optics must be used. The advantage of the continuous laser, however, is in its much higher sensitivity. One requirement of cw-CRD is that the noise introduced by the continuous laser must be much narrower than the line width of the cavity. It is projected that the limitation of continuous laser CRD comes in the form of the shot noise of the cavity. The other advantage of using a continuous source is that it is cheap and widely available, such as the laser diode. However, it is limited by a narrow range of tunable wavelengths.

11.3.3   Sensitivity and Detection Limit

Determining an accurate decay time is critical for precise measurements of the extinction coefficient using CRD. In pulsed-CRD systems, the transverse modes inside the cavity lead to nonexponential decay and results in inaccurate determination of the decay time. Two effects of transverse modes are commonly observed in CRD-S. The first modulates the decay as a result of multiple modes, which form different optical paths inside the cavity. To overcome this issue, it is possible to use a spatial filter for mode matching. The second effect is caused by variations in the quantum efficiency at the detector surface as the laser beam impinges on it. This is overcome by tight focusing of the laser beam on a small surface of the detector (Scherer et al., 1997).

The primary application of CRD spectroscopy is the detection of trace molecules and particles at very low concentrations, which cannot be detected using classical spectroscopic techniques. The detection limit of a CRD system is dictated by the ability to measure precisely the minimum difference between τ_0 and the measured τ, given by the following equation Pettersson et al., 2004:

$$\alpha_{min} = \frac{L}{C \cdot d} \cdot \frac{\sqrt{2} S \tau_0}{\tau_0^2 \sqrt{RT}} \qquad (11.17)$$

where L is the length of the cavity, d is the actual part of the cavity filled with absorber molecules or particles, C is the speed of light, and $S\tau_0$ is the minimum detectable change in the ring-down time for one laser shot ($\tau_0 - \tau$) upon introducing molecules or particles to the cavity, R is the repetition rate and T is the sampling time. The sensitivity of the CRD system can be improved by using mirrors with higher reflectivity, which increases τ_0 and decreases $S\tau_0$, and by carrying out the experiments at higher repetition rates. Currently, a wide range (down to 8.8×10^{-12} cm^{-1} Hz$^{-1/2}$) of sensitivity is reported, which is highly dependent on the experimental setup (Spence et al., 2000).

11.4 APPLICATION OF CRD-AS FOR MEASURING THE EXTINCTION COEFFICIENT (α_{ext}), EXTINCTION CROSS SECTION (σ_{ext}) AND EXTINCTION EFFICIENCY (Q_{ext}) OF AEROSOLS

CRDS has been recently employed for measuring extinction coefficients of laboratory and field aerosols. Sappy et al. pioneered the use of CRD for detecting ambient particles nonresonantly at 532 and 355 nm. They used a highly reflective mirrors at 532 nm ($R = 99.9986\%$) obtaining 107.2 μs for an empty cavity and $R = 99.99\%$ for 355 nm cavity. They were able to detect as low as 200 particles/L of sizes 100 nm and smaller (Sappey et al., 1998). Although this experiment was not quantitative, it was the first to demonstrate the ability of CRD in aerosol studies. Vander Waal and Ticich used pulsed-CRD to study the absorption of soot produced from methane-air flame and to calibrate laser-induced incandescence measurements, which is widely used to measure soot volume fraction (Vander Waal and Ticich, 1999). The first measurements of atmospheric aerosols in the environment were performed by Smith et al. A dual wavelength CRD (532 and 1064 nm) was used during two different days; the first day was hazy with relatively short visual range, while the second day was clear. On both days, clear air measurements were used as the baseline (only Rayleigh scattering contributed to the extinction). Aerosols were the major contributor to the extinction on the hazy day, while on the clear day their contribution was significantly less pronounced. In addition, the extinction was measured for the ambient air using different single stage impactors for size selection (Figure 11.5) (Smith and Atkinson, 2001). Recently, this system was validated in laboratory experiments by measuring the extinction coefficient of different aerosols and was combined with nephelometer for measuring the scattering coefficient in order to directly measure the SSA (ω) (James G. Radney, 2009). Using simultaneous measurements at 510.6 and 578.2 nm, Thompson et al. monitored the change in atmospheric optical extinction coefficient during a wildfire and local fireworks event (Thompson et al., 2002). Baynard et al. designed a pulsed-CRD-AS for field measurements capable of measuring simultaneously and independently the extinction of atmospheric aerosols at 4 different wavelengths (1064, 532, 355, and 683 nm). They used this system to measure the extinction due to aerosols at the New England Air Quality Study–Intercontinental Transport and Chemical Transformation 2004 (NEAQS-ITCT 2004). In this campaign, they separated the contribution of absorbing gases such as O_3 and N_2O at 532 nm from the extinction of aerosols. They achieved that by measuring periodically the sample passing either directly to the CRD or through a filter that removes the aerosols (Baynard et al., 2007).

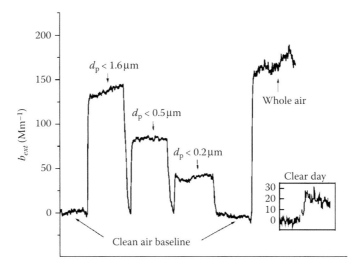

FIGURE 11.5 CRD measurements at 532 nm of the extinction of ambient aerosols coming out of different impactor stages on a hazy day. The inset represents the measurements of the extinction in a clear day for clean air (no particles exist) baseline and whole air (including particles). (Reproduced from *The Analyst*, 2001, 126, p. 1219. Copyright 2001, The Royal Society of Chemistry. With permission.)

Strawa et al. (2003) were the first to use a cw-CRD for aerosol studies. Using diode lasers at 690 and 1550 nm, they measured a minimum extinction coefficient for both wavelengths of about 1.5×10^{-8} cm^{-1} (better sensitivity could be achieved with higher reflectivity mirrors). By placing a scattering detector perpendicular to the cavity, they measured the extinction coefficient and the scattering coefficient and directly extracted the SSA (which is the ratio between scattering efficiency and the total extinction, a parameter often used in climate calculations). This instrument was improved for the uses in airborne aerosol optical properties measurements and is called Cadenza (Hallar et al., 2006). Lang Yona et al. used a single mode continuous laser to determine the extinction efficiencies (Q_{ext}) as a function of size of different types of model aerosols to retrieve the complex refractive index. The system was operated at 400 Hz with a detection limit of 6.67×10^{-10} cm^{-1} (Lang-Yona et al., 2009). It was later deployed to measure the optical properties of organic secondary aerosols formed from Hom Oak emissions in a plant chamber. (Lang-Yona et al., in preparation) It was found that these fresh aerosols do not absorb and are good scatterers of solar radiation.

Bulatov et al. used a pulsed dye laser at 620 nm to study laboratory-generated nonabsorbing NaCl and $CuCl_2 \cdot H_2O$ aerosols. The measured extinction coefficients were compared to Mie scattering calculations, but there was no attempt to determine the complex refractive index (Bulatov et al., 2002). They also measured the extinction coefficient of size-selected Rhodamine 640 aerosols (a strongly absorbing dye at 615 nm). This was the first use of CRD to measure optical properties of absorbing organic aerosols (other than soot) (Bulatov et al., 2006). Lack et al. applied CRD to derive aerosol absorption of laboratory-generated aerosols and compared the obtained absorption with photoacoustic spectroscopy measurements. They used Nigrosin as a model of strongly absorbing aerosols and highly spherical particles (Lack et al., 2006). Moosmuller et al. used CRD to measure very low extinction in the atmosphere and laboratory environments (Moosmuller et al., 2005). Butler et al. have used CRD for studying single micro-sized aerosols particles (Butler et al., 2007; Miller and Orr-Ewing, 2007). Rudić et al. used CRD to study the extinction large of water droplets of 30–70 μm at 560 nm. They were able to measure the oscillations in the extinction efficiency as a function of the size parameters and to determine the size of the particles. In this study, they needed to introduce a correction factor for the forward scattering in order to have a better match between the experimental and theoretical value of the losses per pass in the CRD (Rudic et al., 2007). This list of studies shows the wide array of applications that CRD-AS has been used for in both field and laboratory settings in order to determine the optical properties of single-compound or complex aerosols.

11.4.1 MEASURING THE COMPLEX REFRACTIVE INDEX (m) OF AEROSOLS BY CRD-AS

The complex refractive index (m) of aerosols can be determined by measuring the extinction efficiency (Q_{ext}) of the aerosols as a function of the size parameter (x), and comparing the measurements to theoretical predictions from Mie theory or from T-Matrix calculations. As previously shown, the relation between the extinction efficiency and size parameter depends on the complex refractive index, and therefore, by measuring Q_{ext} as a function of size parameter it is possible to retrieve the complex refractive index of the aerosols by fitting the Mie theory curve. The fit is done by optimizing the complex refractive index until the calculated curve has the best match to the measured points. To record the size dependent extinction, the CRD-AS system should be integrated with aerosol source generation or sampling, a differential mobility analyzer (DMA) for selection of a narrow size distribution and a condensation particle counter (CPC) to measure the concentration of the particles. Figure 11.6 represents a scheme of this setup for determining the complex refractive index.

In laboratory experiments, an aqueous solution (50–500 mg L^{-1}) of the compound of interest is nebulized using a constant output atomizer with dry, particle-free pure nitrogen, generating a polydisperse distribution of droplets. Following a conditioning bulb and drying in a silica gel column, the dry polydisperse aerosol passes through a neutralizer to obtain an equilibrium charge distribution on the particles. Next, a size-selected monodisperse aerosol is generated with an electrostatic classifier (Differential Mobility Analyzer (DMA)). The size-selected monodisperse aerosol flow is directed then to the CRD-AS cell. To maintain clean mirrors, a small purge flow of dry particle-free nitrogen (0.05 SLM) is introduced in front of each mirror to prevent mirror contamination by deposition of the aerosols. The particles exit the cavity and their concentration is determined by a CPC.

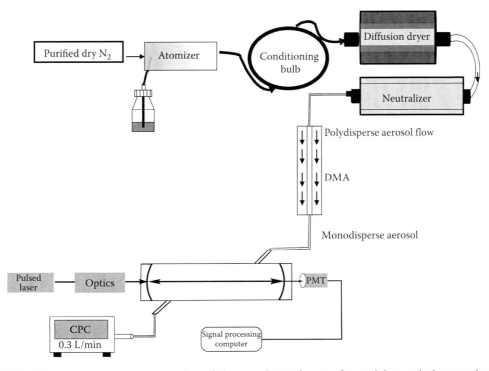

FIGURE 11.6 A schematic representation of the experimental setup for studying optical properties of aerosols and determining the complex refractive index of various aerosols.

The actual length of the cavity filled with particles (*d*) ranges from 50 to 90 cm. In such a design, the particle losses are negligible and can be determined by measuring the particle number density after the DMA and at the exit of the CRD cell.

Petterson et al. demonstrated the use of pulsed laser (532 nm) CRD to study polystyrene spheres (PSS) and dioctyl sebacate (DOS) aerosols (Pettersson et al., 2004). These experiments were the first attempt to obtain quantitative measurements of known optical properties of nonabsorbing aerosols and to compare them to Mie calculations. Good agreement in the scattering cross section and the refractive index (within 5% error) was reported (Pettersson et al., 2004). These were followed by experiments in which the complex refractive index was determined for unknown absorbing and nonabsorbing aerosols by a retrieval algorithm (Riziq et al., 2007; Spindler et al., 2007; Dinar et al., 2008). Dinar et al. used this method to determine the complex refractive index of aerosols composed of complex organic matter that was extracted from biomass burning and pollution aerosols (Dinar et al., 2006). The retrieval algorithm for single-component particles compares the measured extinction efficiency with the extinction efficiency calculated using a Mie scattering subroutine for homogeneous spheres developed by Bohren and Huffman (Bohren and Huffman, 1983), while simultaneously varying the real and imaginary refractive indices of the particles. The algorithm converges to the appropriate refractive by minimizing the "merit function" χ^2/N^2, where χ^2 is

$$\chi^2 = \sum_{i=1}^{N} \frac{\left(Q_{\text{ext measured}} - Q_{\text{ext calculated}} \right)_i^2}{\varepsilon_i^2} \tag{11.18}$$

N is the number of particle sizes, and ε is the estimated error in the measurement (taken as the standard deviation) (Press et al., 1992). The algorithm does not require an initial guess for the real and imaginary parts of the refractive index. Rather, it scans through all possible physical values of the indices and progressively increases the resolution of the search until it finds the absolute minimum in the merit function within the desired precision. Figure 11.7 presents the retrieved complex refractive indices of ammonium sulfate $(NH_4)_2SO_4$, sodium chloride (NaCl) and glutaric acid aerosols at 532 nm. All these aerosols are considered nonabsorbing aerosols in the visible and UV-range. In addition, the retrieved complex refractive index of Nigrosin (a model material for strongly absorbing aerosols) is also presented. The circles in each curve represent the experimental values of the Q_{ext} measured for different sizes of each aerosol type. The solid lines represent the best fit of the Mie theory using the retrieval algorithm. These experiments demonstrate that the CRD method can be applied for determining the complex refractive index of aerosols by comparing the measurements to a calculated Mie theory prediction. This approach lays two very strong assumptions: that the particles are spherical (or nearly spherical) and that the chemical composition of the aerosols is constant throughout the size distribution.

11.4.2 HOMOGENOUSLY MIXED AEROSOLS

The experiments above used single component aerosols. Further experiments were performed to determine the refractive index of complex aerosols, such as homogenously mixed aerosols, to test optical mixing rules calculations that are often used in atmospheric modeling for calculating refractive index of aerosols. A number of methods for calculating the radiative properties of aerosols of internally mixed composition (i.e., homogenously mixed components) are used in climate models. For example, a growth function estimated from measurements or from Mie calculations may be applied to describe the change in scattering coefficient as aerosol water content increases (Bates et al., 2006). Alternatively, Mie scattering calculations are employed explicitly during the simulation

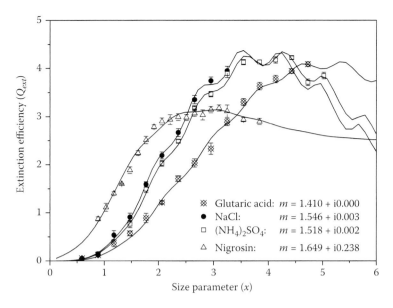

FIGURE 11.7 The measured and calculated extinction efficiency versus the size parameter of glutaric acid, NaCl, $(NH_4)_2SO_4$, and Nigrosin at 532 nm. The complex refractive indices were determined by fitting the Mie theory curve to the measured Q_{ext}.

or in a look-up table fashion, using mixing rules to calculate the effective refractive indices of the mixture or assuming a core-plus-shell configuration (Jacobson, 2002; Erlick, 2006). The mixing rules currently in use will be described. The first rule refers to molar refraction and absorption (Stelson, 1990; Tang, 1997; Born and Wolf, 1999; Jacobson, 2002). For the molar refraction (absorption) mixing rule, it is assumed that the total molar refraction (absorption) of a mixture is given by the linear average of the molar refraction (absorption) of the components in the mixture weighted by their molar volumes, that is

$$R_{tot} = V_{tot} \frac{n_{tot}^2 - 1}{n_{tot}^2 + 2} = \chi_1 R_1 + \chi_2 R_2 \tag{11.19}$$

$$A_{tot} = V_{tot} K_{tot} = \chi_1 A_1 + \chi_2 A_2 \tag{11.20}$$

where R_{tot}, V_{tot}, n_{tot}, A_{tot}, and K_{tot} are the molar refraction, molar volume, real part of the refractive index, molar absorption, and imaginary part of the refractive index of the mixture, respectively, and χ_i, R_i, and A_i are the molar fraction, molar refraction, and molar absorption of the components, respectively. The molar refraction of the components, molar absorption of the components and total molar volume are given by

$$R_i = \frac{M_i}{\rho_i} \frac{n_i^2 - 1}{n_i^2 + 2}, \quad A_i = \frac{M_i}{\rho_i} k_i, \quad V_{tot} \frac{M_{tot}}{\rho_{tot}} \tag{11.21}$$

where M is the molecular weight and ρ is the density.

Second, volume-weighted linear average of the refractive indices is often used, that is, a "linear" mixing rule (d'Almeida and Shettle, 1991). The "linear mixing rule" assumes that the total real and imaginary refractive indices of the mixture are given by the linear average of the indices of the components weighted by their volume fractions:

$$n_{tot} = f_1 n_1 + f_2 n_2 \tag{11.22}$$

$$k_{tot} = f_1 k_1 + f_2 k_2 \tag{11.23}$$

where f_i is the volume fraction of the components.

The third mixing rule used is the Maxwell–Garnett rule in which one or more of the components (usually the undissolved and/or absorbing components) are deemed "inclusions," while the rest of the components comprise a "homogeneous matrix" (Bohren and Huffman, 1983; Chylek et al., 1984). The inclusions are assumed to be small (dipoles), spherical, randomly distributed throughout the aerosol, and dilute, such that the effective dielectric constant of the mixture is given by

$$\varepsilon_{tot} = \varepsilon_{matrix} + \frac{3 f_{incl} \varepsilon_{matrix} (\varepsilon_{incl} - \varepsilon_{matrix})}{\varepsilon_{incl} + 2\varepsilon_{matrix} - f_{incl} (\varepsilon_{incl} - \varepsilon_{matrix})} \quad n_{tot} = (\varepsilon_{tot})^{0.5} \tag{11.24}$$

where ε_{tot}, ε_{incl}, and ε_{matrix} are the complex dielectric constants of the mixture, the inclusions, and the matrix, respectively, f_{incl} is the volume fraction of the inclusions, and n_{tot} is the complex refractive index of the mixture. Finally, the dynamic effective medium approximation mixing rule is an example of a higher order or extended effective medium approximation as compared to the Maxwell–Garnett mixing rule, where the inclusions are allowed higher order effects than the electric dipole (Chylek et al., 2000; Jacobson, 2006). The size or size distribution of the inclusions themselves must be specified.

For testing such models, three different approaches were taken in a set of laboratory experiments using CRD-AS:

1. Two nonabsorbers (NaCl and glutaric acid) are mixed together.
2. A weakly absorbing component (Suwannee River Fulvic acid (SRFA)) (a model material of humic-like substances (HULIS)) is mixed with nonabsorbing material (($NH_4)_2SO_4$) (Dinar et al., 2007, 2008; Lang-Yona et al., 2010).
3. A nonabsorbing compound (($NH_4)_2SO_4$) is mixed with a strongly absorbing compound (Rhodamine 590 (strong absorber at 532 nm)).

Figure 11.8 shows the Q_{ext} of mixtures of nonabsorbing aerosols generated by mixing NaCl and glutaric acid with molar ratios 1:1 and 1:2, and of pure aerosols of each compound as a function of size parameter. The experimental results are shown as circles with a standard deviation determined by repeated measurements for each aerosol type. The solid lines represent the retrieved best fit for Mie theory (Riziq et al., 2007).

The experiment was carried out using a pulsed-CRD aerosol spectrometer employing a 532 nm Nd:YAG laser. The retrieved complex refractive indices (m) of the pure aerosols and those with a variety of mixed ratios were used to test the performance of the different mixing rules and to find out which one provides the best match to experimental measurements through comparison of the merit function (χ^2/N^2) as a result of each mixing rule. For such mixtures it was shown that the linear mixing rule provides the lowest merit function (0.14) for each mixture compared to other methods. This is attributed to the good mixing of the soluble compounds resulting in good agreement with the simple linear mixing rule.

The second type of mixture (nonabsorbing ($NH_4)_2SO_4$ and weakly absorbing SRFA) was also tested by a similar approach described above. These mixed aerosols imitate aerosols composed of inorganic and organic mixtures often detected in urban and polluted areas. The experiment was

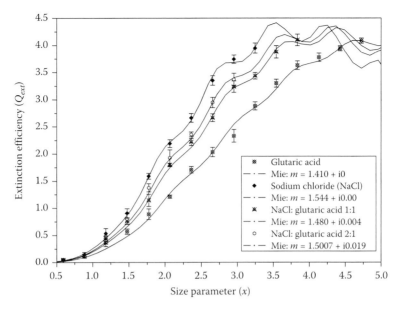

FIGURE 11.8 The measured and calculated extinction efficiency versus the size parameter of glutaric acid, NaCl, and their homogenous mixture at mass ratios 1:1 and 1:2.

carried out at 390 nm using an optical parametric oscillator (OPO), since SRFA has higher absorption in the UV region as implicit in the imaginary part of its refractive index ($m = 1.602 + i0.098$ at 390 nm, while $RI = 1.634 + i0.004$ at 532 nm) (Dinar et al., 2008). Figure 11.9 shows the measured Q_{ext} as a function of size parameter, of aerosols composed of a mixture of $(NH_4)_2SO_4$ and SRFA at 1:1 ratio. The solid line represents the fit of the Mie curve obtained by applying the linear mixing

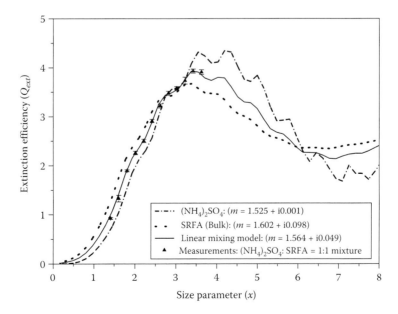

FIGURE 11.9 The measured and calculated extinction efficiency versus the size parameter $(NH_4)_2SO_4$ and SRFA and their mixture at mass ratio a 1:1. The Mie curve correspond to the effective refractive indices calculated with the volume-weight "linear mixing" rule at 390 nm wavelength.

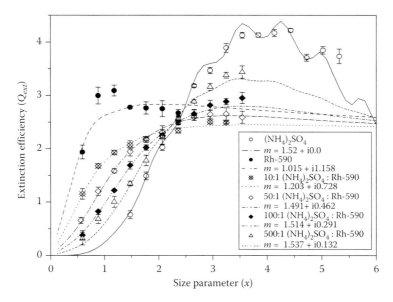

FIGURE 11.10 The measured and calculated extinction efficiency (Q_{ext}) as a function of size parameter obtained for different mixtures of ammonium sulfate and Rhodamine 590 with molar ratios (10:1, 50:1, 100:1, and 500:1, respectively).

rule on this mixture. The dashed and dotted lines represent the calculated Mie curves of $(NH_4)_2SO_4$ and SRFA, respectively.

Herein again, the simple linear mixing rule provides the best match to the experimental Q_{ext} values, but all of the mixing rules tested provide reasonable results; for a 1:1 mixture of $(NH_4)_2SO_4$ and SRFA, the merit function value for the linear mixing rule was very close to the merit function of other mixing rules.

The third type of aerosols mixture is composed of nonabsorbing and strongly absorbing molecules (Panne et al., 2001; Poschl, 2005; Karar et al., 2006; Riziq et al., 2007; Shah et al., 2007; Lang-Yona et al., 2010). Mixtures of different ratios of $(NH_4)_2SO_4$ and Rhodamine 590 were studied as a model of this aerosol type using pulsed-CRD at 532 nm. Figure 11.10 represents the measured Q_{ext} as a function of size parameter of these mixed aerosols and the retrieved complex refractive index (m). Unlike the previous mixtures, the merit function was considerably higher for all concentration ratios. Moreover, for low volume fractions of the absorber (Rhodamine-590), the Mie scattering subroutine for coated spheres was proved to perform better than the mixing rules. This surprising result was explained by the different solubility of the components in the mixture which can lead to formation of coated rather than mixed aerosols. Alternatively, the coated sphere model might represent better the interaction between the electromagnetic fields of the constituents in the case of a strong absorber with low volume fraction even if they are truly homogeneously mixed. For higher volume fractions of Rhodamine-590 in the mixture, the extended effective medium approximation provides the best agreement with measurements. The success of the extended effective medium approximation for high volume fractions of the strong absorber is related to the fact that the extended effective medium approximation retains the highest order representation of the electromagnetic fields, allowing the highest order description of the interaction between internal scattering and absorption.

11.4.3 Coated Aerosols

The coating of aerosols by a material different than the core material is a prevalent process in the atmosphere that can occur in several different ways. It can happen through condensation of semivolatile molecules on primary aerosols (e.g., condensation of organic molecules on soot particles) or by heterogeneous chemical reactions of the aerosol surface (e.g., photochemistry, radical reactions and oxidation) (Ellison et al., 1999; Posfai et al., 1999; Vaida et al., 2000; Grassian, 2001; Guazzotti et al., 2001; Laskin et al., 2003; Falkovich et al., 2004; Laskin et al., 2005; Zhang et al., 2005). The coatings can be either organic or inorganic, depending on the creation process. These coatings can alter the optical properties of the aerosols and therefore, it is crucial to understand how they can change the optical properties of the resulting aerosol (Lesins et al., 2002).

CRD-AS was used to study the effect of coating by measuring the Q_{ext} of coated aerosols as a function of size parameter. Primary aerosols were first generated and size selected, as described previously, and then directed into a heated reservoir saturated with the vapor of the species of interest. Following the coating process of these core aerosols, the aerosols were size-selected by a second DMA to select the thickness of the coated material. Finally, the extinction of these aerosols was measured by CRD-AS. These experiments were carried out with two types of samples:

1. Aerosols with the same core diameter but different thicknesses of the coating shell.
2. Aerosols that have a different core diameter but same shell thickness.

Computed extinction efficiencies of coated aerosols were compared with the experimental results from the CRD-AS. Several calculation methods were used: the subroutine DMiLay (Toon and Ackerman, 1981) and a version of bhcoat (Bohren and Huffman, 1983) adapted for Matlab by C. Mätzler (Toon and Ackerman, 1981; Bohren and Huffman, 1983). These codes compute the scattering parameters for a plane electromagnetic wave incident on a stratified dielectric sphere, that is, a spherical core coated by a spherical shell. These codes are exact implementations of Mie theory, where the core and shell retain their individual refractive indices during the computation. DMiLay is a double precision code designed to accurately handle large size parameters, thin shells, and large imaginary refractive indices, to avoid the ill-conditioning common to other core plus shell codes, although bhcoat produced the same results to the same number of decimal places reported here.

There are three different types of coatings that are atmospherically relevant and have recently been experimentally and theoretically examined.

1. Coating of nonabsorbing aerosols by another nonabsorbing compound (e.g., coating of sea salt aerosols by saturated organic molecules as in a NaCl aerosol coated by glutaric acid).
2. Nonabsorbing aerosols coated by weakly absorbing molecules.
3. Strongly absorbing aerosols coated by nonabsorbing aerosols or weakly absorbing molecules (e.g., soot particles coated by other weakly or nonabsorbing species). This is a very important process in atmosphere, as it has been shown that coatings can significantly enhance the absorption of aerosols, resulting in dramatic change in the RB (Jacobson, 2001; Glen Lesins and Lohmann, 2002; Bond and Bergstrom, 2006; Adler et al., 2010; Lack et al., 2009; Xue et al., 2009).

Here, we will introduce the experiment and theoretical calculations that were conducted to study the third type of coating. The setup and routine were also used to study the other two types of coating. The model molecule used as the core is Nigrosin and was coated by glutaric acid (Riziq et al., 2008). Figure 11.11 presents the Mie curves of Nigrosin (black line) and glutaric acid (red line) aerosols. In addition, it presents the Q_{ext} as a function of the size parameter of different diameters of Nigrosin coated by a constant thin layer (20 nm) of glutaric acid (green dots). The theoretical calculations of this type of coating are also presented as the green line. The Q_{ext} of constant core diameter and increasing shell diameter (thick coatings) is also plotted (blue dots) with the theoretical fit (blue line). The Mie scattering subroutine successfully predicted the Q_{ext} of the thin coating, but over predicted the thick coating by 5–12%. These inconsistencies for the thick coating can be explained by either a change in the sphericity of the generated aerosols due to the thick coating or from

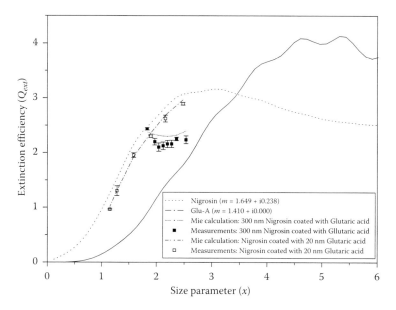

FIGURE 11.11 The measured and calculated extinction efficiency of Nigrosin, glutaric acid, and Nigrosin coated by different thicknesses of glutaric acid, and different sizes of Nigrosin coated by constant thickness of 20 nm of glutaric acid.

changes in the dielectric constant of the coating near the interface between the core and coating. However, no theoretical representations of such physical effects that we tested completely resolved the discrepancy.

More recently, the effect of coating on of soot particles by organic species was also examined using pulsed-CRD (Adler et al., 2010). In addition, Lack et al. studied the absorption and extinction enhancement due coating of absorbing polystyrene spheres (APSS) by oleic acid using cavity ring-down aerosol extinction spectrometer (CRD-AES) to measure the extinction and photoacoustic absorption spectrometer (PAS) to measure the absorption. They measured the enhancement factor of extinction and the absorption (E_{ext}, E_{Abs} respectively) which is the ratio of the extinction and absorption cross-section (σ_{ext}, σ_{Abs}) of the coated particles to that of the uncoated particle. These measurements showed very good agreement within 5% with the values expected from Mie theory and coated sphere Mie theory (Lack et al., 2009).

The other two types of coatings were recently investigated. Polystyrene Spheres Latex (PSL) coated with glutaric acid was used as a model for nonabsorbing/nonabsorbing coating and PSL coated by β-carotene was chosen as a model for nonabsorbing core coated by weakly absorbing shell. β-carotene is weakly absorbing at 532 nm and stable at high temperature during the coating process. The Q_{ext} of the laboratory-generated aerosols as a function of the size parameter was measured for the different coatings using a new cw-CRD system at 532 nm wavelength. The results were compared to calculations using the same Mie scattering subroutine. The Mie scattering subroutine successfully predicted the aerosol extinction for both thin and thick coatings for the nonabsorbing/nonabsorbing coating; however, for the PSL coated by β-carotene, the differences were up to 10% for only a thin coating of shell thickness of 40 nm.

1.4.4 RECENT ADVANCES IN CRD-AS

The CRD-AS technique has become an important method for studying aerosol optical properties, either through measurement of extinction or by combining it with other optical techniques to determine optical parameters such as the SSA (ω) and the Angström exponent (å). The simplest setup combines several CRD-ASs that can be applied to make parallel measurements of optical properties on the same aerosol sample. For example, several studies have reported an arrangement of CRD-ASs recording the optical extinction at two or more different wavelengths operating simultaneously (Smith and Atkinson, 2001; Baynard et al., 2007). Such a combination can also be used to estimate the Angström exponent ($\mathring{a}$), which describes the spectral dependence of aerosol optical properties and the wavelength of the incident light (λ) and is given by

$$\mathring{a} = -\frac{\log(\sigma_{ext1}/\sigma_{ext2})}{\log(\lambda_1/\lambda_2)} \qquad (11.25)$$

The Angström exponent is inversely related to the average size distribution of aerosols; that is, the smaller the aerosol geometric size, the larger the Angström exponent. Therefore, the Angström exponent is a useful parameter for estimating the particle size of atmospheric aerosols and the wavelength dependence of aerosol optical properties.

Tandem CRD-ASs of the same wavelength were used to study the relative humidity (RH) dependence of the aerosol extinction. Usually, the RH dependence of scattering is determined by using two nephelometers, of which one operates at low RH (<40%), and is considered as the reference, and the other at high RH (>85%). Together, they measure the scattering ratio between the reference and the high RH ($f_{\sigma(scat)}$) (RH, RH_{ref}), which is given by

$$f_{\sigma(scat)}(RH, RH_{ref}) = \frac{\sigma_{scat}(RH)}{\sigma_{scat}(RH_{ref})} \qquad (11.26)$$

Baynard et al. were the first to use a tandem CRD-AS for this application (Baynard et al., 2006; Garland et al., 2007). Figure 11.12 presents an illustration of this combination. They used the second harmonic of a Nd:YAG laser (532 nm) split between the two CRD systems. The first CRD was dry and the second was humidified. Particles generated by the atomizer were dried and size-selected using DMA. After exiting the first CRD, the particles were exposed to 95% RH by passing through a temperature control humidifier that consisted of a water vapor preamble membrane

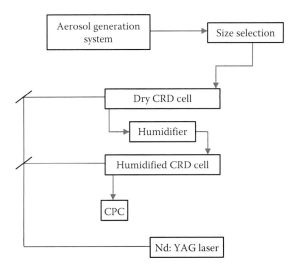

FIGURE 11.12 A schematic representative of the experimental setup for studying optical growth factor of aerosols using CRD-AS. (Reproduced from *Journal of Geophysical Research*, 2007, 112, p. 2. Copyright 2007, The American Geophysical Union. With permission.)

surrounded by liquid water. The humidified particles entered the second CRD with a typical RH 80%. Finally, the particles exited the humidified CRD and enter the CPC to determine their concentration. Using this setup, they studied the variation of $f_{\sigma(ext)}(RH, RH_{ref})$ with respect to particle size, composition and mixing state of NaCl, $(NH_4)_2SO_4$ and the internal mixture of soluble organic compounds with NaCl and $(NH_4)_2SO_4$. They found that the atmospheric $f_{\sigma(ext)}(RH, RH_{ref})$ is very sensitive to the aerosol composition, for example, organic/inorganic, so, for example, increasing the inorganic ratio increases the $f_{\sigma(ext)}(RH, RH_{ref})$. In addition, it is sensitive to the aerosol size; however, the mixing state (internally or externally) does not affect the $f_{\sigma(ext)}(RH, RH_{ref})$ and can therefore be neglected. Similar conclusions were obtained in their expanded study of other organic molecules mixed with $(NH_4)_2SO_4$ aerosols. A recent study of organic compounds measuring the optical growth factor of the polyfunctional aromatic acids (lignin combustion products) for the pure compounds mixed with ammonium sulfate showed similar behavior to the other organic compounds previously studied. It should be noticed that these molecules are less water soluble as compared to the previous organic compound used by Garland and coworkers (Garland et al., 2007; Beaver et al., 2008).

New optical combinations of CRD were developed in order to measure the scattering in addition to the extinction. Such a combination allows a direct measurement of the SSA (ω), which is the ratio between the scattering to the extinction coefficients. Strawa et al. (2003) introduced such an instrument using cw laser as the light sources. They used two cw diode lasers at 690 nm and 1550 nm. The CRD system has a three-mirror configuration, which is different from the two-mirror arrangement described previously and has the advantage that the reflected beam from the CRD mirror does not couple the laser beam back to the laser source. In this design, one of the walls of the cavity is made of glass (BK-7) to allow optical access for the scattering detectors. The scattered light also rings down exponentially similar to the CRD signal and the scattering coefficient can be determined by

$$\sigma_{scat} = \left(\frac{I_{scat}}{I_{rd}} \right) \frac{(1-R)}{(1+R)L} K \tag{11.27}$$

where I_{scat} and I_{rd} are the scattered and the ring down signals, R is the reflectivity of the mirror, L is the effective path length and K is a calibration constant. This was the first airborne CRD and was further modified and improved using a Lambertian diffuser to increase the acceptance angle of the scattering detector from 10–170° to 5–175°. Based on this novel method, Thompson et al. developed the aerosol albedometer, which is a combination of CRD and integrating sphere nephelometer that is capable of direct measurement of the SSA by determining α_{ext} using the CRD and α_{scat} by the integrating sphere (Thompson et al., 2008). The CRD mirrors were connected to the integrating sphere and the scattered light from the probe CRD beam was collected by the integrating sphere to a separate detector. Both methods have the main advantage of measuring the scattering and extinction of the same sample volume, which eliminates errors that arise from using separate instruments that can lead to differences in sample losses and humidity.

The synergistic combination of CRD spectrometer with a PAS for studying the optical properties of aerosols was recently demonstrated (Lack, 2006). The PAS is an extraordinary technique for measuring the absorption with high accuracy compared to traditional techniques, such as the filter-based technique (Moosmuller et al., 1998; Arnott et al., 2003; Schmid, 2006; Moosmuller et al., 2009). In addition, this technique offers high sensitivity that can be comparable to CRD making it very useful in atmospheric studies. This combination enables measurements simultaneously of both the extinction and the absorption of aerosols and, therefore, the SSA can be also determined at a single aerosol size. By using this method, therefore, it is not necessary to measure an entire size range and hence the measurement is faster and can provide more information about different aerosols at different size ranges. Lack et al. (2006) used pulsed-CRD to calibrate their aerosols' PAS using ozone molecules at 532 nm. They retrieved the complex refractive index of absorbing aerosols (Nigrosin) by measuring the extinction using the CRD and the absorbing using the calibrated PAS. In addition, they used the system to study the enhanced absorption and extinction by the coating of absorbing aerosols by nonabsorbing molecules, which is a common process in atmosphere (Lack et al., 2009).

11.5 SUMMARY

It is not surprising that CRD-AS has become a leading technique in the studies of optical properties of aerosols in laboratory or field applications. The capability and performance of this technique in field measurements have been proven in several laboratories and campaigns. In recent years, CRD has become a very promising technique for quantitative and accurate measurements for aerosol study at high sensitivity in field studies as well. The combination of the CRD with other techniques is becoming more reliable leading to improvements in accuracy and in the accessibility to measure simultaneously various optical parameters that are relevant for atmospheric modeling. In addition, the combination of the CRD with other techniques that can measure composition and physical properties such as mass spectrometry will be effective to provide *in situ* data that will be useful in climate modeling.

ACKNOWLEDGMENTS

This work was partially supported by the Israel Science Foundation (Grants 1527/07 and 196/08). Yinon Rudich acknowledges financial support by the Helen and Martin Kimmel Award for Innovative Investigation. We thank Caryn Erlick and Jeremy Seltzer for critical reading of the manuscript.

REFERENCES

Adler, G., A. A. Riziq, C. Erlick, and Y. Rudich. 2010. Effect of intrinsic organic carbon on the optical properties of fresh diesel soot. *Proceedings of the National Academy of Sciences of the United States of America* 107: 6699–6704.

Andreae, M. O. and A. Gelencser, 2006. Black carbon or brown carbon? The nature of light-absorbing carbonaceous aerosols. *Atmospheric Chemistry and Physics Discussions* 6: 3419–3463.

Arnott, W. P., H. Moosmuller, P. J. Sheridan, J. A. Ogren, R. Raspet, W. V. Slaton, J. L. Hand, S. M. Kreidenweis, and J. L. Collett. 2003. Photoacoustic and filter-based ambient aerosol light absorption measurements: Instrument comparisons and the role of relative humidity. *Journal of Geophysical Research-Atmospheres* 108(D1): 1–15.

Atkinson, D. B. 2003. Solving chemical problems of environmental importance using cavity ring-down spectroscopy. *Analyst* 128(2): 117–125.

Atkinson, D. B. and J. W. Hudgens, 1997. Chemical kinetic studies using ultraviolet cavity ring-down spectroscopic detection: Self-reaction of ethyl and ethylperoxy radicals and the reaction $O-2 + C_2H_5->C_2H_5O_2$. *Journal of Physical Chemistry A* 101(21): 3901–3909.

Atkinson, D. B. and J. W. Hudgens, 1999. Rate coefficients for the propargyl radical self-reaction and oxygen addition reaction measured using ultraviolet cavity ring-down spectroscopy. *Journal of Physical Chemistry A* 103(21): 4242–4252.

Atkinson, D. B. and J. W. Hudgens, 2000. Chlorination chemistry. 2. Rate coefficients, reaction mechanism, and spectrum of the chlorine adduct of allene. *Journal of Physical Chemistry A* 104(4): 811–818.

Atkinson, D. B., J. W. Hudgens, et al., 1999. Kinetic studies of the reactions of IO radicals determined by cavity ring-down spectroscopy. *Journal of Physical Chemistry A* 103(31): 6173–6180.

Bates, T. S., T. L. Anderson, et al., 2006. Aerosol direct radiative effects over the northwest Atlantic, northwest Pacific, and North Indian Oceans: Estimates based on *in situ* chemical and optical measurements and chemical transport modeling. *Atmospheric Chemistry and Physics* 6: 1657–1732.

Baynard, T., R. M. Garland, et al., 2006. Key factors influencing the relative humidity dependence of aerosol light scattering. *Geophysical Research Letters* 33(6), doi:10.1029/2005GL024898.

Baynard, T., E. R. Lovejoy, et al., 2007. Design and application of a pulsed cavity ring-down aerosol extinction spectrometer for field measurements. *Aerosol Science and Technology* 41(4): 447–462.

Beaver, M. R., R. M. Garland, et al., 2008. A laboratory investigation of the relative humidity dependence of light extinction by organic compounds from lignin combustion. *Environmental Research Letters* 3(4): 1–8.

Bellouin, N., O. Boucher, J. Haywood, and M. S. Reddy, 2005. Global estimate of aerosol direct radiative forcing from satellite measurements. *Nature* 438(7071): 1138–1141.

Berden, G., R. Peeters, et al., 2000. Cavity ring-down spectroscopy: Experimental schemes and applications. *International Reviews In Physical Chemistry* 19(4): 565–607.

Bohren, C. F. and D. R. Huffman, 1998. *Absorption and Scattering of Light by Small Particles*. Wiley VCH, New York, NY.

Bohren, C. F. and D. R. Huffman, 1983. *Absorption and Scattering of Light by Small Particles*. Wiley, New York, NY.

Bond, T. C. and R. W. Bergstrom, 2006. Light absorption by carbonaceous particles: An investigative review. *Aerosol Science and Technology* 40(1): 27–67.

Born, M. and E. Wolf, 1999. *Principles of Optics*, 7th ed. Cambridge University Press, Cambridge.

Brown, S. S. 2003. Absorption spectroscopy in high-finesse cavities for atmospheric studies. *Chemical Reviews* 103(12): 5219–5238.

Brown, S. S., H. Stark, et al., 2002a. Cavity ring-down spectroscopy for atmospheric trace gas detection: Application to the nitrate radical (NO_3). *Applied Physics B-Lasers and Optics* 75(2–3): 173–182.

Brown, S. S., H. Stark, S. J. Ciciora, R. J. McLaughlin, and A. R. Ravishankara, 2002b. Simultaneous *in situ* detection of atmospheric NO_3 and N_2O_5 via cavity ring-down spectroscopy. *Reviews in Science Instrumentation* 73(9): 3291–3301.

Bulatov, V., Y. H. Chen, et al., 2006. Absorption and scattering characterization of airborne microparticulates by a cavity ringdown technique. *Analytical and Bioanalytical Chemistry* 384(1): 155–160.

Bulatov, V., M. Fisher, et al. 2002. Aerosol analysis by cavity-ring-down laser spectroscopy. *Analytica Chimica Acta* 466(1): 1–9.

Busch, K. W. and M. A. Busch (Eds.), 1999. Cavity-ring-down spectroscopy, an ultratrace absorption measurement technique. *ACS Symposium Series*. Washington, DC: American Chemical Society.

Butler, T. J. A., J. L. Miller, et al., 2007. Cavity ring-down spectroscopy measurements of single aerosol particle extinction. I. The effect of position of a particle within the laser beam on extinction. *Journal of Chemical Physics* 126(17): 1–7.

Campargue, A., L. Biennier, et al., 1999. High resolution absorption spectroscopy of the nu(1) = 2–6 acetylenic overtone bands of propyne: Spectroscopy and dynamics. *Journal of Chemical Physics* 111(17): 7888–7903.

Campargue, A., S. Kassi, et al., 2006. CW-cavity ring down spectroscopy of the ozone molecule in the 6625–6830 cm (−1) region. *Journal of Molecular Spectroscopy* 240(1): 1–13.

Chen, W. T., R. A. Kahn, et al., 2008. Sensitivity of multiangle imaging to the optical and microphysical properties of biomass burning aerosols. *Journal of Geophysical Research-Atmospheres* 113(D10), doi:10.1029/2007JD009414.

Cheskis, S., I. Derzy, et al., 1998. Cavity ring-down spectroscopy of OH radicals in low pressure flame. *Applied Physics B-Lasers and Optics* 66(3): 377–381.

Chow, J. C., J. G. Watson, et al., 2009. Aerosol light absorption, black carbon, and elemental carbon at the Fresno Supersite, California. *Atmospheric Research* 93(4): 874–887.

Chylek, P., V. Ramaswamy, et al., 1984. Effect of graphitic carbon on the albedo of clouds. *Journal of The Atmospheric Sciences* 41(21): 3076–3084.

Chylek, P., G. Videen, D. J. W. Geldart, J. S. Dobbie, and H. C. W. Tso, 2000. Effective medium approximations for heterogeneous particles, in *Light Scattering by Nonspherical Particles, Theory, Measurements and Application*. Academic Press, New York, NY.

Cox, A. J., A. J. DeWeerd, et al., 2002. An experiment to measure Mie and Rayleigh total scattering cross sections. *American Journal of Physics* 70(6): 620–625.

Crunaire, S., J. Tarmoul, C. Fittschen, A. Tomas, B. Lemoine, and P, Coddeville. 2006. Use of cw-CRDS for studying the atmospheric oxidation of acetic acid in a simulation chamber. *Applied Physics B-Lasers and Optics* 85(2–3): 467–476.

d'Almeida, G. A., P. Koepke, and Shettlee, E. P. 1991. *Atmospheric Aerosols, Global Climatology and Radiative Characteristics*. Deepak Pub., Hampton, VA.

Decesari, S., Fuzzi, S., M. C. Facchini, M. Mircea, L. Emblico, F. Cavalli, W. Maenhaut, et al., 2006. Characterization of the organic composition of aerosols from Rondônia, Brazil, during the LBA-SMOCC 2002 experiment and its representation through model compounds. *Atmospheric Chemistry and Physics* 6(2): 375.

Dinar, E., T. F. Mentel, et al., 2006. The density of humic acids and humic like substances (HULIS) from fresh and aged wood burning and pollution aerosol particles. *Atmospheric Chemistry and Physics* 6: 5213–5224.

Dinar, E., A. A. Riziq, C. Spindler, C. Erlick, G. Kiss, and Y. Rudich, 2008. The complex refractive index of atmospheric and model humic-like substances (HULIS) retrieved by a cavity ring down aerosol spectrometer (CRD-AS). *Faraday Discussions* 137: 279–295.

Dinar, E., I. Taraniuk, E. R. Graber, T. Anttila, T. F. Mentel, and Y. Rudich, 2007. Hygroscopic growth of atmospheric and model humic-like substances. *Journal of Geophysics Research* 112(D5), doi:10.1029/2006 JD007442.

Ellison, G. B., A. F. Tuck, et al., 1999. Atmospheric processing of organic aerosols. *Journal of Geophysical Research-Atmospheres* 104(D9): 11633–11641.

Erlick, C. 2006. Effective refractive indices of water and sulfate drops containing absorbing inclusions. *Journal of the Atmospheric Sciences* 63(2): 754–763.

Falkovich, A. H., G. Schkolnik, et al., 2004. Adsorption of organic compounds pertinent to urban environments onto mineral dust particles. *Journal of Geophysical Research-Atmospheres* 109(D2), doi:10.1029/2003JD003919.

Fawcett, B. L., A. M. Parkes, D. E. Shallcross, and A. J. Orr-Ewing, 2002. Trace detection of methane using continuous wave cavity ring-down spectroscopy at 1.65 μm. *Physical Chemistry Chemical Physics* 4(24): 5960–5965.

Finlayson-Pitts, B. J. and J. Pitts, Jr. 1999. *Chemistry of the Upper and Lower Atmosphere*. Academic Press, CA.

Garland, R. M., A. R. Ravishankara, et al., 2007. Parameterization for the relative humidity dependence of light extinction: Organic-ammonium sulfate aerosol. *Journal of Geophysical Research-Atmospheres* 112(D19), doi:10.1029/2006JD008179.

Gelencser, A. 2004. *Carbonaceous Aerosol*. Springer, Honolulu.

Glen Lesins, P. C. and U. Lohmann, 2002. A study of internal and external mixing scenarios and its effect on aerosol optical properties and direct radiative forcing. *Journal of Geophysical Research-Atmospheres* 107: 1–5.

Grassian, V. H. 2001. Heterogeneous uptake and reaction of nitrogen oxides and volatile organic compounds on the surface of atmospheric particles including oxides, carbonates, soot and mineral dust: implications for the chemical balance of the troposphere. *International Reviews in Physical Chemistry* 20(3): 467–548.

Guazzotti, S. A., J. R. Whiteaker, et al., 2001. Real-time measurements of the chemical composition of size-resolved particles during a Santa Ana wind episode, California USA. *Atmospheric Environment* 35(19): 3229.

Hallar, A. G., A. W. Strawa, B. Schmid, E. Andrews, J. Ogren, P. Sheridan, R. Ferrare, et al., 2006. Atmospheric radiation measurements aerosol intensive operating period: Comparison of aerosol scattering during coordinated flights. *Journal of Geophysical Research-Atmospheres* 111(D5), doi:10.1029/2005JD006250.

Hoffer, A., A. Gelencser, et al., 2006. Optical properties of humic-like substances (HULIS) in biomass-burning aerosols. *Atmospheric Chemistry and Physics* 6: 3563–3570.

Howie, W. H., I. C. Lane, et al., 2000. The near ultraviolet spectrum of the FCO radical: Re-assignment of transitions and predissociation of the electronically excited state. *Journal of Chemical Physics* 113(17): 7237–7251.

IPCC, S., D. Qin, M. Manning, R. B. Alley, T. Berntsen, N. L. Bindoff, Z. Chen, et al., 2007. Climate change 2007: Synthesis report. In *Climate Change 2007: The Physical Science Basis. Contribution of Working Group I to the Fourth Assessment Report of the Intergovernmental Panel on Climate Change.* (Eds.) S. Solomon, D. Qin, M. Manning, Z. Chen, M. Marquis, K. B. Averyt, M. Tignor, H. L. Miller. Cambridge University Press, Cambridge.

Jacobson, M. C., H. C. Hansson, et al., 2000. Organic atmospheric aerosols: Review and state of the science. *Reviews of Geophysics* 38(2): 267–294.

Jacobson, M. Z. 2001. Strong radiative heating due to the mixing state of black carbon in atmospheric aerosols. *Nature* 409(6821): 695–697.

Jacobson, M. Z. 2002. Analysis of aerosol interactions with numerical techniques for solving coagulation, nucleation, condensation, dissolution, and reversible chemistry among multiple size distributions. *J. Geophys. Res.* 107(D19): 4366.

Jacobson, M. Z. 2006. Effects of externally-through-internally-mixed soot inclusions within clouds and precipitation on global climate. *Journal of Physical Chemistry A* 110(21): 6860–6873.

Jongma, R. T., M. G. H. Boogaarts, et al., 1995. Trace gas-detection with cavity ring down spectroscopy. *Review of Scientific Instruments* 66(4): 2821–2828.

Karar, K., A. Gupta, et al., 2006. Characterization and identification of the sources of chromium, zinc, lead, cadmium, nickel, manganese and iron in Pm10 particulates at the two sites of Kolkata, India. *Environmental Monitoring and Assessment* 120(1): 347.

Kaufman, Y. J., 2002. A satellite view of aerosols in the climate system. *Nature* 419(6903): 215.

Kaufman, Y. J., I. Koren, et al., 2005. The effect of smoke, dust, and pollution aerosol on shallow cloud development over the Atlantic Ocean. *Proceedings of the National Academy of Sciences of the United States of America* 102(32): 11207–11212.

Kesselmeier, J. and M. Staudt, 1999. Biogenic volatile organic compounds (VOC): An overview on emission, physiology and ecology. *Journal of Atmospheric Chemistry* 33(1): 23–88.

Kim, D. Y. and V. Ramanathan, 2008. Solar radiation budget and radiative forcing due to aerosols and clouds. *Journal of Geophysical Research-Atmospheres* 113(D2), doi:10.1029/2007JD008434.

Koren, I., Y. J. Kaufman, et al., 2004. Measurement of the effect of Amazon smoke on inhibition of cloud formation. *Science* 303(5662): 1342–1345.

Koren, I., L. A. Remer, et al., 2007. On the twilight zone between clouds and aerosols. *Geophysical Research Letters* 34(8), doi:10.1029/2007GL029253.

Kraus, D., R. J. Saykally, et al., 2002. Cavity-ringdown spectroscopy studies of the B-2 Sigma(+) <- X-2 Sigma(+) system of AlO. *Chemphyschem* 3(4): 364–366.

Kulmala, M., L. Laakso, et al., 2004. Initial steps of aerosol growth. *Atmospheric Chemistry and Physics* 4: 2553–2560.

Lack, D. A., C. D. Cappa, et al., 2009. Absorption enhancement of coated absorbing aerosols: Validation of the photo-acoustic technique for measuring the enhancement. *Aerosol Science and Technology* 43(10): 1006–1012.

Lack, D. A., E. R. Lovejoy, T. Baynard, A. Pettersson, and A. R. Ravishankara, 2006. Aerosol absorption measurement using photoacoustic spectroscopy: Sensitivity, calibration, and uncertainty developments. *Aerosol Science and Technology* 40(9): 697–708.

Lang-Yona, M., Y. Rudich, et al., 2009. Complex refractive indices of aerosols retrieved by continuous wave-cavity ring down aerosol spectrometer. *Analytical Chemistry* 81(5): 1762–1769.

Lang-Yona, N., A. Abo-Riziq, C. Erlick, E. Segre, M. Trainic, and Y. Rudich, 2010. Interaction of internally mixed aerosols with light. *Physical Chemistry Chemical Physics* 12, 21–31.

Laskin, A., D. J. Gaspar, et al., 2003. Reactions at interfaces as a source of sulfate formation in sea-salt particles. *Science* 301(5631): 340–344.

Laskin, A., M. J. Iedema, et al., 2005. Direct observation of completely processed calcium carbonate dust particles. *Faraday Discussions* 130: 453–468.

Lesins, G., P. Chylek, et al., 2002. A study of internal and external mixing scenarios and its effect on aerosol optical properties and direct radiative forcing. *Journal of Geophysical Research* 107: 4094, doi:10.1029/2001JD000973.

Lohmann, U. and J. Feichter, 2001. Can the direct and semi-direct aerosol effect compete with the indirect effect on a global scale. *Geophysical Research Letters* 28(1): 159.

Lund Myhre, C. E. and C. J. Nielsen, 2004. Optical properties in the UV and visible spectral region of organic acids relevant to tropospheric aerosols. *Atmospheric Chemistry and Physics Discussion* 4(3): 3013.

Maria, S. F., L. M. Russell, et al., 2004. Organic aerosol growth mechanisms and their climate-forcing implications. *Science* 306(5703): 1921–1924.

Mazurenka, M. I., B. L. Fawcett, et al., 2003. 410-nm diode laser cavity ring-down spectroscopy for trace detection of NO_2. *Chemical Physics Letters* 367(1–2): 1–9.

Miller, J. L. and A. J. Orr-Ewing, 2007. Cavity ring-down spectroscopy measurement of single aerosol particle extinction. II. Extinction of light by an aerosol particle in an optical cavity excited by a cw laser. *Journal of Chemical Physics* 126(17): 174303.

Moosmuller, H., W. P. Arnott, et al., 1998. Photoacoustic and filter measurements related to aerosol light absorption during the Northern Front Range Air Quality Study (Colorado 1996/1997). *Journal of Geophysical Research-Atmospheres* 103(D21): 28149–28157.

Moosmuller, H., R. K. Chakrabarty, et al., 2009. Aerosol light absorption and its measurement: A review. *Journal of Quantitative Spectroscopy & Radiative Transfer* 110(11): 844–878.

Moosmuller, H., R. Varma, et al., 2005. Cavity ring-down and cavity-enhanced detection techniques for the measurement of aerosol extinction. *Aerosol Science and Technology* 39(1): 30–39.

Newman, S. M., W. H. Howie, et al., 1998. Predissociation of the A(2)Pi(3/2) state of IO studied by cavity ring-down spectroscopy. *Journal of the Chemical Society-Faraday Transactions* 94(18): 2681–2688.

O'Keefe, A. and D. A. G. Deacon, 1988. Cavity ring-down optical spectrometer for absorption measurements using pulsed laser sources. *Reviews in Science Instrumentation* 59: 2544–2555.

Panne, U., R. E. Neuhauser, et al., 2001. Analysis of heavy metal aerosols on filters by laser-induced plasma spectroscopy. *Spectrochimica Acta Part B: Atomic Spectroscopy* 56(6): 839.

Parkes, A. M., B. L. Fawcett, et al., 2003. Trace detection of volatile organic compounds by diode laser cavity ring-down spectroscopy. *Analyst* 128(7): 960–965.

Paul, B., J. J. Scherer, et al., 1997. Cavity ringdown measures trace concentrations. *Laser Focus World* 33(3): 71–80.

Pettersson, A., E. R. Lovejoy, C. A. Brock, S. S. Brown, and A. R. Ravishankara, 2004. Measurement of aerosol optical extinction at 532 nm with pulsed cavity ring down spectroscopy. *Journal of Aerosol Science* 35(8): 995–1011.

Poschl, P. U. 2005. Atmospheric aerosols: Composition, transformation, climate and health effects. *Angew. Chem. Int. Ed.* 44(46): 7520.

Posfai, M., J. R. Anderson, et al., 1999. Soot and sulfate aerosol particles in the remote marine troposphere. *Journal of Geophysical Research-Atmospheres* 104(D17): 21685–21693.

Pradhan, M., R. E. Lindley, R. Grilli, I. R. White, D. Martin, and A. J. Orr-Ewing, 2008. Trace detection of C2H2 in ambient air using continuous wave cavity ring-down spectroscopy combined with sample pre-concentration. *Applied Physics B-Lasers and Optics* 90: 1–9.

Press, W. H., S. A. Teukolsky, W. T. Vetterling, and B. R. Flannery, 1992. *Numerical Recipes in C (2nd ed.): The Art of Scientific Computing*. Cambridge University Press, Cambridge.

Radney, J. G., M. H. Bazargan, M. E. Wright, and D. B. Atkinson, 2009. Laboratory validation of aerosol extinction coefficient measurments by a field-deployable pulsed cavity ring-down transmissometer. *Aerosol Science & Technology* 43: 71–80.

Ramanathan, V. and G. Carmichael, 2008. Global and regional climate changes due to black carbon. *Nature Geoscience* 1(4): 221–227.

Ramanathan, V., C. Chung, et al., 2005. Atmospheric brown clouds: Impacts on South Asian climate and hydrological cycle. *Proceedings of the National Academy of Sciences of the United States of America* 102(15): 5326–5333.

Ramanathan, V., P. J. Crutzen, et al., 2001. Atmosphere—Aerosols, climate, and the hydrological cycle. *Science* 294(5549): 2119–2124.

Riziq, A. A., C. Erlick, E. Dinar, and Y. Rudich, 2007. Optical properties of absorbing and non-absorbing aerosols retrieved by cavity ring down (CRD) spectroscopy. *Atmospheric Chemistry and Physics* 7(6): 1523–1536.

Riziq, A. A., M. Trainic, C. Erlick, E. Segre, and Y. Rudich, 2008. Extinction efficiencies of coated absorbing aerosols measured by cavity ring down aerosol spectrometry. *Atmospheric Chemistry and Physics* 8(6): 1823–1833.

Robinson, A. L., N. M. Donahue, et al., 2007. Rethinking organic aerosols: Semivolatile emissions and photochemical aging. *Science* 315(5816): 1259–1262.

Romanini, D., A. A. Kachanov, N. Sadeghi, and F. Stoeckel, 1997. CW cavity ring down spectroscopy. *Chemical Physics Letters* 264(3–4): 316–322.

Romanini, D. and K. K. Lehmann, 1995. Cavity ring-down overtone spectroscopy of Hcn, (Hcn)-C-13 and (Hcn)-N-15. *Journal of Chemical Physics* 102(2): 633–642.

Rudic, S., R. E. H. Miles, et al., 2007. Optical properties of micrometer size water droplets studied by cavity ringdown spectroscopy. *Applied Optics* 46(24): 6142–6150.

Rudich, Y. 2003. Laboratory perspectives on the chemical transformations of organic matter in atmospheric particles. *Chemical Reviews* 103(12): 5097–5124.

Rudich, Y., N. M. Donahue, et al., 2007. Aging of organic aerosol: Bridging the gap between laboratory and field studies. *Annual Review of Physical Chemistry* 58: 321–352.

Saathoff, H., K. H. Naumann, et al., 2003. Coating of soot and $(NH_4)_{(2)}SO_4$ particles by ozonolysis products of alpha-pinene. *Journal of Aerosol Science* 34(10): 1297–1321.

Sappey, A. D., E. S. Hill, et al., 1998. Fixed-frequency cavity ringdown diagnostic for atmospheric particulate matter. *Optics Letters* 23(12): 954–956.

Scherer, J. J., J. B. Paul, et al., 1995. Cavity ringdown laser-absorption spectroscopy and time-of-flight mass-spectroscopy of jet-cooled copper silicides. *Journal of Chemical Physics* 102(13): 5190–5199.

Scherer, J. J., J. B. Paul, et al., 1997. Cavity ringdown laser absorption spectroscopy: History, development, and application to pulsed molecular beams. *Chemical Reviews* 97(1): 25–51.

Scherer, J. J., J. B. Paul, et al., 1995. Cavity ringdown laser-absorption spectroscopy of the jet-cooled aluminum dimer. *Chemical Physics Letters* 242(4–5): 395–400.

Scherer, J. J. and D. J. Rakestraw, 1997. Cavity ringdown laser absorption spectroscopy detection of formyl (HCO) radical in a low pressure name. *Chemical Physics Letters* 265(1–2): 169–176.

Schmid, T. 2006. Photoacoustic spectroscopy for process analysis. *Analytical and Bioanalytical Chemistry* 384(5): 1071–1086.

Schnaiter, M., H. Horvath, et al., 2003. UV–VIS–NIR spectral optical properties of soot and soot-containing aerosols. *Journal of Aerosol Science* 34(10): 1421–1444.

Scholz, S. M., R. Vacassy, et al., 1998. Mie scattering effects from monodispersed ZnS nanospheres. *Journal of Applied Physics* 83(12): 7860–7866.

Schwarz, J. P., J. R. Spackman, et al., 2008. Coatings and their enhancement of black carbon light absorption in the tropical atmosphere. *Journal of Geophysical Research-Atmospheres* 113(D3), doi:10.1029/2007 JD009042.

Seinfeld, J. H. and S. N. Pandis, 2006. *Atmospheric Chemistry and Physics*. John Wiley & Sons Inc., New York, NY.

Shah, M. H., N. Shaheen, et al., 2007. Characterization of selected metals in airborne suspended particulate matter in relation to meteorological conditions. *Journal of the Chemical Society of Pakistan* 29(6): 598–604.

Smith, J. D. and D. B. Atkinson, 2001. A portable pulsed cavity ring-down transmissometer for measurement of the optical extinction of the atmospheric aerosol. *Analyst* 126(8): 1216–1220.

Sokolik, I. N. and O. B. Toon, 1996. Direct radiative forcing by anthropogenic airborne mineral aerosols. *Nature* 381(6584): 681–683.

Sokolik, I. N., O. B. Toon, et al., 1998. Modeling the radiative characteristics of airborne mineral aerosols at infrared wavelengths. *Journal of Geophysical Research-Atmospheres* 103(D8): 8813–8826.

Spence, T. G., C. C. Harb, et al., 2000. A laser-locked cavity ring-down spectrometer employing an analog detection scheme. *Review of Scientific Instruments* 71(2): 347–353.

Spindler, C., A. Abo Riziq, et al., 2007. Retrieval of aerosol complex refractive index by combining cavity ring down aerosol spectrometer measurement with full size distribution information. *Aerosol Science & Technology* 41: 1011–1017.

Stelson, A. W. 1990. Urban Aerosol refractive-index prediction by partial molar refraction approach. *Environmental Science & Technology* 24(11): 1676–1679.

Strawa, A. W., R. Castaneda, T. Owano, D. S. Baer, and B. A. Paldus, 2003. The measurement of aerosol optical properties using continuous wave cavity ring-down techniques. *Journal of Atmospheric and Oceanic Technology* 20(4): 454–465.

Tang, I. N. 1997. Thermodynamic and optical properties of mixed-salt aerosols of atmospheric importance. *Journal of Geophysical Research* 102.

Thompson, J. E., N. Barta, et al., 2008. A fixed frequency aerosol albedometer. *Optics Express* 16(3): 2191–2205.

Thompson, J. E., B. W. Smith, et al., 2002. Monitoring atmospheric particulate matter through cavity ring-down spectroscopy. *Analytical Chemistry* 74(9): 1962–1967.

Toon, O. B. and T. P. Ackerman, 1981. Algorithms for the calculation of scattering by stratified spheres. *Applied Optics* 20(20): 3657–3660.

Vaida, V., A. F. Tuck, et al., 2000. Optical and chemical properties of atmospheric organic aerosols. *Physics and Chemistry of the Earth Part C-Solar-Terrestial and Planetary Science* 25(3): 195–198.

Vallance, C. 2005. Innovations in cavity ringdown spectroscopy. *New Journal of Chemistry* 29(7): 867–874.

Vander Wal, R. L. and T. M. Ticich, 1999. Cavity ringdown and laser-induced incandescence measurements of soot. *Applied Optics* 38(9): 1444–1451.

Wheeler, M. D., S. M. Newman, et al., 1998. Cavity ring-down spectroscopy. *Journal of the Chemical Society-Faraday Transactions* 94(3): 337–351.

Wheeler, M. D., A. J. OrrEwing, et al., 1997. Predissociation lifetimes of the A(2)Sigma(+) v = 1 state of the SH radical determined by cavity ring-down spectroscopy. *Chemical Physics Letters* 268(5–6): 421–428.

Xue, H. X., A. F. Khalizov, et al., 2009. Effects of coating of dicarboxylic acids on the mass-mobility relationship of soot particles. *Environmental Science & Technology* 43(8): 2787–2792.

Yu, T. and M. C. Lin, 1994. Kinetics of the C6H5 + O-2 reaction at low-temperatures. *Journal of the American Chemical Society* 116(21): 9571–9576.

Zakowicz, W. 2002. On the extinction paradox. *Acta Physica Polonica A* 101(3): 369–385.

Zhang, Q., M. R. Canagaratna, et al., 2005. Time- and size-resolved chemical composition of submicron particles in Pittsburgh: Implications for aerosol sources and processes. *Journal of Geophysical Research-Atmospheres* 110(D7), doi:10.1029/2004JD004649.

12 Laser-Induced Fluorescence Spectra and Angular Elastic Scattering Patterns of Single Atmospheric Aerosol Particles

R. G. Pinnick, Y. L. Pan, S. C. Hill, K. B. Aptowicz, and R. K. Chang

CONTENTS

12.1 Introduction 298
    12.1.1 Overview..298
    12.1.2 Single-Particle Diagnostic Techniques..298
    12.1.3 Approach to this Chapter..298
12.2 Measurements of Fluorescence Spectra of Atmospheric Aerosol...................298
    12.2.1 Experimental Setup for Measurement of Fluorescence Spectra of
           Single Aerosol Particles...299
    12.2.2 Measurement Sites...300
    12.2.3 Time-Series Measurements of Atmospheric Aerosol Fluorescence Spectra 300
    12.2.4 Cluster Analysis of Atmospheric Aerosol Fluorescence Spectra 301
    12.2.5 Compositions of Atmospheric Aerosol that Might Contribute to
           Fluorescence Spectra Clusters..302
    12.2.6 Implications of Similarities in LIF Spectra of Atmospheric Aerosol at
           Different Sites..305
    12.2.7 Utility of Single-Particle Fluorescence Spectroscopy to Study OCAs....................306
    12.2.8 Limitations of the PFS Technique 306
12.3 Measurements of Angular Scattering Characteristics of Single
Atmospheric Aerosol Particles 307
    12.3.1 The Attractiveness of Single-Particle Angular Elastic Scattering and
           Some Difficulties in Interpreting the Results 307
    12.3.2 Experimental Setup for Angular Scattering Measurements...................307
    12.3.3 TAOS Measurements of Atmospheric Aerosol 309
    12.3.4 Visual Classification of Scattering Patterns and
           Possible Particle Morphologies..309
           12.3.4.1 Spherical Particle Patterns 310
           12.3.4.2 Perturbed Sphere Patterns...311
           12.3.4.3 Swirl Patterns..312
           12.3.4.4 Fiber-Like Particles..312
           12.3.4.5 Particles of Complex Structure 312
    12.3.5 Frequency-of-Occurrence of Pattern Classes and Implications for
           Atmospheric Aerosol 312

12.3.6 TAOS of Atmospheric Aerosol Using Improved Ellipsoidal Mirror
        and Different Geometry...313
12.4 Summary 313
References...315

12.1 INTRODUCTION

12.1.1 Overview

Aerosols are ubiquitous in the Earth's atmosphere. Their impacts are wide-ranging: atmospheric radiative forcing, cloud microphysics, atmospheric visibility, and inhalation effects on human health, to name a few. Atmospheric aerosols have a multitude of natural and anthropogenic sources including gas-to-particle reactions, biological emission, emissions of spores and pollens, dust storms, biomass combustion, wave action in the oceans, industrial emissions, vehicular traffic, home heating and cooking, and volcanic eruptions. Techniques for rapidly analyzing single aerosol particles, even if they are not as definitive as those used on bulk samples, can be useful for better understanding the sources, chemistry, and fate of atmospheric aerosol. Single-particle techniques provide for: (1) inferences concerning the composition of single particles that have only about a picogram of mass, even if they occur in small concentration, (2) rapid measurement of the temporal and spatial variability of specific aerosol species, particularly those that occur in low concentration, and (3) information on the morphology and internal structure of single particles. In this chapter, we focus on the use of two relatively new techniques for characterization of single particles in atmospheric aerosol: laser-induced-fluorescence (LIF) (sometimes called intrinsic fluorescence or native fluorescence) and two-dimensional angular elastic scattering.

12.1.2 Single-Particle Diagnostic Techniques

Some single-particle techniques being developed for characterization of airborne particles are as follows: (1) Aerosol laser-ablation mass spectroscopy provides elemental composition and masses of laser-ablated molecular fragments of single aerosol particles,[1–6] (2) Laser-induced breakdown spectroscopy (LIBS) yields atomic emission and plasma emission spectra that can be used to determine elemental composition of single particles,[7–12] (3) LIF of aerosol particles measured in one or two broadband wavelength channels[13–21] and spectrally dispersed fluorescence measurements[22–31] can be used to classify organic carbon aerosols (OCAs), and (4) elastic-angular scattering can be used to classify morphologies of some single particles.[18,32–37] Although fluorescence and elastic-angular scattering provide less information than mass spectrometry, these techniques are simpler experimentally, the signals increase linearly with interrogation laser power, and both are relatively nondestructive.

12.1.3 Approach to this Chapter

In this chapter we summarize current findings regarding the fluorescence and angular scattering characteristics of single atmospheric aerosol particles at geographic sites located in Maryland, Connecticut, and New Mexico. The results presented we believe to be are definitive but represent only a tiny sample of aerosol particles at a small number of sampling sites. Additional measurements are needed to investigate more fully the potential diagnostic power of the fluorescence and elastic-angular scattering techniques for characterization of atmospheric aerosols.

12.2 MEASUREMENTS OF FLUORESCENCE SPECTRA OF ATMOSPHERIC AEROSOL

Measurements of the fluorescence spectra of individual atmospheric aerosol particles are sparse. It appears likely that, with rare exceptions fluorescence measurements are only practical for particles containing organic carbon compounds, because of the extremely weak fluorescence signal of most nonorganic material in aerosol particles that have only a few picograms of mass. Early measurements of the fluorescence of atmospheric OCA were made only in one or two (rarely three) spectral bands.[16,17,19,38–41] These measurements provide information on the relative concentration of fluorescent aerosol but little quantitative information concerning particle composition. In parallel to these early efforts, a more capable system to detect the LIF spectra of single particles was developed,[22–26,42,43] with the first measurements of LIF spectra of atmospheric OCAs only recently reported.[27,28]

12.2.1 Experimental Setup for Measurement of Fluorescence Spectra of Single Aerosol Particles

A number of prototypes for measurement of single-airborne-particle LIF spectra have been developed over the last 15 years with continually improved capability.[22,25,28,31,43] A recent prototype capable of atmospheric aerosol measurements[28] is depicted by the schematic and the photograph in Figure 12.1. In this prototype, referred to as the Particle Fluorescence Spectrometer (PFS), a virtual impactor concentrator (nominal sample rate of 330 L/min) is used to concentrate particles in the 2–10 μm

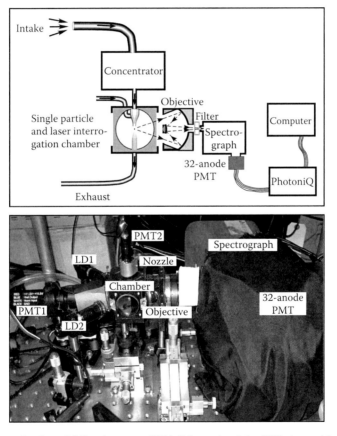

FIGURE 12.1 **(See color insert following page 206.)** Schematic of the PFS. Aerosol is sampled through a virtual impactor and a focusing nozzle inlet, which concentrates super-micron particles into a flowing laminar jet. As single particles within the aerosol jet are drawn through the sampling volume, they are probed with a pulsed UV laser, which excites fluors within the particles. The fluorescence emitted is collected by a reflective objective, focused onto a spectrograph slit, and detected with a 32-anode PMT detector. This arrangement permits rapid measurement of single-particle fluorescence spectra, with aerosol nominal sample rates of about 100 L/min for supermicron particles. (From Pan, Y.L. et al. 2007. *J. Geophys. Res. Atmos.*, 112: doi:1029/2007JD008741. With permission.)

diameter range. The minority outlet flow of the concentrator (nominally 1 L/min) is fed to the PFS inlet nozzle forming a highly focused, laminar, cylindrical aerosol jet within the relatively small optical chamber. Particles in the jet are detected by elastic-scattering signals from two, highly focused, intersecting, different wavelength (650 and 685 nm) diode laser beams, which are used to trigger a pulsed probe laser and the detection system. Fluorescence in particles is excited by a Nd:YLF probe laser frequency-quadrupled to a 263 nm wavelength and having a 2 mm beam diameter, 0.05 mJ energy per pulse, and 10 ns pulse length. The detector, a Hamamatsu model H7260 32-anode Photomultiplier tube (PMT), combined with a data aquisition system specially designed for our LIF spectroscopy (VTech, now Vertilon), has high data acquisition rate, minimal storage requirements, but retains sufficient spectral resolution (15 nm) for fluorescence emission. This PFS prototype can: (1) measure fluorescence spectra of bacterial particles with sizes of 1–10 μm diameter; (2) measure spectra at rates of many thousands of particles per second; (3) measure each particle's elastic scattering, which can be used to estimate particle size; and (4) provide a time stamp for each particle's arrival.

12.2.2 MEASUREMENT SITES

To date limited measurements of aerosol LIF spectra have been made and only at three geographic sites located in Maryland, Connecticut, and New Mexico, USA. These sites are in regions with very different regional climate. Adelphi, MD, USA (39°N latitude, elevation 75 m), is located in the Baltimore–Washington metroplex. This site is a highly populated urban area with moderately high precipitation (101 cm per year), and having large deciduous forests with tall trees and other vegetation. Aerosol was sampled through the (18 m high) roof of the US Army Research Laboratory, Harry Diamond Building in Adelphi, MD. New Haven, CT, USA (41.2°N latitude, elevation 25 m) is in the Atlantic Coastal region of the United States. New Haven is a moderately populated city (New Haven–Meridan metropolitan area population was 542,000 in 2000) located in the heavily industrialized northeastern United States about 10 km from Long Island Sound, 130 km from New York City, and 240 km from Boston, MA. The annual average temperature is 11°C. The New Haven area has moderately high precipitation (134 cm per year) and large deciduous forests. Las Cruces, NM (32.2°N latitude; elevation 1200 m) is in the Chihuahuan Desert of the southwestern United States and northern Mexico. This site is a moderately populous urban area (metropolitan–area population about 193,000) with relatively low precipitation (about 25 cm per year). The average annual temperature is 18°C. The urban metroplex of El Paso, Texas—Ciudad Juarez, Mexico is located 80 km to the south. The site has relatively little vegetation, except in the nearby irrigated Mesilla Valley along the Rio Grande River (6 km west). Outside of a small region along the Rio Grande, and near isolated riparian areas, there are essentially no trees and relatively little vegetation; dusty conditions are common.

Measurements were made during spring (February 26 to April 1, 2003) in MD, during fall (October 24–25, 2006) in CT, and during winter (January 22–23, 2007) in NM. The MD site is located near the Washington Beltway with moderately traveled streets located within 150 m. The CT and NM sites are located on university campuses (Yale University and New Mexico State University) with moderately traveled city streets located less than 50 m away. At the CT and NM sites the virtual-impactor inlet for the PFS intake passes through a window or wall positioned about 2 m above ground level into a laboratory where the PFS is mounted on an optical table. At the MD site the inlet passes through the roof of a four-story building.

12.2.3 TIME-SERIES MEASUREMENTS OF ATMOSPHERIC AEROSOL FLUORESCENCE SPECTRA

Several million spectra were measured in a variety of atmospheric conditions at the three sites. A sample of PFS LIF spectra is presented in Figure 12.2. Shown are the spectra for a series of 1000 consecutively arriving particles measured on October 23, 2006 in CT, over a 16-min period. Spectra of the small fraction of particles that have the largest fluorescence emission dominate this compressed

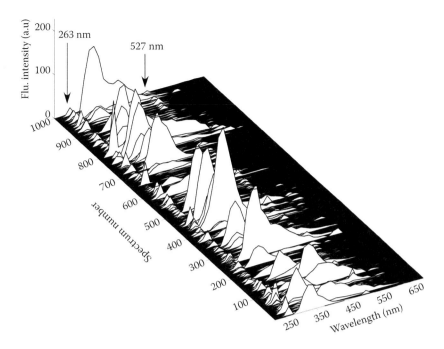

FIGURE 12.2 Fluorescence spectra for a series of 1000 consecutively arriving particles measured during October 23, 2006 in CT, over a 16-min period. (From Pan, Y.L. et al. 2007. *J. Geophys. Res. Atmos.*, 112: doi:1029/2007JD008741. With permission.)

view of the data. In an effort to categorize the LIF spectra from ambient aerosols, an unstructured hierarchical cluster analysis was performed on the data.

12.2.4 CLUSTER ANALYSIS OF ATMOSPHERIC AEROSOL FLUORESCENCE SPECTRA

To perform an unstructured hierarchical analysis, each spectrum was treated as a large multidimensional vector. Only the anodes which correspond to the fluorescence signal (typically from around 300 nm to 600 nm in wavelength) were used in the analysis. The spectra were unit normalized, so that the sum of the intensities in the fluorescent range equaled 1. This allowed the use of the dot product as a means to compare how similar two spectra were to each other—identical spectra would have a dot product of 1, and as the similarity decreased the dot product would also decrease. The hierarchical clustering was then performed by combining the spectra with the largest dot products. The dot products of every pair of spectra are compared, and the pair with the highest dot product is combined into a new cluster, which is then assigned the averaged and renormalized spectra of the two spectra being combined. The process is then repeated, until the largest dot product found is below a set threshold.

We found that even though we took ambient aerosol spectra from three different locations, the final cluster templates were similar (see Figure 12.3); the main difference between the different sites was the percentage of aerosol particles that fall within each cluster. This interesting result suggests that the cluster templates may be sufficiently robust to be applied to different geographic locations having different regional climate. The computation of these cluster templates is a computationally intensive process, because the dot product of every possible pair of spectra must be calculated. The number of spectra that can be clustered using a nonoptimized Fortran code is limited by computer memory on our PC to a few tens of thousand. We investigated other cluster analysis methods, and were able to write an algorithm, which is able to handle about an order of magnitude more spectra, and currently is limited by memory constraints. This new method is based around the k-means algorithm.[44] The k-means algorithm is an optimization algorithm where one starts with a set of clusters, and continues to move data from one cluster to another to maximize the similarity between the data and the cluster

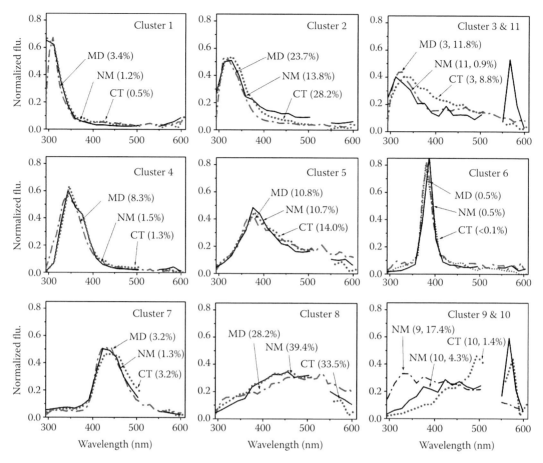

FIGURE 12.3 Cluster templates found for LIF spectra of atmospheric aerosols in Adelphi, MD; Las Cruces, NM; and New Haven, CT. (From Pan, Y.L. et al. 2007. *J. Geophys. Res. Atmos.*, 112: doi:1029/2007JD008741. With permission.)

centers. Because this method depends on some initial clustering as a starting point and to determine how many clusters are solved for, we used a quick hierarchical clustering to seed the *k*-means optimization algorithm. In our quick hierarchical clustering, a random spectrum is chosen, and all spectra that have a dot product higher than a set threshold with that chosen spectrum is combined into that cluster. The process continues until all spectra have been put into a cluster. The *k*-means algorithm is then used to optimize the clusters. Because the initial clustering tends to create many more clusters than are necessary, the final step is to go back through all of the clusters and combine any clusters whose means have a dot product above a set threshold. This process proved to be quite good at finding clusters, while being able to handle a much larger set of spectra using less computational time.

A sample of fluorescence spectra populating the cluster templates is shown in Figure 12.4 for the NM data set measured on January 22–23, 2007. This data set had 58,260 spectra, with 10,157 (17%) having fluorescence above threshold, subject to cluster analyses. Cluster 8 is the most populated cluster, having 39% of fluorescent particles, followed by clusters 9, 2, and 5.

12.2.5 COMPOSITIONS OF ATMOSPHERIC AEROSOL THAT MIGHT CONTRIBUTE TO FLUORESCENCE SPECTRA CLUSTERS

As noted above, we expect that nearly all particles having measurable fluorescence will have some component of organic carbon. We know from previous work that organic compounds can constitute

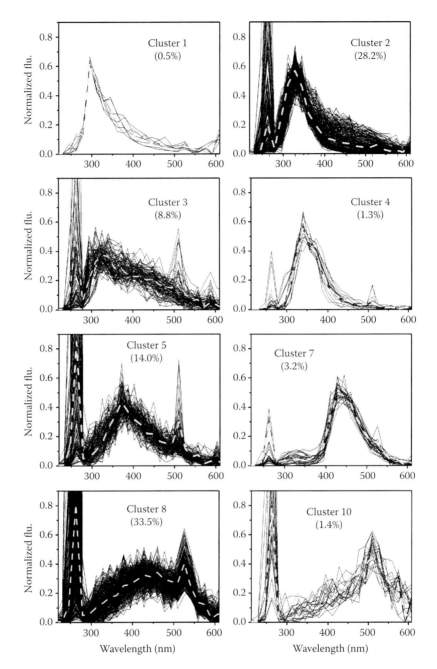

FIGURE 12.4 Illustration of some of the LIF spectra that populate templates of the clusters shown in Figure 12.3 for the NM data set. (From Pan, Y.L. et al. 2007. *J. Geophys. Res. Atmos.*, 112: doi:1029/2007JD008741. With permission.)

a significant but highly variable fraction of aerosol particles and has been reported to occur as alkanes, alkenes, carboxylic acids, ketones, phenols, furans, terpenoids, PAHs, monocyclic and polycyclic aromatic polyacids, alkanoic acids, monocarboxylic acids, lignans, cellulose, humic and fulvic acids, humic-like substances (HULIS), pollens, pollen antigens, bacteria, bacterial spores, viruses, and fungal spores.[45–88] Collectively these cited findings reveal that our knowledge of OCAs in the Earth's atmosphere is rapidly expanding but still limited. Further, many of these finding are site-specific and may only be valid for certain seasons or atmospheric conditions. In the majority of the cited studies the analyses of OCAs were performed on aerosol sample collections because the techniques used require significant mass of aerosol for analysis. However, atmospheric aerosol particles generally do not have uniform composition. Using collections of particles, no definitive inferences can be made about the composition of individual particles. On the other hand, the PFS technique, which rapidly analyzes single particles, and sorts them according to their LIF spectra, provides the potential for a useful diagnostic for their classification.

Cluster 1's spectrum (peaking near 308 nm) rises sharply toward shorter wavelengths but appears to be cutoff by the filters used to block the 266 (or 263) nm laser scattering. Particles with these spectra comprise, on average, 3.4% (MD), 0.5% (CT), and 1.2% (NM), of the fluorescent-particle fraction and have relatively high fluorescence intensity. Compounds having one aromatic ring or an aromatic ring with a small degree of additional conjugation could be present in particles that fall into this cluster. The aromatic amino acid tyrosine has a similar spectrum,[23] as would pure proteins that contain tyrosine but no tryptophan. In proteins that contain both tyrosine and tryptophan, the tyrosine fluorescence is typically very weak because it transfers absorbed energy to tryptophan, which then fluoresces. Compounds that have been found in OCA and that may contribute are the benzoic acids, dibenzoic acids, phenols, and benzaldehydes.[89]

Cluster 2's spectrum peaks at 317 nm and also appears to be cut by the filter. This spectral type occurs frequently (on average 24% in MD, 28% in CT, and 14% in NM) and has relatively high fluorescence intensity. Particles in this cluster may include single-ring aromatics as well as single-ring aromatics having additional conjugated bonds. Double-ring aromatics such as napthalene and its derivatives, or heterocyclic compounds, such as tryptophan in proteins may be present in particles that populate this cluster.

Cluster 3's spectrum appears to be bimodal with a peak near 321 nm and a very broad shoulder between 400 and 500 nm. It occurs in a significant fraction of fluorescent particles in MD and CT (12% in MD, 9% in CT), but not in NM. The cluster corresponds to relatively low intensity. It is similar to spectra of some bacterial samples,[23] especially ones that have not been washed well. The spectrum appears to be a mixture of compounds. The broad hump is similar to that of some mixtures of humic acids or HULIS.

Cluster 4's spectrum (peaking near 340 nm) occurs on an average in 8% of fluorescent particles in MD, but only 1.3% in CT and 1.5% in NM. Particles in this cluster tend to have a moderate fluorescence intensity. This spectrum is similar to that of pure tryptophan and is characteristic of bacteria and tryptophan-containing proteins. Other double-ring aromatic compounds may have similar spectra. Hill et al.[23] found that cigarette side-smoke has a similar spectrum. Pinnick et al.[27] measured fluorescence of pure tryptophan test particles and found 96% of them combined with this cluster. Bacteria or bacterial spores grown in a fluorescent medium and not washed well may not combine with this cluster.[23] The concentration of particles in this cluster is about 90/m[3] in CT and NM, about two orders of magnitude less than that of total bacterial cells measured by Tong and Lighthart[90] at a rural site near Corvallis, Oregon. The fact that bacterial cells smaller than 3 μm likely dominate the concentration in Corvallis likely account for most of this difference.

Cluster 5's spectrum (peaking near 350–360 nm, but with a long tail extending to 600 nm) has similarity to spectra of collections of marine aerosol.[91] It may include bacterial or other biological particles, but the peak emission is further to the red than for laboratory-grown bacteria we have seen. Populations in this cluster are more uniform (11% MD; 14% CT; and 11% NM) across the three sites than any other cluster. The small peak around 532/527 nm is due to leakage of light from the Nd:YAG/Nd:YLF laser and should be ignored. Particles in this cluster have relatively small fluorescence intensity.

Cluster 6's spectrum (peaking near 380 nm) is strikingly narrow. This feature is rare. However, the feature is quite distinct and is not believed to be noise. The fluorescence intensity is relatively strong.

Cluster 7's spectrum (peaking near 430 nm) is similar to the spectra reported for cellulose,[92] although any fluorescence from completely pure cellulose should be very weak. Aromatic and polycyclic aromatic compounds, nicotinamide adenine dinucleotides, and humic acids could also be candidate compounds. The fluorescence intensity is generally the strongest of all clusters. We note that atmospheric cellulose was found in significant concentration and to be a tracer for plant debris in Europe.[80,93] On average only 3.2% of fluorescent particles populate the cluster in MD, 3.2% in CT, and 1.3% in NM.

Cluster 8's spectrum, with its very broad hump, is similar to some spectra recorded for fulvic or humic acids[64,94,95] or HULIS.[96,97] A complex distribution of aromatic and polycyclic aromatic compounds could look similar. Generally this cluster has the most fluorescent particles (on an average of 28% in MD, 33% in CT, and 39% in NM), but generally corresponds to the lowest fluorescence intensity of any cluster. Humic acids in atmospheric particles occur primarily in particles smaller than the 3 μm-diameter cutoff used in our analysis, a finding which is apparently consistent with the small fluorescence signal measured here. However, size cannot be related directly to fluorescence intensity, because the quantum efficiency of the fluorophors is not known.

Cluster 9's spectrum is somewhat similar to cluster 3, which may be thought of a mixture of the spectra of (and maybe the compounds contributing to) clusters 2 and 8. However, cluster 9 appears more humic-like, as the spectrum is relatively flat between 340 nm and 500 nm. This cluster appeared only in NM, where on an average it was populated with 17% of fluorescent particles.

Cluster 10's spectrum is similar to that of cluster 8 (humic-like), but having a prominent 570 nm peak.

Cluster 11's spectrum is the cluster in NM that is most similar to cluster 3. However, its large spike near 575 nm does not occur in cluster 3. We do not have suggestions for organic carbon (OC) materials that would explain the sharp peaks near 380 nm (cluster 6) and 575 nm (cluster 10 and NM cluster 11), and conjecture that some inorganic materials may be responsible.

Many combinations of fluorophors in atmospheric aerosol could have LIF spectra similar to most of the clusters. Determining which of these possible combinations occur in single atmospheric particles will require more research. For example, selected distributions of polycyclic aromatic hydrocarbons (PAHs) and of aged PAHs may be assembled, which could have spectra of any clusters other than clusters 1 and 2. Also, no single spectral template looks exclusively bacterial. Bacteria can occur in complex mixtures/agglomerates in airborne particles. Bacteria may be strong contributors to the spectra of clusters 2, 3, and 4, and may be less strong contributors to cluster 5.

12.2.6 IMPLICATIONS OF SIMILARITIES IN LIF SPECTRA OF ATMOSPHERIC AEROSOL AT DIFFERENT SITES

Most of the fluorescent particles can be clustered into a few spectral types that appear to be quite distinct. These clusters are sufficiently robust that most of the cluster types appear independently in measurements made in three different geographical locations having two very distinct climates. The seven clusters that are most similar at the three sites (clusters 1, 2, 4, 5, 6, 7, and 8) have the majority of all particles (77% in MD, 68% in NM, and 81% in CT). Clusters 1, 4, 5, 6, and 7 are remarkably similar at the three sites, and account for 10–20% of the fluorescent particles. The spectra in clusters 2 and 8 account for 52% of fluorescent particles in MD, 53% in NM, and 62% in CT.

Some factors contributing to the similarities between clusters at the different cities are:

1. Long-range atmospheric transport and turbulent mixing tend to homogenize the ambient aerosol. For example, forest fire smoke can be transported from Alaska to Nova Scotia, Canada[98]; dust from Asia has been measured across the United States[99]; and Saharan dust can travel across the Atlantic Ocean to the United States.
2. In biomass combustion emissions, both the primary and secondary organic aerosol, and the new compounds generated as these aerosols age, may be similar, regardless of geographic

locale. Humidity, temperature, and atmospheric turbulent structure may affect the formation of aerosol particles, partitioning of combustion products between gas and particle phases, and processing of particles.

3. The predominant fluorophors in biological materials (tryptophan, NADH or similar compounds, and flavins) are preserved across species. Some of the main fluorors in plant cell walls (ferulates), and wood (lignans, sinapyl alcohols, etc.) are common in diverse environments.

12.2.7 UTILITY OF SINGLE-PARTICLE FLUORESCENCE SPECTROSCOPY TO STUDY OCAs

Organic carbon in OCAs range from small molecules such as oxalic acid or phenols, to viable bacteria, fungi, and spores (each of which is composed of many thousands of different kinds of molecules), and to highly complex mixtures of decomposition products of biological materials. Sources of organic carbon in aerosol are diverse and include natural wildfires and home heating,[100] residential wood and coal burning,[101] vehicle engine emissions,[101] cooking,[74,102] cattle feed lots,[78] microbial decay of plant matter,[57,103] atmospheric oxidation of soot,[72] biomass combustion,[63] microbial and biochemical degradation of organic debris in the soil,[73] sewage wastewater treatment plants,[69] surfactant matter on the ocean surface aerosolized by wave action,[91] and wind action on plants and soil.[58,80,81] Even so, the sources, composition, concentration, sizes, morphology, and degree of internal mixing of OCAs are not well understood because the OC component of aerosols is so complex and partially volatile, and because sampling and analyzing such an enormous range of molecules is difficult. The PFS technique could provide a diagnostic tool to better understand the sources, diurnal and seasonal variability, atmospheric transport and evolution, and ultimate fate of OCAs.

12.2.8 LIMITATIONS OF THE PFS TECHNIQUE

The ability to measure single particle LIF spectra is particularly useful for studying a minority population of (fluorescing) particles entrained in a dominant background of nonfluorescent particles. However, the PFS technique has several key, probably insurmountable, limitations.

1. The LIF spectra of individual atmospheric particles cannot be unambiguously interpreted, except in some special cases where there is sufficient information about potential sources of particles and/or where the spectra have distinct signatures. Although more information could be gained by adding one or more additional probe lasers, for example, at 355 nm or 351 nm,[22–24,31] fluorescent compounds in atmospheric aerosols can probably never be uniquely identified solely by fluorescence spectral signatures because fluorescence spectral features are typically broad, and because OCAs in the atmosphere are generally complex mixtures.
2. Single-particle fluorescence signals are weak. The laser intensity required to measure 1 μm particles (about 30 MW cm^2) is approaching the plasma breakdown threshold, which is of the order of 100 MW cm^2 for organic particles.[104] In theory, a longer pulse-length probe laser should allow significantly smaller particles to be measured, but our experience with a 125 ns pulse-length laser[105] suggests that increasing the pulse length is of limited value, because of photodegradation of tryptophan and other fluorophors, and possibly because of triplet shelving of excited electrons. In theory, a tighter sample volume, better collection optics, and a detector with better signal-to-noise might take this small-size limit for particles that fluoresce similar to *Bacillus subtilis* down to 0.5 μm, and possibly even a little smaller, but measuring fluorescence spectra of significantly smaller particles may not be feasible.
3. A third limitation is that many OC compounds (e.g., aliphatic alcohols, ketones, carboxylic acids) do not have sufficient conjugation of bonds to have strong fluoresence when excited at 266 nm.

12.3　MEASUREMENTS OF ANGULAR SCATTERING CHARACTERISTICS OF SINGLE ATMOSPHERIC AEROSOL PARTICLES

12.3.1　THE ATTRACTIVENESS OF SINGLE-PARTICLE ANGULAR ELASTIC SCATTERING AND SOME DIFFICULTIES IN INTERPRETING THE RESULTS

Single-particle angular elastic scattering appears promising for characterization of individual atmospheric particles for these reasons: (1) scattering signals are strong, (2) scattering patterns are highly sensitive to particle morphology and internal structure, (3) the illumination of particles results in small, perhaps negligible, changes that might be caused by loss of volatile components, (4) very good correlations between measured and computed scattering have been demonstrated for some particles, and (5) capabilities to calculate scattering are now available for highly complex particles. In particular, the attractiveness of elastic scattering techniques derives partly from the enormous progress that has been made in calculations of scattering by nonspherical, highly inhomogeneous, and ever larger and more complex particles, as numerical methods continue to be improved, and as computer capabilities increase. The moment method (MOM), developed by Jack Richmond and others in the early 1960s[106,107] has been used very extensively. The simplest form of the MOM is often termed the Purcell-Pennypacker method,[108] or more recently the discrete-dipole method (DDA), a name it was given in the late 1980s by researchers who seem to have been unaware of the MOM. The finite-difference time-domain method[109] has continued to be improved and extended to more complex problems and various nonlinear problems. The T-matrix method[110] has continued to be extended and applied to more complex particles. There is a beauty in the mathematics and other aspects of these forward scattering calculational methods, which has a strong appeal to many of us, and has led to a plethora of MS and PhD dissertations.

Because of these attractive features, the use of angularly resolved elastic light scattering to characterize individual airborne particles has been investigated for more than 30 years.[18,32–37,111–115,129]

In spite of the attractions of angular elastic scattering, the science and engineering required for obtaining useful shape and compositional information from angular scattering patterns appears to be in its infancy. Obtaining morphology of atmospheric particles from elastic scattering is subject to a myriad of problems, except in cases where either the shapes are very simple, and/or the particles to be distinguished occur in only a small number of simple shapes. Neither case appears to apply in studying atmospheric particulate matter. Except for spherical homogeneous particles, it is unclear what particle characteristics can be determined from scattering patterns. Unique relations between scattering patterns and particle morphologies (3-D complex refractive index distributions) do not exist except when additional information is provided, for example, in cases where it is known that certain particles are homogenous. Definitively determining particle characteristics from scattering patterns has been recognized to be an important problem for several decades, but relatively little progress has been made. Even the far simpler problem of determining from the scattering something about particle shape, for example, is it a spherical, fiber-like, a more complex particle, is highly nontrivial. As a demonstration of the rudimentary state of extracting morphology from scattering patterns, we have resorted to a visual classification approach to these problems (see Section 12.4).

12.3.2　EXPERIMENTAL SETUP FOR ANGULAR SCATTERING MEASUREMENTS

Here we focus on a technique referred to as two-dimensional angular optical scattering (TAOS).[36,37] TAOS uses a pulsed green laser (532 nm) to illuminate single particles as they traverse one focal point of an ellipsoidal mirror. The angularly scattered light is collected by the mirror and sensed by a gated image-intensified charge coupled device (ICCD) camera positioned beyond the second focal point of the mirror.

A TAOS prototype is shown schematically in Figure 12.5a. To sample atmospheric aerosol, air was drawn through a duct protruding through the roof of our laboratory. Since we wanted to sample

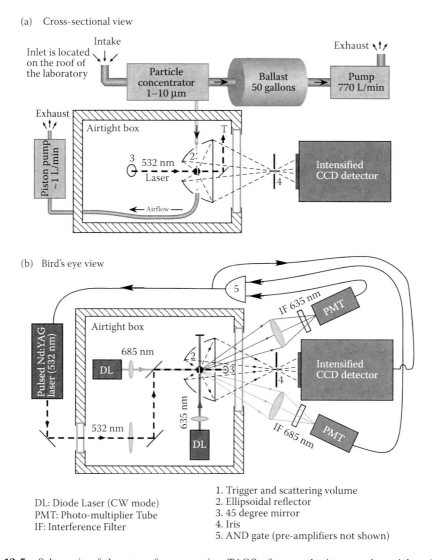

FIGURE 12.5 Schematic of the setup for measuring TAOS of atmospheric aerosol particles. Aerosol is sampled by a virtual impactor concentrator and the minority outlet flow of the concentrator fed to an airtight box through an aerodynamic focusing nozzle (a); single particles flowing from the nozzle pass through the intersection of crossed diode laser beams and are subsequently interrogated by a pulsed green laser (b). (From Aptowicz, K.B. et al. 2006. *J. Geophys. Res.—Atmospheres*, 111: D12212, doi:10.1029/2005JD006774. With permission.)

particles having a large range of sizes, an air-to-air virtual impactor concentrator (Dycor model XMX) was used to increase the concentration of larger particles (within a size range of 2–20 μm). Air from the duct was drawn at a rate of 770 L/min into the concentrator, and the concentrated particles were drawn from the minor exit flow of ~1 L/min. Smaller particles undergo almost no concentration, but pass unimpaired through the device. The minor flow is drawn under slight negative pressure (about 2 mbar) into a cubical airtight aluminum box of 46 cm on a side through a conically shaped aerodynamic nozzle to achieve a focused laminar aerosol jet. The aerosol nozzle is positioned slightly above (~0.5 cm) the scattering "focal volume" where the scattering measurement is made. The jet of aerosol exits the airtight box through an eduction tube positioned about 1 cm below the scattering focal volume.

Particles in the aerosol jet traverse the focus of an ellipsoidal mirror (Opti-Forms Inc., Model E64-3). Custom holes were drilled in the ellipsoid mirror to gain access to the focal volume, as is evident in Figure 12.5a. Light originating from a particle at the focal point of this mirror is reflected through a quartz window to the second focal point (and iris) located outside the box. The light rays from this "virtual" particle located at the iris are then detected by a 1024×1024 two-dimensional ICCD detector (Andor Technology, Model iStar DH734–25F-03). By taking into account the geometry of the system, every pixel of the detector can be matched with a unique scattering angle (polar angle θ and azimuth angle ϕ). This geometry collects a large solid angle ($75° < \theta < 135°, 0° < \phi < 360°$) of scattered light; however, it suffers severely from the off-axis aberration coma that leads to uncertainty in labeling the scattering angle. A ray-tracing analysis of the system shows that for slight misalignment, the error in angle varies approximately as 1 degree for every 10 μm of particle misalignment away from the mirror's focal point.[116] The particle-laden aerosol jet has a diameter of ~400 μm, which leads to severely distorted scattering patterns if particles were probed throughout the entire jet.

To mitigate this error, a cross-beam trigger system[24] is incorporated into the optical train, as shown in Figure 12.5b. The two continuous wave (CW) TEM00 diode lasers (Microlaser Systems), emitting light at 635 nm (power 25 mW) and 685 nm (40 mW), are focused with 7.5 cm and 5 cm focal length lenses to spot sizes 25 and 13 μm, respectively. PMT (Hamamatsu model H6780-02) attached to long (16 cm) working distance microscope objectives detect the scattered light as the particle traverses the two beams. An interference filter is placed in front of each PMT so that it is only detecting the scattered light from a single diode-laser beam. The output from each PMT is amplified and passes to a discriminator to determine if a preset pulse-height threshold is met. The output of each analyzer is fed to a logic AND gate. For coincident scattering events the AND gate output TTL pulse triggers both the pulsed laser source, a frequency-doubled Nd:YAG laser (Spectra Physics model X-30), and the ICCD camera. The coincidence pulse also switches off the CW diode trigger beams, eliminating these unwanted signals from illuminating the ICCD during the "on" time. The crossed-diode-laser trigger system described above effectively reduces the errors due to off-axis aberrations, such as coma mentioned above, by limiting the sample volume to the volume defined by the crossed diode lasers (an ellipsoid with about 300 μm^2 cross section), which is considerably smaller than the aerosol jet.

12.3.3 TAOS MEASUREMENTS OF ATMOSPHERIC AEROSOL

The TAOS prototype described above was used to collect approximately 6000 angular scattering patterns of ambient aerosol particles starting at 3 PM on October 6 and ending at 9 AM on October 7, 2004. Aerosol was sampled at the same MD site described above used for fluorescence measurements. From a statistical perspective this data set is indeed meager. We have no evidence that the patterns would be repeated on different days, different seasons, or during different meteorological conditions. Further measurements are needed to ascertain the generality of the patterns.

Twenty atmospheric aerosol TAOS patterns, captured sequentially, are shown in Figure 12.6. The intensity of the patterns is adjusted for ease of viewing. As is evident, there is a high degree of particle-to-particle variability in the patterns. Some of the patterns are recognizable; for example, images 1, 8, 15, and 20 are likely from spheres. Some patterns appear to arise from perturbed spheres, as suggested by their similarity to sphere-type patterns, whereas some patterns have similarities to those of fibers. Still other patterns have either blotchy islands, or swirl-like patterns of variable intensity and do not readily suggest particular particle shapes.

12.3.4 VISUAL CLASSIFICATION OF SCATTERING PATTERNS AND POSSIBLE PARTICLE MORPHOLOGIES

To examine the scattering pattern characteristics, 2525 patterns (about 500 patterns from five different scattering intensity ranges) were sorted visually. Examples of some patterns that fit into the

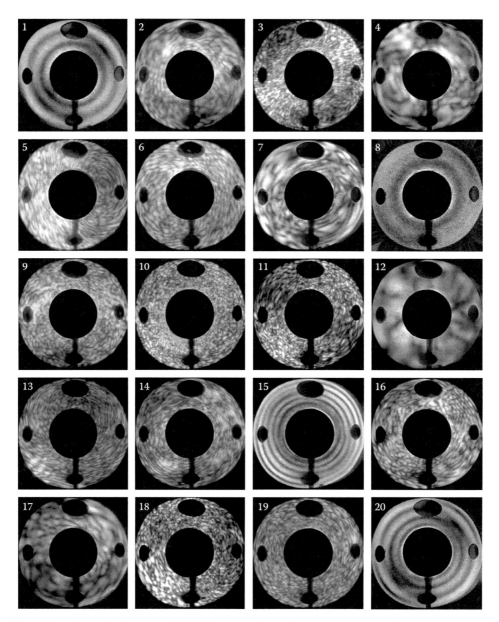

FIGURE 12.6 A sample of sequential TAOS patterns for single atmospheric aerosol particles measured on October 6–7, 2004 at Adelphi, MD. (From Aptowicz, K.B. et al. 2006. *J. Geophys. Res.—Atmospheres*, 111: D12212, doi:10.1029/2005JD006774. With permission.)

five classes into which we sort are shown in Figure 12.7. This figure is a collection of 15 patterns representing five different classes (denoted sphere, perturbed sphere, swirl, fiber, and complex structure) spanning three size ranges. Below are descriptions of the different pattern classes as well as possible characteristics of the aerosol particles.

12.3.4.1 Spherical Particle Patterns
TAOS patterns resulting from particles believed to be spherical, as defined by their characteristic ring-like structure, are shown in the first row of Figure 12.7. Candidate particles for these spherical patterns include: (1) both primary and secondary (formed by atmospheric gas-to-particle reactions)

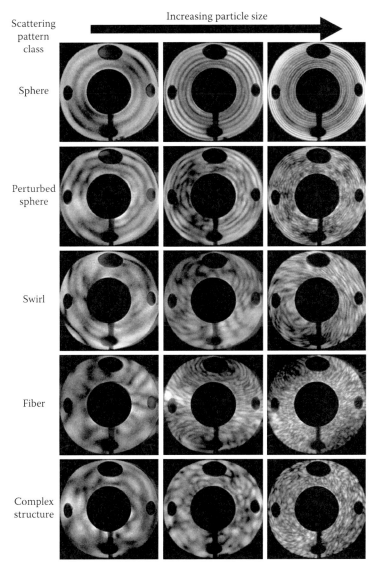

FIGURE 12.7 Example of TAOS patterns for some of the particles measured during October 6–7, 2004 at Adelphi, MD. The TAOS patterns were visually classified into the five categories shown. (From Aptowicz, K.B. et al. 2006. *J. Geophys. Res.—Atmospheres*, 111: D12212, doi:10.1029/2005JD006774. With permission.)

organic carbon particles including single-ring, double-ring, and polycyclic aromatic hydrocarbons, carboxylic acids, organic polymers, humic acids, fulvic acids and HULIS[27,56,63,64,72,73,75,78,117]; (2) aqueous mixtures of inorganic salts including sulfates, nitrates, chlorides, sulfites, and carbonates,[118–120] methane sulfonic acid,[121] sea salt aerosol[118]; (3) some coal and oil fly-ash particles,[122] amorphous carbonaceous tar balls[123]; and (4) silicate and iron oxide spheres.[124]

12.3.4.2 Perturbed Sphere Patterns

TAOS patterns in the second row of Figure 12.7 have a deformed or broken ring-like structure suggestive of a slightly perturbed spherical particle, for example, a slightly rough sphere, a sphere having small particles on its surface, a slightly nonspherical particle (either homogeneous or of mixed composition), or a sphere containing small inclusions.[125,126] Possible atmospheric particles for these patterns may include (1) multicomponent particles, such as organic carbon with sulfates, organic carbon with nitrates, organic carbon with crustal material, organic carbon with black carbon inclusions, organic carbon with sulfuric acid, water-soluble organic carbon with nonwater soluble organic carbon, acid sulfate particles with black carbon inclusions, metal inclusions in sulfuric acid, silicate fly ash, sulfuric acid mixed with crustal material, and smoke from biomass burning and internal combustion engines,[3,122,124,127] (2) neutralized acid sulfates,[128] and (3) nearly spherical pollens or spores.[124]

12.3.4.3 Swirl Patterns

The TAOS patterns in the third row of Figure 12.7, labeled "swirl" patterns seem to be composed of broken or twisted rings. This category differs from the other four categories in that it is defined not by a particle type, but by a type of scattering pattern. The patterns have some similarity to those for droplets with inclusions (perhaps having inclusions larger than those for perturbed spheres) or deformed droplets,[126] multiplet particles,[34] and flake particles.[35] Candidate particles that may produce the patterns are: (1) the mixtures suggested for the perturbed sphere class but with more non-homogeneity, (2) crystalline-like leaf surface waxes dislodged by the wind or by rubbing motions of leaves against each other,[103] (3) multiplet particles, (4) sea-salt crystals,[124] (5) silica shards produced by the combustion of coal,[130] and (6) crystalline particles of quartz or clay minerals.[124]

12.3.4.4 Fiber-Like Particles

The TAOS patterns in the fourth row of Figure 12.7 are characteristic of fibers,[34–36] or doublet particles of similar size.[34] These patterns might be attributable to fiber-like particles (possibly ammonium nitrate crystals),[119] rod-like bacterial endospores,[131] and tire debris (owing to the heavy traffic in the MD locale), or possibly by doublet particles (particles stuck together) of the same or different composition but with similar size.

12.3.4.5 Particles of Complex Structure

The final row of Figure 12.7 shows TAOS patterns comprised primarily of islands where the orientation of the islands does not appear to be strongly correlated to any particular direction. These patterns look similar to those captured previously for aggregates,[36] and to kaolin, Arizona road dust, and to dried droplets containing ammonium sulfate, sodium chloride, or bovine albumin.[116] They also have similarity to scattering patterns of some pollens.[132] The test particle results suggest that the particles corresponding to these patterns likely have a complex structure. Candidate atmospheric particles that may produce these patterns are: (1) mineral dust of soil origin, (2) biological particles with complex morphology that are injected directly into the atmosphere including fragments of skin, leaves, bark, pollens, plant spores, algae, and fungi,[132,133] (3) particles composed of water soluble materials (e.g., salts) that have formed from droplets that dried in such a way that the resulting dry particles were not spherical, (4) agglomerates of particles of the same composition (e.g., aggregated mineral dust of soil origin),[124,134] carbonaceous chain aggregates[128] including aggregate soot particles from oil-fired or coal-fired power plants,[124,135] and diesel engine emissions,[136] and (5) agglomerates of particles of different composition that may include sulfates, nitrates, quartz, clay minerals, organic carbon, black carbon.[124]

12.3.5 Frequency-of-Occurrence of Pattern Classes and Implications for Atmospheric Aerosol

To investigate the frequency-of-occurrence of the various scattering patterns as a function of particle size, subsets of the data ensemble were sorted according to particle-scattering intensity integrated over all measured angles. The result (Table 12.1) reveals a marked decrease in the fraction of spheres (from 42% to <0.1%) as particle size increases from nominally 0.5 to 5 μm (light scattering diameter). The fraction of perturbed spheres also decreases with increasing size, although not as much (from 17% to 5%). The fraction of swirl-like and fiber-like patterns do not vary much with

TABLE 12.1
Frequency-of-Occurrence of Ambient Aerosol Scattering Pattern Class Types for Several Particle Sizes

	Total Particles Analyzed	Spheres (%)	Perturbed Spheres (%)	Swirls (%)	Fiber (%)	Complex Structure (%)
Intensity ~0.5 (nominally 0.5 μm)	460	42	17	26	5	10
Intensity ~1 (nominally 1 μm)	523	42	23	22	5	9
Intensity ~2 (nominally 2 μm)	505	21	24	26	4	25
Intensity ~4 (nominally 3 μm)	527	6	13	32	6	43
Intensity ~8 (nominally 5 μm)	510	<0.1	5	18	6	71

Note: Most particles in the light gray region are likely formed by gas-to-particle reactions in the atmosphere; most particles in the dark gray region are likely injected directly into the atmosphere. Particle scattering patterns were measured during October 6–7, 2004 at Adelphi, MD.

size. On the other hand, there is a significant increase in the number of complex-structure particle patterns with increasing size; from 10% to 71%.

These findings are consistent with the notion that micron-sized particles are primarily liquid, and are formed mainly by heterogeneous nucleation gas-to-particle reaction processes in the atmosphere,[72,73,118,127,137–140] whereas most super-micron particles are directly injected into the atmosphere from a variety of sources,[118,133,140] or may be formed by agglomeration of smaller particles in the atmosphere.

12.3.6 TAOS OF ATMOSPHERIC AEROSOL USING IMPROVED ELLIPSOIDAL MIRROR AND DIFFERENT GEOMETRY

More recently, TAOS measurements have been made using a diamond-machined ellipsoidal mirror with improved imaging characteristics and with different illumination and collection geometry. Instead of the green probe laser being directed along the axis of the mirror (as Figure 12.5), the probe laser was directed perpendicular to the axis, allowing for both forward and backward scattering of single particles to be measured. A sample of a typical sphere-like scattering pattern, a perturbed sphere pattern, a swirl pattern, and a complex structure pattern of atmospheric aerosol particles measured in Las Cruces, NM on January 21, 2007, around 11:07 PM local time are shown in Figure 12.8. The pattern labeled ICCD image refers to the 1024 × 1024 pixel raw image. Each image is then projected onto the spherical coordinate system centered on the scattering particle and forward and backward scattering patterns are generated. These patterns represent a small subset of approximately 10,000 patterns collected between 8:28 PM on January 21 and ending at 1:48 AM on January 22, 2007. More detailed analyses of these data are ongoing.

12.4 SUMMARY

LIF spectra and angular elastic scattering patterns of single micron-sized atmospheric aerosol particles can be measured in near real time. These capabilities provide a means to study the diurnal and seasonal variability, the concentration variability, the compositional variability, and the morphological variability of atmospheric aerosol. Although many compounds can have similar LIF spectra, and although LIF spectra are relatively broad and lack sharp distinguishing features, such as those occurring in Raman or IR spectra; single-particle LIF spectroscopy is clearly useful for looking for increases in biological materials in the atmosphere. Typically, only about one percent of atmospheric aerosol exhibit fluorescence spectra similar to those of bacteria.

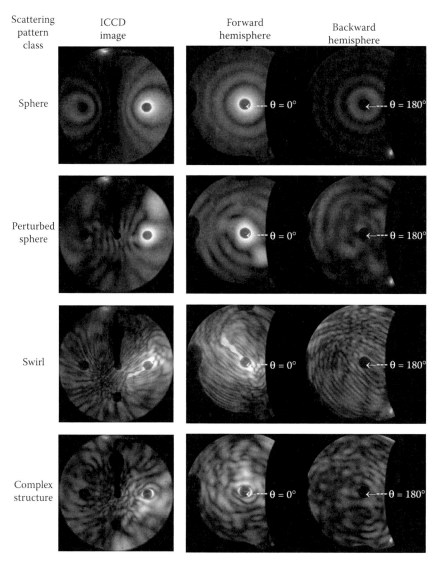

FIGURE 12.8 Sample TAOS patterns collected using different scattering geometry. The TAOS patterns were visually classified into the four categories previously mentioned in the text. Each ICCD image is reconstructed into the forward and backward scattering hemispheres as shown. The patterns were measured in Las Cruces, NM on January 21, 2007 around 11:07 pm local time.

The utility of single-particle TAOS patterns is less clear. If nothing else, these patterns might at least offer the potential to differentiate between spherical and nonspherical or inhomogeneous particles. More research is required in developing automated methods for recognizing particle shapes and classes. Also, it appears that much more work needs to be done in determining the particle types in which bacteria or other cells or materials occur in the natural atmosphere. It has been demonstrated[115] that TAOS from laboratory generated bacterial spores compare favorably with calculated TAOS patterns of rod-like structures with end caps. However, the range of bacteria-containing particle types in the earth's atmosphere is not known. One avenue to explore that appears promising is to measure for each particle, both the LIF spectrum and angular scattering pattern, select the particles with bacteria-like LIF spectra, and investigate to what extent the angular scattering patterns of this subset of bacteria-like particles can be classified.

REFERENCES

1. Prather, K.A., Nordmeyer, T., and Salt, K. 1994. Real-time characterization of individual aerosol-particles using time-of-flight mass-spectrometry. *Anal. Chem.*, 66: 1403–1407.
2. Noble, C.A. and Prather, K.A. 2000. Real-time single particle mass spectrometry: A historical review of a quarter century of the chemical analysis of aerosols. *Mass Spectrom. Rev.*, 19: 248–274.
3. Lee, S.-H., Murphy, D.M., Thomson, D.S., and Middlebrook, A.M. 2002. Chemical components of single particles measured with particle analysis by laser mass spectrometry (PALMS) during the Atlanta Supersite Project: Focus on organic/sulfate, lead, soot, and mineral particles. *J. Geophys. Res., [Atmos.]*, 107, 4003, doi:10.1029/2000JD000011.
4. Murphy, D.M., Middlebrook, A.M., and Warshawsky, M. 2003. Cluster analysis of data from the particle analysis by laser mass spectrometry (PALMS) instrument. *Aerosol Sci. Technol.*, 37: 382–391.
5. Spencer, M.T. and Prather K.A. 2006. Using ATOFMS to determine OC/EC mass fractions in particles. *Aerosol Sci. Technol.*, 40(8): 585–594.
6. Moffet, R.C., de Foy, B., Molina, L.T., Molina, M.J., and Prather, K.A. 2008. Measurement of ambient aerosols in northern Mexico City by single particle mass spectrometry. *Atmos. Chem. Phys.*, 8(16): 4499–4516.
7. Radziemski, L., Loree, T., Cremers, D., and Hoffman, N. 1983. Time-resolved laser-induced breakdown spectrometry of aerosols. *Anal. Chem.*, 55: 1246–1252.
8. Hahn, D.W. 1998. Laser-induced breakdown spectroscopy for sizing and elemental analysis of discrete aerosol particles. *Appl. Phys. Lett.*, 72: 2960–2962.
9. Hahn, D.W. and Lunden, M.M. 2000. Detection and analysis of aerosol particles by laser-induced breakdown spectroscopy. *Aerosol Sci. Technol.*, 33: 30–48.
10. Samuels, A.C., DeLucia, F.C., McNesby, K.L., and Miziolek, A.W. 2003. Laser-induced breakdown spectroscopy of bacterial spores, molds, pollens, and protein: Initial studies of discrimination potential. *Appl. Opt.*, 42(30): 6205–6209.
11. Hohreiter, V. and Hahn, D.W. 2005. Dual-pulse laser induced breakdown spectroscopy: Time-resolved transmission and spectral measurements, *Spectrochim. Acta Part B—Atomic Spectrosc.*, 60(7–8): 968–974.
12. Hahn, D.W. 2009. Laser-induced breakdown spectroscopy for analysis of aerosol particles: The path toward quantitative analysis. *Spectroscopy*, 24(9): 26–33.
13. Pinnick, R.G., Hill, S.C., Nachman, P., Pendleton, J.D., Fernandez, G.L., Mayo, M.W., and Bruno, J.G., 1995. Fluorescence particle counter for detecting airborne bacteria and other biological particles. *Aerosol Sci. Technol.*, 23: 653–664.
14. Hairston, P.P., Ho, J., and Quant, F.R. 1997. Design of an instrument for real-time detection of bioaerosols using simultaneous measurement of particle aerodynamic size and intrinsic fluorescence. *Aerosol Sci. Technol.*, 28: 471–482.
15. Reyes, F.L., Jeys, T.H., Newbury, N.R., Primmerman, C.A., Rowe, G.S., and Scanchez, A. 1999. Bio-aerosol fluorescence sensor. *Field Anal. Chem. Technol.*, 3: 240–248.
16. Seaver, M., Eversole, J.D., Hardgrove, J.J., Cary, W.K. Jr., and Roselle, D.C. 1999. Size and fluorescence measurements for field detection of biological aerosols. *Aerosol Sci. Technol.*, 30: 174–185.
17. Eversole, J.D., Hardgrove, J.J., Cary Jr., W.K., Choulas, D.P., and Seaver, M. 1999. Continuous, rapid biological aerosol detection with the use of UV fluorescence: Outdoor test results. *Field Anal. Chem. Technol.*, 3: 249–259.
18. Kaye, P.H., Barton, J.E., Hirst. E., and Clark, J.M. 2000. Simultaneous light scattering and intrinsic fluorescence measurement for the classification of airborne particles. *Appl. Opt.*, 39: 3738–3745.
19. Ho, J. 2002. Review: Future of biological aerosol detection. *Anal. Chim. Acta*, 457: 125–148.
20. Kaye, P.H., Stanley, W.R., Hirst, E., Foot, E. V., Baxter, K.L., and Barrington, S.J. 2005. Single particle multichannel bio-aerosol fluorescence sensor. *Opt. Express*, 13: 3583–3593.
21. Sivaprakasam, V., Huston, A.L., Scotto, C., and Eversole, J.D. 2004. Multiple UV wavelength excitation and fluorescence of bioaerosols. *Opt. Express*, 12: 4457–4466.
22. Pinnick, R.G., Hill, S.C., Nachman, P., Videen, G., Chen, G., and Chang, R.K. 1998. Aerosol fluorescence spectrum analyzer for rapid measurement of single micrometer-sized airborne particles. *Aerosol Sci. Technol.*, 28: 95–104.
23. Hill, S.C., Pinnick, R.G., Pan, Y.L., Holler, S., Chang, R.K., Bottiger, J.R., Chen, B.T., Orr, C.-S., and Feather, G. 1999. Real-time measurement of fluorescence spectra from single airborne biological particles. *Field Anal. Chem. Technol.*, 3: 221–239.
24. Pan, Y.L., Holler, S., Chang, R.K., Hill, S.C., Pinnick, R.G., Niles, S., and Bottiger, J.R. 1999. Single-shot fluorescence spectra of individual micrometer-sized bioaerosols illuminated by a 351- or 266-nm ultraviolet laser. *Opt. Lett.*, 24: 116–118.

25. Pan, Y.L., Cobler, P., Rhodes, S., Potter, A., Chou, T., Holler, S., Chang, R.K., Pinnick, R.G., and Wolf, J.P. 2001. High-speed, high-sensitivity aerosol fluorescence spectrum detection using a 32-anode photo-multiplier tube detector. *Rev. Sci. Instr.*, 72: 1831–1836.

26. Pan, Y.L., Hartings, J., Pinnick, R.G., Hill, S.C., Halverson, J., and Chang, R.K. 2003. Single-particle fluorescence spectrometer for ambient aerosols. *Aerosol Sci. Technol.*, 37: 628–639.

27. Pinnick, R.G., Hill, S.C., Pan, Y.L., and Chang, R.K. 2004. Fluorescence spectra of atmospheric aerosol at Adelphi, Maryland, USA: Measurement and classification of single particles containing organic carbon. *Atmos. Environ.*, 38: 1657–1672.

28. Pan, Y.L., Pinnick, R.G., Hill, S.C., Rosen, J.L., and Chang, R.K., 2007. Single-particle laser-induced-fluorescence spectra of biological and other organic-carbon aerosols in the atmosphere: measurements at New Haven, CT and Las Cruces, NM, USA. *J. Geophys. Res. Atmos.*, 112: doi:1029/2007JD008741.

29. Pan, Y., Chang, R.K., Hill, S.C., and Pinnick, R.G. 2008. Using single-particle fluorescence to detect bioaerosols, *Opt. Photon. News*, 19(9), 30–33.

30. Huang, H.C., Yong-Le, P., Hill, S.C., Pinnick, R.G., and Chang, R.K. 2008. Real-time measurement of dual-wavelength laser-induced fluorescence spectra of individual aerosol particles, *Opt. Express*, 16(21): 16523–16528.

31. Pan, Y.L., Pinnick, R.G., Hill, S.C., and Chang, R.K. 2009. Particle-fluorescence spectrometer for real-time single-particle measurements of atmospheric organic carbon and biological aerosol. *Environ. Sci. Technol.*, 43: 429–434.

32. Marshall, T.R., Parmenter C.S., and Seaver, M. 1976. Characterization of polymer latex aerosols by rapid measurement of 360 degree light-scattering patterns from individual particles. *J. Colloid Interface Science*, 55(3): 624–636.

33. Wyatt, P.J., Schehrer, K.L., Phillips, S.D., Jackson, C., Chang, Y.J., Parker, R.G., Phillips, D.T., and Bottiger, J.R. 1988. Aerosol-particle analyzer, *Appl. Opt.*, 27(2): 217–221.

34. Kaye, P.H., Hirst E., Clark, J.M., and F. Micheli, J.M. 1992. Airborne particle-shape and size classification from spatial light-scattering profiles. *J. Aerosol Sci.*, 23(6): 597–611.

35. Kaye, P.H. 1998. Spatial light-scattering analysis as a means of characterizing and classifying non-spherical particles. *Meas. Sci. Technol.* 9: 141–149.

36. Pan, Y.L., Aptowicz, K.B., Hart, M., Eversole, J.D., and Chang, R.K. 2003. Characterizing and monitoring respiratory aerosols by light scattering. *Opt. Lett.*, 28(8): 589–591.

37. Aptowicz, K.B., Pinnick, R.G., Hill, S.C., Pan, Y.L., and Chang, R.K. 2006. Optical scattering patterns from single urban aerosol particles at Adelphi, Maryland, USA: A classification relating to particle morphologies. *J. Geophys. Res.—Atmospheres*, 111: D12212, doi:10.1029/2005JD006774.

38. Eversole, J.D., Cary, W.K., Scotto, C.S., Pierson, R., Spence, M., and Campillo, A.J. 2001. Continuous bioaerosol monitoring using UV excitation fluorescence: Outdoor test results. *Field Anal. Chem. Technol.*, 15: 205–212.

39. Snyder, A.P., Maswadeh, W.M., Tripathi, A., Eversole, J., Ho, J., and Spence, M. 2004. Orthogonal analysis of mass and spectral based technologies for the field detection of bioaerosols. *Anal. Chim. Acta*, 513: 365–377.

40. Sivaprakasam, V., Huston, A., Lin, H.B., Eversole, J., Falkenstein, P., and Schultz, A. 2007. Field test results and ambient aerosol measurements using dual wavelength fluorescence excitation and elastic scatter for bioaerosols. In Fountain III, A.W. (Ed.), *Chemical and Biological Sensing* VIII, *Proc. of SPIE* 6554(65540R-1), doi:10.1117/12.719326.

41. Gabey, A.M., Gallagher, M.W., Whitehead, J., and Dorsey, J. 2009. Measurements of coarse mode and primary biological aerosol transmission through a tropical forest canopy using a dual-channel fluorescence aerosol spectrometer. *Atmos. Chem. Phys. Discuss.*, 9: 18965–18984.

42. Nachman, P., Chen, G., Pinnick, R.G., Hill, S.C., Chang, R.K., Mayo, M.W., and Fernandez, G.L. 1996. Conditional-sampling spectrograph detection system for fluorescence measurements of individual airborne biological particles. *Appl. Opt.*, 35: 1069–1076.

43. Chen, P., Nachman, P., Pinnick, R.G., Hill, S.C., and Chang, R.K. 1996. Conditional firing aerosol fluorescence spectrum analyzer for individual airborne particles with pulsed 266-nm excitation. *Opt. Lett.*, 21: 1307–1309.

44. Hartigan, J.A. 1975. *Clustering Algorithms*. Wiley, New York, NY, Ch. 4.

45. Mueller, P.K. 1982. Atmospheric particulate carbon observations in urban and rural areas of the United States. In Wolff, G.T. and Klimisch, R.L. (Eds.), *Particulate Carbon, Atmospheric Life Cycle*. Plenum Press, New York, NY, pp. 343–370.

46. Shah, J.J., Johnson, R.L., Heyerdahl, E.K., and Huntzicker, J.J. 1986. Carbonaceous aerosol at urban and rural sites in the United States. *J. Air Pollut. Control Assoc.*, 36: 254–257.

47. Huntzicker, J.J., Heyerdahl, E.K., McDow, S.R., Rau, J.A., Griest, W.H., and MacDougall, C.S., 1986. Combustion as the principal source of carbonaceous aerosol in the Ohio river valley. *J. Air Pollution Control Assoc.*, 36: 705–709.

48. Larson, S.M., Cass, G.R., and Gray, H.A. 1989. Atmospheric carbon particles and the Los Angeles visibility problem. *Aerosol Sci. Technol.*, 10: 118–130.

49. White, W.H. and Macias, E.S. 1989. Carbonaceous particles and regional haze in the western United States. *Aerosol Sci. Technol.*, 10: 111–117.

50. Novakov, T. and Penner, J.E. 1993. Large contribution of organic aerosols to cloud condensation nuclei concentrations. *Nature*, 365: 823–826.

51. Rogge, W.F., Hildemann, L.M., Mazurek, M.A., Cass, G.R., and Simoneit, B.R.T. 1993. Sources of fine organic aerosol: Particulate abrasion products from leaf surfaces of urban plants. *Environ. Sci. Technol.* 27: 2700–2710.

52. Rogge, W.F., Mazurek, M.A., Hildemann, L.M., and Cass, G.R. 1993. Quantification of urban organic aerosols at a molecular level: Identification of abundance and seasonal variation. *Atmos. Environ.* 27A: 1309–1330.

53. Lighthart, B. and Mohr, A.J. (Eds). 1994. *Atmospheric Microbial Aerosols*. Chapman & Hall, New York.

54. Madelin, T.M. and Madelin, M.F. 1995. Biological analysis of fungi and associated molds. In Cox, C.S. and Wathes, C.M. (Eds.), *Bioaerosols Handbook*. CRC Press, Boca Raton, FL, pp. 361–386.

55. Kunit, M. and Puxbaum, H. 1996. Enzymatic determination of the cellulose content of atmospheric aerosols. *Atmos. Environ.*, 30: 1233–1236.

56. Saxena, P. and Hildeman, L. 1996. Water-soluble organics in atmospheric particles: A critical review of the literature and application of thermodynamics to identify candidate compounds. *J. Atmos. Chem.*, 24: 57–109.

57. Havers, N., Burba, P., Lambert, J., Klockkow, D. 1998. Spectroscopic characterization of humic-like substances in airborne particulate matter. *J. Atmos. Chem.*, 29: 45–54.

58. Lighthart, B. 1997. The ecology of bacteria in the alfresco atmosphere. *FEMS Microbiol. Ecol.*, 23: 263–274.

59. Lighthart, B. and Tong, Y. 1998. Measurements of total and culturable bacteria in the alfresco atmosphere using a wet-cyclone sampler. *Aerobiologia*, 14: 325–332.

60. Seinfeld, J.H. and Pandis, S.N. 1998. *Atmospheric Chemistry and Physics: From Air Pollution to Climate Change*. Wiley, New York.

61. Larson, S.M., Cass, G.R., and Gray, H.A. 1989. Atmospheric carbon particles and the Los Angeles visibility problem. *Aerosol Sci. Technol.*, 10: 118–130.

62. Hitzenberger, R., Berner, A., Giebl, H., Kromp, R., Larson, S.M., Rouc, A., Koch, A., Marischka, S., and Puxbaum, H. 1999. Contribution of carbonaceous material to cloud condensation nuclei concentrations in European background (Mt. Sonnblick) and urban (Vienna) aerosols. *Atmos. Environ.*, 33: 2647–2659.

63. Zappoli, S., Andracchio, A., Fuzzi, S., Facchini, M.C., Gelencser, A., Kiss, G., Krivacsy, Z., et al. 1999. Inorganic, organic and macromolecular components of fine aerosol in different areas of Europe in relation to their water solubility. *Atmos. Environ.*, 33: 2733–2743.

64. Krivacsy, Z., Kiss, Gy., Varga, B., Galambos, I., Sarvari, Zs., Gelencser, A., Molnar, A. et al. 2000. Study of humic-like substances in fog and interstitial aerosol by size-exclusion chromatography and capillary electrophoresis. *Atmos. Environ.*, 34: 4273–4281.

65. Krivacsy, Z., Gelencser, A., Kiss, G., Meszaros, E., Molnar, A., Hoffer, A., Meszaros, T. et al. 2001. Study on the chemical character of water soluble organic compounds in fine atmospheric aerosol at the Jungfraujoch. *J. Atmos. Chem.*, 39: 245–259.

66. Gelencser, A., Sallai, M., Krivacsy, Z., Kiss, G., and Meszaros, E. 2000. Voltammetric evidence for the presence of humic-like substances in for water. *Atmos. Res.*, 54: 157–165.

67. Bauer, H., Kasper-Giebl, A., Zibuschka, F., Hitzenberger, R., Kraus, G.F., and Puxbaum, H. 2002a. Determination of the carbon content of airborne fungal spores. *Anal. Chem.*, 74: 91–95.

68. Bauer, H., Kasper-Giebl, A., Loflund, M., Giebl, H., Hitzenberger, R., Zibuschka, F., and Puxbaum, H. 2002b. The contribution of bacteria and fungal spores to the organic carbon content of cloud water, precipitation and aerosols. *Atmos. Res.*, 64: 109–119.

69. Bauer, H., Fuerhacker, M., Zibuschka, F., Schmid, H., and Puxbaum, H. 2002c. Bacteria and fungi in aerosols generated by two different types of wastewater treatment plants. *Water Res.*, 36: 3965–3970.

70. Fraser, M.P., Yue, Z.W., Tropp, R.J., Kohl, S.D., and Chow, J.C. 2002. Molecular composition of organic fine particulate matter in Houston, TX. *Atmos. Environ.*, 36: 5751–5758.

71. Chen, L.-W.A., Doddridge, B.G., Dickerson, R.R., Chow, J.C., and Henry, R.C. 2002. Origins of fine aerosol mass in the Baltimore-Washington corridor: Implications from observation, factor analysis, and ensemble air parcel back trajectories. *Atmos. Environ.*, 36: 4541–4554.

72. Decesari, S., Facchini, M.C., Matta, E., Mircea, M., Fuzzi, S., Chughtai, A.R., and Smith, D.M., 2002. Water soluble organic compounds formed by oxidation of soot. *Atmos. Environ.*, 36: 1827–1832.

73. Gelencser, A., Hoffer, A., Krivacsy, Z., Kiss, G., Molnar, A., and Meszaros, E. 2002. On the possible origin of humic matter in fine continental aerosol. *J. Geophys. Res.*, 107: ACH2 1–5.

74. Herckes, P., Hannigan, M.P., Trenary, L., Lee, T., and Collett, J.L. 2002. Organic compounds in radiation fogs in Davis, California. *Atmos. Res.*, 64: 99–108.

75. Kiss, G., Varga, B., Galambos, I., and Ganszky, I. 2002. Characterization of water-soluble organic matter isolated from atmospheric fine aerosol. *J. Geophys. Res.*, 107, 8339: ICC 1–8.

76. Limbeck, A., Handler, M., Neuberger, B., Klatzer, B., and Puxbaum, H. 2005. Carbon-specific analysis of humic-like substances in atmospheric aerosol and precipitation samples. *Anal. Chem.* 77(22): 7288–7293.

77. Robinson, A.L., Subramanian, R., Donahue, N.M., Bernardo-Bricker, A., and Rogge, W.F. 2006. Source apportionment of molecular markers and organic aerosols-1. Polycyclic aromatic hydrocarbons and methodology for data visualization. *Environ. Sci. Technol.*, 40(24): 7803–7810.

78. Rogge, W.F., Medeiros, P.M., and Simoneit, B.R.T. 2006. Organic marker compounds for surface soil and fugitive dust from open lot dairies and cattle feedlots. *Atmos. Environ.*, 40(1): 27–49.

79. Feczko, T., Puxbaum, H., Kasper-Giebl, A., Handler, M., Limbeck, A., Gelencser, A., Pio, C., Preunkert, S., and Legrand, M. 2007. Determination of water and alkaline extractable atmospheric humic-like substances with the TU Vienna HULIS analyzer in samples from six background sites in Europe. *J. Geophys. Res.*, 112(D23): D23S10.

80. Sanchez-Ochoa, A., Kasper-Giebl, A., Puxbaum, H., Gelencser, A., Legrand, M., and Pio, C. 2007. Concentration of atmospheric cellulose: A proxy for plant debris across a west–east transect over Europe. *J. Geophys. Res. Atmos.*, 112(D23): D23S08.

81. Elbert, W., Taylor, P.E., Andreae, M.O., and Poschl, U. 2007. Contribution of fungi to primary biogenic aerosols in the atmosphere: Wet and dry discharged spores, carbohydrates, and inorganic ions. *Atmos. Chem. Phys.*, 7: 4569–4588.

82. Kotianova, P., Puxbaum, H., Bauer, H., Caseiro, A., Marr, I. L., and Cik, G. 2008. Temporal patterns of *n*-alkanes at traffic exposed and suburban sites in Vienna. *Atmos. Environ.*, 42(13): 2993–3005.

83. Bauer, H., Schueller, E., Weinke, G., Berger, A., Hitzenberger, R., Marr, I.L., and Puxbaum, H. 2008. Significant contributions of fungal spores to the organic carbon and to the aerosol mass balance of the urban atmospheric aerosol. *Atmos. Environ.*, 42(22): 5542–5549.

84. Deguillaume, L., Leriche, M., Amato, P., Ariya, P.A., Delort, A.M., Poschl, U., Chaumerliac, N., Bauer, H., Flossmann, A.I., and Morris, C.E. 2008. Microbiology and atmospheric processes: Chemical interactions of primary biological aerosols. *Biogeosciences*, 5(4): 1073–1084.

85. Winiwarter, W., Bauer, H., Caseiro, A., and Puxbaum, H. 2009. Quantifying emissions of primary biological aerosol particle mass in Europe. *Atmos. Environ.*, 43(7): 1403–1409.

86. Lukacs, H., Gelencser, A., Kiss, G., Horvath, K., and Hartyani, Z. 2009. Quantitative assessment of organosulfates in size-segregated rural fine aerosol. *Atmos. Chem. Phys.*, 9(1): 231–238.

87. Heald, C.L. and Spracklen, D.V. 2009. Atmospheric budget of primary biological aerosol particles from fungal spores. *Geophys. Res. Lett.* 36: (L09806).

88. Wiedinmyer, C., Bowers, R.M., Fierer, N., Horanyi, E., Hannigan, M., Hallar, A.G., McCubbin, I., and Baustian, K. 2009. The contribution of biological particles to observed particulate organic carbon at a remote high altitude site. *Atmos. Environ.*, 43(28): 4278–4282.

89. Seinfeld, J.H. and Pandis, S.N. 2006. *Atmospheric Chemistry and Physics: From Air Pollution to Climate Change*, 2nd edn. Wiley, New York.

90. Tong, Y. and Lighthart, B. 1999. Diurnal distribution of total and culturable atmospheric bacteria at a rural site. *Environ. Sci. Technol.*, 30: 246–254.

91. Oppo, C., Bellandi, S., Degli Innocenti, N., Stortini, A.M., Loglio, G., Schiavuta, E., and Cini, R. 1999. Surfactant components of marine organic matter as agents for biogeochemical fractionation and pollutant transport via marine aerosols. *Marine Chem.*, 63: 235–253.

92. Olmstead, J.A. and Gray, D.G. 1997. Fluorescence spectroscopy of cellulose, lignin and mechanical pulps: A review. *J. Pulp Paper Sci.*, 23: J571–J581.

93. Puxbaum, H. and Tenze-Kunit, M. 2003. Size distribution and seasonal variation of atmospheric cellulose. *Atmos. Environ.*, 37: 3693–3699.

94. De Souza Sierra, M.M., Giovanela, M., and Soriano-Sierra, E.J., 2000. Fluorescence properties of well-characterized sedimentary estuarine humic compounds and surrounding pore waters. *Environ. Technol.*, 21: 979–988.

95. Klapper, L., McNight, D.M., Fulton, J.R., Blunt-Harris, E.L., Nevin, K.P., Lovley, D.R., and Hatcher, P.G. 2002. Fulvic acid oxidation state detection using fluorescence spectroscopy. *Environ. Sci. Technol.*, 36: 3170–3175.

96. Ouatmane, A., D'Orazio, V., Hafidi, M., and Senesi, N. 2002. Chemical and physicochemical character-ization of humic acid-like materials from composts. *Compost Sci. Utilization*, 10: 39–46.

97. Chen, J., LeBoeuf, E.J., Dai, S., and Gu, B. 2003. Fluorescence spectroscopic studies of natural organic matter fractions. *Chemosphere*, 50: 639–647.

98. Duck, T.J., Firanski, B.J., Miller, D.B., Goldstein, A.H., Allan, J., Holzinger, R., Worsnop, D.R. et al. 2007. Transport of forest fire emissions from Alaska and the Yukon Territory to Nova Scotia during summer 2004. *J. Geophys. Res.*, 112: D10S44, doi:10.1029/2006JD007716.

99. VanCuren, R.A. and Cahill, T.A. 2002. Asian aerosols in North America: Frequency and concentration of fine dust. *J. Geophys. Res.*, 107(D24): 4804, doi:10.1029/2002JD002204.

100. Mazurek, M.A. and Cass, G.R. 1991. Biological input to visibility-reducing aerosol particles in the remote arid southwestern United States. *Environ. Sci. Technol.*, 25: 684–694.

101. Dasch, J.M. and Cadle, S.H. 1989. Atmospheric carbon particles in the Detroit urban area: Wintertime sources and sinks. *Environ. Sci. Technol.*, 10: 236–248.

102. Robinson, A.L., Subramanian, R., Donahue, N.M., Bernardo-Bricker, A., and Rogge, W.F. 2006. Source apportionment of molecular markers and organic aerosol. 3. Food cooking emissions. *Environ. Sci. Technol.*, 40 (24): 7820–7827.

103. Rogge, W.F., Mazurek, M.A., Hildemann, L.M., and Cass, G.R. 1993. Quantification of urban organic aerosols at a molecular level: Identification of abundance and seasonal variation. *Atmos. Environ.*, 27A: 1309–1330.

104. Pinnick, R.G., Chylek, P., Jarzembski, M., Creegan, E., Srivastava, V., Fernandez, G., Pendleton, J.D., and Biswas, A. 1988. Aerosol-induced laser breakdown thresholds: Wavelength dependence. *Appl. Opt.*, 27: 987–996.

105. Hill, S.C., Pinnick, R.G., Niles, S., Fell, N.F., Pan, Y.L., Bottiger, J., Bronk, B.V., Holler, S., and Chang, R.K. 2001. Fluorescence from airborne microparticles: Dependence on size, concentration of fluoro-phores, and illumination intensity. *Appl. Opt.*, 40: 3005–3013.

106. Richmond, J.H. 1965. Scattering by a dielectric cylinder of arbitrary cross section shape. *IEEE Trans. Antennas Propagation*, AP-13: 334–341.

107. Harrington, R.F. 1968. *Field Computation by Moment Methods* (Harrington: 1968) [reprinted by the IEEE Press Series on Electromagnetic Wave Theory, 1993 and 2001].

108. Lakhtakia, A. 1992. General theory of the Purcell–Pennypacker scattering approach and its extension to bianisotropic scatterers. *Astrophys. J.*, 394: 494–499.

109. Yee, K.S. 1966. Numerical solution of initial boundary value problems involving Maxwell's equations in isotropic media. *IEEE Trans. Antennas Propagation*, AP-14: 302–307.

110. Waterman, P.C. 1971. Symmetry, unitarity and geometry in electromagnetic scattering. *Phys. Rev. D.*, 3: 825–839.

111. Gucker, F.T., Tuma, J., Lin, H.-M., Huang, C.-M, Ems, S.C., and Marshall, T.R. 1973. Rapid measure-ment of light-scattering diagrams from single particles in an aerosol stream and determination of latex particle size. *J. Aerosol Sci.*, 4: 389–404.

112. Bartholdi, M., Salzman, G.C., Hiebert, R.D., and Kerker, M. 1980. Differential light scattering photom-eter for rapid analysis of single particles in flow. *Appl. Opt.*, 19(10): 1573–1581.

113. Hirst, E. and Kaye, P.H. 1996. Experimental and theoretical light scattering profiles from spherical and nonspherical particles. *J. Geophys. Res.*, 101(D14): 19,231– 19,235.

114. Dick, W.D., Ziemann, P.J., Huang, P.F., and McMurry, P.H. 1998. Optical shape fraction measurements of submicrometre laboratory and atmospheric aerosols. *Measure. Sci. Technol.*, 9(2): 183–196.

115. Auger, J.C., Aptowicz, K.B., Pinnick, R.G., Pan, Y.L., and Chang, R.K. 2007, Angularly resolved light scattering from aerosolized spores: Observations and calculations. *Opt. Lett.*, 32: 3358–3360.

116. Aptowicz, K.B. 2005. Angularly-resolved elastic light scattering of micro-particles. Ph.D. thesis, 28pp., Yale University, New Haven, CT.

117. Mukai, A. and Ambe Y. 1986. Characterization of humic acid-like brown substance in airborne particu-late matter and tentative identification of its origin. *Atmos. Environ.*, 20: 813–819.

118. Prospero, J.M., Charlson, R.J., Mohnen, V., Jaenicke, R., Delany, A.C., Moyers, J., Zoller, W., and Rahn, K., 1983. The atmospheric aerosol system. An overview. *Rev. Geophys.*, 21(7): 1607–1629.

119. Kopcewicz, B., Nagamoto, C., Parungo, R., Harris, J., Miller, J., Sievering, H., and Rosinski, J. 1991. Morphological studies of sulfate and nitrate particles on the east coast of North America and over the North Atlantic Ocean. *Atmos. Res.*, 26: 245–271.

120. Charlson, R.J. and Wigley, T.M.L. 1994. Sulfate aerosol and climate change, *Sci. Am.*, 270(2): 28–35.

121. Qian, G.-W. and Ishizaka, Y. 1993. Electron microscope studies of methane sulfonic acid in individual aerosol particles. *J. Geophys. Res.*, 98(C5): 8459–8470.

122. Mamane, Y., Miller, J.L., and Dzubay, T.G. 1986. Characterization of individual fly ash particles emitted from coal- and oil-fired power plants. *Atmos. Environ.*, 20(11): 2125–2135.

123. Posfai, M., Gelencsér, A., Simonics, R., Arató, K., Li, J., Hobbs, P.V., and Buseck, P.R. 2004. Atmospheric tar balls: Particles from biomass and biofuel burning. *J. Geophys. Res.*, 109: D06213, doi:10.1029/2003JD004169.

124. Ebert, M., Weinbruch S., Rausch A., Gorzawski G., Helas G., Hoffmann P., and Wex H. 2002. Complex refractive index of aerosols during LACE 98 as derived from the analysis of individual particles. *J. Geophys. Res.*, 107(D21): 8121, doi:10.1029/2000JD000195.

125. Videen, G., Sun, W., Fu, Q., Secher, D.R., Greenaway, R.S., Kaye, P.H., Hirst, E., and Bartley, D., 2000. Light scattering from deformed droplets and droplets with inclusions. II Theoretical treatment. *Appl. Opt.*, 39(27): 5031–5039.

126. Secker, D.R., Kaye, P.H., Greenaway, R.S., Hirst, E., Bartley, D.L., and Videen, G. 2000. Light scattering from deformed droplets and droplets with inclusions. *Appl. Opt.*, 39(27): 5023–5030.

127. Lim, H.-J. and Turpin, B. J. 2002. Origins of primary and secondary organic aerosol in Atlanta; Results of time-resolved measurements during the Atlanta supersite experiment, *Environ. Sci. Technol.*, 36(21): 4489–4496.

128. Sheridan, P.J., Schnell, R.C., Kahl, J.D., Boatman, J.F., and Garvey, D.M. 1993. Microanalysis of the aerosol collected over south-central New Mexico during the ALIVE field experiment, May–December 1989. *Atmos. Environ.*, Part A, 27: 1169–1183.

129. Ashkin, A. and Dziedzic, J.M. 1980. Observation of light-scattering from nonspherical particles using optical leviation. *Appl. Opt.*, 19(5): 660–668.

130. Rietmeijer, J.M. and Janeczek, J. 1997. An analytical electron microscope study of airborne industrial particles in Sosnowiec, Poland. *Atmos. Environ.*, 31(13): 1941–1951.

131. Shaffer, B.T. and Lighthart, B. 1997. Survey of culturable airborne bacteria at four diverse locations in Oregon: Urban, rural, forest, and coastal. *Microb. Ecol.*, 34: 167–177.

132. Surbek, M., Esen, C., and Schweiger, G. 2009. Elastic light scattering on single pollen: Scattering in a large space angle. *Aerosol Sci. Technol.*, 43(7): 679–684.

133. Jaenicke, R. 2005. Abundance of cellular material and proteins in the atmosphere. *Science* 308: 73.

134. Pinnick, R.G., Fernandez, G., Hinds, B.D., Bruce, C.W., Schaefer, R.W., and Pendleton, J.D. 1985. Dust generated by vehicular traffic on unpaved roadways: Sizes and infrared extinction characteristics. *Aerosol Sci. Technol.*, 4: 99–121.

135. Chylek, P., Ramaswamy, V., Cheng, R., and Pinnick, G. 1981. Optical properties and mass concentration of carbonaceous smokes. *Appl. Opt.*, 20(17): 2980–2985.

136. Huang, X.-F., Yu, J.Z., He, L.-Y., and Yuan, Z. 2006. Water soluble organic carbon and oxalate in aerosols at a coastal urban site in China: Size distribution characteristics, sources, and formation mechanisms. *J. Geophys. Res.*, 111: D22212, doi:10.1029/2006JD007408.

137. Weiss, R.E., Waggoner, A.P., Charlson, R.J., and Ahlquist, N.C. 1977. Sulfate aerosol: Its geographic extent in the midwestern and southern United States. *Science*, 195: 979–981.

138. Odum, J.R., Hoffmann, T., Bowman, F., Collins, D., Flagan, R.C., and Seinfeld, J.H. 1996. Gas/particle partitioning and secondary organic aerosol yields. *Environ. Sci. Technol.*, 30: 2580–2585.

139. Novakov, T., Hegg, D.A., and Hobbs, P.V. 1997. Airborne measurements of carbonaceous aerosols on the east coast of the United States. *J. Geophys. Res.*, 102(D25): 30,023–30,030.

140. Sullivan, R.C. and Prather, K.A. 2005. Recent advances in our understanding of atmospheric chemistry and climate made possible by on-line aerosol analysis instrumentation. *Anal. Chem.*, 77: 3861–3886.

13 Femtosecond Spectroscopy and Detection of Bioaerosols

Luigi Bonacina and Jean-Pierre Wolf

CONTENTS

13.1 Introduction 321
13.2 Backward Enhancement 322
    13.2.1 Application to Standoff Detection 324
13.3 Pump–Probe Spectroscopy to Identify Bacteria and Biomolecules 325
    13.3.1 The PPD Scheme 325
    13.3.2 Liquid-Phase Results 326
    13.3.3 PPD Results on Bioaerosols 328
13.4 Pump–Probe Measurement of Particle Size: Ballistic Trajectories in Microdroplets 330
13.5 Femtosecond LIBS in Bioaerosols 331
13.6 Toward a Coherent Identification of Bacteria in Air 333
    13.6.1 Optimal Dynamic Discrimination of Spectrally Identical Molecules 333
    13.6.2 Recent Technical Advancements 336
13.7 Conclusions 337
References 338

13.1 INTRODUCTION

Rapid detection and identification of pathogenic aerosols from potential bioterrorism release and epidemic spreads are urgent safety issues. In order to efficiently protect populations, bioaerosol detection devices have to be *very fast* (typ. minutes) and *very selective* to discriminate pathogenic from nonpathogenic particles and minimize false alarm rates. This very difficult task recently initiated major research and development efforts.

Biochemical identification procedures such as polymerase chain reaction (PCR),[1–4] antibiotic resistance determination,[5,6] or matrices of biochemical microsensors[7,8] are selective, but slow (at least some hours). On the other hand, optical techniques provide information in "real time" but until now lack in specificity. In particular, several optical systems, based on fluorescence[9–13] and/or elastic scattering[14,15] have been developed to distinguish bio- from nonbio-aerosols. The most advanced experiments address *individual* aerosol particles, whose fluorescence is spectrally analyzed.[12,16,17] The major drawback of these approaches resides, however, in the presence of frequent false alarms triggered by other organic aerosols, such as diesel particles or cigarette smoke.[16,18] Figure 13.1 shows, as an example, the similitude in the fluorescence spectra of diesel fuel, soot particles, the amino acid tryptophan (Trp) and *Bacillus subtilis*, which is commonly used as a biosimulant for *Bacillus anthracis*. The major contribution in the UV-Vis fluorescence (around 340 nm) in biological particles is due to the amino acid Trp. The longer wavelength tail of the fluorescence is attributed to the emission of nicotinamide adenine dinucleotide (NADH; around 450 nm) and flavins (riboflavin (RBF), flavin mononucleotide FMN, and flavin adenine dinucleotide

321

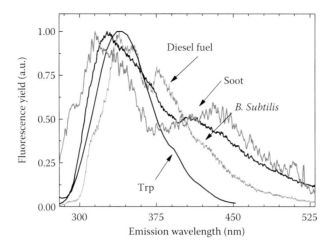

FIGURE 13.1 Comparison of the fluorescence spectra of Tryptophan, *B. Subtilis* and Diesel fuel. (Reprinted from Courvoisier, F., Boutou, V., Guyon, L. et al. 2006. *Journal of Photochemistry and Photobiology A—Chemistry* 180: 300–306. With permission from Elsevier.)

FAD around 560 nm).[16] Due to the interference with (polycyclic aromatic hydrocarbons (PAHs)-containing soot and diesel, the identification of bioaerosols in a background of traffic-related particles (typical of urban conditions) is therefore extremely difficult. It is even more elusive to expect discriminating different types of bacteria.[9] An interesting approach was reported combining optical and biochemical analyses[11]: a fluorescence/scattering device is used to sort "online" bioaerosols from other particles, which can be subsequently chemically analyzed *in situ*.

Optical techniques are also attractive as they can provide information remotely. The light detection and ranging (Lidar) technique[20] allows mapping aerosols in 3D over several kilometers, similar to an optical Radar. Lidars are able to detect the release and spread of potentially harmful plumes (such as pathogen releases from terrorists or legionella from cooling towers) at large distance and thus allow taking measures in time for protecting populations or identifying sources. So far, Lidar detection of bioaerosols has been demonstrated either using elastic scattering[20] or UV-LIF (laser induced fluorescence).[20,21] However, the distinction between bio- and nonbio-aerosols was either impossible (elastic scattering only) or unsatisfactory for LIF-Lidars (interference with pollens and organic particles like traffic-related soot or PAHs).

These inherent difficulties motivate our interest in exciting the fluorescence with ultrashort laser pulses in order to access and exploit specific molecular dynamical features. Recent experiments using coherent control and multiphoton ultrafast spectroscopy have shown the ability to discriminate between molecular species that have similar one-photon absorption and emission spectra.[22,23] Two-photon excited fluorescence (2PEF) and pulse-shaping techniques should allow for selective enhancement of the fluorescence of one molecule versus another that has similar spectra. Optimal dynamic discrimination (ODD)[24] of similar molecular agent provides the basis for generating optimal signals for detection.

13.2 BACKWARD ENHANCEMENT

One of the most prominent feature of nonlinear processes in aerosol particles is strong localization of the emitting molecules within the particle, and subsequent backward enhancement of the emitted light.[25,26] This unexpected behavior is extremely attractive for remote detection schemes, such as Lidar applications. Localization is achieved by the nonlinear processes themselves, which typically involve the nth power of the internal intensity $I^n(\mathbf{r})$ ($\mathbf{r}$ stands for position inside the

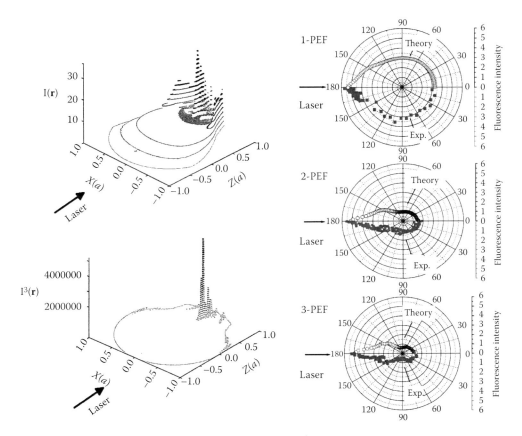

FIGURE 13.2 Backward enhanced MPEF from spherical microparticles. Molecular excitation within drop-lets, proportional to $I^n(\mathbf{r})$ with $n = 1$ and 3 (left). Angular distribution of MPEF emission for 1, 2, and 3 photon excitation (right). (Reprinted from Hill, S. C., Boutou, V., Yu, J. et al. 2000. *Physical Review Letters* 85: 54–57. Copyright 2000, American Physical Society. With permission.)

particle—Figure 13.2). The backward enhancement can be qualitatively understood by the reci-procity (or "time reversal") principle: re-emission from regions with high $I^n(\mathbf{r})$ tends to return toward the illuminating source by essentially retracing the direction of the incident beam that gave rise to the focal points. This backward enhancement has been observed for both spherical and nonspheri-cal[47] microparticles.

More precisely, we investigated, both theoretically and experimentally, incoherent multiphoton processes involving $n = 1$ to 5 photons.[25] For $n = 1$, 2, 3 (at 800 nm incident wavelength), multi-photon excited fluorescence (MPEF) occurs in bioaerosols because of the absorptions of amino acids (Trp, tyrosine), NADH, and flavins. The strong anisotropic MPEF emission was demonstrated on individual water microdroplets containing Trp, RBF, or other synthetic fluorophors in ethanol.[25–27] MPEF angular distribution for the one- (400 nm), two- (800 nm), and three-photon (1, 2 μm) excita-tion show that the fluorescence emission is maximal in the direction toward the exciting source. The directionality of the emission is dependent on the increase of n, because the excitation process involves the nth power of the intensity $I^n(\mathbf{r})$. The ratio $R_f = P(180°)/P(90°)$ increases from 1.8 to 9 when n changes from 1 to 3 with P being the emitted light power. For 3PEF, fluorescence from aerosol microparticles is therefore mainly backwards emitted, which is ideal for Lidar experiments.

At higher intensities, significant ionization occurs in water itself, involving $n = 5$ photons. The growth of the plasma is also a nonlinear function of $I^n(\mathbf{r})$. We showed that both localization and backward enhancement strongly increases with the order n of the multiphoton process, exceeding

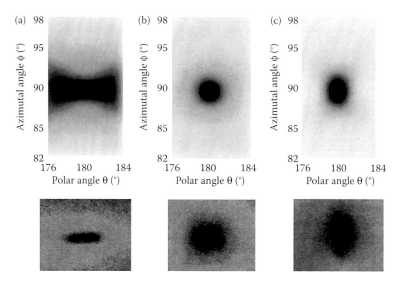

FIGURE 13.3 Backward enhanced 2PEF in spheroidal microdroplets. *S*imulations (upper panels) using ray-tracing and experimental measurements (lower panels) using a CCD camera capturing 2PEF in the backward direction from (a) oblate, (b) spherical, and (c) prolate micro droplets. (From Kasparian, J., Boutou, V., Wolf, J. P. et al. 2008. *Applied Physics B—Lasers and Optics* 91: 167–171. With kind permission from Springer+Science Media.)

$R_f = P(180°)/P(90°) = 35$ for $n = 5$.[28] Notice that the light emitted by the plasma has the potential of providing information about the aerosols composition, as shown in Section 13.5.

Backward enhancement has been observed for both spherical and nonspherical[27,29] microparticles, such as 2PEF from dye doped spheroids.[30] Within aspect ratios ranging from 0.8 to 1.2, backward enhancements of similar values as for spheres are obtained, although the round shape of the backward fluorescence image (Figure 13.3, lower panels) is somewhat affected by the shape of the particle. In particular, multi foci are produced within the spheroid. Due to the large aspect ratios considered, exact Lorentz–Mie calculations could not be performed. Novel ray-tracing approaches were therefore developed in order to determine both the intensity distribution (and its square $I^2(\mathbf{r})$) and the re-emission efficiencies. Although limited to rather large droplets (50 μm in our case), the ray-tracing approach satisfactorily reproduced the experimental data, as shown in the upper panels of Figure 13.3.

13.2.1 Application to Standoff Detection

This unique backward emission behavior allowed us to demonstrate the first MPEF Lidar detection of biological aerosols[29,31] using the "Teramobile" system.* The bioaerosol particles, consisting of 1 μm size water droplets containing 0.03 g/L RBF were generated at a distance of 50 m from the Teramobile system. RBF was excited by two photons at 800 nm and emitted a broad fluorescence around 540 nm. The broad fluorescence signature was clearly observed from the particle cloud (typ. 10^4 p/cm³), with a range resolution of a few meters.

Primarily, MPEF-Lidar is advantageous as compared to linear LIF-Lidar for the following reasons: (1) MPEF is enhanced in the backward direction, and (2) the transmission of the atmosphere is much higher for longer wavelengths. For example, if we consider the detection of Trp with 3-PEF,

---

* The Teramobile (http://www.teramobile.org) is the first femtosecond-terawatt laser-based Lidar. It was developed by an European consortium, now formed by the Universities of Iena, Berlin, Lyon, Geneva, and the Ecole Polytechnique (Palaiseau).

the transmission of the atmosphere is typically 0.6 km^{-1} at 270 nm, whereas it is 3×10^{-3} km^{-1} at 810 nm (for a clear atmosphere, depending on the background ozone concentration). This compensates the lower 3-PEF cross-section compared to the 1-PEF cross-section at distances larger than a couple of kilometers. The most attractive feature of MPEF is, however, the possibility of using pump–probe techniques, as described in Section 13.3, in order to discriminate bioaerosols from background interferents such as traffic-related soot or PAHs.

13.3 PUMP–PROBE SPECTROSCOPY TO IDENTIFY BACTERIA AND BIOMOLECULES

13.3.1 THE PPD SCHEME

As stated in Section 13.1, a major drawback inherent to LIF approaches resides in the lack of selectivity, because UV–Vis fluorescence is incapable of discriminating different molecules with similar absorption and fluorescence signatures. While mineral and carbon black particles do not fluoresce significantly, aromatics and PAHs from organic particles and diesel soot strongly interfere with biological fluorophors such as amino acids.[27,49] The similarity between the spectral signatures of PAHs and biological molecules under UV–Vis excitation seen in Figure 13.1 relies on the fact that similar π-electrons from carbonic rings are involved. Therefore, PAHs (such as napthalene) exhibit absorption and emission bands similar to those of amino acids like tyrosine or Trp. Some shifts are present because of differences in specific bonds and the number of aromatic rings, but the broad featureless nature of the bands renders them almost indistinguishable.

In order to discriminate these fluorescent molecules, we developed the double-pulse excitation scheme represented in Figure 13.4, which is based on the time-resolved observation of the competition between excited state absorption (ESA) into higher lying excited states and fluorescence into the ground state. Such an approach makes use of two physical processes beyond that exploited for standard linear fluorescence spectroscopy: (1) the dynamics in the intermediate pumped state (S_1), and (2) the coupling efficiency to higher lying excited states (S_n).

As sketched in Figure 13.4, a first femtosecond pump pulse (at 270 nm for Trp and PAHs, 405 nm in the case of flavin molecules), resonant with the first absorption band of the fluorophores, coherently excites them from the ground state S_0 to a set of vibronic levels $S_1\{v'\}$. The

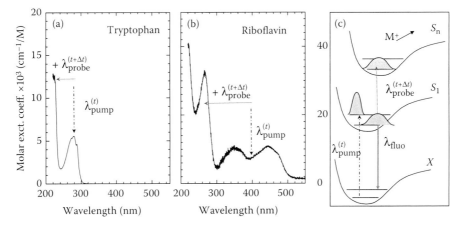

FIGURE 13.4 Absorption spectra of tryptophan (a) and riboflavin (b). (c) PPD scheme in Trp, flavins, and polycyclic aromatics. The pump pulse brings the molecules in their first excited state S_1. The S_1 population (and therefore the fluorescence) is depleted by the second pump pulse. (Courvoisier, F., Boutou, V., Guyon, L. et al. 2006. *Journal of Photochemistry and Photobiology A—Chemistry* 180: 300–306. With permission from Elsevier.)

vibronic excitation relaxes by internal energy redistribution to lower {v} modes. Fluorescence relaxation to the ground state occurs within a lifetime of several nanoseconds. Meanwhile, a second 810 nm femtosecond "re-pump" pulse is used to transfer part of the $S_1\{v\}$ population to higher lying electronic states S_n. The depletion of the S_1 population under investigation depends on both the molecular dynamics in this intermediate state and the transition probability to S_n. The relaxation from the intermediate excited state may be associated with different processes, including charge transfer, conformational relaxation,[32,33] and intersystem crossing with repulsive $\pi\sigma^*$ states.[34,35] S_n states are both autoionizing[35] and relax radiationless[36] into S_0. By varying the temporal delay Δt between the UV–Vis and the IR pulses, the dynamics of the internal energy redistribution within the intermediate excited potential hypersurface S_1 is explored. The S_1 population and the fluorescence signal is therefore depleted as a function of Δt. As different species have distinct S_1 hypersurfaces, discriminating signals can be obtained.

13.3.2 LIQUID-PHASE RESULTS

Figure 13.5a shows the PPD dynamics of S_1 in Trp as compared to diesel fuel and napthalene in cyclohexane, one of the most abundant fluorescing PAHs in diesel. While fluorescence depletion δ,

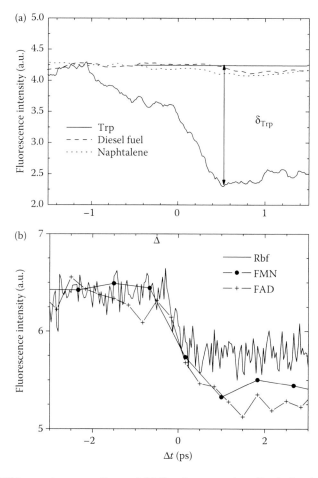

FIGURE 13.5 (a) PPD experiment on Trp and PAHs, demonstrating discrimination capability between the amino acid and other aromatic molecules. (b) Similar results obtained in flavins. (Reproduced from Courvoisier, F., Bonacina, L., Boutou, V. et al. 2008. *Faraday Discussions* 137: 37–49. With permission from the Royal Society of Chemistry.)

which is defined as $\delta = (P_{undepleted} - P_{depleted})/P_{undepleted}$ with P fluorescence power, reaches as much as 50% in Trp, diesel fuel and naphthalene appear almost unaffected (within a few percent), at least on these timescales.[37] This remarkable difference allows efficient discrimination between Trp and organic species, although they exhibit very similar linear excitation/fluorescence spectra (Figure 13.1). Two reasons might be invoked to understand this difference: (1) the intermediate state dynamics is predominantly influenced by the NH- and CO-groups of the amino acid backbone and (2) the ionization efficiency is lower for the PAHs. Further electronic structure calculations are required to better understand the process, especially on the higher lying S_n potential surfaces.

Fluorescence depletion has been obtained as well for RbF, FMN, and FAD (Figure 13.5b). However, the depletion in this case is only about 15% (with a maximum pulse intensity of 5×10^{11} W/cm^2 at 810 nm). We then repeated the experiment but exciting the flavins at 270 nm, as for Trp. Flavins indeed absorb in both spectral regions (Figure 13.4). The second excitation step by the 810 nm pulse then promotes the molecules to the excited states around 200 nm. The fluorescence depletion observed with the 270 nm excitation reached 35%. This is an indication that the branching ratio to autoionization in the 270 nm band is much lower than around 200 nm. To get the depletion while exciting at 405 nm, two photons at 810 nm probably have to be used. This is confirmed by intensity dependences, which show a quadratic dependence in I_{810} in this latter case. A model based on rate equations was recently developed in order to quantitatively explain the PPD behavior in Trp and RbF.[38]

To get closer to the application of detecting and discriminating bioagents from organic particles, we applied PPD spectroscopy to live bacteria ($\lambda_1 = 270$ nm and $\lambda_2 = 810$ nm), such as *Escherichia coli*, *Enterococcus*, and *B. subtilis*. Artefacts due to preparation methods have been discarded by using a variety of samples, that is, lyophilized cells and spores, suspended either in pure or in biologically buffered water (i.e., typically 10^7–10^9 bacteria per cc). The bacteria containing solutions replaced the Trp or flavin containing solutions of the previous experiments. The observed pump–probe depletion results are remarkably robust (Figure 13.6), with similar depletion values for all the bacteria tested (results for *Enterococcus*, not shown in Figure 13.6, are identical), although the Trp microenvironment within the bacteria proteins is very different from water.

On the other hand, the very similar depletion behavior for all bacteria and Trp also shows the limitations of PPD spectroscopy in the present configuration: biomolecules can be distinguished from other aromatics but PPD is unable to discriminate two different bacteria in solution. As

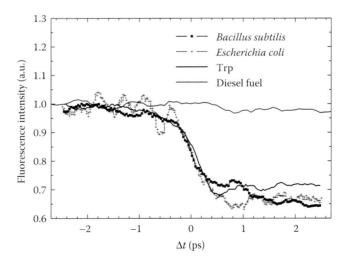

FIGURE 13.6 Discrimination between bacteria and diesel fuel using PPD ultrafast spectroscopy. (Reprinted from Courvoisier, F., Boutou, V., Guyon, L. et al. 2006. *Journal of Photochemistry and Photobiology A—Chemistry* 180: 300–306. With permission from Elsevier.)

described in Section 13.6, this fundamental issue can be addressed by employing coherent excitation schemes and ODD.

13.3.3 PPD RESULTS ON BIOAEROSOLS

After the experiments with liquids, we applied the PPD scheme to bioaerosols and specifically to Trp- and/or FMN-containing water microdroplets (Figure 13.7). The droplets radius was about 25 μm, which is larger than the size of single bacteria (1 μm) or even bacteria clusters (typically 10 μm), but still constitutes an acceptable model. The laser intensities at 810 and 270 nm for Trp (respectively 405 nm for FMN) that excited the microparticles were similar to the intensities used for the liquid-phase experiments. Depletion is as much as 80% for Trp and 60% for FMN. The peak in Figure 13.7b shows the cross-correlation of both laser pulses, which indicates the time resolution of the experiment.

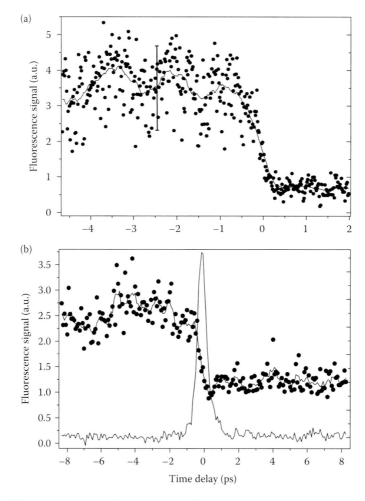

FIGURE 13.7 PPD spectroscopy in bioaerosols: (a) Trp-containing microdroplets, (b) FMN-containing microdroplets. Depletion is much as 80% for Trp and 60% for FMN. The peak in (b) shows the cross-correlation of the two laser pulses, indicating the experimental time resolution. (Courvoisier, F., Bonacina, L., Boutou, V. et al. 2008. *Faraday Discussions* 137: 37–49. With permission from the Royal Society of Chemistry.)

The most impressive outcome of these measurements is the very high PPD efficiency as compared to depletion ratios in liquids. The depletion factor δ reaches indeed 80% for Trp droplets and 60% for FMN droplets (to be compared to 50% and 15% in liquids, respectively). Some tentative explanations could be invoked for this unexpectedly high efficiency, but the definitive reason is not clear yet: (1) the spatial overlap between pump and probe pulses might be enhanced by the shape of the droplet, (2) the spherical shape induces hot spots inside the droplet where intensities are up to 100 times higher than the incident one,[26,27] but the total hot spot volume is rather small, (3) there might be some surface effects (orientation of molecules on the surface) that could enhance two-photon absorption.

We finally repeated the measurements on 20 μm water droplets containing typically 100 live bacteria (*E. coli*). As shown in Figure 13.8, the depletion factor δ is again greatly enhanced as compared to bacteria in bulk water: 60% depletion in the microdroplet and 20% in solution with $\lambda_1 = 270$ nm and $\lambda_2 = 810$ nm excitation.

This experiment is also interesting for field applications, as bacteria and viruses are efficiently transmitted by droplets of saliva (coughing, breathing, speaking, etc.).

We propose to use the unique discrimination capability of PPD as a basis for a novel selective bioaerosol-detection technique that avoids interference from background (traffic-related) organic particles in air. For instance, let us consider a mixture of N_B bacteria and N_D diesel particles. The excitation shall consist of a PPD sequence (270 nm and 810 nm pulses). The fluorescence power P emitted by the mixture shall be measured as the second laser (at 810 nm) is alternately switched on and off (denoted as P_{on} and P_{off}). Without the second laser pulse, the fluorescence cross-sections are $\sigma_D(I_{810} = 0) = \sigma_D$ and $\sigma_B(I_{810} = 0) = \sigma_B$, while with the probe laser the cross-sections are reduced by a factor R_B (= 0.2) and R_D (= 0.98) for the biological and the diesel particles, respectively. This differential procedure allows determining the concentration of the two types of particles

$$N_B = \frac{R_D P_{off} - P_{on}}{I_{270}\sigma_B(R_D - R_B)} \quad \text{and} \quad N_D = \frac{R_B P_{off} - P_{on}}{I_{270}\sigma_D(R_B - R_D)}$$

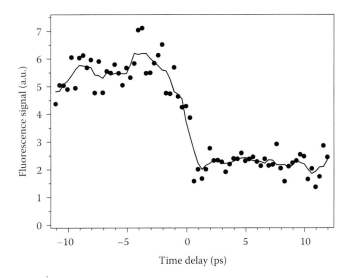

FIGURE 13.8 PPD in 20 μm droplets containing about 100 *E. coli* bacteria. (Reproduced from Courvoisier, F., Bonacina, L., Boutou, V. et al. 2008. *Faraday Discussions* 137: 37–49. With permission from the Royal Society of Chemistry.)

the method's performance strongly depends on the difference $R_B - R_D$ between the two species to be discriminated, which is large in our case. Notice that in order to precisely quantify the concentration of the diesel particles, quantitative knowledge of the cross sections is required, which might be difficult to determine because of the variety of possible organic particles.

13.4 PUMP–PROBE MEASUREMENT OF PARTICLE SIZE: BALLISTIC TRAJECTORIES IN MICRODROPLETS

The very small spatial extension of femtosecond pulses (15 fs corresponds to 3.4 μm in water, that is, the equator of a 0.6 μm droplet) can be used for measuring the size of microparticles.[39,40]

An experimental scheme for this consists in creating an optical correlator between two ultrashort pulses, centered at wavelengths $\lambda_1 = 1200$ nm, and $\lambda_2 = 600$ nm, which circulate on ballistic orbits (see Figure 13.9). 2PEF is then recorded as a function of the time delay between the two pulses in order to quantify the path length traveled within the particle. With this method, the size of droplets up to 670 μm could be precisely measured.[40]

The time delay between the two pulses gives access to ballistic pathlengths. If a measured pathlength differs from an integer multiple of the orbital roundtrip, it indicates the contribution of cross-correlation signals associated with pulses traveling on other trajectories than morphological-dependent resonances (MDRs). In order to better understand this behavior, the impact parameter (indicated by a in Figure 13.10) of the laser onto the droplet was modified, so that evanescent coupling in surface modes could be distinguished from refractive penetration onto inner ballistic trajectories. A detailed analysis of the results showed that for large impact parameters, only surface modes contributed, while for smaller ones, light bullets traveled on rainbow trajectories. The sizing application of this method was applied for a wide range of sizes, but could not be performed experimentally for droplets smaller than 100 μm, due to experimental limitations.

In order to address Lidar applications of the pump–probe approach for measuring remotely both the size and composition of atmospheric aerosols, time-resolved Lorentz–Mie calculations have

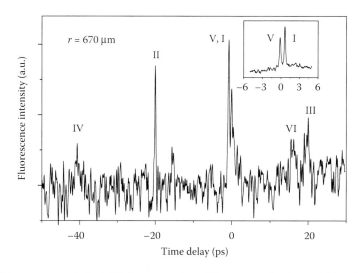

FIGURE 13.9 Measurement of the ballistic trajectories of femtosecond pulses within microdroplets using 2PEF pump–probe technique. Each peak of the series I, II, III, IV corresponds to a roundtrip of the traveling pulse. V and VI correspond to other types of trajectories than MDRs, such as rainbow. (Reprinted from, Wolf, J. P., Pan, Y. L., Turner, G. M. et al. 2001. *Physical Review A* 64: 023808. Copyright 2001, American Physical Society. With permission.)

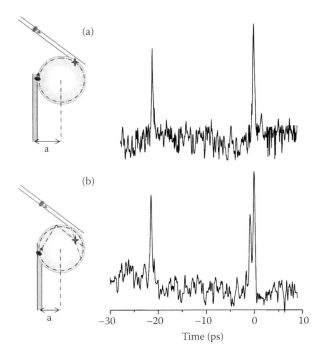

FIGURE 13.10 Control of the ballistic trajectories used by the traveling pulse by modifying the impact parameter: (a) evanescent coupling (b) evanescent + refractive coupling. (Reprinted from Wolf, J. P., Pan, Y. L., Turner, G. M. et al. 2001. *Physical Review A* 64: 023808. Copyright 2001, American Physical Society. With permission.)

been carried out (50 fs pulses).[39] In a pump–probe 2PEF Lidar experiment, the composition would be addressed by the excitation/fluorescence signatures and the size by the time delay between the two exciting pulses. The high contrast between time-resolved peaks obtained by the numerical simulations shows that measurements should be feasible for submicron size particles.

13.5. FEMTOSECOND LIBS IN BIOAEROSOLS

Thanks to the high-intensity/energy ratio provided by femtosecond laser pulses, aerosols are not significantly deformed during the induced light emission. Therefore, even higher order nonlinear processes such as (LIBS) and plasma emission is enhanced in the backward direction as shown in Section 13.2.[25,28] The resulting "nanoplasma" is, for instance, highly localized at the focal line within the droplet and the associated plasma lines can be efficiently recorded in a Lidar arrangement.

Although nanosecond-laser LIBS (nano-LIBS) has already been applied to the study of bacteria,[41,42] femtosecond lasers open new perspectives in this respect. The plasma temperature is indeed, much lower in the case of femtosecond excitation, which strongly reduces the blackbody background and interfering lines from excited N_2 and O_2 air molecules. This allows performing time-gated detection with very short delays, and thus observing much richer and cleaner spectra from biological samples. This crucial advantage is shown in Figure 13.11, where the K line emitted by a sample of *E. coli* is clearly detected in femto-LIBS and almost unobservable under ns-laser excitation.[43]

The low thermal background in fs-LIBS allowed the recording of 20–50 lines for each bacterial sample considered (*Acinetobacter, E. coli, Erwinia, Shewanella,* and *B. subtilis*). The results are promising, as can be seen in Figure 13.12, where significant difference between *E. coli* and *B.*

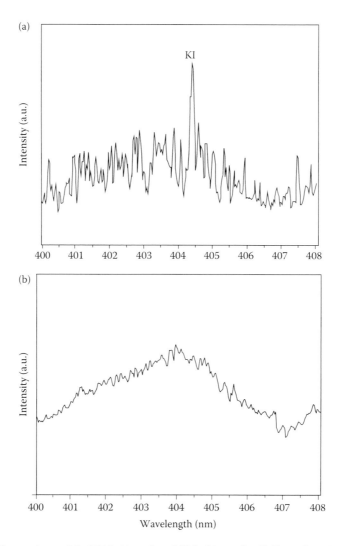

FIGURE 13.11 Comparison of fs-LIBS (a) and ns-LIBS (b) on the K line of *Escherichia Coli*. (From Baudelet, M., Guyon, L., Yu, J. et al. 2006. *Journal of Applied Physics* 99: 084701. Copyright 2006, American Institute of Physics. With permission.)

subtilis are observed for the Li line intensity (also observed for the Ca line). This difference can be understood by the typical difference of the cell wall structure between a Gram-positive and a Gram-negative bacterium. The ratio of these lines as compared to Na for example constitutes an "all optical Gram test."

Low temperature plasma is not the only advantage of fs-LIBS: the ablation process itself seems different. fs-LIBS acts more as a direct bond-breaking and evaporation process than as thermal evaporation. Evidence of this particular ablation process is in the observation of not only atomic and ionic lines, but also molecular signatures such as CN or C_2.[44] It was shown in particular that these molecular species are directly ablated from the sample, and not created by recombination of C atoms or ions with nitrogen from the air (which occurs for ns excitation). Obtaining molecular signatures in addition to trace elements is a significant improvement of the LIBS method. The presence of CN molecules is, for instance, a good indicator for a biological material.

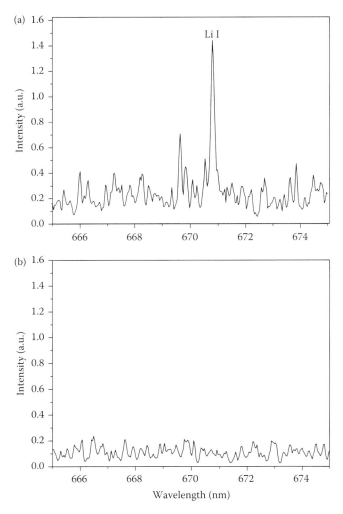

FIGURE 13.12 The "all optical Gram test" : *B. subtilis* (a) vs *E. coli* (b). (Reprinted from Baudelet, M., Guyon, L., Yu, J. et al. 2006. *Journal of Applied Physics* 99: 084701. Copyright 2006, American Institute of Physics. With permission.)

13.6 TOWARD A COHERENT IDENTIFICATION OF BACTERIA IN AIR

13.6.1 OPTIMAL DYNAMIC DISCRIMINATION OF SPECTRALLY IDENTICAL MOLECULES

As illustrated in Section 13.3, PPD spectroscopy has proven itself capable to discriminate bio-aerosols from other organics. However, it is unable to discriminate one type of biological aerosol from another (Figure 13.6). A possible reason is that the averaged dynamics in the excited states of the biofluorophors (embedded in complex proteins) are quite similar in all living organisms. A natural approach is therefore to extend the PPD technique to coherent control, where the amplitude and phase of every spectral component of the exciting pulses are shaped in order to fit at best the potential hypersurfaces of the molecules addressed. The method is then extremely sensitive to the details of these hypersurfaces, which provides unprecedented selectivity.[22,23] For this, a large number of parameters (corresponding to the amplitude and phase of each spectral component within the exciting laser pulses) has to be controlled. This "pulse-shaping" technique is usually performed by introducing a liquid crystal (LC) array in the Fourier plane between two gratings

(4*f* arrangement). In 1992, the concept of "optimal control" was introduced,[45,46] in which a feedback loop optimizes the laser pulse characteristics to reach the desired target most efficiently. Excellent results have been obtained using coherent control schemes in atomic and molecular systems (mostly in gaseous phase).[47] Recent theoretical analysis yields the general conclusion that quantum systems differing even infinitesimally in structure may be distinguished by means of their dynamics when acted upon by a suitably shaped, ultrafast control field.[48] Such molecular ODD can in principle achieve dramatic levels of control, and hence provide a valuable testing ground to probe the fundamental selectivity limits of quantum control despite finite laser resources. We recently demonstrated in a joint experiment with the group of Professor H. Rabitz in Princeton, the control of system selectivity for two structurally similar flavin molecules: aqueous-phase FMN and RBF. Although we chose these specific molecules for this study, the results should be broadly applicable to control of systems whose static spectra show essentially indistinguishable features associated with the degrees of freedom desired for manipulation. Considering their very similar structure (Figure 13.13a) it is not surprising that the two molecules are nearly identical by the static spectroscopic methods[49]: the flavins' electronic spectroscopy is primarily associated with their common chromophore ($\pi \rightarrow \pi^*$ type transitions localized on the isoalloxazine ring), and is influenced indirectly, and only very slightly, by the chemical moieties (H versus $PO(OH)_2$) on the terminal side chains.[50] Therefore, the spectra in Figure 13.13b are practically indistinguishable throughout the entire visible and far UV, including the region of the pump–pulse wavelength at 400 nm. Consequently, achieving any dynamical specificity must rely on the underlying flavin vibronic structure.

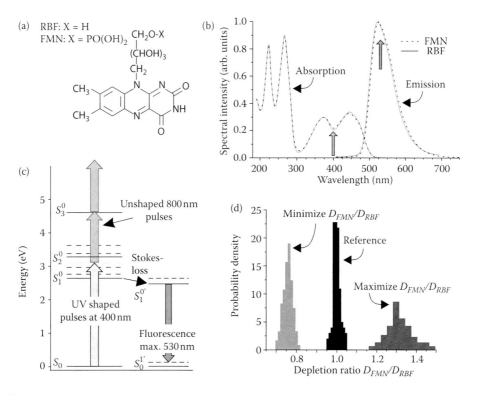

FIGURE 13.13 **(See color insert following page 206.)** RBF and FMN are structurally (a) and spectroscopically (b) quite similar. A shaped UV pulse coordinated with a time-delayed IR pulse enables selectivity (c) and is able to both maximize and minimize the fluorescence depletion ratio by ±28% (d). Arrows at 400 nm and 530 nm in (b) locate regions of UV control and collected fluorescence, respectively. (Reprinted from Roth, M., Guyon, L., Roslund, J. et al. 2009. *Physical Review Letters* 102: 253001. Copyright 2009, American Physical Society. With permission.)

This application of ODD may appear to be inordinately complex considering upwards of 100 vibrational degrees of freedom for each flavin, control in a polarizable solvent, and a thermal population of many low-frequency vibrational modes. Yet, rather than acting as a hindrance, this system complexity may be exploited to achieve high excitation specificity even with modest control resources.

Our particular implementation of ODD relies on a two-stage control process very similar to the PPD scheme described in Section 13.3. First, a specially shaped UV pulse coherently transfers ground-state population into a vibrational progression of the flavin electronic states S_1 or S_2. Tailoring of the UV spectral phase amplifies minute differences in each system's vibronic structure to create unique dynamical wave packets (Figure 13.13c). Regardless of the specific shape of this preparation pulse, the tailored wave packets of both systems eventually evolve to the same final, de-cohered state when this shaped pulse acts alone. Consequently, any distinction between the two systems is lost in the incoherent, spectrally identical fluorescence signals. However, coordinated application of a second, time-delayed IR pulse on a coherent timescale disrupts the carefully created vibronic excitations and results in an additional excitation that is dependent upon the precise structure, position, and coherence of the tailored wave packets generated by the first pulse. Thus, the action of dual control pulses provides the means to dynamically interrogate the two statically identical systems and thereby produce a discriminating difference in their respective depleted fluorescence signals. The optimal UV pulse was found by closed-loop optimization of the fitness function J for a fixed delay Δt:

$$J(\Delta t) = \frac{\delta_{\text{FMN}}(\Delta t)}{\delta_{\text{RBF}}(\Delta t)} + 0.05\delta_{\text{RBF}}(\Delta t) \tag{13.1}$$

where $\delta_{\text{FMN}}(\Delta t)/\delta_{\text{RBF}}(\Delta t) = 1$ indicates no recognizable discrimination from an unshaped UV pulse and δ is the depletion factor introduced in Section 13.3.2. The first term of $J(\Delta t)$ aims to maximize the depletion ratio of FMN over that of RBF, and the last term prevents the modulation from time-shifting the UV pulse to follow the IR pulse. An analogous expression, with reversal of the roles of RBF and FMN, was used to minimize the depletion ratio of the two systems. This fitness function was optimized at $\Delta t = 500$ fs using a 50-dimensional UV spectral phase (maintaining constant UV pulse energy) under the guidance of a genetic algorithm (GA).

Repeated application of the closed-loop optimization procedure yielded a family of uniquely shaped UV control pulses, each one associated to the same degree of dynamic discrimination. Although this collection of UV pulses possesses considerable temporal variability, each member is equally successful at dynamically distinguishing the two systems. This set of discriminating laser pulses permits exploiting ODD as a novel means for optical detection in demanding circumstances. To demonstrate such capability, the fluorescence signal from a flavin mixture generated with the nth UV–IR control pulse pair is related to the fractional flavin contents by the following equation:

$$F_{\text{mix},n} = c_{\text{RBF}} F_{\text{RBF},n}^{\text{d}} + c_{\text{FMN}} F_{\text{FMN},n}^{\text{d}}, \tag{13.2}$$

where $F_{\text{RBF(FMN)},n}^{\text{d}}$ is the depleted fluorescence signal of a pure solution of RBF (FMN) and $0 \leq c_{\text{FMN(RBF)}} \leq 1$. These standardizing fluorescence signals are recorded for 53.7 and 52.6 µM pure, separate solutions of RBF and FMN, respectively, utilizing the previously determined nth optimal pulse. For the purpose of concentration determination, both flavins are combined in a single cell, and the resulting fluorescence $F_{\text{mix,n}}$ from the nth optimized pulse is obtained with 1 min of signal averaging (1 kHz system operation). Although two distinct ODD pulses could successfully determine the individual fractional contents of the two flavins, an increase in the number of distinct interrogating pulses decreases the extraction error. Accordingly, the retrieval equation is overspecified by measuring the mixture fluorescence at delay times of $\Delta t = 250, 375, 400, 500, 600,$ and

700 fs for the two optimized pulses that maximize and minimize the depletion ratio $\delta_{FMN}(\Delta t)/\delta_{RBF}(\Delta t)$. A typical result from this linear least squares procedure is $c_{RBF} = 0.35 \pm 0.04$ (18.8 μM) and $c_{FMN} = 0.68 \pm 0.05$ (35.8 μM) when the known fractional contents were $c_{RBF} = 0.33$ and $c_{FMN} = 0.67$. No attempt was made to push the limit of detectability for this proof-of-concept selectivity control experiment, and these signals were easily measured for concentrations approaching the physiologically significant micromolar level.

Although the laser resources exploited for discriminating the nearly identical flavins consisted of a modest ~3.5 nm of UV bandwidth and ~10 nm of IR bandwidth, dramatic selectivity was achieved with optimal UV pulses, whereas the static spectra in Figure 13.13b appear nearly identical, subtle differences exploited by control are nonetheless profound and allow a flavin discrimination of 12σ (Figure 13.13d).[51] System complexity (e.g., high vibrational state density, thermal population, solvent-induced line broadening) effectively amplifies the control field capabilities and compensates for the constraint of limited bandwidth, thus making dramatic levels of control possible even in the weak-field limit, a result potentially important for achieving quantum control in numerous demanding applications. Additionally, this successful demonstration of ODD specificity bolsters the positive theoretical projections for quantum control and its potential for yielding high selectivity amongst tightly competitive channels in a variety of practical applications.

13.6.2 Recent Technical Advancements

Most of the coherent control techniques rely on the use of 1D or 2D spatial light modulators (SLM) based on LC arrays. Despite some promising results in the UV [3], LC modulators are limited to the visible and NIR, while most important applications of coherent control in organic chemistry and biology require excitation in the UV. For instance, absorption bands of amino acids, proteins, and nucleic acids in DNA–RNA all lie in the 200–300 nm region. The preceding example of Section 13.6.1 together with a few recent approaches for direct femtosecond pulse shaping in the near UV made use of acousto-optic,[52–54] while indirect schemes are essentially based on frequency mixing of shaped pulses in the visible and NIR.[55,56] Although encouraging, these techniques still suffer either of low throughput (some percent) due to diffraction losses or of insufficient spectral bandwidth (typ. 10% of the central wavelength). Group velocity dispersion in the crystal usually further reduces the flexibility of the output waveforms.

Considering the broad absorption features of organic molecules in solution and the fast decoherence time of their vibronic excitations, it is highly desirable to have at disposal an UV pulse shaper with no strict bandwidth limitations, based for instance on reflective elements, such as deformable mirrors[57] or micro-electro-mechanical-systems (MEMS) micromirror arrays. While plain deformable mirrors usually lack in spectral resolution, MEMS appear as an appealing solution for these requirements, in particular if used to shape ultrabroad sources that can be obtained by filamentation in rare gas cells. This approach was recently followed to generate pulses at 800 nm as short as 5 fs, by recompressing the broadened pulse with chirped mirrors.[58] Laser filaments[58,59] arise in the nonlinear propagation of ultrashort, high-power laser pulses in transparent media. They result from a dynamic balance between Kerr-self-focusing and defocusing by self-induced plasma. These spatiotemporal solitonic structures are able to generate an extraordinary broad supercontinuum by self-phase modulation and four wave mixing, spanning from the UV to the IR.[59] We have recently conducted a first series of broadening experiments at 400 nm using filamentation in order to obtain a broadband laser source for flavins (and NADH) excitation. In this case, filamentation was produced in an Ar-gas filled cell (7 bars) of 100 cm length. While the incoming laser bandwidth was only 3 nm (0.5–1 mJ, 150 fs), it reaches 15–20 nm (full-width at half-maximum) after filamentation.

MEMS shapers are ideal for tailoring pulses of such a large bandwidth in the UV region of the spectrum. With their pioneering work, Hacker et al. demonstrated the aptness of a 2D MEMS device from Fraunhofer IPMS for pulse-shaping applications at 400 nm.[60] Within this approach, shaping of the UV–Vis pulses is carried out in the experimental arrangement shown in Figure 13.14b, which

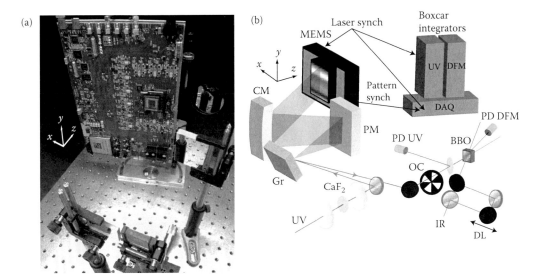

FIGURE 13.14 (See color insert following page 206.) (a) Picture of the shaper setup. We recognize the Fraunhofer MEMS chip, placed in the middle right of the driving board. (b) Experimental scheme. Gr: grating; PM: plane mirror; CM: cylindrical mirror; DL: temporal delay line; OC: optical chopper. PD UV/DFM: detection photodiodes, DAQ: acquisition card. The OC is synchronized to half the repetition rate of the laser to reject every second UV pulse. (From Rondi, A., Extermann, J., Bonacina, L. et al. 2009. *Applied Physics B—Lasers and Optics* 96. With kind permission from Springer + Science Media.)

acts as a $4f$ zero dispersion compressor, with the mirror array placed in the Fourier plane. The pulses are first dispersed by a reflection grating; successively a cylindrical lens focuses the spectrum on the mirror array. The phase modulation is accomplished by varying the overall optical path length of the different components, acting on the pistons of the mirror elements. Note that the unique spectral restriction of the apparatus is represented by the maximum travel of the mirror pistons, which limits the maximum phase shift at a given wavelength $\Delta\Phi(\lambda)$. A few research groups worldwide have exploited the 2D features of the same MEMS chip for shaping NIR pulses in diffraction mode,[61] as well as for shaping two beams (NIR and UV–VIS) simultaneously.[62] We very recently reported a custom design for broadband femtosecond pulse shaping in the deep-UV around 266 nm[63]: we characterized the device in this wavelength region, addressed a series of technical details related to its operation, and demonstrated its capability of re-compressing self-phase modulation (SPM) broadened UV pulses with a closed-loop approach based on a genetic algorithm.[64] On the other hand, the strong diffraction inherent to the two-dimensional pixellated nature of the device should be taken into account because it results the principal factor at the origin of the intensity losses of the setup.

We are presently realizing a new linear MEMS device allowing both phase and amplitude shaping of pulses in a wide spectral region, ranging from 250 to 900 nm. With respect to the Fraunhofer device, an additional degree of freedom is added to the mirrors, in that the intensity attenuation is performed by tilting the mirrors in the vertical plane.[65]

13.7 CONCLUSIONS

Femtosecond spectroscopy opens new ways for the optical detection and identification of microorganisms in water and in air. Its unique capability of distinguishing molecules that exhibit almost identical absorption and fluorescence signatures is a key feature for identifying bacteria in a background of urban aerosols. The technique can also be applied for the remote detection of microorganisms in air, if a nonlinear Lidar based configuration is used, as for the Teramobile system. A more difficult task will be the distinction of one bacteria species from another, and in particular the identification of pathogen from nonpathogen bioaerosols. A possible way to reach this difficult goal might be coherent control and ODD. Significant technical improvements are currently developed, which should provide a definitive answer on the potential of coherent excitation schemes in the UV spectral region. A complementary approach can be provided by fs-LIBS, as the wealth of emission lines under femtosecond excitation might allow us to target some biological process that is characteristic from one type of bacteria.

REFERENCES

1. Belgrader, P., Benett, W., Hadley, D. et al. 1999. Infectious disease—PCR detection of bacteria in seven minutes. *Science* 284: 449–450.
2. Ho, J. 2002. Future of biological aerosol detection. *Analytica Chimica Acta* 457: 125–148.
3. Makino, S. I., Cheun, H. I., Watarai, M. et al. 2001. Detection of anthrax spores from the air by real-time PCR. *Letters in Applied Microbiology* 33: 237–240.
4. Tenover, F. C. and Rasheed, J. K. 1999. Genetic methods for detecting antibacterial and antiviral resistance genes. In *Manual of Clinical Microbiology*, Ed. P. R. Murray, E. J. Baron, M. A. Pfaller, F. C. Tenover, and R. H. Yolken. Herndon, VA: American Society for Microbiology.
5. De, B. K., Bragg, S. L., Sanden, G. N. et al. 2002. Two-component direct fluorescent-antibody assay for rapid identification *Bacillus anthracis. Emerging Infectious Diseases* 8: 1060–1065.
6. Pourahmadi, F., Taylor, M., Kovacs, G. et al. 2000. Toward a rapid, integrated, and fully automated DNA diagnostic assay for *Chlamydia trachomatis* and *Neisseria gonorrhoeae. Clinical Chemistry* 46: 1511–1513.
7. Francois, P., Bento, M., Vaudaux, P. et al. 2003. Comparison of fluorescence and resonance light scattering for highly sensitive microarray detection of bacterial pathogens. *Journal of Microbiological Methods* 55: 755–762.
8. Hagleitner, C., Hierlemann, A., Lange, D. et al. 2001. Smart single-chip gas sensor microsystem. *Nature* 414: 293–296.
9. Eversole, J. D., Cary, W. K., Scotto, C. S. et al. 2001. Continuous bioaerosol monitoring using UV excitation fluorescence: Outdoor test results. *Field Analytical Chemistry and Technology* 5: 205–212.
10. Luoma, G. A., Cherrier, P. P., and Retfalvi, L. A. 1999. Real-time warning of biological-agent attacks with the Canadian Integrated Biochemical Agent Detection System II (CIBADS II). *Field Analytical Chemistry and Technology* 3: 260–273.
11. Pan, Y. L., Boutou, V., Bottiger, J. R. et al. 2004. A puff of air sorts bioaerosols for pathogen identification. *Aerosol Science and Technology* 38: 598–602.
12. Pan, Y. L., Hartings, J., Pinnick, R. G. et al. 2003. Single-particle fluorescence spectrometer for ambient aerosols. *Aerosol Science and Technology* 37: 628–639.
13. Reyes, F. L., Jeys, T. H., Newbury, N. R. et al. 1999. Bio-aerosol fluorescence sensor. *Field Analytical Chemistry and Technology* 3: 240–248.
14. Kaye, P., Hirst, E., and Wang, T. Z. 1997. Neural-network-based spatial light-scattering instrument for hazardous airborne fiber detection. *Applied Optics* 36: 6149–6156.
15. Pan, Y. L., Aptowicz, K. B., Chang, R. K. et al. 2003. Characterizing and monitoring respiratory aerosols by light scattering. *Optics Letters* 28: 589–591.
16. Hill, S. C., Pinnick, R. G., Niles, S. et al. 1999. Real-time measurement of fluorescence spectra from single airborne biological particles. *Field Analytical Chemistry and Technology* 3: 221–239.
17. Pan, Y. L., Cobler, P., Rhodes, S. et al. 2001. High-speed, high-sensitivity aerosol fluorescence spectrum detection using a 32-anode photomultiplier tube detector. *Review of Scientific Instruments* 72: 1831–1836.
18. Pinnick, R. G., Hill, S. C., Pan, Y. L. et al. 2004. Fluorescence spectra of atmospheric aerosol at Adelphi, Maryland, USA: Measurement and classification of single particles containing organic carbon. *Atmospheric Environment* 38: 1657–1672.
19. Cheng, Y. S., Barr, E. B., Fan, B. J. et al. 1999. Detection of bioaerosols using multiwavelength UV fluorescence spectroscopy. *Aerosol Science and Technology* 30: 186–201.
20. Weitkamp, C. 2005. *LIDAR—Range-Resolved Optical Remote Sensing in the Atmosphere.* New York: Springer-Verlag.
21. Immler, F. and Schrems, O. 2004. Measurements of aerosol and cirrus clouds in the UTLS by a shipborne lidar. *22nd International Laser Radar Conference (Ilrc 2004)*, Matera, Italy, Vols. 1 and 2, 561: 415–418.

22. Brixner, T., Damrauer, N. H., Kiefer, B. et al. 2003. Liquid-phase adaptive femtosecond quantum control: Removing intrinsic intensity dependencies. *Journal of Chemical Physics* 118: 3692–3701.

23. Brixner, T., Damrauer, N. H., Niklaus, P. et al. 2001. Photoselective adaptive femtosecond quantum control in the liquid phase. *Nature* 414: 57–60.

24. Li, B. Q., Rabitz, H., and Wolf, J. P. 2005. Optimal dynamic discrimination of similar quantum systems with time series data. *Journal of Chemical Physics* 122: 154103.

25. Boutou, V., Favre, C., Hill, S. C. et al. 2002. Backward enhanced emission from multiphoton processes in aerosols. *Applied Physics B—Lasers and Optics* 75: 145–152.

26. Hill, S. C., Boutou, V., Yu, J. et al. 2000. Enhanced backward-directed multiphoton-excited fluorescence from dielectric microcavities. *Physical Review Letters* 85: 54–57.

27. Pan, Y. L., Hill, S. C., Wolf, J. P. et al. 2002. Backward-enhanced fluorescence from clusters of microspheres and particles of tryptophan. *Applied Optics* 41: 2994–2999.

28. Favre, C., Boutou, V., Hill, S. C. et al. 2002. White-light nanosource with directional emission. *Physical Review Letters* 89: 035002.

29. Mejean, G., Kasparian, J., Yu, J. et al. 2004. Remote detection and identification of biological aerosols using a femtosecond terawatt lidar system. *Applied Physics B—Lasers and Optics* 78: 535–537.

30. Kasparian, J., Boutou, V., Wolf, J. P. et al. 2008. Angular distribution of non-linear optical emission from spheroidal microparticles. *Applied Physics B—Lasers and Optics* 91: 167–171.

31. Kasparian, J., Rodriguez, M., Mejean, G. et al. 2003. White-light filaments for atmospheric analysis. *Science* 301: 61–64.

32. Callis, P. R. and Vivian, J. T. 2003. Understanding the variable fluorescence quantum yield of tryptophan in proteins using QM-MM simulations. Quenching by charge transfer to the peptide backbone. *Chemical Physics Letters* 369: 409–414.

33. Vivian, J. T. and Callis, P. R. 2001. Mechanisms of tryptophan fluorescence shifts in proteins. *Biophysical Journal* 80: 2093–2109.

34. Dedonder-Lardeux, C., Jouvet, C., Perun, S. et al. 2003. External electric field effect on the lowest excited states of indole: *ab initio* and molecular dynamics study. *Physical Chemistry Chemical Physics* 5: 5118–5126.

35. Steen, H. B. 1974. Wavelength dependence of quantum yield of fluorescence and photoionization of indoles. *Journal of Chemical Physics* 61: 3997–4002.

36. Iketaki, Y., Watanabe, T., Ishiuchi, S. et al. 2003. Investigation of the fluorescence depletion process in the condensed phase; application to a tryptophan aqueous solution. *Chemical Physics Letters* 372: 773–778.

37. Courvoisier, F., Boutou, V., Wood, V. et al. 2005. Femtosecond laser pulses distinguish bacteria from background urban aerosols. *Applied Physics Letters* 87: 063901.

38. Courvoisier, F., Boutou, V., Guyon, L. et al. 2006. Discriminating bacteria from other atmospheric particles using femtosecond molecular dynamics. *Journal of Photochemistry and Photobiology A—Chemistry* 180: 300–306.

39. Mees, L., Wolf, J. P., Gouesbet, G. et al. 2002. Two-photon absorption and fluorescence in a spherical micro-cavity illuminated by using two laser pulses: Numerical simulations. *Optics Communications* 208: 371–375.

40. Wolf, J. P., Pan, Y. L., Turner, G. M. et al. 2001. Ballistic trajectories of optical wave packets within microcavities. *Physical Review A* 64: 023808.

41. Dixon, P. B. and Hahn, D. W. 2005. Feasibility of detection and identification of individual bioaerosols using laser-induced breakdown spectroscopy. *Analytical Chemistry* 77: 631–638.

42. Morel, S., Leone, N., Adam, P. et al. 2003. Detection of bacteria by time-resolved laser-induced breakdown spectroscopy. *Applied Optics* 42: 6184–6191.

43. Baudelet, M., Guyon, L., Yu, J. et al. 2006a. Femtosecond time-resolved laser-induced breakdown spectroscopy for detection and identification of bacteria: A comparison to the nanosecond regime. *Journal of Applied Physics* 99: 084701.

44. Baudelet, M., Guyon, L., Yu, J. et al. 2006b. Spectral signature of native CN bonds for bacterium detection and identification using femtosecond laser-induced breakdown spectroscopy. *Applied Physics Letters* 88: 053901.

45. Judson, R. S. and Rabitz, H. 1992. Teaching lasers to control molecules. *Physical Review Letters* 68: 1500–1503.

46. Warren, W. S., Rabitz, H., and Dahleh, M. 1993. Coherent control of quantum dynamics—The dream is alive. *Science* 259: 1581–1589.

47. Dantus, M. and Lozovoy, V. V. 2004. Experimental coherent laser control of physicochemical processes. *Chemical Reviews* 104: 1813–1859.

48. Li, B. Q., Turinici, G., Ramakrishna, V. et al. 2002. Optimal dynamic discrimination of similar molecules through quantum learning control. *Journal of Physical Chemistry B* 106: 8125–8131.

49. Heelis, P. F. 1982. The photophysical and photochemical properties of flavins (isoalloxazines). *Chemical Society Reviews* 11: 15–39.

50. Sikorska, E., Khmelinskii, I., Komasa, A. et al. 2005. Spectroscopy and photophysics of flavin-related compounds: Riboflavin and iso-(6,7)-riboflavin. *Chemical Physics* 314: 239–247.

51. Roth, M., Guyon, L., Roslund, J. et al. 2009. Quantum control of tightly competitive product channels. *Physical Review Letters* 102: 253001.

52. Coudreau, S., Kaplan, D., and Tournois, P. 2006. Ultraviolet acousto-optic programmable dispersive filter laser pulse shaping in KDP. *Optics Letters* 31: 1899–1901.

53. Pearson, B. J. and Weinacht, T. C. 2007. Shaped ultrafast laser pulses in the deep ultraviolet. *Optics Express* 15: 4385–4388.

54. Roth, M., Mehendale, M., Bartelt, A. et al. 2005. Acousto-optical shaping of ultraviolet femtosecond pulses. *Applied Physics B—Lasers and Optics* 80: 441–444.

55. Nuernberger, P., Vogt, G., Selle, R. et al. 2007. Generation of shaped ultraviolet pulses at the third harmonic of titanium–sapphire femtosecond laser radiation. *Applied Physics B—Lasers and Optics* 88: 519–526.

56. Schriever, C., Lochbrunner, S., Optiz, M. et al. 2006. 19 fs shaped ultraviolet pulses. *Optics Letters* 31: 543–545.

57. Zeek, E., Maginnis, K., Backus, S. et al. 1999. Pulse compression by use of deformable mirrors. *Optics Letters* 24: 493–495.

58. Hauri, C. P., Kornelis, W., Helbing, F. W. et al. 2004. Generation of intense, carrier-envelope phase-locked few-cycle laser pulses through filamentation. *Applied Physics B–Lasers and Optics* 79: 673–677.

59. Berge, L. and Skupin, S. 2005. Self-channeling of ultrashort laser pulses in materials with anomalous dispersion. *Physical Review E* 71: 016602.

60. Hacker, M., Stobrawa, G., Sauerbrey, R. et al. 2003. Micromirror SLM for femtosecond pulse shaping in the ultraviolet. *Applied Physics B—Lasers and Optics* 76: 711–714.

61. Stone, K. W., Milder, M. T. W., Vaughan, J. C. et al. 2007. Spatiotemporal femtosecond pulse shaping using a MEMS-based micromirror SLM. *Ultrafast Phenomena XV* 88: 184–186.

62. Abe, T., Wang, G., and Kannari, F. 2008. Femtosecond pulse shaping on two-color laser superposition pulse using a MEMS micromirror SLM. *2008 Conference on Lasers and Electro-Optics & Quantum Electronics and Laser Science Conference*, 1–9: 2832–2833.

63. Rondi, A., Extermann, J., Bonacina, L. et al. 2009. Characterization of a MEMS-based pulse-shaping device in the deep ultraviolet. *Applied Physics B—Lasers and Optics* 96: 757–761.

64. Bonacina, L., Extermann, J., Rondi, A. et al. 2007. Multiobjective genetic approach for optimal control of photoinduced processes. *Physical Review A* 76: 023408.

65. Waldis, S., Webert, S. M., Noell, W. et al. 2008. Large linear micromirror array for UV femtosecond laser pulse shaping. *2008 IEEE/LEOS International Conference on Optical Mems and Nanophotonics, 11–14 August, Freiburg, Germany*, pp. 39–40.

14 Light Scattering by Fractal Aggregates

C. M. Sorensen

CONTENTS

14.1 Introduction 341
14.2 Fractal Aggregates 342
14.3 Light Scattering 344
    14.3.1 Polarization 344
    14.3.2 Cross Sections 345
        14.3.2.1 The Differential Scattering Cross Section 345
        14.3.2.2 The Total Scattering Cross Section 345
    14.3.3 The Rayleigh–Debye–Gans Approximation 346
14.4 The Structure Factor 347
    14.4.1 The Scattering Wave Vector 347
    14.4.2 The Fundamental Equation for the Structure Factor 348
        14.4.2.1 Small q Behavior of the Structure Factor: The Guinier Regime 350
    14.4.3 The Structure Factor for Fractal Aggregates 350
14.5 Light-Scattering Cross Sections 351
    14.5.1 Rayleigh Scattering 351
    14.5.2 Fractal Aggregate Light-Scattering Cross Sections 352
        14.5.2.1 Fractal Aggregate Differential Scattering Cross Section 352
        14.5.2.2 Fractal Aggregate Absorption Cross Section 353
        14.5.2.3 Fractal Aggregate Total Scattering Cross Section 353
        14.5.2.4 Fractal Aggregate Albedo 354
        14.5.2.5 The Meaning of the RDG Fractal Aggregate Approximation 354
14.6 Scattering from an Ensemble of Aggregates 355
    14.6.1 Effects of Polydispersity 355
    14.6.2 The Tyndall Effect 357
14.7 Optical Particle Sizing 358
    14.7.1 The Optical Structure Factor 359
14.8 Tests of the Validity of RDG Fractal Aggregate Light Scattering 360
14.9 Conclusion 362
Acknowledgments 362
References 362

14.1 INTRODUCTION

Particulate matter, such as aerosols, are a nonequilibrium state that with time evolves, via aggregation, to form larger particles. Aggregation with coalescence, like one might observe in a cloud of water droplets, leads to bigger, dense particles. Aggregation without coalescence, like one might observe in a cloud of solid, carbonaceous soot particles, leads to bigger, ramified (as opposed to dense) particles. Three decades ago, we learned that these ramified aggregates could be quantitatively described as fractal aggregates.[1]

Fractal aggregates are very common in both natural and technical settings. Perhaps the most common is soot from various combustion processes that occur on the face of the Earth, due to either human or natural activity. Other examples include any solid particle aerosol that has been allowed sufficient time for the aggregation processes to have effect, such as fumed SiO_2.

The purpose of this chapter is to review the light-scattering properties of fractal aggregates. Since fractal aggregates are common, we desire to know and understand their physical properties; and light scattering is an important physical characteristic. With light scattering we can study or monitor aggregation kinetics in the lab or in industrial settings. Light scattering is also very important for understanding the effects of aerosols on the Earth's climate. In 2001, this author presented a review of light scattering by fractal aggregates.[2] Another review of how fractal aggregates scatter light, that contains some of the material covered in the previous review but to a lesser extent, is presented here. This new review also includes advances made in the intervening nine years.

14.2 FRACTAL AGGREGATES

Fractals are scale invariant objects. This means that they appear the same upon a change of scale; they are self-similar. Once we were taught how to look for fractal aggregates by Mandelbrot,[3] we have come to realize that they abound in nature. Examples include trees, coastlines, mountain ranges, clouds and the subject of this chapter, fractal aggregates.

Fractal aggregates are clusters of particles with a self-similar structure over a finite range of length scales. The particles composing the aggregates are called "primary particles" or "monomers." Ideally, they are spherical with point contacts, but this description is an approximation. Regardless, we take their radius as a, and this sets the lower limit to the fractal scaling range. The upper limit is well described by the radius of gyration R_g, which is a root-mean-square radius that quantifies the overall size of the aggregate. A consequence of self-similarity is that the number of primary particles N in the aggregate scales as a power law with the reduced size of the aggregate as

$$N = k_0 \left(\frac{R_g}{a} \right)^{D_f}$$

(14.1)

In Equation 14.1, D_f is the fractal dimension and k_0 is the scaling prefactor. Equation 14.1 may be considered as the defining relation for fractal aggregates. One can say that the "mass" N scales with the linear size R_g with the dimensionality D_f, which is typically a noninteger. Figure 14.1 shows an ensemble of fractal aggregates captured from a flame–soot aerosol. Figure 14.2 shows a schematic drawing of a fractal aggregate illustrating Equation 14.1.

A second consequence of scale invariant self-similarity is that the pair correlation function of the primary particles is a power law that can be expressed as

$$g(r) \sim r^{D_f - d} \, h\left(\frac{r}{R_g} \right)$$

(14.2)

Here d is the spatial dimension and $h(x)$ is the cutoff function that describes how the aggregate ends at its perimeter. To preserve the power law, $h(x) = 1$ for $x \ll 1$, and to cut-off the power law, $h(x)$ must decrease faster than any power law for $x \gg 1$. Such a function that has proven to be adequate for fractal aggregates is the stretched exponential

$$h(x) = \exp(-x^\gamma)$$

(14.3)

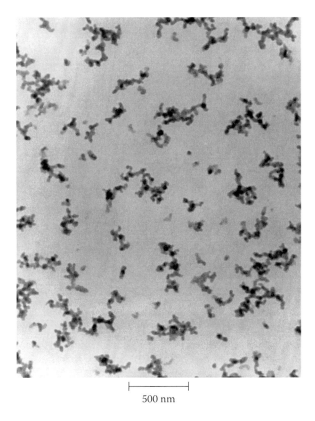

FIGURE 14.1   Soot fractal aggregates with fractal dimension $D_f = 1.78 \pm 0.07$.

Various studies have shown that $\gamma \sim 2$–2.5 (see Refs. [4]–[7]). Aerosol particles almost always aggregate via random Brownian motion; aggregates meeting and sticking together. Such a process is called diffusion limited cluster aggregation (DLCA).[8,9] In three-dimensional space, DLCA leads to fractal aggregates with $D_f = 1.78 \pm 0.1$ and $k_0 = 1.3 \pm 0.2$.[10–12] These numbers can be considered universal, since they have been determined from experiments on both aerosols and colloids, from simulations, which show an underappreciated anticorrelation between the fit values of D_f and k_0,[12] and a simple analytical theory.[13] The prefactor k_0 for some combustion aerosols has been measured

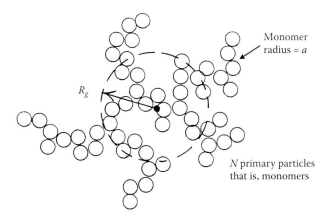

FIGURE 14.2   Schematic diagram of a fractal aggregate showing the monomer or primary particle size, a, the aggregate radius of gyration, R_g, and the number of monomers, N.

to be as large as 2.5^{14} and this possible break from universality appears legitimate and warrants more attention. There also are situations where aggregates are found with larger fractal dimensions as a result of processes following the aggregation processes, such as shear restructuring or cloud processing. More recently, hybrid superaggregates with $D_f = 1.8$ on short length scales and $D_f = 2.6$ on larger length scales have been discovered.[15–17]

In colloids, the probability of cluster–cluster sticking upon meeting can be made very small and then a new type of aggregation ensues, reaction limited cluster aggregation (RLCA). Although RLCA is common in colloids, it does not seem to have been observed in aerosols. RLCA leads to $D_f = 2.05$–2.15 with $k_0 = 0.94$ for the lower value.[7]

14.3 LIGHT SCATTERING

14.3.1 POLARIZATION

The polarization of light is the direction of its electric field vector. Maxwell's laws enforce that this is perpendicular to the direction of propagation; hence, there are two independent polarizations. Natural light that is emitted by the sun, light bulbs, and so on. have equal amounts of each of these two independent polarizations. Such light is sometimes called "unpolarized" or "randomly" polarized. Lasers are very often, but not always, polarized. A typical laboratory setup positions a laser shining its beam in the horizontal direction with the polarization in the vertical direction.

To understand the effect of scattering on polarization, consider a light wave traveling in the positive z direction incident upon a particle at the origin as drawn in Figure 14.3. The propagation direction is described by the incident wave vector $\vec{k}_i$ with magnitude $|\vec{k}_i| = 2\pi/\lambda$ where λ is the wavelength of the light in the medium. The incident light is polarized along the vertical x axis. The y–z plane is horizontal and is called the "scattering plane." Light is scattered in the direction of the scattered wave vector $\vec{k}_s$. We consider only elastic scattering; hence,

$$|\vec{k}_s| = |\vec{k}_i| \tag{14.4}$$

Because these magnitudes are equal, we represent each simply by k.

For particles small compared to the wavelength, the polarization of the scattered light is in the direction of the projection of the incident polarization at the particle, which is in the direction $\hat{x}$, onto a plane perpendicular to $\vec{k}_s$, plane p in Figure 14.3. In vector notation, this is equivalent to

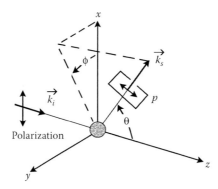

FIGURE 14.3 Geometry of scattering for an incident light wave traveling along the positive z-axis with propagation direction defined by the incident wave vector k_i and with polarization in the vertical x direction. Light is scattered from the small, spherical particle to the detector in the direction of the scattering wave vector k_s. The scattered polarization is in the direction of the projection of the incident polarization at the scatterer onto the plane p, which is perpendicular to the detector direction k_s.

the double cross product, $(\hat{k}_s \times \hat{x}) \times \vec{k}_s$, where $\hat{k}_s$ is the unit vector in the scattering direction. With either the projection or the double cross product, it can be shown that the scattered intensity obeys the proportionality

$$I_{sca} \propto 1 - \cos^2 \phi \sin^2 \theta. \tag{14.5}$$

Most laboratory experiments are confined to the scattering plane; hence, $\phi = 90°$. Then the angle θ is the scattering angle; $\theta = 0$ is forward scattering. An incident vertical polarization projects completely onto plane p of Figure 14.3 with no angular dependence. This is corroborated by Equation 14.5 with $\phi = 90°$. Horizontally polarized incident light would scatter out with horizontal polarization with intensity proportional to $\cos^2 \theta$, shown in Equation 14.5. Randomly polarized light would be a simple linear combination of these two polarizations weighted by their intensities.

14.3.2 CROSS SECTIONS

Cross sections describe the extent of scattering. As the name implies, their units always include an area (m^2). The concept is that the cross section for scattering is the effective area the incident wave sees and hence, blocks its path. The result of this blocking is either scattering, absorption, or both.

14.3.2.1 The Differential Scattering Cross Section

The differential scattering cross section describes the power scattered, $P_{sca}(W)$, in a given direction. The direction has a small spread in angles, a differential of a solid angle, $d\Omega$ (steradian). For an incident irradiance (often called wrongly as the intensity) I_o (W/m^2), the power scattered per solid angle is

$$\frac{P_{sca}}{\Omega} = \frac{dC_{sca}}{d\Omega} I_o \tag{14.6}$$

In Equation 14.6, $dC_{sca}/d\Omega$ is the differential scattering cross section with units of m^2/steradian.
The scattered irradiance is the scattered power per unit area of detection

$$I_{sca} = P_{sca}/A \tag{14.7}$$

The solid angle subtended by the detector a distance r from the scatterer is

$$\Omega = A/r^2 \tag{14.8}$$

Thus, from Equations 14.6 through 14.8, we obtain

$$I_{sca} = I_o \frac{dC_{sca}}{d\Omega} \frac{1}{r^2} \tag{14.9}$$

This equation contains the well-known $1/r^2$ dependence due to the three-dimensional geometry of space.

14.3.2.2 The Total Scattering Cross Section

The total scattering cross section describes the total scattering in all directions. Thus, it is found by the integration of the differential cross section over the complete solid angle

$$C_{sca} = \int_{4\pi} \frac{dC_{sca}}{d\Omega} d\Omega \tag{14.10}$$

The units of C_{sca} are m². This integral must include polarization effects (see above). The differential element $d\Omega$ in three-dimensional Euclidean space is

$$d\Omega = d\cos(\theta)\, d\varphi \qquad (14.11)$$

(See Figure 14.3 for the definition of the polar coordinates θ and φ.) The total power scattered from the incident beam of irradiance I_0 is

$$P_{sca} = I_0\, C_{sca}. \qquad (14.12)$$

14.3.3 THE RAYLEIGH–DEBYE–GANS APPROXIMATION

The exact description of light scattering from any particle requires solution of the Maxwell equations for the interaction of the electromagnetic wave with that particle. Perhaps the best known example of this is the Lorenz–Mie solution for scattering from a dielectric sphere.[18–20] Maxwell's equations are applied to the particular case of spherical boundary conditions to yield an infinite series of special functions related to that symmetry. Such an approach is useless for fractal aggregates, because the aggregate has no such symmetry.

Lessons can be learned, however, from the Lorenz–Mie solution. Studies from this laboratory[21–24] have shown that the Lorenz–Mie solution is based on simple diffraction from the sphere. Alternately, diffraction is the result of the wave nature of light and scattering is a result of both diffraction and the electromagnetic interaction of the light with the particle. If that electromagnetic interaction is small, diffraction dominates. The criterion for diffraction to dominate is that the so-called phase-shift parameter, ρ, defined as

$$\rho = 2kR(m - 1), \qquad (14.13)$$

is less than one. In Equation 14.13, k is the wave number and equals $2\pi/\lambda$, where λ is the light wavelength, R is the radius of the particle and m is the particle's refractive index. Use of the diffraction result to describe light scattering is often called the Rayleigh–Debye–Gans (RDG) approximation.

The RDG approximation can be made somewhat tenable for fractal aggregates with the following argument.[2] We can consider an aggregate as a dispersion of the primary particle optical material in vacuum. We then apply the Maxwell–Garnet effective medium theory, which is good for low and uniform density dispersions. The second condition is not satisfied by fractal aggregates, but remembering this caveat, we proceed. The Maxwell–Garnet theory allows one to calculate the effective index of refraction of the average medium that is composed of the primary particles and the vacuum as

$$(m - 1)_{eff} = f_v(m - 1). \qquad (14.14)$$

For a fractal aggregate, the volume fraction of monomers is approximately

$$f_v = \frac{Na^3}{R_g^3}. \qquad (14.15)$$

This combined with the fundamental equation for fractal aggregates, Equation 14.1, yields

$$(m - 1)_{eff} = R_g^{D_f - 3} a^{3 - D_f} \, (m - 1) \qquad (14.16)$$

where we set $k_0 = 1$. Substitution of Equations 14.13 into 14.16 finds the fractal aggregate phase-shift parameter as

$$\rho = 2k R_g^{D_f - 2} a^{3 - D_f} \, (m - 1) \qquad (14.17)$$

For a given optical wave number k and monomer of size a and refractive index m, Equation 14.17 shows that for $D_f < 2$ the phase-shift parameter gets *smaller* with increasing aggregate size. Thus, the condition for RDG to hold can be ultimately satisfied. On the other hand, for $D_f > 2$ increasing R_g increases the cluster phase-shift parameter, taking it further away from the regime in which the RDG theory can be successful. Equation 14.17 must be viewed with some caution because of the misuse of the Maxwell–Garnet theory.

14.4 THE STRUCTURE FACTOR

14.4.1 THE SCATTERING WAVE VECTOR

We consider a scalar electromagnetic wave incident upon a scattering element, or scatterer, at $\vec{r}$ as shown in Figure 14.4. We choose a scalar wave, because the development that follows is not affected by polarization. We make the scatterer point-like so that it scatters isotropically throughout space. The incident field at $\vec{r}$ in complex notation is

$$E(\vec{r}) = E_0 \exp(i\vec{k}_i \cdot \vec{r}) \tag{14.18}$$

where $i = \sqrt{-1}$ and $\vec{k}_i$ is the incident wave vector. The field scatters toward the detector, at $\vec{R}$, in the direction $\vec{k}_s$, the scattered wave vector. Under the assumption of elastic scattering, Equation 14.4, the field at the detector is

$$E(\vec{R}) \sim E(\vec{r})\exp(i\vec{k}_s \cdot (\vec{R} - \vec{r})) \tag{14.19}$$

Substitution of Equation 14.18 yields

$$E(R) \sim E_0 \exp(i\vec{k}_s \cdot \vec{R})\exp(i(\vec{k}_i - \vec{k}_s) \cdot \vec{r}) \tag{14.20}$$

We have dropped the equality for the proportionality, because we do not know and do not need to know the strength of the scattering element at $\vec{r}$. The second term of Equation 14.20 shows that the phase at the detector is a function of the position of the scattering element and the vector

$$\vec{q} = \vec{k}_i - \vec{k}_s \tag{14.21}$$

This vector $\vec{q}$ is called the *scattering wave vector*. Its direction is in the scattering plane from k_s to $\vec{k}_i$, as shown in Figure 14.5. Figure 14.5, the elasticity condition Equation 14.4, and simple

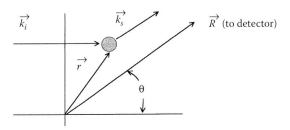

FIGURE 14.4 Diagram of an incident wave with propagation wave vector $\vec{k}_i$ scattering from a point-like scatterer at $\vec{r}$ the scattered wave leaves $\vec{r}$ in the direction of $\vec{k}_i$, the scattered wave vector, toward the detector at great distance $\vec{R}$.

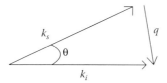

FIGURE 14.5   The scattering wave vector and its association with and the scattering angle θ.

trigonometry yield the magnitude of q to be

$$q = 2k \sin(\theta/2) \tag{14.22a}$$

$$= (4\pi/\lambda) \sin(\theta/2) \tag{14.22b}$$

where θ is the scattering angle.

The importance of q is that it describes how the phase at the detector, whatever that might be, depends on the position of the scatterer and the angle of scattering. Our approach below involves envisioning any scattering object, as composed of a great many point-like scatterers at various $\vec{r}$. The object is represented as a system of scatterers.[25] An important characteristic of q is that its inverse, q^{-1}, represents the length scale, or the probe length, of the scattering experiment. This follows from the second term in Equation 14.20, which can now be written as

$$E_{sca}(\vec{q}, \vec{r}) \sim E_0 \exp(i\vec{q} \cdot \vec{r}) \tag{14.23}$$

where E_{sca} is the amplitude of the scattered wave. Equation 14.23 shows that if the variation of r is small compared to q^{-1}, the scattered field does not significantly change; whereas if r varies greatly relative to q^{-1}, the scattered field changes significantly. Thus, q^{-1} represents a length scale to be compared to length scales of the scatterer; this comparison determines the scattered field.

14.4.2   THE FUNDAMENTAL EQUATION FOR THE STRUCTURE FACTOR

To determine the scattering from a material object of finite extent, we consider that object to be composed of a multitude of point-like scatterers. We call this concept a "system of scatterers."[25] Here we assume that the object meets the RDG criterion. Then, the scattering from each individual scatterer of the object is so weak that they do not sense each other's scattered fields. Hence, each scatterer sees only the incident field. Each scatterer then sends (scatters) a wave out to the detector with phase information given by Equation 14.23. Then the total field at the detector due to the system of N scatterers is the sum over N scatterers

$$E(q) \sim E_0 \sum_{i}^{N} \exp(i\vec{q} \cdot \vec{r}_i) \tag{14.24}$$

Then the intensity of the scattered wave is the square of the amplitude, which for complex amplitudes is the product of the amplitude and its complex conjugate

$$I(q) \sim E_0^2 \sum_{i}^{N} \exp(i\vec{q} \cdot \vec{r}_i) \sum_{j}^{N} \exp(-i\vec{q} \cdot \vec{r}_j) \tag{14.25}$$

Since we must often deal with ensembles of scatterers with random orientations, we have dropped the vector notation on the left-hand side of Equation 14.25.

The structure factor, $S(q)$, is proportional to the scattered intensity $I(q)$ and defined with the normalization $N^{-2}I(q)$; thus

$$S(q) = N^{-2} \sum_i^N \sum_j^N \exp(i\vec{q} \cdot (\vec{r}_i - \vec{r}_j)) \tag{14.26}$$

We regain the equality, because this defines the structure factor.* This normalization ensures $S(0) = 1$.

To convert the sums to integrals in Equation 14.26, we write the density function of the system of scatterers as

$$n(\vec{r}) = \sum_i^N \delta(\vec{r} - \vec{r}_i) \tag{14.27}$$

where $\delta(\vec{r})$ is the Dirac delta function. Then

$$\sum_i^N \exp(i\vec{q} \cdot \vec{r}_i) = \int \exp(i\vec{q} \cdot \vec{r}) n(\vec{r}) d\vec{r} \tag{14.28}$$

From this and Equation 14.26 we write the structure factor as

$$S(q) = N^2 \iint n(\vec{r}) n(\vec{r}') \exp(i\vec{q} \cdot (\vec{r} - \vec{r}')) d\vec{r} \, d\vec{r}' \tag{14.29}$$

Changing variables to $\vec{r}$ and $\vec{u} = \vec{r} - \vec{r}'$ we obtain

$$S(q) = \iint n(\vec{r}) n(\vec{r} - \vec{u}) \exp(i\vec{q} \cdot \vec{u}) d\vec{r} \, d\vec{u} \tag{14.30}$$

Now recognize that the integral over $\vec{r}$ is a convolution of the density with itself, which we write as

$$g(\vec{u}) = \int n(\vec{r}) n(\vec{r} - \vec{u}) d\vec{r} \tag{14.31}$$

Then

$$S(q) = \int g(\vec{u}) \exp(i\vec{q} \cdot \vec{u}) du \tag{14.32}$$

We find that the structure factor is the Fourier transform of the self convolution of the object's density function.

The function $g(\vec{u})$ of Equations 14.31 and 14.32 is well known in the scattering literature[27–29] and beyond, not so much as the self convolution but as the density autocorrelation function of the object. This function answers the question of how the density correlates with itself. It is a joint probability which, given a scatterer at an arbitrary position, determines the probability of finding another scatterer a position u away. Its importance transcends scattering theory and for condensed matter includes the thermodynamics of the system. Equation 14.32 implies that $S(q)$ and $g(\vec{u})$ are Fourier transform pairs.

Under the assumption of isotropy, $S(\vec{q}) = S(q)$ and $g(\vec{u}) = g(u)$, and the solid angle integration can be performed on Equation 14.32 to yield

$$S(q) = 4\pi \int g(u)[(\sin qu)/qu]u^2 \, du \tag{14.33}$$

---

* Some authors use a N^{-1} normalization to define $S(q)$.

14.4.2.1 Small q Behavior of the Structure Factor: The Guinier Regime

The sine function in the integrand of Equation 14.33 can be expanded for small qu. Then Equation 14.33 becomes two integrals, the first of which is unity by the normalization of $g(u)$, so that we obtain

$$S(q) \approx 1 - \frac{q^2}{6} \int u^2 g(u)\, du \tag{14.34}$$

One can show that the radius of gyration for an object is given by

$$R_g^2 = \frac{1}{2} \int u^2 g(u)\, du \tag{14.35}$$

Thus,

$$S(q) \approx 1 - \frac{1}{3} q^2 R_g^2. \tag{14.36}$$

Equation 14.36 is the Guinier equation,[26] good in the regime $qR_g < 1$ for all shapes. It is very useful for measuring particle size as described by the radius of gyration. The radius of gyration is a root-mean-square radius. For a sphere $R_g = \sqrt{3/5}R$.

14.4.3 THE STRUCTURE FACTOR FOR FRACTAL AGGREGATES

Substitution of the correlation function for fractal aggregates, Equations 14.2 and 14.3, into Equation 14.33 yields the structure factor for fractal aggregates. The key question is what is the value of the exponent γ in the cut-off function? Much of the older literature assumes $\gamma = 1$, that is, an exponential cut-off.[27,28] With this, one finds

$$S(q) = \frac{\sin[(D_f - 1)\tan^{-1}(q\xi)]}{(D_f - 1)\, q\xi(1 + q^2\xi^2)^{(D_f - 1)/2}} \tag{14.37}$$

where

$$\xi^2 = \frac{2R_g^2}{D_f(D_f + 1)} \tag{14.38}$$

When $D_f = 2$, the exponential structure factor simplifies to the Fisher–Burford[29] form

$$S(q) = \left(1 + \frac{2q^2 R_g^2}{3D_f}\right)^{-D_f/2} \tag{14.39}$$

This remains a good approximation when $D_f \approx 2$, such as 1.8 or 2.1 as found for DLCA and RLCA, respectively. Because of its simplicity, the Fisher–Burford form has been used a lot; but it is wrong. Surprisingly, it fits data from polydisperse systems well, under the assumption of no polydispersity of the aggregates. This is erroneous; aggregates form due to aggregation, which leads to polydispersity. More correct fits are obtained with more correct structure factors integrated over the polydispersity[30] as described below.

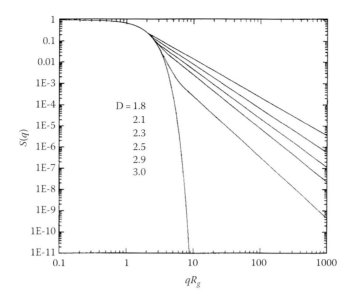

FIGURE 14.6 Double log plot of the structure factor for an aggregate with a Gaussian cut off for its density correlation function.

Current science[4-7] supports a Gaussian cut-off with $\gamma = 2$ as a better descriptor of the fractal correlation function. Then

$$S(q) = {}_1F_1\left[\frac{D_f}{2}, \frac{3}{2}; -\frac{(qR_g)^2}{D_f}\right]$$ (14.40)

where ${}_1F_1$ is the Kummer or hypergeometric function. Figure 14.6 shows a plot of the Gaussian structure factor for different D_f values.[2] Note the small hump near $qR_g \approx 2$.

14.5 LIGHT-SCATTERING CROSS SECTIONS

14.5.1 RAYLEIGH SCATTERING

We assume that the monomers (primary particles) are Rayleigh scatterers. The conditions for Rayleigh scattering are[18-20]

$$ka < 1$$ (14.41a)

$$|m|\, ka < 1$$ (14.41b)

In Equation 14.41b $m = n + ik$ is the complex refractive index of the particle. If we deal with optical wavelengths of say $\lambda = 500$ nm and reasonable refractive indices, these conditions imply a particle radius of $a < 80$ nm. The Rayleigh differential scattering cross section for the particle is

$$\frac{dC_{sca}}{d\Omega} = k^4 a^6 F(m)$$ (14.42)

where

$$F(m) = \left|\frac{(m^2 - 1)}{(m^2 + 2)}\right|^2$$ (14.43)

Absorption in the RDG regime is given by

$$C_{abs} = 4\pi k a^3\, E(m) \tag{14.44}$$

where

$$E(m) = \mathrm{Imag}\!\left[\frac{(m^2 - 1)}{(m^2 + 2)}\right] \tag{14.45}$$

Imag means to take the imaginary part.

14.5.2 Fractal Aggregate Light-Scattering Cross Sections

Under the RDG approximation, the individual monomers of the fractal aggregate act as point scatterers. Thus, they are the scatterers in our system of scatterers, above. Since they do not interact, the total scattering from the aggregate is determined by the cross section of each individual monomer adding with the others with the proper accounting for the phase differences of the scattered waves from each particle at the detector. This proper accounting is taken care of by the structure factor; recall the structure factor contained solely and completely in the phase information.

14.5.2.1 Fractal Aggregate Differential Scattering Cross Section

The total differential cross section for a fractal aggregate of Rayleigh primary particles under the RDG approximation is

$$\frac{dC_{sca}^{agg}}{d\Omega} = N^2 \frac{dC_{sca}^{m}}{d\Omega} S(q) \tag{14.46}$$

where "agg" and "m" mean "aggregate" and "monomer," respectively.

This result is sketched in Figure 14.7 as the scattered intensity, which is directly proportional to the differential cross section, as a function of q on a double logarithmic plot. There we see a q independent Rayleigh regime at small q with an N^2 dependence on the number of monomers in the aggregate. In this regime, the length scale of the scattering process q^{-1} is too big to see the aggregate as anything but an unresolved blob. As q increases to $q \sim R_g^{-1}$, $I(q)$ starts to decline as the Guinier

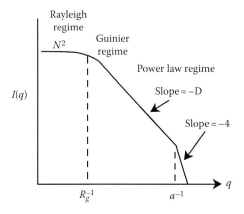

FIGURE 14.7 Double log plot of the scatter light intensity as a function of q for a fractal aggregate made of N spherical particles of radius a. The aggregate has a radius of gyration of R_g and a fractal dimension D_f.

regime is entered. Note, this is when q^{-1} is small enough to begin to resolve the overall aggregate size. It is in this regime that the size of the aggregate can be measured. As q increases further, the power-law regime is entered described by

$$I(q) \sim C(qR_g)^{-D_f} \tag{14.47}$$

Thus, the slope of the log–log plot is $-D_f$ and this provides an opportunity for measurement of the fractal dimension. The best value for the coefficient is $C = 1.0 \pm 0.1$.[2] Finally, a second bend is found at $q \sim a^{-1}$, where the scattering begins to resolve the monomer size, a. Thereafter, the slope is typically -4, the Porod scattering from a sphere.[31,32]

14.5.2.2 Fractal Aggregate Absorption Cross Section

For absorption, the particle independence and nonmultiple scattering conditions of RDG imply the simple result

$$C_{\text{abs}}^{\text{agg}} = N C_{\text{abs}}^{\text{m}} \tag{14.48}$$

The absorption of the aggregate is just the sum of the independent absorptions of its monomers.

14.5.2.3 Fractal Aggregate Total Scattering Cross Section

The total scattering cross section is formally related to the differential cross section by Equation 14.10. The integral must account for the polarization. Under the RDG approximation, the polarization law for the aggregate is the same as it is for the monomers of which it is composed. Then for the common experimental situation of light vertically polarized relative to the scattering plane, the factor of Equation 14.5 must be included to yield

$$C_{\text{sca}} = \iint \frac{dC_{\text{sca}}}{d\Omega} \, (1 - \cos^2 \varphi \sin^2 \theta) \, d(\cos \theta) \, d\varphi \tag{14.49}$$

Substitution of the expression for the differential cross section of a fractal aggregate, Equation 14.46, leads to an integral over the structure factor, which leads to algebraically elaborate results. Instead, we follow the simpler approach of Dobbins and Megaridis[33] who used the small qR_g, Guinier form, universal to all structure factors, and a generalization based on the Fisher–Burford structure factor to obtain

$$C_{\text{sca}}^{\text{agg}} = N^2 C_{\text{sca}}^{\text{m}} G(kR_g), \tag{14.50}$$

where

$$C_{\text{sca}}^{\text{m}} = \frac{8\pi}{3k^4 a^6} F(m), \tag{14.51}$$

the Rayleigh total scattering cross section for a sphere of radius a, and

$$G(kR_g) = \left[1 + (4/3D_f)k^2 R_g^2 \right]^{-D_f/2} \tag{14.52}$$

Dobbins and Megaridis compared Equations 14.50 through 14.52 to both simulation data for total scattering of Mountain and Mulholland[4] and their own porous sphere model with success. We remark that Koylu and Faeth,[14] Kazakov and Frenklach[34] and Mulholland and Choi[35] have considered the problem of the total scattering cross section for an aggregate, but the resulting expressions are more complex than those above.

14.5.2.4 Fractal Aggregate Albedo

A useful quantity is the albedo, which is the ratio of scattering to total extinction

$$\omega = \frac{C_{\text{sca}}}{(C_{\text{sca}} + C_{\text{abs}})} \tag{14.53}$$

For a particle or aggregate with a real refractive index (no imaginary part), $C_{\text{abs}} = 0$ and hence, $\omega = 1$.

For an aggregate of particles with complex refractive index, the use of Equations 14.48 and 14.50 in Equation 14.53 yields

$$\omega^{\text{agg}} = \left(1 + \frac{3}{2}\left(\frac{E(m)}{F(m)}\right) ka^{-3}\left(NG(kR_g)\right)^{-1}\right)^{-1} \tag{14.54}$$

where $G(kR_g)$ is given by Equation 14.52. This albedo is plotted for soot in Figure 14.8. The calculations were performed[2] for $\lambda = 500$ nm and $3(E/F)/2k_0 = 1$, which is a typical combination for carbonaceous soot. Figure 14.8 shows that initially $\omega^{\text{agg}} \sim R_g$, but this quickly saturates near $R_g = 100$ nm. In a previous review, Equation 14.54 was shown to give a good comparison to the soot albedo data of Mulholland and Choi[35] attesting to the veracity of G as well.

14.5.2.5 The Meaning of the RDG Fractal Aggregate Approximation

Attention to the development above uncovers the meaning of the RDG Fractal Aggregate Approximation.

First, the mathematical machinery of RDG is the Fourier transform, the physical motivation of the Fourier transform is addition of waves from a source. The addition of waves from a

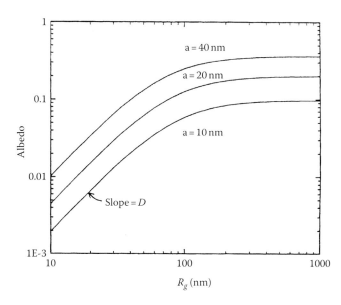

FIGURE 14.8 The albedo for a soot-like fractal aggregate calculated with Equation 14.54 with $D_f = 1.75$, $E/Fk_0 = 2/3$, $\lambda = 500$ nm and various monomer radii a. (From Sorensen, C.M. 2001. *Aerosol Sci. Tech.* 35:2648–2687. With permission.)

source is diffraction, and since we implicitly assumed the far field, Fraunhofer diffraction, a subject in sophomore physics. Simply said, RDG is diffraction from whatever object you are scattering from.

Second, in RDG fractal aggregate light-scattering theory, the electromagnetism is added at the monomer level as Rayleigh scattering. This gives a magnitude to the scattered waves that are subsequently added together via the laws of diffraction.

Third, the electromagnetism follows the diffraction sequentially; they are not convoluted together. Such convolution of the wave and electromagnetic natures of light is the complete and correct description of light's interaction with matter and is the basis for complete theories such as the Lorenz–Mie theory. It is the weak coupling between the monomers of the aggregate that allows the deconvolution of the electromagnetism and the diffraction.

14.6 SCATTERING FROM AN ENSEMBLE OF AGGREGATES

14.6.1 EFFECTS OF POLYDISPERSITY

Since aggregates are the result of aggregation, we expect that any ensemble of aggregates would be polydisperse. We need to understand the effect of polydispersity on the shape of the structure factor measured with light-scattering experiments, and with this, quantify polydispersity in terms of a measurable size distribution.[5,36] We also need to understand what the mean size we measure is, in terms of moments of the size distribution.

In general, the effective structure factor for an ensemble of aggregates can be written as

$$S_{\text{eff}}(q) = \frac{\int N^2 n(N)\, S[qR_g(N)]\, \mathrm{d}N}{\int N^2 n(N)\, \mathrm{d}N} \tag{14.55}$$

In Equation 14.55, $n(N)$ is the size distribution, that is, the number of clusters per unit volume with N monomers per cluster. One procedure to follow would be to fit data numerically using a single cluster structure factor above and an appropriate size distribution with fit variables, such as D_f and the size distribution parameters of most probable size, distribution width, and so on. Here we describe an analysis that relies on physical sense and simplicity.

We represent the single aggregate structure factor by its Rayleigh, Guinier, and power-law limits

$$S(q) = 1 \quad qR_g < 1 \tag{14.56a}$$

$$= 1 - \frac{q^2 R_g^2}{3} \quad qR_g \sim 1 \tag{14.56b}$$

$$= C\,(qR_g)^{-D_f} \quad qR_g > 1 \tag{14.56c}$$

The value of C in Equation 14.56c is 1.0 ± 0.1 for the Gaussian structure factor.[2]

The size distribution is completely described by its moments. We define the ith moment of the size distribution as

$$M_i = \int N^i n(N)\, \mathrm{d}N \tag{14.57}$$

Now substitute Equations 14.56 into 14.55 to find the ensemble equivalent of Equations 14.56

$$S_{eff}(q) = 1 \quad qR_{g,z} < 1 \tag{14.58a}$$

$$= \left(1 - \frac{q^2 R_{g,z}^2}{3}\right) \quad qR_{g,z} \sim 1 \tag{14.58b}$$

$$= C_p C(qR_{g,z})^{-D_f} \quad qR_{g,z} > 1 \tag{14.58c}$$

where $R_{g,z}$ is the z-average[28] radius of gyration given by

$$R_{g,z}^2 = a^2 k_0^{-2/D_f} \frac{M_{2+2/D}}{M_2} \tag{14.59}$$

Equation 14.58b says that a Guinier analysis of scattered light from an ensemble of particles yields the z-average radius of gyration, Equation 14.59 might be confusing at first, but consider Equation 14.1 rewritten as $R_g = a (N/k_0)^{1/D_f}$. With this we see that the z-average radius of gyration is one which is weighted by the moments of the size distribution. Further, note that for $D_f \approx 2$ the moments are ca. M_3/M_2, high moments that stress the large N end of the size distribution. This is not surprising, because light scatters more from larger particles than smaller ones; the light-scattering process weighs the large end of the size distribution.

Notable results in Equation 14.58c are that the power law at large q retains the exponent—D_f; thus the fractal dimension measurement is unaffected by simple polydispersity (see, however, Martin and Ackerson[37]). A change occurs in that the coefficient of the power law is modified by the polydispersity of the ensemble. The modifying factor C_p is significantly different than unity, so that the use of a single cluster structure factors on scattering data would yield erroneous results. Fortunately, the result opens an opportunity to measure, to some degree, the polydispersity of the ensemble.

We call C_p the polydispersity factor; it is given by[36]

$$C_p = \frac{M_1}{M_2} \left(\frac{M_{2+2/D_f}}{M_2}\right)^{D_f/2} \tag{14.60}$$

Given a size distribution, the polydispersity factor C_p can be calculated. It is well established that an aggregating system develops a self preserving, scaling distribution given by[38–40]

$$n(N) = As^{-2} x^{-\tau} e^{-\alpha x} \tag{14.61}$$

where A and α are constants, x is the relative size

$$x = \frac{N}{s}, \tag{14.62}$$

and s is a mean size. Notice that the size in this context is the aggregation number, N. The exponent τ is a measure of the width of the distribution with large τ implying a broad distribution. This scaling form is valid when $x > 1$, the small x form being different. Since scattering strongly weighs the large end part of the distribution, that is, $x > 1$, the small x part has little effect on the properties of scattering from an ensemble of aggregates and hence can be ignored.

The log normal distributions are frequently used in the literature. However, we have shown[41] that these distributions yield erroneous values for distribution moments higher than the second when compared to the exact scaling distribution. For the second moment and lower, the distributions

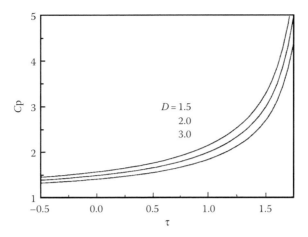

FIGURE 14.9 The polydispersity factor of Equation 14.60 plotted versus the scaling exponent τ for three different fractal dimensions. (From Sorensen, C.M. and Wang, G.M. 1999. *Phys. Rev. E* 60:7143–7148. With permission.)

agree well. Since scattering involves higher moments, such as $M_{2+2/D} \approx M_3$ for $D \approx 2$, it is erroneous to use the log normal distribution for light-scattering analysis.

In Figure 14.9, C_p is shown as a function of the width parameter τ for a variety of fractal dimensions D. In particular, for DLCA in the continuum regime it is expected and well verified that $\tau = 0$ and $D_f = 1.75$ to 1.8. From Figure 14.9 we find $C_p = 1.53$. We have applied Equation 14.60 successfully for measuring the size distribution exponent.[36]

In summary, the optical structure factor is a very capable method for yielding R_g, D_f, and the scaling distribution polydispersity exponent τ. Under the RDG approximation, the particle refractive index need not be known. If data beyond $qR_g \approx$ are not available, the D_f measurement can only be considered qualitative.

14.6.2 THE TYNDALL EFFECT

The Tyndall effect is the increased scattering as an ensemble of particles aggregates. We assume that the system neither gains nor loses particulate mass, that is, the total number of monomers (also called "primary particles") is constant, but the particles come together and form larger particles. Here we will take these larger particles as fractal aggregates. We also require no expansion or contraction of the system to maintain constant volume. Let n be the number of clusters per unit volume in the system. Then the scattered intensity has the proportionality

$$I(q) \sim nN^2 \, S_{\text{eff}}(q) \tag{14.63}$$

With the use of Equation 14.56, we can write

$$I(q) \sim nN^2, \quad qR_g < 1 \tag{14.64a}$$

$$I(q) \sim nN^2 (qR_g)^{-D_f}, \quad qR_g > 1. \tag{14.64b}$$

The product nN is the total number of monomers in the system, N_m, which is conserved. From Equation 14.1, $NR_g^{-D_f}$ is a constant. With these facts, Equations 14.64 become

$$I(q) \sim N_m N, \quad qR_g < 1 \tag{14.65a}$$

$$I(q) \sim N_m, \quad qR_g > 1. \tag{14.65b}$$

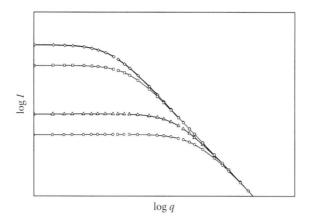

FIGURE 14.10 Sketch of the evolution with time of the scattered intensity versus q on a log–log plot as the system aggregates. The enhanced scattering in the Rayleigh regime is the Tyndall effect for fractal aggregates. (From Sorensen, C.M. 2001. *Aerosol Sci. Tech.* 35:2648–2687. With permission.)

We see that with aggregation the scattered intensity in the Rayleigh regime increases as the number of monomers in the aggregates N increases, but the power-law regime stays constant during aggregation. This behavior is sketched in Figure 14.10 and shown with an aggregating aerosol in Figure 14.11.[42]

14.7   OPTICAL PARTICLE SIZING

One of the major motivations for understanding how aggregates scatter and absorb light is to allow *in situ* light-scattering and absorption measurements of particle size, morphology, and number density. Light scattering is noninvasive, remote, and as good as or better than any other more "direct" method, such as transmission electron microscopy (TEM). Reciprocal space is just as good as real space.

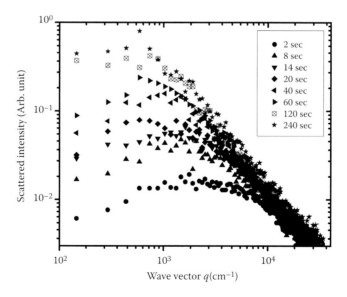

FIGURE 14.11 Experimental light-scattering data from a dense, aggregating, soot aerosol demonstrating the Tyndall effect. The legend indicates time after the aerosol was created by explosion of a fuel rich acetylene/ oxygen mixture.

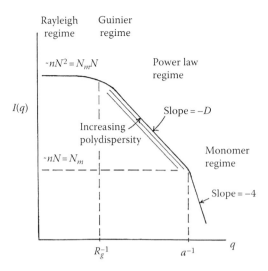

FIGURE 14.12 Double log plot of the scatter light intensity as a function of q for a group of fractal aggregates made of spherical particles of radius a with average monomer number N. The aggregates have a z-average radius of gyration of $R_{g,z}$ and a fractal dimension D_f. The group has n aggregates per unit volume and the total number of monomers in the system is $N_m = nN$. (From Sorensen, C.M. 2001. *Aerosol Sci. Tech.* 35:2648–2687. With permission.)

14.7.1 THE OPTICAL STRUCTURE FACTOR

Scattered intensity versus angle is the experimental method used, but the angle is best converted to q which, under the RDG approximation, is reciprocal or Fourier space of the scattering object. We call the measurement of I vs. q plotted log–log an optical structure factor analysis.[43] From the optical structure factor, one can determine the aggregate average R_g and D_f, the polydispersity of the aggregate size distribution, and the monomer size a, if short wavelengths or large a is available. These facts are outlined schematically in Figure 14.12, which is the ensemble-scattering analogue of Figure 14.7.

An example from some of our studies[2] is given in Figure 14.13 for scattering from a titania aerosol. This aerosol was created by the hydrolysis of titanium tetraisopropoxide vapor in humid air and then

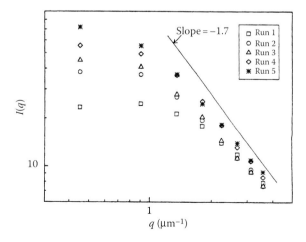

FIGURE 14.13 Optical structure factor analysis of an aggregating TiO_2 aerosol. Later times are larger run numbers.

contained in a 6 L scattering chamber. The incident light had a wavelength of 488 nm. Inspection of Figure 14.13 shows that with increasing age of the aerosol as indicated by run number, the bend in the optical structure factor where the slope of I vs. q goes from zero to negative, progresses to smaller q. A cardinal rule is that a change in slope on these log–log plots implies a length scale. In this case, the length scale is the overall aggregate size, and since $R \sim q^{-1}$, this is a direct observation, albeit qualitative, of the aggregate size increasing with time. Also seen is a significant power-law regime with a slope implying $D_f \approx 1.7$. The $I(q = 0)$ limit (the Rayleigh regime) increases with run time. This is the Tyndall effect that as a system coarsens, and it scatters more light. On the other hand, the power-law regime in Figure 14.13 approaches a constant intensity as described above.

Guinier analysis of the data in Figure 14.13 is shown in Figure 14.14. We have found this works best by inverting the Guinier result, Equation 14.58b, to

$$\frac{I(0)}{I(q)} \approx 1 + \frac{q^2 R_{g,z}^2}{3} \tag{14.66}$$

Plotting the inverse, normalized scattered intensity versus q^2 yields linear graphs with a slope equal to $R_{g,z}^2/3$.

14.8 TESTS OF THE VALIDITY OF RDG FRACTAL AGGREGATE LIGHT SCATTERING

In a previous review of this subject,[2] the author made a comprehensive analysis of work testing the validity of the RDG description of fractal aggregate scattering and absorption up to 1999. The general conclusion was that the neglect of the monomer–monomer scattering interaction, referred to as internal multiple scattering in that study, may lead to errors in the cross sections in the 10–20% range both positive and negative. For small ka, the scattering can be greater than RDG predictions. On the other hand, if the imaginary refractive index is significant, absorption can cause screening of the aggregate interior with concomitant less scattering than the RDG prediction. A fractal dimension $D_f > 2$ also causes less actual scattering analogous to the decreased scattering of spheres relative to the Rayleigh regime, once out of the Rayleigh regime. Comparison to calculations by

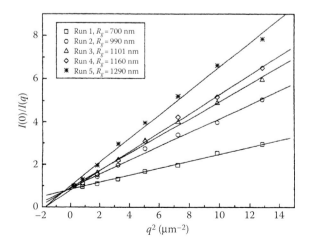

FIGURE 14.14 Titania aerosol of Figure 14.13 plotted for a Guinier analysis ala Equation 14.66. Note that the bulk of the data extend beyond the Guinier regime.

Farias et al.[44] indicated that a phase-shift parameter for fractal aggregates could be defined and RDG theory was good to 10% when

$$\rho = 2kR_g(m-1) < 3. \qquad (14.67)$$

where m is the real part of the index of refraction of the monomers.

The analysis of experimental tests not only outlined all the problems in the optical measurements but also more problems in comparative measurements, most notably perturbative sampling of the aerosol and measurement of properties under the TEM. Experiments found no major discrepancies for measuring R_g, N, and D_f, but both the optical and TEM methods had significant uncertainties.

The overall conclusion of that review was that for $D_f \approx 1.78$ and ka on the order of <0.5, RDG was good to 10%.

Subsequently, Wang and Sorensen[45] tested the RDG theory with light-scattering measurements on TiO_2 and SiO_2 aerosols that aggregated to yield $D_f \approx 1.8$ aggregates. Note, that the TiO_2 has a large refractive index of 2.6. They dealt with polydisperse aerosols, measured the polydispersity with light-scattering methods, and compared scattering RDG, predicted and experimentally measured cross sections. The agreement was within the experimental errors of ca. 10%.

Chakrabarty et al.[46] compared the RDG theory predictions to data from a soot aerosol with $D_f = 1.7$. They used a tandem DMA set up to achieve a quasi-monodisperse distribution of aggregate sizes centered at a mobility radius of 400 nm. This eliminated the problem of the size distribution and the higher moment weighting of light scattering. Agreement between theory and experiment was good up to 10–25% depending on the value of the soot refractive index, another notorious experimental uncertainty.

Van-Hulle et al.[47] compared the RDG results to both the discrete dipole approximation (DDA)[48,49] and the "rigorous solution" (RS) of Xu et al.[50] Their theoretical aggregates were soot with $D_f = 1.8$

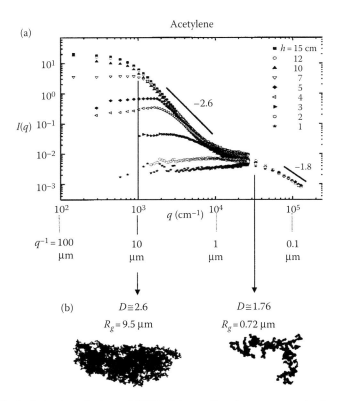

FIGURER 14.15 Optical structure factor and TEM pictures of soot superaggregates in and obtained from a laminar diffusion acetylene flame.

with $ka = 0.15$ and $N = 64$, 128, and 256. DDA and RS agreed well with each other for both absorption and total scattering cross sections and with RDG for absorption cross sections. However, RDG was 60 to 120% in error for the scattering.

Yon et al.[51] compared the RDG theory to more exact calculations that included monomer–monomer electromagnetic interactions, that is, internal multiple scattering. They created $D_f = 1.8$ aggregates for the calculations with $N = 64$, 128, and 256 and varied the monomer size up to $ka = 0.9$. They found that for small ka the scattering at $q = 0$ was ca. 5% greater than predicted by RDG (see above) but fell to as much as 20% smaller at the largest ka. The absorption cross section was good up to 10%. The predictions of total scattering, Equations 14.50–14.52, were verified but the best fit to RDG required $D_f = 2.24$. Most unexpected, however, was that the large qR_g regime yielded fractal dimensions of 2.34 ± 0.15 instead of the fractal dimension of 1.8. These large values of D_f were not explained and stand in strong contrast to many experiments that have yielded the correct fractal dimension when compared to nonscattering methods.

Finally, I will stretch the envelope a bit. Studies in our labs over the past several years[15–17] have uncovered the existence of soot superaggregates; aggregates of one morphology composed of aggregates of another morphology. Figure 14.15 shows both optical structure factor measurements and TEM pictures of soot from a heavily sooting, laminar diffusion, acetylene flame. Both indicate that the soot has a fractal dimension of $D_f = 2.6$ for length scales in the range of ca. 1–10 μm and a fractal dimension of $D_f = 1.8$ for submicron scales. We ask: how is it that the analysis of the light-scattering data with RDG scattering theory agrees well with the TEM measurements? Does the RDG theory work for 10 μm, $D_f = 2.6$ superaggregates?

We conclude in much the same way we did in 1999. It appears that the RDG formulation of scattering and absorption by fractal aggregates is accurate to ca. 10%.

14.9 CONCLUSION

The fundamental aspects of RDG fractal aggregate light scattering have been laid out above in what the author hopes is a straightforward and accessible manner. Despite the author's failings, the theory's simplicity speaks for itself. Although approximations that allow this simplicity cause concern for the theory's validity, the theory has proven extremely useful in describing the otherwise complex problem of how fractal aggregates scatter and absorb light. Its use for the diagnostics of aerosols (and colloids) is no less suspect than many other alternative diagnostics. Finally, I will end in the same way I did in 2001 with my first review of this subject[2] by reminding the reader that the scattering wave vector q (not θ) is the physically motivated independent variable for all scattering experiments. Moreover, this is by and large a geometric universe and hence, logarithmic axes more readily uncover features in data than linear ones. Plot your data double log, scattered intensity versus q.

ACKNOWLEDGMENTS

Since 2001 my work has benefited from the pleasant and stimulating collaboration with my students M. Beavers, M. Berg, R. Dhaudhabel, D. Fry, C. Gerving, S.E. Gilbertson, J. Hubbard, W.G. Kim, T. Mokhtari, F. Pierce, and R. Pyle and my good friend and colleague Amit Chakrabarti. My work has been supported by NSF and NASA.

REFERENCES

1. Forrest, S.R. and Witten, T.A. 1979. Long-range correlations in smoke-particle aggregates. *J. Phys. A* 12:L109–L117.
2. Sorensen, C.M. 2001. Light scattering from fractal aggregates. A review. *Aerosol Sci. Tech.* 35:2648–2687.
3. Mandelbrot, B. 1983. *The Fractal Geometry of Nature*. Freeman, San Francisco, CA.

4. Mountain, R.D. and Mulholland, G.W. 1988. Light-scattering from simulated smoke agglomerates. *Langmuir* 4:1321–1326.

5. Nicolai, T., Durand, D., and Gimel, J.-C. 1994. Static structure factor of dilute solutions of polydisperse fractal aggregates. *Phys. Rev. B* 50:16357–16363.

6. Cai, J., Lu, N., and Sorensen, C.M. 1995. Analysis of fractal cluster morphology parameters: Structural coefficient and density autocorrelation function cutoff. *J. Colloid Interface Sci.* 171: 470–473.

7. Lattuada, M., Wu, H., and Morbidelli, M. 2003. A simple model for the structure of fractal aggregates. *J. Colloid Interface Sci.* 268:106–120.

8. Meakin, P. 1983. Formation of fractal clusters and networks by irreversible diffusion-limited aggregation, *Phys. Rev. Lett.* 51:1119–1122.

9. Kolb, M., Botet, R., and Jullien, R. 1983. Scaling of kinetically growing clusters. *Phys. Rev. Lett.* 51:1123–1126.

10. Jullien, R. and Botet, R. 1987. *Aggregation and Fractal Aggregates*. World Scientific, Singapore.

11. Meakin, P., Fractal Aggregates 1988. *Adv. Colloid & Interface Sci.* 28:249–331.

12. Sorensen, C.M. and Roberts, G. 1997. The prefactor of fractal aggregates. *J. Colloid Interface Sci.* 186:447–452.

13. Sorensen, C.M. and Oh, C. 1998. Divine proportion shape invariance and the fractal nature of aggregates. *Phys. Rev.* E58:7545–7548.

14. Koylu, U.O. and Faeth, G.M. 1995. Optical properties of overfire soot in buoyant turbulent diffusion flames at long residence times. *Trans. ASME* 116:152–159.

15. Sorensen, C.M., Kim, W., Fry, D., and Chakrabarti, A. 2003. Observation of soot superaggregates with a fractal dimension of 2.6 in laminar acetylene/air diffusion flames. *Langmuir* 19:7560–7563.

16. Kim, W., Sorensen, C.M., and Chakrabarti, A. 2004. Universal occurrence of soot aggregates with a fractal dimension of 2.6 in heavily sooting laminar diffusion flames. *Langmuir* 20:3969–3973.

17. Kim, W., Sorensen, C.M., Fry, D., and Chakrabarti, A. 2006. Soot aggregates, superaggregates and gel-like networks in laminar diffusion flames. *J. Aerosol Sci.* 37:386–401.

18. van de Hulst, H.C. 1981. *Light Scattering by Small Particles*. Dover, New York, NY.

19. Kerker, M. 1969. *The Scattering of Light and Other Electromagnetic Radiation*. Academic Press, New York, NY.

20. Bohren, C.F. and Huffman, D.R. 1983. *Absorption and Scattering of Light by Small Particles*. John Wiley & Sons, New York, NY.

21. Sorensen, C.M. and Fischbach, D.E. 2000. Patterns in Mie scattering. *Opt. Commun.* 173:145–153.

22. Sorensen, C.M. and Shi, D. 2000. Guinier analysis for homogeneous dielectric spheres of arbitrary size. *Opt. Commun.* 178:31–36.

23. Sorensen, C.M. and Shi, D. 2002. Patterns in the ripple structure in Mie scattering. *JOSA* 19:122–125.

24. Berg, M.J., Sorensen, C.M., and Chakrabarti, A. 2005. Patterns in Mie scattering: Evolution when normalized by the Rayleigh cross section. *Appl. Opt.* 44:7487–7493.

25. Oh, C. and Sorensen, C.M. 1999. Scaling approach for the structure factor of a generalized system of scatterers. *J. Nanopart. Res.* 1:369–377.

26. Guinier, A., Fournet, G., Walker, C.B., and Yudowitch, K.L. 1955. *Small Angle Scattering of X-Rays*. Wiley, New York, NY.

27. Teixeira, J. 1986. Experimental methods for studying fractal aggregates, in *On Growth and Form, Fractal and Non-Fractal Patterns in Physics*, H.E. Stanley and N. Ostrowski, (Eds), Nijhoff, Dordrecht. 145–165.

28. Martin, J.E. and Hurd, A.J. 1987. Scattering from fractals. *J. Appl. Cryst.* 20:61–78.

29. Fisher, M.E. and Burford, R.J. 1967. Theory of critical-point scattering and correlations I. The Ising model. *Phys. Rev.* 156:583–622.

30. Sorensen, C.M., Cai, J., and Lu, N. 1992. Test of static structure factors for describing light scattering from fractal soot aggregates. *Langmuir* 8:2064–2069.

31. Porod, G. 1951. Die Röntgonkleinwinkelstreuung von Dichtgepacken Kolloiden Systemen. *Kolloid Z.* 124:83–114.

32. Jullien, R. 1992. From Guinier to fractals. *J. Phys. I* (France) 2:759–770.

33. Dobbins, R.A. and Megaridis, C.M. 1991. Absorption and scattering of light by polydisperse aggregates. *Appl. Opt.* 30:4747–4754.

34. Kazakov, A. and Frenklach, M. 1998. Dynamic modeling of soot particle coagulation and aggregation; implementation with the method of moments and application to high-pressure laminar premixed flames. *Combust. Flame* 114:484–501.

35. Mulholland, G.W. and Choi, M.Y. 1998. Measurement of the mass specific extinction coefficient for acetylene and ethene smoke using the large agglomerate optics facility, *Twenty-Seventh Symposium (International) on Combustion*, The Combustion Institute, 1515–1522.

36. Sorensen, C.M. and Wang, G.M. 1999. Size distribution effect on the power law regime of the structure factor of fractal aggregates. *Phys. Rev. E* 60:7143–7148.

37. Martin, J.E. and Ackerson, B.J. 1985. Static and dynamic light scattering by fractals. *Phys. Rev.* A31:1180–1182.

38. Friedlander, S.K. and Wang, C.S. 1966. The self-preserving particle size distribution for coagulation by Brownian motion. *J. Colloid Interface Sci.* 22:126–132.

39. Wang, C.S. and Friedlander, S.K. 1967. The self-preserving particle size distribution for coagulation by Brownian motion. *Jour. Coll. Interface Sci.* 24:170–179.

40. van Dongen, P.G.J. and Ernst, M.H. 1985. Dynamic scaling in kinetics of clustering. *Phys. Rev. Lett.* 54:1396–1399.

41. Sorensen, C.M., Cai, J., and Lu, N. 1992. Light-scattering measurements of monomer size, monomers per aggregate, and fractal dimension for soot aggregates in flames. *Appl. Opt.* 31:6547–6557.

42. Dhaubhadel, R., Chakrabarti, A., and Sorensen, C.M. 2009. Light scattering study of aggregation kinetics in dense, gelling aerosols. *Aerosol Sci. Technol.* 43:1053–1062.

43. Gangopadhyay, S., Elminyawi, I., and Sorensen, C.M. 1991. Optical structure factor measurements for soot particles in a premixed flame. *Appl. Opt.* 30:4859–4864.

44. Farias, T.L., Koylu, U.O., and Carvalho, M.G. 1996. Range of validity of the Rayleigh–Debye–Gans theory for optics of fractal aggregates. *Appl. Opt.* 35:6560–6567.

45. Wang, G.M. and Sorensen, C.M. 2002. Experimental test of the Rayleigh–Debye–Gans theory for light scattering by fractal aggregates. *Appl. Opt.* 41:4645–4651.

46. Chakrabarti, R.K., Moosmuller, H., Arnott, W.P., Garro, M.A., Slowik, J.G., Cross, E.S., Han, J.-H., Davidovits, P., Onasch, T.B., and Worsnop, D.R. 2007. Light scattering and absorption by fractal-like carbonaceous chain aggregates: Comparison of theories and experiment. *Appl. Opt.* 46, 6990–7006.

47. Van-Hulle, P., Weill, M.-E., Talbaut, M., and Coppalle, A. 2002. Comparison of numerical studies characterizing optical properties of soot aggregates for improved EXSCA measurements. *Part. Part. Syst. Charact.* 19:47–57.

48. Draine, B.T. 1988. The discrete-dipole approximation and its application to interstellar graphite grains. *Astrophys.* 333:848–872.

49. Draine, B.T. and Flatau, P.J. 1994. Discrete-dipole approximation for scattering calculations. *J. Opt. Soc. Am.* A. 11:1491–1499.

50. Xu, Y.L. and Gustafson, A.S. 1999. Comparison between multisphere light-scattering calculations. Rigorous solution and discrete dipole approximation. *Astrophys. J.* 513:894–909, and references therein.

51. Yon, J., Roze, C., Girasole, T., Coppalle, A., and Mees, L. 2008. Extension of RDG-FA for scattering prediction of aggregates of soot taking into account interactions of large monomers. *Part. Part. Syst. Charact.* 25:54–67.

Section IV

UV, X-ray, and Electron Beam Studies

15 Aerosol Photoemission

Kevin R. Wilson, Hendrik Bluhm, and Musahid Ahmed

CONTENTS

15.1 Introduction 367
15.2 Photoelectric Charging of Aerosols.. 370
    15.2.1 Aerosol Adsorption and Desorption Kinetics....................................... 371
    15.2.2 Circular Dichroism in Aerosol Photoemission 373
    15.2.3 Probe Molecule Aerosol Photoemission... 375
    15.2.4 Summary 377
15.3 Synchrotron-Based Aerosol Photoemission 378
    15.3.1 Vacuum UV Photoemission of Aerosols using Velocity Map Imaging 378
    15.3.2 Photoemission from Biological Nanoparticles 379
    15.3.3 Angle-Resolved Photoemission from Nanoparticles............................. 382
    15.3.4 Nanoparticle Photoemission in Interstellar Dust Clouds..................... 387
    15.3.5 Threshold Photoemission from Multicomponent Aerosol..................... 387
    15.3.6 Summary: Threshold Aerosol Photoemission and VMI 389
    15.3.7 XPS of Submicron Particles 389
    15.3.8 XPS of Liquid Aerosols.. 391
15.4 Summary and Outlook 394
Acknowledgments... 395
References.. 395

15.1 INTRODUCTION

Photoelectron spectroscopy is based on the photoelectric effect[1] where the energy of an incident photon ($h\nu$) is transferred to a valence or core electron of an atom, leading to the emission of an electron if the incident photon energy is larger than the binding energy (BE) of the electron in the atom. Using energy-dispersive detectors, the kinetic energy (KE) of the emitted electron is measured. From the difference of the incident photon energy and the measured KE, the BE of the electron within the atom can be determined. The BE for a core-level electron is sensitive to the chemical bonds to neighboring atoms ("chemical shifts") and can be used to probe the oxidation state of a given atom or different functional groups. The inelastic mean free path (IMFP) of photoelectrons in a condensed sample is on the order of a few Angstroms to many nanometers and depends both on the KE and the chemical composition of the sample. Figure 15.1a shows a general functional form for electron escape lengths versus KE based on an empirical formula from Seah and Dench.[2] It is the short IMFP of electrons that makes photoemission spectroscopy a surface-sensitive probe of chemical composition. On the other hand, electrons are also scattered by gas molecules. Since gases densities are a small fraction (10^{-6}) of the condensed phase, the IMPF in a gas is about 10^6 times longer in gases than in solids. The electron IMPF depends on KE and pressure as shown in Figure 15.1b for nitrogen (a proxy for air) and water vapor.[3–6] Figure 15.1b shows that the IMPF of electrons at kinetic energies above 30 eV is much shorter than that below 30 eV. Spectroscopic techniques that measure electrons at higher kinetic energies are therefore restricted to low-pressure environments or have to utilize differential pumping schemes that limit the path length of the electrons in the

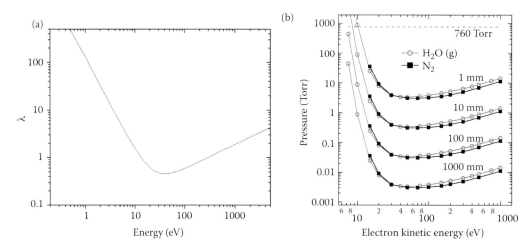

FIGURE 15.1 (a) Electron IMFP (in nanometers and shown as λ on *Y* axis) versus electron KE reproduced from Seah and Dench.[2] (b) Electron IMFP in a gas as a function of electron KE and pressure. The plot shows equi-IMFP lines for both nitrogen and water vapor. (Data are taken from Blanco, F. Garcia, G. 2003. *Phys. Rev. A* 67:022701; Garcia, G.; Roteta, M.; Manero, F. 1997. *Chem. Phys. Lett.* 264:589–595; Hoffman, K. R.; Dababneh, M. S.; Hsieh, Y. F.; Kauppila, W. E.; Pol, V.; Smart, J. H.; Stein, T. S. 1982. *Phys. Rev. A* 25:1393–1403. Munoz, A.; Oller, J. C.; Blanco, F.; Gorfinkiel, J. D.; Limao-Veira, P.; Garcia, G. 2007. *Phys. Rev. A* 76:052707. With permission.)

high-pressure region. Techniques that measure low KE electrons, on the other hand, can operate even at atmospheric pressure.

This chapter focuses on photoemission studies of substrate-free nanoparticles, droplets, and aerosols. Since there is a vast literature on clusters (i.e., very small nanoparticles), we focus our discussion on particle sizes in the range of 20 nm–200 μm. While the distinction between large clusters and small nanoparticles is rather arbitrary, this chapter focuses only on applications of photoelectron spectroscopy to particles that are stable at atmospheric pressure and room temperature, thereby excluding the wide range of clusters and nanoparticles that can be formed at low temperatures in molecular beam expansions. As an introduction to aerosol photoemission, we review only a small portion of the broader literature on photoelectron spectroscopy. A few select examples from clusters, single particle, and liquid jet spectroscopy are used to illustrate in broad terms what can be learned from photoelectron spectroscopy when it is applied to complex samples such as aerosol particles.

Bjorneholm and coworkers[7] pioneered the study of clusters using core-level spectroscopy, most notably x-ray photoelectron spectroscopy (XPS). This group recently published a review article[7] detailing almost two decades of work that began with studies of small rare gas clusters,[8] and has evolved to more complex studies of hydrogen-bonded clusters[9,10] (e.g., water) and molecular solids (NaCl). XPS spectra are used to examine changes in electronic structure that occur when small molecular aggregates are formed from monomeric species. In general, photoemission spectroscopy has revealed that molecules or atoms that reside at a cluster surface have unique electronic features that are, in general terms, intermediate between gas and bulk phase properties. This is because surface atoms or molecules lack the full coordination of species inside the cluster, which leads to unique electronic properties and in some cases enhanced chemical reactivity. Furthermore, these unique surface states evolve as a function of cluster size, since the fraction of surface molecules increases with decreasing aggregate size. These results suggest that any fundamental understanding of submicron aerosol properties requires experimental methods that distinguish bulk from surface electronic states to reveal how surface properties evolve with particle size.

Photoemission from semiconductors and insulators is often complicated by the buildup of electrical charge produced by ionizing radiation. Rühl and coworkers[11] used single nanoparticles suspended in an electrodynamic trap to examine fundamental particle charging mechanisms. A single ~500 nm SiO_2 nanoparticle was exposed to soft x-rays, and the resulting primary photoelectrons, Auger electrons, and secondary electrons are analyzed. It was observed that the number of electrons emitted per absorbed photon changed with excitation photon energy, while electron emission decreased with the number of elementary charges on the particle. When the silica particle is coated with a 40-nm gold layer, fast primary photoelectron emission is suppressed, while the yield of secondary electrons increased considerably compared to bare silica nanoparticles due to multiple ionization.[12] An inelastic loss mechanism on the rough gold surface was used to explain this phenomenon. These trapping experiments show that photoemission can be used to examine how various charging mechanisms depend on the detailed surface structure of a nanoparticle with core–shell morphologies. As discussed later in this chapter, measuring the change in particle charge state upon photoexcitation is a sensitive way to examine a variety of surface localized properties occurring on nanometer-sized aerosols suspended in a carrier gas.

While photoemission from gas and condensed phase systems has been studied for a number of years, the recent combination of micron-sized liquid jets with synchrotron radiation has taken this field in exciting new directions. Early work has been reviewed extensively by Faubel and Winter.[13] Initial work focused on the electronic structure of liquid water[14] has recently expanded to studies of solvation in aqueous solutions (alkali halide salts,[15,16] surfactants,[17,18] DNA bases[19]). The alkali halide work shows how photoemission can be used to gain insights into the solvation of anions at the surface of a liquid jet. Other studies have focused on surface versus bulk solvation in solutions comprised surfactants. In related work, Saykally and coworkers[20–23] have combined liquid jets with near-edge x-ray absorption spectroscopy to examine the electronic structure of water and aqueous solutions. These measurements have implications for many important atmospheric processes such as acid rain formation, stratospheric ozone depletion, and aerosol formation; all processes that depend, in part, on the way gas-phase species are accommodated and then solvated at the surface of a liquid droplet or solid particle.

There are a number of early ultraviolet (UV) photoemission studies of aerosols conducted at atmospheric pressure. Schmidt-Ott and coworkers[24] measured unexpectedly large photoemission quantum yields from 4 to 10 nm Ag and Au nanoparticles. These quantum yields were determined to be 10–100 times greater than macroscopic samples of the same composition. A variety of theoretical efforts focusing on nanoparticle geometry have been used to semiquantitatively model this enhanced photoemission.[25,26] Photoemission from soot nanoparticles generated in a flame has been reported in the literature.[27]

Within the field of aerosol chemistry, a number of groups have developed electron emission techniques for particle size and composition analysis. For example, Ziemann and McMurry[28] examined the chemical properties of submicron aerosol particles using secondary electron yields produced by high-energy (100–600 eV) electron bombardment. The overall secondary electron yield[29] was found to be a sensitive function of the chemical composition of multicomponent aerosol particles. In a related study, Ziemann et al.[28] used electron impact charging of aerosols. They found that the saturation charge on an aerosol was a linear function of particle diameter and sensitive to the particle shape (e.g., sphere vs cube). The development of commercial particle instruments, which are based on sizing particles by electrical mobility, benefited enormously from these fundamental studies of nanoparticle charging mechanisms.

Finally, aerosol photoemission[30] is the basis for a number of real-time, field-based particle detectors. These devices operate at ambient conditions and use UV light to photoelectrically charge nanometer-sized aerosol particles, which are then detected by an electrometer. These instruments[30–33] are especially sensitive to nanometer-sized combustion aerosols comprised of molecules with low ionization energies, such as polycyclic aromatic hydrocarbons (PAHs). These instruments are important tools in urban environments to track combustion sources and to assess human exposure to combustion particulates.[34]

It is now widely accepted that aerosol particles play critical roles in, for example, planetary atmospheres, combustion reactions, and heterogeneous catalysis. This is in part due to the variety of heterogeneous processes that may occur when small particles are suspended in a gas. Photoelectron spectroscopy is a powerful tool for probing the electronic structure of surfaces and as such its application to aerosols is expected to further our understanding of nanoparticle surface chemistry. Aerosol photoemission is still a young field, partly due to the challenges of detecting photoelectrons from particles surrounded by a carrier gas. This chapter aims to highlight various approaches to aerosol photoemission as well as outlines the kind of information that can be obtained using photoelectron-based methods. The chapter first focuses on the use of photoelectric charging techniques to measure surface composition, desorption kinetics, nanoparticle chirality, and interfacial polarity and solvation. Next, a number of synchrotron-based techniques are presented to show how classical gas-phase and surface science techniques have been adapted to new studies of aerosol surfaces. The chapter closes with a discussion of future prospects and challenges for aerosol photoemission.

15.2 PHOTOELECTRIC CHARGING OF AEROSOLS

In this section, we focus on the application of atmospheric pressure photoemission for the detection and chemical characterization of aerosols. The development of the first photoelectron-based aerosol detectors was pioneered by a number of research groups[30,31,33,35] interested in the detection of PAH molecules on soot particles. PAHs are suspected carcinogens and are emitted in copious amounts in urban environments from vehicles and factories. The basic experimental concept is shown in Figure 15.2. Particles entrained in a carrier gas are passed through a photoelectric charging cell where the aerosol is irradiated with either a low-pressure mercury lamp, with primary wavelengths of 185 nm (6.7 eV) and 254 nm (4.9 eV), or a UV laser. When the ionization energy of molecules that are condensed at the particle surface falls within this energy range, they absorb the UV photons and emit photoelectrons. Since the particles themselves are smaller than the mean free path of electrons, the probability of back diffusion of the photoelectrons to the particle surface is small, thus yielding an overall net positive charge on the particle. This change in particle charge state is then detected by a conductive filter connected to a picoammeter. In some cases, electrostatic precipitators are used to ensure that photoelectrons or negative ions are extracted out of the aerosol flow to prevent partial particle reneutralization and to enhance detection efficiency. Charging efficiencies

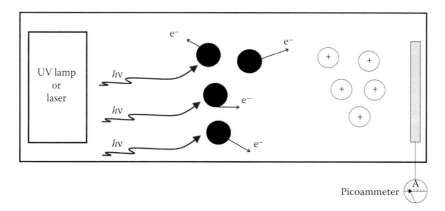

FIGURE 15.2 A schematic of a photoelectric charging cell. Aerosol particles, entrained in a carrier gas, pass into the cell and are irradiated with UV light from a laser or Hg Lamp. Upon absorption, the particles emit photoelectrons, which leave the particle with a net positive charge. The change in particle charge state is then measured by a picoammeter connected to a conductive filter.

as high as 65–85% for particles between 50 and 200 nm can be achieved, with overall detection limits on the order of 1 ng m^{-3} for particle-phase PAHs.[30]

Although the underlying physics that govern photoexcitation and the subsequent ejection of a photoelectron from a nanometer-sized object is rather complex, PAH detectors based on the principle of photoelectric charging are commercially available (EcoChem Analytics) and are routinely deployed in urban environments to quantify the concentration of combustion aerosols.[36,37] A detailed discussion of these field measurements is beyond the scope of this chapter; however, there are a number of studies that use photoelectric charging to probe in new ways the surface composition, chemistry, and physical properties of aerosol particles. These studies are the main focus of this section, where we discuss the current status of the field and also illustrate new ways for probing the surfaces of nanometer-sized particles using the simple photoelectric approach outlined in Figure 15.2.

Niessner et al.[38] used photoelectric charging to examine the ionization energy for a number of PAH molecules absorbed to nanometer-sized NaCl and carbon seed aerosols. In this study, a tunable laser with 3–6 µJ pulse energies was used instead of a UV lamp to charge the particles. The authors then examined how the photoelectric yield varied with PAH coverage, substrate, molecular structure, and photon wavelength. Size-selected particles (~50 nm) were passed through a temperature-controlled oven to condense PAH molecules onto the surface of seed aerosol (NaCl and carbon). The resulting surface coverage was determined using an electrostatic classifier and a screen diffusion battery. Niessner and coworkers showed that measurable photoelectric signals could be obtained for particles covered by as little as 0.1 monolayer equivalent PAH coverage, thereby illustrating the exquisite sensitivity that can be obtained simply by measuring the change in particle charge state upon photoionization.[38]

Photoionization spectra can be obtained by measuring the total change in charge state of the aerosol as a function of laser wavelength. The analysis of these spectra yields ionization energies for various surface-bound PAH molecules. The ionization onsets for the seven PAH molecules, measured by Niessner et al.[38] fall between 5.0 and 5.4 eV. For example, the photoionization spectra of perylene adsorbed on a 30-nm carbon particle is shown in Figure 15.3. The ionization onset is found to be ~5.5 eV (i.e., ~225 nm), which is 1.4 eV below the ionization energy of gas-phase perylene (6.90 eV). The reduction in ionization energy that occurs in going from the gas to solid phases is generally attributed to the electrostatic stabilization of the cation by neighboring polarizable molecules.

This energy shift, or polarization energy, is largest for fully coordinated molecules embedded in a bulk organic crystal. However, at a surface, the ionization energy should be in between the gas phase and a fully coordinated molecule in the crystal, due to the reduction in the number of neighboring molecules at an interface. For instance, Sato et al.[39] reported an ionization energy for solid perylene of 5.2 eV (238.4 nm), which appears slightly lower than the onset of the photoelectric yield observed by Niessner et al.[38] This difference is most likely due to the reduction of the polarization energy of a perylene molecule residing at the surface with submonolayer coverage. In fact, shifts in ionization energy can be used, as presented in the following sections, to examine the microscopic changes in electronic structure of an aerosol interface that occur upon the addition of an adsorbate, such as water.

15.2.1 Aerosol Adsorption and Desorption Kinetics

Hueglin et al.[40] and Steiner et al.[41] used photoelectric charging to measure the adsorption kinetics and thermodynamics of PAHs bound to an aerosol surface. This is the "aerosol analog" to temperature-programmed desorption used for measuring adsorbate thermodynamics in classical surface science. The measurement of adsorption kinetics and energetics is important for understanding how organic vapors are partitioned between the gas and aerosol phases in the atmosphere.

The experiment[40] consists of first coating aerosol seed particles (e.g., NaCl or carbon) with perylene, and then exposing the aerosol stream to a rapid temperature jump and measuring the desorption rate of perylene using a photoelectric detector. The photoelectric charging is proportional

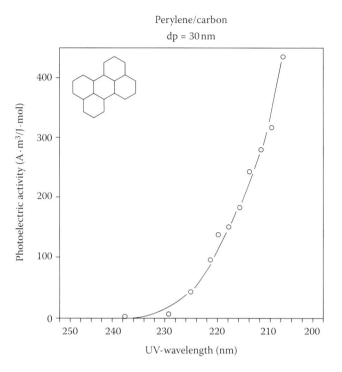

FIGURE 15.3 Photoionization spectra of perylene (submonolayer) adsorbed onto a 30-nm carbon particle. (Reproduced from Niessner, R.; Robers, W.; Wilbring, P. 1989. *Anal. Chem.* 61:320–325. Copyright 1989 American Chemical Society. With permission.)

to the surface concentration of perylene. If the desorption process is first order, then the rate of desorption of molecule A from a surface as a function of temperature (T) is given by an Arrhenius-type equation

$$[A](t) = [A]_0 e^{-k_d t}$$

where,

$$k_d = v e^{-E_d/RT},$$

v is the preexponential factor, and E_d is the desorption activation enthalpy. The desorption enthalpy can be obtained by measuring k_d as a function of temperature. To obtain k_d, desorption profiles like the one shown in Figure 15.4 are obtained by measuring the change in particle photoemission as a function of time.[40] Shown in Figure 15.5 is an Arrhenius plot for desorption of perylene from a 90-nm NaCl aerosol. The desorption activation enthalpy, obtained by Hueglin et al.[40] is 122.9 kJ/mol, similar to thermodynamic measurements for the same system reported by Steiner et al.[41] using the same technique. These results for the perylene/NaCl system are very different than the desorption kinetics measured for perylene absorbed onto diesel and carbon seed particles. Both Steiner et al.[41] and Hueglin et al.[40] find that perylene desorption from these carbon interfaces is far more complex and cannot be modeled using a single value for the desorption enthalpy and a simple Arrhenius expression, as in the case of NaCl. Instead, the desorption kinetics can be only modeled by assuming two distinct surface sites with different activation enthalpies. These results illustrate that photoelectric charging can be used to obtain adsorbate-surface energetics in nanometer-sized particles, which have

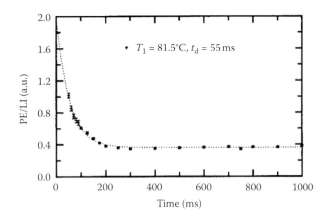

FIGURE 15.4 Desorption of perylene from polydisperse NaCl particles at 81.5°C. The desorption kinetics are measured by monitoring the photoelectric (PE) signal as a function of time. (Reproduced from Hueglin, C.; Paul, J.; Scherrer, L.; Siegmann, K. 1997. *J. Phys. Chem. B* 101:9335–9341. Copyright 1997 American Chemical Society. With permission.)

important ramifications for understanding the partitioning of organic material between the gas and particle phases, both in the troposphere as well as in combustion exhaust streams.

15.2.2 CIRCULAR DICHROISM IN AEROSOL PHOTOEMISSION

Photoelectric detection has also been used to examine more fundamental interactions of electromagnetic radiation with aerosol particles. While circular dichroism (CD) is a well-known effect in both photoabsorption and photoemission spectra of chiral molecules, observing this effect in a randomly oriented ensemble of free nanoparticles suspended in a carrier gas is more exotic and was first reported by Paul et al.[42] These experiments utilize the photoelectric charging scheme described

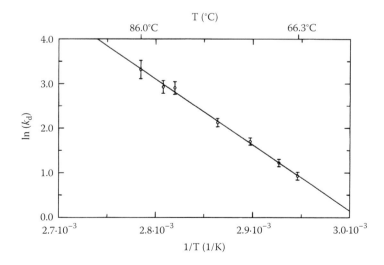

FIGURE 15.5 The desorption rate constant as a function of reciprocal temperature. This Arrhenius plot is used to compute desorption activation energies for perylene-coated NaCl particles. (Reproduced from Hueglin, C.; Paul, J.; Scherrer, L.; Siegmann, K. 1997. *J. Phys. Chem. B* 101:9335–9341. Copyright 1997 American Chemical Society. With permission.)

above and shown in Figure 15.2. Circularly polarized laser light (193.4 nm, 6.4 eV) is directed into a photoelectric charging cell and the change in particle charge state is measured as a function of laser polarization. Particles comprised of pure enantiomers are formed, size-selected, and illuminated with right- or left-hand circularly polarized light. The resulting change in particle charge state for each helicity is then measured for each enantiomeric nanoparticle. The asymmetry in photoelectric charging[42] is defined as

$$A = \frac{I_{PE}^{LCP} - I_{PE}^{RCP}}{I_{PE}^{LCP} + I_{PE}^{RCP}},$$

where I_{PE} is the photoelectric charging intensity for left circularly polarized (LCP) light and right circularly polarized (RCP) light circularly polarized light. For nonchiral particles or a racemic mixture, $A = 0$. Shown in Figure 15.6 is the asymmetry for pure particles comprised of $R(+)$ or $S(-)$ acetyl-cylopentadienyl-carbonyl-triphenylphosphine-iron as a function of particle size.[43] The asymmetry in photoelectric charging is a strong function of particle size. For particle sizes smaller than 20 nm A approaches 0 (i.e., no CD). For diameters larger than 50 nm, there is clear evidence for strong CD in both entaniomeric particles. It should be noted that there are large numbers (10^3–10^6) of chiral molecules in these particles and that the observed nanoparticle CD is not simply due to molecular chirality, but rather must arise from long-range order of the nanoparticle crystal lattice or

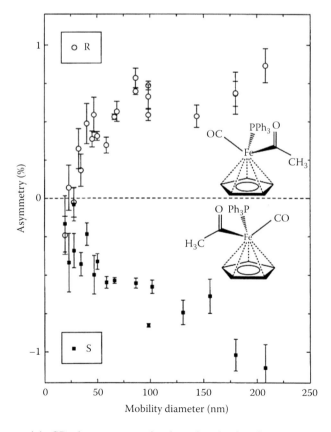

FIGURE 15.6 Nanoparticle CD: the asymmetry in photoelectric charging versus aerosol size. (Reprinted from Kasper, M.; Keller, A.; Paul, J.; Siegmann, K.; Siegmann, H. C. 1999. *J. Electron Spectrosc. Relat. Phenom.* 98–99:83–93. Copyright 1999 Elsevier. With permission.)

surface. It is noted by the authors[42,43] that the particles are randomly oriented in the photoemission cell and that various checks were performed to rule out particle alignment or other asymmetries in the experiment that might be responsible for these effects. Furthermore, no such asymmetry is observed when s or p linearly polarized light is used to charge the particles.[42]

A second example of nanoparticle CD was reported by the same group.[42] In this case, both *R*- and *S*- (1,1′-binapthyl 2,2′-diyl hydrogen phosphate) enantiomers showed the same effect, with smaller particles (<20 nm) exhibiting little asymmetry in photoelectric charging. The authors suggest that the lack of CD for small particles in these cases might be due to a fundamental change in the phase or nanoparticle structure. They speculate[42] that smaller particles may be disordered or amorphous, and therefore exhibit inversion symmetry, which could help explain the size dependence in CD photoemission. While this is a plausible explanation, further work is clearly needed to understand how molecular chirality induces nanoparticle chirality. In any case, a general and ongoing challenge for aerosols scientists is to determine the phase of an aerosol (liquid, solid, crystalline amorphous) produced in laboratory sources (nucleation, atomization) or in the atmosphere (e.g., combustion and secondary organic aerosol).[44] Phase may play an important role in heterogeneous reactions, thermodynamics, and microphysics. The authors[42] suggest that there are many particle types in the atmosphere that could be chiral, including viruses, spores, and pollens. CD photoemission spectroscopy might find an application in the rapid detection and online characterization of these important bioaerosols.

15.2.3 Probe Molecule Aerosol Photoemission

In recent years, Woods and coworkers[45–47] have pioneered a method which combines photoelectric charging with solvatochromic probe molecules to examine the solvation environment of aerosol surfaces. The absorption maximum of a solvatochromic molecule in a bulk solution is sensitive to solvent polarity. Woods and coworkers[46] measure the polarity of an aerosol using a solvatochromic probe molecule deposited on the particle surface. Through the combination of photoionization energy measurements with solvatochromic shifts, trends in the surface polarity of aerosol particles can be determined.[46]

Woods et al. use tunable lasers instead of UV lamps to photoionize the particles in a conventional photoelectric charging cell. This makes possible the use of various laser pump–probe schemes to ionize the aerosol via one or two photon absorption. An optical parametric oscillator is used to scan the photon energy in order to measure ionization thresholds. Woods and coworkers[46] can also measure excited state spectra using photoelectric charging action spectroscopy. In these experiments, the probe dye molecule, on the surface of the aerosol, is excited to the first electronic state and subsequently ionized by a second UV laser.

The probe molecule is a laser dye, such as Coumarin 314, and is placed on the aerosol surface by co-atomizing the dye in a NaCl solution. This technique produces 75 nm NaCl particles with as few as 50 dye molecules on the surface. Woods et al.[46] studied the surface solvation environment of NaCl aerosol by measuring the solvatochromic shifts of Coumarin 314 as a function of relative humidity (RH). Shown in Figure 15.7 is an example of their data. There is a clear blue shift in the electronic spectra of dye with increasing RH. For Coumarin 314 the absorption maximum red shifts with increasing solvent polarity, since the S_1 excited state has a larger dipole moment than the ground state. The blue shift with increasing RH is counterintuitive, since water is a polar solvent, and thus one would expect a red shift with increasing water coverage on the NaCl surface. Similar counterintuitive changes in the Coumarin ionization energy were observed. Shown in Figure 15.8 are photoionization spectra of Coumarin 314 as a function of RH. There is a ~0.2 eV blue shift in the ionization onset in going from 20% to 65% RH. If water played a direct role in solvating the dye molecule, the ionization energy would red shift with increasing water coverage on the aerosol surface. The authors[46] conclude that the addition of water to the aerosol surface in fact induces a more complex and indirect change in the polarity of the aerosol surface rather than

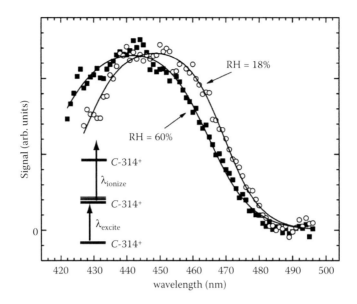

FIGURE 15.7 The two color electronic spectra of Coumarin 314 adsorbed onto a NaCl aerosol surface. There is a blue shift in the band maxima with increasing RH. (Reproduced from Woods, E.; Morris, S. F.; Wivagg, C. N.; Healy, L. E. 2005. *J. Phys. Chem. A* 109:10702–10709. Copyright 2005 American Chemical Society. With permission.)

direct solvation of the probe molecule by water molecules. They conclude that water changes the polarity of the interface by reducing the electric field of the NaCl surface.[46] This could occur if water alters the surface topology of the NaCl nanoparticle through enhanced ion mobility. As a consequence, the effective electric field experienced by the Coumarin molecule would decrease, leading in turn to a decrease in surface polarity, a blue shift in ionization energy, and an increase in the S_1 excited state transition energy.

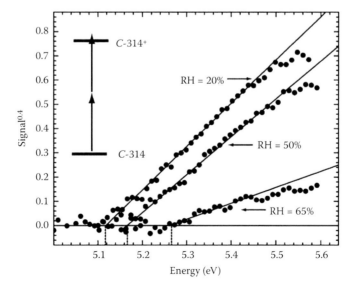

FIGURE 15.8 Two color photoionization spectra of Coumarin 314 on an NaCl aerosol surface. The ionization threshold of the probe molecule shifts to higher photon energies with increased RH. (Reproduced from Woods, E.; Morris, S. F.; Wivagg, C. N.; Healy, L. E. 2005. *J. Phys. Chem. A* 109: 10702–10709. Copyright 2005 American Chemical Society. With permission.)

Woods and coworkers[45] use their probe molecule photoemission technique to examine the hygroscopic growth of surfactant-coated aerosol particles, in particular how surfactant-coated aerosols (sodium dodecyl sulfate (SDS) on potassium iodide (KI)) inhibit the uptake of water.[45] As shown previously the electronic structure of the probe molecule, Coumarin 314, is sensitive to the local polarity of the surface, which is in turn a function of the water coverage of the aerosol surface. As shown in Figure 15.9, the relative photoelectric charging efficiency is a complex function of RH for the SDS/KI aerosol system. Clear evidence of discontinuous changes in photoelectric charging efficiency is observed as a function of RH. The authors attribute these changes to phase transitions. The largest photoelectric charging efficiency is observed at very low RH (<5%) where little water is absorbed to the surface of the salt aerosol, which contains patches of dry SDS surfactant. As the RH is increased, an inflection point in the charging efficiency is observed at RH >5%, followed by a monotonic decrease in charging efficiency until about 40% RH. This decrease in charging is attributed to the capture of the photoelectrons by greater amounts of adsorbed water on the particle surface. This intermediate region, as shown in Figure 15.9, is assigned to a particle covered by a thin aqueous solution film of SDS on the KI particle, prior to deliquescence. For RH >40%, the authors see a sharp decrease in photoelectric charging efficiency and attribute this loss to the formation of a liquid aerosol with a micelle structure.[45]

15.2.4 SUMMARY

Photoelectric charging by UV photons is one of the few methods currently available to interrogate the surfaces of submicron aerosol particles under atmospheric conditions and in real time. Although in traditional surface science measurements, surface charging of insulating samples during photoemission is an experimental problem, the change in particle charge state for an aerosol produces a

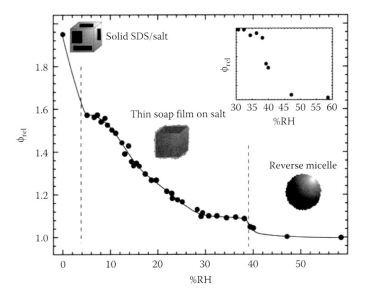

FIGURE 15.9 Relative photoelectric charging efficiency (φ_{rel}) as a function of RH for potassium iodide particles coated with an SDS surfactant. Discontinuous changes (shown as dashed lines) in φ_{rel} versus RH are observed and assigned to phase transitions whose structures are shown in the figure. (Inset) An expanded view of φ_{rel} from 30% to 60% RH. (Reproduced from Woods, E.; Kim, H. S.; Wivagg, C. N.; Dotson, S. J.; Broekhuizen, K. E.; Frohardt, E. F. 2007. *J. Phys. Chem. A* 111:11013–11020. Copyright 2007 American Chemical Society. With permission.)

clear signature of the aerosol surface, which can be used to study desorption kinetics, surface solvation, and polarization. Currently, photoelectric charging has only been implemented using low photon energies to probe molecules with low ionization thresholds (e.g., 6.5 eV for Coumarin 314). However, new approaches could be envisioned that utilize higher photon energies in order to access the ionization energies of a much larger array of molecular species that might comprise an aerosol surface. Most large hydrocarbons have gas-phase ionization energies below 11 eV; the surface ionization energy is expected to be even lower. There are currently a number of commercially available VUV lamps that operate in this energy range (e.g., Kr at 10.0 eV and Xe at 8.4 eV). For reference, the ionization energies of N_2, O_2, and H_2O are 15.5 eV, 12.0 eV, and 12.6 eV, respectively; hence, these species will not directly interfere with particle charging. Using higher photon energies might require more careful engineering of the electrostatic precipitators to filter out any gas-phase ion or electron signals that might interfere with the direct detection of the particles. Although it might be difficult to use high photon energies for ambient particle detection due to the many gas-phase species that might have ionization energies below 11 eV, such an approach might be useful for fundamental laboratory studies of absorption and desorption kinetics, and water uptake, on a wide variety of aerosol particles containing nonaromatic chromophores.

15.3 SYNCHROTRON-BASED AEROSOL PHOTOEMISSION

Recently, there has been new interest in applying synchrotron radiation to study photoemission in micron- and nanometer-sized aerosol particles.[11,12,48–55] Soft x-ray light sources, such as the Advanced Light Source, produce continuously tunable radiation from 7 to 2000 eV with photon fluxes between 10^{12} and 10^{16} photon s^{-1}. Furthermore, synchrotrons produce small (10–500 μm) photon spot sizes, which are ideally suited for interrogating beams of aerosol particles or trains of droplets as described below.

In the photoelectric charging experiments described above, the change in particle charge state is measured. However, the high pressures in the photoelectric charging cell prevent a detailed analysis of the photoelectron KE as well as the angular distributions of photoelectrons ejected from the aerosol surface. Measuring the KE and angular distributions of photoelectrons can provide important information about the electronic structure of the surface, critical in understanding chemical reactivity and polarization effects that naturally arise when a molecule is absorbed at an interface. Angular and KE analysis in aerosol photoemission requires new experimental approaches to deliver nanoparticles to the detection regions of electron energy analyzers.

This section begins with the application of VUV photoemission for investigating the polarization energy of biological nanoparticles. Next, we discuss the use of vacuum UV photoemission for the examination of the angular distribution of threshold electrons emitted from the beams of nanoparticles. Finally, we describe the application of XPS to the study of nanoparticles and micron-sized droplets, where chemical information can be obtained from solid and liquid interfaces.

15.3.1 VACUUM UV PHOTOEMISSION OF AEROSOLS USING VELOCITY MAP IMAGING

Velocity map imaging (VMI) allows the simultaneous measurement of the energy and angular distributions of charged particles originating from a photoionization event.[56,57] VMI is generally used to investigate the gas-phase photoionization dynamics of atoms, molecules, and clusters. A VMI spectrometer consists of a set of flat plate electrostatic optics and a dual multichannel plate detector coupled to a fast phosphor screen. Photoelectron images are recorded using a video camera. The overall experimental arrangement is shown in Figure 15.10. The electrostatic optics are biased in such a way as to focus all electrons with the same momentum in the plane of the detector to the same radial position on the phosphor detector. The angular distribution of the photoelectrons is preserved in the VMI spectrometer. An example of a VMI photoelectron image of gas-phase xenon is shown

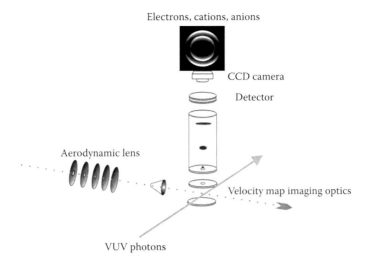

Electrons, cations, anions

CCD camera

Detector

Aerodynamic lens

Velocity map imaging optics

VUV photons

FIGURE 15.10 Experimental setup for a VMI spectrometer. Aerosol is sampled from atmospheric pressure into vacuum via a set of aerodynamic lenses that focuses the particles into a narrow beam. Synchrotron radiation intersects the particle beam at 90° in the interaction region of the VMI spectrometer. The photoelectron images are recorded using a multichannel plate detector coupled to a phosphor screen.

in Figure 15.11. Zero KE photoelectrons appear as a spot in the center of the detector, while electrons with larger kinetic energies appear as concentric rings on the detector. The recorded image is actually a two-dimensional (2D) projection of the nascent 3D velocity distribution. Established tomographic techniques are used to reconstruct the 3D photoelectron distribution.[58] The photoelectron velocity (i.e., KE) distributions are obtained by performing an angular integration of the reconstructed images as a function of the radius. Two clear rings are observed in the reconstructed Xe velocity map image, which correspond to the two 5p spin orbit states of the atom. Also shown in Figure 15.11 are photoelectron angular distributions orientated around the polarization axis of the photon beam. The angular pattern corresponds to the symmetry of the orbital from which the photoelectron originates. In general, VMI is a highly sensitive way (4π collection efficiency) to obtain low-energy photoelectron spectra (<10 eV).

The VMI spectrometer is ideally suited to studying photoionization dynamics of gas-phase molecules, which is accomplished by passing a molecular beam between the extraction optics of the spectrometer. Adapting this approach to aerosol particles requires replacing the molecular beam nozzle with an aerodynamic lens to form particle beams. A focused nanoparticle beam is formed by passing aerosols through a set of aerodynamic lenses (ADL),[59,60] yielding a beam diameter of ~1.0 mm[53] in the interaction region of the spectrometer located 25 cm from the exit aperture of the ADL. Since this particle beam has low divergence, it can be steered through a series of differentially pumped chambers into the interaction region of the VMI spectrometer. Synchrotron radiation intersects the nanoparticle beam at 90°, forming a detection plane perpendicular to the electron detector as shown schematically in Figure 15.10. During a typical experiment,[55] the average particle flux into the spectrometer is ~2.5×10^7 particles s^{-1} with an average particle beam velocity of 100 m s^{-1}. For a 1×1-mm photon–particle interaction volume, an average of 1–20 photons interact with each particle.

15.3.2 Photoemission from Biological Nanoparticles

Some of the first experiments combined aerosol beams with VMI, in order to understand the electronic structure of nanoparticles comprised of biological molecules.[54] Amino acids are the building

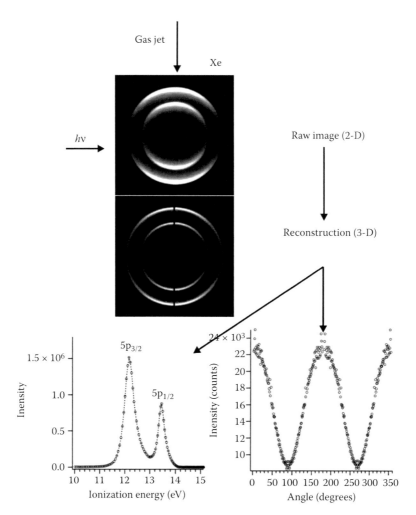

FIGURE 15.11 Velocity map images of gas-phase xenon showing how raw images are reconstructed and analyzed to obtain a photoelectron spectrum and angular distributions.

blocks of larger protein molecules whose electrical conductivity and electron transfer processes have been subject to long-standing theoretical and experimental interest.[61] Furthermore, there is also a growing need to understand the interaction of VUV and soft x-ray light with solid amino acids in the context of astrobiology.[62] This is because amino acids are thought to be formed on the surfaces of meteorites, comets, and space dust.[63,64] Very recently, a theoretical model suggested that prebiotic chemistry can occur within the cavities and interstitial voids of dust aggregates when VUV light radiation impinges on molecular material trapped within the grain.[65] Recent experiments indicate that VUV and soft x-ray irradiation of amino acid thin films initiate polymer formation;[66] that is, peptides are formed from amino acids, a possible precursor to life.

At the Chemical Dynamics Beamline of the Advanced Light Source, biological nanoparticles were introduced into a VMI spectrometer and interrogated with VUV radiation.[54] Photoelectron images are collected over the energy range of 7.5–15.5 eV for an amino acid glycine (gly) and a polypeptide phenylalanine–glycine–glycine (phe–gly–gly). Reconstructed photoelectron images of nanophase glycine taken at three photon energies are shown in Figure 15.12. The direction of the particle beam is from the left to right, whereas the synchrotron beam propagates from the top to the bottom of Figure 15.11. There is a band that energetically corresponds to the valence band of glycine, with an additional sharp ring that corresponds to photoelectrons originating from residual

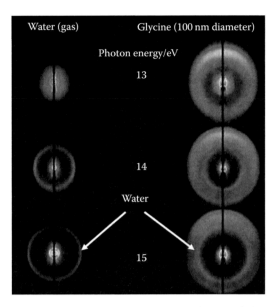

FIGURE 15.12 **(See color insert following page 206.)** Velocity map images of 100 nm glycine particles recorded at photon energies of 13, 14, and 15 eV (right). Images of the gas-phase components (water) that travel with the aerosol beam are shown at the left.

gas-phase H_2O in the particle beam. The intense spot in the middle of the image arises primarily from inelastically scattered photoelectrons (zero KE electrons). On the left side of Figure 15.12 are shown corresponding images with the nanoparticle component removed from the beam showing the gas-phase contribution to the photoelectron signal.

Electron KE distributions are obtained by integrating over the reconstructed images. The electron energy distribution is broad and featureless. While the shapes of the photoelectron spectra by themselves do not provide much chemical information, the falling edge at the maximum KE release can be used to extract the ionization onset of nanophase gly and phe–gly–gly. These values are plotted in Figure 15.13 for glycine, compared with gas-phase water and oxygen as a function of incident photon energy. It is clear in all cases that there is a linear relationship between the maximum KE release and incident photon energy. Extrapolation to zero KE yields the ionization energy, which is determined to be 7.6(±0.2), 7.5(±0.2), 12.1(±0.2), and 12.5(±0.2) eV for nanophase gly, phe–gly–gly, and gas-phase O_2 and H_2O, respectively. This would suggest that the photoelectron imaging technique, as implemented here, indeed provides correct threshold ionization energies for the solid biological nanoparticles.[54]

The polarization energy of an organic solid is the difference in ionization energies of the gaseous and solid phases, as described in the previous section. Sato and coworkers[39] studied a variety of molecules, both in the gas phase and the solid state, using UV photoelectron spectroscopy. They observed that the peak position of the solid spectrum correlates very well with the gas-phase spectrum, although there is an overall shift in the ionization onset toward lower energies. The polarization energy provides a glimpse of important physical chemical properties of the organic solid state, for example, the energetics of charge-carrier formation and intermolecular bonding.

Using thermal vaporization of similar biological nanoparticles in conjunction with tunable VUV light, Wilson et al.[53] measured the gas-phase ionization energies of gly (9.3 ± 0.1 eV) and phe–gly–gly (9.1 ± 0.1 eV). Combining these results with the nanophase ionization energies yields polarization energies of 1.7 ± 0.2 eV and 1.6 ± 0.2 eV for gly and phe–gly–gly, respectively. The polarization energy of glycine, measured in this experiment, may be used to estimate the molecular polarizability (χ) if the crystal structure of the biological nanoparticle is known. This first-order approximation

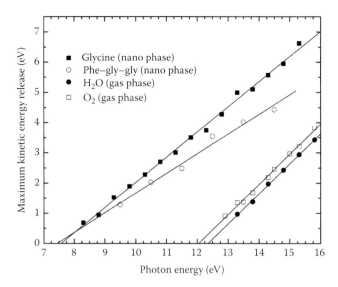

FIGURE 15.13 Photoelectron KE release versus incident photon energy for gly (■), phe–gly–gly (●), gas-phase O_2 (■), and gas-phase H_2O (●). (Reproduced from Wilson, K. R.; Peterka, D. S.; Jimenez-Cruz, M.; Leone, S. R.; Ahmed, M. 2006. *Phys. Chem. Chem. Phys.* 8:1884–1890. With permission of the PCCP Owner Societies.)

does not include the effects of higher multipoles or the interaction of induced dipoles. X-ray powder diffraction measurements[54] revealed that the structure of the glycine particles is crystalline, with a $P2_1/n$ space group. At room temperature, the unit cell comprises four molecules and has a volume of 309.9 Å^3.[67] Using the formalism of Sato,[39] χ for glycine is determined to be 4.7 ± 0.3 Å^3 (31.9 ± 1.9 a.u). This is the first experimental determination of χ for glycine; however, theoretical estimates of χ have yielded values that are below the experimentally determined ones. This could be due to either an inaccurate force field used in models, or the oversimplified empirical model used in the interpretation of the experimental values, where higher multipoles as well as induced dipole interactions are neglected.

These experiments illustrate that VMI of nanoparticles can be used to probe into the electronic structure of fragile biological molecules. A key component in nanoscale electronic or optoelectronic devices is the junction between a metal and an organic molecule, monolayer, or film. Hybrid interfaces between metals and organic molecules are key structural elements in state-of-the art dye sensitized solar cells, single molecule wires and switches, chemical sensors, field-effect transistors, electroluminescent devices, and light emitting diodes. Hence, there is immense interest in probing the electronic structure in metal–biological systems, and future experiments where metal nanoparticles are coated with biological and organic molecules could allow the VMI probing of metal–molecule contacts as individual components isolated in a particle beam.

15.3.3 ANGLE-RESOLVED PHOTOEMISSION FROM NANOPARTICLES

Wilson et al.[55] used VMI of nanoparticle beams to examine how the angular distribution of photo-electrons depends on particle size. Monodisperse particle beams are formed using a commercial differential mobility analyzer. Shown in Figure 15.14 are the raw and reconstructed photoelectron images of a polydisperse size distribution of NaCl aerosol recorded at a photon energy of 12 eV.[55] Relative to both the direction of the incoming photon and particle beams (shown in Figure 15.14) there is a clear enhancement of photoelectron intensity emanating from the side of the particle beam directly illuminated by the incoming synchrotron radiation.

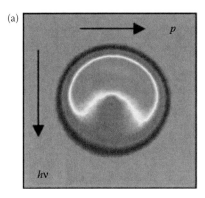

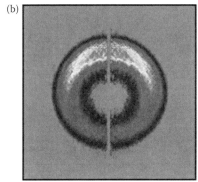

FIGURE 15.14 **(See color insert following page 206.)** (a) Raw and (b) reconstructed photoelectron images of NaCl (polydisperse size distribution) obtained at a photon energy of 12 eV. Arrows show the photon (*h*v) and particle (*p*) beam propagation directions. (Reproduced from Wilson, K. R.; Zou, S. L.; Shu, J. N.; Rühl, E.; Leone, S. R.; Schatz, G. C.; Ahmed, M. 2007. *Nano Lett.* 7:2014–2019. Copyright 2007 American Chemical Society. With permission.)

A particle photoelectron spectrum, shown in Figure 15.15, is obtained by radially integrating the reconstructed image. It is found that despite the angular asymmetry, the photoelectron KE distributions in the forward and backward hemispheres are identical within the instrumental resolution. The peak of the particle photoelectron band, which originates mainly from anion p levels (Cl 3p),[68] appears at an energy of 10.2 eV, consistent with previous spectra recorded for NaCl surfaces as shown in Figure 15.15.[68,69] Furthermore, the ionization threshold of the NaCl nanoparticles was determined to be 8.2 ± 0.1 eV, also in good agreement with previous thin film studies (8.1 eV).[70] No size-dependent change in the particle-phase photoelectron spectrum or ionization onset was observed, which indicates that for the particle sizes (50–500 nm) investigated, no modification of the bulk-phase electronic structure occurs. The close agreement of the photoelectron spectrum with previous thin film studies suggests that particle charging does not influence the absolute KE axis in these experiments. Sample charging and radiation damage often complicate the photoemission measurements of insulating surfaces at synchrotron light sources.

To investigate the asymmetry in angular photoemission, a differential mobility analyzer is used to produce a beam of monodisperse particles.[55] Raw velocity map images are recorded as a function of both particle size and photon energy as shown in Figure 15.16. The orientation of the photon beam (*h*v) and particle beams (*p*) are shown in Figure 15.16. Zero KE electrons appear as a bright spot in the center of the image. There is a clear depletion of signal in the forward hemisphere with respect to the incoming photon beam. The magnitude of this asymmetry (α) can be quantified by taking the ratio of the integrated intensities in the forward and backward hemispheres. Equal intensity in the forward and backward hemispheres ($\alpha = 1$) corresponds to isotropic photoemission, while

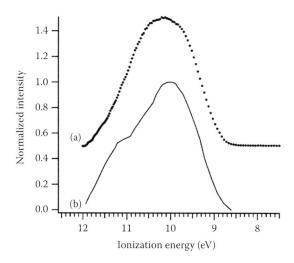

FIGURE 15.15 NaCl photoelectron spectra: (a) nanoparticles and (b) thin film[68] recorded at photon energies of 12 and 75 eV, respectively. (Reproduced from Wilson, K. R.; Zou, S. L.; Shu, J. N.; Rühl, E.; Leone, S. R.; Schatz, G. C.; Ahmed, M. 2007. *Nano Lett.* 7:2014–2019. Copyright 2007 American Chemical Society. With permission.)

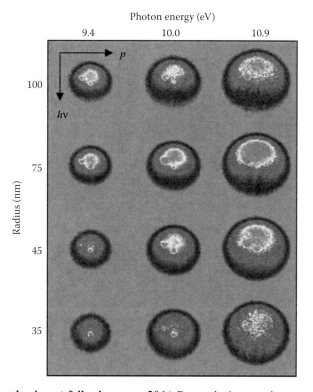

FIGURE 15.16 **(See color insert following page 206.)** Raw velocity map images of size-selected NaCl aerosol as function of photon energy. The arrows at the top of the figure indicate the photon (*h*v) and particle (*p*) beam propagation directions. (Reproduced from Wilson, K. R.; Zou, S. L.; Shu, J. N.; Rühl, E.; Leone, S. R.; Schatz, G. C.; Ahmed, M. 2007. *Nano Lett.* 7:2014–2019. Copyright 2007 American Chemical Society. With permission.)

an asymmetry ratio of 0.5 indicates that more electrons are emitted in the backward direction along the axis of the incoming photon beam. It is found that α ($h\nu$ = 10.9 eV) is particle size dependent as shown in Figure 15.17 where α is plotted versus R, the particle radius. Plotting α versus R^{-1} yields a straight line, as shown in Figure 15.17 (inset). A linear extrapolation of this plot predicts an asymmetry ratio of 1 (i.e., isotropic photoemission) at $R = 24 \pm 4$ nm.

Two characteristic lengths govern photoemission in nanoparticles; the absorption length of the incident photons, and electron IMFP of the photoelectrons. The photoabsorption length in NaCl at 10 eV is ~14 nm.[55] For low-energy electrons (<10 eV), the escape lengths are uncertain owing to complex electron–phonon interactions in the solid. Nevertheless, these escape lengths may range from a few to several hundred nanometers and are clearly expected to be on the order of the dimensions of the NaCl nanoparticles used in this study. For particle sizes that are larger than both the photoabsorption and electron escape lengths, the photoelectron images become more asymmetric, indicating that only a fraction of the total particle volume emits photoelectrons. The linear relationship of α with R^{-1} suggests that photoemission asymmetry is scaling as the surface-to-volume ratio of the aerosol.

Ahmed and coworkers[55] developed a model based on Mie theory, to further examine how the confined nanoparticle geometry influences both the photoabsorption and electron migration steps in aerosol photoemission. The internal electromagnetic field, E, inside a spherical nanoparticle is computed to map out the regions of the particle that have a high photo-excitation probability (see Figure 15.18). For particles that are 100 nm and larger, the electric field amplitude is strongly peaked on the side of the particle which is directly illuminated by the incoming photon beam, indicating that there is a strong preference for photoelectrons to be emitted in the backward direction in this case. The photoexcitation probability is more uniformly distributed in the particle as the particle size decreases, as observed for the 5 nm radius particle in Figure 15.18.

By combining the size-dependent photoexcitation probability with a standard model for photoelectron emission, the size dependence of the experimental asymmetry parameter can be directly simulated. The agreement between model and experiment is satisfactory, as shown in Figure 15.17.

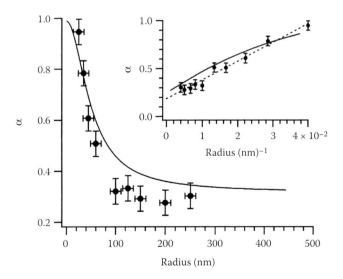

FIGURE 15.17 Experimental asymmetry (α) parameter ($\bullet$) as a function of particle radius (R) measured at a photon energy of 10.9 eV. (Inset) α versus inverse radius fit with a line (dashed). Results from model simulation are shown as solid lines. (Reproduced from Wilson, K. R.; Zou, S. L.; Shu, J. N.; Rühl, E.; Leone, S. R.; Schatz, G. C.; Ahmed, M. 2007. *Nano Lett.* 7:2014–2019. Copyright 2007 American Chemical Society. With permission.)

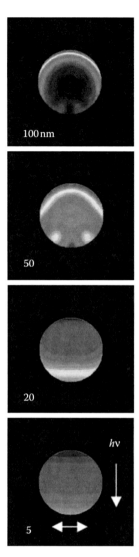

FIGURE 15.18 **(See color insert following page 206.)** Mie theory calculations of the internal electric field amplitude as a function of particle radius. (Reproduced from Wilson, K. R.; Zou, S. L.; Shu, J. N.; Rühl, E.; Leone, S. R.; Schatz, G. C.; Ahmed, M. 2007. *Nano Lett.* 7:2014–2019. Copyright 2007 American Chemical Society. With permission.)

Any differences between the model and experimental data are likely due to the approximation of NaCl particles as spheres, which is necessary in order to apply Mie theory. It is expected that more sophisticated computational treatments of the photoexcitation probability (e.g., using a nonspherical nanoparticle in the model) would yield a better agreement with the experimental results.

Combining nanoparticle beams with VMI provides a new way to probe the surfaces of condensed matter using already established techniques designed for gas-phase atoms, molecules, and clusters. The nanoparticles in these experiments are continuously replenished, which eliminates significant sample charging, contamination, and radiation damage that often complicates traditional photoelectron measurements of macroscopic insulating samples. This technique could be particularly useful for probing the electronic structure of nanomaterials in the absence of perturbations produced by substrates, with potential new applications in characterizing environmental and aerosol surfaces as well as the photophysics of interstellar grains.

15.3.4 Nanoparticle Photoemission in Interstellar Dust Clouds

The angular distributions of photoelectrons emitted from nanoparticles may have consequences for the alignment of dust grains in interstellar clouds.[71–75] It has been observed that starlight passing through an interstellar dust cloud is polarized, suggesting that particles within the cloud exhibit some overall net alignment. If interstellar dust particles are irregular, then a unidirectional radiation field that impinges on a cloud could produce a net torque on the grains and therefore particle alignment.[76] There could be a number of radiation induced "forces" to align particles that include absorption, scattering, hydrogen desorption, and photoemission.

A number of authors[72–74] have considered asymmetric photoemission by dust grains, induced by UV and visible radiation, to examine the physical origins of grain alignment in interstellar dust clouds. For example, Weingartner and Draine[76] considered VUV photoemission from carbonaceous particles and silicate grains. Using a computational model, they obtained results for the asymmetric production of photoelectrons in small particles, which are similar to the ones described above for NaCl photoemission.[76] Shown in Figure 15.19 is the asymmetry factor computed by Weingartner and Draine[76] for carbonaceous and silicate particles as a function of particle size at a photon energy of 11 eV. The computed asymmetry factor is qualitatively similar to that observed for NaCl shown in Figure 15.17. There are differences in the absolute magnitude of the asymmetry factor, which is a sensitive function of the optical properties of the modeled particles. In general, Weingartner and Draine conclude that processes connected with asymmetric particle photoemission might be important to consider in interstellar regions, where the grain electrical potential is low and the radiation spectrum is shifted toward higher energies.

15.3.5 Threshold Photoemission from Multicomponent Aerosol

The KE resolution afforded by VMI is sufficient to examine the surface composition of more complex particle types. The electronic structure and heterogeneous chemistry of aerosols of mixed composition, in particular inorganic/organic aerosol, are of particular interest in molecular electronic applications as well as in atmospheric chemistry. While detecting the bulk composition of such particles is routine using aerosol mass spectrometry,[77] examining the surface composition

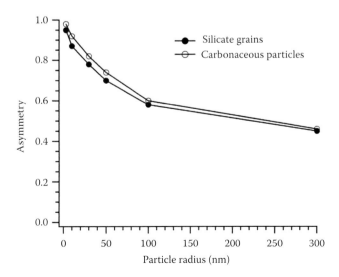

FIGURE 15.19 Computed asymmetry in photoemission as a function of particle radius for carbonaceous particles and silica grains at a photon energy of 11.0 eV. (Adapted from Weingartner, J. C.; Draine, B. T. 2001. *Astrophys. J.* 553:581–594. With permission.)

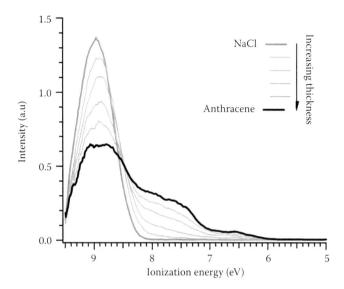

FIGURE 15.20 The evolution of photoelectron spectra as a function of anthracene coating thickness on a NaCl seed aerosols. The uncoated NaCl particle exhibits a band at ionization energy of 9.0 eV, while contributions of anthracene appear at lower ionization energies.

remains challenging. Shown in Figure 15.20 is a simple example of how VMI can be used to study an aerosol surface comprised of two distinct species. NaCl particles are formed using an atomizer and passed through an oven containing solid anthracene. By changing the oven temperature, the NaCl particles can be coated with varying amounts of anthracene. To analyze the chemical composition of the particles, they are introduced into a VMI spectrometer through an aerodynamic lens.

Shown in Figure 15.20 is a set of VMI photoelectron spectra measured as a function of anthracene oven temperature at photon energy 12 eV. The particle spectra evolve from pure NaCl to that of a mixed particle, which is comprised of both NaCl and anthracene. The two phases in the particle can be clearly distinguished, with anthracene exhibiting two bands at ionization energies of 6.5 and 7.8 eV. These bands are well separated from the main NaCl band located at 9 eV. As the anthracene coating increases, the NaCl band at 9 eV, which contains some contributions from both species, decreases, while the low ionization energy bands assigned to anthracene grow in magnitude. The thickness of the anthracene coating is estimated from the change in average particle diameter using a differential mobility analyzer. The change in photoelectron signal, for ionization energies between 6 and 8 eV, is plotted as a function of estimated anthracene thickness in Figure 15.21. The anthracene thickness is only a rough estimate since there are large errors in estimating the change in average particle size for a polydisperse particle distribution. Furthermore, the change in anthracene thickness is computed assuming that the NaCl particles are spherical and that the coverage of anthracene is uniform. Nevertheless, the photoelectron signal increases at low coverage, reaching a plateau around an estimated film thickness of ~20–30 nm. This plateau suggests that anthracene has completely covered the NaCl surface and that the thickness of the film has exceeded the photoelectron escape length. This is consistent with IMFPs of ~10 nm for 1–3 eV KE photoelectrons in pure NaCl particles.

This preliminary data illustrates how VMI can be used to study more complex aerosol surfaces containing organic coatings. For instance, many important reactions might occur only within a thin organic coating on the aerosol surface, for example, heterogeneous aerosol oxidation in the atmosphere. Furthermore, bulk probes of aerosol composition might be unable to distinguish between surface and bulk-phase-mediated chemistries or processes.

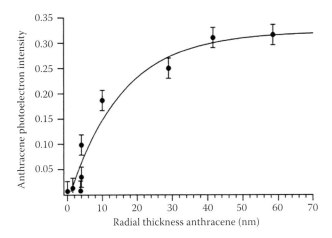

FIGURE 15.21 Photoelectron intensity, obtained by integrating the 6.5 and 7.8 eV bands of anthracene (Figure 15.20), as a function of anthracene coating thickness on NaCl nanoparticles.

15.3.6 SUMMARY: THRESHOLD AEROSOL PHOTOEMISSION AND VMI

Threshold photoemission of particle beams using VUV synchrotron radiation is a promising technique for real-time surface characterization of aerosol particles. VMI can be easily implemented within existing molecular beam spectrometers and, when combined with commercial particle sizing equipment, opens up many new opportunities for studying fundamental processes that occur at the surfaces of submicron aerosol particles. Furthermore, pump–probe studies of nanoparticle dynamics might be feasible using ultrafast laser systems. In this respect, the techniques described above are indeed universally applicable to various aerosol species and can be easily implemented in laboratory settings to examine, for example, the ultrafast electron dynamics of free quantum dots isolated in the gas phase without the perturbation of a substrate.

In the remaining sections of this chapter we describe the application of synchrotron-based XPS of free aerosols. This is a relatively recent development that provides new opportunities for the detailed analysis of aerosol surfaces, including micron-sized liquid droplets, using the elemental, chemical, and surface sensitivity of XPS.

15.3.7 XPS OF SUBMICRON PARTICLES

The heterogeneous chemistry of aerosol particles is determined by their surface properties, that is, the chemical composition and speciation at the aerosol surface. Ideally, the surface properties of aerosol particles are investigated in real time, without prior collection of the aerosols on a substrate, which may alter their properties. XPS is one of the most widely used techniques for the investigation of surfaces. Owing to the short mean free path (several Angstrom to several nanometers) of electrons with typical energies in the 10–1000 eV KE range (see Figure 15.1a), XPS is inherently surface sensitive. In order to investigate a wide range of model aerosol particles with direct relevance to processes in the atmosphere or environment, including nitrates, sulfates, soot, salts, and PAHs, Mysak et al.[78] recently performed proof-of-principle experiments that combined an aerodynamic lens with an ambient pressure XPS (APXPS). This approach could be used in the future to probe the surface of particles collected directly from the ambient environment. Through the use of a differentially pumped electrostatic lens system,[79–83] APXPS allows photoelectron spectroscopy experiments at elevated pressures in the Torr range, that is, at much higher pressures than the ultrahigh vacuum conditions in a conventional XPS experiment. Figure 15.22 shows the principle layout of the experiment. Using an ADL, a gas–particle stream (in this case, SiO_2 particles in N_2) with a beam waist of 0.2 mm is produced. The particle stream then passes in front of the entrance aperture (0.3 mm

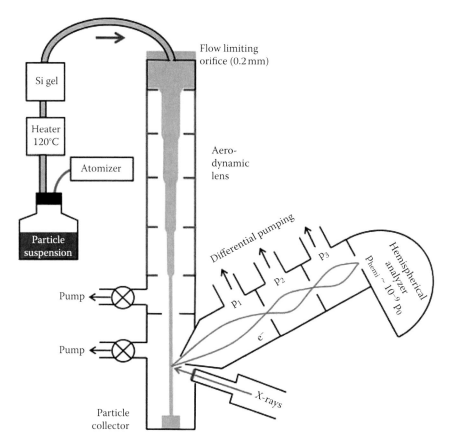

FIGURE 15.22   Schematic drawing of a combined aerodynamic lens/ambient pressure XPS setup. For details see the text.

diameter) of the differentially pumped lens system of the APXPS spectrometer at a distance of ~0.6 mm, and it is irradiated by incident x-rays from the undulator beamline 11.0.2 of the Advanced Light Source in Berkeley, CA.

The particle concentration (7×10^5 cm^{-3}) and the average size (270 nm) of the SiO$_2$ particles at the inlet of the ADL are measured by a scanning mobility particle sizer (SMPS). The flow rate of the carrier gas through the flow limiting orifice at the entrance to the ADL was 5 cm^3 s^{-1}, resulting in a particle flow rate of 3.5×10^6 s^{-1}. Typical particle velocities in an ADL for 270 nm particles are about 150 m/s.[77] From the given particle velocity and particle concentration as well as the incident x-tray beam diameter (0.2 mm) and the acceptance area of the electron analyzer (diameter ~0.3 mm), Mysak et al.[78] estimated that an average of only 3–4 particles are in the probed volume at any given time, resulting in small but detectable XPS signals (see Figure 15.23). The Si2p and O1s peaks are shifted to higher BE (compared to the literature values) by about 4 eV, which was attributed by the authors to charging of the particles in the atomizer and potentially also in the x-ray beam. From the relative Si2p/O1s signal intensity ratio, the measured photon flux, and the literature values for the photoionization cross sections,[84] the Si–O stoichiometry was determined to be 0.45 ± 0.15, that is, in agreement with the expected stoichiometry of 0.5.

The measurements above show that the combination of ADL and APXPS provides a means to quantitatively determine the chemical composition of unsupported aerosol particles. The signal levels at present are not yet sufficient to make it a routine tool for aerosol surface studies. The signal can be increased either by increasing the concentration of particles in the stream or by increasing the sensitivity of detection. The upper concentration of aerosols in the stream is, however, limited

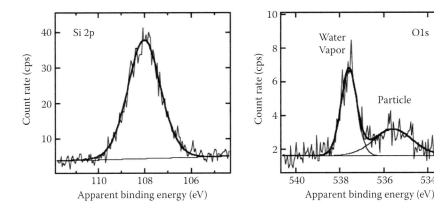

FIGURE 15.23   Si 2p and O 1s spectra measured from a particle stream of 270-nm-diameter SiO$_2$ particles. The incident photon energies are 310 eV (Si2p) and 735 eV (O1s).

by the onset of coagulation. Next-generation APXPS instruments promise an order-of-magnitude improvement in the signal levels. One open question, however, is the detection of volatile species at the surface of the aerosol particles, which potentially could evaporate while passing the differential pumping stages of the ADL.

15.3.8   XPS OF LIQUID AEROSOLS

APXPS has also been used to study the surface of liquid aerosols. The properties of liquid–vapor interfaces influence the abundance and reactivity of trace gas molecules that are important for many heterogeneous processes in the atmosphere and environment, such as uptake (e.g., HNO$_3$, HCl, N$_2$O$_5$)[85] and release (e.g., halogen radicals[86]) of atmospheric trace gases. Till date, little is known about the concentration of solution-phase species at the liquid–vapor interface, which may significantly differ from the bulk solution concentration and is an important quantity in the modeling of heterogeneous reactions at liquid–vapor interfaces.[87,88]

APXPS has been used to study liquid–vapor interfaces, beginning with Hans Siegbahn and coworkers' early designs in the 1970s.[89,90] In those experiments, the effective pressure limit is 1 Torr. The development of synchrotron-based APXPS instruments that use differentially pumped electrostatic lens systems have increased operating pressures to above 5 Torr, allowing neat water surfaces, in equilibrium with their vapor, to be investigated (the vapor pressure of water at the triple point is 4.6 Torr). A number of different approaches have been developed over the past decades to prepare liquid surfaces inside a measurement chamber. Hans Siegbahn and coworkers have used wires,[89] rotating trundles,[91] and disks[92] that are continuously moved through a liquid reservoir and are thereby coated by a thin liquid layer that is then investigated using XPS. Another approach is taken by Winter, Faubel, and coworkers. They perform XPS measurements on liquid surfaces using a jet with a small diameter (~10 μm).[93,94] The jets are expanded into a measurement chamber with a working pressure of 10^{-5} Torr. Due to the low chamber pressure in these experiments, measurements under realistic partial pressures of water, for example, are not possible.

For the investigation of liquid aerosols, Starr et al.[95] developed an instrument that combines APXPS and a vibrating orifice aerosol generator (VOAG). VOAGs represent an established technique in aerosol science to produce a droplet train with droplets of stable size and spacing, and have been used for some time in atmospheric science experiments to measure the uptake kinetics of gases by liquids.[96,97] Droplets in the train are stable over long distances (up to 40 cm at $p < 60$ Torr[97] and about a meter at $p < 20$ Torr; Worsnop, D. R., personal communication), and allows one to vary the exposure time of trace gases to the liquid surface. Millisecond droplet/gas exposure times can be achieved under typical operating conditions. A schematic of a combined VOAG/APXPS setup is

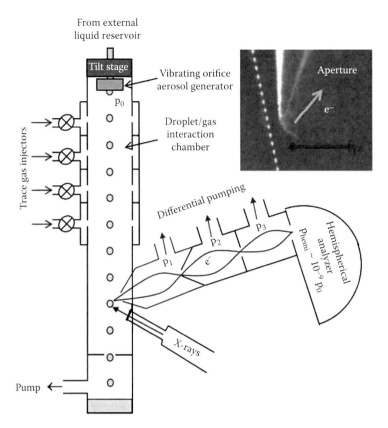

FIGURE 15.24 Schematic drawing of the combined ambient pressure photoemission/droplet train setup. Liquid from an external reservoir is pumped through an orifice (diameter 30–75 μm). The orifice is vibrated with frequencies in the 20–100 kHz range, which leads to the formation of evenly sized and spaced droplets that traverse the vacuum chamber. The droplets are collected at the bottom of the chamber in a temperature-controlled reservoir. Incident photons irradiate the droplets and gas phase in the space between the droplets in front of the entrance aperture of a differentially pumped electrostatic lens system that transfers the photoelectrons to the entrance aperture of a hemispherical analyzer. The inset shows a stroboscopic photograph of a droplet train prepared from a 40% ethanol solution in front of the entrance aperture of the differentially pumped lens system (diameter 0.3 mm). (Reprinted from Starr, D. E.; Wong, E. K.; Worsnop, D. R.; Wilson, K. R.; Bluhm, H. 2008. *Phys. Chem. Chem. Phys.* 10:3093–3098. PCCP Owner Societies. With permission.)

shown in Figure 15.24. This setup is currently under construction in our laboratory. Liquid from an external reservoir is forced through an orifice (30–75 μm diameter) mounted on a piezoelectric element. By oscillating the orifice using the piezoelectric element, the liquid jet is broken up into regular droplets with a diameter that depends on both the flow rate and the driving frequency of the piezo. The droplet train passes through a flow tube in front of the entrance aperture of the electrostatic lens system of the APXPS endstation, where incident photons irradiate the droplets in front of the aperture. The emitted electrons are collected by the differentially pumped lens system. For kinetic measurements, trace gases can be injected at different positions along the axis of the flow tube. The chemical change in the droplet surface, upon the uptake of trace gases, is then probed using APXPS. The droplets are collected at the bottom of the chamber through a differentially pumped orifice. Since the incident photon beam not only irradiates the droplet surface, but also the gas in front of the droplets and in the gaps between them, gas-phase XPS peaks are also observed in the spectra. The binding energies of gas-phase and condensed species differ from each other,[98] allowing unique surface spectral features to be identified and analyzed in the XPS spectrum. In addition, the composition of the gas phase can be determined from the gas-phase XPS peaks at partial pressures above ~0.05 Torr.

The inset in Figure 15.24 shows a stroboscopic photograph of a droplet train, prepared from a 40% ethanol solution, passing in front of the entrance aperture (diameter 0.3 mm) of the differentially pumped electrostatic lens system. The background vapor pressure is 4.3 Torr. The orifice diameter d_o of the VOAG is 50 μm. From the known flow rate F (2.2 ml/min), and the driving frequency f of the piezo element (54 kHz) the nominal droplet diameter D_d is calculated to be 110 μm according to $D_d = (6F/\pi f)^{1/3}$.[97] The droplet velocity v_d can be calculated from $v_d = 4F/\pi d_o^2$ and is found to be 1870 cm/s.[97]

The results of proof-of-principle experiments on mixed methanol/water solutions carried out in a simpler version of the setup are shown in Figure 15.24. Shown in Figure 15.25 are XPS spectra of the methanol solution–vapor interface recorded from droplets with a nominal bulk methanol mole fraction of $\chi = 0.21$. Both molecular dynamics (MD) simulations[99,100] and optical sum frequency generation measurements[101–103] indicate that methanol should be segregated to the vapor–solution interface. Using the dependence of the probing depth in XPS on the KE of the photoelectrons (see Figure 15.1a), depth profiles of the stoichiometry at the solution–vapor interface can be measured.

Figure 15.25 shows C1s and O1s photoemission spectra of the droplets at a nominal droplet temperature of −25°C, with a measured background vapor pressure in the chamber of 2.5 Torr.[95] Both gas-phase and liquid-phase photoemission peaks are visible. The C1s peak in Figure 15.25 of methanol in the solution is shifted by ~1.0 eV to lower BE from the gas-phase methanol peak (the C1s gas-phase peak is split due to vibrational excitations of the C–H stretching mode[104]). In the O1s spectrum (Figure 15.25), the methanol and water gas-phase peaks are separated by 0.85 eV, in good agreement with the literature values.[98] The liquid-phase methanol and water O1s peaks overlap too closely to be separately fit. Using experimentally determined C/O sensitivity factors, a depth-dependent methanol/water stoichiometry was calculated, which is plotted in Figure 15.26. The dotted line in Figure 15.26 corresponds to the molecular ratio of methanol/water (0.26) from the bulk concentration of a $\chi_{liquid} = 0.21$ methanol/water solution. At high kinetic energies (i.e., greater probing depth) the expected bulk value of methanol/water = 0.26 is approached, while the most surface-sensitive measurements show a clear enhancement in the methanol-to-water ratio of about a factor of 3 over the bulk value, which is in agreement with the results of MD simulations.[99]

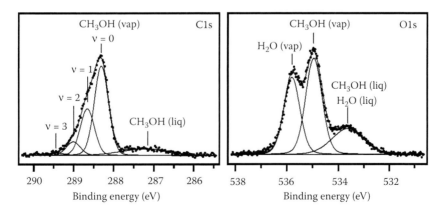

FIGURE 15.25 C1s and O1s XPS spectra of methanol solution droplets (methanol mole fraction 0.21). The C1s and O1s spectra were taken using incident photon energies of 730 eV and 938 eV, respectively, that is, at a photoelectron KE of ~450 eV. Both gas-phase and liquid-phase photoemission peaks are visible in the C1s and O1s spectra. (Reprinted from Starr, D. E.; Wong, E. K.; Worsnop, D. R.; Wilson, K. R.; Bluhm, H. 2008. *Phys. Chem. Chem. Phys.* 10:3093–3098. PCCP Owner Societies. With permission.)

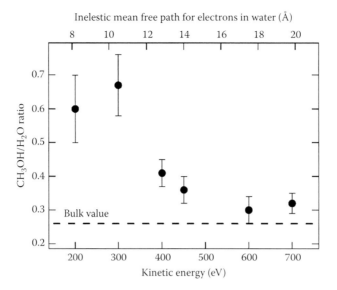

FIGURE 15.26 Methanol/water ratio as a function of photoelectron KE. At the highest kinetic energies the ratio approaches the expected bulk methanol/water ratio for a $\chi_{liquid} = 0.21$ solution. At low kinetic energies the ratio is enhanced in favor of methanol, indicating the segregation of methanol to the solution–vapor interface. (Reprinted from Starr, D. E.; Wong, E. K.; Worsnop, D. R.; Wilson, K. R.; Bluhm, H. 2008. *Phys. Chem. Chem. Phys.* 10:3093–3098. PCCP Owner Societies. With permission.)

The measurements in ref. 95 show that the combination of a droplet train with ambient pressure XPS allows one to determine the surface composition of free liquid aerosols. Future experiments will focus on the uptake kinetics and reaction of trace gases with liquid aerosols.

15.4 SUMMARY AND OUTLOOK

Nanoparticle surfaces play key roles in many fields ranging from atmospheric chemistry and catalysis to combustion. Heterogeneous processes on nanoparticle surfaces depend sensitively on their electronic structure and topology, which can be rather complex for real aerosol particles produced, for example, by combustion sources. Developing new methods for probing the surface chemistry and physics of nanometer-sized objects suspended in a carrier gas remains an experimental challenge. Aerosol photoemission, as illustrated here, is one of the few currently available techniques for achieving such surface sensitivity. UV photoelectric charging, at atmospheric pressure, uses the simple change in particle charge state to measure monolayer desorption kinetics, surface polarity, CD, and polarization energy in free nanoparticles. Many of the techniques highlighted in this chapter are in essence "aerosol analogs" of classical surface science and molecular beam techniques. For instance, rather than depositing a monolayer film on a flat geometric surface, as is done in ultrahigh vacuum surface science, submonolayer adsorbate coverage on nanometer-sized aerosols can be easily achieved and measured using simple devices such as an atomizer, oven, SMPS, UV lamp, and an electrometer. An aerodynamic lens creates beams of particles, which can easily be substituted for conventional molecular beam sources used for measuring photoionization dynamics of atoms, molecules, and clusters.

The use of synchrotron radiation has opened up new opportunities to study VUV and x-ray photoemission from nanometer aerosols and micron-sized droplets. The photon energies and intensities available from these light sources would allow new studies of surface chemistry, which are not limited to molecules with ionization energies below ~6–8 eV. In particular, XPS studies using variable photon energies yield chemically resolved depth profiles of the surface and subsurface of a droplet or particle. Such detailed information is essential in revealing how a liquid or solid interface controls heterogeneous reactivity.

Finally, we anxiously await the future use of ultrafast extreme UV and x-ray lasers to probe the electronic structure and dynamics in free nanoparticles using aerosol photoemission. In fact, free electron lasers and fourth-generation synchrotron light sources may require developing new ways of probing condensed-phase material, perhaps as nanoparticle beams, to eliminate the sample damage that is caused when a high-intensity light pulse collides with a stationary macroscopic sample. By combining aerosol photoemission with novel light sources, we anticipate gaining a richer insight into the complex chemistry and physics of aerosol interfaces; ranging from fundamental studies of size-dependent electronic properties to technologically important measurements of surface reactivity, essential for understanding the chemistry of the atmosphere as well as for the development of new catalytic nanomaterials.

ACKNOWLEDGMENTS

The authors acknowledge the support of the Director, Office of Energy Research, Office of Basic Energy Sciences, Chemical Sciences Division of the U.S. Department of Energy under contract No. DE-AC02-05CH11231.

REFERENCES

1. Einstein, A.; 1905. Generation and transformation of light. *Ann. Phys.* 17:132–148.
2. Seah, M. P.; Dench, W. A. 1979. Quantitative electron spectroscopy of surfaces: A standard data base for electron inelastic mean free paths in solids. *Surf. Interface Anal.* 1:2–11.
3. Blanco, F.; Garcia, G. 2003. Improvements on the quasifree absorption model for electron scattering. *Phys. Rev. A* 67:022701.
4. Garcia, G.; Roteta, M.; Manero, F. 1997. Electron scattering by N_2 and CO at intermediate energies: 1–10 keV. *Chem. Phys. Lett.* 264:589–595.
5. Hoffman, K. R.; Dababneh, M. S.; Hsieh, Y. F.; Kauppila, W. E.; Pol, V.; Smart, J. H.; Stein, T. S. 1982. Total-cross-section measurements for positrons and electrons colliding with H_2, N_2, and CO_2. *Phys. Rev. A* 25:1393–1403.
6. Munoz, A.; Oller, J. C.; Blanco, F.; Gorfinkiel, J. D.; Limao-Vieira, P.; Garcia, G. 2007. Electron-scattering cross sections and stopping powers in H_2O. *Phys. Rev. A* 76:052707
7. Bjorneholm, O.; Ohrwall, G.; Tchaplyguine, M. 2009. Free clusters studied by core-level spectroscopies. *Nucl. Instrum. Methods Phys. Res. Sect. A* 601:161–181.
8. Hergenhahn, U.; Barth, S.; Ulrich, V.; Mucke, M.; Joshi, S.; Lischke, T.; Lindblad, A.; Rander, T.; Ohrwall, G.; Bjorneholm, O. 2009. 3p valence photoelectron spectrum of Ar clusters. *Phys. Rev. B* 79:155448.
9. Abu-Samha, M.; Borve, K. J.; Winkler, M.; Harnes, J.; Saethre, L. J.; Lindblad, A.; Bergersen, H.; Ohrwall, G.; Bjorneholm, O.; Svensson, S. 2009. The local structure of small water clusters: imprints on the core-level photoelectron spectrum. *J. Phys. B At. Mol. Opt. Phys.* 42:1–5.
10. Lindblad, A.; Bergersen, H.; Pokapanich, W.; Tchaplyguine, M.; Ohrwall, G.; Bjorneholm, O. 2009. Charge delocalization dynamics of ammonia in different hydrogen bonding environments: Free clusters and in liquid water solution. *Phys. Chem. Chem. Phys.* 11:1758–64.
11. Grimm, M.; Langer, B.; Schlemmer, S.; Lischke, T.; Becker, U.; Widdra, W.; Gerlich, D.; Flesch, R.; Rühl, E. 2006. Charging mechanisms of trapped element-selectively excited nanoparticles exposed to soft X-rays. *Phys. Rev. Lett.* 96:1–4.
12. Graf, C.; Langer, B.; Grimm, M.; Lewinski, R.; Grom, M.; Rühl, E. 2008. Investigation of trapped metallo-dielectric core–shell colloidal particles using soft X-rays. *J. Electron Spectrosc. Relat. Phenom.* 166:74–80.
13. Winter, B.; Faubel, M. 2006. Photoemission from liquid aqueous solutions. *Chem. Rev.* 106:1176–1211.
14. Winter, B.; Aziz, E. F.; Hergenhahn, U.; Faubel, M.; Hertel, I. V. 2007. Hydrogen bonds in liquid water studied by photoelectron spectroscopy. *J. Chem. Phys.* 126:124504.
15. Winter, B.; Faubel, M.; Vacha, R.; Jungwirth, P. 2009. Behavior of hydroxide at the water/vapor interface. *Chem. Phys. Lett.* 474:241–247.

16. Brown, M. A.; Winter, B.; Faubel, M.; Hemminger, J. C. 2009. Spatial distribution of nitrate and nitrite anions at the liquid/vapor interface of aqueous solutions. *J. Am. Chem. Soc.* 131:8354–8355.

17. Winter, B.; Weber, R.; Schmidt, P. M.; Hertel, I. V.; Faubel, M.; Vrbka, L.; Jungwirth, P. 2004. Molecular structure of surface-active salt solutions: Photoelectron spectroscopy and molecular dynamics simulations of aqueous tetrabutylammonium iodide. *J. Phys. Chem. B* 108:14558–14564.

18. Winter, B.; Weber, R.; Hertel, I. V.; Faubel, M.; Vrbka, L.; Jungwirth, P. 2005. Effect of bromide on the interfacial structure of aqueous tetrabutylammonium iodide: Photoelectron spectroscopy and molecular dynamics simulations. *Chem. Phys. Lett.* 410:222–227.

19. Slavicek, P.; Winter, B.; Faubel, M.; Bradforth, S. E.; Jungwirth, P. 2009. Ionization energies of aqueous nucleic acids: Photoelectron spectroscopy of pyrimidine nucleosides and *ab initio* calculations. *J. Am. Chem. Soc.* 131:6460–6467.

20. Uejio, J. S.; Schwartz, C. P.; Duffin, A. M.; Drisdell, W. S.; Cohen, R. C.; Saykally, R. J. 2008. Characterization of selective binding of alkali cations with carboxylate by x-ray absorption spectroscopy of liquid microjets. *Proc. Natl. Acad. Sci. USA* 105:6809–6812.

21. Smith, J. D.; Saykally, R. J.; Geissler, P. L. 2007. The effects of dissolved halide anions on hydrogen bonding in liquid water. *J. Am. Chem. Soc.* 129:13847–13856.

22. Smith, J. D.; Cappa, C. D.; Messer, B. M.; Drisdell, W. S.; Cohen, R. C.; Saykally, R. J. 2006. Probing the local structure of liquid water by x-ray absorption spectroscopy. *J. Phys. Chem. B* 110:20038–20045.

23. Cappa, C. D.; Smith, J. D.; Messer, B. M.; Cohen, R. C.; Saykally, R. J. 2007. Nature of the aqueous hydroxide ion probed by X-ray absorption spectroscopy. *J. Phys. Chem. A* 111:4776–4785.

24. Schmidt-Ott, A.; Schurtenberger, P.; Siegmann, H. C. 1980. Enormous yield of photoelectrons from small particles. *Phys. Rev. Lett.* 45:1284–1287.

25. Chen, Q. Y.; Bates, C. W. 1986. Geometrical factors in enhanced photoyield from small metal particles. *Phys. Rev. Lett.* 57:2737–2740.

26. Muller, U.; Burtscher, H.; Schmidt-Ott, A. 1988. Photoemission from small metal spheres—A model calculation using an enhanced 3-step model. *Phys. Rev. B.* 38:7814–7816.

27. Mitchell, J. B. A.; Rebrion-Rowe, C.; Legarrec, J. L.; Taupier, G.; Huby, N. 2002. X-ray synchrotron radiation probing of an ethylene diffusion flame. *Combust. Flame* 131:308–315.

28. Ziemann, P. J.; Kittelson, D. B.; McMurry, P. H. 1996. Effects of particle shape and chemical composition on the electron impact charging properties of submicron inorganic particles. *J. Aerosol Sci.* 27:587–606.

29. Ziemann, P. J.; McMurry, P. H. 1998. Secondary electron yield measurements as a means for probing organic films on aerosol particles. *Aerosol Sci. Tech.* 28:77–90.

30. Burtscher, H.; Scherrer, L.; Siegmann, H. C.; Schmidt-Ott, A.; Federer, B. 1982. Probing aerosols by photoelectric charging. *J. Appl. Phys.* 53:3787–3791.

31. Burtscher, H.; Reis, A.; Schmidt-Ott, A. 1986. Particle charge in combustion aerosols. *J. Aerosol Sci.* 17:47–51.

32. Burtscher, H.; Schmidt-Ott, A. 1984. Surface enrichment of soot particles in photoelectrically active trace species. *Sci. Total Environ.* 36:233–238.

33. Burtscher, H.; Schmidt-Ott, A.; Siegmann, H. C. 1988. Monitoring particulate-emissions from combustions by photoemission. *Aerosol Sci. Tech.* 8:125–132.

34. Bluhm, H.; Siegmann, H. C. 2009. Surface science with aerosols. *Surf. Sci.* 603:1969–1978.

35. Niessner, R.; Hemmerich, B.; Wilbring, P. 1990. Aerosol photoemission for quantification of polycyclic aromatic-hydrocarbons in simple mixtures adsorbed on carbonaceous and sodium chloride aerosols. *Anal. Chem.* 62:2071–2074.

36. Qian, Z. Q.; Siegmann, K.; Keller, A.; Matter, U.; Scherrer, L.; Siegmann, H. C. 2000. Nanoparticle air pollution in major cities and its origin. *Atmos. Environ.* 34:443–451.

37. Sakai, R.; Siegmann, H. C.; Sato, H.; Voorhees, S. 2002. Particulate matter and particle-attached polycyclic aromatic hydrocarbons in the indoor and outdoor air of Tokyo measured with personal monitors. *Environ. Res.* 89:66–71.

38. Niessner, R.; Robers, W.; Wilbring, P. 1989. Laboratory experiments on the determination of polycyclic aromatic hydrocarbon coverage of submicrometer particles by laser induced aerosol photoemission. *Anal. Chem.* 61:320–325.

39. Sato, N.; Seki, K.; Inokuchi, H. 1981. Polarization energies of organic solids determined by ultraviolet photoelectron spectroscopy. *J. Chem. Soc. Faraday Trans. II* 77:1621–1633.

40. Hueglin, C.; Paul, J.; Scherrer, L.; Siegmann, K. 1997. Direct observation of desorption kinetics with perylene at ultrafine aerosol particle surfaces. *J. Phys. Chem. B* 101:9335–9341.

41. Steiner, D.; Burtscher, H. K. 1994. Desorption of perylene from combustion, NaCl and carbon particles. *Environ. Sci. Technol.* 28:1254–1259.

42. Paul, J.; Dorzbach, A.; Siegmann, K. 1997. Circular dichroism in the photoionization of nanoparticles from chiral compounds. *Phys. Rev. Lett.* 79:2947–2950.

43. Kasper, M.; Keller, A.; Paul, J.; Siegmann, K.; Siegmann, H. C. 1999. Photoelectron spectroscopy without vacuum: Nanoparticles in gas suspension. *J. Electron. Spectrosc. Relat. Phenom.* 99:83–93.

44. Cappa, C. D.; Lovejoy, E. R.; Ravishankara, A. R. 2008. Evidence for liquid-like and nonideal behavior of a mixture of organic aerosol components. *Proc. Natl. Acad. Sci. USA* 105:18687–18691.

45. Woods, E.; Kim, H. S.; Wivagg, C. N.; Dotson, S. J.; Broekhuizen, K. E.; Frohardt, E. F. 2007. Phase transitions and surface morphology of surfactant-coated aerosol particles. *J. Phys. Chem. A* 111:11013–11020.

46. Woods, E.; Morris, S. F.; Wivagg, C. N.; Healy, L. E. 2005. Probe molecule spectroscopy of NaCl aerosol particle surfaces. *J. Phys. Chem. A* 109:10702–10709.

47. Woods, E.; Wivagg, C. N.; Chung, D. 2007. Linear solvation energy parameters for model tropospheric aerosol surfaces. *J. Phys. Chem. A* 111:3336–3341.

48. Mysak, E. R.; Wilson, K. R.; Jimenez-Cruz, M.; Ahmed, M.; Baer, T. 2005. Synchrotron radiation based aerosol time-of-flight mass spectrometry for organic constituents. *Anal. Chem.* 77:5953–5960.

49. Northway, M. J.; Jayne, J. T.; Toohey, D. W.; Canagaratna, M. R.; Trimborn, A.; Akiyama, K. I.; Shimono, A.; et al. 2007. Demonstration of a VUV lamp photoionization source for improved organic speciation in an aerosol mass spectrometer. *Aerosol Sci. Tech.* 41:828–839.

50. Shu, J. N.; Wilson, K. R.; Ahmed, M.; Leone, S. R. 2006. Coupling a versatile aerosol apparatus to a synchrotron: Vacuum ultraviolet light scattering, photoelectron imaging, and fragment free mass spectrometry. *Rev. Sci. Instrum.* 77:043106.

51. Shu, J. N.; Wilson, K. R.; Ahmed, M.; Leone, S. R.; Graf, C.; Rühl, E. 2006. Elastic light scattering from nanoparticles by monochromatic vacuum-ultraviolet radiation. *J. Chem. Phys.* 124:034707.

52. Shu, J. N.; Wilson, K. R.; Arrowsmith, A. N.; Ahmed, M.; Leone, S. R. 2005. Light scattering of ultrafine silica particles by VUV synchrotron radiation. *Nano Lett.* 5:1009–1015.

53. Wilson, K. R.; Jimenez-Cruz, M.; Nicolas, C.; Belau, L.; Leone, S. R.; Ahmed, M. 2006. Thermal vaporization of biological nanoparticles: fragment-free vacuum ultraviolet photoionization mass spectra of tryptophan, phenylalanine–glycine–glycine, and, beta-carotene. *J. Phys. Chem. A* 110:2106–2113.

54. Wilson, K. R.; Peterka, D. S.; Jimenez-Cruz, M.; Leone, S. R.; Ahmed, M. 2006. VUV photoelectron imaging of biological nanoparticles: Ionization energy determination of nanophase glycine and phenylalanine–glycine–glycine. *Phys. Chem. Chem. Phys.* 8:1884–1890.

55. Wilson, K. R.; Zou, S. L.; Shu, J. N.; Rühl, E.; Leone, S. R.; Schatz, G. C.; Ahmed, M. 2007. Size-dependent angular distributions of low-energy photoelectrons emitted from NaCl nanoparticles. *Nano Lett.* 7:2014–2019.

56. Eppink, A.; Parker, D. H. 1997. Velocity map imaging of ions and electrons using electrostatic lenses: Application in photoelectron and photofragment ion imaging of molecular oxygen. *Rev. Sci. Instrum.* 68:3477–3484.

57. Parker, D. H.; Eppink, A. 1997. Photoelectron and photofragment velocity map imaging of state-selected molecular oxygen dissociation/ionization dynamics. *J. Chem. Phys.* 107:2357–2362.

58. Heck, A. J. R.; Chandler, D. W. 1995. Imaging techniques for the study of chemical-reaction dynamics. *Annu. Rev. Phys. Chem.* 46:335–372.

59. Liu, P.; Ziemann, P. J.; Kittelson, D. B.; Mcmurry, P. H. 1995. Generating particle beams of controlled dimensions and divergence. 1. Theory of particle motion in aerodynamic lenses and nozzle expansions. *Aerosol Sci. Tech.* 22:293–313.

60. Liu, P.; Ziemann, P. J.; Kittelson, D. B.; Mcmurry, P. H. 1995. Generating particle beams of controlled dimensions and divergence. 2. Experimental evaluation of particle motion in aerodynamic lenses and nozzle expansions. *Aerosol Sci. Tech.* 22:314–324.

61. Gray, H. B.; Winkler, J. R. 2003. Electron tunneling through proteins. *Q. Rev. Biophys.* 36:341–372.

62. Meierhenrich, U. J.; Caro, G. M. M.; Bredehoft, J. H.; Jessberger, E. K.; Thiemann, W. H. P. 2004. Identification of diamino acids in the Murchison meteorite. *Proc. Natl. Acad. Sci. USA* 101:9182–9186.

63. Bernstein, M. P.; Dworkin, J. P.; Sandford, S. A.; Cooper, G. W.; Allamandola, L. J. 2002. Racemic amino acids from the ultraviolet photolysis of interstellar ice analogues. *Nature* 416:401–403.

64. Caro, G. M. M.; Meierhenrich, U. J.; Schutte, W. A.; Barbier, B.; Segovia, A. A.; Rosenbauer, H.; Thiemann, W. H. P.; Brack, A.; Greenberg, J. M. 2002. Amino acids from ultraviolet irradiation of interstellar ice analogues. *Nature* 416:403–406.

65. Cecchi-Pestellini, C.; Saija, R.; Iati, M. A.; Giusto, A.; Borghese, F.; Denti, P.; Aiello, S. 2005. Ultraviolet radiation inside interstellar grain aggregates. I. The density of radiation. *Astrophys. J.* 624:223–231.

66. Kaneko, F.; Tanaka, M.; Narita, S.; Kitada, T.; Matsui, T.; Nakagawa, K.; Agui, A.; Fujii, K.; Yokoya, A. 2005. Chemical evolution of amino acid induced by soft X-ray with synchrotron radiation. *J. Electron. Spectrosc. Relat. Phenom.* 144–147:291–294.

67. Destro, R.; Roversi, P.; Barzaghi, M.; Marsh, R. E. 2000. Experimental charge density of alpha-glycine at 23 K. *J. Phys. Chem. A* 104:1047–1054.

68. Wertheim, G. K.; Rowe, J. E.; Buchanan, D. N. E.; Citrin, P. H. 1995. Valence-band structure of alkali-halides determined from photoemission data. *Phys. Rev. B.* 51:13675–13680.

69. Poole, R. T.; Jenkin, J. G.; Liesegang, J.; Leckey, R. C. G. 1975. Electronic band-structure of alkali-halides .1. Experimental parameters. *Phys. Rev. B.* 11:5179–5189.

70. Taylor, J. W.; Hartman, P. L. 1959. Photoelectric effects in certain of the alkali halides in the vacuum ultraviolet. *Phys. Rev.* 113:1421–1435.

71. Draine, B. T. 2003. Interstellar dust grains. *Ann. Rev. Astron. Astrophys.* 41:241–289.

72. Watson, W. D. 1972. Heating of interstellar H I clouds by ultraviolet photoelectron emission from grains. *Astrophys. J.* 176:103–110.

73. Watson, W. D. 1973. Photoelectron emission from small spherical-particles. *J. Opt. Soc. Am.* 63:164–165.

74. Weingartner, J. C.; Draine, B. T. 2001. Photoelectric emission from interstellar dust: Grain charging and gas heating. *Astrophys. J. Supp. Ser.* 134:263–281.

75. Weingartner, J. C.; Draine, B. T.; Barr, D. K. 2006. Photoelectric emission from dust grains exposed to extreme ultraviolet and X-ray radiation. *Astrophys. J.* 645:1188–1197.

76. Weingartner, J. C.; Draine, B. T. 2001. Forces on dust grains exposed to anisotropic interstellar radiation fields. *Astrophys. J.* 553:581–594.

77. Jayne, J. T.; Leard, D. C.; Zhang, X. F.; Davidovits, P.; Smith, K. A.; Kolb, C. E.; Worsnop, D. R. 2000. Development of an aerosol mass spectrometer for size and composition analysis of submicron particles. *Aerosol Sci. Tech.* 33:49–70.

78. Mysak, E. R.; Starr, D. S.; Wilson, K. R.; Bluhm, H. 2010. *Rev. Sci. Instrum.* 81(1):016106.

79. Ogletree, D. F.; Bluhm, H.; Lebedev, G.; Fadley, C. S.; Hussain, Z.; Salmeron, M. 2002. A differentially pumped electrostatic lens system for photoemission studies in the millibar range. *Rev. Sci. Instrum.* 73:3872–3877.

80. Ogletree, D. F.; Bluhm, H.; Hebenstreit, E. D.; Salmeron, M. 2009. Photoelectron spectroscopy under ambient pressure and temperature conditions. *Nucl. Instrum. Methods Phys. Res., Sect. A* 601:151–160.

81. Bluhm, H.; Havecker, M.; Knop-Gericke, A.; Kiskinova, M.; Schlogl, R.; Salmeron, M. 2007. *In situ* x-ray photoelectron spectroscopy studies of gas–solid interfaces at near-ambient conditions. *MRS Bull.* 32:1022–1030.

82. Salmeron, M.; Schlogl, R. 2008. Ambient pressure photoelectron spectroscopy: A new tool for surface science and nanotechnology. *Surf. Sci. Rep.* 63:169–199.

83. Bluhm, H. 2010. Photoelectron spectroscopy under humid conditions. *J. Electron Spectrosc. Relat. Phenom.* 177:71–84.

84. Yeh, J. J.; Lindau, I. 1985. Atomic subshell photoionization cross-sections and asymmetry parameters—1 less-than-or-equal-to Z less-than-or-equal-to 103. *Atom. Data. Nucl. Data.* 32:1–155.

85. Vandoren, J. M.; Watson, L. R.; Davidovits, P.; Worsnop, D. R.; Zahniser, M. S.; Kolb, C. E. 1990. Temperature-dependence of the uptake coefficients of HNO_3, HCl, and N_2O_5 by water droplets. *J. Phys. Chem. A* 94:3265–3269.

86. Knipping, E. M.; Lakin, M. J.; Foster, K. L.; Jungwirth, P.; Tobias, D. J.; Gerber, R. B.; Dabdub, D.; Finlayson-Pitts, B. J. 2000. Experiments and simulations of ion-enhanced interfacial chemistry on aqueous NaCl aerosols. *Science* 288:301–306.

87. Ghosal, S.; Hemminger, J. C.; Bluhm, H.; Mun, B. S.; Hebenstreit, E. L. D.; Ketteler, G.; Ogletree, D. F.; Requejo, F. G.; Salmeron, M. 2005. Electron spectroscopy of aqueous solution interfaces reveals surface enhancement of halides. *Science* 307:563–566.

88. Jungwirth, P.; Tobias, D. J. 2002. Ions at the air/water interface. *J. Phys. Chem. B* 106:6361–6373.

89. Fellner-Feldegg, H.; Siegbahn, H.; Asplund, L.; Kelfve, P.; Siegbahn, K. 1975. ESCA applied to liquids .4. wire system for ESCA measurements on liquids. *J. Electron Spectrosc. Relat. Phenom.* 7:421–428.

90. Siegbahn, H. 1985. Electron-spectroscopy for chemical-analysis of liquids and solutions. *J. Phys. Chem. A* 89:897–909.

91. Siegbahn, H.; Svensson, S.; Lundholm, M. 1981. A new method for ESCA studies of liquid-phase samples. *J. Electron Spectrosc. Relat. Phenom.* 24:205–213.

92. Moberg, R.; Bokman, F.; Bohman, O.; Siegbahn, H. O. G. 1991. Esca studies of phase-transfer catalysts in solution—Ion-pairing and surface-activity. *J. Am. Chem. Soc.* 113:3663–3667.

93. Winter, B.; Weber, R.; Hertel, I. V.; Faubel, M.; Jungwirth, P.; Brown, E. C.; Bradforth, S. E. 2005. Electron binding energies of aqueous alkali and halide ions: EUV photoelectron spectroscopy of liquid solutions and combined *ab initio* and molecular dynamics calculations. *J. Am. Chem. Soc.* 127:7203–7214.

94. Winter, B.; Weber, R.; Widdra, W.; Dittmar, M.; Faubel, M.; Hertel, I. V. 2004. Full valence band photoemission from liquid water using EUV synchrotron radiation. *J. Phys. Chem. A* 108:2625–2632.

95. Starr, D. E.; Wong, E. K.; Worsnop, D. R.; Wilson, K. R.; Bluhm, H. 2008. A combined droplet train and ambient pressure photoemission spectrometer for the investigation of liquid/vapor interfaces. *Phys. Chem. Chem. Phys.* 10:3093–3098.

96. Gardner, J. A.; Watson, L. R.; Adewuyi, Y. G.; Davidovits, P.; Zahniser, M. S.; Worsnop, D. R.; Kolb, C. E. 1987. Measurement of the mass accommodation coefficient of SO_2 (G) on water droplets. *J. Geophys. Res. Atmos.* 92:10887–10895.

97. Worsnop, D. R.; Zahniser, M. S.; Kolb, C. E.; Gardner, J. A.; Watson, L. R.; Vandoren, J. M.; Jayne, J. T.; Davidovits, P. 1989. Temperature-dependence of mass accommodation of SO_2 and H_2O_2 on aqueous surfaces. *J. Phys. Chem.* 93:1159–1172.

98. Siegbahn, K.; Nordling, C.; Johansson, G.; Hedman, J.; Hedén, P. F.; Jamrin, K.; Gelius, U.; et al. 1969. *ESCA Applied to Free Molecules*. North-Holland, Amsterdam.

99. Chang, T. M.; Dang, L. X. 2005. Liquid–vapor interface of methanol–water mixtures: A molecular dynamics study. *J. Phys. Chem. B* 109:5759–5765.

100. Partay, L.; Jedlovszky, P.; Vincze, A.; Horvai, G. 2005. Structure of the liquid–vapor interface of water–methanol mixtures as seen from Monte Carlo simulations. *J. Phys. Chem. B* 109:20493–20503.

101. Superfine, R.; Huang, J. Y.; Shen, Y. R. 1991. Nonlinear optical studies of the pure liquid vapor interface—Vibrational-spectra and polar ordering. *Phys. Rev. Lett.* 66:1066–1069.

102. Stanners, C. D.; Du, Q.; Chin, R. P.; Cremer, P.; Somorjai, G. A.; Shen, Y. R. 1995. Polar ordering at the liquid-vapor interface of N-alcohols (C_1–C_8). *Chem. Phys. Lett.* 232:407–413.

103. Wolfrum, K.; Graener, H.; Laubereau, A. 1993. Sum-frequency vibrational spectroscopy at the liquid air interface of methanol—Water solutions. *Chem. Phys. Lett.* 213:41–46.

104. Wiklund, M.; Jaworowski, A.; Strisland, F.; Beutler, A.; Sandell, A.; Nyholm, R.; Sorensen, S. L.; Andersen, J. N. 1998. Vibrational fine structure in the C is photoemission spectrum of the methoxy species chemisorbed on Cu(100). *Surf. Sci.* 418:210–218.

16 Elastic Scattering of Soft X-rays from Free Size-Selected Nanoparticles

Harald Bresch, Bernhard Wassermann, Burkhard Langer, Christina Graf, and Eckart Rühl

CONTENTS

16.1 Introduction 401
16.2 Experimental 402
16.3 Results and Discussion 403
    16.3.1 Energy Dependences of Scattered Light Intensity in the Regime of Inner-Shell
    Excitation: Surface Properties and Size Distributions of Nanoparticles..................404
    16.3.2 Angle-Resolved Elastic Light Scattering in the Regime of Inner-Shell
    Excitation and Surface Roughness 409
16.4 Conclusion 414
Acknowledgments..415
References..415

16.1 INTRODUCTION

Growing interest in research on nanoparticles is to a great extent motivated by their numerous applications in fields of life science, chemistry, technology, as well as materials and environmental research [1–5]. Focus of attention are custom-made nanoparticles, since their properties can be engineered according to the needs of applications. Powerful synthesis approaches of such nanoparticles reach from plasmas and flames [6,7] to colloidal chemistry [8–10]. This requires a variety of analysis techniques, which are commonly used to determine the particle size and size distribution, as well as to optimize synthesis conditions. This includes electron microscopy, such as transmission and scanning electron microscopy (TEM, SEM), where the dry nanoparticles are studied [11]. Alternatively, scanning probe microscopy (SPM) has been used [12]. Static and dynamic light scattering (DLS) are standard techniques for probing the size of nanoparticles in solution [13], where swelling, aggregation as well as bulky ligand shells can explain different sizes and size distributions compared to studies using electron microscopy [14]. In addition, the advanced analysis of properties of nanoparticles requires spectroscopic approaches, which permit to determine not only their size and size distribution, but also their surface and bulk properties. Conventional approaches include infrared, visible, ultraviolet, and Raman spectroscopy [15–17]. Further studies made use of soft x-rays, where the electronic structure [17–21] and small angle scattering [22] were probed. Spectroscopic work on nanoparticles has been combined with optical microscopy approaches [23] or more advanced techniques, such as near-field microscopy [24], stimulated emission-depletion (STED)-microscopy [25], tunnel- and force microscopy [12], as well as tip-enhanced approaches

401

[26]. All these methods have in common that they require depositing the samples on a substrate. Thus, their intrinsic properties may be modified due to particle–substrate or particle–particle interactions as well as radiation damage, especially if high-energy radiation or intense photon fields are used. Such aspects become increasingly important with decreasing particle size, since the surface-to-volume ratio increases and deposited nanoparticle samples undergo readily severe radiation damage if they are, for example, exposed to x-rays [27]. Therefore, experimental approaches have been developed more recently, with which either single trapped nanoparticles were studied [28–30] or alternatively, a beam of free size-selected nanoparticles has been utilized more recently [31–35]. The latter approach is subject of the present work, where emphasis is put on elastic light scattering in the soft x-ray regime, similar to previous studies [35].

Already the very first scattering measurements performed on free, size-selected silica (SiO_2) nanoparticles in the soft x-ray regime clearly showed that simple Mie simulations yield distinct differences compared to the experimental results, if the optical constants of macroscopic condensed matter were used to simulate the angle-resolved elastic light-scattering patterns [35]. This result was seeking a profound quantitative assignment, being supported by a reliable model. In this work, a suitable model for simulating the experimental results is reviewed. This is based on a rough and graded surface, that is, the surface properties of the nanoparticles are taken into account. Further, the comparison with scattering from solvent coated or soaked silica nanoparticles is included and the role of surface contaminations is discussed, where soft x-rays are a sensitive probe for surface properties. Particle size distributions of the samples, obtained from the analysis of Mie scattering spectra, are compared to those, which were obtained from electron microscopy. Perspectives for improving the present model by extending the existing theories to the x-ray regime are discussed in detail.

We also report novel results gained more recently, since our previous work [35] has been published. The diffuse scattering part, which was previously considered as being purely reflective, is described in the current context by contributions from the interior of the nanoparticles. In addition, the perturbations due to diffuse scattering are applied to the specular unperturbed Mie intensity, and not simply added to the incident intensity, as was reported before (cf. [35]). In this way, it becomes possible to reproduce important details of the angular dependence of the scattered light at large scattering angles, which were not considered in our previous work [35].

16.2 EXPERIMENTAL

The experimental approach described in this work is a nanoparticle beam, which is used for the preparation of single, size-selected nanoparticles in an ultrahigh vacuum surrounding. The experimental setup is schematically shown in Figure 16.1. It is similar to previous work [35]. The nanoparticle samples consist of amorphous silica (SiO_2). Silica nanoparticles are used for the present work as a simple and robust model system for elastic light scattering. They are prepared by the standard synthesis by Stöber et al. [36]. This approach is well known to yield spherical nanoparticles of variable size in narrow size distributions of the order of a few percent [34,35]. After synthesis the samples are characterized by electron microscopy. The particles are kept in dispersions using either water or ethanol. Ethanol is preferred, since it dries off the nanoparticles more rapidly than water when they are transferred into the gas phase and subsequently in ultrahigh vacuum via a nanoparticle beam. Aerosol formation is accomplished by using an atomizer (TSI 3076). The aerosol phase contains both, neat solvent droplets as well as those containing silica nanoparticles. The solvent is dried off by using a diffusion drier and the size distribution is controlled by a differential mobility analyzer (DMA) (TSI 3080L), which is coupled to a condensation particle counter (TSI 3022A). This yields typically up to 10^7 nanoparticles per cm^3 with a typical gas flow of 1–2 L/min. The control of the nanoparticles in the gas phase at ambient pressure in nitrogen as a carrier gas allows us to verify that single, isolated nanoparticles are formed and to avoid the formation of aggregates, which are easily formed if the sample concentration in the solvent is too high

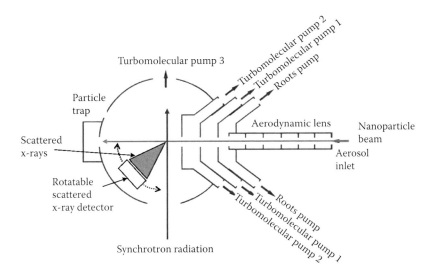

FIGURE 16.1 Schematic diagram of the experimental setup showing the transfer of the nanoparticles into the high-vacuum system, where the focused beam is crossed by monochromatic synchrotron radiation and angle resolved elastic light scattering is measured. Details on the transfer of the nanoparticles into the gas phase are not included (see [39] for more details).

(cf. [34]). If there are pre-made nanoparticles in the solvent, the full beam of nanoparticles can be used for the experiments, since the DMA lowers the particles flux due to its limited transmission and does not narrow the size distribution further. The transfer of the free nanoparticles into the vacuum is accomplished by a critical nozzle of 200 μm diameter (cf. Figure 16.1). It is followed by a series of aerodynamic lenses, which focus the beam to ca. 400 μm in diameter in the scattering center [35]. This region is separated from the aerodynamic lens system by differential pumping stages, so that the pressure in the scattering center is of the order of 10^{-8} mbar. Typical time periods for the transfer from solution into the scattering region are of the order of 10 s, depending on the flow rate of the carrier gas and the length of the tubing guiding the particles to the critical nozzle. Scattering of monochromatic soft x-rays (80–600 eV) from free nanoparticles is performed at the UE49/2-PGM1 and UE52-SGM undulator beamlines at the electron storage ring BESSY-II (Berlin, Germany) [37,38]. These beamlines deliver typically $\sim 10^{12}$ soft x-ray photons for the experiments at a typical energy resolving power $E/\Delta E \approx 5 \times 10^3$. Detection of x-ray photons is accomplished by a channelplate detector that is rotated around the scattering center, similar to previous work (cf. Figure 16.1 and [35]).

16.3 RESULTS AND DISCUSSION

We discuss in the following results from elastic light scattering recorded in the soft x-ray regime (80 eV ≤ E ≤ 600 eV), where emphasis is put on the regime of resonant Si 2p-excitation ($E \approx 100$ eV). This is specifically advantageous for characterizing nanoparticles in the size regime between 50 and 500 nm by elastic light scattering, since the Mie size parameter $x = 2\pi R/\lambda$ is significantly increased by a factor of ~50 compared to the visible regime. Here, λ is the wavelength of the incident radiation (λ ≈ 10 nm) and R is the particle radius. Intense forward scattering is found for $x > 1$, which is indicative for Mie scattering, where λ < R. Elastic light scattering in forward direction was shown to be sensitive to nanoparticles probed by vacuum ultraviolet radiation [33]. The present work goes beyond previous contributions [33,34], where emphasis was put on vacuum ultraviolet radiation in the energy regime below 20 eV. The focus of this study is the soft x-ray regime (80–600 eV), where element-selective excitation of free nanoparticles is used.

16.3.1 Energy Dependences of Scattered Light Intensity in the Regime of Inner-Shell Excitation: Surface Properties and Size Distributions of Nanoparticles

Figure 16.2 shows results from spectroscopic studies on size-selected silica nanoparticles, where elastically scattered soft x-rays are recorded at a fixed scattering angle. Three samples were studied with diameters of 150, 200, and 250 nm. Note that for simplicity we here use only their nominal size, which is determined by a rounded value of the average size, as obtained from electron microscopy. We will elaborate the issue of particle size further below. The spectra are recorded at fixed scattering angles, as indicated in Figure 16.2. The photon energy is scanned in the Si 2p-regime (85–130 eV) [39]. This gives rise to distinct spectral features of SiO_2, which are due to Si 2p--excitation. Even the raw experimental data shown in Figure 16.2 as thick black lines exhibit significant signatures, which sensitively depend on the particle size and scattering angle.

The energy dependencies of the scattered light are simulated by using a Mie scattering model, where the Bohren–Huffman code [40] was used, as documented in previous publications [41–43]. In addition, the program Mie plot was used [44]. Evidently, this light-scattering model is suitable, yielding in the near-edge regime agreement with the experimental results, if the optical constants of single crystals are used [45]. However, there are some significant deviations from the experimental results, if one takes a closer look into the details of the simulations. These occur primarily below and above the absorption edge at about 105 eV, respectively (see Figure 16.2). They become specifically important, if the scattered light intensity is weak in the Si 2p-regime, as is observed for large scattering angles (cf. Figure 16.2c). This requires on the one hand rationalizing the validity of the light-scattering model and on the other hands, finding possible reasons that are related to surface and bulk properties as well as surface contaminations of the nanoparticles.

It is evident that especially the latter is likely to occur, since the preparation of the nanoparticles takes place within a short time period, in which they are transferred from the liquid phase into high vacuum, where they are investigated.

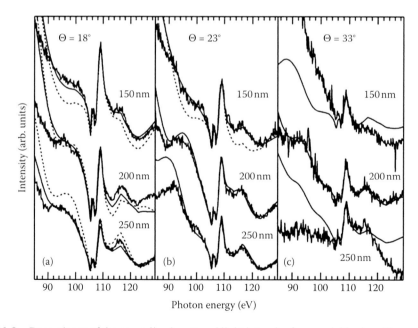

FIGURE 16.2 Dependence of the normalized scattered light intensity from variable size silica nanoparticles on the incident photon energy in the regime of the Si 2p-excitation. Experimental results (thick full lines) are compared to simulated data. Dashed lines: simulations without surface coverage of ethanol, thin full line ethanol coverage on the nanoparticles. Three different particle sizes are studied: 150, 200, and 250 nm. The scattering angle is set to 18° (a), 23° (b), and 33° (c), respectively. See the text for more details.

Figure 16.2a contains such considerations, where two different models were employed to assign the experimental results recorded at 18° scattering angle. They address the importance of surface coatings on the nanoparticles, which are likely to be caused by the solvent sticking on the particle surfaces. These results are also included in Figure 16.2a (thin lines and dashed lines, respectively).

Significant deviations between experiment and model are already observed for the measurements at $\theta \approx 18°$. They become even more significant at $\theta = 33°$, where the scattered light intensity is significantly lower than at small scattering angles. Thus, any disturbances of the weak signal appear as significant deviations from the Mie model (cf. Figure 16.2c).

As one can see from Figure 16.2a the inclusion of several ethanol layers on the surface of nanoparticles significantly improves the agreement with the experimental results, as compared to the neat silica surfaces with empty pores. Ethanol is the solvent, which incompletely evaporates. Furthermore, one expects that the solvent would also stick in the bulk interior of the nanoparticles, since they are porous and therefore soaked by the solvent. The bulk of the nanoparticles are, however, not sensitively probed by elastic light scattering.

Systematic studies on estimating the particle properties were performed, where various parameters, such as surface porosity, deposited layers of the solvent, and ethanol filled pores of silica are systematically varied. Figure 16.3 shows a systematic study on the porosity of the nanoparticles at 18° scattering angle. A 2 nm thick surface layer of a 150 nm nanoparticle is used to vary the amount of silica from 100% to 0%, where the latter corresponds to a 146 nm particle. It becomes clear that surface porosity just leads to a slight shift of the spectral features. Thus, a rough surface cannot easily be probed by such spectroscopic experiments, where the scattering angle is kept constant. Further modeling studies were performed, where the depth of the rough surface layer was varied up to 20 nm, reflecting changes in optical constants due to porosity and remaining solvent. This situation is, however, not backed by earlier transmission electron microscopy studies, which indicate that the thickness of a rough surface layer should be not larger than 2 nm [46]. This is in agreement with the results shown in Figure 16.4, where both the entire nanoparticle with a nominal size of 150 nm from the samples under study is shown (Figure 16.4a) as well as a fraction of the same particle (Figure 16.4b). Clearly, the surface roughness is of the order of 2 nm. A significantly enhanced thickness of a rough surface layer would be unrealistic, as has been modeled in [39]. Silica nanoparticles are porous materials which are soaked by the solvent. Therefore, it is known that silica nanoparticles undergo swelling, which affects their size depending on the degree of drying [47,48]. The time period of particle transfer from the liquid phase into the scattering region is short, so that it appears

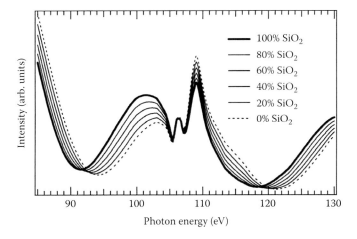

FIGURE 16.3 Scaled influence of a 2 nm porous surface layer on the scattered light spectra of nanoparticles with $D = 150$ nm taken at a scattering angle of $\theta = 18°$.

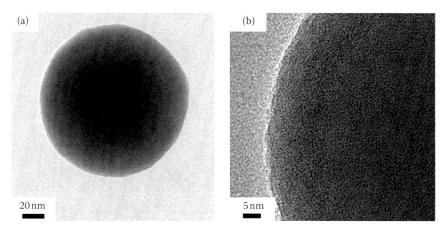

FIGURE 16.4   High-resolution transmission electron microscopy images of a nominally 150 nm silica nano-particle. (a) Image of the entire nanoparticle; (b) close-up look on the surface of this particle.

to be likely that the solvent cannot efficiently evaporate from the pores yielding a drying gradient toward the surface. For elastic light scattering, it is most important to model the surface properties correctly. Therefore, systematic simulations have been performed, which consider a contribution of the solvent reaching up to 10% in a 10 nm surface layer as well as a variation of the thickness of the surface layer at constant mixing ratio of the solvent [39]. Both simulations provide significant changes in scattered light intensity, especially if the surface layer has a thickness of 4–16 nm. These changes become considerably smaller, if one takes a particle size distribution of 10% into account, as it reflects the real sample [34,35]. Thus, the variation of the optical constants due to porosity and filled pores of the sample appears to be considerably less important than the simulation of monodisperse particles would suggest. Specifically, one observes for the regime of the Si 2p-continuum (110–130 eV) that the scattered light intensity at 18° scattering angle is increased by the above mentioned corrections of the particle properties, implying better agreement with the experimental results (cf. Figures 16.2 and 16.3). However, in the pre-edge regime between 85 and 95 eV, the changes are not significant and do not improve the agreement with the experimental results. Thus, one can conclude that drying gradients in porous nanoparticles affect the scattered light intensity. This shows up most clearly for monodisperse particles, but cannot be fully verified for size distributions.

Another important issue are surface layers of the solvent ethanol or other impurities drying on the surface of the nanoparticles upon preparation in the beam. These simulations have been performed on the basis of the Scatmech library [49], where the index of refraction was obtained from [50]. The results on the influence of an ethanol surfaced coating are shown in Figure 16.5, indicating that the Si 2p-near-edge features vanish, if the surface layer is thicker than 2 nm. This appears to be in agreement with the experimental results, which clearly show that the SiO_2 features are dominating the spectra shown in Figure 16.2. Thick surface coatings evidently get lost upon quick transfer into the high-vacuum surroundings. They would lead to strong oscillations of the scattered light intensity, where a minimum is expected to occur near 110 eV, where the near-edge features of Si 2p excitation are observed. The maxima of these oscillations are located at 97 and 124 eV, respectively, where a small size dependence of the surface layer is derived from the model (see Figure 16.5).

The question remains, if one can estimate the thickness of the sticking ethanol layer from the experimental results. This appears to be possible, if one looks at the results shown in Figure 16.6 for three different nanoparticle sizes. There, the difference between the experimental results and the simulated spectra of neat silica nanoparticles is shown. This yields, besides imperfect subtraction near the near-edge features, broad oscillations of the same frequency as modeled for ethanol surface coatings, which are quite similar in frequency to the results shown in Figure 16.5. Therefore, it is assumed that solvent sticking to the nanoparticle surface gives rise to these low amplitude features,

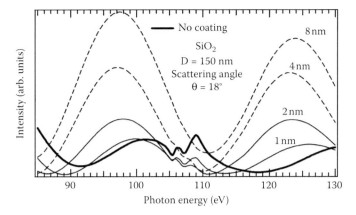

FIGURE 16.5　Influence of the thickness of an ethanol surface layer coating silica nanoparticles ($D = 150$ nm, scattering angle: $\theta = 18°$).

which are estimated to be at most a few monolayers. These contributions are also considered by the simulations shown in Figure 16.2, where an improved agreement with the experimental results has been obtained from inclusion of surface coatings of the solvent. This also implies that other impurities, which may stick on the surface of the nanoparticles do not show significant contributions to the scattered light spectra in the Si 2p-regime.

In summary, it can be concluded, that via energy scans elastic light scattering from free nanoparticles at constant scattering angle in the regime of core level excitation provides sensitively and element-selectively information on the properties of the nanoparticle surface. The classical Mie formalism can be successfully used for simulating the experimental results for spherical particles in order to receive information on the surface properties of the nanomaterials as well as the surface coverage by solvents or adsorbates. This goes beyond previous work of elastic light scattering from macroscopic surfaces [45]. Thus, elastic light scattering from free nanoparticles exhibits with its surface sensitivity a unique capability as a tool for surface analysis of nanoscopic building blocks.

Besides surface properties the particle size distribution is another important property of nanoparticle samples. After optimization colloidal chemistry approaches yield narrow size distributions

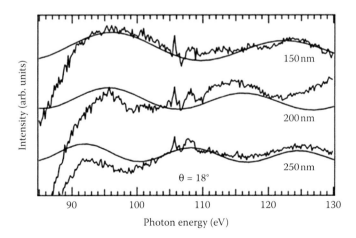

FIGURE 16.6　Comparison of the experimental scattering spectra (noisy curves) and the simulated ones (cf. Figure 16.2) for three different nominal particle sizes of 150, 200, and 250 nm, where exclusively contributions from the solvent are visualized: Contributions from neat silica have been subtracted from the experimental curves showing only contributions from ethanol. These results should be compared to Figure 16.5, where an ethanol layer covering the nanoparticles is simulated. See the text for further details.

of nanoparticles. Centrifugation removes only large aggregates which may have been formed during the synthesis. TEM (transmission electron microscopy) is the standard approach to determine the size distribution as well as aggregate formation. As a result, controlled particle distributions with a polydispersity of ≤12% are commonly achieved [34,35]. This also applies to the samples under study. Note that considerably narrower size distributions of <1% have been obtained using a micro-emulsion-based synthesis route [51], which goes beyond this study.

The nanoparticle size distribution obeys the well-known formula of log-normal distribution, which is given by

$$I(D) = \frac{1}{\sigma\sqrt{2\pi}} \cdot e^{-\frac{1}{2}\left[\frac{\ln(D/\bar{D})}{\sigma}\right]^2}, \tag{16.1}$$

where I denotes the abundance, σ the standard deviation, D the diameter of the nanoparticles, and $\bar{D}$ is the mean value of the size distribution. It is known for amorphous silica, that the particle diameter obtained from TEM experiments is lower than from dynamical light scattering, where the particles are studied in a solvent [47]. On the one hand, this is due to shrinking of the particles upon drying in TEM studies. On the other hand in solution adsorbed species and swelling of the nanoparticles contribute to increased diameters. The difference between both approaches can reach up to 10% [47]. In addition, aggregates of nanoparticles may be formed upon preparation using the nanoparticle beam approach, if the concentration of nanoparticles in the liquid phase is not properly adjusted [34].

Elastic light scattering along with the nanoparticle beam technique is another approach to study the particle size and its size distributions. The resulting energy or angular dependence of the nanoparticle target for a given size distribution is obtained from the weighted sum of the angular distributions of the scattered light. Figure 16.7 shows the scattered light intensity in the Si 2p-regime. The simulated spectra undergo slight changes with increasing number of particles in an assumed size distribution of $\sigma = 12\%$, where the particle size is chosen to be $D = 150$ nm and the scattering angle is set to 23°. The number of particles forming the size distribution is varied from 5 to 100. It is clearly seen, that spectral changes only occur, if a small number of particles is measured. In

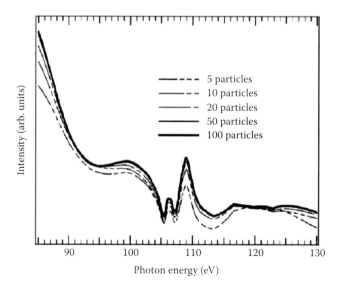

FIGURE 16.7 Scattering intensity for a nanoparticle with $D = 150$ nm, taken at $\theta = 23°$, for different numbers of particles in a distribution of 12% width. See the text for further details.

contrast, there are no more changes if more than 50 particles are considered. It was also found that for a small number of particles the results were quite different by changing the distribution from log-normal to normal, whereas for a large number of particles, that is >50, this change in size distribution had practically no effect, where changes of the order of ±1% were derived. These considerations are of importance for data acquisition times, while scanning the photon energy using the nanoparticle beam, where the experimental spectra shown in this work correspond to the sum over a large number of nanoparticles.

The fits to the experimental spectra considering the size distribution of silica nanoparticles were tested by using spectral scans at three fixed scattering angles of 18°, 23°, and 33°, respectively. This yields the average size and the width of the size distribution of the nanoparticles under study. The particles with a nominal size of 150 nm are best described by an average size of 147 nm and a width of the size distribution of $\sigma = 12\%$, which is in agreement with the results from electron microscopy. Similar, the 200 nm particles correspond to an average size of 193 nm and a size distribution of $\sigma = 8\%$. Finally, the 250 nm sample is found to have an average size of 240 nm and a size distribution of $\sigma = 8\%$. These values appear to be somewhat larger than those derived before [35], but one has to take into account that the present analysis makes use of three different scattering angles and appears to be more reliable than previous estimates that were just considering a single scattering angle.

16.3.2 ANGLE-RESOLVED ELASTIC LIGHT SCATTERING IN THE REGIME OF INNER-SHELL EXCITATION AND SURFACE ROUGHNESS

Angle-resolved light-scattering patterns from free nanoparticles have recently been reported in the regimes of vacuum ultraviolet radiation and soft x-rays [33–35]. Specifically, in the soft x-ray regime the experimental results exhibit strong deviations from model calculations, if the classical Mie theory for ideal spheres and the optical constants from condensed matter are used (cf. [45]). Instead of the expected position of a deep minimum at 90°, this feature occurs in the present work at smaller angles of $30° < \theta < 60°$, depending on the photon energy of the incident beam [35]. The experimental results can be presented suitably as a function or the scattering vector $\vec{q}$, where $\vec{q} = \vec{k}_s - \vec{k}_i = 4\pi/\lambda \, \sin(\theta/2)$ is defined as a vector difference between the scattered ($\vec{k}_s$) and the incident ($\vec{k}_i$) wavevectors, and θ denotes the angle between these vectors. These deviations from simple Mie theory can be seen from the results published earlier [35]. They cannot be rationalized by the polydispersity of the sample prepared in the nanoparticle beam. We present in the following an extended model, which goes beyond previous work [35]. It provides a quantitative and a qualitative explanation for the experimental results.

Sorensen and Fischbach pointed out in previous work that the phase shift has to be considered in elastic light scattering [52]. This quantity is given by $\rho = 2kR|N - 1| = 2x|N - 1|$, where k is the wave number, and N is the index of refraction of the sphere [53]. The quantity $x = kR$ is the above mentioned Mie size parameter. The phase shift ρ reflects the phase difference between two beams passing the distance $2R$ in vacuum and in the sphere, respectively. As discussed in Section 16.3.1, the roughness of the surface, as well as a gradient of drying are strongly influencing the patterns of scattered light. Thus, a component of the diffusely reflected light should be added to the intensity of Mie scattering in order to model properly the angle resolved scattering patterns. The analysis of this contribution was already a subject of extensive theoretical studies [54–58]. As it was pointed out in [54,56] in the pure perturbation approach, originating from Rayleigh, the diffusely reflected light component is introduced by taking the product of the Fourier transform of the correlation function $G(\varepsilon, \theta_i, \theta_r)$ and some expressions $f(\theta_i, \theta_r)$ involving only the angles of incidence θ_i and reflection θ_r. The zero-order intensity is expressed in the present case by the sum of transmitted and the specular reflected intensity, as given by Barrick et al. [57]. The analytic form of this correlation function is chosen to be a Gaussian for smoothly curved surfaces, whereas for jagged and edged surfaces with many corners an exponential form is preferred. Note that rigorous, but rather complicated analytic expressions for the resulting scattering cross sections of rough dielectric spheres can be found elsewhere [56], whereas Farias et al. [54] recommend the explicit choice of the correlation function to be based on physical considerations and mathematical convenience. We have selected a simple Gaussian form, since the observed roughness of the nanoparticles is about five times smaller than the wavelength of the incident radiation ($\lambda \approx 12$ nm) (see Figure 16.4). Furthermore, as it was pointed out by Elson et al. [59], the expression for G should reflect the real experimental conditions. Keeping this in mind, we suggest two alternative treatments to obtain from these considerations the analytic form of the correlation function G. In a first approach, we use the fact that for polished glass surfaces typical correlation lengths a are of the order of 0.1–0.2 µm [59], which is similar to the diameter of the nanoparticles under study. The correlation distance of a surface grating consisting of such spheres was given by Pietsch [60] and for the limiting case of adjacent spheres it simply equals the distance $2R$. In the present case of single spheres, attenuation is determined by the optical path difference $\vec{\tau} = 2\vec{R}|N-1|$, which is produced by the light passing between two boundaries and is responsible for the phase shift ρ (see above). The autocovariance function of surface roughness g, as defined in [59], is expressed by the Fourier transform to G

$$g(\vec{k}) = \int d^2\tau\, G(\tau)\, e^{i\vec{k}\vec{\tau}} = \int d^2\tau\, G(\tau)\, e^{i\vec{k}\,2\vec{R}(N-1)} \tag{16.2}$$

This leads to the following form of a Gaussian autocovariance function:

$$g(\vec{k}) \sim \pi\delta^2(2R)^2 \exp\left[-\left(2kR|N-1|\cos\theta_s\right)^2\right] = \pi(2\delta R)^2 \exp\left[-(\rho\cos\theta_s)^2\right], \tag{16.3}$$

where δ is the surface roughness and θ_s is the scattering angle between the direction of the incident light and the scattered light, as indicated in Figure 16.8. As a result, the position of the minimum in the angular dependence of the scattered light intensity becomes a function of the phase shift ρ. The only fit parameter in this approach is the surface roughness δ. For the remaining angular part $f(\theta_i, \theta_r)$ of the surface-scattering function, the simple diffuse ("white particle") scattering formula [53], was used to approximate the integral in Equation 5 of [59]. This expression is based on the formulae developed by Schoenberg for the diffuse sunlight scattering by planets [61]. The differential power scattered in the unit solid angle $d\Omega = \sin\theta_s\, d\theta_s\, d\varphi$ is then given by

$$I_{\text{diff}} = \frac{1}{P_0}\frac{dP}{d\Omega} = k^4\left(\delta 2R|N-1|\right)^2 \exp\left[-\left(\rho\cos\theta_s\right)^2\right] \cdot \left[\sin\theta_s + (\pi-\theta_s)\cos\theta_s\right]. \tag{16.4}$$

The second approach to model the surface roughness δ is simply to use this quantity as a fit parameter in the exponent of correlation function $g(\vec{k})$, as motivated by previous work by Filatova et al. [45]. In this case, the correlation function shown in Equation 16.3 is replaced by an expression which is similar to that published in [45]. It is independent on the correlation length except for small angles of incidence $\theta_i\,(\pi a\theta_i^2/\lambda \ll 1)$

$$g(\vec{k}) \sim \exp\left[-\left(2\delta k\cos\theta_s\right)^2\right]. \tag{16.5}$$

The mean surface roughness δ amounts to ≈ 1.5 nm and the wavevector of the incident light is approximately ≈ 0.4 nm^{-1}. Then, the requirement $k\delta \ll 1$ is not quite fulfilled and the perturbation series in Equation 16.6 in [56] does not necessarily converge. Therefore, care should be taken by using this approach to model the correlation function.

Figure 16.9 shows a comparison of the experimental and modeled results, where the scattered light intensity is plotted on a logarithmic scale as a function of the scattering vector q. Note that the

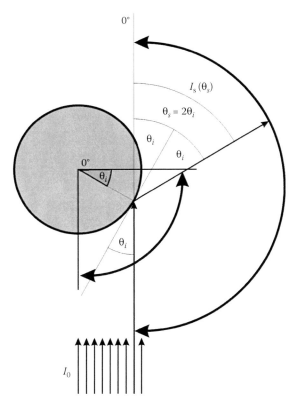

FIGURE 16.8 Optical diagram showing the paths of the incident beam (I_0) and the scattered beam (I_s), where θ_i is the angle of incidence and θ_s is the scattering angle.

distinct and narrow feature near $q \approx 1$ is due to the nanoparticle beam hitting the detector. It defines the scattering angle of 90° and is not a subject of assignment. The fits to the experimental data are performed using a perturbation model for diffuse scattering, which is added to the contribution of Mie scattering

$$I_s = I_{\mathrm{Mie}} + A \cdot I_{\mathrm{diff}}, \tag{16.6}$$

where A is a fit factor of the perturbation strength. The agreement with the experimental results is almost perfect except for the absence of the distinct minimum in the small-angle regime at 0.08–0.1 nm^{-1}. This behavior is observed for all particle sizes and for all four excitation energies. The evolving perturbation strength factors A are summarized as $2R = 150$ nm: $A = 0.2$; $2R = 200$ nm: $A = 0.3$; and $2R = 250$ nm: $A = 0.3$, which indicates that the contribution of Mie scattering is always dominant.

Table 16.1 summarizes the values of surface roughness δ obtained from the fits to experimental angular dependences with different excitation energies and particle sizes. Since δ is expected to be a constant, the observed slight deviation at 107 eV, for example, can be related to the distinct minimum of the near-edge structure (see Figure 16.2). These results indicate that the surface roughness is of the order of 1–2 nm, which is in agreement with results from electron microscopy (cf. Figure 16.4).

Scattering of soft x-rays at large angles is mostly dominated by diffusively scattered light. Thus, it becomes clear that this is the reason for the deviation of the fit from the experimental results using the approach given by Equation 16.6. In addition, one has to consider the contribution to diffusive scattering generated by the transmitted radiation I_{Mie} coming from the inside of the sphere. By

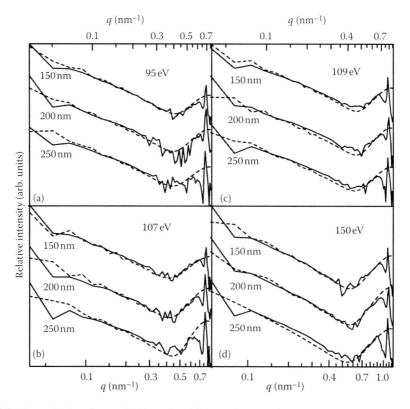

FIGURE 16.9 Perturbation theory fit, Equation 16.6, to the experimental angular dependences on SiO_2 nanoparticles as a function of excitation energy: (a) 95 eV, (b) 107 eV, (c) 109 eV, (d) 150 eV. Calculated curves are indicated by dashed lines, where a Mie model for scattering of soft x-rays from a rough surface of the nanoparticles is considered. Ssee the text for further details.

considering I_{Mie} (and not I_0) as an additional source for diffusive scattering in the surface layer, Equation 16.6 is replaced by

$$I_s = I_{Mie} + I_{Mie} A \exp\left[-\left(2\delta k \cos \theta_s\right)^2\right] \tag{16.7}$$

Fits using Equation 16.7 are shown in Figure 16.10, where the overall agreement with the model is clearly improved. This becomes evident when the range of scattering angles between 70° and 90° is considered, corresponding 0.8–1.0 nm^{-1}.

TABLE 16.1
Surface Roughness of Silica Nanoparticles, as Obtained from Equation 16.5 for Different Excitation Energies and Particle Sizes

Energy (eV)	Surface Roughness (nm)		
	Particle Size: 150 nm	Particle Size: 200 nm	Particle Size: 250 nm
95	1.2	1.0	0.7
107	1.5	1.5	1.5
109	1.2	1.0	0.7
150	1.2	1.0	0.7

Sorensen and Fischbach applied a power law analysis to the plots, where the scattered light intensity is plotted on a logarithmic scale as a function of the dimensionless parameter qR [52]. Note that in the present case the phase shift ρ is slightly larger than unity. This implies that the Rayleigh–Gans (RG) theory for the qR slope analysis [52] should be applied with some care. Nevertheless, all experimental curves in Figures 16.9 and 16.10, exhibit without exception a slope -4 for the qR dependence in a wide range. On the other hand, the long-range slope of the line drawn to the top of the first Mie maximum shows the expected slope of -2, as shown in Figure 16.11. This result is expected from the classical RG theory (see [52] for details). This discrepancy in slope could be easily explained by the strong contribution of diffuse scattering. This is because in both cases, that is for ρ (cf. Equation 16.3) or δk (cf. Equation 16.5), the exponential term in Equation 16.4 can be replaced for angles $\theta > 20°$ by a linear dependence of $(\cos \theta)^2$. The scattering dependence is then dominated by a $q^{-4} \sim (\sin (\theta/2))^{-4}$ factor. Note that this yields the same result as the -4 slope, which is well known for scattering in the bulk, according to the Porod law. Although even the range of the qR^{-4} law is the same for equal phase shifts ρ, the experimental dependence cannot be reproduced with the help of classical Mie theory, delivering the mentioned shift of the minimum at $\theta \approx 90°$.

Some problems with modeling of the experimental results obtained from soft x-ray experiments are arising if one is attempting to extend known models that have been developed for long wavelength radiation into the x-ray regime. This is specifically true, since the index of refraction N of SiO_2 becomes smaller than unity. This causes several expressions in the formalism of [56] to become

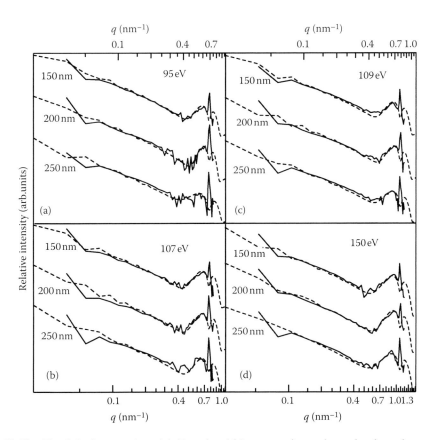

FIGURE 16.10 Fit of the improved model, Equation 16.7, to experimental angular dependences taken on SiO_2 nanoparticles as a function of excitation energy: (a) 95 eV, (b) 107 eV, (c) 109 eV, (d) 150 eV. Calculated curves are indicated by dashed lines. See the text for further details.

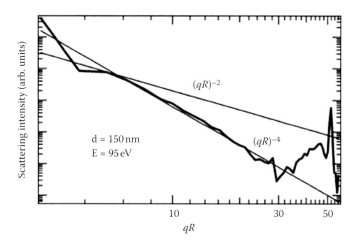

FIGURE 16.11 Scattered light intensity from SiO_2 particles (nominal size: 150 nm) at 95 eV photon energy, where the dimensionless parameter qR is used (cf. [52]). The fit of the qR dependence to slopes of the experimental data is indicated by straight lines. See the text for further details.

complex. After necessary algebraic rearrangements and extension of the calculations into the complex plane, some care should also be taken by considering the application limits (a) through (d) given in part II of [56] to the x-ray region. These are elucidated in the following. We can immediately see that the first criterion (a), requesting the product of the average deviation h from the smooth surface and the wavelength λ to be <<1, is almost satisfied for the energy range under study as well as the particle quality, yielding $hk \approx 0.15$. Criterion (b), requesting the surface slopes being smaller than unity, is satisfied, as well. The entire particle is exposed to the incident x-rays. Thus, rule (c), requesting the illuminated area to be much larger than the correlation length l of the roughness structures, is not violated either. Finally, electron microscopy studies on SiO_2 nanoparticles show that the surface roughness is quite homogeneously distributed and isotropic (see Figure 16.4), which satisfies criterion (d). Therefore, we can conclude that the empirical model presented in this work can be tested in the frame of existing theories of [57,58].

16.4 CONCLUSION

Elastic light scattering of free nanoparticles using tunable soft x-rays is used for analyzing the properties of free nanoparticles, which are prepared in a beam. The nanoparticle beam is a universal preparation scheme for dense optical targets of isolated nanoscopic systems in ultrahigh vacuum surroundings. The combination of this sample preparation with soft x-rays yields specific properties of isolated nanoscopic matter from elastic light scattering. This includes the particle size, size distributions, and surface properties. The present approach appears to be a suitable way of characterizing the properties of nanoparticles in a quantitative way. This can contribute to solid knowledge on optical properties of nanoscopic matter and how much they deviate from microscopic and macroscopic solids. Detailed knowledge on nanoscopic building blocks can contribute for an optimization of such properties with respect to possible applications. On the other hand, it is of interest to use elastic light scattering from free nanoparticles as a novel approach for absolute nanoparticle size determinations. This is of specific interest, since currently different results in particle size and specifically size distributions are obtained from various analysis approaches [14]. Further, the sample surroundings may lead to changes in sample size due to swelling, shrinking or changes in size of the ligand shell, which can also be studied in solution, if hard x-rays are used.

ACKNOWLEDGMENTS

Financial support by the BMBF (contract 05 KS7KEA) and the Fonds der Chemischen Industrie is gratefully acknowledged. We thank the staff of the storage ring BESSY-II (Helmholtz Centre Berlin, Germany) for their continuous support. Finally, we thank Sören Selve, Technische Universität Berlin, Zentraleinrichtung Elektronenmikroskopie (ZELMI) for taking the HRTEM images.

REFERENCES

1. Bruchez, M.; Moronne, M.; Gin, P.; Weiss, S.; and Alivisatos, A. P. 1998. Semiconductor nanocrystals as fluorescent biological labels. *Science* 281: 2013–2016.
2. Graf, C.; Meinke, M.; Gao, Q.; Hadam, S.; Raabe, J.; Sterry, J.; Blume-Peytavi, U.; Lademann, J.; Rühl, E.; and Vogt, A. 2009. Qualitative detection of single submicron and nanoparticles in human skin by scanning transmission X-ray microscopy. *J. Biomed. Opt.* 14: 021015.
3. Burda, C.; Chen, X. B.; Narayanan, R.; and El-Sayed, M. A. 2005. Chemistry and properties of nanocrystals of different shapes. *Chem. Rev.* 105: 1025–1102.
4. Daniel, M. C.; and Astruc, D. 2004. Gold nanoparticles: Assembly, supramolecular chemistry, quantum-size-related properties, and applications toward biology, catalysis, and nanotechnology. *Chem. Rev.* 104: 293–346.
5. Zhu, Y. F.; Hinds, W. C.; Kim, S.; Shen, S.; and Sioutas, C. 2002. Study of ultrafine particles near a major highway with heavy-duty diesel traffic. *Atmos. Environm.* 27: 4323–4335.
6. Meinen, J.; Leisner, T; Khasminskaya, S.; Rühl, E.; and Baumann, W. 2010. The TRAPS apparatus: Enhancing target density of nanoparticle beams in vacuum for X-ray and optical spectroscopy. *Aerosol Sci. Technol.* 44: 316–328.
7. di Stasio, S.; Mitchell, J. B. A.; LeGarrec, J. L.; Biennier, L.; and Wulff, M. 2006. Synchrotron SAXS (*in situ*) identification of three different size modes for soot nanoparticles in a diffusion flame. *Carbon* 44: 1267–1279.
8. Murray, C. B.; Kagan, C. R.; and Bawendi, M. G. 2000. Synthesis and characterization of monodisperse nanocrystals and close-packed nanocrystal assemblies. *Ann. Rev. Mater. Sci.* 30: 545–610.
9. Vogel, R.; Hoyer, P.; and Weller, H. 1994. Quantum-sized PbS, CdS, Ag_2S, Sb_2S_3, and Bi_2S_3 particles as sensitizers for various nanoporous wide-band gap semiconductors. *J. Phys. Chem.* 98: 3183–3188.
10. Graf, C.; and van Blaaderen, A. 2002. Metallodielectric colloidal core-shell particles for photonic applications. *Langmuir* 18: 524–534; Graf, C.; Dembski, S.; Hofmann, A.; and Rühl, E. 2006. A general method for the controlled embedding of nanoparticles in silica colloids. *Langmuir* 22: 5604–5610.
11. Wang, Z. L. 2000. Transmission electron microscopy of shape-controlled nanocrystals and their assemblies. *J. Phys. Chem. B* 104: 1153–1175.
12. Grabar, K. C.; Brown, K. R.; Keating, C. D.; Stranick, S. J.; Tang, S.-L.; and Natan, M. J. 1997. Nanoscale characterization of gold colloid monolayers: A comparison of four techniques. *Anal. Chem.* 69: 471–477.
13. Schmidt, M. 1993. Simultaneous static and dynamic light scattering: Application to polymer structure analysis. In *Dynamic Light Scattering, the Method and Some Applications.* Ed. W. Brown. Clarendon Press, Oxford.
14. Lenggoro, I. W.; Xia, B.; Okuyama, K.; and de la Mora, J. F. 2002. Sizing of colloidal nanoparticles by electrospray and differential mobility analyzer methods. *Langmuir* 18: 4584–4591; Goertz, V.; Dingenouts, N.; and Nirschl, H. 2009. Comparison of nanometric particle size distributions as determined by SAXS, TEM, and analytical ultracentrifuge. *Part. Part. Syst. Charact.* 26: 17–24.
15. Guo, L.; Yang, S.; Yang, C.; Yu, P.; Wang, J.; Ge, W.; and Wong, G. K. L. 2000. Highly monodisperse polymer-capped ZnO nanoparticles: Preparation and optical properties. *Appl. Phys. Lett.* 76: 2901–2903.
16. Dembski, S.; Graf, C.; Krüger, T.; Gbureck, U.; Ewald, A.; Bock, A.; and Rühl, E. 2008. Photoactivation of CdSe/ZnS quantum dots embedded in silica colloids. *Small* 4: 1516–1526.
17. Barglik-Chory, C.; Buchold, D.; Schmitt, M.; Kiefer, W.; Heske, C.; Kumpf, C.; Fuchs, O.; et al. 2003. Synthesis, structure and spectroscopic characterization of water-soluble CdS nanoparticles. *Chem. Phys. Lett.* 379: 443–451.
18. Hamad, K. S.; Roth, R.; Rockenberger, J.; van Buuren, T.; and Alivisatos, A. P. 1999. Structural disorder in colloidal InAs and CdSe nanocrystals observed by X-ray absorption near-edge spectroscopy. *Phys. Rev. Lett.* 83: 3474–3477.

19. McGinley, C.; Borchert, H.; Pflughoefft, M.; Al Moussalami, S.; de Castro, A. R. B.; Haase, M.; Weller, H.; and Möller, T. 2001. Dopant atom distribution and spatial confinement of conduction electrons in Sb-doped SnO_2 nanoparticles. *Phys. Rev. B* 64: 245312.

20. Nowak, C.; Döllefeld, H.; Eychmüller, A.; Friedrich, J.; Kolmakov, A.; Löfken, J. O.; Riedler, M.; et al. 2001. Innershell absorption spectroscopy on CdS: Free clusters and nanocrystals. *J. Chem. Phys.* 114: 489–494.

21. Ethiraj, A. S.; Hebalkar, N.; Kulkarni, S. K.; Pasricha, R.; Urban, J.; Dem, C.; Schmitt, M.; et al. 2003. Enhancement of photoluminescence in manganese-doped ZnS nanoparticles due to a silica shell. *J. Chem. Phys.* 118: 8945–8953.

22. Araki, T.; Ade, H.; Stubbs, J. M.; Sundberg, D. C.; Mitchell, G. E.; Kortright, J. B.; and Kilcoyne, A. L. D. 2006. Resonant soft x-ray scattering from structured polymer nanoparticles. *Appl. Phys. Lett.* 89: 124106.

23. Zhang, B.; Li, Y.; Fang, C.-Y.; Chang, C.-C.; Chen, C.-S.; Chen, Y.-Y.; and Chang, H.-C. 2009. Receptor-mediated cellular uptake of folate-conjugated fluorescent nanodiamonds: A combined ensemble and single-particle study. *Small* 5: 2716–2721.

24. Hillenbrand, R.; and Keilmann, F. 2001. Optical oscillation modes of plasmon particles observed in direct space by phase-contrast near-field microscopy. *Appl. Phys. B* 73: 239–243.

25. Willig, K. I.; Keller, J.; Bossi, M.; and Hell, S. W. 2006. STED microscopy resolves nanoparticle assemblies. *New J. Phys.* 8: 106.

26. Anderson, N.; Bouhelier, A.; and Novotny, L. 2006. Near-field photonics: Tip-enhanced microscopy and spectroscopy on the nanoscale. *J. Opt. A* 8: S227–S233.

27. Döllefeld, H.; McGinley, C.; Almousalami, S.; Möller, T.; Weller, H.; and Eychmüller, A. 2002. Radiation-induced damage in x-ray spectroscopy of CdS nanoclusters. *J. Chem. Phys.* 117: 8953–8958.

28. Grimm, M.; Langer, B.; Schlemmer, S.; Lischke, T.; Widdra, W.; Gerlich, D.; Becker, U.; and Rühl, E. 2004. New setup to study trapped nanoparticles using synchrotron radiation. *AIP Conf. Proc.* 705: 1062–1065.

29. Grimm, M.; Langer, B.; Schlemmer, S.; Lischke, T.; Becker, U.; Widdra, W.; Gerlich, D.; Flesch, R.; and Rühl, E. 2006. Charging mechanisms of trapped element-selectively excited nanoparticles exposed to soft X-rays. *Phys. Rev. Lett.* 96: 066801.

30. Lewinski, R.; Graf, C.; Langer, B.; Flesch, R.; Bresch, H.; Wassermann, B.; and Rühl, E. 2009. Size-Effects in clusters and free nanoparticles probed by soft X-rays. *Eur. J. Phys. Special Topics* 169: 67–72.

31. Shu, J. N.; Wilson, K. R.; Ahmed, M.; and Leone, S. R. 2006. Coupling a versatile aerosol apparatus to a synchrotron: Vacuum ultraviolet light scattering, photoelectron imaging, and fragment free mass spectrometry. *Rev. Sci. Instrum.* 77: 043106.

32. Wilson, K. R.; Zou, S.; Shu, J. N.; Rühl, E.; Leone, S. R.; Schatz, G. C.; and Ahmed, M. 2007. Size-dependent angular distributions of low energy photoelectrons emitted from NaCl nanoparticles. *Nano Lett.* 7: 2014–2019.

33. Shu, J. N.; Wilson, K. R.; Arrowsmith, A. N.; Ahmed, M.; and Leone, S. R. 2005. Light scattering of ultrafine silica particles by VUV synchrotron radiation. *Nano Lett.* 5: 1009–1015.

34. Shu, J. N.; Wilson, K. R.; Ahmed, M.; Leone, S. R.; Graf, C.; and Rühl, E. 2006. Elastic light scattering from nanoparticles by monochromatic vacuum-ultraviolet radiation. *J. Chem. Phys.* 124: 034707.

35. Bresch, H.; Wassermann, B.; Langer, B.; Graf, C.; Flesch, R.; Becker, U. Österreicher, B.; Leisner, T.; and Rühl, E. 2008. Elastic light scattering from free sub-micron particles in the soft X-ray regime. *Faraday Discuss.* 137: 389–402.

36. Stöber, W.; Fink, A.; and Bohn, E. J. 1968. Controlled growth of monodisperse silica spheres in micron size range. *J. Colloid Interface Sci.* 26: 62–69.

37. Sawhney, K. J. S.; Senf, F.; and Gudat, W. 2001. PGM beamline with constant energy resolution mode for U49–2 undulator at BESSY-II. *Nucl. Instrum. Meth. A* 467–468: 466–469.

38. Godehusen, K.; Mertins, H.-C.; Richter, T.; Zimmermann, P.; and Martins, M. 2003. Electron-correlation effects in the angular distribution of photoelectrons from Kr investigated by rotating the polarization axis of undulator radiation. *Phys. Rev. A* 68: 012711.

39. Bresch, H. 2007. Photoionisation von freien Aerosolpartikeln mit Synchrotronstrahlung. Dissertation, Freie Universität Berlin, Berlin, Germany.

40. Bohren, C. F.; and Huffman, D. R. 1983. *Absorption and Scattering of Light by Small Particles.* Wiley-VCH, Weinheim, Germany.

41. Damaschke, N. 2002. Light scattering theories and their use for single particle characterization. Dissertation, Technische Universität Darmstadt, Darmstadt, Germany.

42. Vortisch, H. 2002. Beobachtung von Phasenübergängen in einzeln levitierten Schwefelsäuretröpfchen mittels Raman-Spektroskopie und elastischer Lichtstreuung. Dissertation. Freie Universität Berlin, Berlin, Germany.

43. Leisner, T. 2006. Private Communication. Forschungszentrum Karlsruhe.

44. Laven, P. 2003. Simulation of rainbows, coronas, and glories by use of Mie theory. *Appl. Opt.* 42: 436–444.

45. Filatova, E.; Lukyanov, V.; Barchewitz, R.; André, J.-M.; Idir, M.; and Stemmler, P. 1999. Optical constants of amorphous SiO_2 for photons in the range of 60–3000 eV. *J. Phys. Condens. Matter* 11: 3355–3370.

46. Bogush, G. H.; and Zukoski, C. F. 1991. Uniform silica particle precipitation: An aggregative groth model. *J. Colloid Interface Sci.* 142: 19–34.

47. van Helden, A. K.; Jansen, J. W.; and Vrij, A.. 1981. Preparation and characterization of spherical monodisperse silica dispersions in non-aqueous solvents. *J. Colloid Interface Sci.* 81: 354–368; van Blaaderen, A.; and Kentgens, A. P. M. 1992. Particle morphology and chemical microstructure of colloidal silica spheres made from alkoxysilanes. *J. Non-Cryst. Solids* 149: 161–178.

48. Costa, C. A. R.; Leite, C. A. P.; and Galembeck, F. 2003. Size dependence of stöber silica nanoparticle microchemistry. *J. Phys. Chem. B* 107: 4747–4755.

49. Germer, T. A. 2008. SCATMECH: Polarized light scattering C^{++} class library; available at http://physics.nist.gov/scatmech

50. Henke, B. L.; Gullikson, E. M.; and Davis J. C. 1993. X-ray interactions—photoabsorption, scattering, transmission, and reflection at $E = 50$–30,000 eV, $Z = 1$–92. *At. Data Nucl. Data Tables*, 54: 181–342.

51. van Blaaderen, A.; van Geest, J.; and Vrij, A. 1992: Monodisperse colloidal silica spheres from tetraalkoxysilanes: Particle formation and growth mechanism. *J. Colloid Interface Sci.* 154: 481–501; Arriagada, F.J.; and Osseo-Asare, K. 1999. Synthesis of nanosize silica in a nonionic water-in-oil microemulsion: Effects of the water/surfactant molar ratio and ammonia concentration. *J. Coll. Int. Sci.* 211: 210–220

52. Sorensen, C. M.; and Fischbach, D. J. 2000. Patterns in Mie scattering. *Opt. Commun.* 173: 145–153.

53. van de Hulst, H. C. 1981. *Light Scattering by Small Particles*. General Publishing Company, Toronto, Canada.

54. Farias, G. A.; Vasconcelos, E. F.; Cesar, S. L.; and Maraduduin, A. A. 1994. Mie scattering by a perfectly conducting sphere with a rough surface. *Physica A* 207: 315–322.

55. Drossart, P. 1990. A statistical model for the scattering by irregular particles. *Astrophys. J.* 361: L29–L32.

56. Schiffer, R.; and Thielheim, K. O. 1984. The effect of slight surface roughness on the scattering properties of convex particles. *J. Appl. Phys.* 57: 2437–2454.

57. Barrick, D. E.; and Peake, W. H. 1967. Batelle Memorial Institute, Report No. BAT-197-A-10-3, AD 662 751.

58. Vinogradov, A. V.; Zorev, N. N.; Kozhevnikov, I. V.; Sagitov, S. I.; and Turyanskii, A. G. 1988. X-ray scattering by highly polished surfaces. *Sov. Phys. JETP* 67: 1631–1638.

59. Elson, J. M. 1975. Light scattering from semi-infinite media for non-normal incidence. *Phys. Rev. B* 12: 2541–2542; Elson, J. M.; Rahn, J. P.; and Bennet, J. M. 1983. Relationship of the total integrated scattering from multilayer-coated optics to angle of incidence, polarization, correlation length, and roughness cross correlation properties. *Appl. Opt.* 22: 3207–3219.

60. Pietsch, U. 2002. X-ray and visible light scattering from light-induced polymer gratings. *Phys. Rev. B*: 66: 155430.

61. Schoenberg, E. 1929. Theoretische Photometrie. Chapter 1 in *Handbuch der Astrophysik II*. Eds. Eberhard, G.; Kohlschütter, A.; and Ludendorff, H., Springer, Berlin.

17 Scanning Transmission X-ray Microscopy

Applications in Atmospheric Aerosol Research

Ryan C. Moffet, Alexei V. Tivanski, and Mary K. Gilles

CONTENTS

17.1 Introduction and Synchrotron Background 420
    17.1.1 Storage Rings and Insertion Devices..420
        17.1.1.1 An Electron's Life in the Synchrotron 420
        17.1.1.2 Synchrotron Operation Modes..421
        17.1.1.3 Insertion Devices 422
        17.1.1.4 Beamline Optical Components..422
        17.1.1.5 Beamline Carbon Contamination...422
17.2 The STXM/NEXAFS Technique...423
    17.2.1 STXM Description 423
    17.2.2 Background on NEXAFS Spectroscopy 424
        17.2.2.1 Electronic Transitions Probed..424
        17.2.2.2 Conversion to Optical Density 425
        17.2.2.3 Carbon NEXAFS 427
    17.2.3 Polarization...428
    17.2.4 Aerosol Technique Comparison 429
    17.2.5 Potential for Radiation Damage...430
    17.2.6 Practical Sample Considerations 431
        17.2.6.1 Sample Preparation and Collection 431
        17.2.6.2 Substrates 432
    17.2.7 Data Analysis..432
17.3 STXM/NEXAFS Aerosol Studies 433
    17.3.1 Soot/Black Carbons/Brown Carbons 433
        17.3.1.1 Black Carbon/Soot Single-Energy Images and Spectra 434
        17.3.1.2 C:O Atomic Ratios..436
        17.3.1.3 Identifying Black Carbon/Soot via NEXAFS Spectra 437
        17.3.1.4 Spectral Deconvolution...438
        17.3.1.5 Quantifying the Graphitic Nature: sp^2 Hybridization.............439
        17.3.1.6 Biomass Burn Aerosols...439
        17.3.1.7 Tar Balls..440
        17.3.1.8 Evolution of Aerosol Mixing States..443
    17.3.2 Organic Carbon 446
    17.3.3 Inorganic Species...448
        17.3.3.1 Speciation of Zn-Containing Aerosols 449

17.3.3.2 Fe Speciation..449
17.3.3.3 Marine Boundary Layer Sulfur Speciation 450
17.3.4 Other Studies 452
17.3.4.1 Power Plant Plume Processing: Presence of Organosulfates? 452
17.3.4.2 Ice Nucleation and Cloud Condensation Nuclei 452
17.4 Opportunities for Future Applications and Developments....................................453
17.4.1 Combining Complementary Microscopic and Spectroscopic Techniques..............453
17.4.2 Development of *in situ* Techniques..454
Acknowledgments..455
References...455

17.1 INTRODUCTION AND SYNCHROTRON BACKGROUND

Scanning transmission x-ray microscopy (STXM) combines x-ray microscopy and near edge x-ray absorption fine structure spectroscopy (NEXAFS). This combination provides spatially resolved bonding and oxidation state information. While there are reviews relevant to STXM/NEXAFS applications in other environmental fields[1-4] (and magnetic materials[5]) this chapter focuses on atmospheric aerosols. It provides an introduction to this technique in a manner approachable to non-experts. It begins with relevant background information on synchrotron radiation sources and a description of NEXAFS spectroscopy. The bulk of the chapter provides a survey of STXM/NEXAFS aerosol studies and is organized according to the type of aerosol investigated. The purpose is to illustrate the current range and recent growth of scientific investigations employing STXM-NEXAFS to probe atmospheric aerosol morphology, surface coatings, mixing states, and atmospheric processing.

Scanning transmission x-ray microscopy (STXM) is performed at synchrotron light sources, which produce intense and tunable x-ray beams.[1-5] An exceptional reference is provided by Attwood[6]; hence, only a few basic concepts are presented here. Workshops and introductory short courses are often available to introduce those unfamiliar with the field, and interested readers can consult websites for individual light sources.

17.1.1 STORAGE RINGS AND INSERTION DEVICES

17.1.1.1 An Electron's Life in the Synchrotron

Electrodynamics tells us that a charged particle under the influence of a magnetic force emits radiation. Synchrotrons exploit this by subjecting relativistic electrons (i.e., those traveling at 99.999985% of the speed of light) to magnetic fields, thereby causing them to emit radiation. Electrons produced by an electron gun are contained in an ultrahigh vacuum tube (to minimize their loss via collision with gas species) and are accelerated to relativistic velocities using radio frequency cavities. Electrons are subjected to magnetic forces as they pass through an insertion device resulting in the emission of synchrotron radiation in a cone tangent to the trajectory of the electron. From each insertion device, the photons are emitted and directed down a beamline to an experiment. A spread of photon energies are produced in the insertion device and a monochromator provides additional energy selection. Depending on the monochromator (and gratings), the energy resolution may vary by an order of magnitude or more. Typically, STXMs are on beamlines with resolutions on the order of 20–40 meV (at ~300 eV). At the end of each beamline, there may be one (or more) experiments (endstations) either permanently attached to the beamline or temporarily attached for a specific set of experiments. The schematic in Figure 17.1 shows a synchrotron, an insertion device (undulator), and the primary components of an STXM. More specific details on synchrotrons and insertion devices are available in Attwood[6] as well as on individual synchrotron websites. The following

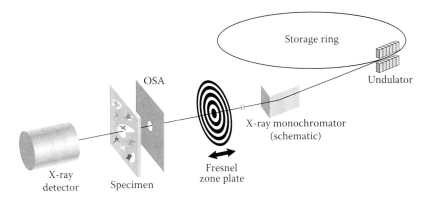

FIGURE 17.1 Schematic of a synchrotron (not to scale) showing a single insertion device (undulator) and beamline with an x-ray monochromator for energy selection. The STXM instrument includes the Fresnel zone plate for focusing, an order sorting aperture to eliminate unfocused and higher-order focused x-rays, a sample, and the detector. The sample is located at the focal point and raster scanned to obtain an image. [Adapted from Maser, J. et al. *Journal of Microscopy* **197**, 68–79 (2000).]

section discusses only the most relevant concepts for atmospheric (or environmental) scientists wishing to understand or pursue STXM/near-edge x-ray absorption fine structure spectroscopy (NEXAFS) experiments at a synchrotron.

17.1.1.2 Synchrotron Operation Modes

Synchrotrons operate in a variety of modes: multibunch, top-off, and two-bunch. In multibunch mode, a set number of electron bunches are in the synchrotron (~300). For most experiments, the timing between these bunches is short (several ~few nanoseconds) compared to the timescale for detection and the synchrotron is treated as a continuous photon source. As electrons pass through an insertion device, they lose energy as synchrotron radiation (emit photons). This energy is then replenished to the electrons using radio frequency cavities. The time between electron bunches is determined by the frequency used to accelerate the electrons. Over the course of several hours (4–12), the electron beam current decreases due to collisions with gas molecules and other loss processes. The decreasing number of electrons within bunches produces a corresponding decrease in photon flux.

When the photon flux is sufficiently low, the electron beam is forcibly lost ("dumped") and the synchrotron is "injected" with new bunches of electrons resulting in an increase in photon flux. In a two-bunch mode, there are only two bunches of electrons circulating in the synchrotron. The time between these two bunches is larger and is often used for experiments in which time is a parameter (e.g., time of flight coincidence experiments). Due to fewer electron bunches in the synchrotron, the photon flux is also correspondingly lower. Several synchrotrons now operate in top-off mode where the injection of new electrons into the synchrotron is quasi-continuous. In this mode, fresh electron bunches are injected into the ring approximately every half-minute to replenish any decrease in electron current. Top-off mode significantly increases the average photon flux available as the beam current remains roughly constant over a period of approximately a week. Additionally, unlike multibunch mode there is no downtime for injecting fresh electron bunches into the synchrotron. Hence, several hours per day previously used for "injection" are recovered as usable experimental time. Due to the higher average photon flux, particular care must be taken to minimize any potential radiation damage to samples. STXM/NEXAFS data on atmospheric aerosols are generally acquired using the multibunch or top-off mode; the low flux during the two-bunch mode would result in prohibitively long data acquisition times.

17.1.1.3 Insertion Devices

To produce synchrotron radiation several types of magnetic structures (discussed in detail in Attwood[6]) are used: wigglers, bending magnets, and undulators. Insertion devices differ in spatial focusing, energy spread, and polarization. Undulators and bending magnets are employed in current STXM instruments. There are a few important differences between these two insertion devices. A bending magnet emits radiation in a wider angle than an undulator. Undulators produce highly collimated, partially coherent radiation with a narrower energy range (frequency spread) having higher photon fluxes than bending magnets. Depending on the insertion device, the contribution from higher harmonics (photons whose frequency is a multiple of the fundamental frequency) may vary significantly. In addition, the broader energy range produced by a bending magnet makes the alignment of some of the STXM optical components more stringent. Another important difference is that the linear polarization of the synchrotron radiation is fixed for a bending magnet and can be rotated for an undulator (which can also produce circularly or elliptically polarized light).

17.1.1.4 Beamline Optical Components

Key optical components between the insertion device and an STXM are indicated in Figure 17.1. The monochromator is an important beamline component for STXM/NEXAFS studies. Depending on its design, there may be entrance and/or exit slits, or it may be slitless. Monochromators may have a single grating or multiple gratings blazed onto a single substrate and the gratings switched depending on the relative importance of energy, flux, or spectral resolution in a particular experiment. A fundamental difference between optics in other energy regions, such as visible or infrared, is that all optical components can absorb x-ray radiation. As reflectivity is enhanced at oblique angles, beamline optics are normally employed in a grazing incidence configuration. Typically, STXMs are coupled with high spectral resolution monochromators that enable functional groups within carbon spectra to be resolved. Depending on the energy region of interest, energy calibration is performed with gaseous CO_2, O_2, or N_2. To maintain photon beam alignment, slits and positioning mirrors (not shown in Figure 17.1) are coupled by a feedback system. Although Figure 17.1 indicates a single end station, some beamlines have a mirror that directs the photon beam into separate experiments. Specific synchrotron websites normally contain datasheets for each beamline providing information about the endstations, insertion devices, energy range, and monochromators.

17.1.1.5 Beamline Carbon Contamination

Since carbon is a major component of atmospheric aerosols, the STXM has filled a unique niche by providing detailed information at the carbon K-edge (290 eV). However, one critical issue that arises in examining carbonaceous samples is that often beamline components (gratings, zone plates, mirrors) are contaminated with carbon. This arises from the materials used to manufacture the beamline components and by the deposition of carbon onto the optical components during the initial building, development, and commissioning of a beamline. Therefore, special measures must be taken to ensure the lowest levels of carbon contamination. Any carbon contamination effectively absorbs photons emitted from the insertion device causing a decreased flux in the energy region where carbon absorbs. At best, the majority of the photon flux is not absorbed and at worst only a fraction (<5%) of the photons are transmitted to the STXM. In this event, it is nearly impossible to obtain quantitative carbon spectra. Although some degree of carbon contamination is always present, selecting a beamline with low levels of contamination is critical for obtaining high-quality (and quantitative) carbon spectra. Over time, within the STXM itself, carbon deposits onto optical elements such as the zone plate. Examining the background absorption spectra (photons detected in the absence of a sample) is critical to ensure that absorption by carbon on beamline optical components does not compromise the measured NEXAFS spectra.

17.2 THE STXM/NEXAFS TECHNIQUE

STXM was developed by Kirz, Jacobsen, Ade, and coworkers[7] at the National Synchrotron Light Source, NSLS, (Brookhaven, USA) during the late 1980s. During the early years a number of developments were also underway at Daresbury (United Kingdom).[8] In the following years, STXM instruments at the Advanced Light Source, ALS (Berkeley, USA), went through several iterations. The Advanced Photon Source (Argonne National Laboratory, Argonne, USA)[9] STXM operates in the higher energy region (0.8 keV–4 keV). To date, the majority of STXM experiments in the soft x-ray region (~100–1000 eV) have been at the ALS or NSLS. Several new instruments are either operational or anticipated in the next few years: Diamond Light Source (United Kingdom),[10] Canadian Light Source (Saskatoon, Canada),[11] BESSY II (Berlin, Germany),[12] Swiss Light Source (Paul Scherrer Institut, Villingen, Switzerland),[13] Elettra (Trieste, Italy),[14] European Synchrotron Radiation Source (France),[15] SOLEIL II (France), Pohang Light Source (Korea),[16] Shanghai (China), Stanford Synchrotron Radiation Laboratory (USA), and a third STXM at the ALS (USA).

17.2.1 STXM DESCRIPTION

The STXM spatial resolution is determined by the x-ray spot size at the focal point. For visual light microscopes (VLMs), the diffraction-limited spatial resolution is of the order of 200 nm. As soft x-rays have much shorter wavelengths (1–10 nm), they can be focused to a smaller spot size (Rayleigh criterion). Rather than focusing using refraction, zone plates focus monochromatic x-rays with a circular diffractive grating (called a Fresnel zone plate). As shown in Figure 17.2, zone plates consist of variable width concentric gold rings (zones) on a thin transparent silicon nitride support. Zone plates are manufactured using lithography to create high-aspect ratio zones. The zone-width (width of the gold zone) decreases with increasing radial distance from the center of the zone plate. The width of the outermost zone determines the largest diffraction angle and plays a critical role in obtaining the best spatial resolution. The spatial resolution depends on the coherence of the x-ray radiation incident on the zone plate and the quality of the zone plate. Spot sizes of 25 nm are routinely achieved, although 15 nm spot sizes are reported.[17] Future improvements in the quality of zone plates are expected to further decrease the spot size. Zone plates are designed such that diffracted light from each of the

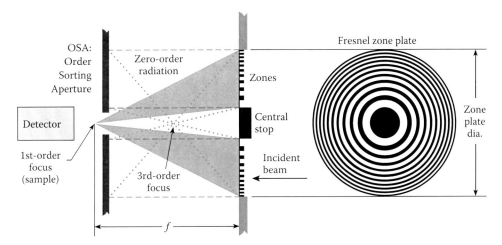

FIGURE 17.2 (Right) Schematic of a Fresnel zone plate showing the central stop and the series of zones having decreasing width with increasing distance from the center. (Left) Schematic of a zone plate (rotated 90°) with the proper positioning of the OSA to prevent third-order diffracted light from impinging on the sample. The sample lies in the focal plane for first-order diffracted light. The zero-order (nondiffracted) light is eliminated by the zone plate central stop.

zones constructively interferes at the focal point. As shown in Figure 17.2, STXM instruments employ zone plates with a central stop to prevent unfocussed incident x-rays from impinging on the sample. Typical zone plates have diameters of ~120–240 μm, with 100s–1000s of zones, central stops with diameters of 50–90 μm, and width of outermost zone ranges from 20 to 50 nm. The zone plate focal length is a function of photon energy and varies from ~1200 μm at 280 eV to ~3000 μm at 700 eV.

Light incident on the zone plate may contain significant levels of higher-order diffracted light from the monochromator that can decrease the spatial resolution and contaminate spectra. To mitigate these effects, an order sorting aperture (OSA) is used. The OSA indicated in Figure 17.2 is a pin hole, 40–60 μm in diameter. When placed at the proper distance between the sample and the zone plate, higher-order diffracted light is eliminated and only first-order diffracted light is transmitted to the sample. The correct alignment of this aperture both in distance between the sample and the zone plate (z) and in the x, y sample plane is important for obtaining quantitative spectra. The importance of the OSA alignment is somewhat dependent on the insertion device used at the beamline. Due to the larger energy spread produced by bending magnets, the possibility of focusing higher harmonics onto the sample is increased and the correct OSA alignment is more critical.

Typically, the sample is held on a set of piezo fine stages operating over a range of ~60 μm. These fine stages are mounted onto a set of coarse stages with a range of motion in 10s of millimeters. Hence, a plate containing multiple samples can be loaded and individual samples selected by positioning the coarse stages. The fine stages, used for the majority of data acquisition, are monitored using laser interferometers to accurately determine their position. Additional information and a more detailed description of microscopes at the ALS and NSLS are provided elsewhere.[18–20]

Images are obtained at a given photon energy by raster scanning the sample through the focal point while measuring the transmitted x-rays. To obtain spatially resolved spectral information, a sequence of images is measured as a function of photon energy. This sequence of images is referred to as a "stack." From this three-dimensional data set, a spectral image is obtained with the spatial resolution determined by the zone plate. As the zone plate, focal length is proportional to the incident x-ray photon energy, during acquisition of a "stack," each time the energy is changed, the distance between the zone plate and the sample (z) changes to keep the sample in focus.

A variety of detection techniques are used in STXM: gas flow proportional counters[21,22]; diode detectors with segmentation[23]; avalanche photodiodes run in pulse-counting mode[20]; fluorescence detectors[15]; charge-coupled-device (CCD) detectors[24]; and differential phase contrast and dark field imaging.[25] To date, most atmospheric aerosol STXM studies have used a detector consisting of a phosphor-coated Lucite pipe coupled to a photomultiplier for single photon counting.[18,26]

17.2.2 BACKGROUND ON NEXAFS SPECTROSCOPY

17.2.2.1 Electronic Transitions Probed

When inner (core) shell electrons absorb a soft x-ray photon, several processes such as excitation to a bound or continuum state can occur. Depending on the type of information desired, for example, surface versus bulk or elemental versus chemical, different experiments measure absorbance, fluorescence, and partial or total electron yields. For low-atomic number (Z) elements, fluorescence yields are small (Stöhr, p. 117).[27] The amount of absorption depends on photon energy, elemental composition, sample thickness, and density. If an electron is completely removed from the atom, ionization occurs. At energies just below the ionization threshold, photon absorption excites inner shell electrons into unoccupied valence orbitals. For a specific element, the sudden increase in absorption is referred to as an absorption edge. Table 17.1 lists elements typically present in environmental samples with absorption edges lying in the soft x-ray region.

Most STXMs operate in the soft x-ray region—an energy range exceptionally well suited for probing natural materials.[28] Several STXMs have extended energy ranges approaching 2000 eV. First row elements possess a single core-level transition in the soft x-ray region: the K-shell or 1s orbital. (Note: Chemists refer to electron orbitals as s, p, or d, while x-ray notation refers to them as K, L, and M, shells

TABLE 17.1
Core Edges Typically Used in the Soft X-ray Region

Binding Energy	Edge
150–500 eV	
B, C, N, Mg, Al, Si	K 1s
S, Cl, K, Ca, Sc, Ti	L_2 $2p_{1/2}$ and L_2 $2p_{3/2}$
Y, Zr, Nb, Mo, Tc, Ru, Rh, Pd, Ag, In, Sn	M_4 $3d_{3/2}$ and M_5 $3d_{5/2}$
500–1000 eV	
O, F	K 1s
V, Cr, Mn, Fe, Co, Ni, Cu	L_2 $2p_{1/2}$ and L_2 $2p_{3/2}$
Sb, Te, I, Xe, Cs, Ba, La, Ce, Pr	M_4 $3d_{3/2}$ and M_5 $3d_{5/2}$
1000–1500 eV	
Na, Mg	K 1s
Zn, Ga, Ge, As, Se	L_2 $2p_{1/2}$ and L_2 $2p_{3/2}$
Nd, Pm, Sm, Eu, Gd, Tb, Dy, Ho, Er, Tm, Yb	M_4 $3d_{3/2}$ and M_5 $3d_{5/2}$
1500–2000 eV	
Al, Si	K 1s
Br, Rb, Sr	L_2 $2p_{1/2}$ and L_2 $2p_{3/2}$
Lu, Hf, Ta, W, Re, Os	M_4 $3d_{3/2}$ and M_5 $3d_{5/2}$

(or edges), respectively.) Environmental samples typically contain carbon, nitrogen, and oxygen, for which the 1s electron binding energies are about 290, 400, and 530 eV, respectively. Hence, in the water window (the region between ~280 and ~525 eV) water is essentially transparent, but carbon and nitrogen absorb. In this window, high contrast images can be obtained when probing systems containing carbon such as aerosols, soil samples, biofilms, or biological systems without interference from water.

The transition metal L-edges are also accessible in the soft x-ray energy range. L-edge excitation probes $3d$ valence orbitals that provide metal oxidation state information.[28] They arise from the dipole-permitted transitions from the $2p$ core level to the $3d$ valence level. For transition metals, probing the L-edge is preferred over the core shell (K-edge), because the outermost (valence shell) electrons are most sensitive to the bonding environment.

Fortuitously, atmospheric aerosols have sizes ideal for STXM measurements. Soot and black carbon particulates are in the fine and accumulation modes. Hence, the most strongly absorbing (in the soft x-ray region) particulates are typically in a size range where they are partially transparent to the incident x-ray photons. In addition, the penetration depth is a few microns at the carbon K-edge (~300 eV) allowing the soft x-rays to penetrate the entire particle volume. Larger particles, mineral dusts, sulfates, and sea salts typically contain less carbon and it is possible to measure carbon spectra on particles with diameters from ~100 nm–10s of microns. In Figure 17.3, the calculated mass absorption in the soft x-ray region is plotted for 100-nm-thick samples of mineral dust, sea salt residue, and secondary aerosols. As seen in Figure 17.3, the soft x-ray region is ideal for probing a variety of common atmospheric aerosol constituents.

17.2.2.2 Conversion to Optical Density

Quantitative analysis is obtained by converting the transmitted intensity signal into absorbance (commonly referred to as optical density, OD, in the soft x-ray energy region), a dimensionless quantity, given by the Beer–Lambert's law:

$$A = \mathrm{OD} = -\ln\left(\frac{I(d)}{I_0}\right) = \rho\mu d = \sigma d, \tag{17.1}$$

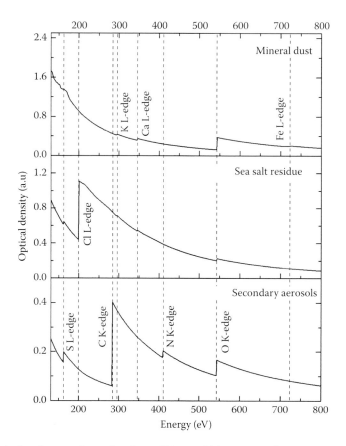

FIGURE 17.3 Calculated mass absorption for a 100-nm-thick sample of common atmospheric aerosols: mineral dust (calculated using the elemental abundance of the earth's crust, $O_{63}Si_{21}Al_6Fe_2Mg_2Ca_2Na_3K$, and the density of SiO_2, 2.6 g/cm^3), sea salt residue (calculated using the formula $Na_{420}Cl_{490}Mg_{50}Ca_{10}K_{10}(SO_4)_{25}(HCO_3)_2$ and the density of NaCl, 2.2 g/cm^3) and secondary aerosols consisting of a mixture of 0.25 mole $(NH_4)_2SO_4$, 0.25 mole $(NH_4)NO_3$, and 0.5 mole palmitic acid, $CH_3(CH_2)_{14}CO_2H$ (where the density was determined from their respective mole fractions).

where I_0 is the incident flux transmitted through a sample free region of the substrate, $I(d)$ is the flux transmitted through the sample, ρ is the sample density, μ is the mass absorption coefficient, d is the sample thickness, and σ is the absorption cross-section. The decrease in flux is due to absorption through the sample column and is a bulk measurement. However, STXM samples are typically thin (80–200 nm) and the surface contribution is significant. One advantage of the conversion into OD is that absorption from a uniform homogeneous substrate (carbon thin film or silicon nitride window) is eliminated. For a sample containing several different chemical components, the OD is given by the sum of the ODs of the individual compounds. A common assumption is that at energies sufficiently far from the ionization threshold, the material may be modeled as a collection of noninteracting atoms. The mass absorption coefficient, $\mu(E)$ (cm^2/g), depends on the incident photon energy, E, and atomic number, Z. For a material composed of a single element, A, the mass absorption coefficient is directly related to the total atomic absorption cross section, σ_A (cm^2/atom) by

$$\mu(E) = \frac{N_A}{Z}\sigma_A,\tag{17.2}$$

where N_A is Avogadro's number. Tabulations of atomic absorption cross-sections are found in Henke et al.[29] and in the X-ray Data Booklet.[30] Transmission can be calculated from these or by using the website maintained by the Center for X-ray Optics (CXRO).[31] Spectra of known elemental composition can be calculated to estimate an appropriate sample thickness. Alternatively, these spectra can be normalized to a specific thickness and used to quantify their contribution in an unknown sample[32] or to derive quantitative composition maps.[32–34]

17.2.2.3 Carbon NEXAFS

Carbon NEXAFS spectra are rich in details and can provide insight into the molecular nature of the carbonaceous material present in aerosols. The book by Stöhr, a critical reference in this field, covers theoretical and experimental considerations related to carbon NEXAFS spectroscopy.[27] Carbon NEXAFS spectra contain features dependent on bonding and unoccupied molecular orbitals. Figure 17.4 schematically illustrates the core-level transitions producing the absorption spectrum. Energy levels of unoccupied valence orbitals provide information on the chemical bonding environment, for example, carbon functional group bonding and oxidation states.

Core-level spectroscopy is a powerful technique, both because it is element specific and because the localized chemical bonding environment has a significant effect on spectral features. Spectra from carbon 1s electrons are in the energy region of ~280–320 eV. At photon energies close to the absorption edge, sharp peaks can be observed in the NEXAFS spectra. These peaks arise from electronic resonance transitions of different functional groups and involve both 1s $\rightarrow \pi^*$ and/or 1s $\rightarrow \sigma^*$ transitions (denoted as π^* and σ^* throughout this chapter). The transition peak width is governed by the Heisenberg uncertainty principle. Therefore, absorption features arising from excitation to short-lived states produce wider peaks in the absorption spectra and those to long-lived states result in sharp and narrow peaks such as the exciton peak observed in highly oriented graphite. Peaks arising from σ^* transitions are usually broader than peaks from π^* transitions and may be superimposed on the photoionization continuum. The peak positions can shift depending on the local coordination environment around different atoms (electron-donating or electron-withdrawing groups). NEXAFS peak positions and intensities are used to determine functional groups and ultimately provide quantitative information on the functional groups present. Since the fine structure in the NEXAFS region above the ionization potential is broad with overlapping σ^* transitions, usually only resonance transitions below the ionization potential are used to quantify chemical composition. Broad peaks can also arise from multiple electronic transitions that are close in energy and overlap with one another.

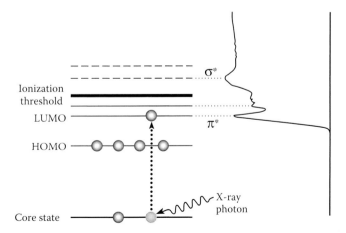

FIGURE 17.4 (Left) Schematic energy-level diagram showing the highest occupied molecular orbital (HOMO), the lowest unoccupied molecular orbital (LUMO), and the ionization potential. Typically, for unsaturated compounds, the LUMO is a π^* molecular orbital, such a transition is indicated. (Right) Example of a resulting NEXAFS spectrum.

Specific peak positions are often noted in the literature; however, caution is necessary when assigning transitions based on specific positions. Urquhart and Ade[35] examined carbonyl π^* transition energies as a function of different bonding environments and observed that transition energies for a particular electronic transition may vary significantly. The transition energy depends on a range of variables that affect the molecular orbital energies. These variables include neighboring electron withdrawing or donating groups, the presence of long-range order, or extensive bond conjugation or resonance. For complex environmental samples, an approach that assigns functional groups based on an energy range[36,37] rather than a specific peak position is preferred. While not exhaustive, Figure 17.5 provides guidance in assigning peaks to specific functional groups. Obtaining NEXAFS spectra at additional edges can provide further confirmation of peak assignments. For example, carbonyl (C=O) groups exhibit a π^* transition at both the carbon and oxygen K-edges (denoted by C $1s \rightarrow \pi^*_{R(C^*=O)R}$ and O $1s \rightarrow \pi^*_{R(C=O^*)R}$, respectively). The presence of these two peaks can be spatially correlated with one another to confirm an assignment.[38] Spectra from standards of known chemical composition can also aid in peak assignments.

17.2.3 POLARIZATION

Although not discussed explicitly in the STXM/NEXAFS aerosol literature, polarization effects may be observed for crystalline and highly structured materials. While a detailed discussion is beyond the scope of this chapter, electronic transitions are polarization dependent and the reader is referred to Stöhr.[27] Polarization dependence arises due to the directionality of chemical bonds and molecular orbitals. For K-shell excitation, the spatial orientation of the molecular orbital relative to the electric vector of the incident radiation determines the angular dependence. In a single bond, the σ^* orbital is localized along the internuclear axis. Double-bonded species, such as C=C or C=O, have both the σ^* as well as a π^* orbital. NEXAFS spectra measured from highly oriented pyrolytic

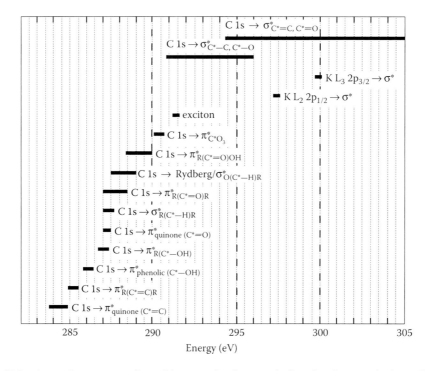

FIGURE 17.5 Approximate range of transition energies for organic functional groups in the carbon K-edge NEXAFS spectra. Potassium L-edge transitions are also indicated.

graphite display strong polarization dependence. Depending on the insertion device employed, different polarizations may be produced; linear, circular, or elliptically polarized.

If sample collection resulted in a preferential orientation of aerosols on the substrate, polarization dependence could be observed. In many atmospheric field campaigns the particle-to-particle spectral variation would make it difficult to ascertain the presence of more subtle polarization effects. Soot, the atmospheric substance most likely to exhibit polarization dependence at the carbon edge, is fortuitously constructed of spherically symmetric spherules. These are the spherules formed by graphitic soot and black carbons consisting of small sheets of graphite, stacked to create symmetric spherules on the order of 15–70 nm in diameter. These spherules then combine in fractal-like structures forming larger particles. Given the spherical symmetry, atmospheric black carbons and soot should not exhibit polarization dependence. However, if during combustion (or other processes) small sheets of oriented graphite are formed, they could exhibit polarization dependence. Evidence for this has been observed when examining changes in carbon bonding upon laser irradiation.[39] Thus so far, little evidence for polarization dependence has been observed in carbonaceous atmospheric aerosols.

17.2.4 AEROSOL TECHNIQUE COMPARISON

Figure 17.6 illustrates the different capabilities of a variety of electron, photon, and ion-imaging techniques used to study atmospheric aerosol particles. By far the most widely used techniques for aerosols are x-ray and electron microscopies; these offer the best combination of chemical information and spatial resolution. Transmission electron microscopy (TEM) offers high spatial resolution and a variety of different chemical detection schemes; however, it is time consuming and not well-suited for analyzing organic species. Scanning electron microscopy (SEM) provides the advantage of high counting statistics but is limited to quantitative elemental analysis of heavy elements ($Z > 11$) with a lower spatial resolution than TEM. Micro-Raman and Fourier transform infrared (FTIR) spectroscopy provide more bonding information, but at the expense of lower spatial resolution.

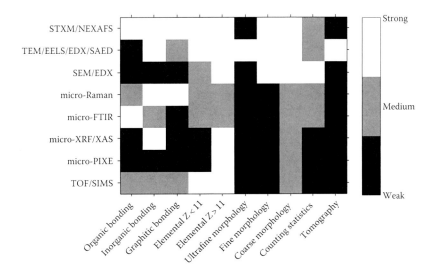

FIGURE 17.6 Analysis of capabilities (x-axis) of imaging techniques (y-axis) used to analyze atmospheric aerosol. The gray scale corresponds to analysis capability which is classified as strong, medium, or weak. Acronyms are as follows: TEM/EELS/EDX/SAED: transmission electron microscopy/electron energy loss spectroscopy/energy dispersive x-ray/selected area electron diffraction, SEM: scanning electron microscopy, FTIR: Fourier transform infrared spectroscopy, XRF/XAS: x-ray fluorescence/x-ray absorption spectroscopy, PIXE: proton-induced x-ray emission and TOF-SIMS: time-of-flight secondary ion mass spectrometry.

Micro x-ray fluorescence (XRF), x-ray absorption (XAS), and proton-induced x-ray emission (PIXE) are typically used for time-resolved bulk analysis of small amounts of particulate matter collected on filters. Secondary ion mass spectrometry coupled with a time-of-flight mass analyzer (TOF-SIMS) can perform elemental analysis while obtaining limited bonding information with a spatial resolution of ~200 nm.

While the spatial resolution of STXM is lower than TEM and SEM, it has several advantages over electron microscopy including (1) chemical specificity, (2) higher energy resolution, (3) lower excitation energy (100–2000 eV versus 50–200 keV), (4) reduced radiation exposure (discussed in more detail in a following section), and (4) no ultrahigh vacuum requirements. Therefore, STXM provides an enhanced chemical sensitivity for obtaining chemical bonding, speciation, and oxidation state information compared to electron microscopy.

Braun et al. examined SLX-25 graphite using STXM/NEXAFS and high-resolution electron energy loss spectroscopy (EELS).[40] As seen in Figure 17.7, several carbon functional groups observed in STXM studies were not apparent in the EELS spectrum (dotted line). In the STXM/NEXAFS spectra shown in Figure 17.7 (solid line), the C 1s → $\pi^*_{R(C^*=C)R}$ peak at 285 eV is sharp and more intense than the 292 eV (C 1s → σ^*) peak, whereas in the EELS spectra, the 285 eV peak is significantly broader and the 292 eV peak is more intense.[40] Katrinak et al.[41] were also unable to observe oxygen functional groups in uncoated soot using EELS, while in contrast, Hopkins et al.[42] observed oxygen in every uncoated soot/soot surrogate examined using STXM/NEXAFS. Hence, the EELS data are inconsistent with the STXM/NEXAFS studies. Braun et al. concluded that EELS was unable to detect carbonyl or carboxyl groups in soot.[40] Hitchcock et al. showed that spectra obtained using TEM-EELS, partially due to the lower energy resolution, exhibit much less structural information than those obtained using STXM/NEXAFS.[43]

17.2.5 POTENTIAL FOR RADIATION DAMAGE

Detailed mechanisms of x-ray damage are uncertain and the topic of current research[44–51]; a journal issue was recently devoted to this topic.[50] However, it is clear that exposure to radiation can change bonding within a sample. Soft x-ray photons are much lower in energy (~300 eV at the carbon K-edge) than the 50–100 keV electrons used for TEM or SEM. Further, it is thought that a significant amount of sample damage results from the dissipation of heat energy from energetic electrons and electron collisions. Generally, x-ray damage from STXM is considered to be at least two orders

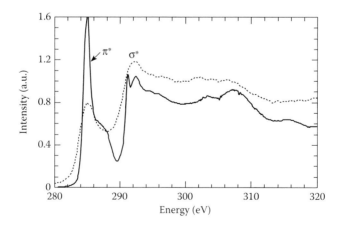

FIGURE 17.7 Spectra of SLX-25 graphite taken using electron energy loss spectroscopy, EELS (dotted line) and STXM/NEXAFS (solid line). The π^* peak at 285 eV is much broader and less intense in the EELS spectra than observed in the NEXAFS spectra. The improved energy resolution of STXM/NEXAFS results in structural features that are not observed in EELS spectra. [Adapted from Braun, A. et al. *Carbon* **43**, 117–124 (2005).]

of magnitude lower than electron microscopy.[52] This potential for less sample damage motivated some of the first STXM experiments on soft matter such as polymers.[53] Several studies have been performed on radiation sensitive systems either by taking images at only a few key energies,[32,54] or with the intention of patterning a sample.[51,55]

Radiation damage is often manifested as a decrease or increase in the intensity of a peak as a function of radiation exposure. For example, Braun et al. examined soot extracts with increasing x-ray exposure and noted an increase in the C 1s → $\pi^*_{C^*O_3}$ (~290.2 eV) peak and a corresponding decrease in the C 1s → $\pi^*_{R(C^*=O)OH}$ 288.5 eV peak.[46] The radiation dose employed, 10^6 Gy (Gy or Gray is the unit of the absorbed radiation dose of ionizing radiation), lies within a normal range for STXM experiments. The authors speculated that atmospheric oxygen present in their STXM experiments (not within a sealed chamber) could have contributed to the spectral changes observed.[46] Attempts by our group to induce bonding changes in the carbon K-edge NEXAFS spectra from mixed composition aerosols containing soot (Mexico City samples), were unsuccessful. This could be due to the fact that the ALS STXMs are contained within an evacuated chamber and backfilled with helium. This is in contrast to the open STXM used by Braun et al. which was purged by helium flow.[46] Nevertheless, systematic examination for changes in bonding with radiation dose is a critical component in obtaining quantitative data. As each electron absorbs x-ray photons, the absorption cross-section increases with increasing atomic number. Hence, absorption increases across a row in the periodic table as well as down.[50] This is consistent with experimental observations that sulfur-containing aerosols were more sensitive to radiation exposure than carbonaceous aerosols.[56] If a sample is radiation sensitive, images can be taken at a few chemically selective energies,[32,54] short dwell times used, or spectra measured using a defocused mode.

17.2.6 PRACTICAL SAMPLE CONSIDERATIONS

17.2.6.1 Sample Preparation and Collection

While the broad categories of STXM aerosol experiments are deferred to Section 17.3, here a few brief remarks are made about samples and substrates. Although STXM measurements require only a single particle, most studies examine a large number of particles. Typically, samples are collected throughout a field campaign. Depending on the research group's preference, samples are sealed and either stored at ambient temperatures or frozen. Atmospheric aerosols are often collected via impaction onto substrates. For maximum time resolution, impaction can be done using a time-resolved aerosol collector (TRAC)[57] or multisample impactor.[58] Time-resolved samples are collected for 2–20 min depending on the location of collection (ground-based site or during flight) and local aerosol concentration. Regardless of lower aerosol concentrations, shorter collection times are often used during flights due to the rapid change in location and local conditions. For studies on cloud drop residuals, TRAC samplers can be located downstream of a counterflow virtual impactor (CVI), followed by a heater to selectively collect particles that formed droplets. Occasionally, STXM substrates are taped to filters in a Micro Orifice Uniform Deposit Impactor (MOUDI) cascade impactor (Micro Orifice Uniform Deposit Impactor, Applied Physics Co., Niwot, CO, USA).

Obtaining samples with good particle coverage (e.g., six or more particles in an 8 μm square region) with distinct individual particles is critical to optimize data acquisition. In the initial stages of a campaign, a few samples are collected and examined with a VLM or an SEM coupled with energy-dispersive x-ray spectroscopy (SEM/EDX) to ensure an appropriate level of sample loading on the substrates and alignment of the impactor nozzle. This prescreening aids in developing protocols for collection under different aerosol loading conditions. Ideal impaction includes a region of the sample, where particles are distinct and separated (nonoverlapping), yet closely spaced so that time spent on acquiring data on the regions that are sparsely loaded is minimized.

The VLM is also used prior to STXM data acquisition. Samples are photographed, and the images calibrated to the STXM coordinates and read into the STXM data acquisition program to directly guide data acquisition. As time allocated on the STXM microscopes is normally limited, this step saves significant time in locating sample regions for potential data acquisition.

A variety of artifacts may influence the measured chemical composition and morphology of collected aerosols. For example, liquid droplets containing salts could dry upon collection and the relative solubility of each salt present will affect the final particle morphology. When using the TRAC or MOUDI impactor, impacted particles are exposed to reduced pressure conditions for up to 6 h during collection with the MOUDI, or 48 h with the TRAC sampler. Hence, a basic assumption is that all semivolatile organics evaporate during collection. In addition, the STXM instrument is routinely either flushed with helium or partially evacuated to ~80 mTorr and backfilled with helium, both of which could result in additional losses of semivolatile components. Therefore, STXM measurements provide a lower limit for organics.

For comparison with atmospheric samples, standards are prepared by a variety of impaction techniques or a gentle contact of the silicon nitride membrane to a fine powder. Liquid standard samples are prepared by micropipetting a dilute solution onto the corner of the membrane and allowing the liquid to dry on the membrane. If wet samples are desired a sandwich surrounding the liquid can be formed by attaching a second silicon nitride window. Although not generally used for atmospheric aerosols, samples prepared by focused ion beam milling (FIB)[59] or microtomed samples embedded in epoxy or sulfur[60] work for STXM. Generally, samples suitable for SEM or TEM also work well for STXM experiments. The most fundamental requirement for STXM measurements is that the absorption lies within the linear regime of Beer–Lambert's law (approximately 30–80% absorption or OD 0.4–1.5 on a natural logarithmic scale).

17.2.6.2 Substrates

Several types of substrates are frequently used to collect atmospheric aerosols (or for standards) for STXM experiments. Silicon nitride windows (Si_3N_4) have a thin membrane (50–100 nm thick) placed on a silicon wafer. The wafer is then etched away to reveal an area containing only the thin membrane film. These have advantages of being relatively optically transparent in the soft x-ray region as well as being a uniform film. One major disadvantage is that they are quite fragile and are easily broken. Hence, the failure rate is high, due to particle impaction when sampling in a region with large, dense particles with a relatively high inertia. Carbon-coated TEM grids have a higher survival rate under these conditions. However, these films are composed of carbon and some oxygen, and absorb in the energy range of interest. In our experience, these films are relatively uniform and with careful normalization can be used successfully in studies on carbonaceous aerosols. Silicon monoxide-coated TEM grids are occasionally used when trying to minimize nitrogen absorption from Si_3N_4 windows. Holey carbon films are another type of coated TEM grid; however, they are probably the least desirable. Often aerosols appear to spread along the holey carbon web and the nonhomogenous thickness of the carbon web may result in a larger uncertainty in the spectral analysis.

17.2.7 Data Analysis

Several software programs are commonly used to analyze STXM data: AXIS 2000,[61] singular value decomposition (SVD) analysis,[62] a cluster analysis program,[63–65] and MATLAB®-based scripts. As aerosol studies examine a large number of spectra from individual particles, or statistically examine specific spectral components within particles, Moffet et al.[66] developed a series of openly available MATLAB scripts.[67] In practice, data analysis often combines several of these methods and software. Regardless of which analysis program is used, several basic steps are required: image alignment, selection of a background region for I_0, and conversion of the transmitted flux data into OD. Beyond these basic steps, a range of techniques are used to extract, map, and quantify spectra from individual pixels as well as individual particles. A detailed description of the analysis is beyond the scope of this chapter and is provided both in individual publications, as well as in references to the software programs.

17.3 STXM/NEXAFS AEROSOL STUDIES

The primary advantages of STXM/NEXAFS over electron microscopies are the ability to distinguish between carbon functional groups, the higher spectral energy resolution of STXM/NEXAFS, the reduced radiation exposure and subsequent sample damage, and (depending on the insertion device) the capacity to change the polarization of the incident photons. Disadvantages include limited access to the STXM, fewer samples can be studied than automated microscopies, and obtaining the full elemental composition is impractical. Soil scientists, marine biologists, astrophysicists, and atmospheric scientists share common interests in naturally occurring carbon, minerals, and their chemical processing. Hence, results from other environmental fields on humic and fulvic acids,[34,68–76] soils and clays,[45,77–85] marine and aquatic samples,[86] natural organic matter,[45,73,87] black carbons,[81,88,89] char,[36] minerals,[77,90] biofilms,[34,91–102] interstellar dust and particles,[60,103–108] and meteorites[105–107,109,110] may be of interest to atmospheric scientists.

Aerosols may be studied on a single-particle basis or results combined with data from larger numbers of particles to provide statistical evidence to support broader conclusions. Typical atmospheric field campaigns employ a suite of measurements examining particle concentrations, size distributions, light absorption and scattering, composition, and meteorology. The results from these measurements can guide sample selection and help focus the scientific questions for STXM investigation. Research on atmospheric aerosols using STXM is currently limited and falls into a few categories; characterizing black carbons and light-absorbing components within aerosols, characterizing organic carbons (OCs), and chemical speciation of a variety of elements other than carbon (S, Fe, Zn).

This section begins with research on black carbons/soot, light-absorbing carbon, and biomass burn particulates. These studies provide the foundation for evaluating changes in carbonaceous aerosol mixing states and atmospheric processing of black carbons. In this section, details on analysis are provided that further illustrate the development (and limitations) of current analysis techniques. The progression has been to probe aerosols in a statistically representative manner by connecting STXM/NEXAFS results with collocated measurements and complementary spectromicroscopy measurements. It also reflects the growth in STXM research from more well-defined laboratory samples (polymers and magnetic samples) to complex and often poorly characterized environmental samples. The next section emphasizes research focused on the organic composition of aerosols and their processing. Both of these sections begin with studies on a handful of particles and grow to encompassing much larger numbers of samples and incorporate STXM measurements more fully within a larger picture. This organization allows the reader to develop an understanding of the method as well as the development of the analysis from a single sample to a statistical analysis. Next, we explore the more limited studies on chemical speciation of sulfur-containing aerosols, and the identification and quantification of oxidation states for elements such as Fe, S, and Zn. Finally, we survey recent studies on cloud droplet residuals and interstitial aerosols. The chapter closes with a brief discussion of recent and future developments of interest to the atmospheric community.

17.3.1 SOOT/BLACK CARBONS/BROWN CARBONS

Although its definition differs across fields, black carbon plays a critical role in nearly every environmental field. STXM work on black carbons explore the chemical inhomogeneities of carbonaceous samples including coals,[111–115] soils,[80,83–85] absorption of chemical species,[88] carbon bonding changes from charring plant biomass,[36] interplanetary dust particles,[105–108,110,116] stardust comet samples,[60,103,104] and marine sediments,[86,89] as well as aerosol studies presented here. Many of the early STXM studies on soot, diesel exhaust, and black carbons focused on nonatmospheric samples. Early work on coal[111–115] showed the potential for using STXM/NEXAFS to examine diesel exhaust particulate. Braun et al. examined diesel exhaust particulates for the purpose of source apportionment.[117] Building on the concept of source apportionment, Vernooij et al. studied wood combustion and diesel combustion samples.[37] However, they concluded that unequivocal source assignment based on NEXAFS spectra was unlikely due to the influence of atmospheric processing. With the eventual goal of monitoring changes in mixing states of aerosols containing black carbons (and light-absorbing carbons), their atmospheric evolution, and *in situ* reactivity, Hopkins et al.[42] sought to identify the ranges in bonding and composition of atmospheric black carbons and the variation in laboratory black carbon surrogates. They observed a significant range in the carbon bonding of these surrogates, but determined that it was much less than that in atmospheric light-absorbing aerosols. This large variation in carbon bonding for light-absorbing atmospheric aerosols was also seen in a study focused on aged tar balls.[38] Laboratory-generated biomass burn particulates were examined for carbon bonding, which was then correlated with measured optical properties.[118] These studies provided the foundation to distinguish soot/black carbons from OCs and to monitor changes in mixing state with particle growth and age.[119] Current, unpublished work on black carbons includes the influence of charring cellulose (with and without added salts) on carbon bonding as well as field studies on atmospheric processing.

The early STXM study of Braun et al. on diesel soot initially pursued source identification,[117] and it clearly indicated the potential of STXM for distinguishing graphitic carbon from hydrocarbons, organic or inorganic carbon, and indirectly motivated studies monitoring the evolution of the mixing states of carbonaceous aerosols. Diesel exhaust was collected in filters, dispersed in acetone, and ultrasonicated and deposited onto silicon nitride membranes.[117] STXM images taken below the carbon absorption edge (<283 eV) showed much lower absorption than images acquired at the peak of the C 1s → $\pi^*_{R(C^*=C)R}$ peak (~285 eV). In higher-energy images (>288 eV), more diffuse particle edges were observed. Although these particulate samples had been mixed in acetone, they clearly illustrated the ability of STXM to distinguish between graphitic and nongraphitic regions of carbonaceous particles with a spatial resolution of ~50 nm.[117] Comparing flame soot and field collected biomass particles, Gilles et al.[120] reported enhanced oxygenated peaks and decreasing graphitic content for some biomass samples. Subsequent STXM work by Braun examined NEXAFS spectra from a range of deposited combustion products: creosote (wood smoke deposits) from chimneys and wood burners, particulates from engines, and an NIST standard reference material (SRM 1648 Urban PM).[121] Spectra from the wood smoke deposits and the NIST Urban particulate matter exhibited minor C 1s → $\pi^*_{R(C^*=C)R}$ peaks and were dominated by C 1s → $\pi^*_{R(C^*-OH)}$ and C 1s → $\pi^*_{R(C^*=O)OH}$ peaks. These initial studies by Gilles and Braun indicated that biomass burn particulates could have significantly different chemical bonding than fossil fuel combustion.

17.3.1.1 Black Carbon/Soot Single-Energy Images and Spectra

With an ultimate objective of identifying black carbons/soot/light-absorbing carbons in atmospheric aerosols using STXM/NEXAFS, Hopkins et al.[42] examined a variety of laboratory surrogates for black carbons. For the black carbon surrogates, specifically flame-generated soot, particles were not visible in images taken at the carbon pre-edge (280 eV). This is in contrast to the pre-edge images of diesel soot, where absorbance (possibly due to the sample preparation or from noncarbonaceous components in the original sample) was observed.[117] The lack of absorbance at energies below the carbon pre-edge (<280 eV) and strong absorbance around 285.4 eV is one of the identifying features for black carbon. Figure 17.8 displays single-energy images of particles collected during the MILAGRO field campaign.[119] These images were recorded at the (a) 278 eV (carbon pre-edge), (b) 285.4 eV C 1s → $\pi^*_{R(C^*=C)R}$, (c) 288.5 eV carboxyl C 1s → $\pi^*_{R(C^*=O)OH}$, and (d) 320 eV at the carbon postedge region. A subtraction map of total carbon shown in Figure 17.8e is obtained by subtracting the pre-edge image (a) from the postedge image (d). The component map shown in (f), discussed in a later section, identifies regions containing elemental carbon (EC) (soot), OC, and noncarbonaceous inorganic (IN) material. The particle indicted with an arrow in (a) shows a stronger absorbance (relative to other particles) at the pre-edge indicating the presence of an element other than carbon, such as metals, mineral dust, salts, and/or sulfur. In (c), this particle displays its strongest absorbance in the 288.5 eV image, indicating a large carboxyl peak. In the total carbon map (e), the

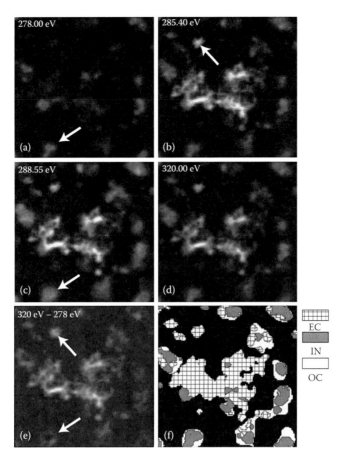

FIGURE 17.8 Single-energy STXM images (6 μm × 6 μm) shown on the same OD scale (0–1.8): (a) 278 eV, (b) 285.4 eV, (c) 288.5 eV, and (d) 320 eV. The difference image, (e), obtained by subtracting the pre-edge (278 eV) from the postedge (320 eV) where each is first converted into OD (shown on OD scale −0.2–1.4). The particle indicated in (a) absorbs at 278 eV (the carbon pre-edge), indicating the presence of noncarbonaceous species (such as inorganic salts); the area surrounding this region shows stronger absorption at 288.5 (c) than at 285.4 (b), indicating an organic coating. The particle indicated in (b) has little or no absorption at the carbon pre-edge (a) and strong absorption at 320 eV (d), 285.4 eV, and in the difference map (e), indicating a carbonaceous (potentially soot) particle. Panel (f) is a map showing the components, EC or soot, inorganic (IN) carbon and OC.

same particle indicated in (a) and (c) appears dark in the center with a brighter surrounding region. Hence, this particle appears to contain a crystalline inorganic inclusion surrounded by OC.

The particle indicated with an arrow in Figure 17.8b does not appear in Figure 17.8a. The image recorded at 320 eV shows a stronger absorbance for this particle compared to the image at 280 eV as a result of high carbon content in the particle. This pattern is consistent with carbonaceous particles containing considerable C=C bonds, and may be an indicator of soot. The difference map in (e) represents the total amount of carbon in the sample. This proves to be a useful tool for a preliminary identification of a particle as a soot/black carbon particle or as a soot/black carbon inclusion. Black carbon has a density of ~1.8 g cm^{-3}, whereas organic compound densities are closer to ~1 g cm^{-3}. Hence, at the carbon K-edge, a 100 nm-thick black carbon particle absorbs nearly a factor of two more strongly than OC simply due to the difference in carbon density. Thus, even without examining the NEXAFS spectra, the first indications that a particle contains soot/black carbon are single-energy images at 280, 285.4, and 320 eV. Particles should have little or no absorbance in the 280 eV

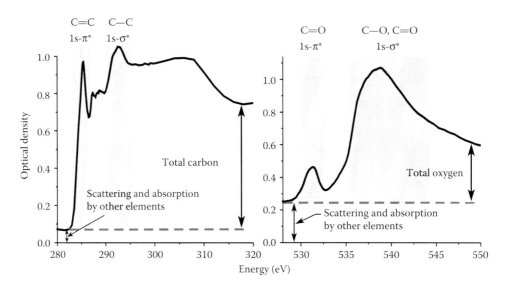

FIGURE 17.9 NEXAFS spectra for flame-generated methane soot taken at the carbon and oxygen K-edges. Spectra are raw data without any pre-edge subtractions or scaling. The small pre-edge absorbance for carbon (OD < 0.1) and strong total carbon absorption (absorption at 320 eV—absorption at 280 eV) is often observed for aerosols consisting primarily of carbon and oxygen. Note that the higher pre-edge at the oxygen edge (~0.25 OD) arises from absorption and scattering of atoms other than oxygen (including carbon).

image and the 285.4 and 320 eV images should have strong absorbance (of similar values). For soot/ black carbons, absorbance should increase more strongly as a function of particle size than organics.

17.3.1.2 C:O Atomic Ratios

Figure 17.9 displays representative carbon and oxygen K-edge spectra for diffuse flame-generated soot using methane for the fuel. The most prominent peaks in the carbon spectrum, C $1s \rightarrow \pi^*_{R(C^*=C)R}$ and $1s \rightarrow \sigma^*_{C^*-C}$ are indicated. The pre-edge backgrounds (280 eV for carbon and 525 eV for oxygen) arise from the photoionization of valence electrons and the weak, but finite, absorption and scattering of other elements present in the sample. The difference between the postedge and pre-edge absorbance represents the total amount of a particular element in a sample. For a two-component carbonaceous system, calculating the C:O ratio is relatively straightforward.[38] These calculations neglect interactions between atoms and assume that at energies sufficiently far away from valence transitions, tabulated atomic cross sections are valid. Absorption edges for other elements are far away and are neglected for these calculations as they would contribute only minor amounts to the observed difference between post-edge and pre-edge absorbance. Therefore, the total number of carbon, x_c, and oxygen, x_o, atoms in the sample is calculated using[38,42]

$$\frac{x_o}{x_c} = \frac{OD_O \mu_C Z_C}{OD_C \mu_O Z_O}, \tag{17.3}$$

where μ_O, Z_O, and μ_C, Z_C are the mass absorption coefficients and atomic masses for oxygen and carbon atoms, respectively. The difference in OD, 320 eV–280 eV, is used to calculate the total carbon. Similarly, the total oxygen can be calculated using pre- and postedge energies of 525 and 550 eV, respectively. The values of OD_O and OD_C are obtained from the difference in the postedge and pre-edge absorbance in the oxygen and carbon NEXAFS spectra measured over the same particle region. For the flame soot, *n*-hexane, ethylene, and methane, C:O ratios were ~85:15, which compares well with the elemental compositions of Akhter et al.[122]

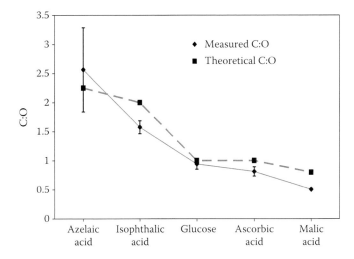

FIGURE 17.10 C:O atomic ratios determined (see text) from NEXAFS spectra of compounds containing carbon and oxygen (neglecting hydrogen). Compounds measured include azelaic acid ($HO_2C(CH_2)_7CO_2H$), isophthalic acid ($C_8H_6O_4$), glucose ($C_6H_{12}O_6$), ascorbic acid ($C_6H_8O_6$), and malic acid ($C_4H_6O_5$). C:O atomic ratios determined from the molecular formula are indicated as theoretical values. Error bars represent the standard deviation taken from four independent measurements.

To further evaluate the approximations used for calculating C:O atomic ratios, aerosols of azelaic acid, isophthalic acid, glucose, ascorbic acid, and malic acid were generated and impacted onto silicon nitride substrates. The NEXAFS spectra were measured at the carbon and oxygen K-edges and the C:O atomic ratios were calculated from the spectra in a blind study. The resulting C:O atomic ratios, as well as that calculated from the chemical formulas, are shown in Figure 17.10. The difference between the measured C:O and calculated atomic ratios was largest for malic acid (−36%) and smallest for glucose (6%). This method is useful for determining the relative amount of oxygen in single particles and in different types of particles within a sample. This method differs from mass spectrometric methods that can determine C:O for organic, inorganic, or both species, but not for soot/black carbons.[123] STXM/NEXAFS determines C:O for the total particle (inorganic + organic + soot/black carbon). For samples containing significant amounts of other elements, determining the C:O ratio is further complicated as their often nonlinear pre-edge and postedge absorbance in the energy region where carbon and oxygen absorb, must be considered.

17.3.1.3 Identifying Black Carbon/Soot via NEXAFS Spectra

The carbon NEXAFS spectra of surrogate soot shown in Figure 17.11 are scaled to a thickness of 100 nm and density of 1.8 g cm^{-3} using carbon mass absorption coefficients.[29] This normalizes the spectra to the same total carbon and allows a more meaningful comparison. (The maximum OD measured for these samples was 1.2; clearly in the linear regime of Beer Lambert's law.) A magic angle spectrum of HOPG is included for comparison. All these spectra display a strong peak at ~285.3 eV, arising from the C 1s $\rightarrow \pi^*_{R(C^*=C)R}$ transition, and a broader peak at 292 eV corresponding to the C 1s $\rightarrow \sigma^*_{C^*-C}$ transition. A valley is observed between these two peaks. This valley varies substantially due to overlapping peaks from multiple transitions and differences in bonding. As seen in Figure 17.5, typical transitions in this energy region include aliphatic C—H groups, ketones or carboxylic acid carbonyl peaks, ethers, and alcohols. Aside from hydrogen, a primary component of black carbon is oxygen. The flame-generated methane soot oxygen K-edge spectra (Figure 17.9) have a pre-edge OD at 525 eV, about a factor of two larger than that observed in the carbon pre-edge. This enhanced absorbance arises from the extended tail of the carbon absorption spectrum as well as atomic scattering. The oxygen spectra contains a distinct peak at ~531 eV, due

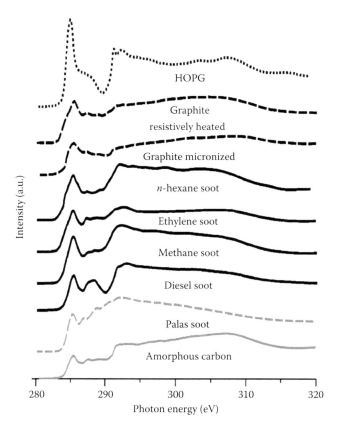

FIGURE 17.11   Carbon K-edge NEXAFS spectra from a variety of laboratory-generated black carbons (soot) and purchased carbonaceous materials. Indications of black carbon are the intense π* peak with a peak intensity similar to the postedge absorbance, a high absorption for the particle size (due to the increased density of carbon), and a valley in the energy region ~286–290 eV. Except for amorphous carbon, all these spectra would indicate the presence of black carbon in atmospheric samples.

to the O 1s → $\pi^*_{C=O^*}$ carbonyl transition. The presence of this peak in the oxygen spectra confirms that the carbonyl group should be present in the carbon spectrum, even if too weak to be resolved. Measuring spectra at multiple edges provides confirmation for spectral peak assignments. Although there are differences in the absorption spectra shown in Figure 17.11, all of these (except perhaps the amorphous carbon) would be suspected of a black carbon/soot origin rather than organic or inorganic species in atmospheric aerosols. The dominant peak, C 1s → $\pi^*_{R(C^*=C)R}$, with an intensity approximately equal to or greater than the absorbance at 320 eV, and the disappearance of the particle in the pre-edges images with a strong absorbance at the postedge, all indicate the possibility of a black carbon or soot.

17.3.1.4   Spectral Deconvolution

More quantitative information is obtained by spectral fitting of the NEXAFS spectra. Fitting routines employ an arctangent function for the ionization step at ~290.3 eV combined with the minimum number of Gaussian, Lorentzian, or Voigt,[124] functions required to fit the observed spectra. Peak energies are set at the peak maxima of distinct spectral features and additional nonresolved peaks may be required to fit the measured spectrum. Spectral fitting programs typically minimize the χ^2 value by adjusting peak positions and peak widths are either fixed, or allowed to vary. For environmental samples, these fits provide a semiquantitative method to compare samples of complex compositions. To ensure that changes in peak intensity are not a function of sample thickness, peak areas are normalized to the integral of the spectrum over the energy range for the particular edge. Studies may employ the approximation that the area under the peak for a particular functional group is representative of the number of bonds. For example, the percentage that a particular functional group contributes to the total carbon in the sample is estimated from the ratio of the integral of an individual peak (π^* or σ^*) to the integral of the spectrum over the energy range 280–320 eV (some researchers integrate over smaller energy ranges). Additional uncertainty arises due to ambiguous peak assignments in the carbon NEXAFS spectra. As indicated in Figure 17.4, multiple transitions may occur in a similar energy region resulting in poorly resolved spectral features. Comparison of transitions occurring in both the C and O (or C and N) NEXAFS spectra can confirm the presence of specific functional groups. While not a rigorous treatment, given the chemical complexity of aerosol and of environmental samples in general, this may be the most reasonable method to compare relative compositions and provide additional insight into the nature of these samples.

17.3.1.5 Quantifying the Graphitic Nature: sp^2 Hybridization

The relative amount of graphitic sp^2 carbon can be estimated from the ratio of the area under the C $1s \rightarrow \pi^*_{R(C^*=C)R^*}$ peak ($A_{C=C}$) to the overall area of the spectrum, for example, in the region between 280 and 310 eV ($A_{280-310}$) according to

$$f_{sp^2} = \frac{A_{C=C}}{A_{280-310}}. \tag{17.4}$$

Similar to Lenardi et al.[125] this can be referenced to the corresponding ratio for the HOPG magic angle spectrum, assuming a 100% abundance of sp^2 carbon using

$$f_{sp^2}^{soot} = \frac{A_{C=C}}{A_{280-310}} \frac{A_{280-310}^{HOPG}}{A_{C=C}^{HOPG}}. \tag{17.5}$$

The area under the spectrum $A_{280-310}$ is obtained by numerical integration of the experimental data, and the area under the π^* peak $A_{C=C}$ can be calculated by fitting this transition to a Gaussian function and integrating the area under the peak.[38,39,42,118] The black carbons and soot in Figure 17.11 are ordered by their graphitic content; highly ordered pyrolytic graphite (top) has the highest amount (100%) and amorphous carbon (bottom) has the least (41%). As noted above, if these spectra were observed in atmospheric samples, with perhaps the exception of amorphous carbon, they would all be identified as black carbon/soot.

This determination of the sp^2 hybridization is considered relative and semiquantitative. Depending on whether a single peak is fit to the area under the curve for the C $1s \rightarrow \pi^*_{R(C^*=C)R}$ peak, or if the entire spectrum is deconvoluted, the sp^2 value may differ by as much as 15%. Some additional uncertainties could arise due to possible polarization effects. Other methods for determining variations in carbon bonding include examining intensity variations[41] in the π^* peak versus the σ^* in EELS spectra or by examining the sp^2/sp^3 ratios by fitting the region below 287 eV as π^* and those between 294 and 301 eV as σ^*.[126] Each of these methods contains its own set of problems; none is perfect. A study combining Raman spectroscopy, another method used to estimate the graphitic nature of carbon, with STXM/NEXAFS could provide an insight into systematic uncertainties in determining carbon bonding.

17.3.1.6 Biomass Burn Aerosols

Aerosols containing biomass burn particulate matter show a wider range in their carbon K-edge spectra than that observed for the black carbon/soot surrogates.[42] Anthropogenic combustion sources often exhibit a higher percentage of sp^2-bonded carbon compared to biomass burn particles.[37,42] They also often exhibit a shoulder at ~284.2 eV attributed to a C=C bond in the ring of quinone (C $1s \rightarrow \pi^*_{quinone(C^*=C)}$). Occasionally, the sharp exciton peak is observed in the carbon spectrum.

Several studies report biomass burn influenced samples with a higher proportion of oxygen-containing functional groups than those produced via anthropogenic combustion or even similarities between biomass burn particulates with humic-like substances or fulvic acids. Keiluweit et al. also recently observed significant changes in carbon bonding upon charring plant biomass.[36]

Hopkins et al. examined biomass burn particulate matter combining SEM/EDX and STXM/NEXAFS measurements produced from about a dozen fuels.[118] The SEM images, elemental composition, and carbon K-edge NEXAFS spectra showed surprising variability. These range from fuels from semiarid regions that contain salts and produce spectra, sp^2 hybridization, and C:O ratios nearly indistinguishable (aside from the K doublet) from methane flame soot. Other samples were dominated by the presence of mixed carbonaceous and inorganic salts or produced a liquid/oily OC with fractal inclusions. Hopkins et al. showed that the sp^2 hybridization was correlated with the measured optical absorbance at visible wavelengths. As Bond[127] has shown, these types of correlations are clearly expected.

Vernooij et al. examined particulates from a village where ambient particulate matter is dominated by wood stove emissions and a sampling site dominated by traffic using TEM and STXM/NEXAFS.[37] Diesel soot contained a C 1s → $\pi^*_{quinone(C^*=C)}$ peak (~284 eV), the C 1s → $\pi^*_{R(C^*=C)R}$ peak (~285 eV), and a shoulder at ~291 eV resulting from the corresponding σ^* transitions. The wood smoke particulates had a signature phenolic peak at ~287 eV with a smaller contribution from the C 1s → $\pi^*_{R(C^*=C)R}$ peak; however, only ~60% of the particulates were identified as wood smoke soot. The range and variation in particulates produced from biomass burning, the dependency on flaming versus smoldering conditions, and the propensity of specific fuels to flame or smolder are all poorly understood and need additional studies.

17.3.1.7 Tar Balls

Tar balls are amorphous carbonaceous spheres, with a structure distinctly different from soot, and have been observed in numerous field campaigns.[128–130] Hand et al. performed extensive single-particle analysis on samples collected at the base of Yosemite National Park during transportation of an air mass impacted by forest fires in Oregon.[131] The highest concentrations of light-absorbing carbon were correlated with periods when SEM/EDX analysis indicated that tar balls dominated the particle samples. Tivanski et al. employed the results from the detailed particle analysis of Hand et al. to guide the selection of samples from specific collection times for STXM experiments. The goal of STXM/NEXAFS in these experiments was to examine the carbon bonding and determine if there were any similarities with soot/black carbons that would account for their absorption of solar radiation in the atmosphere.

In Figure 17.12, three different types of particles observed in this study are shown: the particles within circles are tar balls, the solid-line square surrounds an organic particle, and a dotted-line square indicates an agglomerate. Although not shown in this image, occasionally, particles were observed with a fractal structure of soot (and corresponding NEXAFS spectra). In this image, the organic particle, although larger in diameter, is lighter in color (less absorbing) than the smaller tar balls. To determine if the particles identified as tar balls were spherical, Tivanski et al. combined the two-dimensional STXM/NEXAFS measurements with a model that allowed the determination of a three-dimensional particle shape.[38] The averaged absorbance signal <OD(d)> through a spherical particle of uniform density ρ and diameter d was determined from the following equation:

$$\left\langle \mathrm{OD}(d) \right\rangle = \frac{2}{d} \int_0^{d/2} 2\mu\rho \sqrt{\left(\frac{d}{2}\right)^2 - x^2} \, dx = \frac{\rho\mu d}{2\sin(1)} = A\rho\mu d. \tag{17.6}$$

where x was the radius variable and $A = 1/2 \sin(1) \approx 0.59$ for a perfect sphere. The maximum OD, $\mathrm{OD}_{max}(d) = \rho\mu d$, corresponds to the absorption through the center of the sphere. Thus, for a spherical particle, a plot of the particle's average absorbance versus its maximum absorbance yields a line with

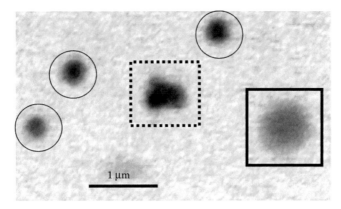

FIGURE 17.12 Single-energy (320 eV) STXM image of representative particles from an aged biomass burn-ing event. Circles indicate tar balls, the solid square is an organic particle, and the dotted square indicates an agglomerate. The tar ball appears darker than the organic particle, indicating increased x-ray absorption.

slope $A = 0.59$ and an intercept of zero. Figure 17.13 displays the measured averaged absorbance versus the maximum absorbance for the C edge at 320 eV. The mean absorbance value for ~7 particles in ~50-nm-wide bins (based on their diameters) is represented by each cross. A fit to the data yielded a slope of 0.59 ± 0.03. Although not shown here, the analogous plot at the O edge at 550 eV yielded a slope of 0.6 ± 0.05. Both of these slopes are in excellent agreement with the predicted slope of 0.59, implying that tar balls are essentially perfect spheres of uniform densities, and that the majority of C and O atoms are homogeneously distributed throughout the tar ball. (A discussion below considers an inhomogeneous surface coating with a thickness similar to the spatial resolution of the STXM/ NEXAFS experiments.) If the particles had been shaped like a flat disc (pancake), the maximum OD would be equal to the averaged OD. Hence, although STXM/NEXAFS is strictly a two-dimensional imaging technique some information on the third dimension is obtained.

The fact that tar balls are perfect spheres of uniform densities with a homogeneous distribution of C and O atoms uniquely allows a quantitative determination of their elemental composition and particle density. Tivanski et al.[38] developed an analytical model that predicts a linear relationship

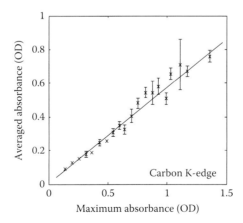

FIGURE 17.13 Averaged absorbance versus the maximum absorbance from the C K-edge spectra at the postedge (320 eV) for tar balls. Each data point represents the mean absorbance value for ~7 particles in ~50-nm-wide bins (based on their diameters). A fit to the data yields a slope of 0.59 ± 0.03. Although not shown here, the analogous plot at the O edge at 550 eV yields a slope of 0.6 ± 0.05. These slopes are in excel-lent agreement with a slope of 0.59 calculated for perfect spheres.

between the averaged total carbon (oxygen) absorbance and tar ball diameter with the slope that depends on atomic carbon to oxygen ratio and the particle density. Using the model and measured size-dependent total carbon and oxygen absorbencies, the atomic ratio of C to O was determined to be 1.2 (55% C, 45% O), with a total density of 0.75 g/cm^3.[38]

One of the primary interests in tar balls is that they are correlated with the absorption of long-wavelength solar radiation. Typically, soot and graphitic carbons are the most strongly absorbing species at long wavelengths in the visible spectrum. NEXAFS spectra of ~150 tar balls were measured and compared to spectra of black carbons and humic-like substances. Figure 17.14 displays a representative tar ball spectrum as well as spectra for soot (measured in the same samples as the tar balls), humic and fulvic acids. Tar ball spectra were unique in that unlike most atmospheric samples, there was almost no particle-to-particle variation in the spectra. As previously reported,[130,131] tar balls were extremely insensitive to radiation exposure. Spectra measured on samples that had been examined via SEM analysis were identical to those measured using very low-radiation STXM scans. A single particle was scanned repeatedly, for a total of ~5 times at normal radiation exposure and the spectra did not vary. Tar ball spectra measured on samples from African biomass burn were identical to those reported by Tivanski et al.[38] Vernooji et al. report tar ball spectra with greatly reduced carboxyl carbonyl intensity.[37] While this discrepancy could be due to the different types of fuel (dried wood versus live biomass), it is unclear if they rigorously determined that the particles they observed were spherical.

The tar ball C edge spectrum contains four sharp resonance transitions at 285.1 (C 1s $\rightarrow \pi^*_{R(C^*=C)R}$, 286.7 (possible transitions: C 1s $\rightarrow \pi^*_{R(C^*=O)R}$, C 1s $\rightarrow \pi^*_{quinone(C^*=O)}$, C 1s $\rightarrow \pi^*_{phenolic(C^*-OH)}$), 288.5 eV C 1s $\rightarrow \pi^*_{R(C^*=O)OH}$ and 289.5 eV (possible functional groups; C 1s $\rightarrow 3p/\sigma^*_{RO(C^*-H)R}$, C 1s $\rightarrow \pi^*_{RR'N(C^*=O)}$ $_{NRR'}$ (carbamide). For tar balls, the intensity of the 288.5 eV and 289.5 eV peaks were strongly correlated. A weak ($R^2 \approx 0.4$) correlation between the peak at 286.7 and the 531.8 eV peak (O 1s $\rightarrow \pi^*_{R(C=O^*)R}$) in the oxygen spectra indicates at least a partial contribution from a ketonic carbonyl. However, a comparison of the spectra shown in Figure 17.14 indicates that tar ball bonding is

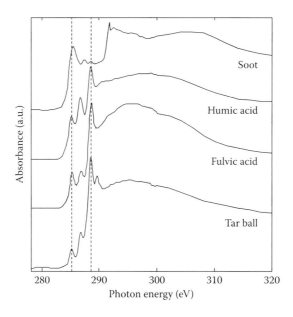

FIGURE 17.14 Spectra of tar balls, soot (measured in the same sample as the tar balls), and humic and fulvic acids. Although tar bars are correlated with long-wavelength absorption in the visible region, these spectra show that the carbon bonding in tar balls is much more similar to humic and fulvic acids than soot or black carbons.

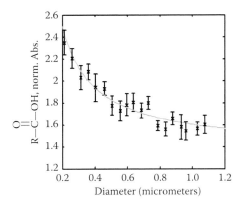

FIGURE 17.15 Carboxylic carbonyl absorbance (normalized to the total carbon) plotted as a function of tar ball diameter. The increased contribution of the carboxyl group at smaller particle sizes indicates a surface enhancement.

different from soot and more similar to humic and fulvic acids. The spectra from these acids and the tar balls all contain relatively small contributions from aromatic carbon (C=C), ketonic carbonyl, and a large contribution from carboxylic carbonyl. Additional insight into the nature of tar balls was obtained by normalizing the functional group intensities to the total carbon and plotting this as a function of diameter. An example of this is shown for the carboxylic carbon in Figure 17.15. If a specific functional group is more concentrated at the surface at smaller particle diameters, its contribution is enhanced, similar to the surface-limited oxidation described in Maria et al.[132] described in the following section. While the carboxylic and O-alkyl-C functional groups display a dependence on particle diameter, the aromatic and ketonic carbonyls showed no dependence on particle diameter. These results indicate the presence of an oxygenated layer at the surface of the tar balls that is thinner than the STXM spatial resolution (35 nm for this study).

17.3.1.8 Evolution of Aerosol Mixing States

Applying the soot identification method of Hopkins et al.[42] Moffet et al. performed an analysis of particles collected from three sampling sites located progressively farther from the Mexico City center: T0, T1, and T2.[119] The sample date was selected to coincide with a time of air-mass transport from the city center toward the peripheral sites. Soot was identified by identifying particles that contained regions with $\geq 35\%$ sp^2 hybridized carbon. Figure 17.16 demonstrates how C K-edge NEXAFS spectra change by selecting only regions having a $\%sp^2$ above a specific value. From Figure 17.16, it is clear that organic coatings are visible below 35% sp^2; at and above 35% sp^2 only particle cores are visible. In Figure 17.11, all soot had C 1s $\rightarrow \pi^*_{R(C^*=C)R}$ peak intensities greater than the absorbance, due to total carbon at 320 eV. In Figure 17.16, below a threshold of 35% sp^2, the absorbance at 285 eV (C 1s $\rightarrow \pi^*_{R(C^*=C)R}$) is less than that at 320 eV. Additionally, both the total carbon and the intensity of the C 1s $\rightarrow \pi^*_{R(C^*=C)R}$ peak rapidly increase above 35% sp^2, consistent with previous measurements of soot spectra. Inorganic, predominantly noncarbonaceous, regions within particles were identified by a pre-edge (278–280 eV) to postedge (320 eV) ratio of 0.6. Organic regions were identified by absorption at the energy of the C 1s $\rightarrow \pi^*_{R(C^*=O)OH}$ peak.[66] The assignment of inorganic regions was confirmed with CCSEM/EDX; as in samples from other sites, the large pre-edge absorbance could be attributed to salts and mineral dust.[86] Each of the identifying components—soot (EC), inorganic (IN) carbon, and OC—were used to label particles in order to quantify trends in particle-type number fractions. These components were used to derive SVD maps of particles from the three sampling sites as shown in Figure 17.17 (T0, T1, and T2 are shown in panels a, b, and c, respectively).

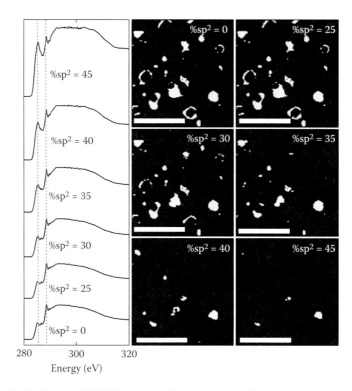

FIGURE 17.16 (Left) Carbon NEXAFS spectra from a Mexico City sample obtained by setting different threshold values for sp^2 hybridization. Vertical gray dotted lines at 285.4 (C 1s $\rightarrow$ $\pi^*_{R(C^*=C)R}$) and 288.6 eV (C 1s $\rightarrow$ $\pi^*_{R(C^*=O)OH}$). (Right) White regions indicate particle regions above the threshold value. Noncarbonaceous, inorganic regions are masked to highlight organic coatings. At $\geq 35\%$ sp^2, the intensity of the 285.4 peak is equal to or greater than the total absorbance at 320 eV, similar to soot/black carbon samples previously reported. White scale bar is 3.2 μm.

Figure 17.17 demonstrates the complex, internal heterogeneous mixtures of components within particles at T0. Farther from the city center, the number fraction of homogenous OCs lacking both soot and inorganic inclusions (Figure 17.17) increased. Although these particles lacked organic inclusions, by examining the sulfur edge it was determined that inorganic sulfate was homogenously mixed with the organic material.

As shown in Figure 17.17, NEXAFS spectra for the OC indicated a relative increase in all carbon functionalities (increased spectral congestion) with the exception of sp^2 hybridized carbon. The decrease in sp^2 with distance from the city center was determined to be either from condensation and/or oxidation processes. All the soot inclusions identified in Mexico City had a strong contribution from COOH groups, indicating possible surface reactions or rapid photochemical oxidation and condensation of gas-phase species. As with the OC, the $\%sp^2$ of the soot inclusions decreased with increasing time and distance from the urban center. However, these heterogeneous processes could not be decoupled from the condensation of gas-phase organics.

Figure 17.18 presents number fractions of particles containing combinations (or absence-NOID) of these three components, EC, OC, and IN, at each sampling site. The increase in OC particles without inorganic or soot inclusions as a function of the distance from the city center was attributed to the growth of ultrafine particles by secondary aerosol condensation.[119] This observation was confirmed by demonstrating that the average size of the homogeneous OC particles increased with distance from the urban center. This study from Mexico City is the first STXM investigation to account for multicomponent particles containing soot, noncarbonaceous inorganic species, and OC. These results clearly indicate that the mixing state of the particles changed with plume age. Such

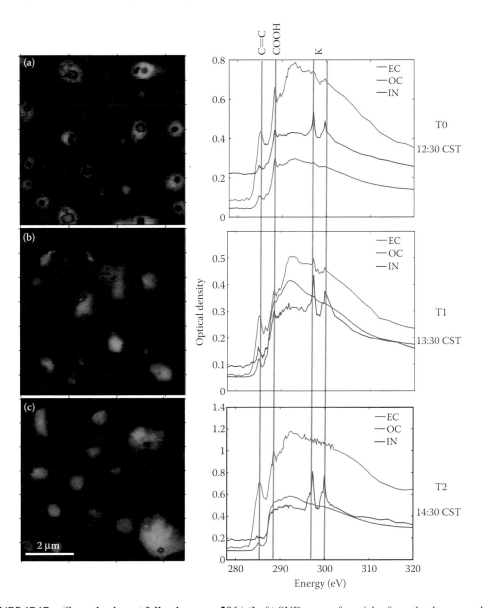

FIGURE 17.17 **(See color insert following page 206.)** (Left) SVD maps of particles from the three sampling sites showing soot (red), inorganic regions (blue), and organic regions (green). Panels a, b, and c refer to sampling sites T0, T1, and T2 located progressively farther from the urban center of Mexico City. Soot was defined as regions with ≥35% sp^2 hybridized carbon, inorganic regions were defined as having a pre-edge (278–280 eV) to postedge (320 eV) ratio of 0.6 and organic regions contained absorption at the COOH peak. (Right) Spectra used to produce SVD maps shown on the left.

changes are expected to strongly influence chemical and physical properties such as hygroscopicity and optical properties.

Knopf et al. used samples collected at the same locations and similar times as those analyzed by Moffet et al.[119] to examine their ice nucleation activity.[133] The particles collected from Mexico City nucleated ice under cirrus cloud temperatures and relative humidities. This finding is notable, because the STXM/NEXAFS showed they have a large organic content. Presumably even dust particles, which normally make efficient ice nuclei, would be coated with organics and thought to

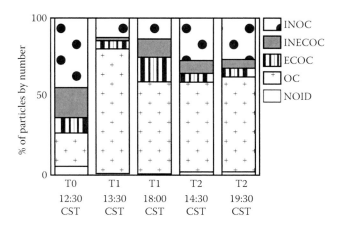

FIGURE 17.18 Percentage of particle types, by number, for samples from the three sites, T0, T1, and T2, located progressively farther from the Mexico City urban center. Particle types are indicated with labels: IN for inorganic, EC for elemental carbon (soot), and OC for organic carbon. Combinations of these are used to indicate particle mixing state. Particles that did not fit into any of these classifications are denoted NOID (not identified). Farther from the city center, the number fraction of homogenous OCs lacking both soot and inorganic inclusions increased.

be rendered inactive with respect to ice nucleation. Previous ice nucleation studies have suggested that organic dominated particles would not serve as ice nuclei under cirrus cloud conditions.

17.3.2 ORGANIC CARBON

Russell et al.[58] were the first to use STXM to map a variety of organic functional groups on individual aerosol particles. The ratios of carboxylic to aliphatic compounds on individual particles indicated surface coatings of shorter chains or more oxygenated groups. This observation indicated that the particles underwent heterogeneous surface oxidation. This study provided a first glimpse of the variable nature of organics in single particles at an unprecedented spatial resolution. Maria et al.[132] demonstrated the utility of STXM/NEXAFS for determining organic aerosol growth processes by exploiting particle-size-resolved organic composition. The size dependence of the functional group mass ratio with respect to either total mass or total carbonaceous mass was used to infer organic growth mechanisms. Total mass was assumed to be proportional to the OD of the region between 278–280 eV. Carbon mass was taken to be proportional to the difference between the absorption at 303–305 eV and 278–280 eV. As illustrated in Figure 17.19, if particles undergo surface-limited oxidation, the ratio of oxidized functional group to total carbon is enhanced for smaller particles due to their larger surface to volume ratio. Conversely, if particles undergo volume-limited oxidation, the spatial distribution of carbonyl groups should be uniform due to fast diffusion of reactants compared to the kinetics of reaction. In the case of a volume-limited reaction, the oxidized functional group to total carbon mass ratio would not be size-dependent. Using this approach for atmospheric samples, surface-limited reactions were found to occur for particles sampled in low relative humidity environments and for particles that likely contained insoluble compounds. Volume-limited reactions were found to occur for particles sampled under high relative humidity conditions. By combining the STXM/NEXAFS results with additional information provided by FTIR and air-mass back trajectories, it was concluded that the oxidation rate used in current models was three times faster than the observations.

Incorporating data from several field campaigns, Takahama et al.[134] demonstrated the first use of spectral cluster analysis to identify individual particle types using the carbon K-edge NEXAFS spectrum. Particle cluster analysis performs an automated grouping of similar spectra based on mathematical criteria. Individual particles were grouped into 14 different types (a-m), shown in 3

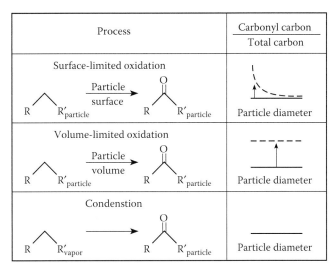

FIGURE 17.19 Particles undergoing surface-limited oxidation contain an oxidized functional group ratio to total carbon that would be enhanced for small particles due to their larger surface-to-volume ratio. If particles underwent volume-limited oxidation, the spatial distribution of carbonyl groups would be uniform due to fast diffusion of reactants compared to the kinetics of reaction. In the case of a volume-limited reaction, the oxidized functional group to total carbon mass ratio would not exhibit size dependence. Particles formed via condensation would also not exhibit size dependence for organic functional groups. [Adapted from Maria, S. F., Russell, L. M., Gilles, M. K., and Myneni, S. C. B. *Science* **306**, 1921–1924 (2004).]

panels in Figure 17.20. The most abundant particle type (a), presumably secondary organic aerosol (SOA), was dominated by the carboxylic peak. This observation highlighted the ubiquity of carboxylic acids in particles sampled over different geographical regions. Furthermore, organic classes indicated in Figure 17.20 (panel 2) had little contribution from the C 1s $\rightarrow \pi^*_{R(C^*=C)R}$ peak. Other particle types were defined based on the abundance of C=C (C 1s $\rightarrow \pi^*_{R(C^*=C)R}$), CO_3 (C 1s $\rightarrow \pi^*_{C^*O_3}$), and K ($L_2\, 2p_{1/2,3/2} \rightarrow \sigma^*$). Particle types with large contributions from C=C most likely contain soot (Figure 17.20, panel 1); within soot particle types, there are different amounts of oxidized functional groups presumably due to differences in atmospheric processing. The presence of K in the particles (Figure 17.20, panel 2) may be indicative of either dust or biomass burning. If particles contained both K and CO_3 (Figure 17.20, panel 3), they were classified as dust based on their similarity with the pine ultisol soil NEXAFS spectrum.[2]

The approaches of Maria et al.[132] and Takahama et al.[134] were combined in the analysis of airborne samples taken as part of the INTEX-B field campaign.[135] INTEX-B covered the western Pacific Ocean as well as California, Oregon, and Washington.[132,134,135] The majority of particles observed during INTEX-B were attributed to SOA formation processes based on the submicron size and dominant contribution from carboxylic acid. The methods of Maria et al.[132] were employed to show that growth of organic aerosol occurs via both condensation and surface-limited processes.[135]

As part of the INTEX-B campaign, a separate study focused on the analysis of dust transported from Asia.[136] Dust plumes observed at Whistler, British Columbia were associated with an increase in sulfate, indicating anthropogenic influences. These plumes occurred with a very low concentration of accumulation mode organics. Half of the NEXAFS spectra sampled during the Asian dust event indicated characteristic peaks for dust (K and CO_3) in addition to a well-defined carboxyl peak. This finding was corroborated by ion chromatography of water-extracted filters, which showed significant concentrations of coarse mode formate (CHO_2^-). The mixing of carboxylic acids and dust within a particle indicates that the mineral dust acts as a condensable surface for organics. The authors concluded that the condensation of organics on large dust particles may diminish their indirect radiative forcing while potentially enhancing direct forcing.

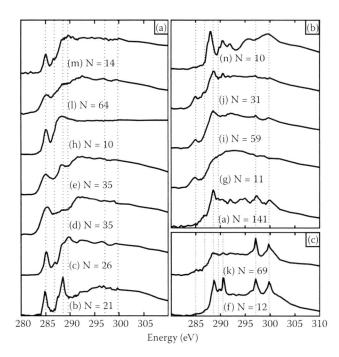

FIGURE 17.20 Spectra derived from k-means clustering of ambient-collected particles from Takahama et al.[134] Here, clusters are organized into panels as follows: (a) soot, (b) organic, and (c) inorganic. Dotted lines indicate 285 eV (C 1s $\rightarrow \pi^*_{R(C^*=C)R}$), 286.8 eV (C 1s $\rightarrow \pi^*_{R(C^*=O)R}$), 288.6 eV (C 1s $\rightarrow \pi^*_{R(C^*=O)OH}$), 289.6 eV (unassigned), 297.1 eV (K L$_2$ 2p$_{1/2}$ $\rightarrow \sigma^*$), and 299.8 eV (K L$_2$ 2p$_{3/2}$ $\rightarrow \sigma^*$) eV. For dust, a line at 290.5 eV was added for the C 1s $\rightarrow \pi^*_{C^*O_3}$ transition.

To analyze particles from Mexico City, Liu et al.[137] followed a method similar to Takahama et al.[134] but used positive matrix factorization (PMF) in addition to k-means clustering. Samples were collected aboard an aircraft on a high-altitude mountain and at an urban site. Three main aerosol types were classified based on their carbon K-edge spectra. By number, soot particles were most abundant (70% of the total number) followed by biomass burning and processed/secondary particles. Biomass burning particles had spectra similar to those of fulvic acid and tar balls.[38] The secondary/processed particle type had a dominant carboxylic acid contribution, consistent with the studies outlined previously. Although distinct differences were seen in the number fractions of different particle types, the PMF factors exhibited approximately equal contributions from three factors, biomass, secondary/processed, and soot. This difference may arise, because while there were more soot particles than organic particles, the organic particles contained more mass.

17.3.3 Inorganic Species

The potential to explore chemical bonding and oxidation states of sulfur and transition metals is of interest to atmospheric and environmental scientists, because they provide insight into branching ratios and chemical processing. Ultimately, aerosol toxicity may be related to specific oxidation states; hence, such studies could be important to understand health effects of aerosols. To date, only a few studies have focused on speciation of metals, such as manganese, iron, or sulfur, in environmental samples.[77,94,96,138] This section discusses STXM/NEXAFS studies characterizing the bonding and oxidation states of metals (Zn and Fe) and sulfur partitioning in the marine boundary layer.

17.3.3.1   Speciation of Zn-Containing Aerosols

Historically, ambient aerosol measurements in Mexico City have indicated elevated Pb, Zn, and Cl levels in some neighborhoods. During the MILAGRO campaign, single-particle mass spectrometry measurements indicated a daily cycle of internally mixed Pb, Zn, and Cl.[139] Combining the chemical evidence with the observation that the daily cycle was suppressed over weekends and holidays, it was concluded that their source was likely to be either industrial smelting and/or garbage burning. Because of the widespread nature of these particles and the increased blood lead levels in children residing in these neighborhoods,[140] the speciation and morphology were pursued using STXM/NEXAFS and CCSEM/EDX. Microscopic information is important, since particle morphology and speciation controls uptake by the human body. CCSEM/EDX measurements indicated that particles containing the characteristic mixture of Pb, Zn, and Cl generally had two morphologies: compact structures and needle-like structures. Based on their tetrahedral morphology and chemical composition, some needle-like particles were hypothesized to be ZnO. As shown in Figure 17.21, STXM/NEXAFS measurements at the Zn L-edge confirmed the presence of needle-like ZnO particles. An additional particle type composed of $Zn(NO_3)_2$ was also observed. Representative Zn L-edge NEXAFS spectra for both ambient particle classes along with standards $(Zn(NO_3)_2 \cdot 6H_2O$ and ZnO) are shown in Figure 17.21. The comparison with the ambient particle spectra with the standard spectra in Figure 17.21 provide clear chemical identification. $Zn(NO_3)_2$ may be formed from the reaction of $ZnCl_2$ with HNO_3. Mass spectrometry measurements showed the replacement of Cl by NO_3 on particles containing Zn, Pb, and Cl. Presumably, $PbCl_2$ undergoes a similar reaction to form $Pb(NO_3)_2$. Measurements at the carbon K-edge confirmed the single-particle mass spectrometry observation that metals were internally mixed with soot, indicating a combustion source. This study illustrates the capacity for transition metal speciation using STXM/NEXAFS, as well as the benefits of combining STXM with other microspectroscopic (CCSEM/EDSX) and mass spectrometric techniques.

17.3.3.2   Fe Speciation

It is estimated that the majority of bioavailable iron in the ocean comes from aerosols. Hence, understanding the origins, transport, and eventual fate of iron is of interest in several environmental fields. Several groups have explored iron speciation in oceans,[101,141] biofilms,[91,93,96,99,142–145] and aerosols,[146] using STXM. Although dust dominates Fe-containing aerosol, anthropogenic/industrial

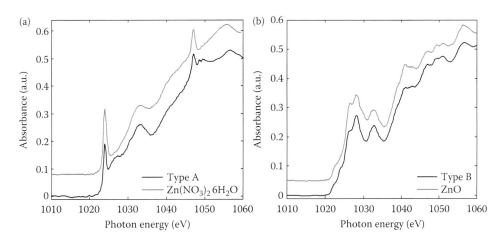

FIGURE 17.21   Zn spectra from ambient particles from Mexico City compared with standard spectra. (a) representative Zn L-edge NEXAFS spectra for particle type A and $Zn(NO_3)_2 6H_2O$ standard and (b) spectra from a second particle type B, and spectra from ZnO standard. Spectra measured for ZnS, $ZnSO_4$, and $ZnCl_2$ did not match either type A or B.

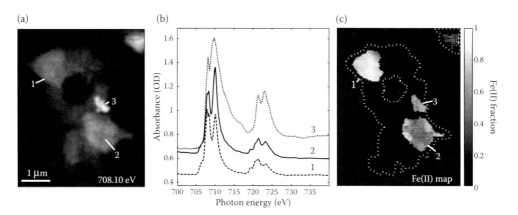

FIGURE 17.22 (a) A single-energy STXM image taken at 708.1 eV (at the peak of the Fe L_3 edge), with numbers indicating bright regions containing Fe. (b) Spectra taken from regions indicated in (a) and (c). Differences in Fe(II) fraction is indicated by changes in the relative heights of the peaks occurring between 705 and 710 eV. (c) Fe(II) fraction indicated by the grayscale with non-Fe-containing regions masked in black and the particle perimeter indicated by the dotted line. Region 1 is Fe(II) rich while regions 2, and 3 contain less Fe(II).

sources can also be important. As demonstrated in Figure 17.22, using the methods of Dynes et al.[96] and Takahama et al.[146] using the methods of Dynes et al. and Takahama et al. spatially resolved Fe speciation for nanometer to micron-sized particles are obtained using STXM at the Fe L-edge. Takahama et al.[146] determined Fe oxidation states using analysis methods established by Dynes et al.[96] and references therein. In Takahama et al.[146] 63 particles (from a variety of samples) indicated Fe (II) fractions from 0 to 0.73. The absence of a strong correlation between Fe (II) fraction and distance from the particle surface indicated that surface reactions did not control Fe speciation. Furthermore, no correlation between OC and Fe (II) fraction was observed, implying that organic ligands did not play a noticeable role in determining Fe speciation. However, a group of spherical particles (collected during the ACE Asia campaign) had both high Fe(II) fraction and OC content; it was hypothesized that these particles were associated with biomass burning.

17.3.3.3 Marine Boundary Layer Sulfur Speciation

Although more bonding information is obtained at the L-edge, there are only a few studies at this edge[147,148] and most sulfur NEXAFS spectra have been measured at the K-edge.[1] During the Marine Stratus Experiment (MASE), dry residue particles from individual cloud droplets and interstitial aerosol were analyzed using an array of complementary microspectroscopic techniques at the Environmental Molecular Science Laboratory (collaboration with A. Laskin) combined with sulfur L-edge STXM/NEXAFS from the ALS. This study was aimed at chemical speciation of particulate sulfur.[56]

CCSEM/EDX provided microscopic particle imaging and quantitative elemental compositions. An analysis of ~10,000 particles indicated two prominent particle classes: sea salt or sulfur-rich (S-rich) particles. S-rich particles were submicron in diameter and spherical. Sea salt particles were cubic-shaped crystals, with diameters larger than 1 μm, surrounded by irregularly shaped residues. TOF-SIMS showed that sea-salt particles contained both $CH_3SO_3^-$ and SO_4^{2-} while S-rich particles were composed of a mixture of H_2SO_4 and $(NH_4)_2SO_4$.

STXM/NEXAFS analysis sought to quantify $CH_3SO_3^-/SO_4^{2-}$ ratios within individual sea-salt particles, because this ratio is indicative of different chemical processes in the marine boundary layer. (Note: the STXM/NEXAFS measurements list only the anion as several cationic species could be paired with these species.) Figure 17.23 shows a representative STXM carbon K-edge image (290.8 eV) indicating two particle classes: circles for S-rich and squares for sea salt. Maps generated

STXM image   Carbon map

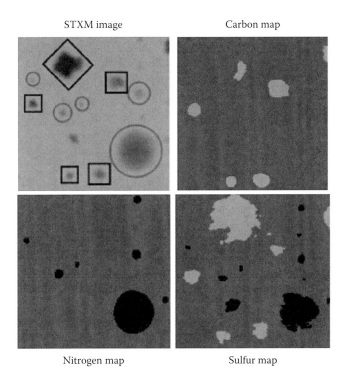

Nitrogen map   Sulfur map

FIGURE 17.23 Single-energy (290.8 eV) STXM image (12 μm×12 μm) of externally mixed sea salt/$CH_3SO_3^-$/ SO_4^{2-} (squares) and H_2SO_4/$(NH_4)_2SO_4$ (circles) particle residuals (left top). Principal component maps at the carbon K-edge (right top), nitrogen K-edge (left bottom), and sulfur L-edge (right bottom). Two types of sulfur spectra were observed: one consistent with H_2SO_4/$(NH_4)_2SO_4$ (measured on nitrogen containing particles) and a the other (measured on carbon-containing sea-salt particles) that is consistent with a mixture of $CH_3SO_3^-$/SO_4^{2-}.

by principle component analysis (PCA) of STXM/NEXAFS spectra taken at the carbon (b), nitrogen (c), and sulfur edges (d) are shown. These maps indicate that particles may contain carbon or nitrogen, but do not contain both. The PCA maps indicate two different types of sulfur; one component associated with carbon (gray) and a different component associated with nitrogen (black). This is consistent with particle compositions of either mixed sea salt/$CH_3SO_3^-$/SO_4^{2-} (gray) or mixed H_2SO_4/ $(NH_4)_2SO_4$ (black). For comparison, reference sulfur L-edge NEXAFS spectra of Na_2SO_4, $(NH_4)_2SO_4$, and CH_3SO_3Na were measured. Sulfur spectra recorded from $(NH_4)_2SO_4$ and Na_2SO_4, were identical to one another with three distinct peaks. However, the CH_3SO_3Na spectrum had no resolvable peaks. Particles containing nitrogen (black) display a sulfur L-edge NEXAFS spectrum indicative of SO_4^{2-} (consistent with mixed H_2SO_4/$(NH_4)_2SO_4$ composition). Particles indicated with gray had sulfur L-edge spectra that were intermediate between $CH_3SO_3^-$ and SO_4^{2-} NEXAFS spectra (consistent with a mixed sea salt/$CH_3SO_3^-$/ SO_4^{2-} composition). Combining the Na_2SO_4 and CH_3SO_3Na spectra would both broaden the resolved peaks and alter relative peak intensities. Using linear combinations of the Na_2SO_4, and CH_3SO_3Na reference spectra, a calibration curve was generated for the peak heights. Spectra from ~100 individual mixed sea salt/$CH_3SO_3^-$/non-sea salt-SO_4^{2-} particles were measured and their $CH_3SO_3^-$/total-SO_4^{2-} ratio determined using the calibration curve. The CCSEM/EDX, TOF-SIMS, and STXM/NEXAFS data are all consistent with one another and indicate enhanced formation of particulate $CH_3SO_3^-$ in sea salt droplets and particles. This unique combination of single-particle measurements enables quantitative assessment of $CH_3SO_3^-$/non-sea salt-SO_4^{2-} ratios in individual particles. On the basis of particle morphology, chemical composition, and chemical bonding, two types of sulfur-containing particles were identified: chemically modified or aged sea salt particles and sulfate particles. Previously, only bulk measurements, which are an average over all particle types, had examined sulfur partitioning of methanesulfonate and sulfate.

17.3.4 OTHER STUDIES

17.3.4.1 Power Plant Plume Processing: Presence of Organosulfates?

Particles collected downwind of a power plant plume were examined using STXM/NEXAFS combined with CCSEM/EDX.[149] NEXAFS spectra of particles at the three locations chosen progressively further from the power plant, P0, P1, and P2, were recorded at the carbon, nitrogen, oxygen, and sulfur edges. Unlike most field-collected aerosol samples, spectra (within each specific location) were extremely homogeneous both within individual particles and from particle to particle. Therefore, at each location, spectra were averaged over ~50 individual particles. All carbon K-edge spectra exhibited a weak peak at 285.15 eV arising from the C 1s $\rightarrow \pi^*_{R(C^*=C)R}$ transition. Three distinct peaks were observed in the sulfur L-edge spectra. STXM/NEXAFS indicated that both carbon and sulfur were homogeneously distributed within individual particles and were not inclusions (e.g., $(NH_4)_2SO_4$) within the particles. Hopkins et al. observed that spectra from $(NH_4)_2SO_4$ and Na_2SO_4 were indistinguishable at the sulfur L-edge.[56] However, ammonium sulfate is widely known to be very sensitive to x-ray exposure (as indicated by the growth of a peak at 401 eV or evaporation in a TEM). However, the nitrogen spectra from the power plant plume samples did not vary with exposure to x-ray radiation. Based on the homogenous distribution of carbon and sulfur and the insensitivity to x-ray exposure, the observed sulfate bonding was tentatively assigned to organosulfates. Organosulfates could be produced via secondary chemistry with organics.[150] While the exact formation mechanisms are currently being investigated, some laboratory evidence exists for the formation of organosulfates and nitroxy organosulfates from the reaction of acidic sulfate particles with isoprene, as well as photooxidation products such as pinonaldehyde and dihydroxyepoxides, under conditions similar to those found in the nocturnal power plant plume investigated in this study.

Using the carbon spectra, the sp^2 hybridization at each location downwind of the power plant plume was estimated using the C 1s $\rightarrow \pi^*_{R(C^*=C)R}$ peak intensity (as discussed previously). The calculated % sp^2 hybridizations were approximately 8%, 4%, and 1.6%, respectively, for the locations P0, P1, and P2. The decrease from 8% to less than 2% indicates a loss of C=C bonds along the plume trajectory. One possible explanation is that a heterogeneous reaction between elevated levels of the NO_3 radicals present in the power plant plume with the C=C bonds leads to the observed loss.

17.3.4.2 Ice Nucleation and Cloud Condensation Nuclei

Many recent aerosol studies have examined how aerosols affect clouds, that is, the aerosol indirect effect on climate. Aerosol indirect effects have a potentially large, but extremely uncertain effect on climate. STXM/NEXAFS was used to analyze particles collected using a CVI in several field campaigns. CVI impactors are designed to sample only cloud droplets or crystals while excluding interstitial aerosol. By sampling cloud droplet residues and comparing their chemical composition with interstitial aerosols, links between particle composition and cloud nucleating ability are studied.

During the ice in clouds experiment—layer clouds (ICE-L), the impact of playa (dry lakebed) dust as ice nuclei was explored in wave clouds consisting primarily of ice.[151] A CVI sampled cloud drops which were subsequently heated and collected. Single-particle mass spectrometers indicated the presence of playa dust in ice clouds while bulk aerosol mass spectrometers indicated that highly oxygenated organics were present. These measurements of clouds over Wyoming implied that desert playa dust advected over the western United States made efficient ice nuclei. Based on the characteristic markers of Na, Ca, K, Cl, and Mg, the single-particle mass spectrometer identified periods of high playa dust ice nuclei. Although the single-particle mass spectrometer measured some minor peaks for organics the presence of organics within the playa dust remained uncertain due to possible matrix effects. The presence of organics within single cloud particle residuals was confirmed based on STXM/NEXAFS measurements of C, K, and Ca edge spectra that identified internally mixed

CO3, K, Ca, and COOH. These findings indicate that the mixture of both dust and organics may contribute to ice nuclei. These results are in contrast with the hypothesis that organics inhibit ice nucleation by occupying active sites where ice nucleation would occur.

CCSEM/EDX and STXM were combined to study mixed phase (ice and liquid) clouds during the indirect and semidirect aerosol campaign (ISDAC) conducted over Barrow, AK. Interstitial aerosol and cloud drop residuals were characterized.[152] Based on the carbon STXM/NEXAFS analysis, samples taken from a period of high ice nucleation contained enhanced fractions of coated carbonate and EC particles (compared to interstitial aerosol). STXM measurements indicated that COOH and CO_3 were internally mixed (consistent with ICE-L and INTEX-B findings).[151,135] By examining the pre-edge to postedge ratio at the carbon K-edge, it was concluded that noncarbonaceous species were enhanced during the CVI sampling periods. SEM/EDX measurements showed enhanced Mg in cloud drop residuals (relative to interstitial aerosols), thus confirming the STXM/NEXAFS results. NEXAFS spectra of interstitial aerosol showed two types of OC: one with relatively high carboxylic acid content, and the other with a larger contribution in the region specific to $R-OCH_2-R$ (the carbon adjacent to the oxygen in ether) groups. OC containing more $ROCH_2R$ was also found in a Siberian biomass burning plume sampled during the same campaign. The enhanced organics in the interstitial aerosol and enhanced Mg in the cloud drop residuals indicate that inorganic species are more efficient condensation nuclei/ice nuclei than organic species.

17.4 OPPORTUNITIES FOR FUTURE APPLICATIONS AND DEVELOPMENTS

17.4.1 COMBINING COMPLEMENTARY MICROSCOPIC AND SPECTROSCOPIC TECHNIQUES

One of the most difficult aspects of studying natural samples is defining the scientific problem narrowly enough for focused experiments. Focusing on a specific scientific problem is simplified when the chemical composition is known, such as polymer or magnetic systems. In contrast, natural samples are incredibly complex with poorly characterized chemical compositions. As only a limited number of samples can be examined using STXM they must be chosen carefully, often based on data and insight obtained from other methods. For example, x-ray fluorescence,[119] SEM,[38] and mass spectrometry,[139] can guide STXM measurements. In addition, atmospheric samples should be statistically representative. Using CCSEM/EDX or single-particle mass spectrometry, to sample thousands to millions of particles provides a statistical analysis for the development of a classification scheme for particle types.[38,56,118,119,139,149] The particle types can then be examined in greater detail using STXM/NEXAFS. These results can then be linked back to SEM/EDX and mass spectrometry data to provide more representative results.

In environmental science, STXM/NEXAFS can be combined with other microscopy and spectroscopy methods to exploit the strengths of each method. As the number of STXM instruments increases, more investigations will combine multiple techniques. TEM has been used in combination with STXM/NEXAFS to examine environmental samples including biofilms,[92] fossil spores,[153] marine sediments,[98] meteorites,[154] biomineralization,[143,155] and nanoparticles in drinking water.[156] Numerous investigations have employed other complementary techniques such as micro-FTIR,[98] Raman imaging,[154] atomic force microscopy (AFM),[54,69,156–158] and TOF-SIMS.[158]

Although very few studies have employed multiple techniques in combination with STXM/NEXAFS on atmospheric aerosol samples,[38,56,118,119,139] a variety of applications are apparent. One limitation of STXM/NEXAFS is that the measured absorbance is the product of particle density times its thickness. Separating these two requires knowledge of the particle shape. For understanding chemical and physical properties, knowing the aerosol density and morphology is useful. By performing AFM and STXM measurements on identical particles, their three-dimensional shape, size, height, density, and total atomic absorbance (mass) can be quantified. The Tivanski research group has used this approach to determine, for the first time, single-particle densities for various atmospherically relevant organic compounds, such as low-molecular-weight dicarboxylic acids and natural fulvic acid.[159]

17.4.2 Development of *in situ* Techniques

Controlled *in situ* studies relevant to aerosols, such as those reported in catalysis[160–162] are certain to develop in the coming years. Ideally, aside from temperature control (heating and cooling), and reactive gas flow, reliable humidity control is desired. This would allow the reactivity of a single particle to be monitored for changes in chemical bonding and oxidation states using STXM/NEXAFS.

Several research groups have been exploring STXM/NEXAFS for studies on water vapor uptake. Ghorai and Tivanski have developed a novel approach based on STXM/NEXAFS for *in situ* studies of the hygroscopic properties of individual submicron aerosols.[163] In particular, several inorganic aerosols (NaCl, NaBr, NaI, and NaNO$_3$) deposited on silicon nitride membranes were characterized using STXM/NEXAFS at varying relative humidities, up to 90%. The amount of water absorbed by a single particle during hydration/dehydration cycle was measured as a function of relative humidity and the measured hygroscopic properties were in excellent agreement with results determined by established techniques. Zelenay et al. recently incorporated an *in situ* reactor to examine the variations in oxygen NEXAFS spectra and changing particle morphology of individual mixed particles of adipic acid/ammonium sulfate as a function of relative humidity.[164] Hence, future studies are expected to include atmospherically relevant aerosols, ambient aerosols, and organic/inorganic mixtures as well as improvements in controlling and measuring the relative humidity. The ability to probe organic/inorganic mixtures is particularly important, due to the extremely limited knowledge of how the type and degree of mixing, and aerosol microstructure influence their hygroscopic properties.

Other potential *in situ* experiments could combine irradiation (with UV or laser light) with STXM to monitor chemical changes or the extent of progression of a reaction with time. Depending on experimental settings, STXM can be used to follow time-resolved chemical changes with a temporal resolution of several milliseconds. Several groups have combined STXM with UV/Visible light irradiation to examine chemical changes upon light exposure[112] or to provide experimental confirmation of the sample heterogeneity and the spatial distribution of different components within the sample.[75]

Although the NSLS cryo-STXM has been in existence for years,[20] and Tzvetkov and Fink recently examined the temperature-dependent phase changes in polymers,[165] cooling techniques for STXM are not fully developed. Thibault et al. imaged a zone plate buried beneath layers of gold by combining elements of scanning microscopy with coherent diffraction resulting in reconstructed images with higher resolution than the focused spot size.[166] Several groups are developing techniques of contrast imaging, fluorescence detection, and tomography for STXM. These techniques along with improved *in situ* capabilities will ultimately enhance the ability to probe environmental samples.

Although a relatively new technique for atmospheric research and environmental science, STXM/NEXAFS applications have grown from examining laboratory samples of soot and mapping functional groups on just a few particles to detailed experiments on hundreds of particles. Initially, a very limited number of STXM instruments were available and the data acquisition and analysis were labor intensive. However, the number of instruments available worldwide is increasing rapidly, and top-off mode has significantly improved data acquisition. The development of software focused on statistical analysis has decreased the time required for data analysis. Combined with other imaging techniques (SEM/EDX, TEM, EELS, AFM), spectroscopic techniques (FTIR, Raman, TOF-SIMS) and technological advances (*in situ* cells, improved temperature and humidity control and measurement) we anticipate a rapid increase both in the number and type of scientific investigations in atmospheric and environmental sciences that employ STXM/NEXAFS.

ACKNOWLEDGMENTS

The authors acknowledge financial support provided by the Atmospheric System Research Program, Office of Biological and Environmental Research (OBER) of the U.S. Department of Energy (DOE). R. C. Moffet acknowledges additional financial support from a Lawrence Berkeley National Laboratory Glenn T. Seaborg Fellowship. A. V. Tivanksi acknowledges The University of Iowa for the financial support. A significant portion of the STXM/NEXAFS particle analysis was performed at beamlines 5.3.2 and 11.0.2 at the ALS at Lawrence Berkeley National Laboratory. The work at the ALS was supported by the Director, Office of Science, Office of Basic Energy Sciences, of the U.S. Department of Energy under Contract No. DE-AC02-05CH11231. Supporting CCSEM/EDX particle analysis was performed in the Environmental Molecular Sciences Laboratory, a national scientific user facility sponsored by the Department of Energy's Office of Biological and Environmental Research at Pacific Northwest National Laboratory. PNNL is operated by the U.S. Department of Energy by Battelle Memorial Institute under contract DE-AC06-76RL0. The authors also acknowledge collaborators who have provided their expertise and insights, as well as samples: P. Ziemann, R. A. Zaveri, S. Wirick, K. R. Wilson, Z. Wang, C. H. Twohy, S. Takahama, P. O. Sprau, J. Smith, V, Shutthanandan, D. K. Shuh, L. M. Russell, K. A. Pratt, K. A. Prather, S. Prakash, L. Muntean, L. T. Molina, H. A. Michelsen, B. D. Marten, D. A. Knopf, T. Kirchstetter, A. Laskin, R. J. Hopkins, T. R. Henn, J. L. Hand, R. Gonzalez, J. Fast, I. J. Drake, and Y. Desyaterik. Additionally, the authors personally acknowledge the spirit of generosity extended by A. P. Hitchcock, C. Jacobsen, A. D. Kilcoyne, G. Mitchell, and T. Tyliszczak in encouraging new users and in advancing the technique.

REFERENCES

1. Myneni, S. C. B. Soft X-ray spectroscopy and spectromicroscopy studies of organic molecules in the environment. *Applications of Synchrotron Radiation in Low-Temperature Geochemistry and Environmental Sciences* **49**, 485–579 (2002).
2. Ade, H. and Urquhart, S. G. *NEXAFS Spectroscopy and Microscopy of Natural and Synthetic Polymers.* World Scientific Publishing, Singapore (2002).
3. Hitchcock, A. P., Morin, C., Heng, Y. M., Cornelius, R. M., and Brash, J. L. Towards practical soft X-ray spectromicroscopy of biomaterials. *Journal of Biomaterials Science, Polymer Edition* **13**, 919–937 (2002).
4. Yoon, T. H. Applications of soft X-ray spectromicroscopy in material and environmental sciences. *Applied Spectroscopy Review.* **44**, 91–122 (2009), DOI: 10.1080/05704920802352531.
5. Ade, H. and Stoll, H. Near-edge X-ray absorption fine-structure microscopy of organic and magnetic materials. *Nature Materials* **8**, 281–290 (2009).
6. Attwood, D. *Soft X-Rays and Extreme Ultraviolet Radiation: Principles and Applications.* Cambridge University Press, New York (1999).
7. Kirz, J., Ade, H., Jacobsen, C. et al. Soft-X-ray microscopy with coherent X-rays. *Review of Scientific Instruments* **63**, 557–563 (1992).
8. G. R. Morrison, S. Bridgwater, M. T. Browne, R. E. Burge, R. C. Cave, P. S. Charalambous, G. F. Foster, A. R. Hare, A. G. et al. Development of x-ray imaging at the Daresbury SRS. *Reviews Scientific Instruments* 60, 2464–2467 (1989). DOI:10.1063/1.1140700.
9. McNulty, I., Paterson, D., Arko, J. et al. The 2-ID-B intermediate-energy scanning X-ray microscope at the APS. *Journal de Physiques. IV France* **104**, 11 (2003).
10. Beelen, T. P. M., Shi, W. D., Morrison, G. R. et al. Scanning transmission X-ray microscopy: A new method for the investigation of aggregation in silica. *Journal of Colloid and Interface Science* **185**, 217–227 (1997).
11. Kaznatcheev, K. V., Karunakaran, C., Lanke, U. D. et al. Soft X-ray spectromicroscopy beamline at the CLS: Commissioning results. *Nuclear Instruments and Methods in Physics Research Section A-Accelerators Spectrometers Detectors and Associated Equipment.* **582**, 96–99 (2007).
12. Mitrea, G., Thieme, J., Guttmann, P., Heim, S., and Gleber, S. X-ray spectromicroscopy with the scanning transmission X-ray microscope at BESSY II. *Journal of Synchrotron Radiation* **15**, 26–35 (2007).

13. Raabe, J., Tzvetkov, G., Flechsig, U. et al. PolLux: A new facility for soft x-ray spectromicroscopy at the Swiss Light Source. *Review of Scientific Instruments* **79**, 113704 (2008).

14. Zangrando, M., Finazzi, M., Zacchigna, M. et al. A multi-purpose experimental station for soft x-ray microscopy on BACH beamline at Elettra. *Soft X-Ray and EUV Imaging Systems II* **4506**, 154–162 (2001).

15. Barrett, R., Kaulich, B., Salome, M., and Susini, J. Current status of the scanning X-ray microscope at the ESRF. *AIP Conference Proceedings* **15**, 507, 458–463 (2000).

16. Shin, H.-J., Chung, Y., and Kim, B. The first undulator beamline at the Pohang Light Source for high-resolution spectroscopy and spectromicroscopy. *Journal of Electron Spectroscopy and Related Phenomena* **101–103**, 985–989 (1999).

17. Chao, W. L., Harteneck, B. D., Liddle, J. A., Anderson, E. H., and Attwood, D. T. Soft X-ray microscopy at a spatial resolution better than 15 nm. *Nature* **435**, 1210–1213 (2005).

18. Kilcoyne, A. L. D., Tyliszczak, T., Steele, W. F. et al. Interferometer-controlled scanning transmission X-ray microscopes at the Advanced Light Source. *Journal of Synchrotron Radiation* **10**, 125–136 (2003).

19. Jacobsen, C., Williams, S., Anderson, E. et al. Diffraction-limited imaging in a scanning-transmission X-ray microscope. *Optics Communication* **86**, 351–364 (1991).

20. Maser, J., Osanna, A., Wang, Y. et al. Soft X-ray microscopy with a cryo scanning transmission X-ray microscope: I. Instrumentation, imaging and spectroscopy. *Journal of Microscopy* **197**, 68–79 (2000).

21. Feser, M., Carlucci-Dayton, M., Jacobsen, C. et al. Applications and instrumentation advances with the Stony Brook scanning transmission X-ray microscope. *Proceedings of SPIE—The International Society for Optical Engineering* **3449**, 19–29 (1998).

22. Rarback, H., Shu, D., Feng, S. C. et al. Scanning x-ray microscope with 75-nm resolution. *Review of Scientific Instruments* **59**, 52–59 (1988).

23. Feser, M., Hornberger, B., Jacobsen, C. et al. Integrating silicon detector with segmentation for scanning transmission X-ray microscopy. *Nucl. Instrum. Methods Phys. Res. Sect. A-Accel. Spectrom. Dect. Assoc. Equip.* **565**, 841–854 (2006).

24. Gianoncelli, A., Morrison, G. R., Kaulich, B., Bacescu, D., and Kovac, J. Scanning transmission x-ray microscopy with a configurable detector. *Applied Physics Letters* **89**, 251117 (2006).

25. Hornberger, B., Feser, M., and Jacobsen, C. Quantitative amplitude and phase contrast imaging in a scanning transmission X-ray microscope. *Ultramicroscopy* **107**, 644–655 (2007).

26. Jacobsen, C., Lindaas, S., Williams, S., and Zhang, X. Scanning luminescence X-ray microscopy: Imaging fluorescence dyes at suboptical resolution. *Journal of Microscopy* **172**, 121–129 (1993).

27. Stöhr, J. *NEXAFS Spectroscopy*. 1st edn, Vol. 25, Springer-Verlag, Berlin (2003).

28. Smith, N. Science with soft x-rays. *Physics Today* **54**, 29–34 (2001).

29. Henke, B. L., Gullikson, E. M., and Davis, J. C. X-ray interactions: Photoabsorption, scattering, transmission, and reflection at E = 50–30,000 eV, Z = 1–92. *Atomic Data and Nuclear Data Tables* **54**, 181–342 (1993).

30. Thompson, A., Attwood, D., Gullikson, E. M. et al. in *LBNL/PUB-490* (Technical and Electronic Information Department, Lawrence Berkeley National Laboratory, Berkeley, CA 94720, 2001).

31. http://www-cxro.lbl.gov/

32. Muntean, L., Planques, R., Kilcoyne, A. L. D. et al. Chemical mapping of polymer photoresists by scanning transmission x-ray microscopy. *Journal of Vacuum Science and Technology B* **23**, 1630–1636 (2005).

33. Hitchcock, A. P., Koprinarov, I., Tyliszczak, T. et al. Optimization of scanning transmission X-ray microscopy for the identification and quantitation of reinforcing particles in polyurethanes. *Ultramicroscopy* **88**, 33–49 (2001).

34. Dynes, J. J., Lawrence, J. R., Korber, D. R. et al. Quantitative mapping of chlorhexidine in natural river biofilms. *Science of the Total Environment* **369**, 369–383 (2006).

35. Urquhart, S. G. and Ade, H. Trends in the carbonyl core (C 1s, O 1s) → p*C=O transition in the near-edge X-ray absorption fine structure spectra of organic molecules. *Journal of Physical Chemistry B* **106**, 8531–8538 (2002).

36. Keiluweit, M. and Kleber, M. Molecular-level interactions in soils and sediments: The role of aromatic p-systems. *Environmental Science and Technology* **43**, 3421–3429 (2009).

37. Vernooij, M. G. C., Mohr, M., Tzvetkov, G. et al. On source identification and alteration of single diesel and wood smoke soot particles in the atmosphere; An X-ray microspectroscopy study. *Environmental Science and Technology* **43**, 5339–5344 (2009).

38. Tivanski, A. V., Hopkins, R. J., Tyliszczak, T., and Gilles, M. K. Oxygenated interface on biomass burn tar balls determined by single particle scanning transmission X-ray microscopy. *Journal of Physical Chemistry A* **111**, 5448–5458 (2007).

39. Michelsen, H. A., Tivanski, A. V., Gilles, M. K. et al. Particle formation from pulsed laser irradiation of soot aggregates studied with a scanning mobility particle sizer, a transmission electron microscope, and a scanning transmission x-ray microscope. *Applied Optics* **46**, 959–977, ISSN: 0003–6935 (2007).

40. Braun, A., Huggins, F. E., Shah, N. et al. Advantages of soft X-ray absorption over TEM-EELS for solid carbon studies—A comparative study on diesel soot with EELS and NEXAFS. *Carbon* **43**, 117–124 (2005).

41. Katrinak, K. A., Rez, P., and Buseck, P. R. Structural variations in individual carbonaceous particles from an urban aerosol. *Environmental Science and Technology* **26**, 1967–1976 (1992).

42. Hopkins, R. J., Tivanski, A. V., Marten, B. D., and Gilles, M. K. Chemical bonding and structure of black carbon reference materials and individual carbonaceous atmospheric aerosols. *Journal of Aerosol Science* **38**, 573–591 (2007).

43. Hitchcock, A. P., Dynes, J. J., Johansson, G., Wang, J., and Botton, G. Comparison of NEXAFS microscopy and TEM-EELS for studies of soft matter. *Micron* **39**, 311–319 (2008).

44. Wang, J., Morin, C., Li, L. et al. Radiation damage in soft X-ray microscopy. *Journal of Electron Spectroscopy and Related Phenomena* **170**, 25–36 (2009).

45. Schäfer, T., Michel, P., Claret, F. et al. Radiation sensitivity of natural organic matter: Clay mineral association effects in the Callovo-Oxfordian argillite. *Journal of Electron Spectroscopy and Related Phenomena* **170**, 49–56 (2009).

46. Braun, A., Kubatova, A., Wirick, S., and Mun, S. B. Radiation damage from EELS and NEXAFS in diesel soot and diesel soot extracts. *Journal of Electron Spectroscopy and Related Phenomena* **170**, 42–48 (2009).

47. Beetz, T. and Jacobsen, C. Soft X-ray radiation-damage studies in PMMA using a cryo-STXM. *Journal of Synchrotron Radiation* **10**, 280–283 (2003).

48. Cody, G. D., Brandes, J., Jacobsen, C., and Wirick, S. Soft X-ray induced chemical modification of polysaccharides in vascular plant cell walls. *Journal of Electron Spectroscopy and Related Phenomena* **170**, 57–64 (2009).

49. Coffey, T., Urquhart, S. G., and Ade, H. Characterization of the effects of soft X-ray irradiation on polymers. *Journal of Electron Spectroscopy and Related Phenomena* **122**, 65–78 (2002).

50. Howells, M. R., Hitchcock, A. P., and Jacobsen, C. J. Introduction: Special issue on radiation damage. *Journal of Electron Spectroscopy and Related Phenomena* **170**, 1–3 (2009).

51. Wang, J., Stover, H. D. H., and Hitchcock, A. P. Chemically selective soft X-ray direct-write patterning of multilayer polymer films. *Journal of Physical Chemistry C* **111**, 16330–16338 (2007).

52. Rightor, E. G., Hitchcock, A. P., Ade, H. et al. Spectromicroscopy of poly(ethylene terephthalate): Comparison of spectra and radiation damage rates in x-ray absorption and electron energy loss. *Journal of Physical Chemistry B* **101**, 1950–1960 (1997).

53. Ade, H. and Hitchcock, A. P. NEXAFS microscopy and resonant scattering: Composition and orientation probed in real and reciprocal space. *Polymer* **49**, 643–675 (2008).

54. Olynick, D. L., Liddle, J. A., Tivanski, A. V. et al. Scanning x-ray microscopy investigations into the electron-beam exposure mechanism of hydrogen silsesquioxane resists. *Journal of Vacuum Science and Technology B* **24**, 3048–3054 (2006).

55. Guay, D., Stewart-Ornstein, J., Zhang, X. R., and Hitchcock, A. P. *In situ* spatial and time-resolved studies of electrochemical reactions by scanning transmission X-ray microscopy. *Analytical Chemistry* **77**, 3479–3487 (2005).

56. Hopkins, R. J., Desyaterik, Y., Tivanski, A. V. et al. Chemical speciation of sulfur in marine cloud droplets and particles: Analysis of individual particles from the marine boundary layer over the California current. *Journal of Geophysical Research-Atmospheres* **113**, D04209 (2008).

57. Laskin, A., Iedema, M. J., and Cowin, J. P. Time-resolved aerosol collector for CCSEM/EDX single-particle analysis. *Aerosol Science and Technology* **37**, 246–260 (2003).

58. Russell, L. M., Maria, S. F., and Myneni, S. C. B. Mapping organic coatings on atmospheric particles. *Geophysical Research Letters* **29** (2002).

59. Brownlee, D., Tsou, P., Aleon, J. et al. Research article—Comet 81P/Wild 2 under a microscope. *Science* **314**, 1711–1716 (2006).

60. Cody, G. D., Ade, H., Alexander, C. M. O. et al. Quantitative organic and light-element analysis of comet 81P/Wild 2 particles using C-, N-, and O-μ-XANES. *Meteoritics and Planetary Science* **43**, 353–365 (2008).

61. Adam Hitchcock Group Home page. http://unicorn.mcmaster.ca/aXis2000.html (accessed August 5, 2010).

62. Koprinarov, I. N., Hitchcock, A. P., McCrory, C. T., and Childs, R. F. Quantitative mapping of structured polymeric systems using SVD analysis of soft X-ray images. *Journal of Physical Chemistry B* **106**, 5358–5364 (2002).

63. Lerotic, M., Jacobsen, C., Gillow, J. B. et al. Cluster analysis in soft X-ray spectromicroscopy: Finding the patterns in complex specimens. *Journal of Electron Spectroscopy and Related Phenomena* **144**, 1137–1143 (2005).

64. Lerotic, M., Jacobsen, C., Schafer, T., and Vogt, S. Cluster analysis of soft X-ray spectromicroscopy data. *Ultramicroscopy* **100**, 35–57(2004).

65. Stony Brook X-ray Microscopy Analysis Software page. http://xray1.physics.sunysb.edu/data/software.php (accessed August 5, 2010).

66. Moffet, R. C., Henn, T. R., Laskin, A., and Gilles, M. K. Automated assay of internally mixed individual particles using X-ray spectromicroscopy maps. Submitted (2010).

67. Mathworks home page. http://www.mathworks.com/matlabcentral/fileexchange/24006 (accessed August 5, 2010).

68. Rothe, J., Denecke, M. A., and Dardenne, K. Soft X-ray spectromicroscopy investigation of the interaction of aquatic humic acid and clay colloids. *Journal of Colloid and Interface Science* **231**, 91–97 (2000).

69. Plaschke, M., Rothe, J., Schafer, T. et al. Combined AFM and STXM *in situ* study of the influence of Eu(III) on the agglomeration of humic acid. *Colloids and Surfaces A—Physicochemical and Engineering Aspects* **197**, 245–256 (2002).

70. Plaschke, M., Rothe, J., Denecke, M. A., and Fanghanel, T. Soft X-ray spectromicroscopy of humic acid europium(III) complexation by comparison to model substances. *Journal of Electron Spectroscopy and Related Phenomena* **135**, 53–62 (2004).

71. Rothe, J., Plaschke, M., and Denecke, M. A. Scanning transmission X-ray microscopy as a speciation tool for natural organic molecules. *Radiochimica Acta* **92**, 711–715 (2004).

72. Plaschke, M., Rothe, J., Altmaier, M., Denecke, M. A., and Fanghanel, T. Near edge X-ray absorption fine structure (NEXAFS) of model compounds for the humic acid/actinide ion interaction. *Journal of Electron Spectroscopy and Related Phenomena* **148**, 151–157 (2005).

73. Schäfer, T., Buckau, G., Artinger, R. et al. Origin and mobility of fulvic acids in the Gorleben aquifer system: Implications from isotopic data and carbon/sulfur XANES. *Organic Geochemistry* **36**, 567–582 (2005).

74. Claret, F., Schafer, T., Rabung, T. et al. Differences in properties and Cm(III) complexation behavior of isolated humic and fulvic acid derived from Opalinus clay and Callovo-Oxfordian argillite. *Applied Geochemistry* **20**, 1158–1168 (2005).

75. Naber, A., Plaschke, M., Rothe, J., Hofmann, H., and Fanghanel, T. Scanning transmission X-ray and laser scanning luminescence microscopy of the carboxyl group and Eu(III) distribution in humic acid aggregates. *Journal of Electron Spectroscopy and Related Phenomena* **153**, 71–74 (2006).

76. Christl, I. and Kretzschmar, R. C-1s NEXAFS spectroscopy reveals chemical fractionation of humic acid by cation-induced coagulation. *Environmental Science and Technology* **41**, 1915–1920(2007).

77. Rothe, J., Kneedler, E. M., Pecher, K. et al. Spectromicroscopy of Mn distributions in micronodules produced by biomineralization. *Journal of Synchrotron Radiation* **6**, 359–361 (1999).

78. Schäfer, T., Hertkorn, N., Artinger, R., Claret, F., and Bauer, A. Functional group analysis of natural organic colloids and clay association kinetics using C(1s) spectromicroscopy. *Journal de Physique IV* **104**, 409–412 (2003).

79. Schmidt, C., Thieme, J., Neuhausler, U. et al. Spectromicroscopy of soil colloids. *Journal de Physique IV* **104**, 405–408 (2003).

80. Schumacher, M., Christl, I., Scheinost, A. C., Jacobsen, C., and Kretzschmar, R. Chemical heterogeneity of organic soil colloids investigated by scanning transmission X-ray microscopy and C-1s NEXAFS microspectroscopy. *Environmental Science and Technology* **39**, 9094–9100 (2005).

81. Lehmann, J., Liang, B. Q., Solomon, D. et al. Near-edge X-ray absorption fine structure (NEXAFS) spectroscopy for mapping nano-scale distribution of organic carbon forms in soil: Application to black carbon particles. *Global Biogeochemical Cycles* **19**, 1, GB1013 (2005).

82. Solomon, D., Lehmann, J., Kinyangi, J., Liang, B. Q., and Schafer, T. Carbon K-edge NEXAFS and FTIR-ATR spectroscopic investigation of organic carbon speciation in soils. *Soil Science Society of America Journal* **69**, 107–119 (2005).

83. Kinyangi, J., Solomon, D., Liang, B. I. et al. Nanoscale biogeocomplexity of the organomineral assemblage in soil: Application of STXM microscopy and C 1s-NEXAFS spectroscopy. *Soil Science Society of America Journal* **70**, 1708–1718 (2006).

84. Wan, J., Tyliszczak, T., and Tokunaga, T. K. Organic carbon distribution, speciation, and elemental correlations within soil micro aggregates: Applications of STXM and NEXAFS spectroscopy. *Geochimica et Cosmochimica Acta* **71**, 5439–5449 (2007).

85. Lehmann, J., Solomon, D., Kinyangi, J. et al. Spatial complexity of soil organic matter forms at nanometre scales. *Nature Geoscience* **1**, 238–242 (2008).

86. Brandes, J. A., Lee, C., Wakeham, S. et al. Examining marine particulate organic matter at sub-micron scales using scanning transmission X-ray microscopy and carbon X-ray absorption near edge structure spectroscopy. *Marine Chemistry* **92**, 107–121 (2004).

87. Schäfer, T., Claret, F., Bauer, A. et al. Natural organic matter (NOM)-clay association and impact on Callovo-Oxfordian clay stability in high alkaline solution: Spectromicroscopic evidence. *Journal de Physique IV* **104**, 413–416 (2003).

88. Yoon, T. H., Benzerara, K., Ahn, S. et al. Nanometer-scale chemical heterogeneities of black carbon materials and their impacts on PCB sorption properties: Soft X-ray spectromicroscopy study. *Environmental Science and Technology* **40**, 5923–5929 (2006).

89. Haberstroh, P. R., Brandes, J. A., Gelinas, Y. et al. Chemical composition of the graphitic black carbon fraction in riverine and marine sediments at sub-micron scales using carbon X-ray spectromicroscopy. *Geochimica et Cosmochimica Acta* **70**, 1483–1494 (2006).

90. Thieme, J., Gleber, S. C., Guttmann, P. et al. Microscopy and spectroscopy with X-rays for studies in the environmental sciences. *Mineralogical Magazine* **72**, 211–216 (2008).

91. Hitchcock, A. P., Morin, C., Tyliszczak, T. et al. Soft X-ray microscopy of soft matter—Hard information from two softs. *Surface Review and Letters* **9**, 193–201 (2002).

92. Lawrence, J. R., Swerhone, G. D. W., Leppard, G. G. et al. Scanning transmission X-ray, laser scanning, and transmission electron microscopy mapping of the exopolymeric matrix of microbial biofilms. *Applied and Environmental Microbiology* **69**, 5543–5554 (2003).

93. Chan, C. S., De Stasio, G., Welch, S. A. et al. Microbial polysaccharides template assembly of nanocrystal fibers. *Science* **303**, 1656–1658 (2004).

94. Toner, B., Fakra, S., Villalobos, M., Warwick, T., and Sposito, G. Spatially resolved characterization of biogenic manganese oxide production within a bacterial biofilm. *Applied and Environmental Microbiology* **71**, 1300–1310 (2005).

95. Benzerara, K., Menguy, N., Lopez-Garcia, P. et al. Nanoscale detection of organic signatures in carbonate microbialites. *Proceedings of the National Academy of Sciences of the United States of America* **103**, 9440–9445 (2006).

96. Dynes, J. J., Tyliszczak, T., Araki, T. et al. Speciation and quantitative mapping of metal species in microbial biofilms using scanning transmission X-ray microscopy. *Environmental Science and Technology* **40**, 1556–1565 (2006).

97. Felten, A., Bittencourt, C., Pireaux, J. J. et al. Individual multiwall carbon nanotubes spectroscopy by scanning transmission X-ray microscopy. *Nano Letters* **7**, 2435–2440 (2007).

98. Schafer, T., Chanudet, V., Claret, F., and Filella, M. Spectromicroscopy mapping of colloidal/particulate organic matter in Lake Brienz, Switzerland. *Environmental Science and Technology* **41**, 7864–7869 (2007).

99. Hunter, R. C., Hitchcock, A. P., Dynes, J. J., Obst, M., and Beveridge, T. J. Mapping the speciation of iron in *Pseudomonas aeruginosa* biofilms using scanning transmission X-ray microscopy. *Environmental Science and Technology* **42**, 8766–8772 (2008).

100. MaClean, L. C. W., Tyliszczak, T., Gilbert, P. et al. A high-resolution chemical and structural study of framboidal pyrite formed within a low-temperature bacterial biofilm. *Geobiology* **6**, 471–480 (2008).

101. Toner, B. M., Santelli, C. M., Marcus, M. A. et al. Biogenic iron oxyhydroxide formation at mid-ocean ridge hydrothermal vents: Juan de Fuca Ridge. *Geochimica Et Cosmochimica Acta* **73**, 388–403 (2009).

102. Dynes, J. J., Lawrence, J. R., Korber, D. R. et al. Morphological and biochemical changes in *Pseudomonas fluorescens* biofilms induced by sub-inhibitory exposure to antimicrobial agents. *Canadian Journal of Microbiology* **55**, 163–178 (2009).

103. Sandford, S. A., Aleon, J., Alexander, C. M. O. et al. Organics captured from comet 81P/Wild 2 by the Stardust spacecraft. *Science* **314**, 1720–1724 (2006).

104. Zolensky, M. E. Report—Mineralogy and petrology of comet 81P/Wild 2 nucleus samples. (vol. 314, pg. 1735, 2006). *Science* **316**, 543–543 (2007).

105. Flynn, G. J., Keller, L. P., Feser, M., Wirick, S., and Jacobsen, C. The origin of organic matter in the solar system: Evidence from the interplanetary dust particles. *Geochimica et Cosmochimica Acta* **67**, 4791–4806 (2003).

106. Keller, L. P., Messenger, S., Flynn, G. J. et al. The nature of molecular cloud material in interplanetary dust. *Geochimica et Cosmochimica Acta* **68**, 2577–2589 (2004).

107. Flynn, G. J., Keller, L. P., Wirick, S., Jacobsen, C., and Sutton, S. R. Analysis of interplanetary dust particles by soft and hard X-ray microscopy. *Journal de Physique IV* **104**, 367–372 (2003).

108. Flynn, G. J., Keller, L. P., Jacobsen, C., and Wirick, S. An assessment of the amount and types of organic matter contributed to the Earth by interplanetary dust. *Space Life Sciences: Steps toward Origin(S) of Life* **33**, 57–66 (2004).

109. Jacobsen, C., Wirick, S., Flynn, G., and Zimba, C. Soft X-ray spectroscopy from image sequences with sub-100 nm spatial resolution. *Journal of Microscopy—Oxford* **197**, 173–184 (2000).

110. Cody, G. D., Alexander, C. M. O., Yabuta, H. et al. Organic thermometry for chondritic parent bodies. *Earth and Planetary Science Letters* **272**, 446–455 (2008).

111. Botto, R. E., Cody, G. D., Kirz, J. et al. Selective chemical mapping of coal microheterogeneity by scanning-transmission X-ray microscopy. *Energy & Fuels* **8**, 151–154 (1994).

112. Cody, G. D., Ade, H., Wirick, S., Mitchell, G. D., and Davis, A. Determination of chemical–structural changes in vitrinite accompanying luminescence alteration using C-NEXAFS analysis. *Organic Geochemistry* **28**, 441–455 (1998).

113. Cody, G. D., Botto, R. E., Ade, H. et al. Inner-shell spectroscopy and imaging of a subbituminous coal— *In-situ* analysis of organic and inorganic microstructure using C(1s)-NEXAFS, Ca(2p)-NEXAFS, and Cl(2s)-NEXAFS. *Energy & Fuels* **9**, 525–533 (1995).

114. Cody, G. D., Botto, R. E., Ade, H. et al. C-NEXAFS microanalysis and scanning-X-ray microscopy of microheterogeneities in a high-volatile a bituminous coal. *Energy & Fuels* **9**, 75–83 (1995).

115. Cody, G. D., Botto, R. E., Ade, H., and Wirick, S. The application of soft X-ray microscopy to the *in-situ* analysis of sporinite in coal. *International Journal of Coal Geology* **32**, 69–86 (1996).

116. Flynn, G. J., Keller, L. P., Jacobsen, C., and Wirick, S. Carbon and potassium mapping and carbon bonding state measurements on interplanetary dust. *Meteoritics & Planetary Science* **33**, A50–A50 (1998).

117. Braun, A., Shah, N., Huggins, F. E. et al. A study of diesel PM with X-ray microspectroscopy. *Fuel* **83**, 997–1000 (2004).

118. Hopkins, R. J., Lewis, K., Desyaterik, Y. et al. Correlations between optical, chemical, and physical properties of biomass burn aerosols. *Geophysical Research Letters* **34**, L18806 (2007).

119. Moffet, R. C., Henn, T. R., Tivanski, A. V. et al. Microscopic characterization of carbonaceous aerosol particle aging in the outflow from Mexico City. *Atmospheric Chemistry and Physics Discussion* **9**, 16993–17033 (2009).

120. Gilles, M. K., Kilcoyne, A. D. L., Tyliszczak, T. et al. Scanning transmission X-ray microscopy imaging of aerosol particles. *EOS Transactions AGU Fall Meeting Supplement* **84**, Abstract A51F-0735 (2003).

121. Braun, A. Carbon speciation in airborne particulate matter with C (1s) NEXAFS spectroscopy. *Journal of Environmental Monitoring* **7**, 1059–1065 (2005).

122. Akhter, M. S., Chughtai, A. R., and Smith, D. M. The structure of hexane soot I: Spectroscopic studies. *Applied Spectroscopy* **39**, 143–153 (1985).

123. Aiken, A. C., Decarlo, P. F., Kroll, J. H. et al. O/C and OM/OC ratios of primary, secondary, and ambient organic aerosols with high-resolution time-of-flight aerosol mass spectrometry. *Environmental Science and Technology* **42**, 4478–4485 (2008).

124. Braun, A., Shah, N., Huggins, F. E. et al. X-ray scattering and spectroscopy studies on diesel soot from oxygenated fuel under various engine load conditions. *Carbon* **43**, 2588–2599 (2005).

125. Lenardi, C., Marino, M., Barborini, E., Piseri, P., and Milani, P. Evaluation of hydrgoen chemisorption in nanostructured carbon films by near edge x-ray absorption spectroscopy. *European Physics Journal B* **46**, 441–447 (2005).

126. Gago, R., Jiménez, I., and Albella, J. M. Detecting with X-ray absorption spectroscopy the modifications of the bonding structure of graphitic carbon by amorphisation, hydrogenation and nitrogenation. *Surface Science* **482–485**, 530–536 (2001).

127. Bond, T. C. and Bergstrom, R. W. Light absorption by carbonaceous particles: An investigative review. *Aerosol Science and Technology* **40**, 1–41 (2006).

128. Li, J., Posfai, M., Hobbs, P. V., and Buseck, P. R. Individual aerosol particles from biomass burning in southern Africa: 2, Compositions and aging of inorganic particles. *Journal of Geophysical Research-Atmospheres* **108** (2003).

129. Pósfai, M., Gelencser, A., Simonics, R. et al. Atmospheric tar balls: Particles from biomass and biofuel burning. *Journal of Geophysical Research-Atmospheres* **109** (2004).

130. Pósfai, M., Simonics, R., Li, J., Hobbs, P. V., and Buseck, P. R. Individual aerosol particles from biomass burning in southern Africa: 1. Compositions and size distributions of carbonaceous particles. *Journal of Geophysical Research-Atmosphere* **108** (2003).

131. Hand, J. L., Malm, W. C., Laskin, A. et al. Optical, physical, and chemical properties of tar balls observed during the Yosemite Aerosol Characterization Study. *Journal of Geophysical Research-Atmospheres* **110** (2005).

132. Maria, S. F., Russell, L. M., Gilles, M. K., and Myneni, S. C. B. Organic aerosol growth mechanisms and their climate-forcing implications. *Science* **306**, 1921–1924 (2004).

133. Knopf, D. A., Wang, B., Laskin, A., Moffet, R. C., and Gilles, M. K. Anthropogenic organic aerosols as potential ice nuclei for cirrus clouds. Submitted for publication (2010).

134. Takahama, S., Gilardoni, S., Russell, L. M., and Kilcoyne, A. L. D. Classification of multiple types of organic carbon composition in atmospheric particles by scanning transmission X-ray microscopy analysis. *Atmospheric Environment* **41**, 9435–9451 (2007).

135. Day, D. A., Takahama, S., Gilardoni, S., and Russell, L. M. Organic composition of single and submicron particles in different regions of western North America and the eastern Pacific during INTEX-B 2006. *Atmospheric Chemistry and Physics* **9**, 5433–5446 (2009).

136. Leaitch, W. R., Macdonald, A. M., Anlauf, K. G. et al. Evidence for Asian dust effects from aerosol plume measurements during INTEX-B 2006 near Whistler, BC. *Atmospheric Chemistry and Physics* **9**, 3523–3546 (2009).

137. Liu, S., Takahama, S., Russell, L. M., Gilardoni, S., and Baumgardner, D. Oxygenated organic functional groups and their sources in single and submicron organic particles in MILAGRO 2006 campaign. *Atmospheric Chemistry and Physics* **9**, 6849–6863 (2009).

138. Tonner, B. P., Droubay, T., Denlinger, J. et al. Soft X-ray spectroscopy and imaging of interfacial chemistry in environmental specimens. *Surface and Interface Analysis* **27**, 247–258 (1999).

139. Moffet, R. C., Desyaterik, Y., Hopkins, R. J. et al. Characterization of aerosols containing Zn, Pb, and Cl from an industrial region of Mexico City. *Environmental Science and Technology* **42**, 7091–7097 (2008).

140. Schnaas, L., Rothenberg, S. J., Flores, M. F. et al. Blood lead secular trend in a cohort of children in Mexico City (1987–2002). *Environmental Health Perspectives* **112**, 1110–1115 (2004).

141. Lam, P. J., Bishop, J. K. B., Henning, C. C., Marcus, M. A., Waychunas, G. A., Fung, I. Y. Wintertime phytoplankton bloom in the subarctic Pacific supported by continental margin iron. *Global Biogeochemical Cycles* **20**, 1, GB1006 (2006) DOI: 10.1029/2005GB002557.

142. Gleber, G., Thieme, J., Niemeyer, J., and Feser, M. Interaction of organic substances with iron studied by O1s spectroscopy—Development of an analysis program. *Journal de Physique IV* **104**, 429–432 (2003).

143. Miot, J., Benzerara, K., Morin, G. et al. Iron biomineralization by anaerobic neutrophilic iron-oxidizing bacteria. *Geochimica et Cosmochimica Acta* **73**, 696–711 (2009).

144. Chan, C. S., Fakra, S., Edwards, D. C., Emerson, D., and Banfield, J. Iron oxyhydroxide mineralization on microbial extracellular polysaccharides. *Geochimica et Cosmochimica Acta* **73**, 3807–3818 (2009).

145. Toner, B. M., Fakra, S. C., Manganini, S. J. et al. Preservation of iron(II) by carbon-rich matrices in a hydrothermal plume. *Nature Geoscience* **2**, 197–201 (2009).

146. Takahama, S., Gilardoni, S., and Russell, L. M. Single-particle oxidation state and morphology of atmospheric iron aerosols. *Journal of Geophysical Research-Atmospheres* **113**, D22202 (2008).

147. Hitchcock, A. P., Tourillon, G., Garrett, R. et al. Inner-shell excitation of gas-phase and polymer thin-film 3-alkylthiophenes by electron energy loss and X-ray photoabsorption spectroscopy. *Journal of Physical Chemistry* **94**, 2327–2333 (1990).

148. Sarret, G., Connan, J., Kasrai, M. et al. Chemical forms of sulfur in geological and archaeological asphaltenes from Middle East, France, and Spain determined by suflur K-, and L-edge X-ray absorption near-edge structure spectroscopy. *Geochimica et Cosmochimica Acta* **63**, 3767–3779 (1999).

149. Zaveri, R. A., Berkowitz, C. M., Brechtel, F. J. et al. Nighttime chemical evolution of aerosol and trace gases in a power plant plume: Implications for secondary organic nitrate and organosulfate aerosol formation, NO_3 radical chemistry, and N_2O_5 heterogeneous hydrolysis. *Journal of Geophysical Research-Atmospheres* **115**, D12304 (2010).

150. Surratt, J. D., Kroll, J. H., Kleindienst, T. E. et al. Evidence for organosulfates in secondary organic aerosol. *Environmental Science and Technology* **41**, 517–527 (2007).

151. Pratt, K. A., Twohy, C. H., Murphy, S. M. et al. Observation of playa salts as nuclei of orographic wave clouds. *Journal of Geophysical Research-Atmospheres* **115**, D15301 (2010) DOI:10.1029/2009JD013606. (2010).

152. Hiranamu, N., Moffet, R. C., Gilles, M. K. et al. Personal communication. (2010).

153. Bernard, S., Benzerara, K., Beyssac, O. et al. Ultrastructural and chemical study of modern and fossil sporoderms by scanning transmission X-ray microscopy (STXM). *Review of Palaeobotany and Palynology* **156**, 248–261 (2009).

154. Amri, C. E., Maurel, M. C., Sagon, G., and Barone, M. H. The micro-distribution of carbonaceous matter in the Murchison meteorite as investigated by Raman imaging. *Spectrochimica Acta Part A, Molecular and Biomolecular Spectroscopy* **61**, 2049–2056 (2005).

155. Benzerara, K., Morin, G., Yoon, T. H. et al. Nanoscale study of As biomineralization in an acid mine drainage system. *Geochimica et Cosmochimica Acta* **72**, 3949–3963 (2008).

156. Kaegi, R., Wagner, T., Hetzer, B. et al. Size, number and chemical composition of nanosized particles in drinking water determined by analytical microscopy and LIBD. *Water Research* **42**, 2778–2786 (2008).

157. Morin, C., Ikeura-Sekiguchi, H., Tyliszczak, T. et al. X-ray spectromicroscopy of immiscible polymer blends: Polystyrene-poly(methyl methacrylate). *Journal of Electron Spectroscopy and Related Phenomena* **121**, 203–224 (2001).

158. Winesett, D. A. and Tsou, A. H. The application of high resolution chemical imaging techniques for butyl rubber blends. *Rubber Chemistry and Technology* **81**, 265–275 (2008).

159. Tivanski, A. V. and Ghorai, S. Personal communication (2010).

160. Drake, I. J., Liu, T. C. N., Gilles, M. et al. An *in situ* cell for characterization of solids by soft x-ray absorption. *Review of Scientific Instruments* **75**, 3242–3247 (2004).

161. de Smit, E., Swart, I., Creemer, J. F. et al. Nanoscale chemical imaging of a working catalyst by scanning transmission X-ray microscopy. *Nature* **456**, U222–U239 (2008).

162. de Smit, E., Swart, I., Creemer, J. F. et al. Nanoscale chemical imaging of the reduction behavior of a single catalyst particle. *Angewandte Chemie International Edition* **48**, 3632–3636 (2009).

163. Ghorai, S. and Tivanski, A. V. Hygroscopic behavior of individual submicron particles studied by X-ray spectromicroscopy. Submitted to *Analytical Chemistry* (2010).

164. Personal communication (2010).

165. Tzvetkov, G., Graf, B., Wiegner, R. et al. Soft X-ray spectromicroscopy of phase-change microcapsules. *Micron* **39**, 275–279 (2008).

166. Thibault, P., Dierolf, M., Menzel, A. et al. High-resolution scanning x-ray diffraction microscopy. *Science* **321**, 379–382 (2008).

18 Electron Beam Analysis and Microscopy of Individual Particles

Alexander Laskin

CONTENTS

18.1 Introduction 463
18.2 Electron Microscopy and Electron Probe Techniques 464
18.3 Applications of Electron Microscopy in Studies of Particle Physical Chemistry 468
    18.3.1 Chemical Characterization of Field-Collected Particles 468
    18.3.2 Hygroscopic Properties of Individual Particles...473
    18.3.3 Ice Nucleation Properties of Individual Particles..477
    18.3.4 Optical Properties of Individual Particles 478
    18.3.5 Laboratory Studies of Gas-Particle Reactions...479
18.4 Application of Complementary Analysis Techniques in Studies of Particle Chemistry......483
18.5 Summary 486
Acknowledgments..486
References..487

18.1 INTRODUCTION

The significance and critical impact of aerosols in a variety of environmental and engineering processes has been recognized in many branches of modern environmental science, including geophysics and climate change, atmospheric chemistry and environmental catalysis, nanoscience and nanotoxicology, pharmacology, and drug delivery. For instance, the scattering of light by atmospheric aerosol has a direct effect on visibility and the Earth's radiation balance. Additionally, aerosol's impact on climate includes modification of cloud properties and precipitation developments as a result of aerosol particles acting as cloud condensation nuclei (CCN) and ice nuclei (IN). Aerosols affect air quality and atmospheric chemistry through their heterogeneous gas-particle reactions that modify particle composition and properties and alter concentrations of gas-phase species. Submicron-sized aerosol particles, either commercially engineered or of anthropogenic pollution origin, often contain inflammatory, carcinogenic, and mutagenic compounds and may have adverse effects on human health upon inhalation. Conversely, the ability of small particles to deeply penetrate into human lungs and interact directly with alveolar cells provides an efficient and quick means of delivery of aerosolized drugs, which are widely used in medical applications.

Environmental aerosols comprise a complex mixture of particles of many different sizes, origins, and physical and chemical properties, which also exhibit diverse distributions of chemical species among different particles (external mixing) and within the same individual particles (internal mixing). Chemical, morphological (size and shape), and phase data of individual particles are of key importance for understanding the formation and reaction mechanisms of aerosols, their possible atmospheric history, source apportionment, potential toxicity, or biocompatibility. Accurate assessment of the environmental impact of airborne particles requires fundamental understanding of the relationship between their composition and chemical and physical properties. However, obtaining comprehensive information on particle composition and properties is inherently challenging because of substantial variations in the composition of individual particles. As a result, a comprehensive characterization of atmospheric particles can be only obtained using a combination of complementary analytical methods discussed in detail in this book.

This chapter provides an overview of the most recent applications of electron microscopy (EM) and microprobe techniques in research aimed to address different aspects of particle environmental chemistry. Microanalysis of particles provides information on size, morphology, phase, and the composition of particles, including the lateral distribution of chemical species within individual particles. The qualitative and quantitative data obtained in these studies are essential for evaluating hygroscopic and optical properties of particles, understanding their origin, possible formation processes, their atmospheric aging, reactivity, and transformations. As will be discussed, data obtained in these studies provide important insights on the relationship between particles' composition and their environmental impacts.

The present chapter is not intended to convey a comprehensive review of the EM and associated microspectroscopy techniques, but to embrace recent applications of these techniques in studies of particle environmental and physical chemistry. This chapter is organized into five sections. Following the introduction, Section 18.2 provides the basic description of the EM-based methods of particle analysis. The specific applications are presented in Section 18.3, which include examples drawn from the most recently published works over the last decade. The topics of the presented studies include chemical characterization of field-collected and laboratory-generated particles, hygroscopic properties of particles, ice nucleation propensity of particles, optical properties of particles, and heterogeneous gas-particle reaction chemistry. Section 18.4 discusses synergism and the application of complementary particle analysis techniques. A summary is provided in Section 18.5.

For more detailed information on the fundamentals of the EM-based microanalytical techniques and specific aspects of particle microanalysis, the reader is referred to a number of additional reviews and book chapters (Anderson and Buseck, 1998; Buseck and Anderson, 1998; Injuk et al., 1998; De Bock and Van Grieken, 1999; Fletcher et al., 2001; Poelt et al., 2002; Burleson et al., 2004; Szaloki et al., 2004; Laskin et al., 2006a; Leppard, 2008).

18.2 ELECTRON MICROSCOPY AND ELECTRON PROBE TECHNIQUES

The unique advantage of electron beam-based methods in particle analysis is that they combine two analytical tasks that inherently complement one another: (1) imaging of particles with sufficient resolution to visualize their morphology, and (2) microspectroscopy analysis of particles that provides information on the local chemical composition, including lateral heterogeneity within individual particles. Figure 18.1 illustrates a particle positioned on the substrate and irradiated with an electron beam along with basic signals produced from the interaction of the incident electrons and the specimen (particle) material. These signals include secondary electrons (SE) ejected mostly from particle surface, backscattered primary electrons (BSE) and transmitted primary electrons (TE), and characteristic x-rays emitted from particle microvolume.

Scanning electron microscopes (SEMs) equipped with detectors for either energy-dispersive analysis (EDX) or wavelength-dispersive analysis of x-rays (WDX) are most commonly used for particle analysis (De Bock and Van Grieken, 1999; Injuk et al., 1998; Laskin et al., 2006a; Szaloki et al., 2004, and references therein). In SEM, the electron beam is focused down to 1–2 nm size, which can be scanned across the specimen to provide an image or stopped at one position for microanalysis. Detection of SE, BSE, and TE signals or their digital mix provides a means of particle imaging, whereas EDX microanalysis is used to detect and quantify the elemental composition

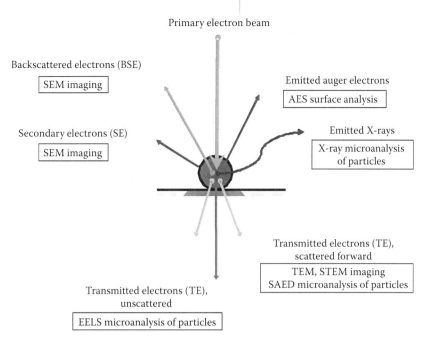

Primary electron beam

Backscattered electrons (BSE)

SEM imaging

Emitted auger electrons

AES surface analysis

Secondary electrons (SE)

SEM imaging

Emitted X-rays

X-ray microanalysis
of particles

Transmitted electrons (TE),
scattered forward

TEM, STEM imaging
SAED microanalysis of particles

Transmitted electrons (TE),
unscattered

EELS microanalysis of particles

FIGURE 18.1 Schematic diagram of various signals used in particle analysis by different techniques and approaches of EM electron-based microspectroscopy.

of particles. Figure 18.2 illustrates capabilities of the SE, BSE, and TE modes of imaging demonstrated for a soot particle with Ca-containing fly-ash inclusions deposited on a grid-supported thin film substrate. A low atomic number carbonaceous material of soot has especially low scattering power. As a result, it produces SE signals, but almost no backscattering occurs. Therefore, fractal soot chains are clearly seen in the SE mode, but are nearly invisible in the BSE mode. In contrast, because of very low BSE yields from the substrate carbon film and from the soot chain, contrast-enhanced BSE signals from the higher atomic number Ca-containing fly-ash inclusions makes them remarkably visible in the BSE mode. The TE mode of imaging is not common for SEM instruments and involves the detection of transmitted primary electrons scattered forward after interaction with the particle. The TE mode of imaging can be implemented only over particles deposited on thin film substrates and requires special arrangement of an annular solid-state electron detector below the sample (Laskin et al., 2006a). Due to the high electron transparency of the thin film, even low forward scattering of electrons by particles produce sharp contrast in the TE signal above the thin film background and results in a largely black-and-white particle image featuring its two-dimensional projection area. Combined together, the three modes of SEM imaging enable comprehensive characterization of particle morphology, including the visualization of the form and the location of possible inclusions within individual particles. Many modern SEMs employ computer-controlled (CC SEM/EDX) particle analysis, which allows automated measurements of basic particle morphology (size and aspect ratio) and elemental composition over a large number of individual particles. In this case, digitally mixed TE/BSE imaging is the method of choice for the automated search of particles and accurate positioning of the electron beam prior to the EDX analysis (Laskin et al., 2006a).

The energy of primary electrons striking the particle is partially absorbed by the material, which results in subsequent emission of x-rays characteristic of the particle composition. Depending on its energy, the electron beam typically penetrates into a volume of a few μm^3 inside the specimen. For submicron particles, the electron beam penetrates through the entire particle, resulting in lateral scattering of electrons transmitted through the particle sides. These electrons strike the

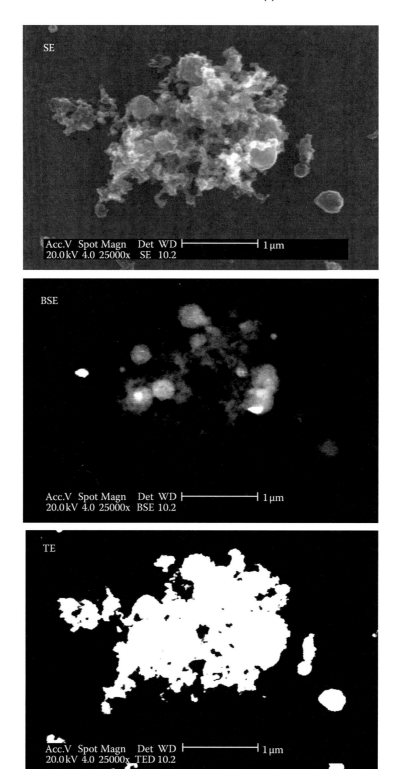

FIGURE 18.2 SEM images of a soot particle with round-shaped Ca-containing fly-ash inclusions. Images were acquired using SE, BSE, and TE signals over the particle supported by a thin film substrate. Size bars on the micrographs depict a 1 μm scale.

nearby substrate and result in a substantial yield of background x-rays. The conventional use of SEM/EDX involves particle samples supported on thick, not electron transparent substrates such as metal foils, polished carbon plates, or polycarbonate filters. In these cases, the characteristic x-rays of submicron particles are difficult to discriminate against the discrete background lines from the substrate and the broad continuum (bremsstrahlung) background. In addition, quantitative analysis of particle composition from acquired x-ray spectra requires corrections for the *ZAF* effects related to atomic number (*Z*)-dependent electron scattering, absorption (*A*), and fluorescence (*F*), as well as the effects of particle size and shape (Armstrong, 1991). Quantitative assessment of these effects is not trivial and requires the application of sophisticated, computer-intensive Monte Carlo-based modeling approaches (e.g., Ro et al., 1999, 2003; Choël et al., 2005, 2007). In practice, the major goal of particle analysis usually is to detect and classify particles into specific particle-type groups present in the sample, which does not necessarily require precise knowledge of particle composition. This can be accomplished using CC SEM/EDX analysis of particles captured on grid-supported thin film substrates (Laskin et al., 2006a). Because of the low background of x-rays emitted from the thin film substrate, the major and minor elemental peaks can be clearly identified in the EDX spectra of particles collected at short acquisition times, including a semiquantitative detection of C, N, and O peaks (Laskin et al., 2003a, 2006a). The approach allows a routine CC SEM/EDX analysis of particles down to 0.1 μm with chemical characterization sufficient for the particle-type classification.

The x-ray signal also can be used to generate element-specific images of particles—elemental maps. Modern SEM/EDX instruments implement elemental mapping in a mode when x-ray spectra are recorded over the entire energy range at each pixel of the SEM image with subsequent reconstruction of maps for any elements of interest. External and internal heterogeneity of individual particles can be effectively visualized using x-ray maps at the submicron level. Again, the use of thin film substrates provides a better contrast and lateral resolution of the obtained elemental maps that can reveal important insights on the chemistry of particles (Krueger et al., 2003a, 2004). Primary electrons of different energies penetrate to different depths of a particle, so that corresponding x-ray spectra recorded at different energies of the incident beam can be used for particle depth profiling and the assessment of its possible core–shell structure (Ro et al., 2001). Additionally, elemental analysis of the particle surface can be performed using a SEM equipped with an appropriate detector for Auger electron microspectroscopy, providing information on oxidation state and chemical bonding environment of elements within the top 2 nm (approximate) outer layer of the particle (Strausser, 1992).

A detailed characterization of the internal structure of particles with subnanometer spatial resolution can be obtained using high-resolution transmission electron microscopy (HR-TEM) (Buseck and Pósfai, 1999; Burleson et al., 2004; Chen et al., 2006; Hays and Vander Wal, 2007; Leppard, 2008; Hower et al., 2008). Compared to SEM, HR-TEM has important advantages for particle analysis because of its higher imaging resolution and applications of unique analytical techniques to probe particle composition. Specifically, the phase and crystalline structures of individual particles can be identified using selected-area electron diffraction (SAED) analysis (Li et al., 2003; Utsunomiya and Ewing, 2003; Chen et al., 2004, 2005, 2006; Utsunomiya et al., 2004; Kojima et al., 2006). Chemical information in HR-TEM can be obtained using electron energy loss spectroscopy (EELS) that provides information on coordination and valence states of elements detected in particles (Chen et al., 2004, 2005, 2006). EELS can be also used to provide high-resolution elemental maps of particles (Pósfai et al., 1999, 2004; Hand et al., 2005). Additionally, HR-TEM options of composition-specific imaging of particles include high-angle annular dark-field (HAADF) imaging (often called Z-contrast imaging) and energy filtered (EF) imaging (elemental mapping) (Utsunomiya and Ewing, 2003; Utsunomiya et al., 2004; Chen et al., 2004, 2005, 2006).

Specific applications of the SEM- and TEM-based techniques in recent studies of particle environmental and physical chemistry will be presented in the following section of this chapter.

18.3 APPLICATIONS OF ELECTRON MICROSCOPY IN STUDIES OF PARTICLE PHYSICAL CHEMISTRY

18.3.1 CHEMICAL CHARACTERIZATION OF FIELD-COLLECTED PARTICLES

The analysis of field-collected particles is, perhaps, the most widespread application of EM and associated microprobe techniques reported in the literature over the last decade. To make microscopy analysis an effective and practical tool for field studies, a few aspects need to be carefully addressed for each study. First, appropriate sampling devices that collect particle samples on substrates suitable for later microscopy analysis need to be employed. In any field study, researchers usually "over" collect samples and later select only a subset for detailed microscopy examination. Therefore, the second important aspect in particle microscopy studies is identifying the most viable samples to undergo extensive microscopy analyses. Typically, field measurements from codeployed *in situ* instrumentation and meteorological records provide guidance for sample selection. Selected samples can be analyzed using a variety of microscopy and microprobe techniques to characterize the size, morphology, phase, and composition of particles. The strategy for these analyses can be generally described as going from relatively quick imaging and the automated assessment of general particle types present in samples using CC SEM/EDX single-particle analysis at a rate of approximately 700 analyzed particles per hour. This is followed by more detailed microscopy studies focused on chemical and physical properties of individual particles, including their internal heterogeneity, which require significant operator time and effort. Below we illustrate the paradigm for these microscopy studies based on a number of recent results reported for different field campaigns.

SEM images shown in Figure 18.3 illustrate different particle types observed in the field-collected samples. In all these studies, particle sampling was carried out using a time-resolved aerosol collector (TRAC) (Laskin et al., 2003a, 2006a). The TRAC device consists of a single jet impactor with a traveling impaction plate on which up to 550 microscopy substrates can be prearranged and replaced automatically over a preset period of time. An important advantage of TRAC-collected samples is their compatibility with an array of microscopy and microspectroscopy techniques. Figure 18.3a,b illustrates two prominent types of biomass smoke particles collected during the Yosemite Aerosol Characterization Study (YACS). The study took place during intensive forest fires in summer 2002 and was focused on the characterization of optical and hygroscopic properties of biomass burning particles (Hand et al., 2005). Organic particles with or without inorganic inclusions (a) and tar ball particles (b) were two of the most abundant particle types observed. Organic particles are largely electron-transparent and, therefore, are usually seen in SEM as dark areas that may or may not contain inorganic inclusions, which are seen as bright specks. The organic nature of the dark particles is inferred from their EDX spectra that show no elements other than C and O, while EDX inspection of the bright specks indicates the presence of K, S, Cl, Si, and Na. Tar balls are recognized by their characteristic spherical morphology. Their EDX and EELS spectra indicate elemental compositions consisting mostly of C and O and trace amounts of N, S, K, Cl, Si, and Na. Tar ball particles are typically observed as individual spheres and their agglomerates with only occasional internal mixing with other particle types. TEM images and EELS maps of the tar balls revealed that their internal structure includes a spherical core and an approximately 30 nm-thin outer oxygenated layer (Hand et al., 2005). The layer was attributed to the atmospheric processing of particles, which resulted in large concentrations of carboxylic carbonyls, oxygen-substituted alkyl (O-alkyl-C), and ketone carbonyl functional groups on the surface of the particles (Tivanski et al., 2007). During the YACS study, real-time optical measurements employed at the sampling site indicated specific time periods of high levels of aerosol light scattering. The CC SEM/EDX analysis indicated that tar balls accounted for approximately 90% of the particles in the samples collected during the time of the light-scattering episode, thereby establishing the correlation of these particles to deteriorating visibility and the specific optical properties measured during their appearance.

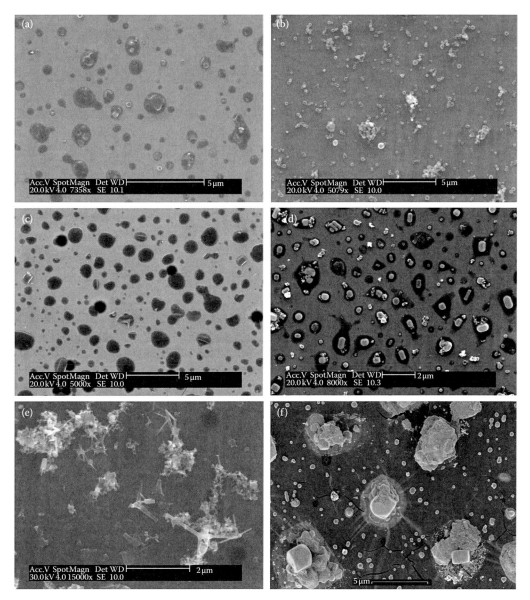

FIGURE 18.3 SEM images illustrating morphology and the internal heterogeneities of particles collected in different field studies. Biomass burning organic particles with inorganic inclusions (a) (From Hand, J. L. et al. 2005. *Journal of Geophysical Research–Atmospheres* 110. Copyright 2005, American Geophysical Union. With permission.) and biomass burning tar ball particles (b) from the YACS 2002 study. (From Hand, J. L. et al. 2005. *Journal of Geophysical Research–Atmospheres* 110. Copyright 2005, American Geophysical Union. With permission.) Homogeneously mixed organic/sulfate particles (c) sampled from the power plant plume during the NAOPEX 2002 experiment. (From Zaveri, R. A. et al. 2010. *Journal of Geophysical Research–Atmospheres* 115.Copyright 2010, American Geophysical Union. With permission.) Primary inorganic particles coated with dark layers of photochemically formed secondary organic material (d) collected in downtown Mexico City during the MILAGRO 2006. (From Moffet, R. C. et al. 2009. *Atmospheric Chemistry and Physics Discussion* 9:16993–17033.Copyright 2009, European Geophysical Union. With permission.) Needle-like particles containing Pb, Zn, and Cl (e) attributed to waste incineration in Mexico City. (From Moffet, R. C. et al. 2008. *Environmental Science and Technology* 42:7091–7097. Copyright 2009, American Chemical Society. With permission.) Atmospherically processed micron-sized sea-salt particles and submicron ammonium sulfate particles (f) collected during the TexAQS 2000 study. (From Laskin, A., M. J. Iedema, and J. P. Cowin. 2002. *Environmental Science and Technology* 36:4948–4955. Copyright 2002, American Chemical Society. With permission.)

Based on the closure analysis of the optical and the particle composition data, a refractive index (RI) of RI = 1.56 + 0.02i specific for the tar balls particles observed in that study was derived.

Figure 18.3a illustrates particles collected on board the G-1 research aircraft during the Nighttime Aerosol–Oxidant Plume Experiment (NAOPEX), which was carried out in New England in July 2002. The experiment was designed to characterize aerosols and trace gases originating from the greater Boston area and their atmospheric evolution during the nighttime. Aircraft flights were conducted to measure downwind characteristics of the urban plume. Using the meteorology data, particle samples collected from different parts of the plume originating from one of the major regional power plants were selected for detailed microscopy analysis. X-ray microanalysis of individual particles indicated the presence of C, O, N, and S, revealing the dominant presence of internally mixed organic and sulfate particles associated with the power plant plume. Additional microspectroscopic analysis of particles using the synchrotron-based scanning transmission x-ray microscopy (STXM) and the near-edge x-ray absorption fine structure spectroscopy (NEXAFS) suggested the presence of organosulfate compounds in particles. STXM/NEXAFS analysis also revealed a decrease of the C=C double bonds in the aerosol organic material sampled along the plume trajectory. The observed loss of C=C bonds was attributed to the heterogeneous chemistry of particles reacting with NO_3 radicals present in the power plant plume (Zaveri et al., 2010).

Figure 18.3d illustrates urban particles sampled in the polluted environment of Mexico City during the Megacities Initiative: Local and Global Research Observations (MILAGRO) experiment conducted in spring 2006. During the study, ground-based aerosol sampling was conducted at three supersites with the goal of characterizing chemical and physical transformations and the ultimate fate of pollution particulates in the urban plume originating in Mexico City and their evolution during regional transport downwind from the city (Moffet et al., 2010). Freshly emitted particles at the city center are mostly black carbon and sulfates, which have been quickly coated in the air with photochemically formed secondary organic materials. As the plume evolves from the city center, the organic mass per particle increases, and the fractions of black carbon and inorganic constituents decrease (Johnson et al., 2005; Adachi and Buseck, 2008; Moffet et al., 2010). The internally mixed nature of primary particles coated with organics is expected to impact their heterogeneous chemistry, optical, and hygroscopic properties. In addition, industrial particulate emissions from large northern point sources in Mexico City result in the frequent appearance of particles with high concentrations of Zn, Pb, and Cl (Johnson et al., 2006). Real-time data obtained using single-particle mass spectrometry were used to select subsets of particle samples for microscopy and microspectroscopy analyses (Moffet et al., 2008). The results indicated that many of the Zn-containing particles had needle-like structures as illustrated Figure 18.3e. Chemical composition of those particles was confirmed as ZnO, $Zn(NO_3)_2 \cdot 6H_2O$, $PbCl_2$, and $Pb(NO_3)_2$, which is indicative of waste incineration particulates that undergo atmospheric heterogeneous processing where chlorides are displaced by nitrates (Moffet et al., 2008).

Figure 18.3f illustrates a particle sample collected during the Texas Air Quality Study (TexAQS) conducted in August 2000. The sample is dominated by two particle types: sea-salt (few micron-sized particles having cubic shaped crystals of NaCl) and ammonium sulfate (submicron particles with nearly spherical shape). CC SEM/EDX analysis made it possible to follow, in detail, the atmospheric aging of individual sea-salt particles detected in a diverse mixture of atmospheric aerosol (Laskin et al., 2003a). Particle samples were collected every 10 min at the measurement site (located on the 62nd floor of a Houston skyscraper) for an entire month of the field deployment. Then, based on the meteorological data, the subset of samples from a particular time period was selected to study particle composition in the air mass that originated from open areas of the Gulf of Mexico and was transported over urban areas of greater Houston. The CC SEM/EDX analysis of these samples indicated that sea-salt particles were plentiful in all analyzed samples. These sea-salt particles were identified based on their characteristic x-ray spectra, and the elemental composition was quantified for each individual particle. The average composition of sea-salt particles was then determined for each of the 20 representative field samples with time intervals of 1–2 h(s) between them. Figure 18.4 indicates the changes in the average composition of sea-salt particles as a function of time. The

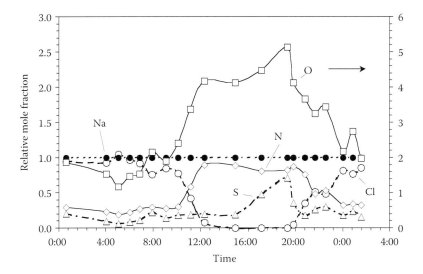

FIGURE 18.4 Diurnal trend in chloride/nitrate replacement in sea-salt particles sampled in Houston, as detected by CC SEM/EDX particle analysis. Markers represent average mole ratios of major elements. Mole ratios of all elements are scaled with respect to that of Na for which the ratio was set equal to 1. Ratios of Na, Cl, N, and S are plotted against the left axis. The ratio of O is plotted against the right axis. Lines connect the symbols of the same element. (From Laskin et al. 2002. *Environmental Science and Technology* 36:4948–4955. Copyright 2002, American Chemical Society. With permission.)

average ratios of detected elements are shown relative to that of Na, for which the ratio was set equal to 1. During the nighttime, the composition of sea-salt particles remained fairly constant and close to that of seawater. Chlorine depletion from particles started around 8 AM and, by noon, chlorine had completely disappeared. Simultaneous with the Cl depletion, sea-salt particles become enriched with N and O, with a nearly 1/3 proportion between them. Observed changes indicate heterogeneous reaction of sodium and magnesium chlorides of sea-salt particles with nitric acid, a typical gas-phase pollutant in urban areas resulting from traffic-related NO_y emissions.

$$NaCl_{(s,aq)} + HNO_{3(g)} \rightarrow NaNO_{3(s,aq)} + HCl_{(g)} \tag{r1}$$

By the end of the day, the air plume changed its direction and speed, which resulted in quick transport of fresh sea-salt particles away from the gulf. A similar diurnal trend of the chloride/nitrate replacement was previously observed using single-particle mass spectrometry measurements over the Los Angeles area (Gard et al., 1998). Both the TexAQS and Los Angeles studies concluded that the key factor determining the extent of chloride/nitrate replacement was the amount of time the particles had spent in the polluted urban atmosphere.

A few years later, another field study showed a complete, irreversible atmospheric processing of crystalline calcium carbonate particles and the formation of highly hygroscopic calcium nitrate particles apparently as a result of the heterogeneous reaction of calcium carbonate-containing mineral dust particles with gaseous nitric acid (Laskin et al., 2005a).

$$CaCO_{3(s)} + 2HNO_{3(g)} \rightarrow Ca(NO_3)_{2(aq)} + H_2O_{(aq)} + CO_{(g)} \tag{r2}$$

In that study, the formation of nitrates in individual calcite and sea-salt particles was followed as a function of time for particle samples collected at Shoresh, Israel. The observed morphology and compositional changes of individual mineral dust particles were consistent with reaction r2. Calcium nitrate particles are exceptionally hygroscopic and deliquesce at a relative humidity (RH) of approximately 11%, lower than the typical atmospheric environment. Transformation of nonhygroscopic dry mineral dust particles into aqueous aerosol may have substantial impact on atmospheric chemistry, light-scattering properties of particles, and their ability to modify clouds.

Often, HR-TEM imaging and associated analytical approaches are used for more focused studies of individual particles sampled in field campaigns. In contrast to SEM, particle analysis using TEM cannot be automated because of the complexity of the method. Therefore, each measurement takes much more time and effort. While the CC SEM is used to provide data over a large number of particles with statistical depth, the HR-TEM is the most powerful tool to study internal structure and the composition of individual particles of specific interest. Figure 18.5 illustrates the application of HR-TEM-based techniques for structural analysis of $K_3Na(SO_4)_2$ particles sampled from ambient air in Lexington, Kentucky (Chen et al., 2006). Panel (a) shows a TEM image featuring four similar crystalline particles approximately 300 nm in size with a

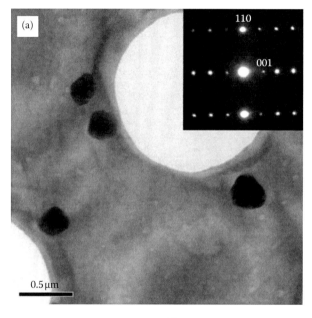

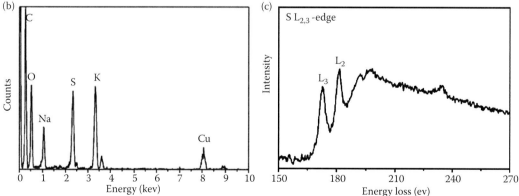

FIGURE 18.5   Comprehensive characterization of $K_3Na(SO_4)_2$ particles using HR-TEM-based methods of analytical microscopy. Panel (a) shows TEM image and crystal structure of particles indicated by SAED pattern (inset). Panel (b) shows the EDX spectrum of the particle indicative of its elemental composition. Panel (c) shows sulfur L-edge EELS spectrum, indicating a S^{6+} valence state revealed from the L-edge peak positions. (From Chen et al. 2006. *Atmospheric Environment* 40:651–663. Copyright 2006, Elsevier. With permission.)

somewhat round morphology. The SAED pattern recorded from one of the particles is shown in the inset. Panel (b) shows an EDX spectrum of the particle indicating the presence of Na, K, S, and O in the particle material (C and Cu are substrate background peaks). Panel (c) shows an S $L_{2,3}$-edge EELS spectrum recorded from one of these particles. Based on EDX and SAED characterization, these particles were identified as $K_3Na(SO_4)_2$ (aphthitalite). In addition, the sulfate (S^{6+}) form of sulfur was independently inferred from the position of the L-edge peaks in EELS spectra. Specifically, peaks at 172 eV (L_3) and 181 eV (L_2) matched peak position characteristics of sulfate standards. Based on the detailed composition, size, and shape data, these particles were apportioned as fly ash that likely originated from the combustion of pelletized biomass fuels used in residential burners (Boman et al., 2004).

Examples of other recent applications of HR-TEM imaging and analysis of field-collected particles include characterization of turbostatic carbon and graphene morphology of soot particles (Hays and Vander Wal, 2007; Vander Wal et al., 2010); microscopic visualization and analysis of metal inclusions in fly ash (Chen et al., 2004, 2005, 2006; Hower et al., 2008) and in secondary organic particles (Adachi and Buseck, 2008); and speciation of external and internal mixing states of urban particles (Utsunomiya et al., 2004), mineral dust particles (Li and Shao, 2009a, 2009b), and biomass-burning particles (Li et al., 2003). In addition, electron tomography using HR-TEM images was applied in order to provide the first three-dimensional imaging of individual soot particles (van Poppel et al., 2005; Adachi et al., 2007).

18.3.2 Hygroscopic Properties of Individual Particles

Hygroscopic transformations of particles occurring in the atmosphere, such as hygroscopic growth, deliquescence (solid-to-liquid), and efflorescence (liquid-to-solid) phase transitions, have a significant effect on aerosol chemistry. Both the kinetics and mechanisms of atmospheric reactions are different for liquid and solid particles. In addition, scattering and absorption of solar radiation and cloud microphysics are also affected by particle hygroscopicity. Laboratory studies of the hygroscopic properties of particles provide quantitative information on the hygroscopic growth factors and characteristic RH values for deliquescence and efflorescence phase transitions. Data obtained in these studies provide important input for computer modeling of atmospheric thermodynamics, transport, heterogeneous chemistry, and climate-forcing effects of environmental aerosols.

The development of environmental SEM (ESEM) (Danilatos, 1993) and environmental TEM (ETEM) (Sharma and Weiss, 1998; Sharma, 2001) instruments made it possible to obtain high-quality microscopy images of samples, while the gaseous environment around the specimen can be controlled by an investigator. Both ESEM and ETEM are relatively new techniques, which only recently have been used to study hygroscopic properties of individual particles. ESEM and ETEM enable imaging of particles in microenvironments (water vapor pressure up to approximately 10 Torr) relevant to real atmospheric conditions. The temperature of the sample can be varied in the range of −20°C to −35°C using either a Peltier stage or a combination of cryogenic cooling and resistive heating. The entire range of RH (1–100% RH) can be established for particle samples by varying both water vapor pressure inside of the chamber and the temperature of the sample. Figure 18.6 demonstrates the first application of ESEM, which visualized the deliquescence phase transition of microscopic NaCl particles (Ebert et al., 2002a). The upper panel shows particles at 75% RH, where water adsorption was first observed in microscopy images. Arrows on the image indicate particle surface areas where initial uptake of water and minor melting of NaCl crystals became visible. The lower panel shows the same particles at 78.3% RH after their complete deliquescence. The observed result was in agreement with the literature values of deliquescence relative humidity (DRH) of 75–76% reported for NaCl particles using different experimental techniques (Martin, 2000 and references therein). Hygroscopic properties of particles composed of other inorganic salts [NH_4NO_3, $(NH_4)_2SO_4$, and Na_2SO_4] measured in that pilot study also concurred with the literature data. A few years later, similar application of ETEM to study phase transitions of test particles composed of

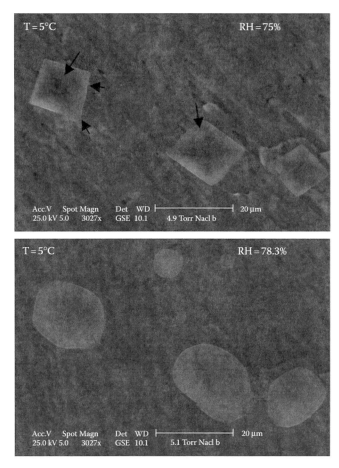

FIGURE 18.6   ESEM images of NaCl particles undergoing deliquescence at 5°C. The upper panel shows solid particles at RH = 75% when first detectable water uptake becomes visible at the sites depicted by arrows. The lower panel shows deliquesced droplet formation at RH = 78.3%. (From Ebert et al. 2002a. *Atmospheric Environment* 36:5909–5916. Copyright 2002, Elsevier. With permission.)

single-component inorganic salts were also reported (Wise et al., 2005). Both studies demonstrated that water uptake on particles, deliquescence, and efflorescence phase transitions could be visualized by ESEM and ETEM with a lateral resolution down to approximately 10 nm, providing the unique opportunity to study hygroscopic properties of particles with a microscopic level of detail. The limit of lateral resolution in these experiments is largely determined by the susceptibility of particles to electron beam damage, rather than the resolution limits of microscopes themselves. Once particles are deliquesced, they become unstable under exposure to the electron beam. To minimize particle damage which also affects data interpretation, relatively coarse magnifications have been employed in ESEM and ETEM studies of particle hydration. Typically, the smallest sizes of particles probed in ESEM and ETEM experiments are on the order of 1 and 0.1 μm, respectively (Laskin et al., 2005b; Wise et al., 2005).

Substantial limits of the ESEM and ETEM techniques are the lack of quantitative measurements of efflorescence relative humidity (ERH) and hygroscopic growth factors. Wetting of the substrate by deliquesced particles can alter the RH at which efflorescence occurs (Ebert et al., 2002a). As a result, ERH values measured in ESEM and ETEM experiments can be significantly higher as compared to ERH reported for airborne particles. Wetting of the substrate also changes the shape of a deliquesced particle (Krueger et al., 2003) and does not allow accurate measurements of the particle size required for the assessment of its growth factors. For this reason, quantitative values of hygroscopic growth factors and ERH based on ESEM and ETEM data are rarely discussed.

An outstanding advantage of the ESEM and ETEM techniques is their ability to visualize hygroscopic properties of complex, mixed-component particles, which enables unique studies of the effect of particle internal composition and mixing state on a particle's ability to take up water. Figure 18.7 illustrates an ESEM experiment that probed hygroscopic transformations of individual sea-salt particles at 5–40% RH. The ESEM image taken at 5.5% RH [panel (a)] indicated a segregation of different salts in vacuum-dried particles—NaCl cubic crystals are seen surrounded by other salts arranged on the NaCl surface. Arrows on the image point at the gaps between segregated salts. With an increase in RH, the outer salts started to take up water and formed a thick, liquid-like layer that filled the gaps between the crystals at 34–39% RH [panels (b) and (c)]. The chemical composition of the outer layer salts was obtained by acquiring SEM/EDX elemental maps over one of the sea-salt particles as shown in Figure 18.8. The observed spatial variations in Na, Cl, Mg, S, and O signals indicate substantial enrichment of the Mg-containing salts in the outer layer and confirm the

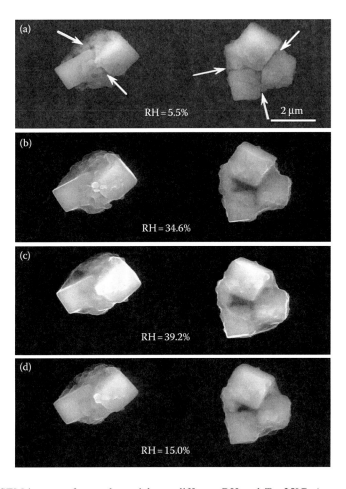

FIGURE 18.7 ESEM images of sea-salt particles at different RH and $T = 25°C$. Arrows on the image (a) point out the gaps between NaCl crystals and other dry salts initially seen at 5.5% RH. (a), (b), and (c) demonstrate consecutive changes in particle morphology as RH increases. Salts present on the surface of NaCl crystals take up water and form a thick outer layer of liquid-like material that grows hygroscopically and fills the gaps between the crystals. (d), taken at a reduced RH of 15%, indicates shrinkage of the outer layer as RH decreases.

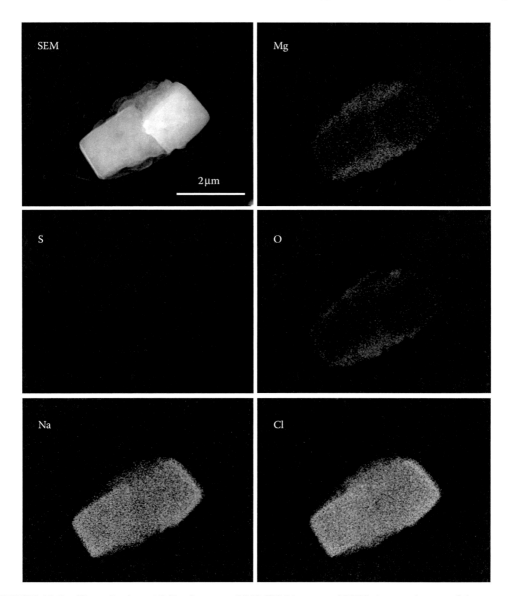

FIGURE 18.8 **(See color insert following page 206.)** SEM image and EDX elemental maps of the sea-salt particle studied in the ESEM experiment shown in Figure 18.7. Segregation of Na- and Mg-containing salts is observed. As a wet sea-salt particle dries out, highly soluble Mg salts crystallize last, forming outer structures on the surface of less soluble NaCl, which crystallizes first.

composition of inner NaCl cubic crystals. The EDX maps are consistent with the thermodynamic predictions of the mineral crystallization sequences during evaporation of seawater. Because of their high water solubility, Mg-containing salts ($MgCl_2 \cdot 6H_2O$, $MgSO_4 \cdot H_2O$, and $KMgCl_3 \cdot 6H_2O$) crystallize last in the sequence (Harvie et al., 1980). Therefore, dry sea-salt particles prepared for the experiment had a multilayered structure with the most soluble Mg-containing salts concentrated in the outer layer. These salts take up water at approximately 30% RH, which is substantially lower than the RH required for the deliquescence of single-component NaCl particles. These microscopic observations suggest that sea-salt particles in the atmospheric environment exist in two major physical forms. At high RH, particles are deliquesced microdroplets, and, at low RH, particle structure comprises a core of NaCl crystal coated with a layer of deliquesced magnesium salts. This conclusion is also supported by similar ETEM observations (Wise et al., 2009) and corroborates with a number of literature reports on the heterogeneous chemistry of sea-salt. Specifically, the experimentally measured kinetics of HNO_3 and SO_2 uptake onto dry sea-salt powders was found to be consistent with the uptake onto liquid surfaces (De Haan and Finlayson-Pitts, 1997).

ESEM and ETEM studies of particle hygroscopic properties have been reported for laboratory-prepared single and mixed salt particles (Hoffman et al., 2004; Matsumura and Hayashi, 2007; Freney et al., 2009; Treuel et al., 2009; Wise et al., 2009), field-collected sea-salt particles (Wise et al., 2007), atmospherically processed (aged) mineral dust particles (Laskin et al., 2005a, 2005b; Shi et al., 2008), and agricultural particles collected at cattle feedlots (Hiranuma et al., 2008). Several studies examined the effect of particle activation at supersaturation values of RH using ESEM. Both morphological and structural changes were observed for individual soot particles after their activation followed by drying (Ebert et al., 2002a; Zuberi et al., 2005). In addition, Shi et al. (2009) reported formation of Fe nanoparticles and an increase in Fe reactivity resulting from activation of selected mineral dust particles.

18.3.3 Ice Nucleation Properties of Individual Particles

One of the least understood properties of aerosol particles is their ability to nucleate ice crystals in the atmosphere. Ice nucleation in the atmosphere occurs by two basic mechanisms: (1) homogeneous freezing of aqueous particles and (2) heterogeneous freezing of water on the surface of selected airborne particles (heterogeneous ice nuclei (HIN)). Heterogeneous ice nucleation proceeds through different modes: deposition mode, where ice nucleates on the particle surface from supersaturated water vapors; immersion freezing mode, which is the freezing of supercooled aqueous droplets induced by the presence of insoluble material inside the droplet; or contact freezing mode, where ice nucleation is induced by collisions of the supercooled droplets and airborne particles (Pruppacher and Klett, 1997). HIN propensity is largely controlled by the surface properties of particles, which can be influenced by many factors, including the chemical composition of the particle and its possible coating; the heterogeneity of the surface chemical composition; possible crystallographic match of surface materials with ice crystals; and particle size, porosity, and morphology.

Understanding the effects of particle morphology and surface chemical heterogeneity on HIN requires detailed microscopic characterization of the ice nucleation process. ESEM has a unique, yet not fully explored, potential to conduct such experiments. Initial results on the application of ESEM for ice nucleation studies have been recently published. These studies demonstrated the application of this technique using silver iodide and model mineral dust particles with well-documented HIN properties (Zimmermann et al., 2007). The same research group extended their studies to examine ice nucleation properties of mineral dust particles common for wind-blown dust in desert areas (Zimmermann et al., 2008). Figure 18.9 shows an image of ice crystals that formed on the surface of illite particles observed in the ESEM experiment. The onset of ice nucleation events was systematically studied as a function of particle temperature and RH over ice (RH_{ice}). The corresponding supersaturation temperature curve for the onset of heterogeneous ice nucleation on illite particles is shown in Figure 18.10. Analogous curves were reported for all nine tested minerals. The reported data on the HIN ability of mineral dust particles covered the range of experimental temperatures from $-10°C$ to $-25°C$ and $RH_{ice} > 110\%$ controlled by a Peltier element. However, understanding the HIN ability of particles relevant to the processes of ice formation in cirrus clouds requires additional experiments at temperatures measuring $-60°C$. An extension of ESEM studies to lower temperatures requires the design and construction of novel, cryogenically cooled sample holders.

Proof-of-principle ESEM experiments examined the onset of ice nucleation by imaging relatively large 100 μm (approximately) ice crystals (see Figure 18.9) that also could be successfully imaged by optical microscopes (e.g., Dymarska et al., 2006; Knopf and Koop, 2006). However, visualization of ice nucleation using ESEM requires imaging of nucleation events on the surface of submicron

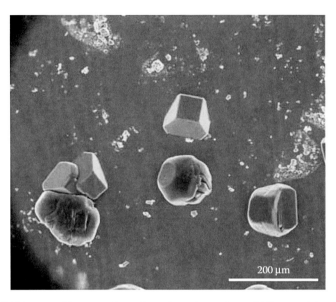

FIGURE 18.9   ESEM image of ice crystals nucleated on the surface of illite particles from supersaturated water vapors. (From Zimmermann et al. 2008. *Journal of Geophysical Research–Atmospheres* 113. Copyright 2008, American Geophysical Union. With permission.)

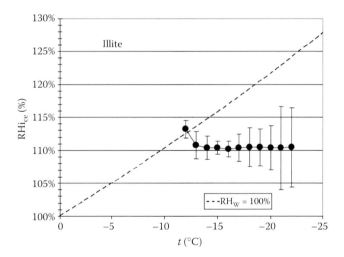

FIGURE 18.10   Supersaturation temperature curves showing the onset conditions of heterogeneous ice nucleation on illite particles reported in ESEM. (From Zimmermann et al. 2008. *Journal of Geophysical Research–Atmospheres* 113. Copyright 2008, American Geophysical Union. With permission.)

particles. Such experiments would provide a fundamental understanding of ice nucleation processes specific to individual particles and their surface chemistry.

18.3.4   OPTICAL PROPERTIES OF INDIVIDUAL PARTICLES

Scattering and absorption of solar light by atmospheric aerosols, known to directly effect climate, may result in both warming and cooling of the atmosphere. The light-scattering particles reflect the visible solar radiation back to space and have a net cooling effect on climate, whereas the light-absorbing aerosol traps this energy in the lower atmosphere and has a warming effect.

Therefore, the magnitude and sign of the direct forcing by aerosols depend on the relative amounts of scattering and absorption, which differ substantially for diverse mixtures of particles.

Optical properties of particles depend on their size, morphology, and RI of the particle matter. In turn, RIs are a function of the chemical composition, phase, and mixing state of particles. Physical and chemical properties of large ensembles of particles can be assessed using EM and electron probe analysis. Based on this characterization, size-selected particles are classified into particle-type groups. The average RI of particle mixture (RI_{mix}) can be calculated based on estimated (RI_i) values for each particle group and a volume approximation for externally mixed particles using Equation 18.1 (Horvath, 1998):

$$RI_{mix} = \frac{\sum_i RI_i V_i}{\sum_i V_i} = \frac{\sum_i n_i V_i}{\sum_i V_i} - i \frac{\sum_i k_i V_i}{\sum_i V_i}. \tag{18.1}$$

In this equation, n_i, and k_i are real and imaginary parts of RI (RI_i), and V_i is the total volume of particles in group i. This approach oversimplifies particle mixture in many ways, and its accuracy distinctly depends on the nature of particles present in the sample. It can be used to provide reasonable estimates of the RI_{mix} values for particle mixtures dominated by inorganic and mineral dust particles for which RI values are well tabulated (Horvath, 1998; Sokolik and Toon, 1999). Such estimates were recently employed in several field studies where practical information on the general trends of optical properties of aerosols in rural and urban air masses were reported (Ebert et al., 2002b, 2004). However, the application of this approach to soot and organic particles is largely unfeasible because of ambiguous data and definitions of particle chemical composition, size, morphology, and generally unknown RI values.

Recently, Alexander et al. (2008) presented an elegant application of the HR-TEM and EELS measurements for determination of optical properties of spherical carbon particles. The mean RI of spherical carbon particles, RI = 1.67–0.27i at a wavelength of λ = 550 nm, was obtained directly from measured EELS spectra. The method used a Kramers–Kronig transformation of EELS spectra of individual particles to obtain corresponding dielectric functions that can be transformed into RI values. The dielectric function calculations assume isotropic material of particles and require accurate measurements of particle thickness. Because of the amorphous nature of carbon particles and their nearly perfect spherical shape, these calculations were relatively straightforward. However, the extension of this method to other types of particles is a recondite process.

18.3.5 LABORATORY STUDIES OF GAS-PARTICLE REACTIONS

Laboratory studies are essential for a fundamental understanding of the atmospheric chemistry of particles and their possible effects on the environment. Over the last decade, EM and microanalysis have been extensively used for chemical characterization of individual particles after their heterogeneous reactions with gas-phase reactants (e.g., Allen et al., 1996; Krueger et al., 2003a, 2003b, 2004; Laskin et al., 2003b, 2005b; Al-Hosney et al., 2005). These microscopy studies provided fundamental information of crucial importance for understanding the reaction mechanisms of atmospheric particles, their possible airborne history, and source apportionment.

An important challenge for laboratory studies of gas-particle heterogeneous chemistry is the ability to capture realistic, atmospherically relevant conditions of RH, temperature, pressure, reaction time, and trace reactive gas concentrations. To address this challenge, a Particle-on-Substrate Stagnation Flow Reactor (PS-SFR) approach has been developed (Liu et al., 2007). In this approach, the reactivity of individual particles is examined by exposing them to trace level concentrations of common atmospheric oxidants under carefully controlled conditions. Detailed microscopic analysis of particle samples is then used to obtain information on particle composition and morphology before and after the reaction. This method provides a unique opportunity to study the chemical reactions of particles of various sizes under different conditions of RH and atmospherically relevant reaction times, as well as the trace concentrations of gas reactants and their mixtures. To date, this approach has been used in the kinetic studies of heterogeneous gas-particle reactions of $NaCl + HNO_3$, $NaCl/MgCl_2 + HNO_3$, and SeaSalt + HNO_3 (Liu et al., 2007); $CaCO_3 + HNO_3$ (Liu et al., 2008); $NaCl + OH$ (Laskin et al., 2006b); and $NaCl/CH_3SO_3Na + HNO_3$ (Laskin and co-workers, unpublished results).

Figure 18.11 shows typical SEM images and EDX spectra of NaCl particles before and after exposure to gaseous HNO_3 in the PS-SFR. Morphological and chemical changes indicative of reactive transformation are clearly seen. The reaction kinetics was followed by quantitative detection of chloride depletion in particles according to the reaction $NaCl + HNO_3 \rightarrow NaNO_3 + HCl$. Figure 18.12 shows values of the experimental uptake coefficients (γ_{net}) determined over a range of 20–80% RH for initially deliquesced particles of $D_p = 0.9$ μm (dry size) composed of NaCl, mixed $NaCl/MgCl_2$, and sea salt, respectively. The overall trend of HNO_3 uptake is in agreement with previously published data (Saul et al., 2006). However, the exact values are somewhat different because γ_{net} is particle-size dependent. The experimental uptake coefficient initially increases as RH decreases from 80%, reaches its maximum at approximately 55% RH, and then decreases at lower RH. This behavior was explained by the variation in chloride concentration in particles and their efflorescence phase transition. Over the RH range from 80% to 55%, a decrease in RH results in an increase in the chloride ion concentration in droplets, leading to a larger reactive uptake. Below the ERH (45–50%), the reactive uptake drops rapidly. However, a sudden "shutoff" in the reactivity was not observed. Considerable HNO_3 uptake onto all three types of particles ($\gamma_{net} = 0.04–0.10$) was measured under quite dry conditions of RH < 30%. It has been suggested that under this humidity, there is still water absorbed on the particle surface. The presence of surface-absorbed water results in

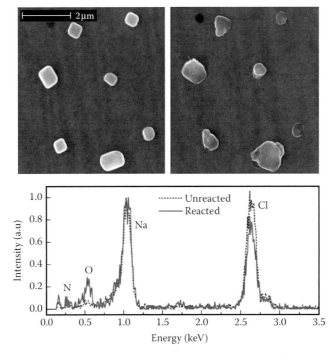

FIGURE 18.11 Top panels: SEM images of NaCl particles before (left) and after (right) exposure to gaseous HNO_3 in the PS-SFR apparatus. Bottom panel: the typical EDX spectra of individual NaCl particle before and after the reaction. (From Liu et al. 2007. *Journal of Physical Chemistry A* 111:10026–10043. Copyright 2007, American Chemical Society. With permission.)

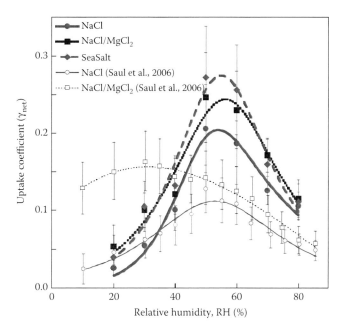

FIGURE 18.12 **(See color insert following page 206.)** Values of initial uptake coefficient γ_{net} as a function of RH for NaCl, mixture of NaCl/MgCl$_2$ ($X_{Mg/Na} = 0.114$), and sea-salt particles. Solid symbols are experimental data from microscopy study with deliquesced particles of dry size $D_p \sim 0.9$ μm. The open symbols represent data from single-particle mass spectrometry study of Saul et al. (2006) for deliquesced particles of dry size $D_p \sim 0.1$ μm. (From Liu et al. 2007. *Journal of Physical Chemistry A* 111:10026–10043. Copyright 2007, American Chemical Society. With permission.)

enhanced ionic mobility and subsequent replenishment of fresh NaCl onto the surface for further exposure.

Figure 18.13 presents γ_{net} values measured as a function of deliquesced particle diameter in PS-SFR experiments along with previously reported data obtained for smaller particles by single-particle mass spectrometry (Tolocka et al., 2004; Saul et al., 2006). The combination of two data sets provided information about the reactivity over a broad range of particle sizes, indicating transition from kinetic ($\gamma_{net} \sim \overline{D_d}$) to diffusion-limited reactivity ($\gamma_{net} \sim 1/\overline{D_d}$). While this behavior could be predicted using the fundamental kinetic and diffusion model of gas-particle reaction (Fuchs and Sutugin, 1970), it was never reported in previous experimental studies, because most methods for studying gas-particle reactivity are limited to narrow ranges of particle sizes or reaction times.

SEM images shown in Figure 18.14 display morphology changes in individual calcium carbonate (CaCO$_3$) and China loess dust particles after their exposure to HNO$_3$ vapor (Laskin et al., 2005b). According to the reaction CaCO$_3$ + 2HNO$_3$ → Ca(NO$_3$)$_2$ + CO$_2$ + H$_2$O, the crystalline solid particles of CaCO$_3$ became enlarged and spherical in shape because of the formation of the deliquesced calcium nitrate product, Ca(NO$_3$)$_2$. The lower panels of Figure 18.14 demonstrate that some of the China loess particles also show identical morphology changes. X-ray mapping of particles that showed a crystalline-to-liquid transformation confirmed their mixed CaCO$_3$/Ca(NO$_3$)$_2$ composition (Krueger et al., 2004). Possible atmospheric effects of these transformations are twofold: first, the resulting hygroscopic particles can act as efficient CCN (Gibson et al., 2006); second, under atmospheric conditions of RH, reacted CaCO$_3$/Ca(NO$_3$)$_2$ particles absorb water efficiently, change their shape and size (Laskin et al., 2005b), and alter their light-scattering and absorption properties.

Figure 18.15 shows experimental uptake coefficients as a function of RH for CaCO$_3$ particles of 0.85 μm size determined in the PS-SFR experiments (Liu et al., 2008). For comparison, data

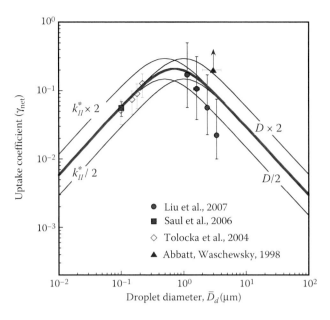

FIGURE 18.13 Experimental values of γ_{net} under RH = 80% as a function of the droplet diameter for NaCl particles measured at 80% RH. Symbols, experimental data; dark solid line, modeling results. (From Liu et al. 2007. *Journal of Physical Chemistry A* 111:10026–10043. Copyright 2007, American Chemical Society. With permission.)

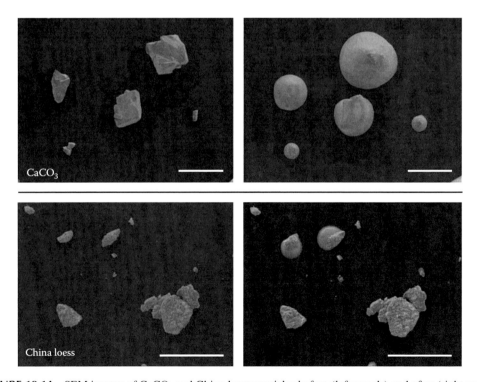

FIGURE 18.14 SEM images of $CaCO_3$ and China loess particles before (left panels) and after (right panels) reaction with gaseous HNO_3 in the presence of water vapor. Size bars on the micrographs depict 5 µm scale. (From Laskin et al. 2005. *Geophysical Research–Atmospheres* 110. Copyright 2005, American Geophysical Union. With permission.)

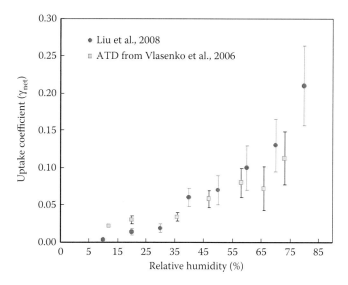

FIGURE 18.15 Experimental uptake coefficient (γ_{net}) measured in the PS-SFR experiments as a function of RH for the HNO_3 reaction with $CaCO_3$ particles of 0.85 μm size. (From Liu et al. 2008. *Journal of Physical Chemistry A* 112:1561–1571. Copyright 2008, American Chemical Society. With permission.)

reported in the literature for the HNO_3 uptake on Arizona test dust (Vlasenko et al., 2006) are also included in Figure 18.15. The PS-SFR results illustrate the significant effects of RH on the experimental uptake (γ_{net}) values, which were found to increase from 0.032 to 0.21 at an RH of 10% and 80%, respectively. The study concluded that the observed reaction enhancement was a combined result of several physicochemical processes mediated by water absorbed on the particle surface, which determines the overall reactivity and kinetics of the HNO_3 uptake on $CaCO_3$.

In summary, the preceding results demonstrate that the combination of EM characterization of particles with the PS-SFR experiment is an efficient approach for studying heterogeneous kinetics. The approach is useful for obtaining kinetic information as a function of particle size, RH, and atmospherically relevant concentrations of gas-phase reactants. Compared to more traditional experiments using flow reactors or cloud chambers, the reaction time accessible in PS-SFR is practically unlimited—a critical advantage that makes it the method of choice for studying slow reactions. However, electron microanalysis techniques cannot probe chemical changes in organic particles. As a result, this approach was used to probe the heterogeneous chemistry of inorganic particles.

18.4 APPLICATION OF COMPLEMENTARY ANALYSIS TECHNIQUES IN STUDIES OF PARTICLE CHEMISTRY

A detailed understanding of particle chemistry often requires a combination of analytical methods and measurements. The application of different analytical methods for aerosol analysis is interdependent, especially when information obtained from one technique frequently triggers and guides subsequent measurements. The paradigm for this approach is illustrated (as follows) using a recent study that focused on the quantitative assessment of methanesulfonate ($CH_3SO_3^-$) and non-sea-salt sulfate (nss-SO_4^{2-}) forms of sulfur in individual marine particles.

The Marine Stratus Experiment (MASE) took place in July 2005 at the Pt. Reyes National Seashore located north of San Francisco. One of the objectives of this study was to compare the composition of particles forming droplets in ambient clouds to those remaining as aerosol particles in cloud interstitial air. Assisted by the meteorology data, the analysis focused on an air mass that originated over the open ocean and then passed through the northern California coast (Hopkins et al., 2008). Unambiguous, qualitative speciation of sulfur-containing compounds and quantitative assessment of the $CH_3SO_3^-/nss\text{-}SO_4^{2-}$ ratios in particles was performed using a combination of three techniques: (a) time-of-flight secondary ionization mass spectrometry (TOF-SIMS) for qualitative molecular speciation of sulfur-containing compounds in individual particles, (b) CC SEM/EDX for a quantitative assessment of the elemental composition of individual particles, and (c) STXM/NEXAFS for a quantitative assessment of different forms of sulfur within individual particles.

Figure 18.16 shows a SEM image of typical pristine marine particles dominated by sea salt and ammonium sulfate aerosol collected in Pt Reyes [panel (a)]. Sea-salt particles are easily recognized in the image as they are typically supermicron in size with cubic shaped NaCl crystals surrounded by irregularly shaped residues of other salts. In contrast, ammonium sulfate particles are submicron in size and spherical. The CC SEM/EDX analysis showed sulfur enrichment and chloride depletion in sea-salt particles. This indicated cloud and chemical processing of sea-salt particles with sulfur compounds, which was consistent with the atmospheric aerosol chemistry expected in the region of the field campaign. For cloud and chemically processed sea-salt particles, chemical reactions of sea-salt components with sulfuric and/or methanesulfonic acids (pervasive species in the marine boundary layer) can be roughly described using the following reactions:

$$2NaCl/MgCl_{2(aq)} + H_2SO_{4(aq,g)} \rightarrow Na_2SO_4/MgSO_{4(aq)} + 2HCl_{(g)} \tag{r4}$$

$$2NaCl/MgCl_{2(aq)} + CH_3SO_3H_{(aq,g)} \rightarrow CH_3SO_3Na/(CH_3SO_3)_2Mg_{(aq)} + 2HCl_{(g)} \tag{r5}$$

These reactions liberate HCl gas into the atmosphere, leaving particles enriched in sulfate and methanesulfonate and depleted in chloride.

CC SEM/EDX analysis provided quantitative data on the S/Na ratios over a statistically significant number of particles. However, no molecular information on the chemical forms of sulfur could be inferred from EDX spectroscopy. Molecular speciation of sulfur-containing compounds was revealed by complementary analysis of individual particles using TOF-SIMS. Chemical forms of sulfur were identified using TOF-SIMS ionic maps produced for selected ions indicative of sulfate and methanesulfonate. The results demonstrated that both methanesulfonate and sulfate are present in sea-salt particles. However, TOF-SIMS experiments are inherently qualitative, providing no quantitative information on the concentrations of different species within particles. Additional approach used STXM/NEXAFS analysis to study chemical bonding information of sulfur and allowed the quantification of $CH_3SO_3^-/SO_4^{2-}$ ratios in individual particles.

Calculation of the $CH_3SO_3^-/nss\text{-}SO_4^{2-}$ ratio relied on a combination of data sets recorded using both STXM/NEXAFS and CC SEM/EDX. Specifically for mixed sea salt/$CH_3SO_3^-/nss\text{-}SO_4^{2-}$ particles, the total S/Na ratio equals the sum of $ss\text{-}SO_4^{2-}/Na$, $nss\text{-}SO_4^{2-}/Na$, and $CH_3SO_3^{2-}/Na$ ratios, as described by Equation 18.2:

$$\frac{\left[\text{total-S}\right]}{\left[\text{Na}\right]} = \underbrace{\frac{\left[ss\text{-}SO_4^{2-}\right]}{\left[\text{Na}\right]}}_{= \, 0.06} + \frac{\left[nss\text{-}SO_4^{2-}\right]}{\left[\text{Na}\right]} + \frac{\left[CH_3SO_3^-\right]}{\left[\text{Na}\right]}. \tag{18.2}$$

The ratio of original sulfate present in sea salt ($ss\text{-}SO_4^{2-}$) to sodium is 0.06 (CRC 1999), leaving two unknowns in this equation. The ratio of $CH_3SO_3^-/\text{total-}SO_4^{2-}$ averaged over a statistically significant number of particles was determined using the STXM/NEXAFS technique. This ratio can be written as follows:

$$\frac{\left[CH_3SO_3^-\right]}{\left[\text{total-}SO_4^{2-}\right]} = \frac{\left[CH_3SO_3^-\right]/\left[\text{Na}\right]}{\left[\text{total-}SO_4^{2-}\right]/\left[\text{Na}\right]} = \frac{\left[CH_3SO_3^-\right]/\left[\text{Na}\right]}{0.06 + \left[nss\text{-}SO_4^{2-}\right]/\left[\text{Na}\right]}. \tag{18.3}$$

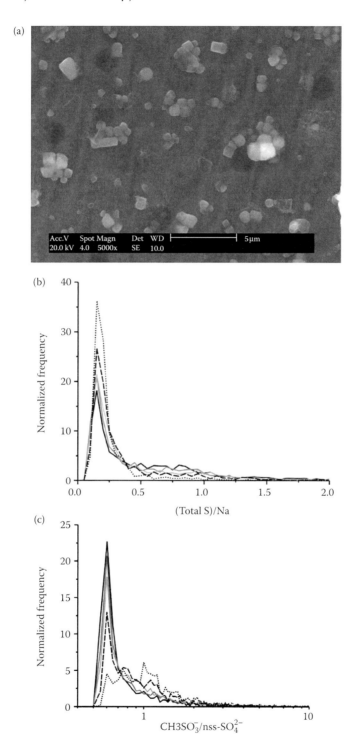

FIGURE 18.16 (a) SEM image of particles collected in Pt. Reyes. Sea-salt particles are large, irregularly shaped particles with NaCl cubic crystal cores. Ammonium sulfate particles are spherical submicron particles. Distribution of (b) total sulfur to sodium ratio and (c) $CH_3SO_3^-$ to nss-SO_4^{2-} ratio present in particles with diameter in the range 0.31–0.5, 0.5–0.79, 0.79–1.26, 1.26–2.00, and >2.00 μm (black, dark gray, light gray, dashed, and dotted lines, respectively). (From Hopkins et al. 2008. *Journal of Geophysical Research–Atmospheres* 113. Copyright 2008, American Geophysical Union. With permission.)

Equations 18.2 and 18.3 contain the same two unknowns. The equations were then solved to determine values for both $[CH_3SO_3^-]/[Na]$ and $[nss\text{-}SO_4^{2-}]/[Na]$.

This analysis was applied to CC SEM/EDX and STXM/NEXAFS data sets and yielded plots presented in the panels (c) and (d) of Figure 18.16. For the first time, size-resolved nss-S/Na and $CH_3SO_3^-/nss\text{-}SO_4^{2-}$ ratios were reported for marine particles (Hopkins et al. 2008). Characteristic ratios of nss-S/Na > 0.10 were obtained for sea-salt particles, with higher values observed for smaller particles indicating more extensive formation of sulfur-containing salts. Characteristic ratios of $CH_3SO_3^-/nss\text{-}SO_4^{2-} > 0.60$ were reported for sea-salt particles of all sizes, with higher values for large particles. This indicates that $CH_3SO_3^-$ salts were likely the dominant form of nss-sulfur in large particles, while SO_4^{2-} was more common in smaller particles.

These results were in qualitative agreement with the schematic of the sulfur–aerosol–climate links in the marine boundary layer mediated by atmospheric oxidation of dimethyl sulfide (DMS)—the major source of sulfur over the oceans. The presence of $CH_3SO_3^-$ in sea-salt particles indicates substantial formation and fast uptake of CH_3SO_3H (reaction r5) presumably resulting from the oxygen addition channel in the DMS oxidation mechanism (von Glasow and Crutzen, 2004). The presence of nss-SO_4^{2-} in sea-salt particles is attributed to reaction r4 with H_2SO_4 formed via the H abstraction channel of DMS oxidation (von Glasow and Crutzen, 2004). The quantitative assessment of the partitioning between the two different forms of sulfur, that is, $CH_3SO_3^-$ and nss-SO_4^{2-}, is important for kinetic modeling. Ratios of $[CH_3SO_3^-]/[Na]$ and $[nss\text{-}SO_4^{2-}]/[Na]$ provide information on the pathways of DMS oxidation in the environment at a given geographic location. The partitioning between these reaction products is important for climate modeling, because it impacts the number and size of CCN produced in the marine atmosphere (von Glasow and Crutzen, 2004).

18.5 SUMMARY

As discussed throughout this chapter, electron beam microscopy and microspectroscopy methods of particle analysis offer unique analytical tools to study the composition and chemistry of environmental particles collected in both field and laboratory experiments. Over the last decade, numerous advances in the instrumentation and analytical methodologies of electron beam techniques have resulted in the development of novel approaches for the fundamental studies of particle physical chemistry. The information obtained in these studies is crucial for evaluating the chemistry and physical properties of particles related to climate change, as well as understanding particle aging and reactivity in the atmosphere. Therefore, microscopy and microspectroscopy studies of individual particles will continue to be a subject of many environmental research projects for years to come.

Obtaining comprehensive information on the chemical composition, physical properties, and atmospheric reactivity of particles is challenging, because no single method of analytical chemistry is capable of providing all the requisite information. For instance, the methods of EM and microanalysis discussed in this chapter can be employed to visualize particle morphology and internal structure at the nanometer scale. However, the analytical power of these techniques in terms of a comprehensive chemical characterization of atmospheric particles is limited to the elemental analysis, which is not informative enough for complex organic particles. A comprehensive characterization of atmospheric particles can only be obtained using a complementary combination of different analytical methods, ranging from microscopic properties of individual particles to an advanced chemical characterization of complex molecules comprising particulate matter. The development of analytical methodologies for comprehensive studies and their applications presents both challenging and exciting opportunities for future research.

ACKNOWLEDGMENTS

The author acknowledges financial support from the Atmospheric Science Program, Office of Biological and Environmental Research (OBER) of the U.S. Department of Energy (DOE); the

Tropospheric Chemistry and the Radiation Sciences programs at the National Aeronautics and Space Administration (NASA); the Laboratory Directed Research and Development funds of Pacific Northwest National Laboratory (PNNL); and operational budget of the Environmental Molecular Sciences Laboratory (EMSL), a national scientific user facility sponsored by the OBER DOE and located at PNNL. PNNL is operated by Battelle Memorial Institute for the DOE under contract no. DE-AC05-76RL01830.

The author also acknowledges his colleagues and collaborators who profoundly influenced the research projects described in this chapter: C. M. Berkowitz, J. P. Cain, J. P. Cowin, Y. Desyaterik, M. J. Ezell, B. J. Finlayson-Pitts, D. J. Gaspar, E. R. Gibson, M. K. Gilles, V. H. Grassian, E. R. Graber, J. L. Hand, R. C. Hoffman, R. J. Hopkins, S. W. Hunt, M. J. Iedema, K. S. Johnson, B. J. Krueger, Y. Liu, W. C. Malm, R. C. Moffet, L. T. Molina, M. J. Molina, K. A. Prather, Y. Rudich, V. Shutthanandan, A. V. Tivanski, T. W. Wietsma, C. Wang, H. Wang, R. A. Zaveri, and B. Zuberi.

REFERENCES

Abbatt, J. P. D. and G. C. G. Waschewsky. 1998. Heterogeneous interactions of HOBr, HNO₃, O₃, and NO₂ with deliquescent NaCl aerosols at room temperature. *Journal of Physical Chemistry A* 102:3719–3725.

Adachi, K. and P. R. Buseck. 2008. Internally mixed soot, sulfates, and organic matter in aerosol particles from Mexico City. *Atmospheric Chemistry and Physics* 8:6469–6481.

Adachi, K., S. H. Chung, H. Friedrich, and P. R. Buseck. 2007. Fractal parameters of individual soot particles determined using electron tomography: Implications for optical properties. *Journal of Geophysical Research–Atmospheres* 112(D14):D14202.

Alexander, D. T. L., P. A. Crozier, and J. R. Anderson. 2008. Brown carbon spheres in East Asian outflow and their optical properties. *Science* 321:833–836.

Al-Hosney, H. A., S. Carlos-Cuellar, S. J. Baltrusaitis, and V. H. Grassian. 2005. Heterogeneous uptake and reactivity of formic acid on calcium carbonate particles: A Knudsen cell reactor, FTIR and SEM study. *Physical Chemistry Chemical Physics* 7:3587–3595.

Allen, H. C., J. M. Laux, R. Vogt, B. J. Finlayson-Pitts, and J. C. Hemminger. 1996. Water-induced reorganization of ultrathin nitrate films on NaCl: Implications for the tropospheric chemistry of sea salt particles. *Journal of Physical Chemistry* 100:6371–6375.

Anderson, J. R. and P. R. Buseck. 1998. Special examples of applications of mineralogy and geochemistry to environmental problems: Atmospheric dust. In *Advanced Minerology*, Vol. 3. Ed. A. S. Marfunin. Berlin: Springer-Verlag, pp. 300–313.

Armstrong, J. T. 1991. Quantitative elemental analysis of individual microparticles with electron beam instruments. In *Electron Probe Quantitation*. Eds. K. F. J. Heinrich and D. E. Newbury, New York, NY: Plenum Press, pp. 261–316.

Boman, C., A. Nordin, D. Bostrom, and M. Ohman. 2004. Characterization of inorganic particulate matter from residential combustion of pelletized biomass fuels. *Energy & Fuels* 18:338–348.

Burleson, D. J., M. D. Driessen, and R. Lee Penn. 2004. On the characterization of environmental nanoparticles. *Journal of Environmental Science and Health A* 39:2707–2753.

Buseck, P. R. and J. R. Anderson. 1998. Analysis of individual airborne mineral particles. In *Advanced Minerology*, Vol. 3. Ed. A. S. Marfunin, Berlin: Springer-Verlag, pp. 292–300.

Buseck, P. R. and M. Pósfai. 1999. Airborne minerals and related aerosol particles: Effects on climate and the environment. *Proceedings of National Academy of Science, USA* 96:3372–3379.

Chen, Y., N. Shah, F. Huggins, and G. Huffman. 2004. Investigation of the microcharacteristics of PM2.5 in residual oil fly ash by analytical transmission electron microscopy. *Environmental Science and Technology* 38:6553–6560.

Chen, Y., N. Shah, F. Huggins, and G. Huffman. 2005. Characterization of ambient airborne particles by energy-filtered transmission electron microscopy. *Aerosol Science and Technology* 39:509–518.

Chen, Y., N. Shah, F. E. Huggins, and G. P. Huffman. 2006. Microanalysis of ambient particles from Lexington, KY, by electron microscopy. *Atmospheric Environment* 40:651–663.

Choël, M., K. Deboudt, and P. Flament. 2007. Evaluation of quantitative procedures for X-ray microanalysis of environmental particles. *Microscopy Research and Technique* 70:996–1002.

Choël, M., K. Deboudt, J. Osán, P. Flament, and R. Van Grieken. 2005. Quantitative determination of low-Z elements in single atmospheric particles on boron substrates by automated scanning electron microscopy-energy-dispersive X-ray spectrometry. *Analytical Chemistry* 77:5686–5692.

Danilatos, G. D. 1993. Introduction to the ESEM instrument. *Microscopy Research and Techniques* 25:354–361.

De Bock, L. A. and R. E. Van Grieken. 1999. Single particle analysis techniques. In *Analytical Chemistry of Aerosols*. Ed. K. R. Spurny, Boca Raton, FL: Lewis Publishers Ltd, pp. 243–275.

De Haan, D. O. and B. J. Finlayson-Pitts. 1997. Knudsen cell studies of the reaction of gaseous nitric acid with synthetic sea salt at 298 K. *Journal of Physical Chemistry A* 101:9993–9999.

Dymarska, M., B. J. Murray, L. Sun, M. Eastwood, D. A. Knopf, and A. K. Bertram. 2006. Deposition ice nucleation on soot at temperatures relevant for the lower troposphere. *Journal of Geophysical Research–Atmospheres.* 111(D4):D04204.

Ebert, M., M. Inerle-Hof, and S. Weinbruch. 2002a. Environmental scanning electron microscopy as a new technique to determine the hygroscopic behaviour of individual aerosol particles. *Atmospheric Environment* 36:5909–5916.

Ebert, M., S. Weinbruch, A. Rausch, et al. 2002b. The complex refractive index of aerosols during LACE 98 as derived from the analysis of individual particles. *Journal of Geophysical Research–Atmospheres* 107(D21):D8121.

Ebert, M., S. Weinbruch, P. Hoffmann, and H. M. Ortner. 2004. The chemical composition and complex refractive index of rural and urban influenced aerosols determined by individual particle analysis. *Atmospheric Environment* 38:6531–6545.

Fletcher, R. A., J. A. Small, and J. H. J. Scott. 2001. Analysis of individual collected particles. In *Aerosol Measurement. Principles, Techniques, and Applications.* Eds. K. Willeke and P. A. Baron, New York: John Wiley & Sons Ltd, pp. 295–363.

Freney, E. J., S. T. Martin, and P. R. Buseck. 2009. Deliquescence and efflorescence of potassium salts relevant to biomass-burning aerosol particles. *Aerosol Science and Technology* 43:799–807.

Fuchs, N. A. and A. G. Sutugin. 1970. *Highly Dispersed Aerosols.* London: Ann Arbor Science Publishers, pp. 1–105.

Gard, E. E., M. J. Kleeman, D. S. Gross, et al. 1998. Direct observation of heterogeneous chemistry in the atmosphere. *Science* 279:1184–1187.

Gibson, E. R., P. K. Hudson, and V. H. Grassian. 2006. Aerosol chemistry and climate: Laboratory studies of the carbonate component of mineral dust and its reaction products. *Geophysical Research Letters* 33(13):L13811.

Hand, J. L., W. C. Malm, A. Laskin, et al. 2005. Optical, physical and chemical properties of tar balls observed during the Yosemite Aerosol Characterization Study. *Journal of Geophysical Research–Atmospheres* 110:D21210.

Harvie, C. E., J. H. Weare, L. A. Hardie, and H. P. Eugster. 1980. Evaporation of seawater—Calculated mineral sequences. *Science* 208:498–500.

Hays, M. D. and R. L. Vander Wal. 2007. Heterogeneous soot nanostructure in atmospheric and combustion source aerosols. *Energy & Fuels* 21:801–811.

Hiranuma, N., S. D. Brooks, B. W. Auvermann, and R. Littleton. 2008. Using environmental scanning electron microscopy to determine the hygroscopic properties of agricultural aerosols. *Atmospheric Environment* 42:1983–1994.

Hoffman, R. C., A. Laskin, and B. J. Finlayson-Pitts. 2004. Sodium nitrate particles: Physical and chemical properties during hydration and dehydration, and implications for aged sea salt aerosols. *Journal of Aerosol Science* 35:869–887.

Hopkins, R. J., Y. Desyaterik, A. V. Tivanski, et al. 2008. Chemical speciation of sulfur in marine cloud droplets and particles: Analysis of individual particles from marine boundary layer over the California current. *Journal of Geophysical Research–Atmospheres* 113(D4):D04209.

Horvath, H. 1998. Influence of atmospheric aerosols upon the global radiation balance. In *Atmospheric Particles. IUPAC Series on Analytical and Physical Chemistry of Environmental Systems*, Vol. 5. Eds. R. M. Harrison and R. Van Grieken, Chichester, UK: John Wiley & Sons Ltd, pp. 543–596.

Hower, J. C., U. M. Graham, A. Dozier, M. T. Tseng, and R. A. Khatri. 2008. Association of the sites of heavy metals with nanoscale carbon in a kentucky electrostatic precipitator fly ash. *Environmental Science and Technology* 42:8471–8477.

Injuk, J., L. A. De Bock, and R. E. Van Grieken. 1998. Structural heterogeneity within airborne particles. In *Atmospheric Particles. IUPAC Series on Analytical and Physical Chemistry of Environmental Systems*, Vol. 5., Eds. R. M. Harrison and R. Van Grieken, Chichester, UK: John Wiley & Sons Ltd, pp. 173–202.

Johnson, K. S., B. Zuberi, L. T. Molina, et al. 2005. Processing of soot in an urban environment: Case study from the Mexico City Metropolitan Area. *Atmospheric Chemistry and Physics* 5:3033–3043.

Johnson, K. S., B. de Foy, B. Zuberi1, et al. 2006. Aerosol composition and source apportionment in the Mexico City metropolitan area with PIXE/PESA/STIM and multivariate analysis. *Atmospheric Chemistry and Physics* 6:4591–4600.

Knopf, D. A. and T. Koop. 2006. Heterogeneous nucleation of ice on surrogates of mineral dust. *Journal of Geophysical Research–Atmospheres* 111(D12):D12201.

Kojima, T., P. R. Buseck, Y. Iwasaka, A. Matsuki, and D. Trochkine. 2006. Sulfate-coated dust particles in the free troposphere over Japan. *Atmospheric Research* 82:698–708.

Krueger, B. J., V. H. Grassian, M. J. Iedema, J. P. Cowin, and A. Laskin. 2003a. Probing heterogeneous chemistry of individual atmospheric particles using scanning electron microscopy and energy-dispersive X-ray analysis. *Analytical Chemistry* 75:5170–5179.

Krueger, B. J., V. H. Grassian, A. Laskin, and J. P. Cowin. 2003b. The transformation of solid atmospheric particles into liquid droplets through heterogeneous chemistry: Laboratory insights into the processing of calcium containing mineral dust aerosol in the troposphere. *Geophysical Research Letters* 30:1148–1151.

Krueger, B. J., V. H. Grassian, J. P. Cowin, and A. Laskin. 2004. Heterogeneous chemistry of individual mineral dust particles from different dust source regions: The importance of particle mineralogy. *Atmospheric Environment* 38:6253–6261. Erratum published in *Atmospheric Environment*, 39, 395 (2005).

Laskin, A., M. J. Iedema, and J. P. Cowin. 2002. Quantitative time-resolved monitoring of nitrate formation in sea salt particles using an automated SEM/EDX single particle analysis. *Environmental Science and Technology* 36:4948–4955.

Laskin, A., M. J. Iedema, and J. P. Cowin. 2003a. Time-resolved aerosol collector for CCSEM/EDX single-particle analysis. *Aerosol Science and Technology* 37:246–260.

Laskin, A., D. J. Gaspar, W. Wang, et al. 2003b. Reactions at interfaces as a source of sulfate formation in sea salt particles. *Science* 301:340–344.

Laskin, A., M. J. Iedema, A. Ichkovich, E. R. Graber, I. Taraniuk, and Y. Rudich. 2005a. Direct observation of completely processed calcium carbonate particles in polluted atmospheric environment. *Faraday Discussions* 130:453–468.

Laskin, A., T. W. Wietsma, B. J. Krueger, and V. H. Grassian. 2005b. Heterogeneous chemistry of individual mineral dust particles with nitric acid: A combined CCSEM/EDX, ESEM, and ICP-MS study. *Geophysical Research–Atmospheres* 110(D10):D10208.

Laskin, A., J. P. Cowin, and M. J. Iedema. 2006a. Analysis of individual environmental particles using modern methods of electron microscopy and X-ray microanalysis. *Journal of Electron Spectroscopy and Related Phenomena* 150:260–274.

Laskin, A., H. Wang, W. H. Robertson, J. P. Cowin, M. J. Ezell, and B. J. Finlayson-Pitts. 2006b. A new approach to determining gas-particle reaction probabilities and application to the heterogeneous reaction of deliquesced sodium chloride particles with gas-phase hydroxyl radicals. *Journal of Physical Chemistry A* 110:10619–10627.

Leppard, G. G. 2008. Nanoparticles in the environment as revealed by transmission electron microscopy: Detection, characterization and activities. *Current Nanoscience* 4:278–301.

Li, J., M. Pósfai, P. V. Hobbs, and P. R. Buseck. 2003. Individual aerosol particles from biomass burning in southern Africa: 2. Compositions and aging of inorganic particles. *Journal of Geophysical Research–Atmospheres* 108(D13):D8484.

Li, W. J. and L. Y. Shao. 2009a. Transmission electron microscopy study of aerosol particles from the brown hazes in northern China. *Geophysical Research–Atmospheres* 114(09):D09302.

Li, W. J. and L. Y. Shao. 2009b. Observation of nitrate coatings on atmospheric mineral dust particles. *Atmospheric Chemistry and Physics* 9:1863–1871.

Lide, D. R. 1999. *Handbook of Chemistry and Physics. A Ready-Reference Book of Chemical and Physics Data*, 80th edn. Boca Raton, FL: CRC Press Inc.

Liu, Y., J. P. Cain, H. Wang, and A. Laskin. 2007. Kinetic study of heterogeneous reaction of deliquesced NaCl particles with gaseous HNO_3 using particle-on-substrate stagnation flow reactor approach. *Journal of Physical Chemistry A* 111:10026–10043.

Liu, Y., E. R. Gibson, J. P. Cain, H. Wang, V. H. Grassian, and A. Laskin. 2008. Kinetics of heterogeneous reaction of $CaCO_3$ particles with gaseous HNO_3 over a wide range of humidity. *Journal of Physical Chemistry A* 112:1561–1571.

Martin, S. T. 2000. Phase transitions of aqueous atmospheric particles. *Chemical Reviews* 100(9): 3403–3453.

Matsumura, T. and M. Hayashi. 2007. Hygroscopic Growth of an $(NH_4)_2SO_4$ aqueous solution droplet measured using an environmental scanning electron microscope (ESEM). *Aerosol Science and Technology* 41:770–774.

Moffet, R. C., Y. Desyaterik, R. J. Hopkins, et al. 2008. Characterization of single aerosol particles containing Pb, Zn, and Cl from an industrial region of Mexico City. *Environmental Science and Technology* 42:7091–7097.

Moffet, R. C., T. R. Henn, A. V. Tivanski, et al. 2010. Microscopic characterization of carbonaceous aerosol particle aging in the outflow from Mexico City. *Atmospheric Chemistry and Physics* 10(3):961–976.

Poelt, P., M. Schmied, I. Obernberger, T. Brunner, and J. Dahl. 2002. Automated analysis of submicron particles by computer-controlled scanning electron microscopy. *Scanning* 24:92–100.

Pósfai, M., J. R. Anderson, P. R. Buseck, and H. Sievering. 1999. Soot and sulfate aerosol particles in the remote marine troposphere. *Journal of Geophysical Research–Atmospheres* 104(D17):21685–21693.

Pósfai, M., A. Gelencser, R. Simonics, et al. 2004. Atmospheric tar balls: Particles from biomass and biofuel burning. *Journal of Geophysical Research–Atmospheres* 109(D6):D06213.

Pruppacher, H. R. and J. D. Klett. 1997. *Microphysics of Cloud and Precipitation*. Dordrecht: Kluwer Press.

Ro, C.-U., J. Osán, and R. Van Grieken. 1999. Determination of low-Z elements in individual environmental particles using windowless EPMA. *Analytical Chemistry* 71:1521–1528.

Ro, C.-U., K.-Y. Oh, J. Osán, et al. 2001. Heterogeneity Assessment in individual $CaCO_3$-$CaSO_4$ particles using ultrathin window electron probe X-ray microanalysis. *Analytical Chemistry* 73:4574–4583.

Ro, C.-U., J. Osán, I. Szalóki, J. de Hoog, A. Worobiec, and R. Van Grieken. 2003. A Monte Carlo program for quantitative electron-induced X-ray analysis of individual particles. *Analytical Chemistry* 75:851–859.

Saul, T. D., M. P. Tolocka, and M. V. Johnston. 2006. Reactive uptake of nitric acid onto sodium chloride aerosols across a wide range of relative humidities. *Journal of Physical Chemistry A* 110:7614–7620.

Sharma, R. 2001. Design and applications of environmental cell transmission electron microscope for *in situ* observations of gas–solid reactions. *Microscopy and Microanalysis* 7:494–506.

Sharma, R. and K. Weiss. 1998. Development of a TEM to study *in situ* structural and chemical changes at an atomic level during gas–solid interactions at elevated temperatures. *Microscopy Research and Technique* 42(4):270–280.

Shi, Z., D. Zhang, M. Hayashi, H. Ogata, H. Ji, and W. Fujiie. 2008. Influences of sulfate and nitrate on the hygroscopic behavior of coarse dust particles. *Atmospheric Environment* 42:822–827.

Shi, Z., M. D. Krom, S. Bonneville, A. R. Baker, T. D. Jickells, and L. G. Benning. 2009. Formation of iron nanoparticles and increase in iron reactivity in mineral dust during simulated cloud processing. *Environmental Science and Technology* 43:6592–6596.

Sokolik, I. N. and O. B. Toon. 1999. Incorporation of mineralogical composition into models of the radiative properties of mineral aerosol from UV to IR wavelength. *Journal of Geophysical Research–Atmospheres* 104:9423–9444.

Strausser, Y. E. 1992. AES: Auger electron spectroscopy. In *Encyclopedia of Materials Characterization*. Eds. R. C. Brundle, C. A. Evans, and S. Wilson, Boston: Butterworth-Heinemann, pp. 310–323.

Szaloki, I., C.-U. Ro, J. Osan, J. De Hoog, and R. Van Grieken. 2004. Speciation and surface analysis of single particles using electron-excited X-ray emission spectrometry. In *X-Ray Spectrometry: Recent Technological Advances*. Eds. K. Tsuji, J. Injuk, and R. Van Grieken, Chichester, UK: John Wiley & Sons Ltd, pp. 570–592.

Tivanski, A. V., R. J. Hopkins, T. Tyliszczak, and M. K. Gilles. 2007. Oxygenated interface on biomass burn tar balls determined by single particle scanning transmission X-ray microscopy. *Journal of Physical Chemistry A* 111:5448–5458.

Tolocka, M. P., T. D. Saul, and M. V. Johnston. 2004. Reactive uptake of nitric acid into aqueous sodium chloride droplets using real-time single-particle mass spectrometry. *Journal of Physical Chemistry A* 108:2659–2665.

Treuel, L., S. Pederzani, and R. Zellner. 2009. Deliquescence behavior and crystallization of ternary ammonium sulfate/dicarboxylic acid/water aerosols. *Physical Chemistry Chemical Physics* 11:7976–7984.

Utsunomiya, S. and R. C. Ewing. 2003. Application of high-angle annular dark field scanning transmission electron microscopy, scanning transmission electron microscopy-energy dispersive X-ray spectrometry, and energy-filtered transmission electron microscopy to the characterization of nanoparticles in the environment. *Environmental Science and Technology* 37:786–791.

Utsunomiya, S., K. A. Jensen, G. J. Keeler, and R. C. Ewing. 2004. Direct identification of trace metals in fine and ultrafine particles in the Detroit urban atmosphere. *Environmental Science and Technology* 38:2289–2297.

Vander Waal, R. L., V. M. Bryg, and M. D. Hays. 2010. Fingerprinting soot (Towards source identification): Physical structure and chemical composition. *Journal of Aerosol Science* 41(1):108–117.

van Poppel, L. H., H. Friedrich, J. Spinsby, S. H. Chung, J. H. Seinfeld, and P. R. Buseck. 2005. Electron tomography of nanoparticle clusters: Implications for atmospheric lifetimes and radiative forcing of soot. *Geophysical Research Letters* 32(24):L24811.

Vlasenko, A., S. Sjogren, E. Weingartner, K. Stemmler, H. W. Gaggeler, and M. Ammann. 2006. Effect of humidity on nitric acid uptake to mineral dust aerosol particles. *Atmospheric Chemistry and Physics* 6:2147–2160.

von Glasow, R. and P. J. Crutzen. 2004. Model study of multiphase DMS oxidation with a focus on halogens, *Atmospheric Chemistry and Physics* 4:589–608.

Wise, M. E., G. Biskos, S. T. Martin, L. M. Russell, and P. R. Buseck. 2005. Phase transitions of single salt particles studied using a transmission electron microscope with an environmental cell. *Aerosol Science and Technology* 39:849–856.

Wise, M. E., E. J. Freney, C. A. Tyree, et al. 2009. Hygroscopic behavior and liquid-layer composition of aerosol particles generated from natural and artificial seawater. *Journal of Geophysical Research–Atmospheres* 114(D03):D03203.

Wise, M. E., T. A. Semeniuk, R. Bruintjes, S. T. Martin, L. M. Russell, and P. R. Buseck. 2007. Hygroscopic behavior of NaCl-bearing natural aerosol particles using environmental transmission electron microscopy. *Journal of Geophysical Research–Atmospheres* 112(D10):D10224.

Zaveri, R. A., C. M. Berkowitz, F. J. Brechtel, et al. 2010. Nighttime chemical evolution of aerosol and trace gases in a power plant plume. *Journal of Geophysical Research–Atmospheres*. 115:D12304.

Zimmermann, F., M. Ebert, A. Worringen, L. Schutz, and S. Weinbruch. 2007. Environmental scanning electron microscopy (ESEM) as a new technique to determine the ice nucleation capability of individual atmospheric aerosol particles. *Atmospheric Environment* 41:8219–8227.

Zimmermann, F., S. Weinbruch, L. Schutz, et al. 2008. Ice nucleation properties of the most abundant mineral dust phases. *Journal of Geophysical Research–Atmospheres* 113(D23):D23204.

Index

Note: n = footnote.

A

A-Train, 230
AAS. *See* Atomic absorption spectroscopy (AAS)
Absorbance. *See* Optical density (OD)
Absorbing polystyrene spheres (APSS), 288
Absorption, 3, 82
    in RDG regime, 352
    spectroscopy, 245
Absorption coefficient, 251, 271–272
    decay time, 277
    measured pressure amplitude, 249
    measurement, 246–247
    measurement error dependence, 248
    PA spectroscopy, 248
Absorption cross-section, 5
    atomic number and, 431
    CDE, 103
    ellipsoidal particles, 7
    optical density, 425–426
    particle volume, 13
Absorption efficiency, 273
    size dependency, 13
Absorption spectroscopy, 245
    CRD, 275
    PA spectroscopy, 248
    riboflavin, 325
    tryptophan, 325
    in UV–VIS spectral range, 244
Acousto-optic modulator (AOM), 278
ADL. *See* Aerodynamic lenses (ADL)
Advanced Light Source (ALS), 423
    Chemical Dynamics Beamline, 380
    STXMs, 431
Advanced very high resolution radiometer (AVHRR), 80
Aerodynamic lenses (ADL), 379, 403
    particle stream, 389
    particle velocities, 390
Aerodynamic particle sizer (APS), 80–81
AERONET. *See* Aerosol Robotic Network (AERONET)
Aerosol albedometer, 290
Aerosol fluorescence spectra measurements, 298
    atmospheric aerosol compositions, 302
    cluster analysis, 301
    cluster templates, 302
    experimental setup, 299
    LIF spectra, 305
    measurement sites, 300
    OCAs study, 306
    PFS, 299, 306
    time-series measurements, 300
Aerosol index, 212
Aerosol interactions and dynamics in atmosphere (AIDA), 247
    aerosol chamber LOPES system, 247
    aerosol vessel, 17
    chamber, 17, 18

Aerosol IR spectra. *See also* Infrared spectroscopy (IR spectroscopy)
    aerosol particle demixing, 45
    CO_2 shell, 44
    core and shell molecules mixing, 44–45
    crystallization kinetics determination, 40, 41
    experimental setups, 36–37
    internal structure influence, 38, 39–40
    mixed CO_2/C_2H_2 particles, 43–44
    multicomponent particles spectral features, 40, 41, 42
    particle composition effect, 43
    particle shape influence, 40, 42
    particle size influence, 37–38, 39
    SO_2/CO_2 aerosol, 44
Aerosol mass spectrometers (AMS), 156
Aerosol particles, 36, 156. *See also* Aerosol(s)
    aggregation, 343
    atmospheric, 82–83
    chemical signatures identification, 228
    in Earth's atmosphere, 156
    hygroscopicity, 156, 157
    inorganic components, 175
    phase transformations, 157
    photoelectric charging, 370
    properties, 156
    reactions, 179
    sample analysis, 201
    scattering cross-section, 213
    single-particle levitation techniques, 156
    sources, 156
Aerosol photoemission, 368, 369. *See also* Aerosols photoelectric charging
    angular and KE analysis, 378
    circular dichroism, 373
    in nanoparticles, 385
    probe molecules, 375
Aerosol photoemission, synchrotron-based, 378
    angle-resolved photoemission, 382
    from biological nanoparticles, 379
    Mie theory, 385, 386
    nanoparticle photoemission, 387
    photoemission asymmetry, 385, 387
    threshold photoemission, 387
    vacuum UV photoemission, 378
    XPS, 389–39
Aerosol Robotic Network (AERONET), 211
    sun photometers, 229, 230
Aerosol spectroscopy, 105–107, 120. *See also* Aerosol IR spectra
    extinction spectra, 112, 117
    filtration method, 107
    FTIR, 106
    gas cell, 105
    glass window, 106
    rotating-brush dust flow generator, 105
    two stage impactor, 107

493

Aerosol(s), 269
    anthropogenic, 269
    carbonaceous, 251
    characterization, 211, 270
    in climate system, 210
    coating, 286–288
    complex refractive index determination, 281
    as condensation nuclei, 270
    dust, 270
    effects, 452, 463
    environmental, 463
    formation processes, 25
    impacts, 298
    inorganic species, 270
    microphysical properties, 210
    mineral dust, 253
    natural, 269
    NEXAFS spectra, 453
    nonabsorbing, 273
    observation techniques, 210
    OC, 270
    optical extinction, 244
    optical properties studies, 281
    phase, 375
    physical and chemical properties, 25
    primary, 270
    RB, 270
    refractive index, 271
    roles, 269, 370
    scattering cross-section, 213
    secondary processes, 270
    sequential TAOS patterns, 310
    shape characterization, 262
    with solar radiation and Mie theory, 271
    sources, 298
    spectral behavior, 251
    surface properties, 389
    surrounding medium, 271
    TAOS measurements, 309, 313
Aerosols photoelectric charging, 370
    adsorption and desorption kinetics, 371
    Arrhenius type equation, 372
    asymmetry in, 374
    CD aerosol photoemission, 373
    photoelectric charging, 371–372
    photoionization spectra, 372
    probe molecule aerosol photoemission,
        377
AFM. See Atomic force microscopy (AFM)
Ag foil, 199, 200. See also Cascade impactors;
        Transmission electron microscopy
        (TEM)—grids
AGB stars. See Asymptotic giant branch stars
        (AGB stars)
Aggregation, 341
    DLCA, 343
    polydispersity, 350
    processes, 344
    RLCA, 344
    scattered intensity and, 358
AIDA. See Aerosol interactions and dynamics in
        atmosphere (AIDA)
ALADIN. See Atmospheric Laser Doppler Instrument
        (ALADIN)

Albedo, 354. See also Single scattering albedo (SSA)
    atmosphere, 210
    single-scattering, 223
    for soot-like fractal aggregate, 354
ALS. See Advanced Light Source (ALS)
Ambient pressure XPS (APXPS), 389, 390. See also
        X-ray photoelectron spectroscopy (XPS)
    liquid aerosols surface study, 391
Ammonium nitrate (AN), 175
    phase transitions, 194
Ammonium sulfate (AS), 158, 175, 470
    CC SEM/EDX analysis, 470, 484
    droplets, 143
    evaporation curves, 158
    hygroscopic growth curves, 158
    hygroscopicity, 172
    hygroscopicity and FWHH, 165
    partial deliquescence behaviors, 168
    phase state, 158
    Raman spectra, 173
    small-particle absorption spectra, 8
    x-ray exposure, 452
Amorphous solid water (ASW), 51
    CO adsorbate bands, 61
    hydrogen adsorption, 62
    infrared bands, 62
    spectrum, 56
AMS. See Aerosol mass spectrometers (AMS)
AN. See Ammonium nitrate (AN)
Angle of incidence, 411
Angle-integrated extinction, 82
Ångström exponent, 253, 289
    absorption, 246, 247
    calculation, 212
    extinction coefficients, 229
    extinction-related particle, 220
Angular scattering coefficient. See Scattering polar
        dependence
Anti-Stokes scattering, 128, 129
AOM. See Acousto-optic modulator (AOM)
APS. See Aerodynamic particle sizer (APS)
APSS. See Absorbing polystyrene spheres (APSS)
APXPS. See Ambient pressure XPS (APXPS)
AS. See Ammonium sulfate (AS)
AS/MA particles, 164. See also Aerosol particles
    elastic light-scattering pattern, 165
    hygroscopicity and FWHH, 165
    Raman spectra, 166, 167, 168
    relative mass changes, 164
    water uptake, 165, 166–167
Asphericity factor, 259
ASW. See Amorphous solid water (ASW)
Asymptotic giant branch stars (AGB stars), 101
    emission spectrum, 117, 118
ATLID. See Atmospheric Lidar (ATLID)
Atmospheric dust loading modeling study, 90
Atmospheric gas-phase oxidants, 179
Atmospheric Laser Doppler Instrument (ALADIN), 233
Atmospheric Lidar (ATLID), 232
Atmospheric processes, 270, 369
    mineral dust aerosol, 79
    supercooled droplets, 40
Atomic absorption spectroscopy (AAS), 195
Atomic force microscopy (AFM), 453

Attenuation, 215, 271
Autoxidation, 180
  conjugated diene hydroperoxides formation, 182
  ozone-induced, 183
  pathways, 184–185
AVHRR. *See* Advanced very high resolution radiometer (AVHRR)
Azimuthal scattering, 258
  asphericity factor, 259
  detector response coefficient of variation, 259
  patterns, 260
  sphericity index, 259
  2D scattering, 254

B

Backscatter Extinction lidar-Ratio Temperature Humidity lidar Apparatus (BERTHA), 224
Backscatter lidar, 213, 214. *See also* Raman lidar
  CALIOP, 227
Backscattered electron image/signal (BEI), 199
Backscattered primary electrons (BSE), 464
  mode of imaging, 465
Backward enhancement, 322
  MPEF, 323
  MPEF-Lidar, 324–325
  reciprocity principle, 323
  standoff detection, 324
  2PEF, 324
Ballistic trajectories
  control, 331
  measurement, 330
  in microdroplets, 330
BC. *See* Black carbon (BC)
BE. *See* Beam expander (BE); Binding energy (BE)
Beam expander (BE), 212
Beam-sensitive particles, 194, 195
Beer–Lambert law, 271
  attenuation, 271
  linear regime, 432
  optical density, 425
BEI. *See* Backscattered electron image/signal (BEI)
BERTHA. *See* Backscatter Extinction lidar-Ratio Temperature Humidity lidar Apparatus (BERTHA)
Binding energy (BE), 367
Biochemical identification procedures, 321
  difficulties, 322
  optical techniques, 322
Black carbon (BC), 245, 270. *See also* Brown carbon (BrC); Organic carbon (OC)
  carbon K-edge NEXAFS spectra, 438
  in climate modeling, 246
  components, 437
  density, 435
  NE XAFS spectra, 437
  single-energy images and spectra, 434
  STXM work, 433
BrC. *See* Brown carbon (BrC)
Brightness temperature (BT), 90
Brown carbon (BrC), 245, 270. *See also* Black carbon (BC); Organic carbon (OC)

BSE. *See* Backscattered primary electrons (BSE)
BT. *See* Brightness temperature (BT)

C

C.H. *See* Clathrate hydrates (C.H.)
C:O ratio, 436–437
Cadenza, 280
CALIOP, 227
Carbon, 422
  NEXAFS spectra, 427
  on sample surfaces, 205–206
Carbon monoxide (CO), 37
Cardinal rule, 360
Cascade impactors, 197
  assembled, 198
  Battelle-type, 198
  Berner-type, 197
  particle separation, 198
  PIXE cascade, 199
Cavity-enhanced Raman scattering (CERS), 136
  composite spectra, 147
  fingerprint of droplet, 145
  from tweezed aqueous droplet, 137
Cavity ring-down aerosol extinction spectrometer (CRD-AES), 288
Cavity ring-down spectroscopy (CRD-S), 270
  advantage, 277
  applications, 275, 279
  CW-CRD spectroscopy, 277–278
  light intensity fluctuations, 275
  methodology, 274
  optical resonator, 275
  pulsed-laser CRD, 275
  sensitivity and detection limit, 278
CC SEM/EDX particle analysis, 465
  S/Na ratios, 484
  sea-salt particles, 484
  tar balls, 468
  TexAQS sample, 470
CCD. *See* Charge-coupled-device (CCD)
CCN. *See* Cloud condensation nuclei (CCN)
CCSEM/EDS. *See* Computer-controlled SEM/EDS (CCSEM/EDS)
CD. *See* Circular dichroism (CD)
CDE. *See* Continuous distribution of ellipsoids (CDE)
Center for X-ray Optics (CXRO), 426
CERS. *See* Cavity-enhanced Raman scattering (CERS)
Charge-coupled-device (CCD), 141, 162, 254, 424
Charge square on sample, 205
Chi-square error, 88
Circular dichroism (CD), 373
  nanoparticle, 374, 375
Clathrate hydrates (C.H.), 73
Clay minerals, 83
  analytic shape simulations, 88
  chi-square error, 88
  experimental IR spectra, 83, 84
  illite IR spectra, 83, 84, 89
  kaolinite IR spectra, 85
  Mie theory simulations, 89–82
  montmorillonite IR spectra, 85, 86

Clay minerals (*Continued*)
  relative integrated area, 88
  simulated IR spectra, 83, 84
  vibrational assignments, 87
CLN. *See* CREST Lidar Network (CLN)
Cloud condensation nuclei (CCN), 156, 270, 463
Clusters, 368
  cluster 8, 302
  spectrum, 304, 305
  templates for LIF spectra, 302
CO. *See* Carbon monoxide (CO)
Collisional cooling cell, 36, 37
Combustion products, 434
  lignin, 290
Cometary dust, 107–108
Complementary techniques, 453
  in particle chemistry studies, 483
Composition-specific imaging, 467
Computer-controlled SEM/EDS (CCSEM/EDS),
    193, 194, 465
Condensable compounds, 101
Condensation particle counter (CPC), 80, 281
Continuous distribution of ellipsoids (CDE), 88, 89, 103
  absorption, 83
  shape model, 92, 93
  spectrum, 115, 116
Continuous wave (CW), 136, 275, 309
  lasers, 139
Continuous wave-cavity ring-down (CW-CRD),
    277–278
Cooling effect, 270. *See also* Aerosol(s)
Cooperative Remote Sensing Science and
    Technology (CREST), 232
Core-level spectroscopy, 427
Coumarin 314. *See* Probe molecule
Counterflow virtual impactor (CVI), 431
  sampling cloud droplet, 452–453
CPC. *See* Condensation particle counter (CPC)
CRD aerosol spectrometry (CRD-AS), 271.
    See also CRD-AS application
  aerosols extinction measurement, 287
  applications, 274
  complex refractive index determination, 281
CRD-AES. *See* Cavity ring-down aerosol extinction
    spectrometer (CRD-AES)
CRD-AS. *See* CRD aerosol spectrometry (CRD-AS)
CRD-AS application, 279. *See also* CRD aerosol
    spectrometry (CRD-AS)
  aerosol studies, 280
  aerosols optical growth factor studies, 289, 290
  coated aerosols, 286–288
  homogenously mixed aerosols, 282–286
  measuring complex refractive index, 281
  optical combinations, 290
  recent advances, 288
  tandem CRD-ASs, 289
CRD-S. *See* Cavity ring-down spectroscopy (CRD-S)
CREST. *See* Cooperative Remote Sensing Science
    and Technology (CREST)
CREST Lidar Network (CLN), 232
  EARLINET, 226
CRH. *See* Crystallization relative humidity (CRH)
Criegee intermediates, 180, 185
Crystallization relative humidity (CRH), 158

CVI. *See* Counterflow virtual impactor (CVI)
CW. *See* Continuous wave (CW)
CW-CRD. *See* Continuous wave-cavity ring-down
    (CW-CRD)
CXRO. *See* Center for X-ray Optics (CXRO)

D

D_2O molecules, isolated, 50
DAS. *See* Data acquisition system (DAS)
Data acquisition system (DAS), 212
DDA approach. *See* Discrete dipole approximation
    approach (DDA approach)
Debris disks, 101, 103
Deduced absorption, 247
Deliquescence relative humidity (DRH), 158, 473
  of AS, 166
  partial deliquescence behaviors, 167, 168
Depolarization ratio, 131
  backscattering linear, 20
  linear particle, 220
  volume, 225
DHS model. *See* Distribution of hollow spheres model
    (DHS model)
Dicarboxylic acids, 163. *See also* Aerosol particles
  hygroscopicity, 164
Difference method (DM), 246, 247, 248
Differential light scattering pattern. *See* Scattering polar
    dependence
Differential mobility analyzer (DMA), 80, 281, 402
  application, 383, 388
  H-TDMA, 161
Diffractometer, 255
Diffusion limited cluster aggregation (DLCA), 343
Dimethyl ether (DME), 74
Dimethyl sulfide (DMS), 486
Dioctyl sebacate aerosols (DOS aerosols), 282
Discrete dipole approximation approach
    (DDA approach), 4, 103, 361, 307.
    See also Moment method (MOM)
Distribution of form factors model (DFF) model,
    104, 116
  IR band profiles, 114–115
  spectrum, 104
Distribution of hollow spheres model
    (DHS model), 104
DLCA. *See* Diffusion limited cluster
    aggregation (DLCA)
DLS. *See* Dynamic light scattering (DLS)
DM. *See* Difference method (DM)
DMA. *See* Differential mobility analyzer (DMA)
DME. *See* Dimethyl ether (DME)
DMiLay, 287
DMS. *See* Dimethyl sulfide (DMS)
Dolomite, 97
DOS aerosols. *See* Dioctyl sebacate aerosols (DOS
    aerosols)
Double-ring aromatics, 304
DRH. *See* Deliquescence relative humidity (DRH)
Dust grains. *See* Dust particles
Dust particles, 101, 103, 108, 213
  in astrophysical environments, 103
  asymmetric photoemission, 387
  cosmic, 104

dust emission spectra, 104
extraterrestrial, 107
interplanetary, 107
interstellar, 387
investigations, 107–108
meteorites, 108
pre-solar grains, 108
properties, 102
Dynamic light scattering (DLS), 401

E

E-AIM. *See* Extended Aerosol Inorganic Model (E-AIM)
EARLINET. *See* European Aerosol Research Lidar Network (EARLINET)
Earth Clouds, Aerosols and Radiation Explorer (EarthCARE), 232, 233
EarthCARE. *See* Earth Clouds, Aerosols and Radiation Explorer (EarthCARE)
EC. *See* Elemental carbon (EC)
EDB. *See* Electrodynamic balance (EDB)
EDB/Raman system, 159, 162
    AC ring electrode, 160
    application, 163, 175, 179
    EDB design, 160
    experiment and theory, 159
    hygroscopic measurements, 160
    morphology and light scattering pattern, 163
    particle mass, 160, 161
    single levitated particles, 161
    solution droplets generation, 159
EDS. *See* Energy-dispersive x-ray spectrometric detection (EDS)
EDX. *See* Energy-dispersive x-ray spectroscopy (EDX)
EDXRF. *See* Energy dispersive x-ray fluorescence (EDXRF)
EELS. *See* Electron energy loss spectroscopy (EELS)
EF imaging. *See* Energy filtered imaging (EF imaging)
Efflorescence relative humidity (ERH), 474
Electrodynamic balance (EDB), 141, 156
    advantage, 156
    applications, 142
    configurations, 160
    hygroscopicity measurement, 158, 159
    SEDB, 158
Electrodynamics, 420
Electron beam-based methods, 464
Electron energy loss spectroscopy (EELS), 430, 467
    application, 479
Electron microscopy (EM), 401, 464
    applications, 468
    approaches, 465
    field-collected particles characterization, 468
    gas-particle reactions laboratory studies, 479–483
    hygroscopic properties, particles, 473
    ice nucleation properties, particles, 477
    optical properties, particles, 478
    studies on SiO_2 nanoparticles, 414
Electron probe microanalysis (EPMA), 200
Electron probe microanalysis with x-ray detection (EPXMA), 193

Elemental carbon (EC), 434
Elemental mapping. *See* Energy filtered imaging (EF imaging)
Ellipsoidal reflectors, 261
EM. *See* Electron microscopy (EM)
Energy dispersive x-ray fluorescence (EDXRF), 196, 202
Energy-dispersive x-ray spectrometric detection (EDS), 193
Energy-dispersive x-ray spectroscopy (EDX), 431, 464, 465
    chemical characterization, 468
    maps, 476
    of sea-salt particles, 475
    spectra of NaCl particle, 480
    spectrum of particle composition, 472, 473
Energy filtered imaging (EF imaging), 467
Environmental SEM (ESEM), 195, 473
    advantage, 475
    of ice crystals, 478
    for ice nucleation studies, 477
    of NaCl particles, 474
    proof-of-principle ESEM experiments, 477–478
    of sea-salt particles, 475
    substantial limits, 474
Environmental TEM (ETEM), 473. *See also* Environmental SEM (ESEM)
    advantage, 475
    application, 473–474
    substantial limits, 474
EPMA. *See* Electron probe microanalysis (EPMA)
EPXMA. *See* Electron probe microanalysis with x-ray detection (EPXMA)
ERH. *See* Efflorescence relative humidity (ERH)
ESA. *See* European Space Agency (ESA); Excited state absorption (ESA)
ESEM. *See* Environmental SEM (ESEM)
ESO. *See* European Southern Observatory (ESO)
ETEM. *See* Environmental TEM (ETEM)
Ethanol, 402, 405
European Aerosol Research Lidar Network (EARLINET), 212, 226, 232
    lidar stations distributions, 227
    observations data, 227
European Southern Observatory (ESO), 102
European Space Agency (ESA), 232
Excitation density, 34, 35
    comparison, 36
    correlation function, 35
    exciton transition moments, 34
    as function of radial coordinate, 38
    local transition moment, 34
Excited state absorption (ESA), 325
Exciton coupling, 8, 27, 28
    effect, 38
Experimental spectra, 202
Extended Aerosol Inorganic Model (E-AIM), 158
Extensive parameters, 220
Extinction. *See* Attenuation
Extinction, total, 271
Extinction coefficient, 117, 244, 272. *See also* Extinction cross section
    determination, 246, 277
    smoke, 225

Extinction cross section, 10, 115, 244, 272
    opposite shape-related trend, 15
    size-bin averaged, 5, 16
Extinction efficiency, 244, 272, 288
    as function of size parameter, 245
    OEC, 246
    vs. size parameter, 274, 283, 285, 286
Extinction paradox, 273
Extinction spectroscopy, 245, 246
    deduced absorption, 247
    extinction spectrum, 21, 89
    LOPES, 247
    measurement error, 248
    mid-infrared, 3
    OEC, 246

F

FAD. *See* Flavin adenine dinucleotide (FAD)
Femtosecond LIBS (fs-LIBS)
    all optical Gram test, 331–332, 333
    in bioaerosols, 331
    low temperature plasma, 332
    and ns-LIBS comparison, 332
FIB. *See* Focused ion beam milling (FIB)
Field stop (FS), 212
Flavin adenine dinucleotide (FAD), 321–322
Flavin mononucleotide (FMN), 321, 334
    PPD spectroscopy, 328
Fluorescence, 141
    relaxation, 326
    spectra, 301
FMN. *See* Flavin mononucleotide (FMN)
Focused ion beam milling (FIB), 431
Fourier transform infrared (FTIR), 80, 106, 175, 429
    advantage, 429
    bands of surface D_2O and HDO, 67
    for D_2O particles, 55
    difference spectra, 59
    of H_2O ice nanoparticles, 55
    sampling of ice aerosol nanocrystals, 52
    spectra at 100 K, 56
    spectra of adsorbed SO_2, 70
    spectral band complex, 66
Fractal aggregates, 342
    absorption cross section, 353
    albedo, 354
    differential scattering cross section, 352–353
    light-scattering cross sections, 352
    pair correlation function, 342
    phase-shift parameter, 346
    primary particles, 342, 343
    radius of gyration, 342
    RDG fractal aggregate approximation, 354
    soot, 343
    stretched exponential, 342
    structure factor, 350
    total scattering cross section, 353
    volume fraction of monomers, 346
Fractals, 342
Fresnel zone plate, 421, 423
    light incident, 424
FS. *See* Field stop (FS)
fs-LIBS. *See* Femtosecond LIBS (fs-LIBS)

FTIR. *See* Fourier transform infrared (FTIR)
Full-width at half-maximum (FWHM), 216
Full-width half-height (FWHH), 142, 163
    hygroscopicity and, 165
    Raman characterization and, 166
FWHH. *See* Full-width half-height (FWHH)
FWHM. *See* Full-width at half-maximum (FWHM)

G

GA. *See* Genetic algorithm (GA)
GALION. *See* GAW Aerosol Lidar Observations
            Network (GALION)
Gaussian
    autocovariance function, 410
    cut-off with, 351
    structure factor, 351
Gaussian random sphere model (GRS model), 115, 116
GAW Aerosol Lidar Observations Network
            (GALION), 232
Genetic algorithm (GA), 335
Geometric regime, 273. *See also* Mie scattering
            regime; Rayleigh scattering
Geostationary operational environmental
            satellites (GOES), 90
Glutaric acid, 171
GOES. *See* Geostationary operational environmental
            satellites (GOES)
Grain growth, 102
GRS model. *See* Gaussian random sphere
            model (GRS model)

H

H-bonds. *See* Hydrogen bonds (H-bonds)
H-TDMA. *See* Hygroscopic Tandem Differential
            Mobility Analyzer (H-TDMA)
HAADF imaging. *See* High-angle annular dark-field
            imaging (HAADF imaging)
Heterogeneous ice nuclei (HIN), 477
Heterogeneous reactions, 145
    changes in particle mass yield, 184
    of organic aerosols, 179
High-angle annular dark-field imaging
            (HAADF imaging), 467
High-resolution infrared radiation Sounder (HIRS), 90
High-resolution transmission electron microscopy
            (HR-TEM), 467
High Spectral Resolution Lidar (HSRL), 214
    multiwavelength, 232
Highest occupied molecular orbital (HOMO), 427
    advantage, 472
    application, 473, 479
    images of silica nanoparticle, 406
    $K_3Na(SO_4)_2$ particles characterization, 472
HIN. *See* Heterogeneous ice nuclei (HIN)
HIRS. *See* High-resolution infrared radiation
            Sounder (HIRS)
Holographic optical tweezers (HOT), 143, 149
HOMO. *See* Highest occupied molecular
            orbital (HOMO)
HOT. *See* Holographic optical tweezers (HOT)
HR-TEM. *See* High-resolution transmission
            electron microscopy (HR-TEM)

HSRL. *See* High Spectral Resolution Lidar (HSRL)
HULIS. *See* Humic-like substances (HULIS)
Humic-like substances (HULIS), 245, 284, 303
Hydrogen bonds (H-bonds), 51
Hygroscopic Tandem Differential Mobility Analyzer (H-TDMA), 161
Hygroscopicity, 156, 157
    aerosol particles assessment, 167–168
    dicarboxylic acid particles, 164
    EDB and SEDB approaches, 158
    and FWHH, 165
    organic coatings effects, 167–168

I

IC. *See* Ion chromatography (IC)
ICCD. *See* Image-intensified charge coupled device (ICCD)
Ice aerosol nanocrystal, 51
    difference spectra, 59–60
    FTIR sampling, 53–54
    FTIR spectra comparison, 55
    FTIR spectra for D_2O particles, 55–56
    H_2O/D_2O nanocrystals preparation, 52–53
    ice subsurface resemblance, 58
    L-defect activity, 64, 65
    Ostwald ripening, 56–57, 59
    proton activity, 63–64, 65
    simulated 4-nm ice particle model, 54
    subsurface spectrum, 57, 59
    surface-based dynamic equilibrium, 69–70
    water clusters computational modeling, 54
    water-molecule vibrational modes sensitivity, 58
    water phases data analysis, 51–52
    weak-acid control, 67–68
    weak adsorbates influence, 59
Ice-defect activity, 69
Ice in clouds experiment-layer clouds (ICE-L), 452
ICE-L. *See* Ice in clouds experiment-layer clouds (ICE-L)
Ice nucleation, 452, 477
    in atmosphere, 477
    ESEM application, 477
    heterogeneous, 477
    ice crystals ESEM image, 478
Ice nuclei (IN), 463. *See also* Dust particles
Ice particles, 13, 50
    formation, 65
    IR spectra of spherical CO, 39
    simulated relaxed 4-nm, 54
Ice surface vibrational spectrum, 50
IDPs. *See* Interplanetary dust particles (IDPs)
IfT. *See* Institute for Tropospheric Research (IfT)
Image-intensified charge coupled device (ICCD), 307
Imaging techniques, 454
    capabilities analysis, 429
IMFP. *See* Inelastic mean free path (IMFP)
Immersion freezing mode, 477
IN. *See* Ice nuclei (IN)
IN material. *See* Inorganic material (IN material)
Incident beam, 131, 411
Incident infrared beam extinction, 3

Indirect and semidirect aerosol campaign (ISDAC), 453
Induced dipole, 28, 128
Inelastic mean free path (IMFP), 367
    electron, 367, 368
Infrared (IR), 79, 101
    extinction spectra of ice spheres, 11
    region, 79, 229
    spectroscopy, 25
Infrared Astronomical Satellite (IRAS), 102
Infrared extinction spectra
    absorption cross-section, 5, 7
    absorption spectrum comparison, 6–7
    extinction cross-section, 5
    infrared optical depth, 5–6
    optical constants derivation, 8–10
    particle shape influence, 7, 13–16
    size distribution retrieval, 16–17
    small-particle absorption spectra comparison, 7–8
    spectral habitus dependency, 10–13
Infrared Space Observatory (ISO), 102
Infrared spectral habitus, 17
    AIDA chamber operation, 17
    cloud particles detection, 18
    comparisons, 19
    homogeneous freezing, 18
Infrared spectroscopy (IR spectroscopy), 3, 25, 44, 45. *See also* Aerosol IR spectra
    aerosol and pellet spectra comparison, 109
    aerosol experiment, 105
    aerosol particle extinction spectra, 26
    astrophysically relevant samples, 107
    experimental setup, 105–107
    free-floating dust particles, 105
    matrix effect, 109
    sample properties, 109
    spectral differences, 27
Infrared Telescope facility (IRTF), 102
Inorganic material (IN material), 434
Inorganic species, 270, 448
    Fe speciation, 449
    marine boundary layer sulfur speciation, 450
    sulfur-containing particles, 451–452
    Zn-containing aerosols speciation, 449
    Zn spectra, 449
Institute for Tropospheric Research (IfT), 211
    laser emission systems, 212
Intensifier, 141
Interfaced approach, 203
    challenges, 203–205
    MRS interfaced with SEM/EDS, 203
Interference filters, 213, 219, 309
Intermediate adsorbates, 50
    role, 59
Interplanetary dust particles (IDPs), 107
Intrinsic fluorescence. *See* Laser induced-fluorescence (LIF)
Inversion, 221
Inversion algorithms, 221, 231–232
    parameters, 222
Ion chromatography (IC), 195
IR. *See* Infrared (IR)

IR band profiles
  agglomeration effect, 113
  form factor distributions, 116
  forsterite samples, 111, 112
  morphological effects on, 110
  rutile samples, 113, 114
  shape effect, 112–113
  size effect, 110–112
  theoretical simulations, 114
IR extinction spectra. *See also* IR spectra
  experimental, 26
  instrument design, 80
  and size distributions measurement, 80
  time-dependent experimental, 41
IR spectra, 1, 3, 36, 102
  of aerosol particles, 37
  astronomical spectra interpretation, 116–119
  CDE, 103
  of CO, 61
  DDA approach, 103
  DFF model, 104
  DHS model, 104
  dust components identification, 102
  of H_2, 62
  influence of particle shape on, 40
  Mie theory, 103
  pellet technique, 104
  statistical models, 103
IR spectroscopy. *See* Infrared spectroscopy
        (IR spectroscopy)
IRAS. *See* Infrared Astronomical Satellite (IRAS)
IRTF. *See* Infrared Telescope facility (IRTF)
ISDAC. *See* Indirect and semidirect aerosol
        campaign (ISDAC)
ISO. *See* Infrared Space Observatory (ISO)

J

Japan Aerospace Exploration Agency (JAXA), 102
JAXA. *See* Japan Aerospace Exploration Agency (JAXA)

K

KE. *See* Kinetic energy (KE)
Kinetic effects, 156
Kinetic energy (KE), 367
Köhler curve, 147–148, 149
Kramers–Kronig integral, 9
Kramers–Kronig relation, 9

L

L-defect activity, 64, 65. *See also* Ice aerosol nanocrystal
  enhanced nanocrystal, 66
  management with SO_2 and H_2S, 70–71
Lambert–Beer equation, 5–6. *See also*
        Beer–Lambert law
Laser diffractometer, 255
  particle trapping, 257
Laser diffractometry. *See* Polar nephelometry
Laser-induced breakdown spectroscopy (LIBS), 298
  fs-LIBS, 331
  fs-LIBS and ns-LIBS comparison, 332
  nano-LIBS, 331

Laser induced-fluorescence (LIF), 298
  cluster templates, 302
  drawback, 325
  implications of similarities, 305
  prototype development, 299–300
  spectra, 303
Laser-induced incandescence measurements, 279
LC. *See* Liquid crystal (LC)
LCP light. *See* Left circularly polarized light (LCP light)
LED. *See* Light-emitting device (LED)
Left circularly polarized light (LCP light), 374
LIBS. *See* Laser-induced breakdown spectroscopy
        (LIBS)
Lidar equation, 214
  Raman, 215–216
Lidar ratio, 211
  height profile, 216
  particle, 220
  urban haze, 221
  variation, 214–215
Lidar technique. *See* Light detection and ranging
        technique (Lidar technique)
LIF. *See* Laser induced-fluorescence (LIF)
Light detection and ranging technique (Lidar technique),
        211, 322
  aerosol characterization, 211
  feature, 211
  instrument setup, 212
  lidar equation, 214
  MPEF, 324
  operation, 232
  signal accumulation, 213
  simple backscatter, 214
  vertically resolved sounding, 211
Light-emitting device (LED), 246
Light scattering, 342, 344, 358
  cross sections, 345
  differential element, 346
  differential scattering cross section, 345
  incident irradiance, 345
  for incident light wave, 344
  phase-shift parameter, 346
  polarization, 344–345
  Rayleigh–Debye–Gans approximation, 346
  scattered irradiance, 345
  total scattering cross section, 345
Light-Scattering cross sections, 351
  fractal aggregate, 352
  Rayleigh scattering, 351
Linoleic acid, 180. *See also* Linolenic acid; Oleic acid
  laser-illuminated light-scattering patterns, 186
  particle images, 186
  particle mass changes, 184
  peak intensity changes, 183
  Raman spectra, 181
Liquid crystal (LC), 333
LO. *See* Longitudinal optical (LO)
Log-normal distribution, 408
Long path extinction spectrometer (LOPES), 247
Longitudinal optical (LO), 58
LOPES. *See* Long path extinction spectrometer (LOPES)
Lowest unoccupied molecular orbital (LUMO), 427
LUMO. *See* Lowest unoccupied molecular
        orbital (LUMO)

M

MA. *See* Malonic acid (MA)
MAARS. *See* Multi-Analysis Aerosol Reactor System (MAARS)
Magnetic structures, 422
Malonic acid (MA), 159
   in AS/MA particles, 164
   hygroscopicity and FWHH, 165
Marine Stratus Experiment (MASE), 450, 483
MARTHA. *See* Multiwavelength–Aerosol–Raman–Temperature–Humidity Apparatus (MARTHA)
MASE. *See* Marine Stratus Experiment (MASE)
Mass absorption coefficient, 426
   C:O ratio, 436
   of forsterite, 111
Mass fraction of solute (*mfs*), 161
Mass transfer processes, 159
Matrix effect, 109
Maxwell–Garnett rule, 284. *See also* Mixing rules
MCP. *See* Multichannel plate (MCP)
MD. *See* Mineral dust (MD); Molecular dynamics (MD)
MDR. *See* Morphological-dependent resonance (MDR)
Megacities Initiative: Local and Global Research Observations (MILAGRO), 470
MEMS. *See* Micro-electro-mechanical-systems (MEMS)
mfs. *See* Mass fraction of solute (*mfs*)
Micro-electro-mechanical-systems (MEMS), 336
   chip, 337
   shaper, 336, 337
Micro Orifice Uniform Deposit Impactor (MOUDI), 431
Micro-Raman spectroscopy (MRS), 193, 203
   advantage, 203
   continuous laser beam, 200
   sparse, 194
Microphysical properties, 3, 210
   boundary layer aerosols, 210
   particles, 221, 230–231
   smoke aerosol, 225
Mid-infrared extinction spectroscopy, 3
Mie scattering model, 404
Mie scattering regime, 273. *See also* Geometric regime; Rayleigh scattering
Mie-scattering theory, 228
Mie simulation code, 82
Mie theory, 10, 80, 103, 271
   application, 82
   implementations, 287
   infrared extinction spectra of ice spheres, 11
   internal electric field amplitude calculations, 386
   inversion schemes, 9–10
   limitations, 83, 103
   radiative transfer model, 90
   simulations, 92
   spectral behavior, 251
MILAGRO. *See* Megacities Initiative: Local and Global Research Observations (MILAGRO)
Mineral dust (MD), 8, 245
Mineral dust aerosol, 79, 253
   atmospheric, 80
   clay minerals, 83
   effect on atmosphere, 79
   mass-specific absorption cross section, 254
   oxide and carbonate minerals, 92

   physical nature, 80
   satellite retrievals, 97
   single scattering albedo, 254
   source, 79
   spectral absorption characteristics, 253
Mixing rules, 282
   dielectric constant of mixture, 283
   linear mixing rule, 282–283
   Maxwell–Garnett rule, 283
   molar absorption, 282
   molar refraction, 282
   total molar volume, 282
Mode number, 134
Mode offset, 150
Moderate resolution imaging spectroradiometer (MODIS), 90
MODIS. *See* Moderate resolution imaging spectroradiometer (MODIS)
Molecular backscatter coefficient, 215
Molecular dynamics (MD), 31, 393
MOM. *See* Moment method (MOM)
Moment method (MOM), 307
Monochromator, 420, 422
Morphological-dependent resonance (MDR), 330
Morphology-dependent resonances. *See* Whispering gallery modes (WGMs)
MOUDI. *See* Micro Orifice Uniform Deposit Impactor (MOUDI)
MPEF. *See* Multiphoton excited fluorescence (MPEF)
MPEF-Lidar, 324–325
MRS. *See* Micro-Raman spectroscopy (MRS)
MRS in parallel with elemental analyses, 202
   interfaced approach, 203
   stand-alone approach, 202
MULIS. *See* Multiwavelength Lidar System (MULIS)
Multi-Analysis Aerosol Reactor System (MAARS), 81
   atomizer, 80
   components, 81
   design, 80
   extinction cell length, 80
   flow conditions, 81
   particle sizing instruments, 80
   volume equivalent diameter, 81–82
Multiangle photometry, 254
Multichannel plate (MCP), 141
Multicomponent particles, 10, 312
   Raman investigation, 163, 164
   spectral features, 40, 41–45
   STXM investigation, 444
Multiphoton excited fluorescence (MPEF), 323, 324–325
Multiwavelength–Aerosol–Raman–Temperature–Humidity Apparatus (MARTHA), 218, 219
   Nd:YAG, 219
   optical parameters, 220
   signal receiver box, 220
   system parameters, 219
   temperature profiles, 219
Multiwavelength HSRL systems (MW-HSRL systems), 232
Multiwavelength Lidar System (MULIS), 224

MW-HSRL systems. *See* Multiwavelength HSRL
           systems (MW-HSRL systems)

N

NADH. *See* Nicotinamide adenine
           dinucleotide (NADH)
nano-LIBS. *See* nanosecond-laser LIBS (nano-LIBS)
Nanoparticle beam, 379, 402
     elastic light scattering with, 408
     nanoparticles aggregates formation, 408
     VMI, 382, 386
Nanoparticles, 401. *See also* Soft x-rays elastic
           scattering
     angle-resolved light-scattering patterns, 409
     angle-resolved photoemission, 382
     ethanol surface layer coating, 406, 407
     into high-vacuum system, 403
     log-normal distribution, 408
     perturbation theory fit, 412
     porosity at 18° scattering angle, 405
     properties analysis, 401
     scattered light intensity, 414
     scattering intensity, 408
     silica, 402
     soft x-ray regime, 403
     surface properties, 402
     surface roughness, 412
     TEM images, 406
nanosecond-laser LIBS (nano-LIBS), 331. *See also*
           Femtosecond LIBS (fs-LIBS)
NAOJ. *See* National Astronomical Observatory of
           Japan (NAOJ)
NAOPEX. *See* Nighttime Aerosol–Oxidant Plume
           Experiment (NAOPEX)
Narrowband sensors, 90
NASA. *See* National Aeronautics and Space
           Administration (NASA)
National Aeronautics and Space Administration (NASA),
           102, 487
National Astronomical Observatory of Japan
           (NAOJ), 102
National Institute for Environmental Studies
           (NIES), 232
National Synchrotron Light Source (NSLS), 423
Native fluorescence. *See* Laser induced-
           fluorescence (LIF)
Nd:YAG. *See* Neodymium:yttrium–aluminum–garnet
           (Nd:YAG)
Near edge x-ray absorption fine structure spectroscopy
           (NEXAFS), 420, 421, 470
     C:O atomic ratios, 437
     carbon K-edge, 438, 440
     for flame-generated methane soot, 436
     sulfur L-edge, 451
     unequivocal source assignment, 433–434
     Zn L-edge, 449
Neodymium:yttrium–aluminum–garnet (Nd:YAG),
           213, 215, 219
New England Air Quality Study–Intercontinental
           Transport and Chemical Transformation
           (NEAQS-ITCT), 279
NEXAFS. *See* Near edge x-ray absorption fine structure
           spectroscopy (NEXAFS)

NEXAFS spectroscopy, 424
     carbon K-edge NEXAFS spectra, 438
     carbon NEXAFS, 427
     electronic transitions probed, 424
     for flame-generated methane soot, 436
     mass absorption coefficient, 426
     OD, 425
     peak positions and intensities, 427
     spectra for OC, 444, 445
Nicotinamide adenine dinucleotide (NADH), 321
NIES. *See* National Institute for Environmental
           Studies (NIES)
Nighttime Aerosol–Oxidant Plume Experiment
           (NAOPEX), 470
NSLS. *See* National Synchrotron Light Source (NSLS)
Nuclepore, 199
Nyquist frequency, 33

O

OC. *See* Organic carbon (OC)
OCA. *See* Organic carbon aerosol (OCA)
OCA method. *See* Optical constants averaging method
           (OCA method)
OD. *See* Optical density (OD)
ODD. *See* Optimal dynamic discrimination (ODD)
OEC. *See* Optical extinction cell (OEC)
Oleic acid, 180. *See also* Linoleic acid; Linolenic acid
     mass loss effect, 185
     oxidation in synthetic seawater droplets, 145
     particle mass changes, 184
     peak intensity changes, 183
     Raman spectra, 181
OPO. *See* Optical parametric oscillator (OPO)
Optical constants averaging method (OCA method), 93
Optical density (OD), 425
Optical extinction cell (OEC), 246
     nephelometer device combination, 348
     single-folded path, 247
Optical levitation, 142–143
     Mie resonances, 146
Optical parametric oscillator (OPO), 277, 285
Optical particle sizing, 358
     cardinal rule, 360
     light scattering, 358
     optical structure factor, 359, 361
Optical properties, 97, 270, 271, 479
     CRD-AS, 274
     Mie theory usage, 80, 82
Optical tweezers, 143
     aqueous droplets chemical composition, 144–145
     comparative thermodynamic measurements,
           148–149
     droplets coagulation, 149
     evaporation experiments, 147–148
     heterogeneous reactions, 145
     mode offset, 150
     Raman measurements, 143–144
Optimal dynamic discrimination (ODD), 322, 333
     implementation, 335
     RBF and FMN, 334
     system complexity, 336
     UV pulse, 335
Order sorting aperture (OSA), 424

Organic carbon (OC), 251, 305, 433, 446
    compounds, 306
Organic carbon aerosol (OCA), 270, 298
Organic solid polarization energy, 381
OSA. *See* Order sorting aperture (OSA)
Ostwald ripening, 56–57, 59
Oxide and carbonate minerals, 92
    CDE simulations, 93, 94, 95, 96
    experimental spectra, 93, 94, 95, 96
    Mie simulations, 93, 94, 95, 96
    optical constant data sources, 96
    quartz experimental IR spectrum, 93

P

PA spectroscopy. *See* Photoacoustic spectroscopy
    (PA spectroscopy)
PAH. *See* Polycyclic aromatic
    hydrocarbon (PAH)
PARAGON. *See* Progressive Aerosol Retrieval
    and Assimilation Global Observing
    Network (PARAGON)
Particle Fluorescence Spectrometer (PFS),
    299, 304
    features, 300
    limitations, 306
Particle Habit Imaging and Polar Scattering
    (PHIPS), 257
    angular light scattering measurement, 257, 258
    hexagonal ice plate 3D model, 258, 259
    probe, 257
    3D model reconstruction principle, 262
Particle mass, 160
    changes in, 184–185
    measurement, 163
    ratio of, 161
Particle morphology, 104, 109
    changes, 475
    characterization, 465
    in spectral feature comparisons, 117
Particle-on-Substrate Stagnation Flow Reactor
    (PS-SFR), 479–483
Particle size distribution, 223, 229, 402
    log-normal distribution, 408
    Mie theory, 80
    monomodal, 229
Particles. *See also* Aerosol(s)
    extinction-to-backscatter ratio, 230
    growth factor, 161
    hygroscopic transformations, 473
    instruments, 369
    lidar ratio height profile, 216
    multicomponent, 312
    optical properties, 479
    phase function, 231
    photoelectron spectrum, 383, 384
    sizing instruments, 80
    trapping, 257
Particles angular scattering characteristics, 307
    atmospheric aerosol TAOS, 309, 313
    experimental, 307
    pattern classes frequency-of-occurrence, 312–313
    scattering patterns visual classification, 309
    single-particle angular elastic scattering, 307

Particles microanalysis, 464
    CC SEM/EDX analysis, 465, 467, 470
    electron beam-based methods, 464
    goal, 467
    HR-TEM, 467
    SEM/EDX, 467
Particulate matter (PM), 193, 341. *See also*
    Aerosol(s); Particles
    analysis and interpretation strategy, 200
    carbonaceous content, 794
    experimental conditions, 200
    point analysis vs. mapping, 201
    sample to result, 195
    samplers, 195
    sampling, 195
    sampling techniques, 195
    spectral library, 202
    substrates, 199
Particulate medium, 244
PAS. *See* Photoacoustic absorption spectrometer (PAS)
PCA. *See* Principle component analysis (PCA)
PCR. *See* Polymerase chain reaction (PCR)
PDA. *See* Photodiode array (PDA)
PE. *See* Polyethylene (PE)
Pellet technique, 104
    advantage, 104
    deficiencies, 104
    IR spectroscopy, 109
PFS. *See* Particle Fluorescence Spectrometer (PFS)
Phase function, 255. *See* Scattering polar dependence
    aerosol particles, 255
    measurement, 256
    normalization, 256
    particle, 231
Phase-shift parameter, 346
Phase transformations, 156, 157, 186
PHIPS. *See* Particle Habit Imaging and Polar
    Scattering (PHIPS)
Photoacoustic absorption spectrometer (PAS),
    288, 290
Photoacoustic spectroscopy (PA spectroscopy),
    248, 251
    cavity geometry, 249
    measured pressure amplitude, 249
    multiwavelength, 250
    physical concept, 249
Photodiode array (PDA), 141
Photoelectric charging cell, 370
Photoelectron spectroscopy, 367, 370
Photoemission, 369
    aerosol, 369, 370
    atmospheric pressure, 370
    spectroscopy, 368
    threshold, 387–388
Photoionization spectra, 370, 371
    Coumarin 314, 375, 376
    perylene, 372
Photomultiplier, 219, 257
Photomultiplier tube (PMT), 141, 213, 309
    H7260 32-anode, 300
    standard, 219
Piezo-electric transducer (PZT), 278
PIXE. *See* Proton-induced x-ray emission (PIXE)
Planck function, 102

PM. *See* Particulate matter (PM)
PM sampling techniques, 195
   to perform bulk analysis, 196–197
   PM2.5 impactor, 196
   samplers and collection, 197
PMF. *See* Positive matrix factorization (PMF)
PMT. *See* Photomultiplier tube (PMT)
PM2.5 impactor, 196
Polar nephelometry, 255
Polarizability tensor, 130–131
Polarization, 344–345, 428
   energy, 371, 381
   incident vertical, 345
   independent, 344
   law, 353
   light, 131, 136, 344
Polychromators, 219
Polycyclic aromatic hydrocarbon (PAH),
      305, 322, 325, 369, 370
   detectors, 371
   ionization onsets, 371
   PPD experiment, 326
Polydispersity, 350
   effects, 355
   factor, 356, 357
Polyethylene (PE), 104
Polymerase chain reaction (PCR), 321
Polystyrene spheres (PSS), 282
Porto notation, 131
Positive matrix factorization (PMF), 448
Pre-solar grains (PS grains), 108
Principle component analysis (PCA), 451
Pristine ice crystals, 258
Probe molecule, 375
   electronic spectra, 376
   photoemission technique, 377
   photoionization spectra, 376
Progressive Aerosol Retrieval and Assimilation Global
      Observing Network (PARAGON), 210
Proton activity, 63–64, 65. *See also* Ice aerosol
      nanocrystal
   activity comparison, 68
   aerosol particle-core, 65
   bare-nanocrystal, 65
   enhanced surface, 66
   FTIR bands of surface D_2O and HDO, 67
   influence of H_2S, 68–69
   weak-acid control, 67
Proton-induced x-ray emission (PIXE), 199, 430
Protoplanetary accretion disks, 101
PS grains. *See* Pre-solar grains (PS grains)
PSS. *See* Polystyrene spheres (PSS)
PS-SFR. *See* Particle-on-Substrate Stagnation Flow
      Reactor (PS-SFR)
Pulsed-laser CRD, 275
   characteristics, 275
   decay time, 276
   ring down curve, 277
   stable cavity, 275
   time constant, 276
   time-dependent light intensity, 276
Pulse-shaping technique, 333
Pump–Probe spectroscopy, 325
   bacteria and biomolecules identification, 325

ballistic trajectories in microdroplets, 330
   bioaerosols, 328
   discrimination between bacteria and diesel fuel, 327
   limitations, 327
   liquid-phase results, 326
   PPD scheme, 325
Purcell factor, 136
PZT. *See* Piezo-electric transducer (PZT)

Q

Quantitative kinetics, 71. *See also* Ice aerosol nanocrystal
   acid-hydrate formation, 72
   ammonia-hydrate formation, 72
   C.H. formation, 73–74
   C.H. transformations, 74
Quantitative particle characterization, 211

R

Radiation budget (RB), 269
   aerosols, 270
   Earth, 210
Radius of gyration, 342, 350, 356
Raman
   effect, 200, 216
   gain, 136
   lidar equation, 215
   transitions, 217
Raman lidar, 211, 215
   advantage, 212
   complex refractive index, 223
   current status, 223
   EARLINET, 226
   GALION, 232
   instrument characteristics, 213
   inversion algorithms, 231
   microphysical particle parameters, 221
   mixed dust from West Africa, 223–226
   molecular backscatter coefficient, 215
   multiwavelength, 218
   optical data, 221
   optical parameters, 220
   outlook, 228
   particle lidar ratio height profile, 216
   Raman effect, 216–218
   Raman lidar equation, 215
Raman lidar instruments, 228
   aerosol particles chemical signatures, 228
   A-Train, 230
   extending measurements into Daytime, 230
   infrared region, 229
   new data products, 231
   particle extinction-to-backscatter ratio, 230
   particle phase function, 231
   particle size distribution, 229
   sun photometers, 230
   ultraviolet region, 228–229
Raman scattering, 128
   amplification, 147
   anti-Stokes, 230
   detector, 201
   intensity, 134
   pure rotational, 216

quartz, 228
   spontaneous, 134, 142, 143
   Stokes, 230
   transitions, 139
   vibrational, 221
Raman spectroscopy, 128
   anisotropic scattering, 131
   depolarization ratio, 131
   energy levels and transitions, 129
   induced dipole, 128
   isotropic scattering, 131
   normalized spontaneous Raman
      spectra, 133
   polarizability tensor, 130–131
   Porto notation, 131
   Raman arrangement, 130
   Raman cross-sections, 129
   Raman shifted wavelengths, 134
   sensitivity, 194
   spontaneous, 128
   spontaneous Raman spectra comparison, 132
   stimulated, 134, 159
   symmetric vibrations, 131
   vibrational frequencies, 134
Raman spectroscopy applications, 141. *See also*
      Raman spectroscopy
   electrodynamic balance, 142
   optical levitation, 142–143
   optical tweezers, 143
   spontaneous Raman scattering, 142
   SRS, 145
Raman spectroscopy instrumentation, 138. *See also*
      Raman spectroscopy
   components, 138
   detectors, 141
   laser sources, 139–140
   microdroplet generation and sampling, 140
   spectral interferences, 141
   spectrometers, 140
   ultrasonic nebulizer, 140
   VOAG, 140
Raman spectrum, 171, 201. *See also* Raman
      spectroscopy
   amorphous C, 200
   change, 143
   complexity, 216
   hematite, 204
RAOS. *See* Reno Aerosol Optics Study
      (RAOS)
Rapid expansion of supercritical solutions setup
      (RESS setup), 37
Rayleigh–Debye–Gans approximation (RDG
      approximation), 346
   fractal aggregate, 354
   RDG description validity, 360–362
Rayleigh–Gans theory (RG theory), 413
Rayleigh scattering, 213, 216, 351
   absorption in RDG regime, 351
   conditions, 351
   differential scattering cross section, 351
   regime, 272. *See also* Geometric regime;
      Mie scattering regime
Rayleigh theory, 4
RB. *See* Radiation budget (RB)

RBF. *See* Riboflavin (RBF)
RCP light. *See* Right circularly polarized light
      (RCP light)
RDG approximation. *See* Rayleigh–Debye–Gans
      approximation (RDG approximation)
Reaction limited cluster aggregation
      (RLCA), 344
Reciprocity principle, 323
Red giant branch stars (RGB stars), 101
Refractive index (RI), 470
   aerosols, 271
   complex, 13, 223, 244, 271
   particle mixture, 479
Refractory carbonaceous particles. *See* Black
      carbon (BC)
Relative humidity (RH), 142, 156, 375, 471
   determination, 148
   Tandem CRD-AS, 289
Reno Aerosol Optics Study (RAOS), 246
Resonant transition dipole coupling. *See*
      Exciton coupling
RESS setup. *See* Rapid expansion of supercritical
      solutions setup (RESS setup)
RG theory. *See* Rayleigh–Gans theory (RG theory)
RGB stars. *See* Red giant branch stars (RGB stars)
RH. *See* Relative humidity (RH)
RI. *See* Refractive index (RI)
Riboflavin (RBF), 321, 324
   absorption spectra, 325
Right circularly polarized light (RCP light), 374
Rigorous solution (RS), 361
RLCA. *See* Reaction limited cluster
      aggregation (RLCA)
RS. *See* Rigorous solution (RS)

S

SA method. *See* Spectral averaging method
      (SA method)
SAED. *See* Selected-area electron diffraction (SAED)
Saharan Mineral Dust Experiments (SAMUM),
      212, 223
Saharan soil samples, 253
SAMUM. *See* Saharan Mineral Dust Experiments
      (SAMUM)
Satellite sensors, 230
   narrowband, 90
   passive, 211
Satellite telescopes, 102
Scaled projection matrix, 29
Scanning EDB (SEDB), 158, 186
   capability, 169
Scanning electron microscopy (SEM),
      193, 401, 429, 464, 465
   CC SEM/EDX particle analysis, 465
   NaCl particles, 480
   particles morphology and internal
      heterogeneities, 469
   pristine marine particles, 484
   sea-salt particles, 475
   soot particle, 466
Scanning mobility particle sizer (SMPS),
      80, 81, 390
Scanning probe microscopy (SPM), 401

Scanning transmission x-ray microscopy
        (STXM), 420, 470
    advantages, 430
    applications and developments, 453
    complementary techniques, 453
    detection techniques, 424
    Fresnel zone plate, 423
    in situ techniques development, 454
    OSA, 424
    spatial resolution, 423
    stack, 424
    technological advances, 454
    zones, 423
    zone-width, 423
Scatter cross-sections, 82
Scattered beam, 411
Scatterer(s), 347
    density function, 349
    system of, 348. *See also* Structure factor
Scattering, 271
    from atmospheric ice crystals, 256
    azimuthal, 257–258
    differential scattering cross section, 345
    intensity for nanoparticle, 408
    monochromatic soft x-rays, 403
    Rayleigh, 351–352
    soft x-rays, 411
    total scattering cross section, 345–346
    two-dimensional, 260–261
    wave vector, 347–348
Scattering angle, 348, 411
Scattering angular dependence
        measurement, 254
    azimuthal scattering, 258
    CCDs, 254
    diffractometer design, 255
    laser diffractometer, 255
    multiple particles, 256
    phase function, 255, 256
    polar nephelometry, 255
    scattering polar dependence, 255
    single particles, 256
    two-dimensional scattering, 260
Scattering diagram. *See* Scattering polar
        dependence
Scattering from aggregates ensemble, 355
    polydispersity effects, 355
    Tyndall effect, 357
Scattering patterns visual classification, 309
    fiber-like particles, 312
    particles of complex structure, 312
    perturbed sphere patterns, 311
    spherical particle patterns, 310
    swirl patterns, 312
Scattering plane, 258, 344
Scattering polar dependence, 255
Scattering theory, classical, 4
Scattering wave vector, 347
    scattering angle and, 348
SDS. *See* Sodium dodecyl sulfate (SDS)
SE. *See* Secondary electrons (SE)
Secondary electrons (SE), 369
    electron beam, 464
    mode of imaging, 465

Secondary organic aerosol (SOA), 447
SEDB. *See* Scanning EDB (SEDB)
Selected-area electron diffraction (SAED), 467
    pattern, 472, 473
SEM. *See* Scanning electron microscopy (SEM)
SEM/EDX, 453
    conventional use, 467
SERS. *See* Surface enhancement Raman
        spectroscopy (SERS)
SFG. *See* Sum-frequency generation (SFG)
SFU. *See* Stacked filter unit (SFU)
Shrinking core model, 71
SID. *See* Small Ice Detector (SID)
Signal receiver unit (SRU), 212
SIMPLe-to-use Interactive Self-modeling Mixture
        Analysis (SIMPLISMA), 202
SIMPLISMA. *See* SIMPLe-to-use Interactive Self-
        modeling Mixture Analysis (SIMPLISMA)
Single aerosol particle measurements, 127
Single particle
    diagnostic techniques, 298
    levitation techniques, 156
Single particles analysis (SPA), 194, 197
    applications, 194
    problems, 201
Single particles Raman spectroscopy applications, 163
    3AN · AS to 2AN · AS, 177
    autoxidation, 180
    comparison to E-AIM predictions, 166
    FWHH analysis, 166
    hygroscopic measurements, 164
    metastable double salts formation, 175
    metastable salts formation, 174
    organic coatings and hygroscopicity, 167–169
    ozone concentration effects, 185
    ozone-processed particles, 180
    ozonolysis, 183
    partial crystallization and deliquescence, 163
    particle mass yield changes, 184
    particles heterogeneous reactions, 179
    Raman spectral features, 177
    water-insoluble organic coating, 169
    water-soluble organic coating, 171
Single scattering albedo (SSA), 245, 253, 271, 280,
        290. *See also* Albedo
    computation, 223
    for dust samples, 254
    MD aerosol, 245
    ranges, 226
    smoke particles, 226
Singular value decomposition (SVD), 432
Size parameters, 222
SLM. *See* Spatial light modulators (SLM)
Small Ice Detector (SID), 259
    experimental scattering patterns, 260
    SID-1, 260
    SID-2, 260
    SID-3, 262
SMPS. *See* Scanning mobility particle sizer (SMPS)
SOA. *See* Secondary organic aerosol (SOA)
Sodium dodecyl sulfate (SDS), 144, 377
Soft x-ray, 401, 402, 424
    core edges, 425
    light sources, 378

photons, 430
region, 403, 423, 425
scattering, 403
Soft x-rays elastic scattering. *See also* Nanoparticles
angle-resolved elastic light scattering, 409–414
experimental, 402
Gaussian autocovariance function, 410
improved model fit, 413
Mie scattering, 411
perturbation theory fit, 412
results and discussion, 403
scattered light intensity, 404–409, 414
surface roughness autocovariance function, 410
Solution ambiguity, 19
distorted spectral habitus approximation, 19
infrared extinction cross-sections approximation, 20
infrared extinction spectra, 19, 20, 21
SPA. *See* Single particles analysis (SPA)
Spatial distribution of scattering. *See* Two-dimensional angular optical scattering (TAOS)
Spatial light modulators (SLM), 336
Spectral averaging method (SA method), 93
Spectroscopic techniques, 149, 367, 454
microscopic techniques combination, 453
single-particle Raman, 159
Sphere diameter, 262
Sphericity index, 259
SPM. *See* Scanning probe microscopy (SPM)
SRFA. *See* Suwannee River Fulvic acid (SRFA)
SRS. *See* Stimulated Raman scattering (SRS)
SRU. *See* Signal receiver unit (SRU)
SSA. *See* Single scattering albedo (SSA)
Stack, 424
Stacked filter unit (SFU), 196
Stand-alone approach, 202
SEM/EDS, 202–203
XRF for bulk elemental concentrations, 202
STED microscopy. *See* Stimulated emission-depletion microscopy (STED microscopy)
Stimulated emission-depletion microscopy (STED microscopy), 401
Stimulated Raman scattering (SRS), 128, 134. *See also* Raman scattering
applications, 145
droplet train, 145
intensity, 136
internal light intensity, 135
Mie resonance, 134, 136
mode number, 134
optical levitation, 146
optical tweezers, 147
Purcell factor, 136
Raman gain, 136
Raman spectra sequence, 135
spectrum, 139
WGMs, 134
Stokes scattering, 128, 129
Strong adsorbates, 50
Structure factor, 347
aggregates ensemble, 355
Fisher–Burford form, 350
fractal aggregates, 350
fundamental equation, 348
Gaussian cut-off, 351

Guinier regime, 350
scattering wave vector, 347
single aggregate, 355
STXM. *See* Scanning transmission x-ray microscopy (STXM)
STXM/NEXAFS, 420
STXM/NEXAFS aerosol studies, 433
aerosol mixing states evolution, 443
IN and CCN, 452
biomass burn aerosols, 439
black carbon, 434, 437
C:O atomic ratios, 436–437
graphitic nature quantification, 439
inorganic species, 448
organic carbon, 446
power plant plume processing, 452
single-energy STXM images, 435
soot, 433
spectral deconvolution, 438
tar balls, 440
STXM/NEXAFS technique, 423
aerosol technique comparison, 429
data analysis, 432
limitation, 453
NEXAFS Spectroscopy, 424
polarization, 428
potential for radiation damage, 430
practical sample considerations, 431
sample preparation and collection, 431
STXM, 423
substrates, 432
Sum-frequency generation (SFG), 52
Sun photometers, 211, 225
AERONET, 230
satellite observations, 231
Surface enhancement Raman spectroscopy (SERS), 200
Surface sites, 51
Suwannee River Fulvic acid (SRFA), 284
SVD. *See* Singular value decomposition (SVD)
Synchrotron, 420, 421
aerosol photoemission, 378
beamline carbon contamination, 422
beamline optical components, 422
electron, 420
insertion devices, 422
operation modes, 421
radiation intersection, 379
radiation use, 394
storage rings and insertion devices, 420

T

TAOS. *See* Two-dimensional angular optical scattering (TAOS)
Tar balls, 440, 468
functional group intensities, 443
spectra, 442
TDL. *See* Tunable diode laser (TDL)
TE. *See* Transmitted primary electrons (TE)
TE modes. *See* Transverse electric modes (TE modes)
TEM. *See* Transmission electron microscopy (TEM)
Teramobile, 324n
TexAQS. *See* Texas Air Quality Study (TexAQS)
Texas Air Quality Study (TexAQS), 470

Time-dependent numerical approach, 32
Time-of-flight secondary ion mass spectrometry
         (TOF-SIMS), 430, 484
Time-resolved aerosol collector (TRAC), 431, 468
Time reversal principle. *See* Reciprocity principle
TM modes. *See* Transverse magnetic modes
         (TM modes)
TO. *See* Transverse optical (TO)
TOF-SIMS. *See* Time-of-flight secondary ion mass
         spectrometry (TOF-SIMS)
TRAC. *See* Time-resolved aerosol collector (TRAC)
Transition dipoles and frequencies
      Hessian matrix calculation, 31
      local environment effects, 31
Transmission electron microscopy (TEM), 358, 401,
         408, 429
      grids, 199, 200
      high-resolution images, 406
      ice nanocrystal preparation, 53
Transmitted primary electrons (TE), 464
      mode of imaging, 465
Transverse electric modes (TE modes), 135, 136
Transverse magnetic modes (TM modes), 135, 136
Transverse optical (TO), 58, 109
Trp. *See* Tryptophan (Trp)
Tryptophan (Trp), 304, 321
      absorption spectra, 325
      fluorescence spectra comparison, 322
      PPD experiment, 326
      PPD scheme, 325
Tunable diode laser (TDL), 18
2D. *See* Two-dimensional (2D)
2PEF. *See* Two-photon excited fluorescence (2PEF)
Two-dimensional (2D), 243, 379
      optical array detector, 141
      scattering, 260
      STXM/NEXAFS, 441
Two-dimensional angular optical scattering (TAOS),
         260, 307
      aerosol measurements, 309
      aerosol nozzle, 308
      cross-beam trigger system, 309
      ellipsoidal mirror and different geometry, 313
      ellipsoidal reflectors, 261
      patterns, 310, 311, 312, 314
      ray-tracing analysis, 309
      scattering plane, 261
      shape characterization, 262
      2D scattering pattern, 261
Two-dimensional scattering. *See* Two-dimensional
         angular optical scattering (TAOS)
Two-photon excited fluorescence (2PEF), 322
      backward enhanced, 324
      pump-probe, 331
Tyndall effect, 357–358, 360

U

Ultrasonic nebulizer, 140
Ultraviolet (UV), 369
      wavelength range, 228–229
Undulators, 420, 421, 422
UNIFAC. *See* Universal functional activity
         coefficient (UNIFAC)

Universal functional activity coefficient
         (UNIFAC), 158
Unsaturated fatty acids, 142
      particle mass changes, 184
      Raman spectra, 180
UV. *See* Ultraviolet (UV)
UV–VIS spectra, 244
      absorption spectroscopy, 245, 248
      aerosol absorption dependence, 246
      carbonaceous aerosol, 251
      complex refractive index, 244
      experimental methods, 246
      extinction cross section, 244
      extinction efficiency, 244
      extinction spectroscopy, 245, 246
      laboratory and aerosol chamber results, 250
      mass-specific cross section, 244
      mineral dust aerosol, 253
      single scattering albedo, 245

V

Vacuum UV photoemission (VUV
         photoemission), 378
Velocity map imaging (VMI), 378
      aerosol surface studies, 388
      gas-phase xenon, 380
      glycine particles, 381
      photoelectron intensity, 389
      raw velocity map images, 383, 384
      spectrometer, 378–379
Vertically resolved lidar observations, 211
Very Large Telescope (VLT), 101–102
Vibrating orifice aerosol generator (VOAG), 140, 391
Vibrational
      exciton, 28
      Hamiltonian, 29
      spectra, 60–61, 61–63
Vibrational exciton model, 27
      dipole-induced dipole coupling, 28–29
      exciton coupling, 27–28
      Hamiltonian matrix elements, 30
      molecular dipole moment operators, 29
      numerical implementation, 32–34
      scaled projection matrix, 29
      vibrational Hamiltonian, 29, 30
Visual light microscopes (VLMs), 423, 431
VLMs. *See* Visual light microscopes (VLMs)
VLT. *See* Very Large Telescope (VLT)
VMI. *See* Velocity map imaging (VMI)
VOAG. *See* Vibrating orifice aerosol generator (VOAG)
VUV photoemission. *See* Vacuum UV photoemission
         (VUV photoemission)

W

Water-soluble organic salts, 174
Water-to-solute ratio (WSR), 161
Wavelength-dispersive analysis of x-rays (WDX),
         464, 465
WDX. *See* Wavelength-dispersive analysis of
         x-rays (WDX)
Weak adsorbates, 50
      interaction, 59

WGMs. *See* Whispering gallery modes (WGMs)
Whispering gallery modes (WGMs), 134
WSR. *See* Water-to-solute ratio (WSR)

X

XAS. *See* X-ray absorption (XAS)
XPS. *See* X-ray photoelectron spectroscopy (XPS)
X-ray absorption (XAS), 430
X-ray fluorescence spectroscopy (XRF
      spectroscopy), 193, 430
   for bulk elemental concentrations, 202
X-ray photoelectron spectroscopy (XPS), 368
   ambient pressure, 389
   C1s and O1s XPS spectra, 393
   liquid aerosols, 391–394
   submicron particles, 389–391
   VOAG/APXPS setup, 391–392

X-ray-powder diffractometry (XRD), 253
XRD. *See* X-ray-powder diffractometry (XRD)
XRF spectroscopy. *See* X-ray fluorescence
      spectroscopy (XRF spectroscopy)

Y

YACS. *See* Yosemite Aerosol Characterization
      Study (YACS)
Yosemite Aerosol Characterization Study
      (YACS), 468

Z

Z-contrast imaging. *See* High-angle annular dark-field
      imaging (HAADF imaging)
Zones, 423
Zone-width, 423

Printed and bound by CPI Group (UK) Ltd, Croydon, CR0 4YY

18/10/2024

01776270-0018